Building a Low-Carbon Society Through Applied Environmental Materials Science

Takaomi Kobayashi
Nagaoka University of Technology, Japan

Published in the United States of America by
IGI Global Scientific Publishing
701 East Chocolate Avenue
Hershey, PA, 17033, USA
Tel: 717-533-8845
Fax: 717-533-8661
E-mail: cust@igi-global.com
Website: https://www.igi-global.com

Copyright © 2025 copyright by IGI Global Scientific Publishing. All rights reserved. No part of this publication may be reproduced, stored or distributed in any form or by any means, electronic or mechanical, including photocopying, without written permission from the publisher.
Product or company names used in this set are for identification purposes only. Inclusion of the names of the products or companies does not indicate a claim of ownership by IGI Global Scientific Publishing of the trademark or registered trademark.

Library of Congress Cataloging-in-Publication Data

Names: Kobayashi, Takaomi, 1958- editor.
Title: Building a low-carbon society through applied environmental
   materials science / edited by Takaomi Kobayashi.
Description: Hershey, PA : Engineering Science Reference, [2025] | Includes
   bibliographical references and index. | Summary: "The significance of
   this book is that technological innovations for a low-carbon society, a
   recycling-oriented society, and symbiosis with nature have become
   essential in recent years. In particular, it is necessary to establish
   technologies that replace petroleum-derived materials with
   biomass-derived or recycled waste materials"-- Provided by publisher.
Identifiers: LCCN 2023049131 (print) | LCCN 2023049132 (ebook) | ISBN
   9798369300039 (h/c) | ISBN 9798369300046 (s/c) | ISBN 9798369300053
   (eISBN)
Subjects: LCSH: Pollution prevention--Materials. | Carbon dioxide
   mitigation. | Green products. | Recycled products.
Classification: LCC TD192 .B85 2024 (print) | LCC TD192 (ebook) | DDC
   628.5--dc23/eng/20240215
LC record available at https://lccn.loc.gov/2023049131
LC ebook record available at https://lccn.loc.gov/2023049132

Vice President of Editorial: Melissa Wagner
Managing Editor of Acquisitions: Mikaela Felty
Managing Editor of Book Development: Jocelynn Hessler
Production Manager: Mike Brehm
Cover Design: Phillip Shickler

British Cataloguing in Publication Data
A Cataloguing in Publication record for this book is available from the British Library.

ll work contributed to this book is new, previously-unpublished material.
he views expressed in this book are those of the authors, but not necessarily of the publisher.
This book contains information sourced from authentic and highly regarded references, with reasonable efforts made to ensure the reliability of the data and information presented. The authors, editors, and publisher believe the information in this book to be accurate and true as of the date of publication. Every effort has been made to trace and credit the copyright holders of all materials included. However, the authors, editors, and publisher cannot assume responsibility for the validity of all materials or the consequences of their use. Should any copyright material be found unacknowledged, please inform the publisher so that corrections may be made in future reprints.

Table of Contents

Preface xv

Chapter 1
Introduction of Environmental Materials and Relevant Technologies 1
  Takaomi Kobayashi, Nagaoka University of Technology, Japan

Chapter 2
Non-Wooden Cellulose Materials Sourced From Plant Wastes 33
  Karla L. Tovar Carrillo, Universidad Autónoma de Ciudad Juárez, Mexico
  Ayano Ibaraki, Nagaoka University of Technology, Japan
  Takaomi Kobayashi, Nagaoka University of Technology, Japan

Chapter 3
Cellulose-Based Functional Fine Particles and Fibers as Environmentally Friendly Materials:
Development and Application 61
  Shoji Nagaoka, Kumamoto Innovative Development Organization, Japan
  Maki Horikawa, Kumamoto Industrial Research Institute, Japan
  Tomohiro Shirosaki, Kumamoto Industrial Research Institute, Japan

Chapter 4
Cellulose Nanofibrils Composite Films 95
  Huixin Jiang, Oak Ridge National Laboratory, USA
  Hannah Snider, Oak Ridge National Laboratory, USA
  Xianhui Zhao, Oak Ridge National Laboratory, USA
  Saurabh Prakash Pethe, University of Tennessee, USA
  Shuvodeep De, Oak Ridge National Laboratory, USA
  Tolga Aytug, Oak Ridge National Laboratory, USA
  Soydan Ozcan, Oak Ridge National Laboratory, USA
  Kashif Nawaz, Oak Ridge National Laboratory, USA
  Kai Li, Oak Ridge National Laboratory, USA

Chapter 5
Environmentally Sustainable Production of Bacterial Nanocellulose in Waste-Based Cell Culture
Media and Applications 119
  Takaomi Kobayashi, Nagaoka University of Technology, Japan

Chapter 6
Production of Sustainable Bioplastics Through Biomass Wastes Valorization to Mitigate Carbon Footprint Emissions 155
Takaomi Kobayashi, Nagaoka University of Technology, Japan
Debbie Dominic, Universiti Sains Malaysia, Malaysia
Nurul Alia Syufina Abu Bakar, School of Industrial Technology, Universiti Sains Malaysia, Malaysia
Siti Baidurah, School of Industrial Technology, Universiti Sains Malaysia, Malaysia

Chapter 7
Chitosan and Its Biomass Composites in Applications 169
Truong Thi Cam Trang, Vietnam National University, Vietnam
Khoa Dang Nguyen, Van Lang University, Vietnam

Chapter 8
Chitosan-Based Hydrogels: Current Strategic Fabrication and Practical Application Perspectives .. 203
Tu Minh Tran Vo, Chulalongkorn University, Thailand
Takaomi Kobayashi, Nagaoka University of Technology, Japan

Chapter 9
Insect Resources for Chitin Biomass 241
Guillermo Ignacio Guangorena Zarzosa, Nagaoka University of Technology, Japan
Takaomi Kobayashi, Nagaoka University of Technology, Japan

Chapter 10
Bioconversion of Waste Materials for the Production of Polylactic Acid to Alleviate Carbon Footprint 269
Debbie Dominic, Universiti Sains Malaysia, Malaysia
Nurul Alia Syufina Abu Bakar, Universiti Sains Malaysia, Malaysia
Siti Baidurah, Universiti Sains Malaysia, Malaysia

Chapter 11
Pectin Materials Sourced From Agriculture Waste: Extraction, Purification, Properties, and Applications 285
Tapanee Chuenkaek, Nagaoka University of Technology, Japan
Tu Minh Tran Vo, Chulalongkorn University, Thailand
Keita Nakajima, Nagaoka University of Technology, Japan
Takaomi Kobayashi, Nagaoka University of Technology, Japan

Chapter 12
Biomass Hydrogel Drug and Ultrasound Delivery Therapy Technology 321
Harshani Iresha, University of Peradeniya, Sri Lanka
Tu Minh Tran Vo, Nagaoka University of Technology, Japan

Chapter 13
ERS Vacuum Fermentation and Drying Bioreactor Contributing to Recycling of Organic
Containing Wastes 351
    Shinichi Shimose, JET Corporation, Japan
    Tomoyuki Katayama, JET Corporation, Japan

Chapter 14
Pollutant Remediation Using Inorganic Polymer-Based Fibrous Composite Adsorbents 369
    Anh Phuong Le Thi, Nagaoka University of Technology, Japan
    Ngan Phan Thi Thu, Nagaoka University of Technology, Japan
    Takaomi Kobayashi, Nagaoka University of Technology, Japan

Chapter 15
Removal of Metal Ions With Biomasses and Bioremediation 399
    Minoru Satoh, National Institute of Technology, Ibaraki College, Japan

Chapter 16
Extracting Technology Upcycling Toward Useful Metallic Materials From Mineral Wastes and
Pollutant Soil by Ultrasound Washing 425
    Tri Phuoc Phan, JET Corporation, Japan

Chapter 17
Assessing the Photocatalytic Performance of Carbon-Based Semiconductors in the Degradation
of Pharmaceutical Wastes 455
    Nursarah Sofea Ismail, Universiti Sains Malaysia, Malaysia
    Ahmad Fadhil Bin Rithwan, Universiti Sains Malaysia, Malaysia
    Sirikanjana Thongmee, Kasetsart University, Thailand
    Siti Fairus Mohd Yusoff, Universiti Kebangsaan Malaysia, Malaysia
    Rohana Adnan, Universiti Sains Malaysia, Malaysia
    Noor Haida Mohd Kaus, Universiti Sains Malaysia, Malaysia

Chapter 18
Geopolymers Prepared From Unused Resources and Their Applications 487
    Yuta Watanabe Nikaido, Tama Chemicals Co. Ltd., Japan
    Sujitra Onutai, Japan Atomic Energy Agency, Japan
    Sirithan Jiemsirilers, Chulalongkorn University, Thailand
    Takaomo Kobayashi, Nagaoka University of Technology, Japan

Chapter 19
Rapid and Easy Colorimetric Detection for Specific Heavy Metal Ions Contaminated in
Environmental Soil 517

 Reiko Wakasugi, National Institute of Technology, Kumamoto College, Japan
 Ryo Shoji, National Institute of Technology, Tokyo College, Japan
 Hitomi Fukaura, Limited Company Sakamoto Lime Industry, Japan
 Yasunori Takaki, Limited Company Sakamoto Lime Industry, Japan
 Hiroyuki Kono, National Institute of Technology, Tomakomai College, Japan

About the Contributors 537

Index 547

Detailed Table of Contents

Preface xv

Chapter 1
Introduction of Environmental Materials and Relevant Technologies 1
    Takaomi Kobayashi, Nagaoka University of Technology, Japan

This chapter describes the importance of building a sustainable society to develop materials related to low-carbon development and to put into practical use recycling technologies using waste materials as raw materials, since people life growth must be achieved while maintaining a balance between society, the economy, and the environment. For this purpose, economic growth is desired while balancing the construction of a recycling-oriented social structure with the conservation of nature. Specifically, the effective utilization of biomass as a carbon-neutral raw material and environmental remediation and conversion to raw materials are effective measures. Against this current background, this chapter also presents a discussion of the recent status of environmental materials and related technologies, especially upcycling technologies, toward building a sustainable society. Three subsections present the following: (1) Biomass waste reuse and biomass functional materials, (2) Green chemical refineries, and (3) Sustainable green technologies and processes using advanced materials.

Chapter 2
Non-Wooden Cellulose Materials Sourced From Plant Wastes 33
    Karla L. Tovar Carrillo, Universidad Autónoma de Ciudad Juárez, Mexico
    Ayano Ibaraki, Nagaoka University of Technology, Japan
    Takaomi Kobayashi, Nagaoka University of Technology, Japan

Cellulose is the most abundant plant waste material, making it a strong candidate to replace petroleum products as a future polymer material. Since cellulose is also abundant in food wastes, upcycling technology to obtain functional materials from these wastes is reviewed from the perspective of resource recycling. Cellulose, which is particularly difficult to handle as a material, has the property of being insoluble in solvents due to its strong cohesive nature. For this reason, recycled cellulose also is discussed as regenerative celluloses. However, compared to these chemically modified celluloses, the utilization of biomass cellulose fibers obtained from plant waste, which are a less scientific process and inexhaustible, enable to contribute to a sustainable society. The current status and technology of unmodified cellulose fibers is presented.Especially, the properties of cellulose hydrogels, which are agglomerated cellulose, and films are introduced in this chapter.

Chapter 3
Cellulose-Based Functional Fine Particles and Fibers as Environmentally Friendly Materials:
Development and Application 61
Shoji Nagaoka, Kumamoto Innovative Development Organization, Japan
Maki Horikawa, Kumamoto Industrial Research Institute, Japan
Tomohiro Shirosaki, Kumamoto Industrial Research Institute, Japan

Due to concerns regarding the effect of the large amount of waste accompanying their consumption on the environment, attention has therefore returned to cellulose and to maximize the potential of cellulose. On the development of cellulose refining, combining, and functionalization technologies, this chapter categorizes and introduces the spherical microbeads and nanofiber and then discusses interfacial hydrophilic microbeads, which are used as raw materials for moisturizing cosmetics. In addition, the fabrication of the semiconductor abrasives, thermally conductive materials, abrasive cleaning agents, and three primary color materials for paints are discussed for development of cellulose composite materials with inorganic materials.

Chapter 4
Cellulose Nanofibrils Composite Films 95
Huixin Jiang, Oak Ridge National Laboratory, USA
Hannah Snider, Oak Ridge National Laboratory, USA
Xianhui Zhao, Oak Ridge National Laboratory, USA
Saurabh Prakash Pethe, University of Tennessee, USA
Shuvodeep De, Oak Ridge National Laboratory, USA
Tolga Aytug, Oak Ridge National Laboratory, USA
Soydan Ozcan, Oak Ridge National Laboratory, USA
Kashif Nawaz, Oak Ridge National Laboratory, USA
Kai Li, Oak Ridge National Laboratory, USA

Nanocellulose derived from biomass is a sustainable, lightweight, and mechanically strong material. Extensive research has been directed towards applying nanocellulose for advanced applications. Nanocellulose-based films are novel and unique for designing functional materials considering their combination of sustainability with new properties. Nanocellulose films can be fabricated through simple strategies and make it easy to access various applications, such as barriers, sensors, energy storage, and so on. In this chapter, we summarized the preparation methods of nanocellulose-based films while focusing on nanocellulose as the film matrix and highlighting some representative applications. Given the sustainability of the nanocellulose and ease of introducing functional groups, we believe that nanocellulose-based films promise great potential for future advanced applications.

Chapter 5
Environmentally Sustainable Production of Bacterial Nanocellulose in Waste-Based Cell Culture
Media and Applications 119
    Takaomi Kobayashi, Nagaoka University of Technology, Japan

Bacterial cellulose has attracted great attention due to the demand for eco–friendly materials and sustainable products. It possesses properties superior to those of plant cellulose and has potential uses in various applications. The wider application of bacterial cellulose depends on the practical considerations such as the scale–up capability and production costs. The high cost of bacterial cellulose production is the main drawback that hinders industrial implementation. The cost–competitiveness can be improved, and bacterial cellulose production can be maximized by the utilization of agricultural and industrial waste and by–products as bacterial cell culture media. This chapter provides an overview of cost–effective culture media for bacterial cellulose production by using agricultural waste as a primary nutrient source. The applications related to nanocellulose, and the future challenges are also described.

Chapter 6
Production of Sustainable Bioplastics Through Biomass Wastes Valorization to Mitigate Carbon
Footprint Emissions 155
    Takaomi Kobayashi, Nagaoka University of Technology, Japan
    Debbie Dominic, Universiti Sains Malaysia, Malaysia
    Nurul Alia Syufina Abu Bakar, School of Industrial Technology, Universiti Sains Malaysia,
        Malaysia
    Siti Baidurah, School of Industrial Technology, Universiti Sains Malaysia, Malaysia

This chapter focuses on the production of bioplastics such as polyhydroxyalkanoates (PHAs) via biomass wastes valorization. The high production costs of PHAs, particularly substrate of the base material, has limited its application. The utilization of biomass wastes from agro-industrial such as molasses, banana trunk juice, palm oil waste effluent, and animal derived chitin from crustacean as an alternative carbon feedstock is explored by many researchers, due to the abundancy, inexpensive, contains high sugar and oil contents. These industrial biomass wastes can be further exploited for lowering the production costs, simultaneously reduce biomass waste accumulation, and accelerate the application of the bioplastics at large scale. This chapter also covers the challenges of utilizing industrial waste as feedstocks for PHA production and its impact on carbon footprint mitigation. These initiatives are in parallel with various Sustainable Development Goals such as number 12, 13, 14, and 15.

Chapter 7
Chitosan and Its Biomass Composites in Applications 169
    Truong Thi Cam Trang, Vietnam National University, Vietnam
    Khoa Dang Nguyen, Van Lang University, Vietnam

Biomass-based materials have been received a lot of attractive attentions due to their renewablity and low-carbon emmsion during processings. Chitosan is a deactylated product from chitin, which is the second most found bio-polymer in nature. Moreover, with the transformation of science and technology as well as the demand of society, chitosan-based composites have been extensively studied to adopt the changes towards sustainable development. Therefore, this chapter summarizes the existing researches of chitosan-based composites in different fields including biomedical applications, degradable food-packaging material and envrionmental remediation.

Chapter 8
Chitosan-Based Hydrogels: Current Strategic Fabrication and Practical Application Perspectives .. 203
   Tu Minh Tran Vo, Chulalongkorn University, Thailand
   Takaomi Kobayashi, Nagaoka University of Technology, Japan

Chitosan holds appeal as an inventing sustainable green material because of the sources of crab and shrimp shell wastes and encompasses applications in the medical, agriculture, and water treatment fields, and the biomass polymer consisting glycoconjugate backbone having amino groups makes it water soluble and can be gelatinized. This chapter highlights the properties of chitosan hydrogels used as several applications. Notably, providing a better understanding of hydrogels and utilization of chitosan-based hydrogels, the preparation methods of chitosan hydrogels involving numerous "green" techniques. The promising properties of chitosan hydrogels are extensively explored in this chapter. Also, the unique properties of chitosan making these hydrogels a powerful bio-friendly material are described in agriculture purpose, applied for slow control release of fertilizer, and in water purification

Chapter 9
Insect Resources for Chitin Biomass 241
   Guillermo Ignacio Guangorena Zarzosa, Nagaoka University of Technology, Japan
   Takaomi Kobayashi, Nagaoka University of Technology, Japan

Among all chitin sources, insects have recently gained the attention of researchers and entrepreneurs. In addition to being supplied primarily from unutilized products of the marine food industry, chitin sources from insect sources are expected to increase in the future as the insect food industry grows. Insect sources of chitin, in particular, are currently the focus of much attention and research, as it is important to know the properties and uses of chitin and how it relates to marine food products. So, this chapter highlights current extraction methods and some characteristics of insect chitin. In addition, some information about insects will be provided to have a holistic view when thinking about insect industrialization and its impact on chitin.

Chapter 10
Bioconversion of Waste Materials for the Production of Polylactic Acid to Alleviate Carbon
Footprint 269
   Debbie Dominic, Universiti Sains Malaysia, Malaysia
   Nurul Alia Syufina Abu Bakar, Universiti Sains Malaysia, Malaysia
   Siti Baidurah, Universiti Sains Malaysia, Malaysia

This chapter presents a comprehensive study on the utilization of various waste sources as feedstocks for the production of polylactic acid (PLA), that is a biodegradable and biocompatible polymer with versatile applications. In the case of PLA, food products containing mainly starch are sources, so that the development of utilizing waste materials instead of relying on such plant resources contributes to securing food resources. This waste approach is discussed in depth in this chapter, including PLA chemical structure, thermal, mechanical, and rheological behaviours and biodegradability. In addition, processing methods of the PLA obtained by bioconversion, as well as its multifaced properties and its applications across various industries are highlighted. Since integration of sustainable waste-PLA strategies provides a deep understanding of PLA, sheds light on its strengths, and identifies its limitations and future prospects, this approach can contribute highly to both environmental protection and a circular economy.

Chapter 11
Pectin Materials Sourced From Agriculture Waste: Extraction, Purification, Properties, and
Applications 285
    Tapanee Chuenkaek, Nagaoka University of Technology, Japan
    Tu Minh Tran Vo, Chulalongkorn University, Thailand
    Keita Nakajima, Nagaoka University of Technology, Japan
    Takaomi Kobayashi, Nagaoka University of Technology, Japan

For abundant citrus fruits, the perspective involves effectively utilizing waste materials such as citrus peels and apple pomace. The pursuit of sustainable pectin sourcing has also driven innovation in extraction techniques. Conventional pectin extraction methods are gradually giving way to more environmentally friendly and green extraction methods. The appeal of pectin lies in its diverse range of applications across various industries. Apart from its traditional uses in the food industry as a gelling agent and stabilizer, pectin has ventured into new frontiers in fields such as packaging, moisturizers, drug delivery, wound healing, and tissue engineering. This chapter provides a comprehensive review of the chemical structure of pectin, its conventional sources, and pectin extraction, with a particular focus on its physical, chemical, and functional properties within the context of various industrial applications.

Chapter 12
Biomass Hydrogel Drug and Ultrasound Delivery Therapy Technology 321
    Harshani Iresha, University of Peradeniya, Sri Lanka
    Tu Minh Tran Vo, Nagaoka University of Technology, Japan

For drug supply systems, there are various applications including the medical field and natural elements are generally polysaccharides like cellulose, chitin and others exhibit as key feature of biocompatibility, biodegradability, non-toxicity, drug-loading ability for regenerated matrix, especially for tissue regeneration and drug release. Highlighting and considering current research with biomass hydrogels of cellulose and chitin and new findings that drug delivery (DD) and drug release (DR) under external stimuli are important and becomes a symbolic area of future research. Therefore, this chapter describes present trends on biomass hydrogel drugs in the DD and DR under external smart stimulation. Among external stimuli, the control of drug release by ultrasonic external stimuli, which can be penetrated from outside the body, is mainly mentioned in this chapter because of its advanced features.

Chapter 13
ERS Vacuum Fermentation and Drying Bioreactor Contributing to Recycling of Organic Containing Wastes 351
Shinichi Shimose, JET Corporation, Japan
Tomoyuki Katayama, JET Corporation, Japan

ERS (Environmental Recycling System), a bioreactor in which microbial fermentation conducts organic waste treatment under mild-temperature heating and vacuum conditions, and in which high-speed fermentation and dehydration drying of wet wastes are carried out in a single reactor unit. The volume and weight reduction of waste containing organic materials and the conversion of the waste into compost, fertilizer, feed, livestock bedding, or biomass fuel are possible. Main features of ERS processing include that there is no emission of drain or bad odors through its processing, that the microorganisms used for fermentation in ERS are those inhabiting the surroundings of the installation site, preventing any disruption to the local microbial environment, and that there is no need to supplement ERS with microorganisms, and that unsorted general waste can be directly introduced into the ERS without sorting. This chapter introduces the current status of organic-containing waste treatment with examples of ERS implementation and also the practical applications.

Chapter 14
Pollutant Remediation Using Inorganic Polymer-Based Fibrous Composite Adsorbents 369
Anh Phuong Le Thi, Nagaoka University of Technology, Japan
Ngan Phan Thi Thu, Nagaoka University of Technology, Japan
Takaomi Kobayashi, Nagaoka University of Technology, Japan

In the prospective of building sustainable development society, environmental remediation technology is important for reconstruction of polluted nature. This chapter highlights new remediation technology of fibrous adsorbents consist of inorganic-polymer composites having several merits on combining both properties of polymer and inorganic components. These materials possess a fibrous structural arrangement, providing them with a substantial surface area that facilitates efficient pollutant capture. Their characters address several topics of a wide range of environmental issues for removal of heavy metals, organic compounds, and gases, as customized to specific pollutant removal requirements. Becasue environmental concerns are increasingly issued, it is clear for decontamination process that fibrous adsorbents have great potential in reducing the negative effects of pollutants on both the environment and human health.

Chapter 15
Removal of Metal Ions With Biomasses and Bioremediation 399
   Minoru Satoh, National Institute of Technology, Ibaraki College, Japan

Eco-friendly technologies using biomasses and bioremediation are presented, since remediation technologies for serious heavy metal pollution are necessary to protect the global environmental. In the perspective of the 3Rs (Recycle, Reuse, Reduce), one of such approaching is used with biomasses, plants and microbe resource derived from living organisms. The removal of metal ions via biomasses and bioremediation, especially with waste biomasses materials, becomes environmentally and friendly technologies in addition with another merit of waste reduction and effective utilization of unused biomass. Thus, bioremediation is gaining attention as a sustainable and effective method of cleaning up environmental pollutants using certain plants and animals. In particular, the method using plants as called phytoremediation, plant roots absorb water and nutrients from contaminated soil, sediment, surface water, and groundwater, removing metals and other toxic chemicals.

Chapter 16
Extracting Technology Upcycling Toward Useful Metallic Materials From Mineral Wastes and
Pollutant Soil by Ultrasound Washing 425
   Tri Phuoc Phan, JET Corporation, Japan

The increasing of industrial waste and soil contamination has raised so many concerns for the environment and the sustainable development of society. Recently, ultrasound (US) technology has emerged as the new tool for enhancing the recovery process of mineral waste and also for the soil washing process, making US highly efficient cleaning soil or extracting pollutants from waste soils. The unique and highly efficient US action is attributed to acoustic cavitation phenomena in ultrasound circumstance, cleaning and extraction can be more efficient than usual one. This chapter reviews the application of high-power US for eliminating pollutants from contaminant soil and US-extracting valuable metals from mineral wastes. Since US washing technology enhances the recovery process from unusual wastes, this also describes an upcycle technology that uses US cleaning to convert them into valuable resources.

Chapter 17
Assessing the Photocatalytic Performance of Carbon-Based Semiconductors in the Degradation
of Pharmaceutical Wastes 455
   Nursarah Sofea Ismail, Universiti Sains Malaysia, Malaysia
   Ahmad Fadhil Bin Rithwan, Universiti Sains Malaysia, Malaysia
   Sirikanjana Thongmee, Kasetsart University, Thailand
   Siti Fairus Mohd Yusoff, Universiti Kebangsaan Malaysia, Malaysia
   Rohana Adnan, Universiti Sains Malaysia, Malaysia
   Noor Haida Mohd Kaus, Universiti Sains Malaysia, Malaysia

This review highlights recent research endeavors in the field, emphasizing the critical role of carbon-based photocatalysts for their superior performance under visible light irradiation and their capacity to efficiently photodegrade pharmaceutical contaminants. The integration of carbon materials with semiconductors has demonstrated remarkable synergistic effects, leading to enhanced photocatalytic efficiency. The exploration of carbon-based semiconductor materials represents a promising avenue for addressing the persistent issue of pharmaceutical waste in the environment, offering a greener and more efficient solution for water treatment.

Chapter 18
Geopolymers Prepared From Unused Resources and Their Applications 487
    Yuta Watanabe Nikaido, Tama Chemicals Co. Ltd., Japan
    Sujitra Onutai, Japan Atomic Energy Agency, Japan
    Sirithan Jiemsirilers, Chulalongkorn University, Thailand
    Takaomo Kobayashi, Nagaoka University of Technology, Japan

As an alumino-silicate material, geopolymers present several of beneficial properties, including high strength, low permeability, high acid resistance, the capacity to immobilize poisonous compounds, and the ability to resist hazardous contaminants. In this chapter, geopolymer materials using raw materials derived from unutilized waste are discussed, especially the case of using metakaolin-based and fly ash-based raw materials, and the preparation and properties of geopolymers and their applications are explained. The main point is that geopolymers are materials that can contribute to the construction of a sustainable society as a technology that can create high value-added products through the effective use of unused waste materials.

Chapter 19
Rapid and Easy Colorimetric Detection for Specific Heavy Metal Ions Contaminated in
Environmental Soil 517
    Reiko Wakasugi, National Institute of Technology, Kumamoto College, Japan
    Ryo Shoji, National Institute of Technology, Tokyo College, Japan
    Hitomi Fukaura, Limited Company Sakamoto Lime Industry, Japan
    Yasunori Takaki, Limited Company Sakamoto Lime Industry, Japan
    Hiroyuki Kono, National Institute of Technology, Tomakomai College, Japan

Among the environmental pollution, soil is making it not only easy for analysis of contamination to remain, but also difficult to clean up. It is therefore necessary to develop a tool that can be used to easily and quickly determine the necessity of soil contamination at the on sites. In this chapter, a method developed to quickly detect soil pollutant species due to four specific types of heavy metals as hexavalent chromium, fluorine, boron, and lead. The detection kit has a sensitivity achieved to the elution standard value stipulated by Japan's Soil Contamination Countermeasures Law. The detection agent consists of a composite of water-absorbing polymer and coloring reagent that specifically reacts with such pollutant. This kit has already been sold under the trade name OCTES® (OCTES, 2014) and can now be used at soil contaminated sites without requiring any particularly difficult operations or electric power sources. This OCTES is expected to be used as a screening method to easily determine the need for detailed soil contamination assessment and purification.

About the Contributors 537

Index 547

Preface

It is with great anticipation that I present *Building a Low-Carbon Society Through Applied Environmental Materials Science*. This book represents a vital contribution to the growing body of work focused on addressing the global need for a low-carbon society. In recent years, technological innovations aimed at creating a recycling-oriented economy and fostering symbiosis with nature have become indispensable, and comprehensive resources that explore these innovations are in high demand.

The chapters in this book explore various topics in materials science where biomass materials and related technologies are directly linked to low-carbon technologies, with an emphasis on promoting industry-academia collaboration in line with the Sustainable Development Goals (SDGs). As areas of materials and technologies that contribute to low-carbon technologies, this book can explore such topics into subjects such as biomass materials for medical applications and cosmetics, food-related recycling innovations, earth coexistence technologies, phytoremediation, inorganic mineral recycling, photocatalytic environmental remediation, and simplified polluted heavy metal analysis. These topics focus on sustainable biomass-derived materials and waste recycled materials technologies. These materials are important in the development of alternatives to materials dependent on a carbonized society and the technologies that use them, are inevitable in building a low-carbon society. They therefore have great potential to transform various industries.

By consolidating cutting-edge information and emerging trends in these fields, this book aims to offer valuable insights for university students, researchers, and corporate engineers alike. It will serve as a guide for understanding the current landscape of materials science in relation to environmental sustainability, providing both a theoretical foundation and practical knowledge to foster advancements in low-carbon technologies.

As we continue to work on the chapters, we are actively seeking contributions from experts who can offer further perspectives on these critical topics. It is my hope that *Building a Low-Carbon Society Through Applied Environmental Materials Science* will inspire innovation and collaboration, paving the way toward a more sustainable and environmentally conscious future.

ORGANIZATION OF THE BOOK

In *Introduction of Environmental Materials and Relevant Technologies*, the Editor sets the stage for a deeper understanding of the environmental materials vital for creating a sustainable society. Kobayashi emphasizes the necessity of balancing economic growth with a recycling-oriented social structure and nature conservation. The chapter examines the utilization of biomass as a carbon-neutral raw material and its role in environmental remediation. With a focus on upcycling technologies, three specific subsections explore biomass waste reuse, green chemical refineries, and sustainable processes using advanced materials—offering a comprehensive overview of the state of environmental technologies as they relate to building a low-carbon future.

Non-Wooden Cellulose Materials Sourced from Plant Wastes as a crucial alternative to petroleum-derived products. In this chapter, the authors outline non-wooden cellulose, which is abundant in plant and food wastes, and highlight challenges and solutions in handling this material due to its strong cohesive nature. The authors review cellulose's abundance in plant and food waste, emphasizing the challenges and solutions in handling this material due to its strong cohesive nature. They discuss regenerative cellulose and unmodified cellulose fibers as sustainable alternatives, with a particular focus on cellulose hydrogels and films. This chapter highlights the current state of unmodified cellulose technology and its potential contributions to a sustainable society.

In *Cellulose-based Functional Fine Particles and Fibers as Environmentally-Friendly Materials: Development and Application,* the authors bring attention to the environmental impact of cellulose waste and the growing efforts to refine and functionalize cellulose for diverse applications. From moisturizing cosmetic microbeads to thermally conductive materials and paint pigments, this chapter categorizes and explores the development of cellulose composite materials with a focus on sustainability and innovation.

Cellulose Nanofibrils Composite Films highlights the emerging field of nanocellulose-based films. These films, derived from biomass, offer a unique combination of sustainability, lightweight properties, and mechanical strength. The authors provide a comprehensive review of fabrication methods and potential applications in energy storage, sensors, and barriers, positioning nanocellulose films as key players in future advanced material solutions.

Environmentally Sustainable Production of Bacterial Nanocellulose in Waste-Based Cell Culture Media and Applications addresses the promising future of bacterial cellulose, a material with superior properties to plant cellulose. The chapter details the cost challenges associated with scaling up bacterial cellulose production and explores solutions through the use of agricultural and industrial waste as bacterial cell culture media. The discussion extends to potential applications of bacterial nanocellulose, presenting a vision for future developments in sustainable materials.

In *Production of Sustainable Bioplastics through Biomass Wastes Valorization to Mitigate Carbon Footprint Emissions,* the authors explore the challenges of producing bioplastics, particularly polyhydroxyalkanoates (PHAs), from biomass waste. They emphasize the potential for reducing production costs and waste accumulation by utilizing agro-industrial by-products. This chapter highlights the impact of these efforts on carbon footprint reduction and aligns them with global Sustainable Development Goals.

Chitosan and its Biomass Composites in Applications investigates the versatility of chitosan, a derivative of chitin, in various fields. From biomedical applications to environmental remediation, the chapter reviews recent advancements in chitosan-based composites, providing insight into how this renewable material supports the shift toward sustainable development in multiple industries.

Chitosan-based Hydrogels: Current Strategic Fabrication and Practical Application Perspectives presents the fabrication techniques and applications of chitosan-based hydrogels. These hydrogels, derived from crustacean shells, have potential uses in agriculture and water treatment. The chapter explores the green techniques used to prepare hydrogels and highlights their promising properties for slow-release fertilizers and water purification, making chitosan a significant player in bio-friendly materials.

Insect Resources for Chitin Biomass focuses on the growing interest in insect-sourced chitin. As the insect food industry expands, researchers are turning their attention to insect chitin as an alternative to marine-sourced chitin. The chapter discusses extraction methods, the characteristics of insect chitin, and its potential applications, while considering the broader implications of insect industrialization.

Bioconversion of Waste Materials for the Production of Polylactic Acid to Alleviate Carbon Footprint offers a detailed analysis of using waste sources as feedstocks for polylactic acid (PLA) production. The chapter examines the potential of PLA as a biodegradable polymer while considering the environmental benefits of waste utilization. Processing methods, properties, and applications of PLA are discussed, positioning this approach as a key driver of environmental protection and circular economy strategies.

In *Pectin Materials Sourced from Agriculture Waste: Extraction, Purification, Properties, and Applications*, the authors present pectin as a versatile material sourced from agricultural waste, such as citrus peels and apple pomace. The chapter covers extraction methods, chemical properties, and the expanding applications of pectin beyond the food industry into fields like drug delivery and tissue engineering. This exploration underscores the growing demand for sustainable materials.

Biomass Hydrogel Drug and Ultrasound Delivery Therapy Technology discusses the potential of biomass hydrogels in drug delivery systems. The chapter delves into recent advancements in using cellulose and chitin hydrogels for tissue regeneration and controlled drug release under external stimuli, with a particular focus on ultrasonic drug delivery. This innovative approach highlights the future of biocompatible and biodegradable materials in the medical field.

In *ERS Vacuum Fermentation and Drying Bioreactor Contributing to Recycling of Organic Containing Wastes*, the authors introduce the Environmental Recycling System (ERS), a bioreactor designed to treat organic waste through microbial fermentation. The chapter highlights the system's ability to reduce waste volume and weight, convert waste into useful products like compost and biomass fuel, and operate without emissions or odors, positioning ERS as a pivotal technology for organic waste recycling.

Extracting Technology Upcycling toward Useful Metallic Materials from Mineral Wastes and Pollutant Soil by Ultrasound Washing focuses on the use of ultrasound (US) technology to recover valuable metals and clean contaminated soils. The acoustic cavitation produced by ultrasound enhances the extraction of pollutants, making it a highly efficient process for soil washing and metal recovery. The chapter also introduces the concept of upcycling, where waste is converted into valuable materials, presenting a sustainable solution to managing industrial waste and soil contamination.

Removal of Metal Ions with Biomasses and Bioremediation explores eco-friendly technologies that utilize biomass and bioremediation for addressing heavy metal pollution. By adopting the 3Rs (Recycle, Reuse, Reduce), biomass materials, including plants and microbes, are used to remove metal ions, thus providing a sustainable approach to waste reduction. The chapter highlights phytoremediation, where plants absorb and neutralize pollutants from soil and water, presenting a cost-effective and environmentally responsible method for remediation.

Pollutant Remediation Using Inorganic Polymer-Based Fibrous Composite Adsorbents delves into the innovative use of fibrous composite adsorbents made from inorganic-polymer materials. These composites blend the advantages of both polymer and inorganic components, allowing for a large surface area that aids in capturing pollutants like heavy metals, organic compounds, and gases. The fibrous structure enhances their decontamination capabilities, making them effective in reducing environmental pollution and improving human health by mitigating the effects of various pollutants.

The chapter, *Assessing the Photocatalytic Performance of Carbon-Based Semiconductors in the Degradation of Pharmaceutical Wastes,* aims to provide a comprehensive assessment of the photocatalytic performance of carbon-based semiconductors in the degradation of pharmaceutical wastes. By exploring recent advances in the synthesis and application of these materials, we evaluate their effectiveness, mechanisms of action, and potential for large-scale implementation in wastewater treatment. Through this exploration, we hope to highlight the critical role of carbon-based semiconductors in advancing sustainable solutions for the mitigation of pharmaceutical pollutants, ultimately contributing to a cleaner and safer environment.

Geopolymers Prepared from Unused Resources and Their Applications focuses on geopolymers derived from aluminosilicate materials such as metakaolin and fly ash, using inorganic waste materials. Despite being derived from waste materials, the resulting geopolymer materials have properties such as high strength, acid resistance, and the ability to immobilize toxic substances, making them useful for a variety of industrial applications, such as general concrete and ceramics. By utilizing waste inorganic resources, geopolymers not only help prevent pollution, but also provide a sustainable technology to create high-value products from unused resources.

The last chapter, *Rapid and Easy Colorimetric Detection for Specific Heavy Metal Ions Contaminated in Environmental Soil* introduces a rapid colorimetric detection method for specific heavy metals, detailing its practicality and effectiveness. This method enables quick on-site analysis, providing a powerful tool for assessing the extent of contamination.

IN CONCLUSION

In addition to biomass materials, the chapters on technological process developments in this book endeavor to provide a comprehensive overview of cutting-edge technologies and strategies designed to address some of the most pressing environmental issues in our lives. The chapters highlight innovative approaches such as inorganic polymer-based fibrous composite adsorbents, biomass-driven bioremediation, ultrasound-assisted pollutant extraction, and the development of geopolymers from unused resources. These contributions not only demonstrate the diversity and potential of new remediation technologies, but also highlight the importance of upcycling waste and contaminants into valuable resources.

As global concerns around pollution and sustainability continue to escalate, the research presented here offers viable pathways for mitigating environmental degradation while simultaneously fostering sustainable development. By merging advancements in materials science, biotechnology, and resource recovery, this book provides a holistic view of how technological innovation can help reconstruct a cleaner, healthier planet.

It is our hope that the work shared in these pages will inspire further research and action, driving a shift toward more sustainable environmental practices. The collective efforts of the contributors reflect the spirit of collaboration and innovation that is essential to solving the complex environmental problems we face. By embracing these approaches, we can look forward to a future where resource management and pollutant remediation work hand in hand to protect both human health and the natural world.

Finally, the networks of some of the contributors summarize the results of research conducted as part of the "Innovative Global Engineer Education Program: Toward Building an Integrated Global Campus through Industry-Academia-Government Collaboration" (2014-2024), supported by the Japanese Ministry of Education, Culture, Sports, Science and Technology (MEXT). In addition, the formation and operation of a consortium network of some contributors from Japan-Thailand-Vietnam-Malaysia were supported by the Japan International Cooperation Agency (JICA) AUN/SEED-Net's "Global Academic-Industry Collaboration and Education Consortium Program (GAICCE) and we would like to thank the "2019-2022" program for its support. We would also like to take this opportunity to thank Nagaoka University of Technology for their cooperation in promoting industry-academia collaboration and global networking.

Takaomi Kobayashi
Nagaoka University of Technology, Japan

Chapter 1
Introduction of Environmental Materials and Relevant Technologies

Takaomi Kobayashi
Nagaoka University of Technology, Japan

ABSTRACT

This chapter describes the importance of building a sustainable society to develop materials related to low-carbon development and to put into practical use recycling technologies using waste materials as raw materials, since people life growth must be achieved while maintaining a balance between society, the economy, and the environment. For this purpose, economic growth is desired while balancing the construction of a recycling-oriented social structure with the conservation of nature. Specifically, the effective utilization of biomass as a carbon-neutral raw material and environmental remediation and conversion to raw materials are effective measures. Against this current background, this chapter also presents a discussion of the recent status of environmental materials and related technologies, especially upcycling technologies, toward building a sustainable society. Three subsections present the following: (1) Biomass waste reuse and biomass functional materials, (2) Green chemical refineries, and (3) Sustainable green technologies and processes using advanced materials.

INTRODUCTION OF ENVIRONMENTAL MATERIALS AND RELATED TECHNOLOGIES

Currently, changes in the global environment are causing global warming and extreme weather. Trends of economic development have emphasized the pursuit of convenience in human life and have development for its own sake, leading to environmental destruction and resource scarcity. Moreover, economic development is spreading across the globe. Flows of people and exchanges across national borders have become a necessity. However, current resource use and consumption patterns are depleting the Earth's finite resources. In addition, the increased stresses of environmental change impose greater demands on the adaptive capacities of nation-states and further diminish the resources available to fulfill other core functions. This convenient life we lead has been supported by the establishment of a high-

DOI: 10.4018/979-8-3693-0003-9.ch001

Copyright ©2025, IGI Global. Copying or distributing in print or electronic forms without written permission of IGI Global is prohibited.

carbon-consumption society that uses energy and which emits vast amounts of carbon. This approach has not allowed us to live in harmony with the Earth. We are facing difficulties related to global warming, extreme weather, environmental pollution, and resource shortages.

As one countermeasure, environmental movements have proposed the concept of green chemistry, a sustainable chemical industry (Anastas & Williamson, 1996; McElroy et al., 2015), which can support ecosystem growth (BenDor et al., 2011). Given this background, UN member states formulated Sustainable Development Goals (SDGs) in 2015. The SDGs led to the "2030 Agenda for Sustainable Development" (United Nations SDGs, 2023), with the following motto: "The Sustainable Development Goals are a blueprint for a better and more sustainable future for all. People aspire to achieve a society that enables them to connect with nature and people." To this end, 17 SDGs have been set forth to eliminate poverty and deprivation (Monkelbaan, 2019) as a plan and goal that "everyone in the world should work together to solve by 2030, in addition to environmental problems (climate change), the many problems we are facing (include) poverty, conflict, human rights issues, and infectious diseases such as new coronaviruses."

Technological developments in chemistry and materials must be achieved for transformation from a high-carbon society to a low-carbon society. Moreover, such a society must coexist with the global environment while upgrading human life. To do so, engineering solutions must be created that surpass current and dominant technologies and which improve, innovate, and invent technologies to achieve sustainability by the benefits of Chemistry and Sustainable Development Goals (ACS, 2023). Such approaches can be expected to have a strong effect on the economy and the society in which we live. It is noteworthy that the economy and society must coexist with the global environment and develop sustainably (Figure 1). In other words, it is important for the development of the economy and society to embrace a viewpoint of global environmental conservation as the foundation on which we live. This foundation has led to the advocacy and rapid popularization of the "sustainable development" concept (Kobayashi & Nakajima, 2021).

Figure 1. Conceptual diagram showing the need for a sustainable society with low-carbon technologies for an economy and society that can coexist with the environment

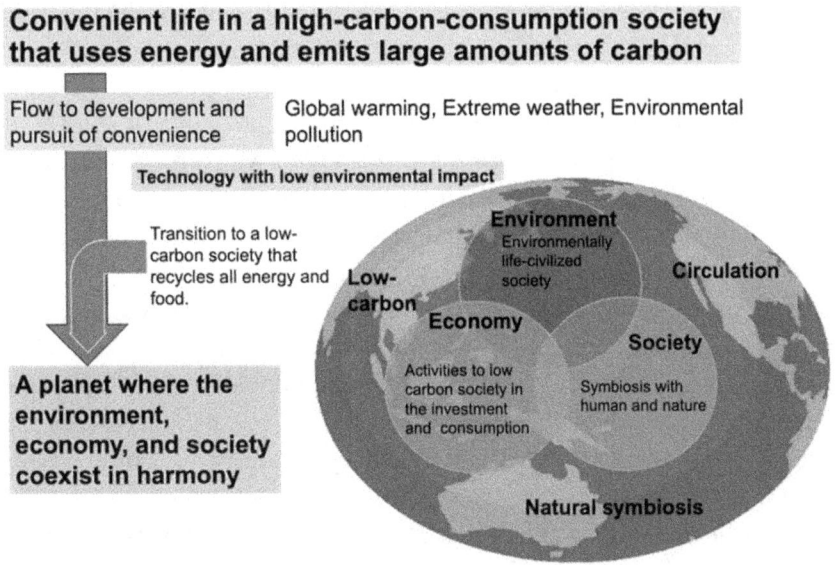

For that sense of social responsibility for current and future chemists, the American Chemical Society supports many SDGs including the following: 2, Zero Hunger; 3, Health and Wellness; 6, Clean Water and Sanitation; 7, Affordable and Clean Energy; 9, Industry; Innovation and Infrastructure; 12, Responsible Consumption and Production; and 13, Climate Change Action. These SDGs can provide opportunities for new chemical research, green and sustainable chemical education, and green and sustainable chemical manufacturing practices (Reyes et al., 2023).

For making our lives sustainable, it is necessary to pursue concepts such as a low-carbon society, a recycling-oriented society, and a society in harmony with nature, in which the environment, economy, and society can be improved together (Ministry of the Environment, Japan, 2007). Here, economic activities bring us closer to a low-carbon society for investment and consumption, as well as the creation of a recycling-oriented economic and social system for coexistence with nature. These activities are also related to the production and management of chemical substances in industry, and are expected to contribute importantly to the control of global environmental pollution. Building a low-carbon society, a recycling-oriented society, and a society in harmony with the global environment are expected to be necessary, with inclusion of the following policies.

Low Carbon Society. To prevent global warming, it is important to reduce emissions of CO_2, an important greenhouse gas, and simultaneously to increase the amount of CO_2 absorbed by forests. The following measures are necessary to realize a low-carbon society.

- Reduce CO_2 emissions from fossil fuel combustion, etc. (saving electricity, reducing fuel use for industrial activities, promoting introduction of renewable energy)
- Increase CO_2 absorption by forests (greening activities, forest conservation, preventing deforestation, etc.)

Although technological progress has enriched people's lives, high economic growth, backed by increased industrial activity, infrastructure development, and other developments are usually regarded as accompanied by increased energy use and its consequent increased greenhouse gas emissions and huge amounts of generated waste. Therefore, it is necessary to popularize the innovative technologies already possessed by economically developed countries to economically developing countries and the rest of the world.

Recycling-Oriented Society. Conventional social systems have given rise to a mass-production, mass-consumption, mass-disposal "use-it-or-lose-it" society in which huge amounts of resources are used to produce goods, which are then discarded after use. The most common methods of disposing of the vast amounts of waste are to burn it and to bury it. A need exists to reduce the wastes of natural resources and energy and to reduce greenhouse gas emissions by reusing once-used materials with commensurate energy consumption for "resources." This achievement is expected to require reduction of natural resource consumption, along with promotion of the reuse and recycling of natural resources. A recurrent cycle of resource use with lower energy consumption must be created so that they can be reused as raw materials or fuel when they are used up. In this way, natural resources can be saved by efficient repetition of the circulation cycle. This efficient cycle conserves natural resources and reduces environmental impacts, as shown in Figure 1.

Natural Symbiotic Society. Biodiversity is a state in which many living organisms adapt and evolve to various environments on Earth, and in which the richness and balance of ecosystems are maintained. Biodiversity is the foundation of a society in harmony with nature. However, as economic development and globalization progress and as burdens on the environment increase, severe difficulties are emerging for the natural environment, such as the deterioration of habitats and growth environments for living creatures and the resulting reduction in biodiversity. A balance must be struck between efforts to build a low-carbon, recycling-oriented society and efforts to build a nature-friendly and symbiotic society.

A balance must exist between efforts to build a low-carbon, recycling-oriented society and efforts to build a natural symbiotic society. In the guidelines for implementing such a society, the establishment of a resource-recycling society and a low-carbon society is important. However, reducing the increase in energy consumption associated with resource recycling is necessary, as is recycling promotion. In addition, the establishment of a "low-carbon society" and the development of innovative technologies are expected to be necessary to achieve this goal. The following three points were identified as basic principles for "Creating a low carbon society" (Paper Recycling Japan, 2022).

(1) Realization of a Carbon Minimum. The key is creation of a social system involving industry and government, with people devoting thorough consideration to minimizing carbon dioxide emissions when making choices and decisions.

(2) Realization of a Simplified Lifestyle with a Sense of Affluence. It is necessary to break away from a uniformity of society that seeks affluence through mass consumption, which was formed mainly in economically developed countries. It is also necessary for producers to transform themselves to satisfy the preferences of consumers.

(3) Realization of Symbiosis with Nature. It is important to secure carbon dioxide sinks, to conserve and restore a rich and diverse natural environment that contributes to adaptation to inevitable global warming in the future, to promote the use of nature-harmonizing technologies, and to ensure that people have opportunities to interact with nature. To realize a society in harmony with nature, a recycling-oriented society through the 3Rs, which are to reduce waste emissions to the greatest extent possible (Reduce) and then reuse (Reuse) and recycle (Recycle) waste in a cyclical manner is an important objective. The remaining waste and other materials can then undergo heat recovery through the introduction of waste power generation and other measures, which can contribute to the reduction of greenhouse gas emissions.

(4) Development of Recycling Products for Low-Carbon Society. The contributions of materials and related technologies to the creation of a low-carbon and recycling-oriented society are considerable. The figure below presents the concept proposed as an alternative to a mass-production, mass-consumption, and mass-disposal society. A society can be built, for which the consumption of natural resources is reduced. The burden on the environment is therefore minimized to the greatest extent possible. Figure 2 portrays the relation of the natural cycle and material circulation in an economy and society.

As an example, one can examine the production and recycling of paper produced from forest resources as pulp material (Paper Recycling Japan, 2022). In 2022, Japan's total production of paper and paperboard was 23.7 million tons, making the country the third largest producer after China and the USA. The amount of wood pulp used annually was 8.1 million tons, whereas 16 million tons were supplied by recovered paper as raw materials for the manufacture of new paper and paperboard. Such recovered paper is generated by residences, offices, shopping malls, grocery stores, printers and bookbinders, carton boxes and corrugated container manufacturers, and other such sources. In the Japanese system, large quantities of paper are collected and given to suppliers, who then deliver the paper as raw material to the mills. Paper recycling is characterized by transforming paper back into new paper: in 2022, approximately 17.9 million tons of paper were collected, of which about 1.8 million tons were exported. Approximately 0.2 million tons were imported. Consequently, domestic recycling amounts to about 16.1 million tons in Japan. The figures indicate that 99% of the recovered paper has been used as raw material for the manufacture of new paper. These systems classify paper into nine types of recovered paper, which are then divided into eight types of paper: liner and corrugated medium, core pare, building board, white paperboard, newsprint, wrapping paper, printing and communication paper, and sanitary paper. Of these paper products, the former group is for paperboard and the latter for paper. The key point for recycling recovered paper is the following basic treatment processes for recycling recovered paper into recovered pulp: defibering, dust removal, dispersing, de-inking, bleaching, washing, and drainage.

As Figure 2 shows, such attempts can reduce the amounts of natural resource input while maintaining production, thereby reducing damage caused by the natural cycle. In other words, it is important to manage resource extraction and environmental impact to avoid the disruptive effects of the cycle. However, after people consume the products, they are discarded as waste. This process can be linked to material recycling to reduce natural resource input. As a result, such a recycling-oriented society can exist in harmony with nature and can halt deterioration of the environment related to global warming, extreme weather, and environmental pollution. Waste that does not fit into this cycle is eventually incinerated or disposed of as landfill.

Figure 2. Illustration of the balance of material circulation in the economy and society and the natural cycle

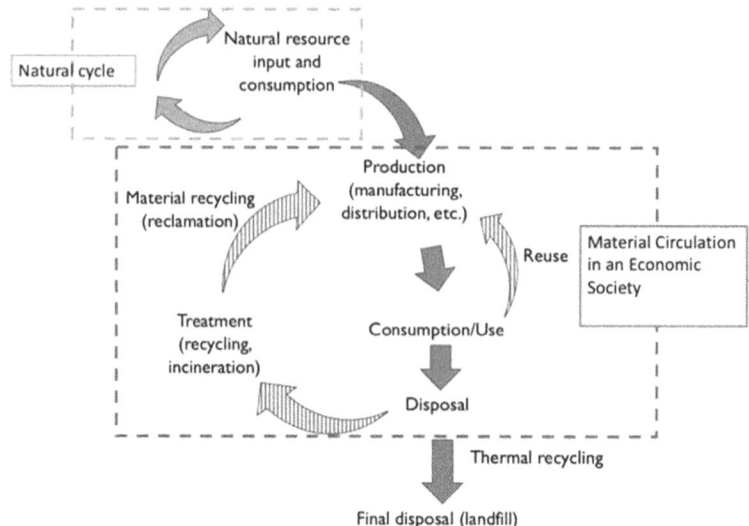

With regard to the recycling of plastics, the Organization of Economic Cooperation and Development (OECD) reported in 2022 that sustainable industrial processes can develop through the exploitation of recycling and conservation of resources (OECD, 2022). The world today generates twice as much plastic waste as it did 20 years ago. Most of it ends up in landfills, incinerated, or leaked into the environment. Only 9% was recycled. The wide variety of plastic products and their diverse uses make their reformulation difficult. Plastic spills continue to occur: in 2019, an estimated 22 Mt of plastic leaked into the environment. The greatest source of leakage (82%) is mismanaged waste, i.e., improperly processed waste. Other sources include microplastic abrasion and loss (12%), littering (5%), and marine activities (1%) (Anthony, 2011). The UN Environmental Programme indicated that action and international cooperation are necessary to reduce current petroleum-based plastic-origin pollution such as microplastic pollution, and reported treatment technologies are crucially important (Andrady, 2011; UN Environmental Programme, 2013). In Honolulu, USA, reducing pollution from current petroleum-based plastics requires action and international cooperation. Treatment technologies are especially important (Honolulu Strategy, 2012). Examples of implementation include the Honolulu Strategy and the four Rs (reduce, reuse, recycle, and recover) for the following goals A, B, and C. Goal A is reduction of the amount and effects of marine debris entering the ocean from land. Goal B is reduction of the amount and effects of sea-based sources of marine debris including solid waste such as lost cargo, and abandoned fishing gear and vessels. Goal C is reduction of marine debris accumulated on shorelines in benthic habitats and in pelagic waters. The Honolulu Strategy is a framework for a comprehensive, global effort to reduce marine debris effects on ecosystems, human health, and economies. Its success requires participation and support at global, regional, national, and local levels and requires the full involvement of intergovernmental organizations and the private sector.

Trends in innovation for environmentally relevant plastics technologies (OECD ENV Working Paper, No199) over the last 30 years demonstrate that the total number of patents rose by a factor of 3.4 during 1990–2017. For improving innovation in plastics prevention and recycling, innovative technologies of the four categories of waste prevention exist (reuse, repair, reduction, and durability), along with waste recycling (sorting, recycling and recyclability), conversion and disposal of waste and leakage removal, and bio-based feedstocks. Among them are chemical recycling of plastics for thermolysis of heterogeneous polymers and solvolysis of homogeneous polymers (Li et al., 2022; Thiounn & Smith, 2022). Thermolysis is concerned with thermal cracking, pyrolysis, gasification, hydrothermal liquefaction, and catalytic pressureless depolymerization, and with eventual conversion to fuels (oil, syngas, gas, diesel, and kerosene) and chemicals such as solvents. By contrast, solvolysis can be used for conversion to monomers via the chemical processes of hydrolysis, alcoholysis, methanolysis, aminolysis, and acidolysis. According to the OECD, plastics emitted 1.8 billion tons of greenhouse gases in 2019, which represent 3.4% of global emissions, 90% of which were attributable to production and conversion from fossil fuels. If such trends continue, it is estimated that greenhouse gas emissions would more than double to 4.3 billion tons by 2026. Therefore, a shift from high-carbon plastics derived from petroleum to low-carbon plastics and conversion to biomass-derived plastics will be necessary.

In addition to these chemical treatments, biodegradable plastics have been used recently as a typical plastic: polylactic acid (PLA) (Komesu et al., 2017; Yee et al., 2022). In this case, the use of biodegradable plastics that break down in nature contributes to alleviation of plastic waste pollution such as microplastic pollution. Nevertheless, concerns persist about competition with food products such as starch as a raw material. Moreover, the overharvesting of naturally occurring resources also affects the environment. Consequently, one direction to take is to develop biomass resources using waste from food products of natural origin. As shown in Figure 2, they have the potential for use in the material recycling loop presented in the figure without any competition.

Development of Innovative Technologies for a Resource-Recycling and Eco-Materials

Eco-Materials. Building an environmentally responsive, recycling-oriented society requires innovative technologies in addition to conventional technologies. Particularly, the effective use of waste materials is an approach that is less resource-depleting and less destructive to nature than the use of competing food and mineral resources. Here, in terms of physical resources, the following points should be regarded as reducing the diffusion of hazardous substances, reducing greenhouse gas emissions, etc. The following four categories are necessary: Prevention of pollution, Reduction of resource and energy consumption, Reduction of landfill and incineration, and Recycling. Achieving reduction of these environmental footprints represents a major challenge for the ecological transition. To achieve this goal, the use of eco-materials has been proposed. The concept of eco-materials (Ishida, 2009; Ito, 1999), here, coined from environmentally friendly materials, is defined as "materials and related material technologies that can be manufactured, used, recycled, and disposed of in a human-friendly manner, with excellent properties and functions and with low environmental impact." These were applied to metallic materials. However, now that the importance of a recycling-oriented society is being reevaluated, this concept should be extended to other materials. In other words, materials and material technologies must undergo a new evolution as the basis for "sustainable" development that can occur in harmony with the global environment. As Figure 3 shows, eco-materials have three characteristics: environmental friendliness, human friendliness, and material properties.

Figure 3. Conceptual diagram of eco-materials. Eco-materials are environmentally friendly, match human life in a friendly manner, retain their properties as materials, and are recyclable

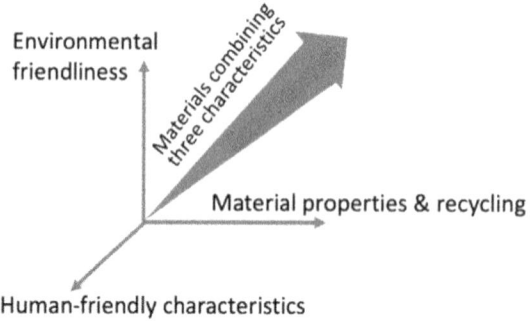

As eco-materials, the following materials can be discarded or disposed of without any environmental impact. Moreover, they contain no hazardous substances, clean the environment, save Earth resource-saving materials, and make effective use of energy. First, it is noteworthy that a need exists for materials that help reduce environmental impacts that occur at each stage of production, use, and disposal. From this perspective, such materials are classified into the following four categories.

(i) "cyclic" materials
(ii) materials for ecological and environmental protection
(iii) materials for society and human health
(iv) materials for energy based on the two main criteria as their sources and functions

Each category includes (i) recycled materials and their efficient use as raw materials, (ii) disposable materials with low environmental impact as material functions, (iii) materials with reduced environmental impact on society and materials with reduced health impact, and (iv) materials with high energy efficiency and green energy materials.

Recycle Materials. Recycling is often associated with lower production costs and carbon emissions, as well as markedly lower energy consumption than the production of goods synthesized from virgin materials. The European Commission (EU) states that natural resources must be used in the most efficient way and without depleting the Earth's resources. As an objective, the **circular economy** decouples economic growth from resource use and boosts sustainable growth while protecting Earth's natural resources. It is preferred that a circular economy use secondary raw materials such as waste to reduce environmental impact (Circular Economy, 2023), thereby helping in efforts to reduce the consumption footprint in the EU over the next decade and to double the rate of circular material use. The commission also proposes bioenergy derived not only from plants, but also from agricultural and municipal waste as sustainable. By securing these raw materials from waste, the commission is committed to promoting the sustainable management of forests, land, and soil, which are the sources of the respective resources

(European Green Deal, 2023). This strategy conserves and restores biodiversity so that the environment can continue to provide many ecosystem services on which humans depend. Additionally, the OECD regulates higher value material loops and waste prevention to enable industrial symbiosis to secure waste resources. In other words, urban areas demand 2/3 of the total energy in OECD countries, emitting about 80% of the greenhouse gases, and creating 50% of the global waste. Therefore, to avoid this situation, the OECD is promoting "circular economy" activities between cities and regions (OECD circular economy). In Japan, the recycling Sub-committee of the Environment Committee of the Industrial Structure Council was established in July 2001 to assess the medium and long-term route of the advancement of the Recycling-Oriented Economic System. The Environmental Protection Agency (EPA) in the U.S. has a recycling program to build an economy in which materials, products, and services stay in circulation as long as possible by following the reduce, reuse, and recycle motto (EPA recycle). The agency reports that learning how reducing, reusing, and recycling is important and then doing so will reduce costs, reduce energy consumption, and conserve natural resources. Recycling is the process of first recovering and processing materials that would otherwise be discarded as garbage, and then transforming them into new products, consequently offering tremendous environmental benefits, particularly in the following four ways: conserve natural resources, climate change for less CO_2 emissions, energy conservation, and waste and pollution reduction. Regarding the recycled content, common products include aluminum cans, car bumpers, carpeting, cereal boxes, comic books, egg cartons, glass containers, laundry detergent bottles, motor oil, nails, newspapers, paper towels, steel products, and trash bags.

However, for plastic waste, the current status is summarized in a report by the Plastic Waste Management Institute (2019). Plastic wastes are divisible into recycled materials and are recycled or un-recycled including processes of mechanical recycling, chemical recycling with solvent-based purification, depolymerization and conversion to simpler chemicals such as hydrocarbons or syngas, followed by energy recovery through combustion of plastic wastes. Moreover, considering the difficulties related to the disposal of plastics, upcycling, referring to the conversion of waste materials into a valuable product has been attractive (Nature, 2019a; Zhao et al., 2022) by including processes of pyrolysis, carbonization or catalytic cracking. However, many factors influence the upcycling ability for conversion into such products by depending on the target application, including the plastic composition, processing history, reaction temperature, and catalyst activity (Tan et al., 2022). Recently, upcycling of chlorinated waste plastics (Xu et al., 2023), contaminated plastics (Wang and Ma, 2023), and mixed plastics (Lepage & Wulff, 2023) have been reported. The latter process implemented compatibilization of polymers, enabling the dynamic crosslinking of molded products.

Material Ecology and Environmental Protection. Ecology, also designated as bioecology, bionomics, or environmental biology, is concerned with the study of the relations between organisms and their environment with pressing problems in human affairs, such as expanding population, food scarcity and environmental pollution including global warming and plant extinction. From an ecological perspective, the reduction of environmental effects related to materials includes, first of all, air pollution, water pollution, noise and vibration, odors, and pollution by toxic substances. To advocate for a clean, productive and well-protected environment, the EPA has delivered good environmental outcomes and knowledge through targeted and timely environmental topics related to climate change, air, drinking water, waste, and others (EPA Environment topics).

Figure 4. Schematic diagram of material sources for environmental materials and related environmental protection technologies

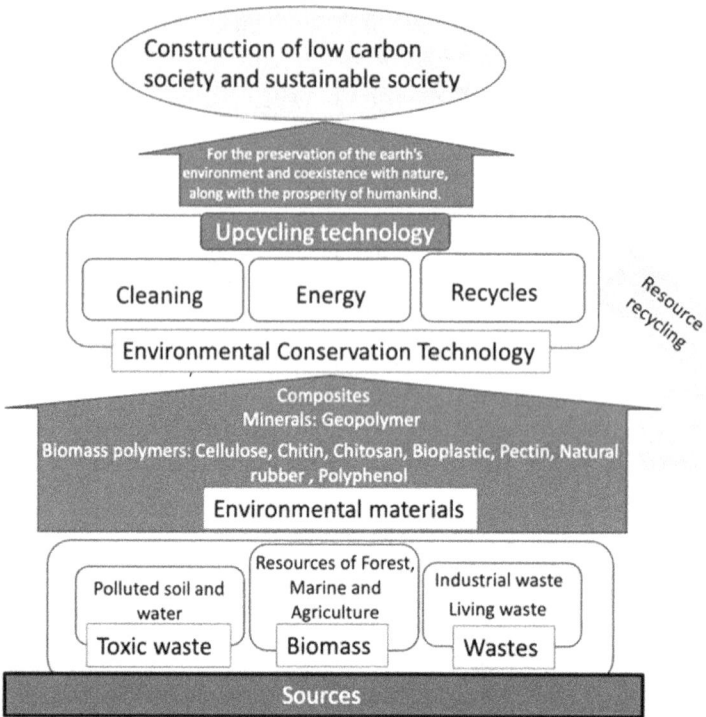

DISCUSSION

Contribution of Environmental Materials and Their Technologies to Develop Sustainable Society

Nature-friendly technologies, which have become necessary for our society and economy, require conversion to a society that is currently dependent on petroleum products into a technological society that supports and contributes to recycling and low carbon emissions. Therefore, materials and related technologies should be developed and deployed in consideration of the coexistence of human society, the economy, and the environment. Environmental materials are materials and related technologies that include elements of eco-materials, but they also accommodate the coexistence of economic, social, and environmental factors through advanced materials and technologies which are friendly to humans and to the Earth. As presented in Figure 4, the concept of environmental materials uses waste, biomass, and pollutants as material resources and uses these materials without discarding them. This usage allows the procurement of materials that can replace, for example, petroleum-derived plastic materials, thereby contributing to a low-carbon society. In addition, technologies supporting the use of garbage and industrial waste generated by human activity as material resources are necessary. Furthermore, environmental conservation technologies that adapt these advanced materials for use in the preservation of the Earth's environment are classified broadly into recycling, energy, and purification technologies, and are categorized according to their coexistence with nature, along with the prosperity of humankind.

Carbon Neutral and Unused Resources and Wastes. For the construction of a sustainable society, the use of waste and unused resources as material resources is a matter that must be considered adequately in future studies. Regarding the procurement of resources for material development, biomass, unused resources, and waste as carbon neutral resources are described as a source of resource procurement in the background of the importance of building a recycling society. The biomass supporting carbon neutral resources includes forest, marine, and agricultural resources. Here, the resources related to carbon neutrality are mainly plant resources, which generate biomass through photosynthesis using CO_2 as a raw material. In other words, biomass materials are released as combustible fuel and as CO_2 in the atmosphere, but plant resources can absorb this CO_2 and increase the amount of resources. Therefore, CO_2, a greenhouse gas, does not increase from the use of plant resources: the CO_2 load on the atmosphere is therefore zero.

Unused resources include inorganic components such as ash, contaminated water, and contaminated soil. For example, fly ash collected in dust collectors from coal ash generated when pulverized coal is burned in thermal power plants is currently used as an additive (Xu & Shi, 2018) for glass (Althoey et al., 2023), geopolymer concrete (Shahab-Ed-Din, 2021), asphalt (Mistry & Roy, 2016), and cement (Fauzi et al., 2016). Most are used as civil engineering materials. In fact, most of them are limited to use as civil engineering materials.

Furthermore, waste includes domestic and industrial wastes. In principle, municipalities are responsible for overseeing the disposal of general waste, whereas operators are responsible for the disposal of industrial waste. Waste from people's daily life is disposed of mainly by combustion or landfill. Currently, however, the rapid growth of general waste is making waste incineration increasingly difficult. It is also becoming more difficult for combustible waste to be landfilled without being incinerated. Here, a waste treatment process that involves the burning of substances included in waste is called incineration. Most facilities are simply incinerators, but in recent years, they have become incineration facilities, commonly designated as waste-to-energy (WTE) facilities, using one of several waste-to-energy technologies such as gasification, pyrolysis, and anaerobic digestion. For example, some electric output capacities (megawatt, MW) by country for municipal waste energy in Europe (2010–2022) are 1067 MW for Germany, 755 MW for the United Kingdom, 572 MW for France, 418 MW for Italy, 413 MW for the Netherlands, 386 MW for Sweden, and 240 MW for Austria (Fernández, 2023). In contrast, as reported by the US Energy Information Administration (EIA), waste-to-energy plants in the United States during 2012–2022 generated 14,000 gigawatt hours (GWh) of electricity each year, with about 60 plants in operation (EIA WTE, 2023). Approximately 90% of the energy produced at WTE plants is fed into the electric grid. The remaining 10% is used as a steam resource, which some WTE facilities send to nearby industrial plants and facilities.

Industrial waste is generated by industrial activities, including all materials that are no longer useful and which are discharged from factories, mines, and other manufacturing processes. Consequently, types of industrial waste include soil, gravel, stone, and concrete from civil engineering projects, scrap metal, oil, solvents, chemicals, and scrap wood from factories, as well as vegetable scraps and food waste from restaurants.

Figure 5. Chemical structures of biosourced polymers of cellulose, chitin, chitosan, pectin, polylactic acid, and polyhydroxybutyrate. Pictures of cellulose source for cotton, sugar cane and agave, shrimp and crab for chitin and citrus for pectin. SEM pictures show bacteria nonocellulose and microbe-producing nanofibers

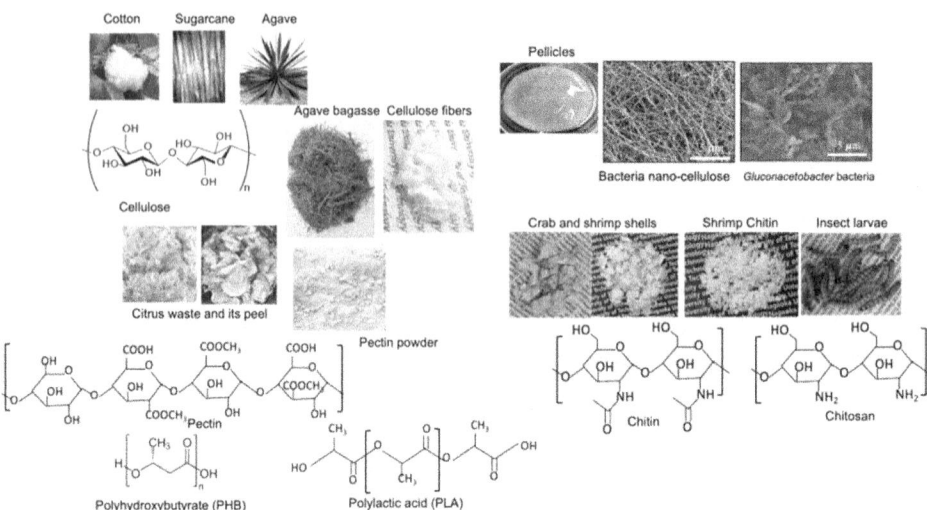

Biomass Materials and Biorefineries. Within this systematized flow of Figure 4, materials derived from biomass resources and related industrial resources include cellulose, chitin, chitosan, pectin, and other polysaccharides for their chemical structures, as shown in Figure 5, and bioplastics such as biocellulose and polyhydroxy alkanoate (PHA) like poly-3-hydroxy butyrate (PHB), and natural rubber materials. Among these, many polysaccharides such as cellulose and chitin are obtained from food wastes. They are promising resources because of the high purity of the target components in waste and unused resources. In fact, cellulose and chitin are the number one and number two biomass polymers on Earth in terms of their reserves. In addition, cotton and natural rubber are producible through agricultural cultivation, and are grown alone by agricultural products, whereas bioplastics such as bacterial cellulose and PHA are producible by cultivating microorganisms (Taokaew et al., 2022). Other promising biomass materials include natural rubber and PLA, which is synthesized from lactic acid (Baidurah et al., 2022). Among these resources for biomass polymers, cellulose is the biomass polymer with the largest reserves. It is expected to be used effectively in the future. However, because of its poor solubility in solvents and difficult handling, its practical use is mainly in the clothing industry, with feedstocks of cotton and other raw materials, and in the paper industry, with feedstocks of wood chips. Cellulose, a non-wood resource, is mainly contained in vegetable resources of agricultural products. Nevertheless, no clear industry has been established, with a route from wood chips to paper. Cellulose deriving from non-wood resources is mainly contained in agro-resources of agricultural products. Actually, food waste is a highly pure cellulosic waste. Bagasse is a typical waste product among agro-resources of food waste. The residue of strained stalks of sugarcane, palm tree seeds, coconut residue, and other fruit residues left over after extraction of food ingredients of sugar essence, oils, and other valuable products. Currently, some bagasse is used mainly as a raw material for paper (Rainey & Covey, 2016) and clothing (Costa et al., 2013), boiler fuel (Dave et al., 2014), gasification (Erlich & Fransson 2011; Ogi et al., 2011), building materials (Li et al.,

2022; de Souza et al., 2017), and livestock feed. Nevertheless, no clear industry has been established, such as that of the huge production stream running from wood chips to paper. For example, most of the palm oil bagasse (empty fruit bunch, EFB) and oil residues of palm oil mill effluent (POME) have been disposed of as portrayed in Figure 6. Long-term abandonment of EFB (c) and POME (e) waste sites, which are residues after oil extraction, has caused the release of the greenhouse gas methane through anaerobic fermentation; methane has become increasingly problematic (Saelor et al., 2017; Said et al., 2021; Sarwani et al., 2019).

Figure 6. Photographs of a palm oil mill in Medan, Indonesia: (a) oil palm peel (mesocarp); (b) transportation of palm kernel oil mill residue (EFB); (c) EFB storage area; (d) EFB; (e) palm oil mill effluent (POME) storage pond

Malaysia and Indonesia are attempting to collect such biogas and to use it as an energy source (Situmeang et al., 2022). However, because of difficulties of collecting gases on site, the trend has been the construction of dedicated plants (Koonaphapdeelert et al., 2020; Pavičić et al., 2022). By contrast, oil palm EFB comprises 20.6–33.5% hemicellulose, 23.7–65.0% cellulose, and 14.1–30.5% lignin (Chang, 2014). Additionally, the cellulose contents are roughly comparable to the cellulose content of wood fiber (38–49%), bamboo (26–43%), and bagasse (32–44%) (Risdianto et al., 2016). Because bagasse wastes such as EFB (Yiin et al., 2019), sugarcane (Mahmud & Anannya, 2021; Nagasone & Kobayashi, 2016a), cassava (Dienl & Anh, 2021), agave (Tovar-Carrillo et al., 2013), and bamboo contain cellulose, a stream of cellulose fiber utilization exists for use as a raw material. Regarding these processes, alkaline treatment, acid-chlorite treatment (bleaching), and acid hydrolysis are some commonly employed processes for the chemical treatments applied to cellulose fibers. Then those are converted to carboxymethylcellulose

(Yimlamai, 2021), nanocellulose (Gond et al., 2021; Varshney et al., 2021) such as nanocrystals (Batista et al., 2022), cellulose nano-fiber (CNF) (Shahi et al., 2020), and hydrogels (Kobayashi, 2015).

However, not many so-called upcycling methods exist to turn plant biomass into new materials via cellulose fiber. Among them, many nano-sized cellulosic materials such as nanofibers and nanocrystalline cellulose have been developed in recent years, especially as reinforcement fillers for polymer composites. After isolating nano-sized cellulose, it is composited with rubber, which dramatically improves the rubber properties and which leads to much research interest for its use as a reinforcing material for composite materials (Favier et al., 1995). Such bio-based substitutes for petroleum-based materials have been reported for CNFs with epoxy (Subbotina et al., 2021), polyvinyl alcohol (Peresin et al., 2010), polyimide (Yeganeh et al., 2023), and others (Oksman et al., 2016; Sharma et al., 2019). These nanomaterials can be tailored using a method that dissolves (unravels) the cellulose aggregates (micrifibrils) originally present in plant cell walls and reduces them to molecular size. The method of defibrillation applied to CNFs generally involves the use of a high-pressure homogenizer, when sugarcane bagasse is used as a source of cellulose (Hongrattanavichit & Aht-Ong, 2020).

Recently, it was reported that cellulose fibers, after purification from several kinds of bagasse, were dissolved from bagasse wastes and cotton cellulose fibers in a dimethylformamide (DMAc)/LiCl solution to produce hydrogel films (Kobayashi, 2015). Those hydrogels were water-swollen. The transparent films with high water content had the properties of cytoaffinity (Nakasone & Kobayashi, 2016b; Tovar-Carrillo et al., 2013), biocompatibility (Nakasone et al., 2016), and applicability for regenerative medicine (Iresha & Kobayashi, 2021; Jiang et al., 2016).

Similarly to cellulose, chitin, an abundant natural material derived primarily from crab and shrimp shells, is a useful biomaterial with beneficial properties such as non-toxicity (Pohling, 2022), biocompatibility (He, 2017), and biodegradability (Makarios-Lahamand & Lee, 1995), but showing difficulty of dissolving chitin because the interaction of OH groups induces a highly crystallized structure. Chitin extraction can be done from crab shells using methods reported by Yen et al. (2008) and Yamaguchi et al. (2003), and with some modifications (Jiang & Kobayashi, 2017). Figure 6 depicts a flow diagram of hydrogel film preparation from crab shell waste via chitin. Briefly, the crab shells (20 g) were washed with large amounts of water, and were treated with 1 N HCl (700 mL) at 26 °C for 24 h to remove mineral components. The waste crab shells were collected from snow crabs (*Genus chinoecetes*) in Teradomari, Niigata, Japan. Then the resultant shells were washed with water until neutral pH was achieved. Then they were treated with 1 N NaOH (700 mL) at 90 °C for 5 h to eliminate proteins. After the shells were washed to neutral pH with distilled water, 1000 mL of ethanol was added. The mixture was heated at 60 °C for 12 h to achieve decolorization. Then, the extracted chitin was obtained after drying in vacuum at 50 °C for 24 h. The chitin hydrogel fabrication was conducted using the following process (Nguyen & Kobayashi, 2020). The obtained chitin was dissolved in DMAc/LiCl solvent for 5 days at room temperature. Then the LiCl content was adjusted to 3, 5, 7, and 10% in DMAc solutions. After the extracted chitin was concentrated, 1 and 2% to the LiCl/DMAc solvent was added. Then the chitin solution was centrifuged at 9000 rpm for 30 min to remove the undissolved parts. Subsequently, the chitin solutions were obtained at 1 and 2% concentrations. Figure 7 presents photographs of the chitin-DMAc solutions prepared to different concentrations of LiCl. For hydrogel fabrication, the resultant chitin solution was cast on a Petri dish of 9.1 cm diameter placed in a plastic container ($12 \times 12 \times 6$ cm^3). Then, 40 mL of distilled water was also placed without contacting the chitin-DMAc solutions in the container. At room temperature, water vapor was induced in the wet-phase inversion process for fabricating the chitin hydrogels. After 24 h, the chitin hydrogel was formed in the Petri dish with DMAc solution.

Among waste biomass derived from plant-based sources with fruits sources, notable examples can be found in citrus waste such as those of oranges, pomelos, lemons, and grapefruits being cultivated worldwide for food industries. Production of these crops has been increasing steadily, reaching a million tons in 2016 (Yadav et al., 2022) and 114 million metric tons in 2020 (FAO, 2022).

Figure 7. Preparation processes of chitin, chitin-DMAc solution containing LiCl from 3 to 10% concentrations and resultant chitin hydrogels

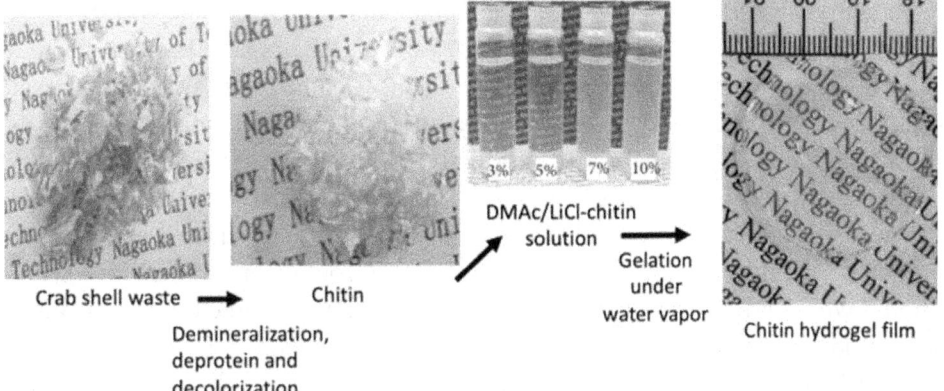

Citrus sources are useful for producing various products such as chemicals, solvents, flavorings, fragrances, paints, cosmetics, and animal feed supplements (Andrade et al., 2014). Additionally, citrus waste is useful for producing bio-based products, including bio-adsorbents, bio-fuels, bio-fertilizers, packaging materials, and activated carbon (Suri et al., 2022) and also for use in producing food packaging films made from citrus peel powders or active compounds extracted from the waste, which has become popular in recent years (Yun & Liu, 2022). One product obtained from citrus fruits is pectin, a polysaccharide consisting of both methylated ester of polygalacturonic acid and 1,4-linked α-D-galacturonic acid (Hoefler, 1991; Panuccio et al., 2022). Those have already attracted attention in the food sector for use as stabilizers during emulsification (Humerez-Flores et al., 2022) and for value-added products as part of a bioeconomic strategy for the agricultural sector (Sarangi et al., 2023), food, biomedical, and pharmaceutical industries (Freita et al., 2021). Because of the gelation properties of pectin, water-soluble conditions polymers have been used widely in food and beverage industries (Hoefler, 1991; Pilgrim et al., 1991). One property of this polysaccharide is that pectin has a surface-active effect as a W/O emulsifier (Leroux et al., 2003) and also supports gelation (Willats et al., 2006) in the presence of calcium ions (Tu Minh et al., 2022). One reason for these diverse application possibilities is that pectin has bioactive properties (Gotoh et al., 2021; Palin & Geitmann, 2012).

Sustainable Green Technologies and Processes with Advanced Materials. Sustainable resources include not only organic compounds such as biomass. Inorganic compounds also have potential. Particularly, unused wastes include large amounts of useful stem materials in industrial waste. The utilization of these materials makes a noteworthy contribution to a sustainable society by linking a network of recycling-oriented materials to their practical use. Particularly, problems related to solid industrial waste generated in the production process remain unresolved in all fields. They are the product of pursuing industrial development without proper management and treatment. For example, a typical mineral industrial waste is copper slag, which is generated at the rate of 2.2 tons of copper slag per ton of refined copper during the copper smelting process (Gorai & Premchand, 2003; Zuo et al., 2022). If not properly managed and treated, this waste continues to accumulate. For that reasons, recycling technologies for this waste stream are extremely important. Its main alternative processing has been as an additive for materials used in the civil engineering field, for example, as an additive for asphalt or as a secondary source of material in place of sand used for cement (Behnood et al., 2015; Siddique et al., 2020). The main components of copper slag are Fe_2O_3, SiO_2, CaO, and Al_2O_3, with iron oxide and silicon oxide accounting for about 80%, and with many other elements such as aluminum oxide, zinc oxide, and copper oxide. Although extraction methods for useful resources such as copper and aluminum have been studied (Gargul et al., 2019; Shi et al., 2020), silica, the main copper slag component, is used only as a building material. It has not been used. For copper slag, no idea to use these rapidly accumulating industrial wastes as a resource exists. However, if silica (SiO_2), which accounts for 1/3 of copper slag, could be extracted efficiently, then it could replace currently used silica not only in industry but also in food, medicine, and cosmetics industries. Especially under current circumstances, the silica source is quartz, existing as sand and ores such as silica sand and silica stone. It is provided by mining from the natural environment. As this example illustrates, when particularly addressing silica as a resource, the procurement of silica from waste sources is important for the construction of a recycling society in the future.

Recently, a technique for upcycling copper slag to silica using ultrasound (US) was reported as a potential source of silica (Tri et al., 2023). The topic is presented in this section.

Figure 8. Picture of copper slag in Onahama Smelting and Refining Co.LTD (Iwaki, Fukushima prefecture, Japan) (left) and experiment flow of US-silica gelation and extraction processes in acidic solution for copper slag waste (Tri et al., 2023)

The copper slag sample, as shown in Figure 8, was donated by Fukushima Prefecture Onahama Smelting (Iwaki city, Fukushima prefecture). The main composition findings were the following: Fe_2O_3 53.5%, SiO_2 23.5%, Al_2O_3 5.64%, CaO 3.79%, MgO 1.41%, CuO 0.98%, ZnO 1.33%, and Na_2O 0.52%. The main component was therefore iron oxide, followed by about 20% silica. A flow chart of the ultrasound silica extraction process from copper slag in acidic water is exhibited in Figure 8. In the glass vessel, HCl, HNO_3, or H_2SO_4 was adjusted to different concentrations. Then the copper slag powder was dispersed in the acidic aqueous solution. Ultrasound was irradiated with 26 kHz frequency with 600 W output. Through the experiments, the ultrasound bath was controlled to 60 °C to avoid a temperature increase caused by ultrasonic heating. During the processing of ultrasonic treatment of copper slag under various acid solution conditions, some cases produced a white gel in the US treatment process, especially in H_2SO_4 cases. In other cases, the leaching solution gelled after the US treating process was stopped. The results are represented by the time variation of the percent silica dissolved in the acid solution (Si LR (%)) and precipitated silica (SW (%)) (Figure 9). The experiment was conducted with ultrasonic irradiation for 120 min.

Figure 9. Time change of Si LR (%) and SW (%) in the leaching solution for (a) HCl 6M, (b) HCl 12M, (c) H_2SO_4 3M, and (d) for H_2SO_4 6M. Sampling was conducted during US exposure from the initial time to 120 min and at 60 min after stopping US until 180 min

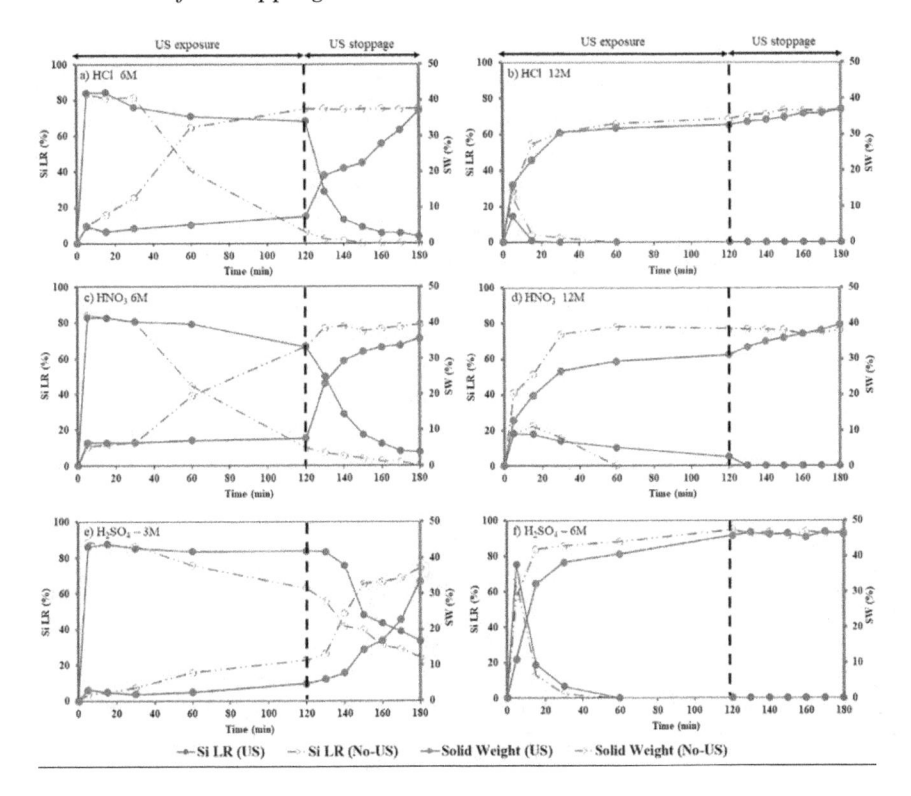

Then irradiation was stopped for 60 min; SiLR% and SW% were measured at each time. Comparison of Si LR% and SW% for each of the 6 M HCl systems shows that the change in Si LR% values indicates that the solubility of silica in US irradiation is about 70–80%. It tends to be higher. By contrast, SW% is less than 10% in each of the US systems. However, when US was stopped at 120 min, corresponding to the decrease in Si LR, the SW value increased considerably to more than 35% of the precipitated whity gels at the end of the 180 min experiment. By contrast, in the case without ultrasonic irradiation (No-US), Si LR % showed a decreasing trend at about 30 min after the start of treatment. The amount of silica dissolved in acidic aqueous solution tended to decrease, whereas SW % increased. In the case of divalent H_2SO_4, the results resembled those of HCl at 3 M, but at 6 M, silica gel sedimentation occurred even when US was manipulated.

The gelatinization and sedimentation of silica occurred with difficulty during ultrasonic irradiation, probably because of the crushing effect of US. However, during sedimentation, iron oxide, which accounts for about 50% of the slag, is aggregated and precipitated. Therefore, there was a strong tendency to obtain silica containing iron oxide when sulfuric acid was used or when US was not used. Finally, after the US was stopped, the precipitated silica gel was collected, washed, and dried. The silica powder (Figure 8) was found to contain 0.04% iron. Silica was recoverable from the copper slag in about 90% yield.

Like copper slag, aluminum hydroxide is discarded as an industrial waste from the anodizing processes of aluminum production. This aluminum (Al) waste has been disposed of by landfilling or other means (Álvarez-Ayuso, 2009; Ribeiro & Labrincha, 2008). Ceramics produced from this waste and incineration fly ash were synthesized into a new structural material using the geopolymer method (Sujitra et al., 2015). The Al waste content in the resultant geopolymer at 40 wt% cured at 80 °C, showing the highest compressive strength of about 40 MPa and also the highest bulk density of about 2.8 g/cm^3. Here, a geopolymer is an inorganic structure resembling cement, produced by mixing alumina silica powder with an alkaline solution. It can be made into ceramics from industrial waste (Davidovits, 2020; Nazari & Sanjayan, 2015), fly ash (Kumar & Kumar, 2011; Sujitra et al., 2016) or clay (Sitthisak et al., 2019) and also from agricultural waste of rice husk ash (Francisco et al., 2014a). Known as "cementless concrete," it is being researched, developed, and put to practical use in the fields of civil engineering and construction as a material that is useful for applications similar to concrete (Figure 10).

Figure 10. Geopolymer sources of Al waste, fly ash, and metakaolin for each geopolymer block and the resulting geopolymers. Pictures of the lower row are for Al waste-fly ash composited geopolymer, porous fly ash geopolymers, and dense metakaolin geopolymer

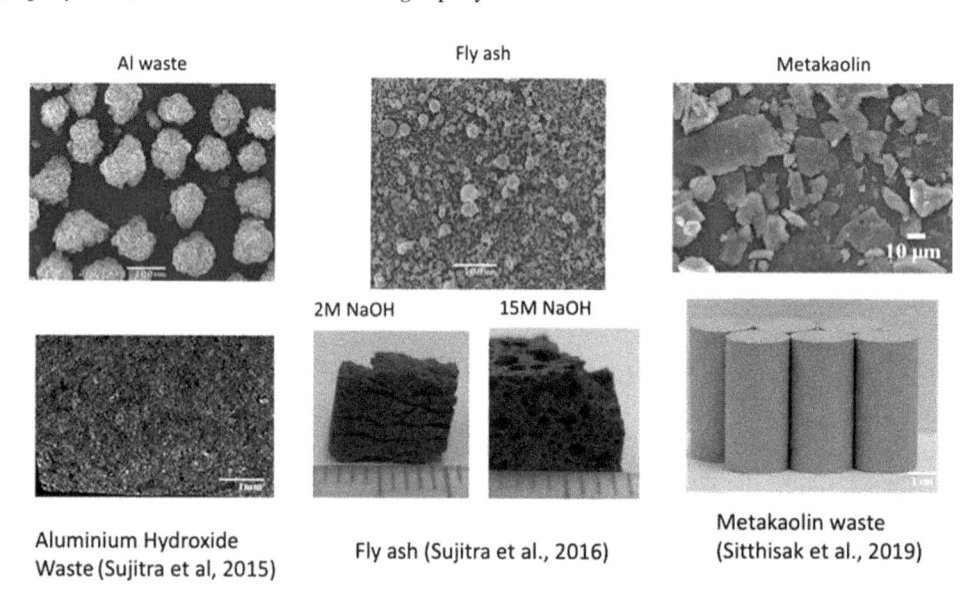

The resulting geopolymers can be made into dense concrete themselves or into porous ceramics by controlling their reaction and fabrication conditions. For example, the fly ash geopolymer in Figure 10 can be heated rapidly using a microwave oven to evaporate water from the slurry and to change the degree of porosity (Sujitra et al., 2016). Pressurization during fabrication along with the preparation process also reduced the amount of alkaline activator and allowed the preparation of dense structures (Sitthisak et al., 2019). Porous geopolymers sourced with fly ash adsorbed lead and copper efficiently (Sujitra et al., 2018). For that reason, they are anticipated for use as adsorbents. Dense geopolymers, which require much energy during conditioning, are anticipated as alternatives to cement. Geopolymers with similar adsorption properties to those of contaminants were reported for cesium (Francisco et al., 2014b) and heavy metals (Francisco et al., 2014c) in inorganic mineral metakaolin-based geopolymers. These properties can also be useful for an immobilizer to confine contaminated heavy metals in the dense matrices (Watanabe et al., 2023) for the production of fly ash geopolymers.

Geopolymers have recently shown promise as an ecological alternative to conventional cement-based materials. Moreover, their use as a composite material for development as a reinforcing additive for organic materials has also been reported, especially by improving their mechanical properties (Davidovits, 2016) with components of fiber (Korniejenko et al., 2016; Ranjbar, 2020), carbon (Růžek et al., 2023), and polymers (Ranjbar, 2016). Composites of geopolymers with other materials can produce composite effects that are not obtainable with geopolymers alone. For these composite materials, examples included self-healing properties of structural materials (Elmesalami, 2022) in civil engineering fields and flexible fiber materials (Sujitra et al., 2019) for use in water remediation.

Environmental Remediation Technology Using Advanced Materials. Although low-carbon technologies that do not pollute the environment are necessary, purification technologies for already polluted soil and water are also important. For example, contaminated soil contains potentially arsenic-contaminated areas, especially in aquifers in India and Bangladesh. These pose potential contamination problems that are specific to the area and which threaten the livelihood of local residents, yet guest soil and other methods are still being used (Abernathy, 2003; Ahmed et al., 2018). Because of strong bonding between contaminated soil and arsenic, soil remediation is difficult to achieve simply by washing with water. Therefore, soil remediation using powerful ultrasound was reported recently (Kobayashi & Tri, 2022).

Figure 11 portrays results of irradiating waste soil containing 9 ppm arsenic discharged from underground tunnel excavation with 1200 W ultrasound in acidic and alkaline water. Under alkaline and acidic conditions, the treatment of waste soil using powerful US grinds the soil, causing it to become finer and leading it to come into contact more easily with alkaline or acid solutions, consequently facilitating the elution of arsenic into the liquid and enabling effective purification.

After cleaning and purifying the soil and extracting contaminants, the next difficulty is disposal of the dissolved contaminants in the supernatant water layer. In the case of arsenic water, it can be treated with a fibrous adsorbent made of limonite minerals complexed with polymers (Le Thi et al., 2023). Similar fibrous adsorbents are visible with zeolite-composite polyethersulfone fibers, which have been adapted for radioactive cesium (Kobayashi et al., 2016; Ohshiro et al., 2017) and for heavy metals (Nakamoto et al., 2017).

Figure 11. Concentration of As residual in the As contaminant soil in (a) supernatant and As residence in the treated soils washed with aqueous (b) NaOH and (c) HCl for using US (300 W– 1200 W). The US exposure was kept for 60 min when experiments (b) and (c) were conducted. The illustration presents US effects of As remediation of the polluted soil

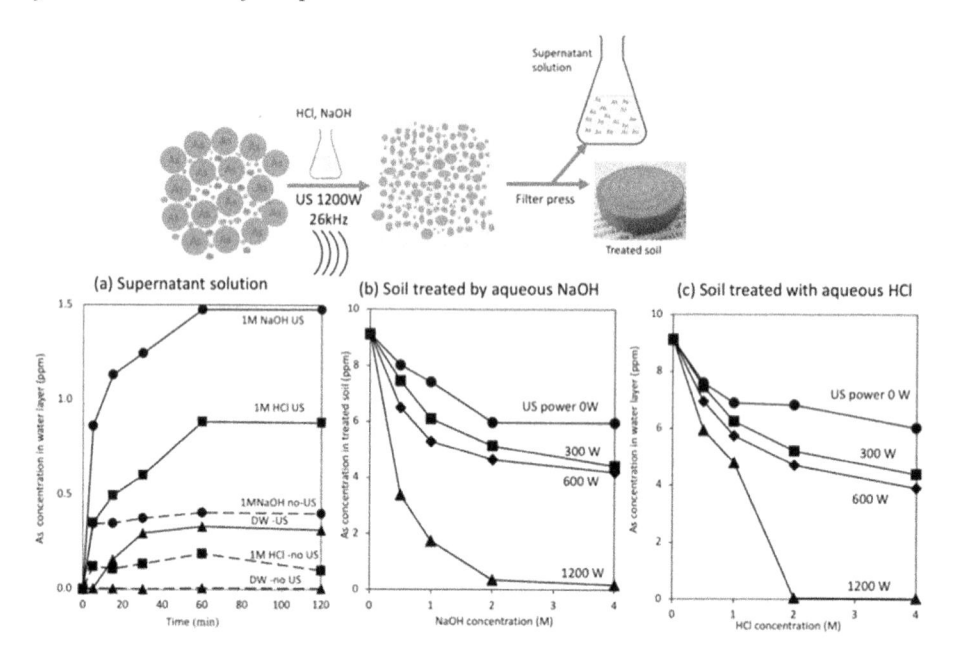

Like arsenic-contaminated water, water remediation processes typically consist of a contamination incident in which significant concentrations of chemical, biological, or radiological agents are found to be present that would adversely affect public health and the environment. Before now, water remediation technologies have been applied using several methods such as coagulation (Teh et al., 2016), filtration (Johnson et al., 2022; Liu et al., 2017), bioremediation (Anekwe & Isa, 2023; Balaet al., 2022), photocatalytic degradation (Kurmar et al., 2021; McCullagh et al., 2011), and adsorption for bacteria (Aryee et al., 2021), pesticides (Bose et al., 2023), and heavy metals (Rajendran et al., 2022). During their remediation processes, the comparative evaluation of both photocatalytic and biodegradation of organic pollutants was particularly emphasized for petrochemical hydrocarbons from various industrial processes released in wastewater (Singh & Borthakur, 2018). Combining those methods of remediation with adsorption (Adeola et al., 2022) was able to remove hazardous organic chemical pollutants from an aqueous medium.

CONCLUSION

This chapter describes the current global situation and efforts to build a sustainable society, with explanations of the contributions to this effort using environmental materials and related advanced technologies. Particularly, the importance of recycling and regenerating of materials using waste and unused resources as starting materials through low-carbon technology is related closely to the conservation of global resources. This chapter also described the importance of maintaining economic development and of improving the lives of human society while living in harmony with nature on Earth by implementing low-carbon technologies.

Specifically, the use of biomass resources and waste as material resources is important because new biomass materials are strong candidates to replace petroleum products. In this case, eco-materials, which are environmentally and human-friendly while retaining their invaluable material functions, are important, but "environmental materials," materials that can be recycled and used as new resources, are also necessary for technological development. Additionally, reusing remediation products of contaminated soil and water using remediation technologies is crucially important. More efficient and nature-friendly technologies must be pursued. The findings presented herein demonstrate that, instead of discarding unused materials and waste, they should be turned into resources and more valuable upcycling must be specifically examined in future studies.

ACKNOWLEDGEMENTS

Author sincerely thank Onahama Smelting and Refining Co. LTD for providing copper slag samples and photographs and also thank Professor Shuji Uchida of Fukushima National College of Technology for his support of related experiments.

REFERENCES

Abernathy, C. O., Thomas, D. J., & Calderon, R. L. (2003). Health Effects and Risk Assessment of Arsenic. *The Journal of Nutrition*, 133(5), 1536S–1538S. DOI: 10.1093/jn/133.5.1536S PMID: 12730460

ACS. https://www.acs.org/sustainability/chemistry-sustainable-development-goals.html

Adeola, A. O., Abiodun, B. A., Adenuga, D. O., & Nomngongo, P. N. (2022). Adsorptive and photocatalytic remediation of hazardous organic chemical pollutants in aqueous medium: A review. *Journal of Contaminant Hydrology*, 248, 104019. DOI: 10.1016/j.jconhyd.2022.104019 PMID: 35533435

Ahmed, M., Matsumoto, M., & Kurosawa, K. (2018). Heavy Metal Contamination of Irrigation Water, Soil, and Vegetables in a Multi-Industry District of Bangladesh. *International Journal of Environmental Research*, 12(4), 531–542. DOI: 10.1007/s41742-018-0113-z

Althoey, F., Zaid, O., Majdi, A., Alsharari, F., Alsulamy, F., & Arbili, M. M. (2023). Effect of fly ash and waste glass powder as a fractional substitute on the performance of natural fibers reinforced concrete. *Ain Shams Engineering Journal*, 6(12), 102247. DOI: 10.1016/j.asej.2023.102247

Álvarez-Ayuso, E. (2009). Approaches for the treatment of waste streams of the aluminium anodising industry. *Journal of Hazardous Materials*, 164(2–3), 409–414. DOI: 10.1016/j.jhazmat.2008.08.054 PMID: 18823703

Anastas, P. T., & Williamson, T. C. (1996). Green chemistry: An overview. ACS (Am Chem Soc). *Symp. Ser.*, 626, 1–17.

Andrade, J. M. M., de Jong, E. V., & Henriques, A. T. (2014). Byproducts of orange extraction: Influence of different treatments in fiber composition and physical and chemical parameters. *Brazilian Journal of Pharmaceutical Sciences*, 50(3), 473–482. DOI: 10.1590/S1984-82502014000300005

Andrady, A. L. (2011). Microplastics in the marine environment. *Marine Pollution Bulletin*, 62(8), 1596–1605. DOI: 10.1016/j.marpolbul.2011.05.030 PMID: 21742351

Anekwe, I. M. S., & Isa, Y. M. (2023). Bioremediation of acid mine drainage – Review. *Alexandria Engineering Journal*, 65, 1047–1075. DOI: 10.1016/j.aej.2022.09.053

Aryee, A. A., Mpatani, F. M., Han, R., Shi, X., & Qu, L. (2021). A review on adsorbents for the remediation of wastewater: Antibacterial and adsorption study. *Journal of Environmental Chemical Engineering*, 9(6), 106907. DOI: 10.1016/j.jece.2021.106907

Baidurah, S., Kobayashi, T., & Aziz, A. A. (2022). PLA Based Plastics for Enhanced Sustainability of the Environment, *Encyclopedia of Materials: Plastics and Polymers*, 2, 511-519. DOI: 10.1016/B978-0-12-820352-1.00175-9

Bala, S., Garg, D., Thirumalesh, B. V., Sharma, M., Sridhar, K., Inbaraj, B. S., & Tripath, M. (2022). Recent Strategies for Bioremediation of Emerging Pollutants: A Review for a Green and Sustainable Environment. *Toxics*, 10(8), 484. DOI: 10.3390/toxics10080484 PMID: 36006163

Behnood, A., Gharehveran, M. M., Gozali, F., & Ameri, A. M. (2015). Effects of copper slag and recycled concrete aggregate on the properties of CIR mixes with bitumen emulsion, rice husk ash, Portland cement and fly ash. *Construction & Building Materials*, 96, 172–180. DOI: 10.1016/j.conbuildmat.2015.08.021

BenDor, T. K., Riggsbee, A. J., & Doyle, M. (2011). Risk and markets for ecosystem services. *Environmental Science & Technology*, 45(24), 10322–10330. DOI: 10.1021/es203201n PMID: 22044319

Bose, S., Kumar, P. S., Rangasamy, G., Prasannamedha, G., & Kanmani, S. (2023). A review on the applicability of adsorption techniques for remediation of recalcitrant pesticides[[Circular economy:]. *Chemosphere*, 313, 13748. https://environment.ec.europa.eu/topics/circular-economy_en. DOI: 10.1016/j.chemosphere.2022.137481 PMID: 36529165

Chang, S. H. (2014). An overview of empty fruit bunch from oil palm as feedstock for bio-oil production. *Biomass and Bioenergy*, 62, 174–181. DOI: 10.1016/j.biombioe.2014.01.002

Costa, S. M., Costa, S. A., Pahl, R., Mazzola, P. G., Marcicano, J. P. P., & Pessoa, A.Jr. (2013). Textile Fiber Produced From Sugarcane Bagasse Cellulose: An Agro-Industrial Residue. *International Journal of Textile and Fashion Technology*, 3(2), 15–28.

Dave, P., Desai, F., & Tailor, H. (2014). Energy Conservation in Bagasse Fired Boiler. *International Journal on Recent and Innovation Trends in Computing and Communication*, 2(9), 2742–2745.

Davidovits, J. (2016). Reinforced Geopolymer Composites: A critical review, *Materials today*. https://www.materialstoday.com/polymers-soft-materials/features/reinforced-geopolymer-composites-a-critical-review/

Davidovits, J. (2020). *Geopolymer Chemistry and Applications* (5th ed.). Institut Géopolymère., www.geopolymer.org

de Souza, P. P. L., Eires, R., & Malheiro, R. (2023). Sugarcane Bagasse as Aggregate in Composites for Building Blocks. *Energies*, 16(1), 398. DOI: 10.3390/en16010398

Dienl, Q., & Anh, T. K. (2021). Nanocellulose Preparation from Cassava Bagasse via Hydrolysis by Sulfuric Acid and Hydrogen Peroxide Medium. *Journal of the Japan Institute of Energy*, 100(8), 135–143. DOI: 10.3775/jie.100.135

EIA WTE. 2023: https://www.eia.gov/todayinenergy/detail.php?id=55900

Elmesalami, N., & Celik, K. (2022). A critical review of engineered geopolymer composite: A low-carbon ultra-high-performance concrete. *Construction & Building Materials*, 346, 128491. DOI: 10.1016/j.conbuildmat.2022.128491

EPA Environment topics: https://www.epa.ie/environmental-topics/

EPA recycle: https://www.epa.gov/recycle

Erlich, C., & Fransson, T. H. (2011). Downdraft gasification of pellets made of wood, palm-oil residues respective bagasse: Experimental study. *Applied Energy*, 88(3), 899–908. DOI: 10.1016/j.apenergy.2010.08.028

European Green Deal. 2023: https://ec.europa.eu/commission/presscorner/detail/%20en/ip_23_3565

FAO. 2022: https://www.fao.org/3/cb6492en/cb6492en.pdf

Fauzi, A., Nuruddin, M. F., Malkawi, B., & Bakri Abdullah, M. M. A. (2016). Study of Fly Ash Characterization as a Cementitious Material. *Procedia Engineering*, 148, 487–493. DOI: 10.1016/j.proeng.2016.06.535

Favier, V., Chanzy, H., & Cavaille, J. Y. (1995). Polymer nanocomposites reinforced by cellulose whisker. *Macromolecules*, 28(18), 6365–6367. DOI: 10.1021/ma00122a053

Francisco, J. L., Sugita, S., & Kobayashi, T. (2014a). Cesium-adsorbent Geopolymer Foams Based on Silica from Rice Husk and Metakaolin. *Chemistry Letters*, 44(1), 128–130.

Francisco, J. L., Sugita, S., Tagaya, M., & Kobayashi, T. (2014b). Geopolymers Using Rice Husk Silica and Metakaolin Derivatives; Preparation and Their Characteristics. *Journal of Materials Science and Chemical Engineering*, 2(5), 35–43. DOI: 10.4236/msce.2014.25006

Francisco, J. L., Sugita, S., Tagaya, M., & Kobayashi, T. (2014c). Metakaolin-Based Geopolymers for Targeted Adsorbents to Heavy Metal Ion Separation. *Journal of Materials Science and Chemical Engineering*, 2, 35–43.

Freitas, C. M. P., Coimbra, J. S. R., Souza, V. G. L., & Sousa, R. C. S. (2021). Structure and Applications of Pectin in Food, Biomedical, and Pharmaceutical Industry: A Review. *Coatings*, 11(8), 922. DOI: 10.3390/coatings11080922

Frnández, L. (2023): https://www.statista.com/statistics/1122082/europe-waste-to-energy-capacity-by-country/

Gargul, K., Boryczko, B., Bukowska, A., Jarosz, P., & Małecki, S. (2019). Leaching of lead and copper from flash smelting slag by citric acid. *Archives of Civil and Mechanical Engineering*, 19(3), 648–656. DOI: 10.1016/j.acme.2019.02.001

Gond, R. K., Gupta, M. K., & Jawaid, M. (2021). Extraction of nanocellulose from sugarcane bagasse and its characterization for potential applications. *Polymer Composites*, 42(10), 5400–5412. DOI: 10.1002/pc.26232

Gorai, B., & Premchand, R. K. J. (2003). Characteristics and utilization of copper slag - a review. *Resources, Conservation and Recycling*, 39(4), 299–313. DOI: 10.1016/S0921-3449(02)00171-4

Gotoh, S., Kitaguchi, K., & Yabe, T. (2021). Involvement of the complex polysaccharide structure of pectin in regulation of biological functions. *Reviews in Agricultural Science*, 9(0), 221–232. DOI: 10.7831/ras.9.0_221

He, M., Wang, X., Wang, Z., Chen, L., Lu, Y., Zhang, X., Li, M., Liu, Z., Zhang, Y., Xia, H., & Zhang, L. (2017). Biocompatible and Biodegradable Bioplastics Constructed from Chitin via a "Green" Pathway for Bone Repair. *ACS Sustainable Chemistry & Engineering*, 5(10), 9126–9135. DOI: 10.1021/acssuschemeng.7b02051

Hoefler, A. C. (1991). CHAPTER 3 - Other Pectin Food Products. In Walter, R. H. (Ed.), *The Chemistry and Technology of Pectin* (pp. 51–66). Academic Press., DOI: 10.1016/B978-0-08-092644-5.50008-X

Hongrattanavichit, I., & Aht-Ong, D. (2020). Nanofibrillation and characterization of sugarcane bagasse agro-waste using water-based steam explosion and high-pressure homogenization. *Journal of Cleaner Production*, 277, 123471. DOI: 10.1016/j.jclepro.2020.123471

Honolulu strategty, 2012: https://wedocs.unep.org/bitstream/handle/20.500.11822/12079/brochure-microplastics.pdf?sequence=1&amp%3BisAllowed=

Humerez-Flores, J. N., Verkempinck, S. H. E., Loey, A. M. V., Moldenaers, P., & Hendrickx, M. E. (2022). Targeted modifications of citrus pectin to improve interfacial properties and the impact on emulsion stability. *Food Hydrocolloids*, 132, 107841. DOI: 10.1016/j.foodhyd.2022.107841

In Sugarcane-Based Biofuels and Bioproducts (Ed.). (2016). *(s):Ian M. O'Hara, Sagadevan G. Mundree.* Wiely., DOI: 10.1002/9781118719862.ch10

Iresha, H., & Kobayashi, T. (2021). Ultrasound-triggered nicotine release from nicotine-loaded cellulose hydrogel. *Ultrasonics Sonochemistry*, 78, 105710. DOI: 10.1016/j.ultsonch.2021.105710 PMID: 34411843

Ishida, H., Ozao, R., Utsumi, T., Shinohara, Y., Halada, K., & Nishimoto, Y. (2009). Trends in eco-materials and products as observed through studies on a web database, eco-MCPS. *Transactions of the Materials Research Society of Japan*, 34(3), 249–252. DOI: 10.14723/tmrsj.34.249

Ito, M., Suzuki, M., & Mifune, N. (1999). Eco-materials in railway-Recycling and chemical analysis. *QR of RTRI*, 40(4), 199–203. DOI: 10.2219/rtriqr.40.199

Jiang, H., & Kobayashi, T. (2017). Ultrasound stimulated release of gallic acid from chitin hydrogel matrix. *Materials Science and Engineering C*, 75, 478–486. DOI: 10.1016/j.msec.2017.02.082 PMID: 28415488

Jiang, J., Tovar-Carrillo, K., & Kobayashi, T. (2016). Ultrasound stimulated release of mimosa medicine from cellulose hydrogel matrix. *Ultrasonics Sonochemistry*, 32, 398–406. DOI: 10.1016/j.ultsonch.2016.04.008 PMID: 27150786

Johnson, S. A., Chen, S., Bolton, G., Chen, Q., Lute, S., Fisher, J., & Brorson, K. (2022). Virus filtration: A review of current and future practices in bioprocessing. *Biotechnology and Bioengineering*, 119(3), 743–761.

Kobayashi, T. (2015). Fabrication of Cellulose Hydrogels and Characterization of Their Biocompatible Films, *Studies in Natural Products Chemistry*, Chapter 1, 45, 8-15, Elsevier.

Kobayashi, T., & Nakajima, L. (2021). Sustainable development goals for advanced materials provided by industrial wastes and biomass sources. *Current Opinion in Green and Sustainable Chemistry*, 28, 100439. DOI: 10.1016/j.cogsc.2020.100439

Kobayashi, T., Ohshiro, M., Nakamoto, K., & Uchida, S. (2016). Decontamination of Extra-Diluted Radioactive Cesium in Fukushima Water Using Zeolite–Polymer Composite Fibers. *Industrial & Engineering Chemistry Research*, 55(25), 6996–7002. DOI: 10.1021/acs.iecr.6b00903

Kobayashi, T., & Tri, P. P. (2022). Effect of High-Power Ultrasound Washing on Arsenic-Polluted Soil. *Journal of Chemical Engineering of Japan*, 55(9), 307–315. DOI: 10.1252/jcej.22we027

Komesu, A., Oliveira, J. A. R., Martins, L. H. S., Maciel, M. R. W., & Filho, R. M. (2017). Lactic acid production to purification: A review. *BioResources*, 12(2), 4364–4383. DOI: 10.15376/biores.12.2.Komesu

Koonaphapdeelert, S., Aggarangsi, P., & Moran, J. (2020). *Biogas Cleaning and Pretreatment, 17-45. Biomethane: Production and Applications*. Springer. DOI: 10.1007/978-981-13-8307-6

Korniejenko, K., Frączek, E., Pytlak, E., & Adamski, M. (2016). Mechanical Properties of Geopolymer Composites Reinforced with Natural Fibers. *Procedia Engineering*, 151, 388–393. DOI: 10.1016/j.proeng.2016.07.395

Kumar, A. S. P., Vo, D., Yaashikaa, P. R., Karishma, S., Jeevanantham, S., Gayathri, B., & Bharathi, V. D. (2021). Photocatalysis for removal of environmental pollutants and fuel production: A review. *Environmental Chemistry Letters*, 19(1), 441–463. DOI: 10.1007/s10311-020-01077-8

Kumar, S., & Kumar, R. (2011). Mechanical activation of fly ash: Effect on reaction, structure and properties of resulting geopolymer. *Ceramics International*, 37(2), 533–541. DOI: 10.1016/j.ceramint.2010.09.038

Le Thi, A. P., Wakasugi, R., & Kobayashi, T. (2023). Suppression of hydrogen sulfide generation via coexistence of anaerobic sludge and goethite-rich limonite/Polyethersulfone composite fibers. *ACS Omega*, 8(38), 35054–35065. Advance online publication. DOI: 10.1021/acsomega.3c04540 PMID: 37779981

Lepage, M. L., & Wulff, J. E. (2023). Mixed plastics upcycled dynamically. *Nature*, 26(April). https://www.nature.com/articles/d41586-023-01352-y PMID: 37100936

Leroux, L., Langendorff, V., Schick, G., Vaishnav, V., & Mazoyer, J. (2003). Emulsion stabilizing properties of pectin. *Food Hydrocolloids*, 17(4), 455–462. DOI: 10.1016/S0268-005X(03)00027-4

Li, H., Aguirre-Villegas, H. A., Allen, R. D., Bai, X., Benson, C. H., Beckham, G. T., Bradshaw, S. L., Brown, J. L., Brown, R. B., Cecon, V. S., Curley, J. B., Curtzwiler, G. W., Dong, S., Gaddameedi, S., García, J. E., Hermans, I., Kim, M. S., Ma, J., Mark, L. O., & Sánchez-Rivera, K. L. (2022). Expanding plastics recycling technologies: Chemical aspects, technology status and challenges. *Green Chemistry*, 24(23), 8899–9002. DOI: 10.1039/D2GC02588D

Li, Y., Chai, J., Wang, R., Zhang, X., & Si, Z. (2022). Utilization of sugarcane bagasse ash (SCBA) in construction technology: A state-of-the-art review. *Journal of Building Engineering*, 56, 104774. DOI: 10.1016/j.jobe.2022.104774

Liu, G., Xiao, M., Zhang, X., Gal, C., Chen, X., Liu, L., Pan, S., Wu, J., Tang, L., & Clements-Croome, D. (2017). A review of air filtration technologies for sustainable and healthy building ventilation. *Sustainable Cities and Society*, 32, 375–396. DOI: 10.1016/j.scs.2017.04.011

Makarios-Laham, I., & Lee, C. T. (1995). Biodegradability of chitin- and chitosan-containing films in soil environment. *Journal of Environmental Polymer Degradation*, 3(1), 31–36. DOI: 10.1007/BF02067791

McCullagh, C., Skillen, N., Adams, M., & Robertson, P. K. J. (2011). Photocatalytic reactors for environmental remediation: A review. *Journal of Chemical Technology and Biotechnology*, 86(8), 1002–1017. DOI: 10.1002/jctb.2650

McElroy, C. R., Constantinou, A., Jones, L. C., Summerton, L., & Clark, J. H. (2015). Towards a holistic approach to metrics for the 21st century pharmaceutical industry. *Green Chemistry*, 17(5), 3111–3132. DOI: 10.1039/C5GC00340G

Md Arif Mahmud, M.A., & Anannya, F.R. Sugarcane bagasse - A source of cellulosic fiber for diverse applications. (2021*Heliyon*78e07771DOI: 10.1016/j.heliyon.2021.e07771 PMID: 34458615

Ministry of the Environment. Japan, 2007: https://www.env.go.jp/policy/hakusyo/zu/h20/html/hj0801010302.html

Mistry, R., & Roy, T. K. (2016). Effect of using fly ash as alternative filler in hot mix asphalt. *Perspectives in Science*, 8, 307–309. DOI: 10.1016/j.pisc.2016.04.061

Monkelbaan, J. (2019). *Governance for the sustainable development goals*. Springer Nature Singapore Pte Ltd., DOI: 10.1007/978-981-13-0475-0

Nakamoto, K., Ohshiro, M., & Kobayashi, T. (2017). Mordenite zeolite—Polyethersulfone composite fibers developed for decontamination of heavy metal ions. *Journal of Environmental Chemical Engineering*, 5(1), 513–525. DOI: 10.1016/j.jece.2016.12.031

Nakasone, K., Ikematsu, S., & Kobayashi, T. (2016). Biocompatibility Evaluation of Cellulose Hydrogel Film Regenerated from Sugar Cane Bagasse Waste and Its in Vivo Behavior in Mice. *Industrial & Engineering Chemistry Research*, 55(1), 3–37. DOI: 10.1021/acs.iecr.5b03926

Nakasone, K., & Kobayashi, T. (2016a). Effect of pre-treatment of sugarcane bagasse on the cellulose solution and application for the cellulose hydrogel films. *Polymers for Advanced Technologies*, 27(7), 973–980. DOI: 10.1002/pat.3757

Nakasone, K., & Kobayashi, T. (2016b). Cytocompatible cellulose hydrog Journal of Environmental Chemical Engineering 5 (2017) 513–525 els containing trace lignin. *Materials Science and Engineering C*, 64, 269–277. DOI: 10.1016/j.msec.2016.03.108 PMID: 27127053

Nazari, A., & Sanjayan, J. G. (2015). Synthesis of geopolymer from industrial wastes. *Journal of Cleaner Production*, 99, 297–304. DOI: 10.1016/j.jclepro.2015.03.003

Nguyen, K. D., Kobayashi, T. (2020). Chitin Hydrogels Prepared at Various Lithium Chloride/ N,N-Dimethylacetamide Solutions by Water Vapor-Induced Phase Inversion, *Journal of Chemistry*, 6645351 | https://doi.org/DOI: 10.1155/2020/6645351

OECD. 2022: https://www.oecd.org/environment/plastic-pollution-is-growing-relentlessly-as-waste-management-and-recycling-fall-short.htm

OECD circular economy: https://www.oecd.org/regional/cities/circular-economy-cities.htm

OECD ENV Working Paper, No199:https://www.oecd-ilibrary.org/docserver/1f6dbd07-en.pdf?expires=1693563348&id=id&accname=guest&checksum=6DB82F3E38D0B18F69C36B1B9FAEC617

Ogi, T., Nakanishi, M., & Fukuda, Y. (2011). Gasification of Empty Fruit Bunch and Bagasse Using an Entrained-flow Mode Reactor. *Journal of Japan Institute of Energy*, 90(9), 886–894. DOI: 10.3775/jie.90.886

Ohshiro, M., Kobayashi, T., & Uchida, S. (2017). Fibrous zeolite-polymer composites for decontamination of radioactive waste water extracted from radio-Cs fly ash. *International Journal of Engineering and Technical Research*, 7(4), 1–6.

Oksman, K., Aitomäki, Y., Mathew, A. P., Siqueira, G., Zhou, Q., Butylina, S., Tanpichai, S., Zhou, X., & Hooshmand, S. (2016). Review of the recent developments in cellulose nanocomposite processing. *Composites. Part A, Applied Science and Manufacturing*, 83, 2–18. DOI: 10.1016/j.compositesa.2015.10.041

Palin, R., & Geitmann, A. (2012). The role of pectin in plant morphogenesis. *Bio Systems*, 109(3), 97–402. DOI: 10.1016/j.biosystems.2012.04.006 PMID: 22554809

Panuccio, M. R., Marra, F., Maffia, A., Mallamaci, C., & Muscolo, A. (2022). Recycling of agricultural (orange and olive) bio-wastes into ecofriendly fertilizers for improving soil and garlic quality. *Resources, Conservation & Recycling Advances*, 15, 200083. DOI: 10.1016/j.rcradv.2022.200083

Paper recycling Japan, 2022: http://www.prpc.or.jp/wp-content/uploads/PAPER-RECYCLING-IN-JAPAN-English.pdf

Pavičić, J., Mavar, K.N., Brkić, V., Simon, K. (2022). Biogas and Biomethane Production and Usage: Technology Development, Advantages and Challenges in Europe, *Energies*, 2, 15(8), 2940. https://doi.org/DOI: 10.3390/en15082940

Peresin, M. S., Habibi, Y., Zoppe, J. O., Pawlak, J. J., Rojas, O. J. (2010). Nanofiber Composites of Polyvinyl Alcohol and Cellulose Nanocrystals: Manufacture and Characterization,

Pilgrim, G. W., Walter, R. H., & Oakenfull, D. G. (1991). CHAPTER 2 - Jams, Jellies, and Preserves. In Walter, R. H. (Ed.), *The Chemistry and Technology of Pectin* (pp. 23–50). Academic Press., DOI: 10.1016/B978-0-08-092644-5.50007-8

Plastic upcycling. (2019a), *Nature Catalysis*, 2, 945–946.

Plastic Waste Management Institute. 2019: https://www.pwmi.or.jp/ei/plastic_recycling_2019.pdf

Pohling, J., Hawboldt, K., & Dave, D. (2022). Comprehensive review on pre-treatment of native, crystalline chitin using non-toxic and mechanical processes in preparation for biomaterial applications. *Green Chemistry*, 24(18), 6790–6809. DOI: 10.1039/D2GC01968J

Rainey, T., & Covey, C. (2016). *Pulp and paper production from sugarcane bagasse*. DOI: 10.1002/9781118719862.ch10

Rajendran, S., Priya, A. K., Kumar, P. S., Hoang, T. K. A., Sekar, K., Chong, K. Y., Khoo, K. S., Ng, H. S., & Show, P. L. (2022). A critical and recent developments on adsorption technique for removal of heavy metals from wastewater-A review. *Chemosphere*, 303, 135146. DOI: 10.1016/j.chemosphere.2022.135146 PMID: 35636612

Ranjbar, N., Talebian, S., Mehrali, M., Kuenzel, C., Metselaar, H. S. C., & Jumaat, M. Z. (2016). Mechanisms of interfacial bond in steel and polypropylene fiber reinforced geopolymer composites. *Composites Science and Technology*, 122, 73–81. DOI: 10.1016/j.compscitech.2015.11.009

Ranjbar, N., Talebian, S., Mehrali, M., Kuenzel, C., Metselaar, H. S. C., & Jumaat, M. Z. (2016). Mechanisms of interfacial bond in steel and polypropylene fiber reinforced geopolymer composites. *Composites Science and Technology*, 122, 73–81. DOI: 10.1016/j.compscitech.2015.11.009

Reyes, K. M. D., Bruce, K., & Shetranjiwalla, S. (2023). Green Chemistry, Life Cycle Assessment, and Systems Thinking: An Integrated Comparative-Complementary Chemical Decision-Making Approach. *Journal of Chemical Education*, 100(1), 209–220. DOI: 10.1021/acs.jchemed.2c00647

Ribeiro, M. J., & Labrincha, J. A. (2008). Properties of sintered mullite and cordierite pressed bodies manufactured using Al-rich anodising sludge. *Ceramics International*, 34(3), 593–597. DOI: 10.1016/j.ceramint.2006.12.005

Risdianto, H., Kardiansyah, H. T., & Sugiharto, A. (2016). Empty fruit bunches for pulp and paper production: The current state in Indonesia. *J. Korea TAPPI*, 48(6), 25–31. DOI: 10.7584/JKTAPPI.2016.12.48.6.25

Růžek, V., Dostayeva, A. M., Walter, J., Grab, T., & Korniejenko, K. (2023). Carbon Fiber-Reinforced Geopolymer Composites: A Review. *Fibers (Basel, Switzerland)*, 11(2), 17. DOI: 10.3390/fib11020017

Saelor, S., Kongjan, P., & O-Thong, S. (2017). Biogas Production from Anaerobic Co-digestion of Palm Oil Mill Effluent and Empty Fruit Bunches. *Energy Procedia*, 138, 17–722. DOI: 10.1016/j.egypro.2017.10.206

Said, M., Sitanggang, A. S., Julianda, R. S., Estuningsih, S. P., & Fudholi, A. (2021). Production of methane as bio-fuel from palm oil mill effluent using anaerobic consortium bacteria, Production of methane as bio-fuel from palm oil mill effluent using anaerobic consortium bacteria. *Journal of Cleaner Production*, 282, 124424. DOI: 10.1016/j.jclepro.2020.124424

Sarangi, P. K., Mishra, S., Mohanty, P., Singh, P. K., Srivastava, P. K., Pattnaik, R., Adhya, T. K., Das, T., Lenka, B., Gupta, V. K., Sharma, M., & Sahoo, U. K. (2023). Food and fruit waste valorisation for pectin recovery: Recent process technologies and future prospects. *International Journal of Biological Macromolecules*, 235, 123929. DOI: 10.1016/j.ijbiomac.2023.123929 PMID: 36882142

Sarwani, M. K. I., Fawzi, M., Osman, S. Z., & Nasrin, A. B. (2019). Bio-Methane from Palm Oil Mill Effluent (POME): Transportation Fuel Potential in Malaysia. *Journal of Advanced Research in Fluid Mechanics and Thermal Sciences*, 63(1), 1–11.

Shah, N., Min, B., Sapkota, B., & Rangari, V. K. (2020). Eco-Friendly Cellulose Nanofiber Extraction from Sugarcane Bagasse and Film Fabrication. *Sustainability (Basel)*, 12(15), 6015. DOI: 10.3390/su12156015

Shahab-Ed-Din, G. M. (2021). *Use of Fly Ash as a Pozzolanic Material in Portland Cement Concrete*. Hassell Street Press.

Sharma, A., Thakur, M., Bhattacharya, M., Mandal, T., & Goswami, S. (2019). Commercial application of cellulose nano-composites – A review. *Biotechnology Reports (Amsterdam, Netherlands)*, 21, e00316. DOI: 10.1016/j.btre.2019.e00316 PMID: 30847286

Shi, G., Liao, Y., Su, B., Zhang, Y., Wang, W., & Xi, J. (2020). Kinetics of copper extraction from copper smelting slag by pressure oxidative leaching with sulfuric acid. *Separation and Purification Technology*, 241, 116699. DOI: 10.1016/j.seppur.2020.116699

Siddique, R., Singh, M., & Jain, M. (2020). Recycling copper slag in steel fibre concrete for sustainable construction. *Journal of Cleaner Production*, 271, 122559. DOI: 10.1016/j.jclepro.2020.122559

Singh, P., & Borthakur, A. (2018). A review on biodegradation and photocatalytic degradation of organic pollutants: A bibliometric and comparative analysis. *Journal of Cleaner Production*, 196, 1669–1680. DOI: 10.1016/j.jclepro.2018.05.289

Sitthisak, P., Wannagon, A., Kobayashi, T., & Sirithan, J. (2019). Reaction mechanisms of calcined kaolin processing waste-based geopolymers in the presence of low alkali activator solution. *Construction & Building Materials*, 221, 409–420. DOI: 10.1016/j.conbuildmat.2019.06.116

Situmeang, R., Mazancová, J., & Roubík, H. (2022). Technological, Economic, Social and Environmental Barriers to Adoption of Small-Scale Biogas Plants: Case of Indonesia. *Energies*, 15(14), 5105. DOI: 10.3390/en15145105

Subbotina, E., Montanari, C., Olsén, P., & Berglund, L. A. (2021). Fully bio-based cellulose nanofiber/epoxy composites with both sustainable production and selective matrix deconstruction towards infinite fiber recycling systems. *Journal of Materials Chemistry. A, Materials for Energy and Sustainability*, 10(2), 570–576. DOI: 10.1039/D1TA07758A

Sujitra, O., Kobayashi, T., Parjaree, T., & Sirithan, J. (2018). Removal of Pb^{2+}, Cu^{2+}, Ni^{2+}, Cd^{2+} from Wastewater Using Fly Ash Based Geopolymer as an Adsorbent. *Key Engineering Materials*, 773, 373–378. DOI: 10.4028/www.scientific.net/KEM.773.373

Sujitra, O., Kobayashi, T., Parjaree, T., & Sirithan, J. (2019). Porous fly ash-based geopolymer composite fiber as an adsorbent for removal of heavy metal ions from wastewater. *Materials Letters*, 236, 30–33. DOI: 10.1016/j.matlet.2018.10.035

Sujitra, O., Sirithan, J., Parjaree, T., & Kobayashi, T. (2015). Aluminium hydroxide waste based geopolymer composed of fly ash for sustainable cement materials. *Construction & Building Materials*, 101, 298–308. DOI: 10.1016/j.conbuildmat.2015.10.097

Sujitra, O., Sirithan, J., Parjaree, T., & Kobayashi, T. (2016). Fast microwave syntheses of fly ash based porous geopolymers in the presence of high alkali concentration. *Ceramics International*, 42(8), 9866–9874. DOI: 10.1016/j.ceramint.2016.03.086

Suri, S., Singh, A., & Nema, P. K. (2022). Current applications of citrus fruit processing waste: A scientific outlook. *Applied Food Research*, 2(1), 100050. DOI: 10.1016/j.afres.2022.100050

Tan, T., Wang, W., Zhang, K., Zhan, Z., Deng, W., Zhang, Q., & Wang, Y. (2022). Upcycling Plastic Wastes into Value-Added Products by Heterogeneous Catalysis. *ChemSusChem*, 19(April), e202200522. Advance online publication. DOI: 10.1002/cssc.202200522 PMID: 35438240

Taokaew, S., Thienchaimongkol, J., Nakson, N., & Kobayashi, T. (2022). Valorisation of okara Waste as an Alternative Nitrogen Source in the Biosynthesis of Nanocellulose. *Chemical Engineering Transactions*, 92, 649–654. DOI: 10.3303/CET2292109

Teh, C. Y., Budiman, P. M., Shak, K. Y. P., & Wu, T. Y. (2016). Recent Advancement of Coagulation–Flocculation and Its Application in Wastewater Treatment. *Industrial & Engineering Chemistry Research*, 55(16), 4363–4389. DOI: 10.1021/acs.iecr.5b04703

Thiounn, T., & Smith, R. (2020). Advances and approaches for chemical recycling of plastic waste. *Journal of Polymer Science*, 58(10), 1347–1364. DOI: 10.1002/pol.20190261

Tovar-Carrillo, K. L., Sugita, S. S., Tagaya, M., & Kobayashi, T. (2013). Fibroblast Compatibility on Scaffold Hydrogels Prepared from Agave Tequilana Weber Bagasse for Tissue Regeneration. *Industrial & Engineering Chemistry Research*, 52(33), 11607–11613. DOI: 10.1021/ie401793w

Tri, P. P., Kobayashi, T., & Uchida, S. (2023). Ultrasound effects on restricted silica gelation during silica extraction from Pyro-Metallurgical copper slag under acidifying conditions. *Ultrasonics Sonochemistry*, 97, 106447. DOI: 10.1016/j.ultsonch.2023.106447 PMID: 37245264

Tu Minh, T. V., Kobayashi, T., & Pranut, P. (2022). Viscoelastic Analysis of Pectin Hydrogels Regenerated from Citrus Pomelo Waste by Gelling Effects of Calcium Ion Crosslinking at Different pHs. *Gels (Basel, Switzerland)*, 8(12), 814. DOI: 10.3390/gels8120814 PMID: 36547338

UN environmental programme. 2013: https://www.unep.org/resources/report/microplastics

United Nations SDGs. https://www.un.org/sustainabledevelopment/sustainable- development-goals/

Varshney, S., Mishra, N., & Gupta, M. K. (2021). Progress in nanocellulose and its polymer based composites: A review on processing, characterization, and applications. *Polymer Composites*, 42(8), 3660–3686. DOI: 10.1002/pc.26090

Wang, M., & Ma, D. (2023). Upcycling contaminated plastics. *Nature Sustainability*, 22(June), 1151–1152. Advance online publication. DOI: 10.1038/s41893-023-01148-y

Watanabe, Y., Sirithan, J., & Kobayashi, T. (2023). Lead Immobilized Fly Ash-Based Geopolymer Ceramics Fabricated by Microwave Quick Cure. *Journal of Chemical Engineering of Japan*, 56(1), 2222780. DOI: 10.1080/00219592.2023.2222780

Willats, W. G. T., Knox, J. P., & Mikkelsen, J. D. (2006). Pectin: New insights into an old polymer are starting to gel. *Trends in Food Science & Technology*, 17(3), 97–104. DOI: 10.1016/j.tifs.2005.10.008

Xu, G., & Shi, G. (2018). Characteristics and applications of fly ash as a sustainable construction material: A state-of-the-art review. *Resources, Conservation and Recycling*, 136, 95–109. DOI: 10.1016/j.resconrec.2018.04.010

Xu, S., Han, Z., Yuan, K., Qin, P., Zhao, W., Lin, T., Zhou, T., & Huang, F. (2023). Upcycling chlorinated waste plastics. *Nature Reviews. Methods Primers*, 3(1), 44. DOI: 10.1038/s43586-023-00227-w

Yadav, V., Sarker, A., Yadav, A., Oktarajifa Mifta, A., Bilal, M., & Iqbal, H. M. N. (2022). Integrated biorefinery approach to valorize citrus waste: A sustainable solution for resource recovery and environmental management. *Chemosphere*, 293, 133459. DOI: 10.1016/j.chemosphere.2021.133459 PMID: 34995629

Yamaguchi, I., Itoh, S., Suzuki, M., Sakane, M., Osaka, A., & Tanaka, J. (2003). The chitosan prepared from crab tendon I: The characterization and the mechanical properties. *Biomaterials*, 24(12), 2031–2036. DOI: 10.1016/S0142-9612(02)00633-6 PMID: 12628822

Yee, V. F., Čučekb, L., Klemeša, J. K., Vujanovicb, A., & Varbanov, P. S. (2022). Life Cycle Assessment Approaches of Plastic Recycling with Multiple Cycles: Mini Review. *Chemical Engineering Transactions*, 4, 85–90. DOI: 10.3303/CET2294014

Yeganeh, Y., Chiewchan, N., & Chonkaew, W. (2023). Cellulose nanofiber/polyimide composites for highly-efficient air filters. *Cellulose (London, England)*, 30(7), 4421–4436. DOI: 10.1007/s10570-023-05131-w

Yen, M., Yang, J., & Mau, J. (2008). Antioxidant properties of chitosan from crab shells. *Carbohydrate Polymers*, 74(4), 840–844. DOI: 10.1016/j.carbpol.2008.05.003

Yiin, C. L, Ho, S., Yusup, S., Quitain, A.T., Chan Y. H., Loy, A.C. M., Gwee, Y. L.(2019). Recovery of cellulose fibers from oil palm empty fruit bunch for pulp and paper using green delignification approach,

Yimlamai, B., Choorit, W., Chisti, Y., & Prasertsan, P. (2021). Cellulose from oil palm empty fruit bunch fiber and its conversion to carboxymethylcellulose. *Journal of Chemical Technology and Biotechnology*, 96(6), 1656–1666. DOI: 10.1002/jctb.6689

Yun, D., & Liu, J. (2022). Recent advances on the development of food packaging films based on citrus processing wastes: A review. *Journal of Agriculture and Food Research*, 9, 100316. DOI: 10.1016/j.jafr.2022.100316

Zhao, X., Korey, M., Li, K., Copenhaver, K., Tekinalp, H., Celik, S., Kalaitzidou, K., Ruan, R., Ragauskas, A. J., & Ozcan, S. (2022, January). Korey. M., Li, K., Copenhaver, K., Tekinalp, H., Celik, S., Kalaitzidou, K., Ruan, R., Ragauskas, A.J., Ozcan, S. (). Plastic waste upcycling toward a circular economy. *Chemical Engineering Journal*, 428, 131928. DOI: 10.1016/j.cej.2021.131928

Zuo, Z., Feng, Y., Dong, X., Luo, S., Ren, D., Wang, W., Wu, Y., Yu, Q., Lin, H., & Lin, X. (2022). Advances in recovery of valuable metals and waste heat from copper slag. *Fuel Processing Technology*, 235, 107361. DOI: 10.1016/j.fuproc.2022.107361

Chapter 2
Non-Wooden Cellulose Materials Sourced From Plant Wastes

Karla L. Tovar Carrillo
Universidad Autónoma de Ciudad Juárez, Mexico

Ayano Ibaraki
Nagaoka University of Technology, Japan

Takaomi Kobayashi
Nagaoka University of Technology, Japan

ABSTRACT

Cellulose is the most abundant plant waste material, making it a strong candidate to replace petroleum products as a future polymer material. Since cellulose is also abundant in food wastes, upcycling technology to obtain functional materials from these wastes is reviewed from the perspective of resource recycling. Cellulose, which is particularly difficult to handle as a material, has the property of being insoluble in solvents due to its strong cohesive nature. For this reason, recycled cellulose also is discussed as regenerative celluloses. However, compared to these chemically modified celluloses, the utilization of biomass cellulose fibers obtained from plant waste, which are a less scientific process and inexhaustible, enable to contribute to a sustainable society. The current status and technology of unmodified cellulose fibers is presented. Especially, the properties of cellulose hydrogels, which are agglomerated cellulose, and films are introduced in this chapter.

INTRODUCTION

Humanity's social life and profit-seeking economic society have prioritized affluence and we now face the collapse of the global environment and are being forced to take countermeasures. Under these circumstances, we are now opening the door to economic vitality rooted in a low-carbon and recycling-oriented society. But contrary to this trend, plenty of wastes is produced due to the increase activity of agricultural, food and paper industry representing a tremendous challenge for the environment. Mean-

DOI: 10.4018/979-8-3693-0003-9.ch002

Copyright ©2025, IGI Global. Copying or distributing in print or electronic forms without written permission of IGI Global is prohibited.

while, industrial waste products are considered as the promising and the suitable material to meet the growing demand for feedstock in terms of resource depletion and conservation of environmental resources (Jawaid & Khalil, 2011). Following this social trend, the conversion of currently underutilized wastes into resources is necessary for the economy to coexist in harmony with nature (Kobayashi, et. al 2022). Reflecting the seriousness of the huge annual disposal of petroleum-based plastics, sustainable green materials are becoming an interesting alternative. Therefore, it is essential to reduce the plastic availability of non-biodegradable petrochemical-based materials, and alternatives such as biomass-based materials are attracting attention. So, it is inevitable that biomass increasingly used in the development of biodegradable films would become value-added bioproducts, and the following approaches can be considered. Natural and non-synthetic polymers are attracting attention to overcome environmental problems, and biomass-derived materials will contribute significantly to a low-carbon society. As a material that can replace synthetic plastics derived from petroleum resources, biomass-derived materials are attracting attention now. Among these sources, as a usual biomass material, polysaccharide is well-known and representative natural polymers like cellulose, which is the most abundant source in the biosphere. In particular, cellulose waste, which has high content of raw materials, is a promising biomass alternative source like food waste and agro-waste. For example, several plants waste of bagasse and agro-waste of bamboo, banana, corn stalks, coir, rice and husk and also oil palm waste are mostly popular sources and the cellulose contents in them are abundant in natural cellulose fiber (Joshi et al., 2004, Jawaid & Abdul Khalil, 2011; Kalia et al., 2009). However, cellulose fiber is found with other materials such as lignin, hemicelluloses and pectin and the percentage in each fiber depend on fiber source, variety, harvest conditions and more (Doree, 1947). For the regenerated cellulose processes from such sources, cellulose has stiffness structure with microfibriles having hydrophilic and hydrophobic parts with complexity structures of the partial crystalline (Swatloski et al., 2002). Therefore, the isolation of cellulose fiber from waste products needs intensive treatment. There are several methods for cellulose obtaining from fibers such as chemical treatments, mechanical treatment, and chemo-mechanical treatment. However, the regenerated cellulose process is limited in dissolving cellulose due to the rigid structure of microfibrils, which have hydrophobic portions and a complex structure of partial crystals (Swatlowski et al., 2002). To date, several studies have been conducted on solvents that dissolve cellulose, especially for solvent systems providing an unstable structure of ether, ester, and acetal derivatives like cellulose acetate, cellulose nitrate, and cellulose xanthate (Heinze & Liebert, 2012). Such regenerated celluloses are fibers that are created by chemically dissolving cellulose and then regenerating it into fibers. These fibers are produced by dissolving cellulose in specific solvents and regenerating by precipitating in an aqueous medium. But there are not many reports on unmodified cellulose (Figure 1). In recent years, there has been a trend to reconsider the use of conventional plastics, which cause oceanic problems such as microplastics (Li, 2007) and require huge amounts of energy to produce, however, this is in contrast to the issue of a low-carbon, recycling-based society. In addition, the recent trend toward energy reduction through lighter weight materials is also driving progress in the study of cellulose-added reinforcing materials. For these reasons, Japan's Ministry of the Environment has been focusing on cellulose nanofibers (hereafter referred to as CNF), a plant-derived material (Japan ministry environment, 2021).

Figure 1. Chemical structure of cellulose

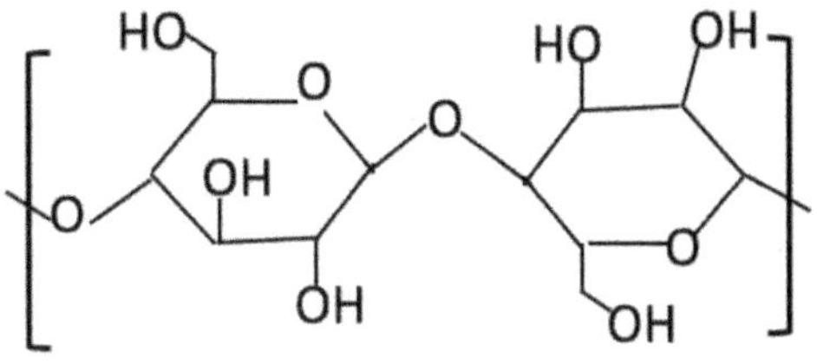

The Ministry of the Environment of Japan is promoting the CNF Performance Evaluation Model Project with the aim of effectively reducing CO_2 emissions through the use of materials that utilize CNF reinforced resin materials (composite resins, etc.) as the foundation of various products, etc., by focusing on CNF (Cai et al. 2014) and micro-crystalline cellulose (MCC) (Majeed et al. 2013). Thus, CNFs and their composites are one of such examples, finding exponentially increasing popularity from a research and development (Kalia, et al. 2011; Trache et al., 2020), but they seem to be still far from being industrially practical at this time.

It has been known that cellulose molecules are natural polymers consisting of linear chains of glucose. Cellulose fiber is a linear, high-molecular-weight homopolysaccharide made up of β-1,4-D-glucose units (Figure 1). Bundles of plant-derived cellulose molecular chains are called cellulose microfibrils (single nanofibers), which have a width of 3-4 nm and are composed of crystalline, associate crystalline, and amorphous portions. In plant tissues, cellulose microfibrils are further bundled into bundles and exist as cellulose fibers of 20 to 40 μm. When cellulose fibers are unraveled to a width of 100 nm or less, which is at the nanoscale, they are called nanocellulose. The technique of defibrillating microfibrils of cellulose fibers at the molecular level is a physical means for nano-cellulose to be insoluble in ordinary solvents. But, in the later year, the direct dissolving process of the regenerated cellulose by using non-derivativing solvents, the cooper salts and concentrated ammonia were effectively cotton dissoluble processes in terms of cuprammonium (Launer & Wilson, 1950) and also alternative ammonia compound as LiCl/DMAc (Striegel, 1997) and metal complexing solvents were used transition metals such as ionic liquid solvents (Heinze & Liebertm, 2012).

On the other hand, sources of cellulose fiber include woody component like pulp and non-woody sources such as agro-products like natural cellulose fibers including cotton and lignin (Batra, S. K, 1985), but the key applications of cellulose fibers include spun yarn, clothing, and fabrics. Among cellulose sources, food waste is currently mostly incinerated or left outdoors after disposal and these food wastes contain less impurities like sugar, lignin, and proteins, and are known to have high cellulose content. Since several approaches to regenerate such wastes to cellulose fibers were reported for scaffolds elaboration from their source, establishment of technology for wastes upcycling toward value-added materials and products becomes important. In these trends using cellulose materials, there is very little movement to convert food waste into ADVANCED MATERIALS and If unused waste and food waste can be utilized as a resource, it is a material that is in line with a sustainable society. For example, the extracted cellulose was used for producing sustainable regenerated textile fibers (Costa et al., 2013). In films, regenerated cellulose fibers attracted more attention to produce the forms used for food and medicine packaging (Klemm et al., 2005) due to their properties of biodegradable, highly transparent, flexible, and low-cost thin gas barrier (Yang et al., 2013). Also, refining and using food-based waste like agave and sugarcane bagasse (Tovar-Carrillo et al., 2014; Nakasone & Kobayashi, 2016) and thinning bamboo agent (Tovar-Carrillo et al., 2013) were reported as cellulose sources since 2013 (Kobayashi, 2015). Therefore, as considered as renewable organic polymeric source was produced in the biosphere (Wooding, 2001), cellulose is a material with great potential for raw material used in the industry widely. Unfortunately, less than 2% was recovered industrially, representing an important waste against disposal problem. So, the re-use of agro-industrial waste is needed, because of the serious economic and environmental problems caused by disposal of these resources.

General Properties of Various Plant Wastes

In order to evaluate several waste fiber properties many studies have been carried out reporting isolation and characterization (Abdul Khalil et al., 2010; Abba et al., 2013; Namvar et al., 2014). Table 1 shows properties of several waste fibers and Figure 2 shows pictures of plant waste and their SEM images for sugarcane, palm leaf, oil palm empty fruit bunch (EFB), coconut and agave.

Table 1. Properties of some waste fibers

Fiber	Fiber length (mm)	Width fiber (μm)	Diameter (μm)	Density g/cm³	Moisture content (%)	UTS (MPa)	Modulus (GPa)
Cotton	3.00-3.50	60.00-110		1.5		500-880	0.05
Jute	3.00-3.60	70.00-120	200	1.45	12	460-533	2.5-13
Coir	2.00-3.00	14.0-17.65	100-450	1.15	10~12	131-175	4~6
Banana	0.90-4.00	80.00-250.00	80-250	1.35	1-~12	529-754	7.7-20.8
Sisal	0.85-1.00	100-300	50-200	1.45	11	568-640	9.4-15.8
Conifer	3.10-3.60	65.00-120		1.5		1100	100
Kraft fiber	1.25-1.60	19.56-20.10		1.54		1000	40
Sunhemp	0.85-1.00	100-250	48	0.673		200-300	2.68
Pineapple	3.00-9.00	20.00-80.00	20-80	1.44		413-1627	34.5
Palmleaf	1.25-1.74	19.58-23.10	240			98.14	2.22
Kenaf	0.66-0.82	17.70-26.70	200	1.47		157.38	12.62
Kusha grass	0.70-0.80	16.03-21.58	390			150.59	5.69

Figure 2. Pictures of plant waste and their SEM images for sugarcane, palm leaf, EFB, coconut and agave

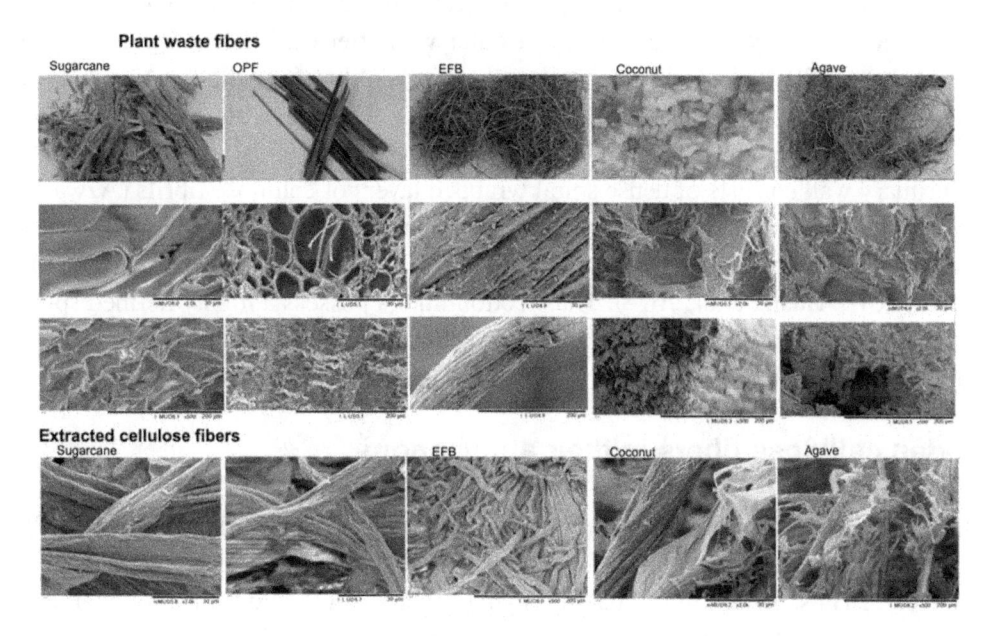

Such waste fibers of plant origin clearly show cellulose microfibrils of about several tens of micrometers, which are aggregates of cellulose molecular chains. The structure differs depending on the waste type. In sugarcane, the cellulose is linearly aggregated, while in agave and EFB, the agglomerated fibrils are already broken up into thick fibers of several hundred μm. This is due to the fact that the fibrillation occurs in high-pressure presses during the squeezing of their extracts. In the fibrous residues remaining after the mechanical chipping in this state, fibrous fibrils are still mixed with impurities such as sugars, lignin, and hemicellulose, and coconut also contains a large amount of oil and fat, which can be observed as slightly flat areas in the SEM photograph.

In contrast, the chemical treatment appears to be enabled to remove such impurities from the cellulose fibers. As seen in SEM pictures, the fiber structure was defibrillated in a twisted form, and eventually, bundles of fine fibers of less than 1 μm were observed to form cellulose fibers of several tens of μm in diameter, consisting of several fibrils that run along the length of the fiber (Kalia et al., 2011; Kalia, et al., 2009; Wooding, 2001).

As summarized in Table I for properties of some waste fibers harvested from different plant source, the structure of cell walls of cellulose, which consist of highly crystalline and regular regions of pure cellulose interspace with small regions of amorphous cellulose. The amorphous cellulose regions usually contain impurities (Kalia et al, 2011; Alberts et al., 2002; Thomas et al, 2011). In addition, the bundles of these cellulose ribbons are called microfibrils. The microfibrils generally vary in width from 0.008 to 0.03 μm and the microfibril bundles are held together by hydrogen bonds formed between the hydroxyl groups of the glucose molecules (Thomas et al., 2011). In the SEM images, extracted cellulose fibers are visible as microfibrils. In the plant wall, the main structural unit of cellulose consists of cellulose microfibrils bonded together by polymer resins like other segments of lignin and hemicellulose. The microfibrils are covalently bonded together by various polymeric sugars and proteins forming fibrils with a diameter around 0.05-0.3 μm and length of 0.15 μm to 20 cm approximately (Wooding, 2001; Nair,

2007; Siro, et. al., 2010). Fibril length and composition is determined by plant species and function as well as waste source for a specific renewable product (Rowell et al., 2000; Satyanarayana, et. al., 2009). The plant cells grow first by forming primary or outer wall after the nucleus has divided during cell division (Osorio et al., 2010). This primary cell wall consists of layered cellulose fibrils. In wood cells, the primary cell wall is formed of matted fibrils resembling matted felt and is usually coated with an outer protective or bonding layer (Batra, 1985; Bledzki & Gassan, 1999). In cotton linters, in contrast to this, the primary wall consists of dense spiral wrapped layers of cellulose fibrils (Azwa et al., 2013) having a waxy cuticle to protect the exposed cell wall. The cotton fibers consist of single individual row of plant cells which are coated by intercellular materials that connect adjoining cell walls together (Cristaldi et al, 2010; Huang et al, 2012). Thus, cotton linters possess a highly defined spiral wrapped pattern of fibrils while the fibrils in wood pulp fibers, which consist mainly of mixture of mature xylem cell, appear almost parallel.

Non-wooden cellulose fibers in their applications.

Since the early 1900s, the need for environment friendly products has been increasing. This has led to a strong trend toward replacing petroleum products with bio-based materials to which the carbon cycle can be adapted. Thus, the use of bio-based polymers, composites, and CNFs, for example, has been gaining momentum. Traditionally, wood pulp has supported the paper industry, which has grown into a large industry (Bajpai, 2014; Reddy, et. al., 2006). However, the supply of wood pulp raw materials is limited, and there is a shift toward the use of recycled pulp. In addition, attempts have been made to convert non-wood pulp into paper (Pardo et al., 2014; Nguong, 2013; Monteiro, et. al., 2011). This is because of the need to protect the global environment from the perspective of resource conservation, non-wood paper is being reevaluated and utilized for various applications. Therefore, utilization of non-wooded materials in the production of paper is one of the most cost-effective and environmentally friendly technologies. In Japan, for example, non-wood paper has been produced and used in the form of Japanese paper since the introduction of paper manufacturing in the United States. Currently, non-wood fibers such as Manila hemp, sisal, flax, and cotton are used for special printing paper such as banknotes, certificates, and certificates, as well as for cigarette paper, teabag paper, electrolytic condenser paper, and various types of filter paper. Non-wood fibers are generally longer and have a greater cell wall thickness than hardwood fibers, making them low-density and bulky, and when made into paper, they have high air permeability (Hara, 1998). In terms of paper strength, such paper has high tearing strength and high strength, and can be used at thinning and low beating. Such non-wooden cellulose fibers offer several properties suitable for diverse industrial applications, for example, in most fibers tensile strengths ranging from 80 MPa for sisal to 938 MPa for ramie. Fiber semi-finished products from non-wood raw materials are widely used in the production of microcrystalline cellulose and cellulose esters in writing, printing, cigarettes, packing, filtering, packaging, sanitary ware, box paper, and more.

Currently, kenaf, ramie, roselle, and flax seed in their polymer matrix composite forms have been studied for their mechanical properties with potential applications in auto parts, mechanical gears, sockets for prostheses, and eco-friendly brakes. The good tensile strength of jute fibers and hemp fibers was utilized in structural applications, construction, sports, domestic products, and auto parts to absorb noises (Abdul Khalil et al., 2010, Abba et al., 2013, Namvar et al., 2014).

Cotton stalks are another important source of raw materials. Cotton is cultivated primarily for textile fibers, and little use is made of the cotton plant stalk. Stalk harvest yield tends to be low and storage can be a problem (Abba et al., 2013; Abdul, el. Al., 2008; Reddy et, al., 2005). Another disadvantage of cotton stalks utilization is uneven distribution of fibers along the stalk induces variable density areas and low resistance to biodegradable agents. The obtained materials exhibited good biocompatibility and bio-integration in the body for cellulose ester (Edgar et al., 2001; Methacanon, et. al., 2010). In the last decade, several studies have reported use of cellulosic natural fibers as a reinforcing agent in biocomposite thermoplastic matrices. Coir, banana, and sisal wastes were used as reinforced polymer composites (Saheb & Jog, 1999, Li et al. 2007). Moreover, the corn husk also was used for biodegradable film, coconut fibers as heat insulator, rice husk in concrete, rice straw and bagasse fiber used as writing and printing papers, and more.

Chemically re-generative treatment of cellulose fibers.

Cellulose is the major component in plants and is an its waste products are excellent for manufacturing many products like regenerate cellulose products of rayon or viscose called commonly, which natural cellulose can be regenerated by xanthogenic acid treatment (Kobayashi et al., 2022; Dungani, et. al., 2014). These are colorless, non-toxic, pure cellulose materials, first trade name cellophane. The raw materials of regenerated cellulose are generally cotton cellulose and wood cellulose, which undergo different manufacturing processes. The process of regeneration is the dissolution treatment of cellulose from solid to solution form, using xanthogenic acid (HS-C(=S) -O-C_2H_5). Such solvation involves the interaction of the hydroxyl groups of the polysaccharide chains of the cellulose with the solvent, which causes the cellulose fibers dissolve. This allowed cellulose to be converted to the xanthogenated form with carbon disulfide in NaOH. The xanthogenated cellulose solution dissolves in the NaOH solution, resulting in a viscous solution as a cellulose solution (Olsson & Westman, 2013; Liu, et. al., 2011; Lloyd, et. al., 1998).

Cellulose solvents are commonly divided in two main categories of derivatizing and non- derivatizing solvents following the key purpose for regenerated cellulose processes (Kobayashi et al., 2023). For the derivatizing process, cellulose is firstly modified example of cellulose acetate, whose process in acetylation changes the hydroxyl groups to acetic anhydride (Sayyed et al., 2019; Lim, et.al., 2021). Historically, the derivatizing process came first with the covalent modification idea. In 1846, cellulose nitrate was reported to prepare a xyloidine by treating starch, sawdust, and cotton with nitric acid. It becomes the nitrocellulose which can be dissolved in ether and ethyl alcohol (Heinze & Liebert, 2012). Viscous also was prepared via dissolution in caustic soda (Wedin et al., 2018). All of these are made from wood pulp or non-wood pulp but high-purity cotton pulp. On the other hand, in the case of bagasse such as plant waste and food waste sticks shown in Figure 2, chemical treatment methods need to be cellulose fiber first.

In order to obtain cellulose from waste fibers chemical treatments are first step. Chemical treatment has been extensively used to remove non-cellulosic compounds in natural fiber, unfortunately this can affect physical and mechanical of natural fibers, implementing with chemical treatment using chlorite bleaching, alkali treatment and acid hydrolysis (Roberts, 1997).

Moreover, approaches on cellulose fibers from rice husk and from jute cellulose (Jahan et al., 2011), have been reported. Examples for the chemical treatments were reported below. The chemical treatment for preparing cellulosic fibers from cellulose plant fibers used sulphuric acid (H_2SO_4) and hydrochloric acid (HCl) were carried out. Also obtaining of cellulose from nanofibers have been reported (Qua et al. 2011; Brinchi et al, 2013). They observed influence on the surface charge and dimensions of the cellulose nanofibers due to chemical treatment with acid hydrolysis. Meanwhile, cellulose fibers were obtained by wet chemical process (Leitner et al. 2007) and, the use of oxygen and peroxide compounds was to remove lignin and other non-cellulosic compounds without degradation of cellulose fibers (Kopania et al, 2012).

Figure 3. Agave fibers treated with different concentrations of NaOCl (Tovar-Carrillo, et al., 2014) and their DMAc/LiCL solution

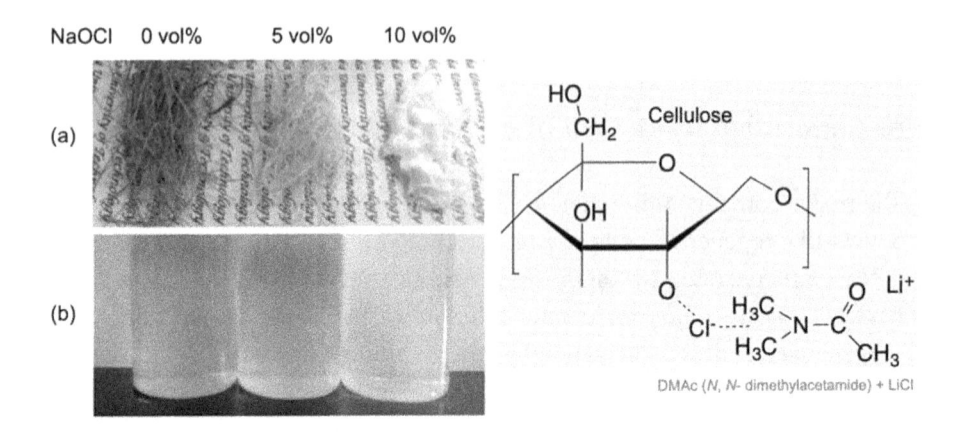

In the cases of food bagasse, several approaches were reported for chemical treatment of agave waste fibers (Tovar-Carrilloet al.,2014), in which lignin was dissolved under alkaline, neutral, or acidic conditions (Nakasone & Kobayashi, 2016b). In the case of acidic method acid sulfite was used with conditions of sulfur dioxide and water at high temperature (Roberts, 1997). The delignification process produced depolymerized cellulose fibers due to acid hydrolysis of glycosidic linkages. Another method consists of alkaline delignification by mixture of sodium sulfide and sodium hydroxide, diminishing crystalline structure of the waste fiber and improving the efficiency of the de-lignification process. During this alkaline process black liquor rich in lignin was obtained and light brown color fibers, bleaching process caused residual lignin from fibers (Figure 3), when used with sodium hypochlorite (NaOCl) (Tovar-Carrillo et al., 2014). In the case of agave bagasse, as shown in Figure 3, the bleaching process with NaOCl changed the color from brown and ochre to white as the amount of bleaching agent increased. After acid and alkali treatment, the non-bleached fiber samples remained slightly brown. The infrared spectra showed that the lignin peaks in the sugarcane sample were almost completely eliminated by this chemical treatment. After the bleaching reaction with chlorine bleach, the sample became white in $NaClO_2$ and gave almost the same spectra as cotton cellulose.

Figure 4 Chemical treatment flow from sugar cane bagasse to cellulose fibers using several chlorine bleaches of NaOCl, NaClO$_2$ and NaClO$_3$ and FT-IR spectra. Sulfuric acid hydrolysis was carried out at 60°C for 1.5 hours, followed by alkali treatment at 10 Wt/V% for 12 hours at 80°C. The resulting non-bleached fibers were then treated with chlorine bleach for 50 hours at 50°C for 3 hours

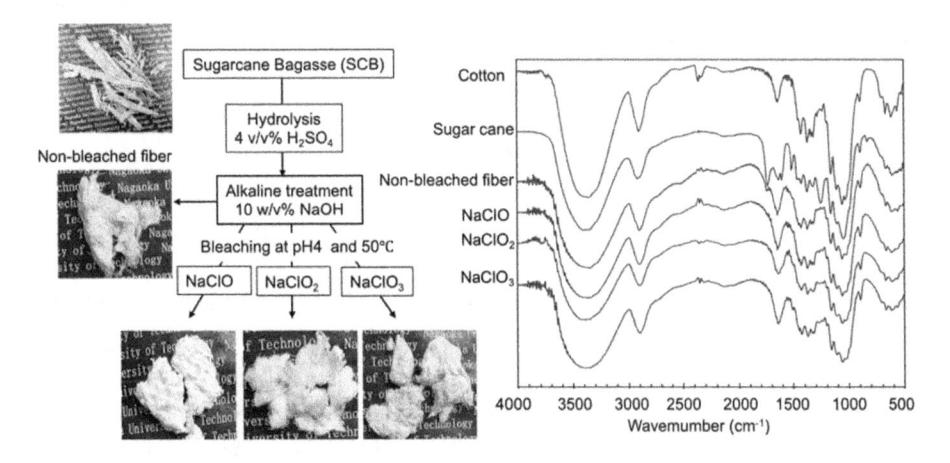

Industrial applications of waste fibers.

Recently, due to necessity of environmentally friendly materials, natural fibers plant waste products offers suitable properties for several industries. The increment of industrial waste products several alternative uses were reported (Izani et al, 2012 & Thiruchitrambalam et al, 2009). Furniture industry increased use of natural fibers as an alternative manufacturing and waste fibers were used for the manufacture of biocomposite advanced (Abdul-Khalil et al, 2012), for example, flax fibers can be used for furniture applications. In other reports, plant fibers such as kenaf, banana, jute and flax has been used as reinforcement. In the last decade, several studies reported use of cellulosic natural fibers as a reinforcing agent in biocomposite thermoplastic matrices; Coir, banana and sisal wastes used as reinforced polymer composites (Saheb & Jog, 1999, Li et al. 2007). Moreover, the corn husk was used for biodegradable film, coconut fibers as heat insulator, rice husk in concrete, rice straw and bagasse fiber used as writing and printing papers, and more.

In the case of automobile industry, waste fibers are the prime candidate for automotive industries. In has been found that by using waste fibers the car weight decreases around 30% and that positive impact on fuel consumption. Recently, biocomposite of polymers reinforced with fibers for manufacturing of seat back, side and door panel, hat rack, boot lining, spare tire lining, business table, dashboard, piller cover panel, instrumental panel and headliner panel have been used (Suddell & Evans, 2005; Bledzki et al, 2006; Holbery & Houston, 2006) among the companies that use those fibers are, Audi, BMW, Fiat, Ford, Renault, Saab, Volvo, Mitsubishi and Peugeot. In addition, Mitsubishi was developed door trim from bamboo composite (Suddell and Evans, 2005; Bledzki et al, 2006; Abdul Khalil, 2012).

Moreover, use of waste fibers as reinforcement of composite materials for packaging increased in the last years. The obtained materials cannot replace completely the properties of synthetic polymers, due to their characteristic hydrophilic behave including permeability to gases and vapor and poor mechanical properties (Johansson et al, 2012; Hirvikorpi et al, 2011 & Siro & Plackett, 2010). However, waste fibers-based materials offers environmental friendly alternative due to their recyclability (Johanson et al, 2012; Majeed et al, 2013). In the last years, several countries have been used waste fibers as composite materials for building applications (Kamble & Behera, 2021; Shakir et al., 2023). Materials such as, doors, panels, roofing sheets and door frames have been elaborated with jute, sisal, and coir. These composite materials produced good insolation and alternative option in low cost. In addition, it has been possible to obtain high transparency films using natural fibers.

Cellulose fibers re-generated from plant wastes

Cellulose solubility. Cellulose solubility is crucial for the obtaining of cellulose materials. Several approaches reported solvent systems for cellulose. Cellulose is hard to dissolve due its stiffness with strong hydrogen bonds linked network. The solvent must be capable to interact with the hydroxyl groups in cellulose fibers, eliminated partially the strong inter-molecular hydrogen-bonding between the polymer chains (Wooding, 2001). However, appropriate solvents systems for cellulose solution obtaining are a key point in order to preserve cellulose fibers properties (Striegel, 1997; Wooding). Table 2 shows several conventional solvent systems for cellulose fibers. Cellulose solubility was reported in acidic condition using sulfuric acid hydrolysis, sulfuric acid destroys the amorphous regions of cellulose fibers (Lindma, et. al, 2010).

Table 2. Conventional and new cellulose solvents

Solvents	Methods and condition	Cellulose	References
NaOH (8-10%)	Direct dissolution 4°C	Treated cellulose, DP 330	Kamide et al. (1984)
NaOH (7-9%)	Direct dissolution	Treated cellulose, DP 330	Yamane et al. (1996)
NaOH (8-9%)	Freeze-thaw	MCC, DP 200	Isogai & Atalla (1998)
NaOH 6%/Urea 12%	Freeze-thaw	Cotton linter DP=690	Zhou & Zhang (2000)
NaOH 7%/Urea 12%	Direct dissolution -10°C	Cotton linter DP=700	Cai & Zhang (2005)
NaOH 9.5%/Thiourea 4.5%	Direct dissolution at -4 °C	Cotton linter DP=620	Ruan et al (2004)
NaOH 9%/PEG 1%	Freeze-thaw	Cellulose powder DP=810	Yan & Gao (2008)
NaOH 12-18% Thiourea 4-6%	Two-step, -2 to 5°C	Cotton linter DP=570 Avicel DP 570	Qi et al (2011)
NMMO	Two-step, 90°C	Cotton linter DP=690	Hon (1997)
DMAc/LiCl	Two-step, room temp	Cotton linter DP=700	Striegel (1997)
DMAc/LiCl	Solvent exchange with water, ethanol, DMAc and DMAc/LiCl	Agave bagasse cellulose Sugar cane bagasse cellulose Commercial cotton products	Tovar-Carrillo et al., 2014 Nakasone et al., 2016 Noguchi and Kobayashi, 2020

DP is degree of polymerization.

On the other hand, alkali systems for cellulose solubility including LiOH and NaOH were reported, although some limitations of NaOH-based aqueous systems were found on dissolving wood pulps fibers (Lindma, et al, 2010; Hon, 1997). In order to obtain cellulose solution, effective destruction of intermolecular hydrogen bonding is essential and critical point for cellulose applications, knowing that intermolecular hydrogen bonding in cellular fiber can be broken by using urea (Tovar-Carrillo, 2014). Unfortunately, urea and NaOH systems could diminish cellulose molecular weight due to the impact on braking intermolecular hydrogen bonding of cellulose structure to allow enhancement water solubility. Alternative method for cellulose solubility was known by using lithium chloride (LiCl)/ *N, N*-dimethylacetamide (DMAc) system (Striegel, 1997). This solvent system has the capability of dissolving cellulose with high polymeric degree like cotton fibers. In LiCl/DMAc system, the activation step is a critical point to allow contact of the solvents with cellulose structure and destroy the hydrogen bonding in the polymer chain by exchange solvent process. In the first step slowed of cellulose fibers in water to lose cellulose structure. The inter-and-intramolecular hydrogen bonds are replaced by hydrogen bonds with water. On the second step, DMAc impedes the re-forming of the inter-and-intramolecular bonds. The solvent system works as solvation of cellulose fibers. Finally, after the activation step, cellulose fibers are ready to dissolve in LiCl/DMAc. It was reported that, concentration of LiCl around 2 and 12% was suitable for cellulose optimal dissolution of cotton and pulp fibers (Tovar-Carrillo et al., 2014). In contrast, LiCl concentrations higher than 12% (12%-15%) results in supersaturated solution and cellulose tended to precipitate. It was found that LiCl concentration in LiCl/DMAc solvent system promote different cellulose fiber arrangements. At lower LiCl concentration the fiber arrangement is mainly lineal, and with the increment of LiCl concentration aggregates shapes were reported (Striegel, 1997; Tovar Carrillo, et al, 2013). In this case, various dissolution mechanisms for cellulose in DMAc/LiCl were proposed as following: 1) $[DMAc_n+Li]^+$ macrocation must exist. 2) In the ion cluster with cellulose, the Cl^- anion is dissociated from the Li^+ cation by intercalating with one or more DMAc molecules. 3) The Li^+ cation interacts with the carbonyl group oxygen of the DMAc molecules. 4) The Cl^- anion disrupting the hydrogen bonds of cellulose can create hydrogen-bond-type interactions with the hydroxyl group hydrogens of cellulose. 5) The macrocation must have weak interactions with the cellulose oxygen. But these processes should be mentioned to be no conclusive evidence. Until today, the interactions between Li^+ cation and the glycosidic oxygen were indicated in the solution system described though semi-empirical MNDO computer models. This type of interaction between cation and various disaccharides in the gas phase (Striegel, 1997; Tovar-Carrillo, 2014).

Film fabrication. Films made from waste-derived cellulose fibers are a much-needed alternative to petroleum-derived plastic films. Regenerated cellulose films prepared from several solutions for example, aqueous alkali/urea solutions exhibited good optical transparency and oxygen barrier properties under dry conditions (Yang et al., 2013). Moreover, amorphous cellulose film was dissolved in regenerated from LiCl (8wt%)/DMAc solution when acetone was coagulated. Cellophane is the mostly produced to cellulose-based film, applying for food packaging due to its transparent, strength and flexibility, making it an attractive natural-based material for food wrapping and adhesive tapping and the derivatives made from wood, cotton, or other sources. This film was obtained by chemically treated processes in alkali solution as called mercerization process (Jewkes et al., 1969) and then, the alkali pulp was continually treated with carbon disulfide to convert to an orange solution known as viscous or cellulose xanthate. To reconvert the viscous solution into cellulose, dilute sulfuric acid and sodium sulfate were prepared in a bath for passing the solution (Venkateshwaran, et. al., 2012; Ververis,et. al., 2004). Various applications were typically coated as nitrocellulose-based waterproof coatings and polyvinylidene chloride-based heat resistance (Wahlang, 2012).

Gelation for cellulose hydrogel films. It was known that cellulose was soluble in DMAc/LiCl, but subsequent material adjustments had to wait until the advent of hydrogel films in 2013 (Tovar-Carrillo et al., 2013). Since cellulose in cellulose-DMAc/LiCl solution (Figure 3) is insoluble in alcohol and water, the presence of these vapors increases the concentration of alcohol and water in the DMAc, causing cellulose to agglomerate and precipitate. A result of such phase inversion from liquid to solid swelling product, hydrogel could be obtained by washing the precipitated transparent gelatinous films having DMAc/LiCl and water (Nakasone and Kobayashi, 2016; Kobayashi, 2018).

Figure 5. Cellulose hydrogel formation via phase inversion process from cellulose liquid solution to gelatinous films. Pictures were hydrogel film of cellulose fibers treated without and with NaOCl bleached agent

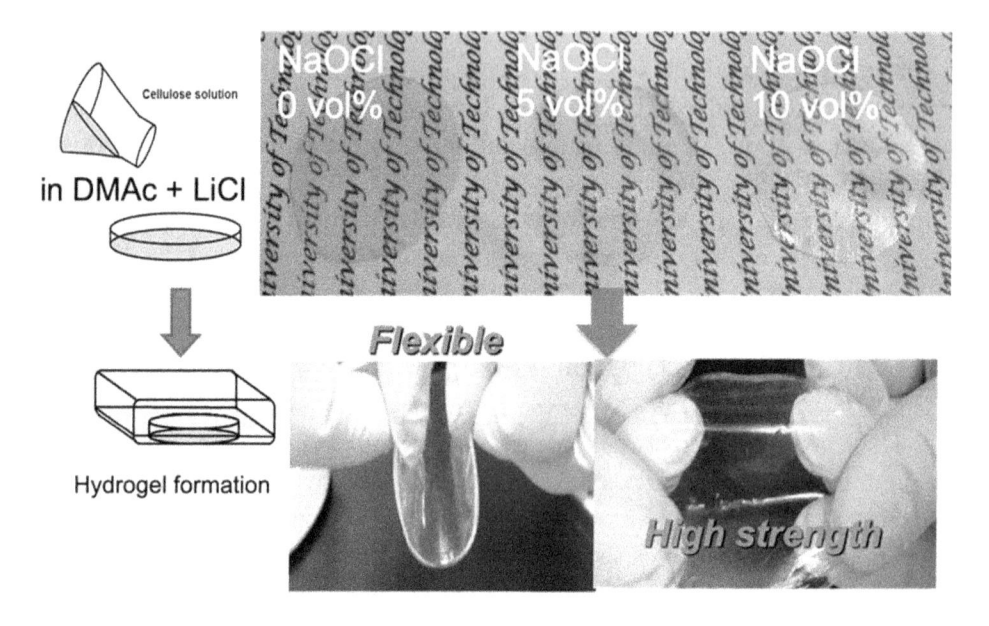

Figure 6 UV-visible absorption spectrum of a film made by heating hydrogel films obtained from NaOCl-treated fibers (as seen in Figure 5) at 150°C-200°C and a photographs of the cellulose films. The absorption spectrum showed that UV light at wavelengths below 350 nm was absorbed by about 60-80%, preventing transmission, while visible light had good transmission performance. The transparency of the cellulose films was ensured, although the film's appearance tended to become slightly cloudy as the heating temperature increased.

Figure 6. Ultraviolet-visible region transmittance spectra of cellulosic films heated at 150°C, 180°C and 200°C applied for hydrogel films and their appearance pictures

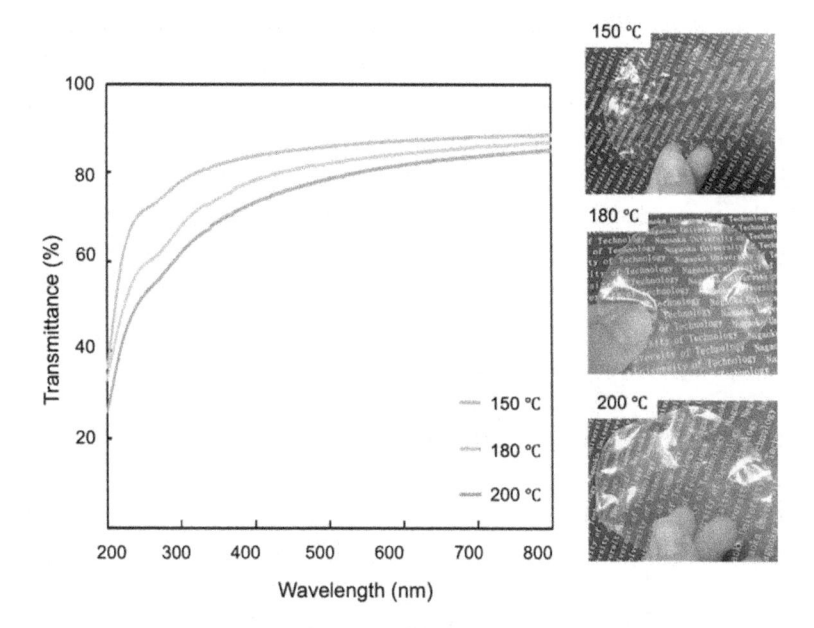

Applications of medical materials elaborated with cellulose obtained from waste fibers.

The main requirement for any material with potential metical application is exhibit long term usage within the body with any adverse or toxic reaction. The material has to be biocompatible for this purpose. Fortunately, the obtained cellulose materials exhibited good biocompatibility and bio-integration in the body for cellulose ester (Edgar et al., 2000). Therefore, cellulose is a suitable candidate to elaborate materials for medical applications due to its nontoxic properties and stable chemically and mechanical as well as completely insoluble under physiological conditions (Bouchard et al. 2006; Dourado et al., 1999) and cellulose polymers are known to have good biocompatibility and wound healing properties like other natural polymers such as chitin, chitosan and alginates (Kimura & Kondo, 2002).

On the other hand, the application of natural polymers for medical applications is limited due to their poor mechanical properties and in some cases, since it is necessary elaborate composite materials combining biopolymers and synthetic polymers (Jiang et al., 2006 ; Henniges et al., 2012). In addition, several materials including cellulose membranes were known to have anti-inflammatory and anti-cancer effects, as reported that hyaluronic acid-carboxymethylcellulose membranes were applied topically for reducing scar formation and surgery (Kim et al., 2013; Kim et al., 2005). Moreover, cellulose membrane has been recognized as a permeation enhancer for drug delivery systems offering a suitable patch material for topical formulations.

Due to cellulose membrane chemical characteristics, it is considered for capsule based controlled drug delivery. Cellulose membranes offered several advantages over conventional capsule materials for uses of controlled release of bulk drug in short time from the membrane (Park et al., 2011; Thombe et al., 1999; Frisbee et al., 2002). Moreover, since the capsule properties were independently modulated without interacting with the core formulation, the obtained capsule was suitable for drug molecules that were difficult and expensive to elaborate, especially sensitive to aqueous environments and elevated temperatures (Kim et al., 2013 ; Kim et al., 2015). But, this provided a smooth, slippery, easily swallowable, and tasteless shell for drug delivery, and these capsules made from regenerated cellulose are commonly used in the commercial fabrication market. Furthermore, cellulosic scaffolds were used for tissue engineering and other medical applications (Digenis et al., 1994; Dahl et al., 1991) and an ideal matrix for large-scale affinity purification procedures (Bussemer & Bodmeier, 2003; Pina & Sousa, 2002), showing chemically inert matrix with excellent physical properties as well as low affinity for nonspecific protein binding suitable for cell adhesion and tissue regeneration.

Cellulose hydrogels for tissue regeneration.

In the last years, several approaches have been conducted to the development of novel hydrogels for tissue regeneration (Langer and Vacanti, 1993) by using natural polymers as elaborated scaffolds for tissue engineering (Sionkowske, 2011; Chen, et al., 2008; Lu et al., 2017). Another reason is that hydrogels are materials that retain large amounts of water, just like living organisms.

Table 3 compares common natural polymers used as scaffolds materials.

Table 3. Summary of commonly employed natural polymers used as scaffolds materials

Main polymer	Properties	Function in wounds	References
Alginate	Wound dressing, Ease application, biodegradability	Wound promotion, Infection control, Haemostatic effect	(Mobed-Miremadi et. Al 2016) Raguvaran et al. 2017)
Chitosan	Wound dressing, In situ gelation, Biodegradability	Healing promotion, Infection control, Anti-inflammatory effect, Haemostatic effect, Delivery of active molecules, Monitoring healing progress	(Balakrishnan et al. 2005, Rocasalbas et. Al. 2013)
Gelatin	In situ gelation, Biodegradability	Delivery of active molecules, Tissue reinforcement, Cell recruitment	(Lee et al. 2014, Jrudi et al. 2015)
Collagen	Wound dressing, In situ gelation, Biodegradability	Delivery of active molecules, Tissue reinforcement, Cell support	(Ribairo et al. 2013, Basu et al. 2017)
Dextran	Skin substitute	Healing promotion, Delivery of active molecules	(Goh et al 2016)
Cellulose	Moisture control	Healing promotion, Cell support	(Wang et al. 2016)
Heparin	Hydrogel sheet, Moisture control	Delivery of active molecules	(Wang et al. 2016)
Glycosaminoglycans	Film dressing, moisture control	Infection control, Tissue reinforcement, Guided tissue regeneration, Promotion of angiogenesis	(Wang et al. 2016, Lu et al. 2017)
Mixed extracellular matrix biopolymers	Sprayable elastic adhesive	Wound closure, Delivery of antimicrobial peptides	(Catanzo et al. 2015, Annabi et al. 2017)

Among them, one strategy has been working with cellulosic polymers, since cellulose natural polymer has easily assimilated in the body due to biocompatibility, hydrophilicity, and biodegradability (Kobayashi & Tovar-Carrillo, 2015). The ultimate purpose of tissue engineering is to replace, repair or enhance the biological function or damage of an organ or tissue. Engineered tissues are produced by using cells that are manipulated through their extracellular environment to allow new tissue formation (Tamada & Ikada, 1993). There are three ways in tissue engineering (Svensson et al. 2005) as followed;

(1) inducing migration of tissue regeneration,
(2) using to encapsulate cells and acting as immuno-isolation barrier,
(3) using as a matrix to support cell growth and cell organization.

Ideals materials of scaffolds should have complex requirements such as no-toxic, biodegradability, appropriate porous structure and mechanical properties. The main characteristic of scaffolds materials is to mimic the extracellular matrix of tissues in the body, both providing support to the cells and having mechanical properties to the tissue. The complex interaction of cells with the extracellular matrix and with neighboring cells is regulated by several reactions. Moreover, extracellular matrix is crucial for cell proliferation, differentiation and death and mediate regeneration and healing process. Base on this, scaffolds materials should provide a stable suitable material for long- term tissue formation without any adverse reaction into the body. Cells do not interact with materials surface directly (Langer, et al. 2006). Cells attach to the absorbed proteins in the scaffold and are important for tissue regeneration.

The implanted material should mimic the environment of the implanted site. After implantation, the first key point is the attachment of water molecules, enhancing protein binding and adsorption (Sionkowske, 2011) and becoming key point of cell adhesion and proliferation, due to cells interact with the adsorbed proteins promoting tissue regeneration. So, hydrogels are suitable scaffolds materials for tissue regeneration for their three dimensional structure. Several approaches have been reported results on tissue regeneration using cellulose hydrogels. Cellulose hydrogels obtained from agave bagasse fibers showed good cyto and biocompatible properties (Tovar-Carrillo, 2013). In addition, cellulose hydrogels exhibited higher cell adhesion number comparing to commercial cell culture assays (Nakasone et al., 2016). Cells adhered to cellulose hydrogels, showing boundaries tightly long shape on the material surface on the first 4 h of cell culture assay (Kobayashi et al., 2015). In the case of skin injuries, severe loss of tissues occurs and scaffolds materials are suitable to intended regenerated tissue due to their three-dimensional structures (Kirker et al., 2002; Raguvaran et al., 2017). Since scaffolds mimic extracellular matrix environment with adequate mechanical and chemical conditions to cell adhesion maintain tissue regeneration for long-termthose, the contribution to mechanical and structural integrity of the surrounded tissue allows nutrients and water transport and facilitated vascularization providing guidance for cell proliferation. In fact, mechanical, strength and stiffness were the most important properties, as well as surface and microstructure (Song et al., 2012; Mishra et al., 2017).

Moreover, scaffolds porosity requirements are dictated by the final application. Pores shape, size and arrangement are important factors to consider. Furthermore, scaffold architecture has an influence on water molecules contact, protein adsorption, cell-cell contact, cell adhesion necessary for cell proliferation and extracellular matrix transport (Gupta, et al., 2008). One important point is cell seeding viability on the scaffold base and source of nutrients and its vascularization ability after implantation to allow cell migration. Scaffolds can be implanted without cells when they are intended to be colonized by cells in short time. Scaffold implantation is determined by surrounded tissue reaction to it, this is mainly a surface manner (Langer et al., 2006; Brumkar et al., 2006). Moreover, protein adsorption is sensitive to mechanical stiffness. Stiffness allows diffusion and permeability of water molecules important for protein adsorption (Amin, et al. 2014). Surface topography also plays an important role on protein adsorption. Nano and micro roughness increase protein adsorption promoting cell adhesion (Tovar-Carrillo, et al., 2013). During the conditioning of the cellulose hydrogel, varying the LiCl in the DMAc changed the cohesiveness of the cellulose segments in the hydrogel, with cellulose oriented side by side at 6 wt% LiCl and a randomly entangled gel at 12 wt% (Figure 7). The growth of fibroblast cells was examined, and it was found that at 6 wt%, the cells grew in the direction of cellulose fiber cohesion, as if they were oriented, and the number of fibroblast cells grew in such a way that they were oriented along the scaffold fibers.

Figure 7. AFM Phase-shift images (upper) of the agave hydrogel films prepared with (a) 6wt% and (b) 12 wt% of LiCl contained DMAc solution and phase-contrast light images (bottom) of their films of (c) 6wt% and (d) 12 wt%, respectively, for 48h in fibroblast cell culture times (Tovar-Carrillo et al., 2013)

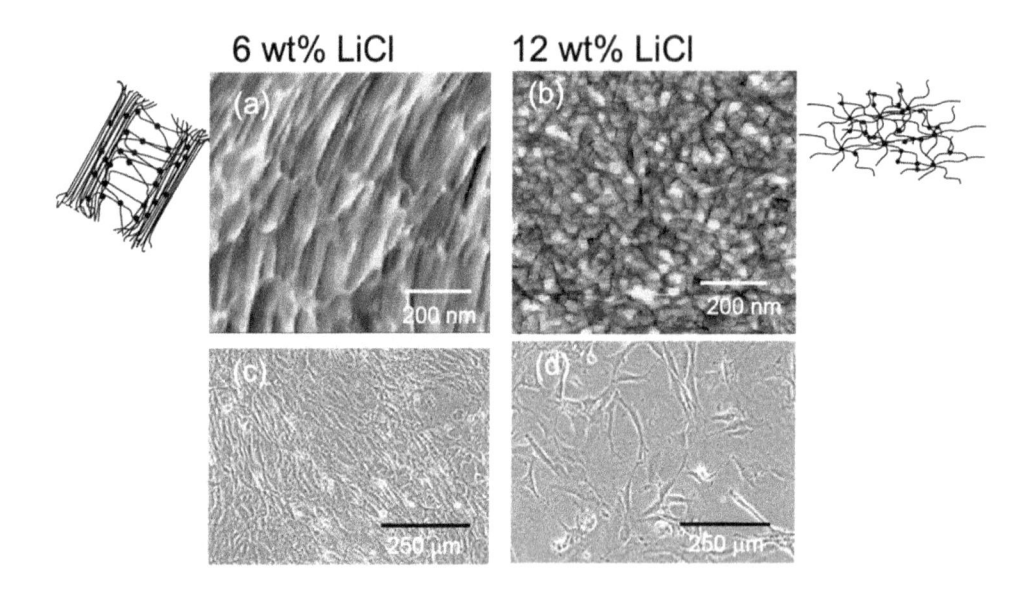

For medical applications, hydrogel and films have been widely used in several fields, such as biomaterials, agriculture, food, and water purification, drug delivery, sensors and smart materials (Gibson, 2005). In pharmaceutical, cellulose has several properties to offer, such as excellent compaction when its blended with other pharmaceutical excipients suitable for oral administration. Cellulose films offered several potentials advantages as a drug delivery excipient (Chen, et al. 2008; Jiang et al., 2016; Iresha & Kobayashi, 2021) and since cellulose films is a low cost, its use provides a substantial environmental advantage compared with other films. Biocompatibility of cellulose hydrogel films regenerated from sugar cane bagasse waste was evaluated by investigating it's *in vivo* behavior in mice (Nakasone at al., 2016). The cellulose hydrogel films were implanted in the intraperitoneal of mice for 4 weeks, showing small influence of the implanted hydrogel films on the growth of mice. It was seen that no inflammation reaction in the intraperitoneal was observed by post-mortem examination, indicating that cellulosic hydrogel films had excellent in biocompatibility.

In drug delivery field, the incorporation of a second component into the film will change the structure and morphology of the network controlling their diffusion properties. It has been reported that cellulose hydrogels provide space and support for cell adhesion and growing, and potential control and function of the engineered tissue in situ, such as; cartilage, bone, muscle, skin, adipose, artery, ligament, tendon, liver, and bladder (Langer & Vacanti, 1993). In addition, cellulose hydrogels can be used in several applications, including drug delivery systems (Sionkowske, 2011), wound healing (Lloyd, et al. 1998). Cellulosic medicines had excellent response to ultrasound trigger to releasing the embedded drug, when ultrasound was exposed to the hydrogel medicines (Figure 8) (Kobayashi, 2023). The enhanced releasing drug by ultrasound was due to soften effect of cellulose hydrogels by breakage hydrogen bonds in the hydrogel under the exposure, but when the ultrasound stopped, the gel fomes self-healing to be original form. Such hydrogel softening and reforming were observed with sono-deviced rheometer (Noguchi & Kobayashi, 2020), causing by hydrogen bond destruction in gels and reforming in the presence and absence of ultrasound.

Figure 8. Illustration of ultrasound response effect of cellulosic hydrogel. Under ultrasound exposure gelation network destroys by hydrogen bond breakage, but, when the exposure stops, the bonds form again

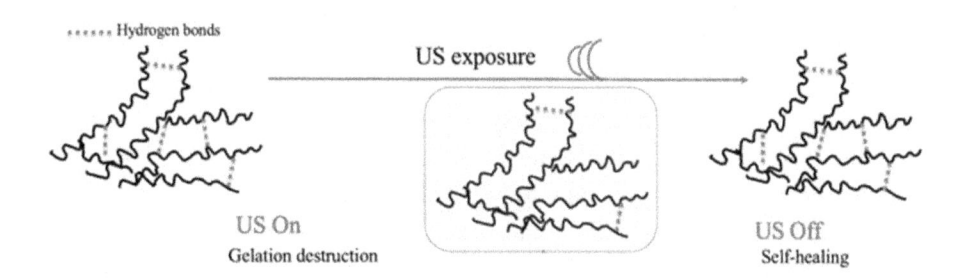

Applications development in current and future.

Materials elaborated from waste fibers can be applied in several fields and industry. In the last years cellulosic materials became a suitable eco-friendly alternative. Regarding to medical field and tissue engineering, cellulose hydrogels are common in non-toxic, water absorbable and reproducibly obtaining flexible. In tissue regeneration, the material interacting with skin cells and enhancing tissue growing at the surface with immunologically inert. It has been found that when the obtaining of the scaffold involves several solvents and compounds traces could remain in the material affecting the biocompatibility of the scaffold. Moreover, cellulosic materials are suitable to implanting in the body (Chandy & Sharma, 1998). In addition, biodegradable nature of cellulose materials became popular, and this property contributes to the construction of a sustainable society as a recyclable material. For example, cellulosic materials can be converted to fertilizer after use. For example, recently, a water-soluble sponge (Figure 9) made from waste cardboard with a binder of agarose was applied to hydroponics. The lettuce cultivation yielded 53 g weight in 42 days, almost equal to the 54 g yield of the commercial urethane sponge. After harvesting, the waste sponge was aerobically fermented in the presence of microorganisms and could be composted.

Figure 9. Hydroponic application and microbial decomposition of cardboard sponges Pictures of lettuce grown on cardboard sponges for 42 days and composted cardboard sponge in microbial presence after harvest

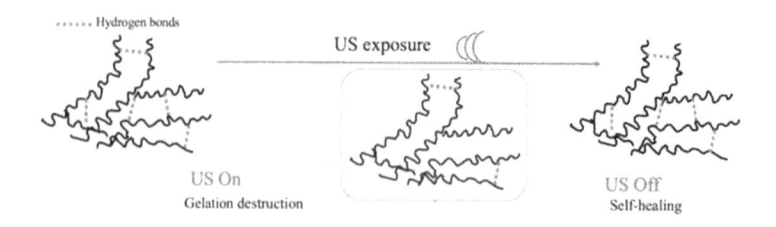

On the other hand, excellent biocompatibility is a practical achievement. In cellulose hydrogel cases, *in vivo* implantation of the hydrogel film in rat model provided excellent results (Nakasone, et al. 2016), showing that non-adverse or inflammation reaction in the peritoneal region after 4 weeks of implantation of the hydrogel film and no cell agglomeration was observed near the hydrogel material. This is because that such cellulose hydrogel is not toxic to the body and showed bio-integration and non-adverse reaction. Several approaches reported results of hydrogels as wound dressings, exudates absorb and moisture retaining for wound healing. In addition, cellulose hydrogels can be applied to solve other needs by regenerative medicines, such as alternative treatments for periodontal and orthodontic, so using for dialyzer membranes for kidney failure (Misha et al. 2017, Meza-valle et al. 2020). Cellulose materials were used as sensing matrix responding to target stimulus, such as pH sensor (Gupta el at. 2008), ammonia sensors (Balakrishnan et al. 2006) and so on. Dialyzer cellulose membranes have the lowest degree of complement with leukocyte activation, inflammatory reactions, and blunting of response of leukocytes (Svensson, et al. 2005), offering an economic alternative for the treatment of this condition. In addition, cellulosic materials had vast applications in medicine detection as sensors (Chen, et al. 2008), pH sensors immobilizing dye on cellulose materials (Svensson, et al. 2005), ammonia sensors (Gibson, 2005), temperature sensors and chemsensors (Sionkowske, 2011). Cellulose materials obtained from waste fibers have a wide number of possibilities on several industrial and medical applications.

CONCLUSION

Due to the growing importance of building a society of sustainability, cellulosic materials made from plant waste have been applied in several industrial sectors in the past few years. In addition, the use of cellulosic fibers, which are abundant as an alternative to petroleum-based plastics, and their industrial diffusion are important factors. In this chapter, cellulose fibers from non-wood plant wastes and un-utilized agro-materials are classified as the main raw materials for obtaining cellulosic materials, and their properties are introduced. The current regenerated cellulose industry was introduced, as well as several papers on the problems of unmodified cellulose fibers and their potential for regenerated cellulose materials. As a distinctive characteristic, the non-toxicity, biocompatibility, and degradability of the obtained cellulosic fibers are described, as well as their economical production, industrialization of recycled cellulose, and environmentally friendly alternative materials. In these cases, cellulosic fibers, have abilities for strong contribution to the future development as advanced materials. We also discussed the economic production and industrialization of recycled cellulose, as well as environmentally friendly alternative materials such as cellulose composites, hydrogels, and films using cellulosic fibers extracted and purified from non-wooden sources.

REFERENCES

Abba, H. A., Nur, I. Z., & Salit, S. M. (2013). Review of agro waste plastic composites production. *Journal of Minerals & Materials Characterization & Engineering*, 1(5), 271–279. DOI: 10.4236/jmmce.2013.15041

Abdul Khalil, H. P. S., Alwani, M. S., Ridzuan, R., Kamarudin, H., & Khairul, A. (2008). Chemical composition, morphological characteristics and cell wall structure of Malaysian oil, palm fibers. *Polymer-Plastics Technology and Engineering*, 47(3), 273–280. DOI: 10.1080/03602550701866840

Abdul Khalil, H. P. S., Amouzgar, P., Jawaid, M., Hassan, A., & Ahmad, F. (2012). New approach to oil palm trunk core lumber material properties enhancement via resin impregnation. *Journal of Biobased Materials and Bioenergy*, 6(3), 299–308. DOI: 10.1166/jbmb.2012.1212

Abdul Khalil, H. P. S., Bhat, A. H., Jawaid, M., Amouzgar, P., Ridzuan, R., & Said, M. R. (2010). Agro-wastes: Mechanical and physical properties of resin impregnated oil palm trunk core lumber. *Polymer Composites*, 31(4), 638–644. DOI: 10.1002/pc.20841

Alberts, B., Johnson, A., Lewis, J., Raff, M., Roberts, K., & Walter, P. (2002). *Molecular biology of the cell* (2nd ed.). Garland Science.

Amin, M. A., & Abdel-Raheem, I. T. (2014). Accelerated wound healing and anti-immflammatory effect of physically crosslinked polyvinyl alcohol-chitosan hydrogel containing honey bee venom in diabetic rats. *Archives of Pharmacal Research*, 37(8), 1016–1030. DOI: 10.1007/s12272-013-0308-y PMID: 24293065

Azwa, Z. N., Yousif, B. F., Manalo, A. C., & Karunasena, W. (2013). A review on the degradability of polymeric composites based on natural fibres. *Materials & Design*, 47, 424–442. DOI: 10.1016/j.matdes.2012.11.025

Bajpai, P. (2014). Recycling and Deinking of Recovered Paper, Elsevier, 1-19, Bajpai, P. (2021). Chapter 2 - Considerations for use of nonwood fiber, 19-24, *Nonwood Plant Fibers for Pulp and Paper*, Elsevier.

Balakrishnan, B., Mohanty, M., Fernandez, A. C., Mohanan, P., & Jayakrishnan, A. (2006). Evaluation of the effect of incorporation of dibutyryl cyclic adenosine monophosphate in an in situ-forming hydrogel wound dressing based on oxidized alginate and gelatin. *Biomaterials*, 27(8), 1355–1361. DOI: 10.1016/j.biomaterials.2005.08.021 PMID: 16146648

Batra, S. K. (1985). Other long vegetables fibers. *In: handbook of Fiber Science and Technology*. Dekker, M. (Ed). Academic Press, San Diego, CA. 727-808.

Bledzki, A. K., Faruk, O., & Sperber, V. E. (2006). Cars from bio-fibres. *Macromolecular Materials and Engineering*, 291(5), 449–457. DOI: 10.1002/mame.200600113

Bledzki, A. K., & Gassan, J. (1999). Composites reinforced with cellulose based fibres. *Progress in Polymer Science*, 24(2), 221–274. DOI: 10.1016/S0079-6700(98)00018-5

Bouchard, J., Methot, M., & Jordan, B. (2006). The effects of ionizing radiation on the cellulose of wood free paper. *Cellulose (London, England)*, 13(5), 601–610. DOI: 10.1007/s10570-005-9033-0

Brinchi, L., Cotana, F., Fortunati, E., & Kenny, J. M. (2013). Production of nanocrystalline cellulose from lignocellulosic biomass: Technology and applications. *Carbohydrate Polymers*, 94(1), 154–169. DOI: 10.1016/j.carbpol.2013.01.033 PMID: 23544524

Brumar, D. R., & Pokharkar, V. B. (2006). Studies on effect of pH on cross-linking of chitosan with sodium tripolyphosphate: A technical note. *AAPS PharmSciTech*, 7, 38–43. PMID: 16796367

Bussemer, T., & Bodmeier, R. (2003). Formulation parameters affecting the performance of coated gelatin capsules with pulsatile release profiles. *International Journal of Pharmaceutics*, 267(1-2), 59–68. DOI: 10.1016/j.ijpharm.2003.07.008 PMID: 14602384

Cai, H. L., Sharma, S., Liu, W. Y., Mu, W., Liu, W., Zhang, X. D., & Deng, Y. L. (2014). Aerogel microspheres from natural cellulose nanofibrils and their applications as cell culture scaffolds. *Biomacromolecules*, 15(7), 2540–2547. DOI: 10.1021/bm5003976 PMID: 24894125

Chandy, T., & Sharma, C. P. (1998). Activated charcoal microcapsules and their applications. *Journal of Biomaterials Applications*, 13(2), 128–157. DOI: 10.1177/088532829801300204 PMID: 9777464

Chen, H., Yuan, L., Song, W., Wu, Z., & Li, D. (2008). Biocompatible polymer materials. Role of protein-surface interactions. *Progress in Polymer Science*, 33(11), 1059–1065. DOI: 10.1016/j.progpolymsci.2008.07.006

Costa, S. M., Mazzola, P. G., Silva, J. C. A. R., Pahl, R., Pessoa, A.Jr, & Costa, S. A. (2013). Use of sugar cane straw as a source of cellulose for textile fiber production. *Industrial Crops and Products*, 42, 189–194. DOI: 10.1016/j.indcrop.2012.05.028

Cristaldi, G., Latteri, A., Recca, G., & Cicala, G. (2010). Composites based on natural fibre fabrics. In Dubrovski, P. D. (Ed.), *Woven Fabric Engineering* (pp. 317–342). InTech Publ.

Doree, C. (1947). *The methods of cellulose chemistry: including methods for the investigation of substances associated with cellulose in plant tissues* (2nd ed.). Van Nostrand Co.

Dourado, F., Mota, M., Pala, H., & Gama, F. M. (1999). Effect of cellulose adsorption on the surface and interfacial properties of cellulose. *Cellulose (London, England)*, 6(4), 265–682. DOI: 10.1023/A:1009251722598

Dungani, R., Abdul Khalil, H. P. S., Sumardi, I., & Suhaya, Y. Sulistyawati, E. (2014). Non-wood renewable materials: Properties improvement and its application. *In: Biomass and Bioenergy: Applications.* Hakeem, K. R., Jawaid, M., Rashid, U. (Eds). Chapter 1. Springer, USA., 1-29.

Edgar, K. J., Buchanan, C. M., Debenham, J. S., Rundquist, P. A., Seiler, B. D., Shelton, D., & Tindall, D. (2001). Advances in cellulose ester performance and application. *Progress in Polymer Science*, 26(9), 1605–1611. DOI: 10.1016/S0079-6700(01)00027-2

Ferdous, T., Ni, Y., Quaiyyum, M. A., Uddin, M. N., & Jahan, M. S. (2021). Non-Wood Fibers: Relationships of Fiber Properties with Pulp Properties. *ACS Omega*, 6(33), 21613–21622. DOI: 10.1021/acsomega.1c02933 PMID: 34471765

Frisbee, S. E., Mehta, K., & McGinity, J. (2002). Processing factors that influence the in vitro and in vivo performance of film-coated drug delivery systems. *Drug Delivery*, 2, 72–76.

Gupta, A., Kumar, R., Upadhyay, N., Surekha, P., & Roy, P. K. (2008). Synthesis, characterization and efficacy of chemically crosslinked PVA hydrogels for dermal wound healing in experimental animals. *Journal of Applied Polymer Science*, 111(3), 1400–1408. DOI: 10.1002/app.28990

Hara, H. (1998). Characteristics and Utilization of Non-wood Pulp and Paper, *Japan. Tappi Journal*, 52(9), 1212–1218. DOI: 10.2524/jtappij.52.1212

Heinze, T., & Liebert, T. (2012). Chapter 10 Celluloses and polyoses/hemicelluloses. In: Henniges, U., Okubayashi, U., Rosenau, T., Potthast, A. (2012). Irradiation of cellulosic pulps: understanding its impact on cellulose oxidation. *Biomacromolecules*, 13(12), 4171–4178. DOI: 10.1021/bm3014457

Holbery, J., & Houston, D. (2006). Natural-fiber-reinforced polymer composites in automotive applications. *Journal of the Minerals Metals & Materials Society*, 58(11), 80–86. DOI: 10.1007/s11837-006-0234-2

Hon, D. N., (1997). *Chemical modification of lignocelluosic materials,* New York, New York, Marcek Dekker, INC.

Huang, Y. H., Fei, B. H., Yu, Y., & Zhao, R. J. (2012). Plant age effect on mechanical properties of MOSO bamboo (*Phyllostachys heterocycle var, Pubescens*) single fibers. *Wood and Fiber Science*, 44, 196–201.

Iresha, H., & Kobayashi, T. (2021). Ultrasound-triggered nicotine release from nicotine-loaded cellulose hydrogel. *Ultrasonics Sonochemistry*, 78, 05710. DOI: 10.1016/j.ultsonch.2021.105710 PMID: 34411843

Izani, M. A. N., Paridah, M. T., Astimar, A. A., Nor, M. Y. M., & Anwar, U. M. K. (2012). Mechanical and dimensional stability proteries of medium-density fibreboard produced from treated oil palm empty fruit bunch. *Journal of Applied Sciences (Faisalabad)*, 12(6), 561–567. DOI: 10.3923/jas.2012.561.567

Jahan, M. S., Saeed, A., He, Z. B., & Ni, Y. H. (2011). Jute as raw material for the preparation of microcrystalline cellulose. *Cellulose (London, England)*, 18(2), 451–459. DOI: 10.1007/s10570-010-9481-z

Jawaid, M. S., & Abdul Khalil, H. P. S. (2011). Effect of layering pattern on the dynamic mechanical properties and thermal degradation of pailm-jute fibers reinforced epoxy hybrid composite. *BioResources*, 6(3), 2309–2322. DOI: 10.15376/biores.6.3.2309-2322

Jewkes, J., Sawers, D., and Stillerman, R., (1969), The soursces of Invention, Mcmillan and CO LTD. 104-117.

Jiang, B., Wu, Z., Zhao, H., Tang, F., Lu, J., Wei, Q., & Zhang, X. (2006). Electron beam irradiation modification of collagen membrane. *Biomaterials*, 27(1), 15–23. DOI: 10.1016/j.biomaterials.2005.05.091 PMID: 16023715

Jiang, H., Tovar-Carrillo, K., & Kobayashi, T. (2016). Ultrasound stimulated release of mimosa medicine from cellulose hydrogel matrix. *Ultrasonics Sonochemistry*, 32, 398–406. DOI: 10.1016/j.ultsonch.2016.04.008 PMID: 27150786

Johansson, C., Bras, J., Mondragon, I., Nechita, P., Plackett, D., Šimon, P., Gregor Svetec, D., Virtanen, S., Giacinti Baschetti, M., Breen, C., & Aucejo, S. (2012). Renewable fibers and bio-based materials for packaging applications a review of recent developments. *BioResources*, 7(2), 2506–2552. DOI: 10.15376/biores.7.2.2506-2552

Joshi, S. V., Drzal, L. T., Mohanty, A. K., & Arora, S. (2004). Are natural fiber composites environmentally superior to glass fiber reinforced composites? *Composites. Part A, Applied Science and Manufacturing*, 35(3), 371–376. DOI: 10.1016/j.compositesa.2003.09.016

Kalia, S., Dufresne, A., Cherian, B. M., Kaith, B. S., Averous, L., Njuguna, J., & Nassiopoulos, E. (2011). Cellulose-based bio and nanocomposites: A review. *International Journal of Polymer Science*, 2011, 1–35. Advance online publication. DOI: 10.1155/2011/837875

Kalia, S., Kaith, B. S., & Kaur, I. (2009). Pretreatments of natural fibers and their applications as reinforcing materials in polymers composites- a review. *Polymer Engineering and Science*, 49(7), 1253–1272. DOI: 10.1002/pen.21328

Kamble, Z., Behera, B. K. (2021). Sustainable hybrid composites reinforced with textile waste for construction and building applications, *Construction and Building Materials*, 284, 17,122800.

Kim, S. M., Eo, M. Y., Kang, J. Y., Myoung, H., Choi, E. K., & Lee, S. K. (2013). Bony regeneration effect of electron-beam irradiation hydroxyapatite and tricalcium phosphate mixtures with 7 to 3 ratio in the calvarial defect model of rat. *Tissue Engineering and Regenerative Medicine*, 9, 24–32.

Kim, S. M., Lee, J. H., Jo, J. A., Lee, J. C., & Lee, S. K. (2005). Development of a bioactive cellulose membrane from sea squirt skin for bone regeneration a preliminary research. *Journal of the Korean Association of Oral and Maxillofacial Surgeons*, 31, 440–453.

Kim, S. M., Park, J. M., Kang, T. Y., Kim, Y. S., & Lee, S. K. (2013). Purification of squirt cellulose membrane from the cystic tunic of Styela clave and identification of its osteoconductive effect. *Cellulose (London, England)*, 20(2), 655–673. DOI: 10.1007/s10570-012-9851-9

Kim, S. M., Woo, K. M., Song, N., Eo, M. Y., Cho, H. J., & Park, J. H. (2015). Electtron beam irradiation to the styela clava derived cellulose membrane. *Polymer*, 39, 1–9.

Kimura, S., & Kondo, T. (2002). Recent progress in cellulose biosynthesis. *Journal of Plant Research*, 115(4), 297–302. DOI: 10.1007/s10265-002-0037-7 PMID: 12582734

Kirker, K., Luo, Y., Nielson, J. H., Shelby, J., & Prestwish, G. D. (2002). Glycosaminoglycan hydrogel films as bio-interactive dressings for wound healing. *Biomaterials*, 43(17), 3661–3671. DOI: 10.1016/S0142-9612(02)00100-X PMID: 12109692

Klemm, D., Heublein, B., Fink, H.-P., & Bohn, A. (2005). Cellulose: Fascinating biopolymer and sustainable raw material. *Angewandte Chemie International Edition*, 44(22), 3358–3393. DOI: 10.1002/anie.200460587 PMID: 15861454

Kobayashi, T. (2018). Chapter 11 Cellulose hydrogels; Fabrication, properties, and their application to biocompatible and tissue engineering, Hydrogels, recent advances, Springer, 297-314.

Kobayashi, T. (2023). *Ultrasound-triggered drug delivery, Advanced and Modern Approaches for Drug Delivery*. Academic press.

Kobayashi, T., Kongklieng, P., & Ibaraki, A. (2022). Encyclopedia of Materials: Plastics and Polymers, 2, 541- 554. DOI: 10.1016/B978-0-12-820352-1.00258-3

Kobayashi, T., & Tovar-Carrillo, K. L. (2015). Fibroblast cell cultivation on wooden pulp cellulose hydrogels for cytocompatibility scaffold method. *Pharmaceutica Analytica Acta*, 6(10), 1–9. DOI: 10.4172/2153-2435.1000423

Kopania, E., Wietecha, J., & Ciechanska, D. (2012). Studies on insolation of cellulose fibres from waste plnat biomass. *Fibres & Textiles in Eastern Europe*, 20, 167–177.

Langer, R., & Vacanti, J. P. (1993). Tissue engineering. *Science*, 260(5110), 920–926. DOI: 10.1126/science.8493529 PMID: 8493529

Langer, S., Botteck, N. M., Bosse, B., Reimer, K., Vogt, P. M., Steinau, H.-U., & Mueller, S. (2006). Effect of polyvinylpyrrolidone-iodine liposoma hydrogel on wound microcirculation in SKH1-Hr hairless mice. *European Surgical Research*, 38(1), 27–34. DOI: 10.1159/000091524 PMID: 16490991

Launer, H. F., & Wilson, W. K. (1950). Preparing cuprammonium solvent and cellulose solutions. *Analytical Chemistry*, 3(3), 455–458. DOI: 10.1021/ac60039a019

Leitner, J., Hinterstoisser, B., Wastyn, M., Keches, J., & Gindl, W. (2007). Sugar beet cellulose nanofibril-reinforced composites. *Cellulose (London, England)*, 14(5), 419–425. DOI: 10.1007/s10570-007-9131-2

Li, X., Tabil, L. G., & Panagrahi, S. (2007). Chemical treatments of natural fiber for use in natural fiber-reinforced composites: A review. *Journal of Polymers and the Environment*, 15(1), 25–33. DOI: 10.1007/s10924-006-0042-3

Lim, X. Z. (2021). Microplastics are everywhere — But are they harmful? *Nature*, 593(7857), 22–25. DOI: 10.1038/d41586-021-01143-3 PMID: 33947993

Lindma, B., Karlstrom, G., & Stigsson, L. (2010). On the mechanism of dissolution of cellulose. *Journal of Molecular Liquids*, 156(1), 76–81. DOI: 10.1016/j.molliq.2010.04.016

Liu, X., Ma, L., Mao, Z., & Gao, C. (2011). Chitosan-based biomaterials for tissue repair and regeneration. *Advances in Polymer Science*, 244, 81–87. DOI: 10.1007/12_2011_118

Lloyd, L. L., Kennedy, J. F., Methacanon, P., Paterson, M., & Knill, C. J. (1998). Carbohydrates polymers as wound management aids. *Carbohydrate Polymers*, 37(3), 315–319. DOI: 10.1016/S0144-8617(98)00077-0

Lu, B., Lu, F., Wu, D., & Lan, G. (2017). In situ reduction of silver nanoparticles by chitosan-L-Glutamic acid/hyaluronic acid: Enhancing antimicrobial and wound-healing activity. *Carbohydrate Polymers*, 173, 556–565. DOI: 10.1016/j.carbpol.2017.06.035 PMID: 28732899

Majeed, K., Jawaid, M., Hassan, A., Bakar, A. A., Abdul Khalil, H. P. S., Salema, A. A., & Inuwa, I. (2013). Potential materials for food packaging from nanoclay/natural fibres filled hybrid composites. *Materials & Design*, 46, 391–410. DOI: 10.1016/j.matdes.2012.10.044

Methacanon, P., Weerawatsophon, U., Sumransin, N., Prahsarn, C., & Bergado, D. T. (2010). Properties and potential application of the selected natural fibers as limited life geotextiles. *Carbohydrate Polymers*, 82(4), 1090–1096. DOI: 10.1016/j.carbpol.2010.06.036

Meza-Valle, K. Z., Saucedo-Acuña, R. A., Rios-Arana, J. V., Lobo, N., Cuevas, J. C., & Tovar-Carrillo, K. L. (2020). Natural film based on pectin and allantoin for wound healing: Obtaining, characterization, and rat model. *BioMed Research International*, 2020, 1–7. DOI: 10.1155/2020/6897497 PMID: 33123582

Mishra, S. K., Mary, D. S., & Kannan, S. (2017). Cooper incorporated microporous chitosan-polyethylene glycol hydrogels loaded with naproxen for effective drug release and anti-infection wound dressing. *International Journal of Biological Macromolecules*, 95, 928–937. DOI: 10.1016/j.ijbiomac.2016.10.080 PMID: 27984151

Monteiro, S. N., Lopes, F. P. D., Barbosa, A. P., Bevitori, A. B., da Silva, I. L. A., & da Costa, L. (2011). Natural lignocellulosic fibers as engineering materials-an overview. *Metallergical and Materials Transactions A*. 42, 2963-2974.

Nair, L. S., & Laurencin, C. T. (2007). Biodegradable polymers as biomaterials. *Progress in Polymer Science*, 32(8-9), 762–767. DOI: 10.1016/j.progpolymsci.2007.05.017

Nakasone, K., Ikematsu, S., & Kobayashi, T. (2016). Biocompatibility Evaluation of Cellulose Hydrogel Film Regenerated from Sugar Cane Bagasse Waste and Its in Vivo Behavior in Mice. *Industrial & Engineering Chemistry Research*, 55(1), 30–37. DOI: 10.1021/acs.iecr.5b03926

Nakasone, K., & Kobayashi, T. (2016a). Effect of pre-treatment of sugarcane bagasse on the cellulose solution and application for the cellulose hydrogel films. *Polymers for Advanced Technologies*, 27(7), 3757–3762. DOI: 10.1002/pat.3757

Nakasone, K., & Kobayashi, T. (2016b). Cytocompatible cellulose hydrogels containing trace lignin. *Materials Science and Engineering C*, 64, 269–27. DOI: 10.1016/j.msec.2016.03.108 PMID: 27127053

Nawar, F., Jawaid, M., Tanir, P. M., Mohamad, R., & Azizi, S. (2014). Potential use of plant fibres and their composites for biomedical applications. *BioResource*, 9, 5688–5706.

Nguong, C. W., Lee, S. N. B., & Sujan, D. (2013). A review on natural fibre reinforced polymer composites. *International Journal of Minerals. Metallurgy and Materials Engineering.*, 7, 52–59.

Noguchi, S., & Kobayashi, T. (2020). Ultrasound response of viscoelastic changes of cellulose hydrogels triggered with Sono-deviced rheometer. *Ultrasonics Sonochemistry*, 67, 105143. DOI: 10.1016/j.ultsonch.2020.105143 PMID: 32446975

Olsson, C., Westman, G. (2013). Direct dissolution of cellulose: Background, means and applications, *Cellulose - Fundamental Aspects,* 143–177. Available at: https://doi.org/.DOI: 10.5772/52144

Osorio, L., Trujillo, E., Van Vuure, A. W., Lens, F., Ivens, J., & Verpoest, I. (2010). The relationship between the bamboo fibre microestructure and mechanical properties. *Proceedings of the 14th European conference on Composite Materials,* June 7-10, Budapest, Hungary. 8, 933-943.

Pardo, M. E. S., Cassellis, M. E. R., Escobedo, R. M., & Garcia, E. J. (2014). Chemical characterization of the industrial residues of the pineapple *(Ananas comosus).Journal of Agricultural Chemistry and Environment*, 3(2), 53–56. DOI: 10.4236/jacen.2014.32B009

Park, J. S., Lee, J. H., Han, C. S., Chng, D. W., & Kim, G. Y. (2011). Effect of hyaluronic acid-carboxymethyl cellulose solution on perineural scar formation after sciatic nerve repair in rats. *Clinics in Orthopedic Surgery*, 3(4), 315–320. DOI: 10.4055/cios.2011.3.4.315 PMID: 22162795

Pina, M. E., & Sousa, A. T. (2002). Application of hydroalcoholic solutions of formaldehyde in preparation of acetylsalicylic acid gastro-resistant capsules. *Drug Development and Industrial Pharmacy*, 28(4), 443–449. DOI: 10.1081/DDC-120003005 PMID: 12056537

Qua, E. H., Hornsby, P. R., Sharme, H. S. S., & Lyons, G. (2011). Preparation and characterization of cellulose nanofibers. *Journal of Materials Science*, 46(18), 6029–6045. DOI: 10.1007/s10853-011-5565-x

Raguvaran, R., Manuja, B. K., Chopra, M., Thakur, R., Anand, T., Kalia, A., & Manuja, A. (2017). Sodium alginate and gum acacia hydrogels of ZnO nanoparticles show wound healing effect on fibroblast cells. *International Journal of Biological Macromolecules*, 96, 185–191. DOI: 10.1016/j.ijbiomac.2016.12.009 PMID: 27939272

Reddy, N., & Yang, Y. (2005). Biofibers from agricultural byproducts for industrial applications. *Trends in Biotechnology*, 23(1), 22–27. DOI: 10.1016/j.tibtech.2004.11.002 PMID: 15629854

Reddy, N., & Yang, Y. (2006). Properties of high-quality long natural cellulose fibers from rice straw. *Journal of Agricultural and Food Chemistry*, 54(21), 8077–8081. DOI: 10.1021/jf0617723 PMID: 17032012

Rowell, R. M., Han, J. S., & Rowell, J. S. (2000). Characterization and factors effecting fiber properties. In Frollini, E., Leao, A. L., & Mattoso, L. H. C. (Eds.), *Natural polymers and agrofibers composites* (pp. 115–134).

Saheb, D. N., & Jog, J. P. (1999). Natural fiber polymer composites: A review. *Advances in Polymer Technology*, 18(4), 351–363. DOI: 10.1002/(SICI)1098-2329(199924)18:4<351::AID-ADV6>3.0.CO;2-X

Satyanarayana, K. G., Arizaga, G. G. C., & Wypych, F. (2009). Biodegradable composites based on lignocellulosic fibers- an-overview. *Progress in Polymer Science*, 34(9), 982–1021. DOI: 10.1016/j.progpolymsci.2008.12.002

Sayyed, A. J., Deshmukh, N. A., & Pinjari, D. V. (2019). A critical review of manufacturing processes used in regenerated cellulosic fibres: Viscose, cellulose acetate, cuprammonium, LiCl/DMAc, ionic liquids, and NMMO based lyocell. *Cellulose (London, England)*, 26(5), 2913–2940. DOI: 10.1007/s10570-019-02318-y

Shakir, M. A., Ahmad, M. I., & Rafatullah, M. (2023). Review on the influencing factors towards improving properties of composite insulation panel made of natural waste fibers for building application, 16 May. https://orcid.org/0000-0002-4590-3153 mrafatullah@usm.my +5 View all authors and affiliations

Sionkowske, A. (2011). Current research on blends of natural and synthetic polymers as a new biomaterial [Review]. *Progress in Polymer Science*, 36(9), 1254–1261. DOI: 10.1016/j.progpolymsci.2011.05.003

Siro, I., & Plackett, D. (2010). Microfibrillated cellulose and new nanocomposite materials: A review. *Cellulose*. 17, 459-494. Sionkowske, A., (2011). Current research on the blends of natural and synthetic polymers as new biomaterials [Review]. *Progress in Polymer Science*, 36, 1254–1261.

Song, A., Rane, A. A., & Christman, K. L. (2012). Antibacterial and cell-adhesive polypeptide and poly(ethylene glycol) hydrogel as a potential scaffold for wound healing. *Acta Biomaterialia*, 8(1), 41–50. DOI: 10.1016/j.actbio.2011.10.004 PMID: 22023748

Striegel, A. M. (1997). Theory and applications of DMAc/LiCl in the analysis of polysaccharides. *Carbohydrate Polymers*, 34(4), 267–273. DOI: 10.1016/S0144-8617(97)00101-X

Suddell, B. C., & Evans, W. J. (2005). Natural fiber composites in automotive applications. *In: natural Fibers, Biopolymers and Biocomposites, Mohanty, A. K., Misra, M., Drzal, L. T. (Eds). CRC Press, USA.* 231-259. DOI: 10.1201/9780203508206.ch7

Svensson, A., Nicklasson, E., Harrah, T., Panilaitis, B., Kaplan, D. L., Brittberg, M., & Gatenholm, P. (2005). Bacterial cellulose as a potential scaffold for tissue engineering of cartilage. *Biomaterials*, 26(4), 419–425. DOI: 10.1016/j.biomaterials.2004.02.049 PMID: 15275816

Swatloski, R., Spear, S. K., Holbrey, J. D., & Rogers, R. D. (2002). Dissolution of cellulose with ionic liquids. *Journal of the American Chemical Society*, 124(18), 4974–4975. DOI: 10.1021/ja025790m PMID: 11982358

Tamada, Y., & Ikada, Q. (1993). Effect of preadsorbed proteins on cell adhesion to polymer surface. *Journal of Colloid and Interface Science*, 155(2), 334–341. DOI: 10.1006/jcis.1993.1044

Thiruchitrambalam, M., Alavudeen, A., Athijayamani, A., Venkasteshwaran, N., and Perumal, A. E., (2009), materials Physics and Mechanics, 8, 165-173.

Thomas, S., Paul, S. A., Pothan, L. A., & Deepa, B. (2011). Natural fibres: structure, properties and applications. In *Cellulose fibers: Bio-and nano-polymer composites* (Vol. 1, pp. 3–42). Springer. DOI: 10.1007/978-3-642-17370-7_1

Thombre, A. G., Cardinal, J. R., DeNoto, A. R., Herbig, S. M., & Smith, K. L. (1999). Asymmetric membrane capsules for osmotic drug delivery: I. Development of a manufacturing process. *Journal of Controlled Release*, 57(1), 55–64. DOI: 10.1016/S0168-3659(98)00100-X PMID: 9863039

Tovar-Carrillo, K. (2014). Thesis: Study on Cellulose hydrogel films regenerated from natural plnat bagasse and their bio and cytocompatible properties for tissue engineering. Nagaoka University of Technology.

Tovar-Carrillo, K., Nakasone, K., Sugita, S., Tagaya, M., & Kobayashi, T. (2014). Effects of sodium hypochlorite on Agave tequilana Weber bagasse fibers used to elaborate cyto and biocompatible hydrogel films. *Materials Science and Engineering C*, 42, 808–815. DOI: 10.1016/j.msec.2014.06.023 PMID: 25063183

Tovar-Carrillo K., Tagaya, M., Kobayashi, T. (2013). Bamboo fibers elaborating cellulose hydrogel films for medical applications. *Journal of Material Science Chemical engineering*, 1, 7-12.

Trache, D., Tarchoun, A. Z., Derradji, M., Hamidon, T. S., Masruchin, N., Brosse, N., & Hussin, M. H. (2020). Nanocellulose: From Fundamentals to Advanced Applications. *Frontiers in Chemistry*, 8, 392. DOI: 10.3389/fchem.2020.00392 PMID: 32435633

Venkateshwaran, N., Elayaperumal, A., & Sathiya, G. K. (2012). Prediction of tensile properties of hybrid-natural fiber composites. *Composites. Part B, Engineering*, 43(2), 793–796. DOI: 10.1016/j.compositesb.2011.08.023

Ververis, C., & Geoghiou, K. (2004). Fiber dimensions, lignin and cellulose content of various plant materials and their suitability for paper production. *Industrial Crops and Products*, 19(3), 245–254. DOI: 10.1016/j.indcrop.2003.10.006

Wahlang, B., Nath, K., Ravindra, U., Chandu, R., & Vijayalaxmi, K. (2012). Precessing and utilization of sugarcane bagasse for functional food formulations. *Proceedings of the International Conference and Exhibition on Food Processing and Technology,* September 22-24, 2012, Hyderabad, India. 106-112.

Wedin, H., Niit, E., Mansoor, Z. A., Kristinsdottir, A. R., de la Motte, H., Jönsson, C., Östlund, Å., & Lindgren, C. (2018). Preparation of viscose fibres stripped of reactive dyes and wrinkle-free crosslinked cotton textile finish. *Journal of Polymers and the Environment*, 26(9), 3603–3612. DOI: 10.1007/s10924-018-1239-y

Wooding, C. (2001). *Regenerated cellulose fibers*. Woodhead Publishing Limited. DOI: 10.1533/9781855737587

Yang, Q., Saito, T., & Isogai, A. (2013). Transparent, flexible, and high-strength regenerated cellulose/saponite nanocomposite films with high gas barrier properties. *Journal of Applied Polymer Science*, 130(5), 3168–3174. DOI: 10.1002/app.39564

Chapter 3
Cellulose–Based Functional Fine Particles and Fibers as Environmentally Friendly Materials:
Development and Application

Shoji Nagaoka
Kumamoto Innovative Development Organization, Japan

Maki Horikawa
Kumamoto Industrial Research Institute, Japan

Tomohiro Shirosaki
Kumamoto Industrial Research Institute, Japan

ABSTRACT

Due to concerns regarding the effect of the large amount of waste accompanying their consumption on the environment, attention has therefore returned to cellulose and to maximize the potential of cellulose. On the development of cellulose refining, combining, and functionalization technologies, this chapter categorizes and introduces the spherical microbeads and nanofiber and then discusses interfacial hydrophilic microbeads, which are used as raw materials for moisturizing cosmetics. In addition, the fabrication of the semiconductor abrasives, thermally conductive materials, abrasive cleaning agents, and three primary color materials for paints are discussed for development of cellulose composite materials with inorganic materials.

INTRODUCTION

Cellulose, a polysaccharide, exists ubiquitously in plants and is an inexhaustible raw material with an annual production far exceeding oil reserves. Since the discovery of cellulose by the French biochemist Anselme Payen in 1838, various techniques for the utilization of cellulose fibers have been developed.

DOI: 10.4018/979-8-3693-0003-9.ch003

Nylon was invented by W. H. Carothers approximately 100 years later (Hermes, 1996). Various petroleum-based synthetic polymers were developed via facile synthetic methods with high production efficiency (Ziegler, 1964; Maddah, 2016; John & Tennant, 1945), following the invention of nylon. These synthetic polymers are commonly utilized for fabricating commodities for daily necessities. Hence, they are produced and consumed in large quantities. The application of cellulose was limited to the manufacture of films, paper, and wood products.

One challenge to utilize celluloses is the processing of cellulose. Its natural form is complex to process as it is resistant to heat, water, or solvents owing to the presence of strong intermolecular hydrogen bonds. Furthermore, the tuning of its reaction is difficult, owing to the three hydroxyl groups in the glucose unit. Conversely, synthetic polymers are easy to process and manufacture with a high production efficiency. Over the years, research on cellulose declined as that on synthetic polymers gained significance.

Synthetic polymers have been produced, consumed, and discarded in large quantities in the 20st century. Mass production of polymers and their waste contributes significantly to global warming due to CO_2 emissions (CO_2 and Greenhouse Gas Emissions website), the depletion of resources (Looney, 2021) and the resulting disasters (Walz et al., 2021). The accumulation of microplastics (MP) pollutes rivers and oceans (Lim, 2021; Bai, et al., 2021; Jambeck, et al., 2015; Xanthos & Walker., 2017), which finally pollute ecosystems through the food chain (Saeedi, 2023). Concepts such as sustainability, sustainable development goals (Brundtland Report, 1987), and carbon neutrality (Carbon Neutral Coalition website, 2023) have been popular in the 21st century as there is a growing awareness on the need to conserve resources. Thus, cellulose-based materials have been gaining renewed attention as alternatives to synthetic polymers. Microcrystalline cellulose (nanocellulose) obtained from wood has been explored as structural or functional materials (Isogai, 2013; Isa et al., 2016; Kondo et al., 2014; Nakagaito & Yano, 2014; Noguchi et al., 2017; Ho et al., 2011). Attempts have been made to replace MP in detergents and cosmetic carriers with environmentally friendly materials, such as cellulose, starch, biodegradable plastics, and silica, to reduce river and ocean pollution (Ministry of the Environmental website, Japanese government; *2020 Current status and future prospects of the fine powder market*, Market research report).

In this chapter, we describe the development of two cellulose-based materials: spherical microbeads and functional nanofibers. We first discuss the development of cellulose spherical microbeads (CSM) packings for liquid chromatography separation (Motozato et al., 1981, 1984; Nagaoka et al., 1994; Hirayama et al., 1995). Next, we explain the development of spherical microbead composites fabricated using cellulose and inorganic materials. The developed composites were used as semiconductor abrasives (Nagaoka et al., 2008), three primary colorants in paints (Nagaoka et al., 2001, 2005a, 2007), and heat-dissipating materials. Skin moisturizers using surface-hydrophilized cellulose microbeads with a focus on MP were also developed.

Composites using conductive polymers have also been fabricated using nanofibers. We have succeeded in increasing the conductivity of a resin (Horikawa et al., 2015, 2018, 2017, 2020) and used it as a heat-shield material absorbing near-infrared (NIR) rays (Yoshida et al., 2020). The developed material was incorporated into a sash that could be used as a heat-insulating interlayer for windows. The performance was verified and the ability of the material to control CO_2 emissions and save energy was confirmed. The practical application of the fabricated composites is also reported.

PREPARATION AND APPLICATION OF CELLULOSE MICROBEADS

Cellulose-based spherical Microbeads (CSM)

In recent years, marine pollution caused by petroleum-derived MP has become a major problem (Lim, 2021; Bai, et al., 2021; Jambeck, et al., 2015; Xanthos & Walker., 2017). The ingestion of MP by marine organisms affects ecosystems. MP are generally defined as extremely small plastics with a diameter < 5 mm and are formed by decomposing, crushing, and fragmenting large plastic fragments via light in the natural environment. They were originally manufactured for personal care products such as face and body wash, and toothpaste (Moore, 2008). Plastic-based MP, also known as microbeads, produced for personal care products are inherently small with sizes between 0.001 and 0.1 mm. It is difficult to recovering MPs from oceans and seas; hence, the influx should be stopped. Environmentally friendly and highly biodegradable plant-based microbeads should be used to replace artificial MP. Thus, cellulose-derived microbeads have become important environment-adaptive materials to replace petroleum-derived microbeads.

This section outlines the preparation of CSM by the spherical particulation method, their characteristics, and applications in continuation of our work in this field.

Spherical Microparticulation of Cellulose Materials

Fine cellulose powder was industrially used as a separation medium in column chromatography for biological components by E. A. Peterson et al. in 1950s (Peterson & Sorber, 1956). However, the crushed material clogged the column, limiting the separation speed and performance. Subsequently, the development of spherical particle technology for preparing spherical microbeads improved their performance as a separating material. Table 1 shows the reported methods for preparing CSM. Cellulose hydrophobic derivatives with hydrophobic groups (acetyl group) (Motozato, 1981) can be dissolved in organic solvents. As another method, cellulose hydrophilic derivatives such as cellulose xanthate (Ohkuma et al., 1986) and cuprammonium (Determann et al.,1968) or rhodan–calcium (Kuga, 1980) cellulose complex can be dissolved in aqueous solvents. Methods for the preparation of spherical microbeads are roughly classified as two categories according to the mechanism of particle formation: 1) O/W and W/O methods using the hydrophobic effect and 2) the W/W method using charge repulsion.

Table 1. Preparation method of cellulose spherical microbeads

Cellulose derivatives	Formation of droplet	Regeneration of cellulose
Cellulose alkylate Y. Motozato et al. 1981 Cellulose—O—C(=O)—CH$_3$	O/W Method Phase separation by hydrophobic effect Evaporation of organic solvent	Saponification by alkali solution Cellulose—O—C(=O)—CH$_3$ $\xrightarrow{\text{Alkali}}$ Cellulose—OH + CH$_3$COOH
Cuprammonium H. Determann et al. 1968 Cellulose$_2$[Cu(NH$_3$)$_4$]$_n$	W/O Method Phase separation by hydrophobic effect Evaporation of organic H$_2$O	Acid treatment Cellulose$_2$[Cu(NH$_3$)$_4$]$_n$ $\xrightarrow{H^+}$ Cellulose—OH + nCu^{2+} + nNH$_4^+$
Calcium rhdanate S. Kuga et al. 1980 Cellulose [SCN$^-$]$_2$Ca^{2+}(H$_2$O)$_2$	W/O Method Phase separation by hydrophobic effect Evaporation of organic H$_2$O	Cellulose [SCN$^-$]$_2$Ca^{2+}(H$_2$O)$_2$ $\xrightarrow{H^+}$ Cellulose—OH + nCa^{2+} + 2SCN$^-$
Cellulose xanthate (Viscose) Okuma et al. 1986 Cellulose—O—C(=S)—SNa	W/W Method Phase separation by Electric Repulsion	Heating and acid treatment Cellulose—O—C(=S)—SNa $\xrightarrow{\Delta}$ Cellulose—OH + CS$_2$ + Na$^+$ $\xrightarrow{H^+}$ Cellulose—OH + CS$_2$ + Na$^+$

Various applications of cellulose microbeads have been reported: chromatographic separation packings (Motozato, 1981), coloring materials (Nagaoka et al., 2005a, 2007), cosmetic materials (Nagaoka et al., 2005c), abrasives for Si semiconductors (Nagaoka et al., 2008), pyrogenic adsorbents (Sakata et al., 2007), and drug carriers (Shi et al., 2021).

Preparation of CSM by Suspension Evaporation Method

The preparation of CSM from hydrophobized cellulose by the suspension evaporation method was developed by a group at Kumamoto University in the mid-1980s (Motozato et al., 1981, 1984). Cellulose was dissolved in organic solvents via hydrophobization. The hydroxyl group of cellulose is protected by a hydrophobic group such as an acetyl or butyryl group. An organic solvent (methylene chloride) with a boiling point less than that of water and immiscible in water was selected to dissolve the hydrophobized cellulose. The solution was added to an aqueous viscous medium with stirring to form droplets of hydrophobized cellulose, followed by gradual evaporation of the organic solvent to form solid spheres. The hydrophobic groups were eliminated to obtain CSM as shown in Figure 1(a). Figures 1(b–d) show the synthesis of natural polysaccharides (pullulan (Motozato at al., 1986), glucomannan (Hirayama et al., 1987), and chitosan (Adachi et al., 1999) or synthetic polymers (polyamino acids (Hirayama, 1990) and polybutadiene (Nagaoka et al., 2005b)) via this method. CSM can be used as adsorbents and separation materials. The synthesis of hydrophobized cellulose from cellulose triacetate (TAC) by the suspension evaporation method is as follows: 1) TAC was dissolved in methylene chloride; 2) the solution was mixed with an aqueous solution using polyvinyl alcohol as a thickening agent; 3) the mixture was stirred at 40 °C for 20 h; 4) TAC microbeads obtained as residues via filtration were washed; 5) the microbeads were dispersed via stirring in an aqueous sodium hydroxide (NaOH) solution (5M) containing a predetermined amount of methanol; and 6) the solution was allowed to stand at 60 °C for 10–12 h to obtain deacetylated CSM via hydrolysis. The particle size could be varied from 2 to 700 μm by changing the stirring speed during the formation of spherical particulates, and the concentration of TAC and polyvinyl alcohol. Porosity could be controlled by adding a porosifying agent, a hydrophobic substance with a boiling point greater than that of water and low compatibility with hydrophobized cellulose during the formation of spherical particulates. CSM could be crosslinked using epichlorohydrin to change the characteristics of the fine pores after hydrolysis.

Figure 1. Polysaccharide spherical microbeads prepared using suspension evaporation method a) Cellulose microbeads, b) Pullulan microbeads, c) Glucomannan microbeads, d) Chitosan microbeads

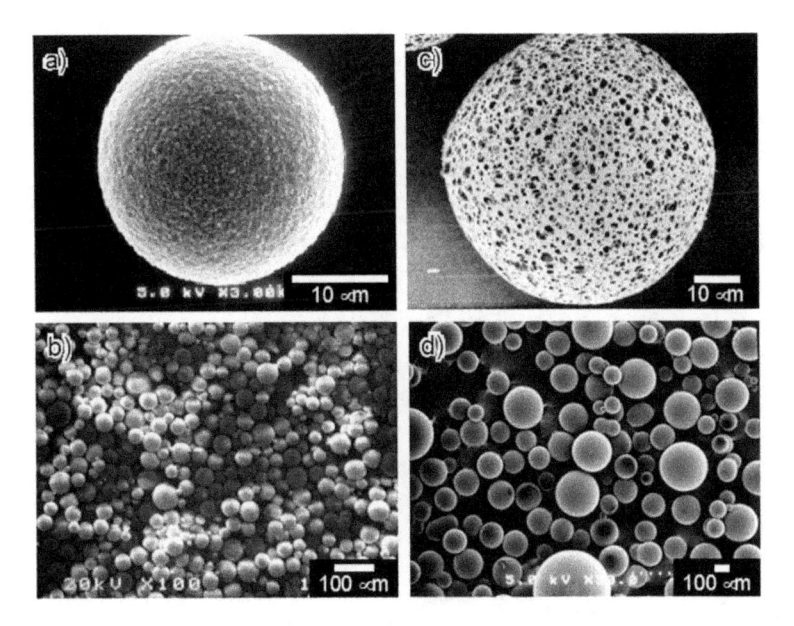

The liquid chromatography column was packed with CSM prepared by varying the amount of the porosity agent. Monodisperse pullulan was used as a standard sample. The molecular weight and separation width changed with an increase in the amount of the porosifying agent because of exclusion limits. Thus, the particle pore size and porosity could be controlled (Hirayama at al., 1990).

CSM by Viscose Phase Separation Particulation Method

The viscose phase separation method is a W/W method to form microspheres. A viscous aqueous solution of cellulose xanthate and polyacrylic acid was mixed and droplets were generated by electric repulsion; the droplets solidified to form CSM (Ohkuma at al., 1986; Nagaoka et al., 2001, 2005c). Aqueous solutions of cellulose xanthate and sodium polyacrylate with a predetermined concentration and calcium carbonate powder were mixed and stirred at room temperature to obtain cellulose xanthate droplets. CSM were formed within 30 min at a temperature of 80 °C and the heating was continued for a further 30 min.

The reaction scheme in Table 1 shows the partial elimination of xanthate, CSS^- as carbon disulfide. This promoted hydrogen bonding and solidified the droplets as spheres. The microbeads obtained by filtration were soft, owing to the presence of water. The xanthate groups were eliminated by immersing the beads in an aqueous solution of hydrochloric acid (HCl) to form hard microbeads. The particle size was controlled from 5 to 200 µm by optimizing the ratio of viscose solution to sodium polyacrylate solution or the molecular weight of sodium polyacrylate. The particle size distribution of the samples was narrower than that formed by other techniques.

Figure 2. Particle size distribution and XRD patterns of cellulose microbeads a) Cellulose microbeads obtained by suspension evaporation method b) Cellulose microbeads obtained by viscose phase separation method

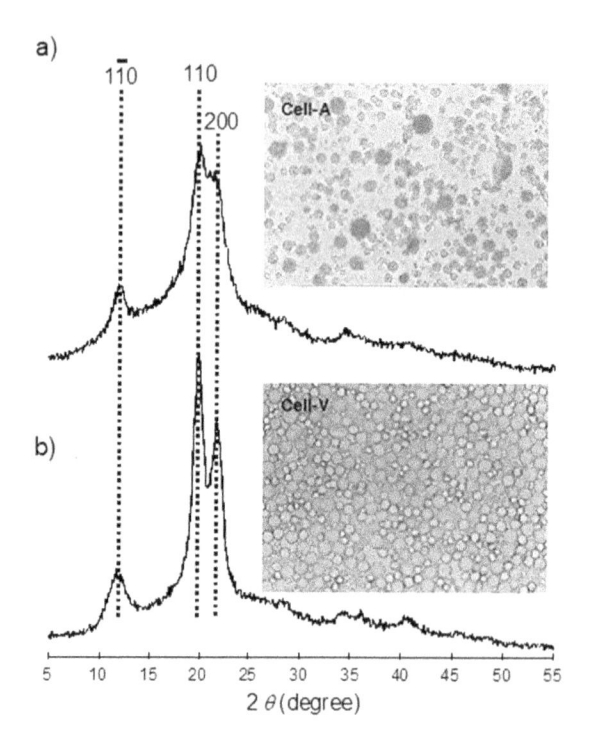

Figure 2 shows the optical micrographs and X-ray diffraction (XRD) patterns of the microbeads prepared using the (a) suspension evaporation method and (b) viscose phase separation method. The microbeads obtained via the viscose phase separation had a narrower particle size distribution than those obtained via suspension evaporation. The XRD pattern of the microbeads obtained via viscose phase separation indicated peaks of 11(−)0, 110, and 200 planes with a typical Type II conformation. In contrast, the microbeads obtained via suspension evaporation had unclear peaks of 11(−)0 and 200 planes, indicating low crystallinity. The microbeads obtained via viscose phase separation had high mechanical strength. The crystallinity of CSM influences the retention of the shape and reactivity of microbeads in oxidation reactions, such as the 2,2,6,6-tetramethylpiperidine 1-oxyl radical (TEMPO)-based oxidation reaction and surface chemical modification.

CELLULOSE/INORGANIC COMPOSITE SPHERICAL MICROBEADS

Cellulose/TiO$_2$ Composite Spherical Microbeads

The viscose phase particulation separation is suitable for the preparation of composite microbeads with inorganic materials. The distribution of inorganic particles in the CSM composite depends on the surface charge of the inorganic particles (Nagaoka et al., 2001, 2007). Cellulose/inorganic composite microbeads were prepared by dispersing inorganic particles in an aqueous solution of cellulose xanthate. The solution was added to an aqueous sodium polyacrylate solution according to the process described previously to form droplets containing inorganic particles. The composite CSM were obtained by eliminating the CSS$^-$ group via heating and acid treatment (Nagaoka et al., 2005c). The preparation of CSM composites with TiO$_2$ used two types of TiO$_2$ nanoparticles (TiO$_2$-A:21 nm and TiO$_2$-B:7 nm). Figures 3(a) and (b) show the electrophoretic zeta potentials of TiO$_2$-A (−29.1 mV) and TiO$_2$-B (−19.1 mV), respectively. Figure 3(a') and (b') show cross-sectional scanning electron microscopy (SEM) images of composite CSM (Cell/TiO$_2$-A and Cell/TiO$_2$-B). TiO$_2$-A was distributed on the surface of Cell/TiO$_2$-A composite microbeads and was absent inside the microbeads, indicating the formation of core–shell microbeads. TiO$_2$-A was excluded from both the negatively charged cellulose xanthate and polyacrylic acid phases during the formation of composite CSM. Consequently, cellulose xanthate spherical domains solidified at the microsphere interface. In contrast, TiO$_2$-B was encapsulated inside the cellulose microbeads. Although the surface of TiO$_2$-B is negatively charged (−19.1 mV), the potential width extends to the positive direction as shown in Figure 3(b). This was probably because TiO$_2$-B solidified without being excluded from the cellulose xanthate phase.

Figure 3. Zeta-potential of TiO$_2$ particles (TiO$_2$-A (a) and TiO$_2$-B (a)) used for preparation of composite spherical microbeads and cross-sectional SEM images of composite CSM prepared using TiO$_2$-A and TiO$_2$-B (Cell/TiO$_2$-A (a') and Cell/TiO$_2$-B (b')) a) TiO$_2$-A: size, approximately 21 nm, b) TiO$_2$-B: size, approximately 7 nm

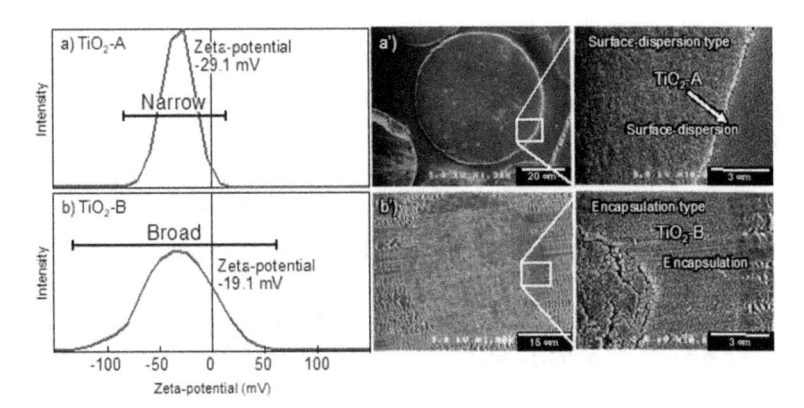

Various Types of Cellulose/Inorganic Composite Spherical Microbeads

Figure 4. SEM images of cellulose spherical microbeads composited with inorganic particles. a) Basic cellulose microbeads, b) Cellulose/TiO$_2$, c) Cellulose/Fe$_2$O$_3$, d) Cellulose/SiC, e) Cellulose/h-BN, f) Cellulose/Al$_2$O$_3$

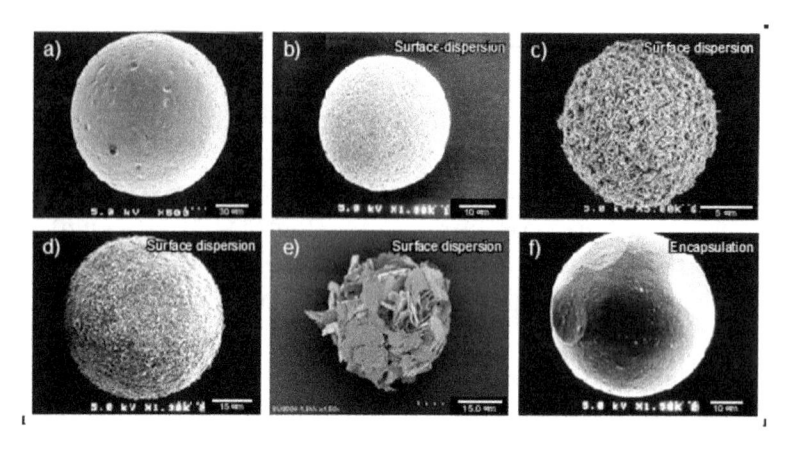

Figure 4 shows the SEM images of composite CSM prepared from various inorganic particles and cellulose. Figures 4(c–e) show that Fe$_2$O$_3$, SiC, and *h*-BN particles are distributed on the surface of microbeads because they have the same surface properties as TiO$_2$-A. However, Figure 4(f) shows Al$_2$O$_3$ particles with their potential width extending to the positive direction, similar to TiO$_2$-B, confirming their distribution inside the microbeads. The viscose phase separation can also be used for combining multiple inorganic pigments. Blue, red, yellow, and green composite CSM can be prepared as coloring materials for cosmetics, such as makeup, lipstick, and coating materials. In addition, TiO$_2$-based CSM composites and inorganic pigments can be combined to be used as outdoor paint for environmental purification. The cellulose content is protected by inorganic pigments and chalking is suppressed, owing to their photocatalytic activity (Nagaoka et al., 2001, 2005a, 2007). Composite microbeads (CM) using hard inorganic powders such as diamond, SiC, ceria, alumina, and zirconia have been used as abrasives with unique polishing properties for semiconductor components, particularly for hard materials such as silicon wafers, quartz, and SiC (Nagaoka et al., 2008). Additionally, we have successfully prepared core-shell spherical microbeads using *h*-boron nitride (*h*-BN) as a thermal conductive material, as shown in Figure 4(e). These core-shell spherical microbeads, shown in Figure 4(e), exhibit remarkable thermal conductivity properties due to the incorporation of *h*-BN as the thermal conductive material.

Cellulose/h-Boron Nitride (h-BN) Core-shell Microbeads Providing High Thermal Conductivity

We then shifted our research focus to *h*-BN as a thermally conductive material. In this section, we covered the following aspects: 1) Core-shell spherical microbeads were used as thermally conductive fillers; 2) we developed these cellulose/*h*-BN core-shell spherical microbeads by viscose phase separation particulation method; and 3) Cellulose/*h*-BN core-shell microbeads were hybridized with epoxy resin by compression molding.

The cellulose/*h*-BN core–shell spherical microbeads was obtained via viscose phase separation from *h*-BN (average particle size = approximately 8 µm) and cellulose xanthate (Nagaoka et al., 2016). The core–shell CSM were hybridized with epoxy resin via compression molding. The *h*-BN-encapsulated cellulose/*h*-BN composite microbeads was obtained by adding *h*-BN particles to the cellulose xanthate phase. Alternately, core–shell cellulose/*h*-BN composite microbeads were obtained with BN dispersed on the microbead surface by the addition of *h*-BN particles to the polyacrylic acid phase, as shown in Figure 4(e). This indicated that BN encapsulation or *h*-BN-surface dispersion can be selected according to the formulation method.

The obtained *h*-BN-surface-dispersed composite microbeads were investigated as thermally conductive fillers for a heat-dissipating resin sheet. The composite CSM were mixed with an epoxy compound and a curing agent. The sheet was fabricated using a matrix of epoxy resin, 1,6-*bis*(2,3-epoxypropoxy) naphthalene, and a filler of cellulose/*h*-BN core–shell CSM or raw *h*-BN. 1-cyanoethyl-2-ethyl-4-methylimidazole was used as the hardening agent. The epoxy resin, hardening agent, and filler were mixed and defoamed under reduced pressure using a centrifugal stirrer. The mixture was placed in a molding die of volume 1.0 cm^3 and heated at 120 °C under a pressure of 2000 kg cm^{-2} using heating and compression molding equipment.

Figures 5(a) and (b) show thermal conductivity as a function of *h*-BN (particle size: approximately 8 µm) content in *h*-BN/epoxy resin composite sheets filled with core–shell CSM and pristine *h*-BN, respectively. The thermal conductivity of the composite sheet with the unmodified *h*-BN filler did not increase unless the *h*-BN content exceeded the percolation threshold of 30 vol.%, in agreement with the percolation theory (Yorifuji & Ando, 2011; Hauser et al., 2008; He et al., 2007). However, the composite sheet filled with core–shell CSM (BN/Cell-1 and BN/Cell-2) showed a sharp increase in thermal conductivity in both thickness and in-plane directions as the *h*-BN content increased. Figures 5(a–b) indicate that the thermal conductivities of composite sheets filled with core–shell CSM were greater than those filled with pristine *h*-BN. Composite sheets with 27 vol.% of pristine *h*-BN sheet showed a thermal conductivity of 1.2 W m^{-1} K^{-1} and 2.0 W m^{-1} K^{-1} in the thickness and in-plane directions, respectively. In contrast, the sheet filled with 32.5 vol.% core–shell CSM showed a thermal conductivity of 4.65 W m^{-1} K^{-1} and 7.77 W m^{-1} K^{-1} in the thickness and in-plane directions, respectively, approximately four times greater than the sheet filled with pristine *h*-BN. The cross-sectional SEM images of the composite sheets filled with pristine h-BN (upper SEM image) and core–shell CSM (lower SEM image) are presented in Figure 5, respectively. Also, the cross-sectional energy dispersive X-ray (EDX) mapping of the composite sheets filled with core-shell CSM is depicted in the photograph on the right-side of SEM images.

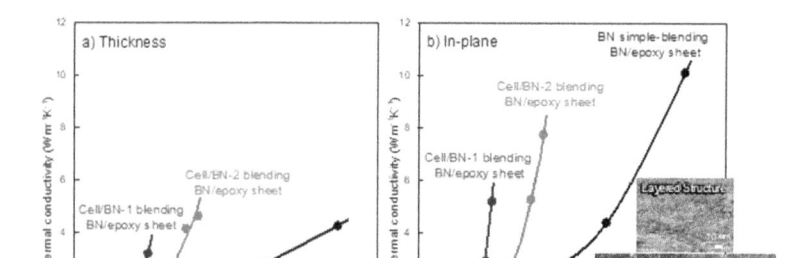

Figure 5. Thermal conductivity for h-BN content in h-BN/epoxy resin composite sheets.

a) Thickness direction, b) In-plane direction

(): BN/epoxy resin composite sheet containing Cell/BN-1 microbeads
(): BN/epoxy resin composite sheet containing Cell/BN-2 microbeads
(): BN/epoxy resin sheet (BN simple-blended)
Note. Cell/BN-1: Cellulose microbeads containing 27.5wt% of h-BN
Note. Cell/BN-2: Cellulose microbeads containing 44.9wt% of h-BN
Note. Upper photograph in Figure 5: cross-sectional SEM image of the composite epoxy resin sheet fabricated blending bare h-BN
Note. Lower photograph in Figure 5: cross-sectional SEM image and its EDX mapping of the composite epoxy resin sheet fabricated blending Cell/BN-1
Note. Red dots: Carbon, Green dots: Boron nitride

The upper SEM image in Figure 5 shows that the composite sheet fabricated using bare *h*-BN had a layered structure. However, the fabricated composite sheet using Cell/BN-1 formed a stone wall structure. We confirmed an increase in the contact area between the shell layers of the composite sheet due to compression. The *h*-BN particles on the boundaries of the core–shell microbeads are distributed due to the presence of nitrogen atoms, as shown in EDX mapping in Figure 5. This distribution generated a stone wall structure with three-dimensionally stacked microbeads in the sheet.

We attributed the large difference in conductivity to the anisotropic structure of *h*-BN. However, the difference between the directions was smaller when core–shell CSM were used as the filler, indicating that the anisotropy of crystalline *h*-BN was suppressed as *h*-BN particles surrounded the cellulose spheres. Furthermore, the changes in thermal conductivity with changing *h*-BN content were outside the percolation theory. The core–shell CSM-based thermally conductive sheets exhibited significantly higher thermal conductivities than those fabricated with pristine *h*-BN. Thus, favorable thermally conductive pathways were formed in the insulating resin and highly thermally conductive sheets could be produced at a low cost.

Hydrophilization of CSM

Western countries such as the United States, Canada, the United Kingdom, France, Sweden, and New Zealand have banned the manufacture and sale of petroleum-derived microbeads used in wash-off cosmetics (Xanthos & Walker, 2017); similar measures will be implemented worldwide in the future. Hence, naturally derived cellulose microbeads are currently attracting attention as an alternative raw material for cosmetics in Western countries and also in Japan and China, where the use of petroleum-derived microbeads has not been prohibited. Cellulose microbeads can be incorporated into creams, gels, or liquid cosmetics and used as moisturizing agents or dispersants. CSM roll on the skin when used in cosmetics. This rolling effect improves the spreading of cosmetics and impart scrubbing effects. CSM prepared via viscose phase separation have a narrow particle size distribution. This makes them promising raw materials for cosmetics. However, their absorption and dispersion in water are insufficient to be used in moisturizing cosmetics, owing to the high crystallinity derived from the cellulose type II structure (Nagaoka et al., 2005c). The intermolecular hydrogen bonds can be dissociated and crystallinity can be lowered by converting the hydroxyl to carboxyl groups. Water can percolate into the cellulose chains, which will enhance their absorption and dispersion in water.

Dicarboxylic Acid Esterified CSM

One end of succinic or glutaric acids, two dicarboxylic acids, can be linked via esterification to CSM. The esterification process is described here: 1) CSM were dispersed in acetone, and potassium hydroxide (KOH) was added to the solution; 2) the mixture was stirred at 50 °C for a predetermined time; 3) dicarboxylic acid anhydride was dissolved in acetone and added dropwise; 4) the reaction mixture was stirred at 50 °C for a predetermined time until a precipitate was formed; 5) the solid precipitate was filtered, washed thoroughly with acetone and water, and lyophilized to obtain esterified cellulose microbeads (Figure 6(a)) (Nagaoka et al., 2005c). The introduction of succinic acid, a naturally derived dicarboxylic acid similar to cellulose, to CSM imparted moisturizing and smooth properties. As they spread when applied to the skin, these products have been blended in lotions and marketed.

Figure 6. Reaction scheme of succinylation (a) and TEMPO-catalyzed oxidation (b) of cellulose

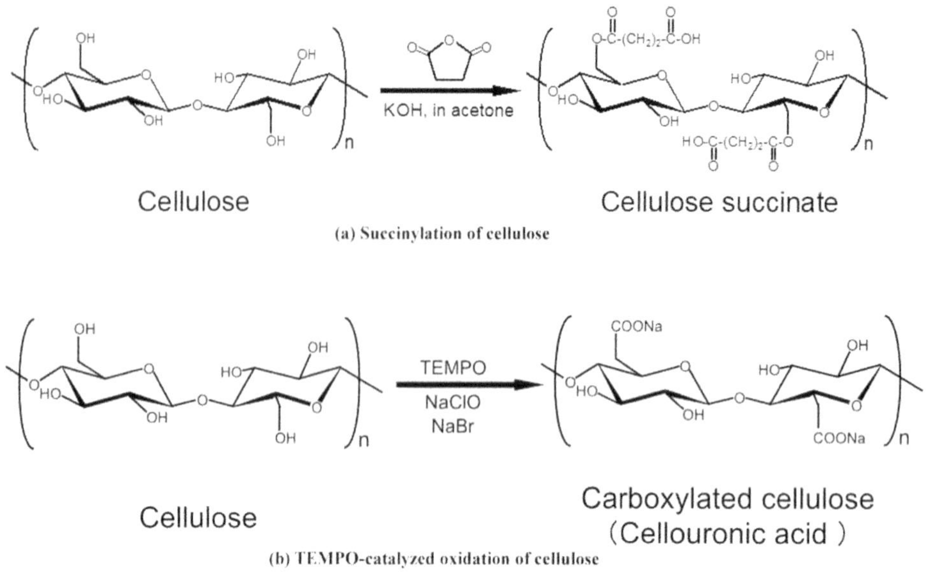

(a) Succinylation of cellulose

(b) TEMPO-catalyzed oxidation of cellulose

TEMPO Oxidized CSM and Application to Cosmetic Materials

Succinic acid-modified CSM are not stable as succinyl groups are released from CSM through a self-cyclization reaction (Nagaoka et al., 2005c). After immersing CSM in a pH 5.4 buffer solution for 70 days, the pH of the CSM suspension decreased from 5.5 to 5.25. This suggests the detachment of succinyl groups from CSM. Therefore, we developed carboxylated CSM with a stable structure. Carboxyl groups were directly bonded to pyranose rings, the structural units of cellulose devoid of ester bonds. An aqueous solution of cellulose treated with TEMPO, sodium bromide (NaBr), and sodium hypochlorite (NaClO) selectively removed the 6-hydroxymethyl group of cellulose to form a sodium salt of the carboxyl group (Figure 6(b)). This reaction is also used to produce cellulose nanofibers. Cellulose nanofibers of diameter approximately 4 nm can be obtained when natural cellulose such as pulp is used as a raw material.

The catalytic oxidation by TEMPO proceeded in water at normal temperature and pressure with a reaction time of approximately 2 h. The reaction consumed NaClO, a food additive. It is expected to be one of the highest production methods. The 6-hydroxymethyl group of cellulose is replaced by a carboxyl group to form cellouronic acid; it is decomposed by bacteria to CO_2 and water as naturally occurring alginic acid (Isogai & Kato, 1998). Although TEMPO is harmful when reduced, it exists as a harmless and stable nitrogen radical in the air (Kato et al., 1998). The presence of generated carboxyl groups is confirmed to be controllable by adjusting the amount of added NaClO during the TEMPO-mediated oxidation of CSM. The number of generated carboxyl groups can be optimized to control the amount of water absorbed by microbeads. The oxidation of CSM by TEMPO is catalyzed in the range 20–25 °C. CSM with an average particle size of 10–200 μm were dispersed in water; TEMPO and NaBr were added to the solution with stirring, followed by the addition of the required amount of NaClO aqueous solution. The cellulose Type II crystal structure inside the microbeads was affected due to the sudden addition of NaClO also reducing the yields, owing to dissolution or failure to maintain the particle size. Therefore, the dropping speed should be maintained appropriately. The mixture was stirred for 2 h until the pH was in the range 10–10.5. The reaction was later neutralized with HCl, filtered and the residue was washed with water to remove TEMPO, NaBr, and NaCl to obtain the target microbeads. However, the microbeads fused on direct vacuum or freeze drying. Therefore, it was necessary to remove the maximum water by washing with ethanol before drying.

Figure 7 shows optical micrographs of the obtained carboxylated CSM with an average particle size of 50 μm in dry and wet conditions. The spherical shape was maintained. An increase in the number of carboxyl groups increased the swelling and absorption in water. Figure 7(c) indicates that CSM with a carboxyl group of 2.1 mmol g^{-1} swelled two- and eight-fold in diameter and volume, respectively. Moreover, they absorbed 55 times more water than their weight. The absorption of water was enhanced as 5–40 mol% of cellulose on the surface and inner region of carboxylated CSM was converted to cellulose acid. The crystalline structure of non-cellouron-oxidized cellulose in the core was attributed to the maintenance of the spherical shape, whereas the cellouron-oxidized outer shell absorbed water. As the carboxylated CSM were highly dispersed in water, no sedimentation occurred even after a month or more, whereas noncarboxylated CSM showed a tendency to sediment. Moreover, carboxylated CSM rolled when applied on a surface, owing to their spherical shape. Therefore, they would provide a smooth texture without damaging the skin when blended with cosmetics and pharmaceuticals.

Figure 7. Optical micrograph of carboxylated cellulose spherical microbeads (a): COOH = 0 mmolg^{-1}, (b): COOH = 1.1 mmolg^{-1}, (c): COOH = 2.1 mmolg^{-1} Upper row: dry state, lower row: swollen state in water

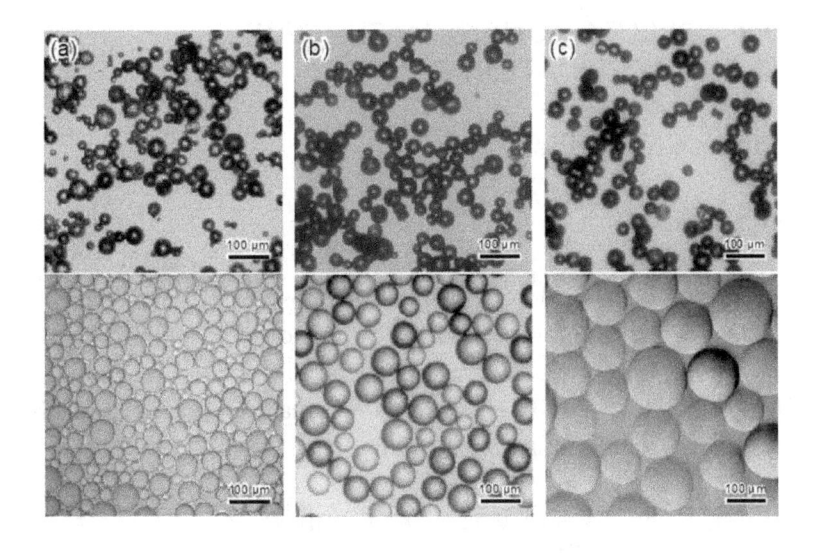

Summary of First Half of This Section

Cosmetic materials for human skin are discharged into rivers and oceans as waste in large quantities after being removed from the body. The annual production of cellulose, a natural polysaccharide derived from plants, far exceeds the production from oil reserves, making it an inexhaustible raw material. A particulation technology for obtaining finer and monodispersed cellulose particles is required to enhance their utilization, as a substitute for artificial polymers. Nanospherical cellulose particles are promising nano-materials, owing to their increased specific surface area, light transmittance, and functionality. Many researchers around the world have been developing nanofiber and nanocrystal manufacturing technologies and applications for nano-miniaturization of cellulose, and many research reports have been published (Isogai, 2013; Isa et al., 2016; Kondo et al., 2014; Nakagaito & Yano, 2014; Noguchi et al., 2017; Ho et al., 2011).

We also developed an application of cellulose nanomaterials for light-shielding glass as will be described in the latter part of the chapter. On the other hand, while there are relatively few development cases of cellulose nanospherical particles in general, we have succeeded in particulating nano-sized cellulose particles with high yields (Sekimoto et al., 2016).

The particulate formation process of cellulose has been studied and progressed despite many challenges, such as the cellulose dissolution technologies in water and solvents. Recently, the use of petroleum-derived microplastics has been restricted worldwide. Hence, cellulose microbeads are in demand as their substitutes. In the future, further progress in particulate technology is expected, and various research institutes, including the authors, will compete to accelerate research and development.

APPLICATION OF CELLULOSE NANOMATERIALS: DEVELOPMENT AND APPLICATION OF

Conductive Cellulose Nanofibers

Cellulose Nanomaterials

In the first half of this chapter, the development of cellulose-based microbeads, that is, various microbeads materials, was reviewed. In the second half of this chapter, we introduce the applications of cellulose nanofiber (CNF). Recently, we have developed the conductive CNF as functionalized cellulose (Horikawa et al., 2015, 2017, 2018, 2020). These composite materials have entered the stage of practical application; here, we present the recent research results.

Conductive Polymer, Poly(3,4-ethylenedioxythiophene)/ polystyrene Sulfonate (PEDOT/PSS)

Indium, a rare metal, is in demand as indium tin oxide, which is used as a transparent electrode material in photovoltaic devices (ITO), such as solar cells and organic light-emitting diodes. Recently, conductive polymers have been used as alternatives to indium tin oxide. However, these materials require an appropriate level of conductivity, workability, and stability. As polypyrrole, polyaniline, and poly(3,4-ethylenedioxythiophene) (PEDOT) possess the required parameters, they have been extensively studied for practical applications. In particular, it is known that PEDOT possesses the highest conductivity. PEDOT was developed by HC Starck AG (formerly Bayer AG) in Germany in 1985. The PEDOT/PSS is obtained by doping PEDOT with polystyrene sulfonate (PSS). The conductivity of the material has been exploited to develop antistatic materials, hole-injection layers for organic thin-film solar cells, and conductive films for touch panels. As PEDOT/PSS is an aqueous dispersion, it can easily form a coating film under atmospheric pressure. The conductivity of PEDOT/PSS dispersion increases when polar molecules are added as secondary dopants (Takano et al., 2012). The addition of substances with polar groups, such as DMSO, NMP, ethylene glycol (EG), DMF, and alcohol promotes the orientation of PEDOT/PSS. In addition, thermal drying increases the crystallized region and improves conductivity. The addition of a secondary dopant changes PEDOT from a benzoid to a quinoid structure, facilitating electron delocalization. Ouyang et al. reported that the addition of ethylene glycol as a secondary dopant to PEDOT/PSS facilitated the formation of a quinoid structure due to the polar hydroxyl group, which increased conductivity from 0.4 to 200 S cm^{-1}(Ouyang et al., 2004).

PEDOT/Cellulose Sulfate (CS) using Cellulose Sulfate as Dopant

We focused on sulfated polysaccharides with hydroxyl groups as dopants and adopted cellulose with rigid chains of linear β-1,4-linked glucopyranose residues, because PEDOT molecules are presumed to line around the cellulose rigid chains and be easy to crystallize (Horikawa et al., 2015). The arrangement of PEDOT along the rigidly ordered structure of sulfated cellulose during the formation of an ionic complex (Figure 8) will suppress the disturbance in the conduction path in the crystal lattice. In addition, cellulose has polar hydroxyl groups, such as EG, which may induce the crystallization of PEDOT and the formation of quinoid structures, which may improve its conductivity. Moreover, in contrast to PSS, in PEDOT, there is no absorption due to functional groups such as aromatic rings. Therefore, the transmittance in the ultraviolet (UV) region increases the positive influence of the transparent conductive films.

Figure 8. Structure of PEDOT/CS

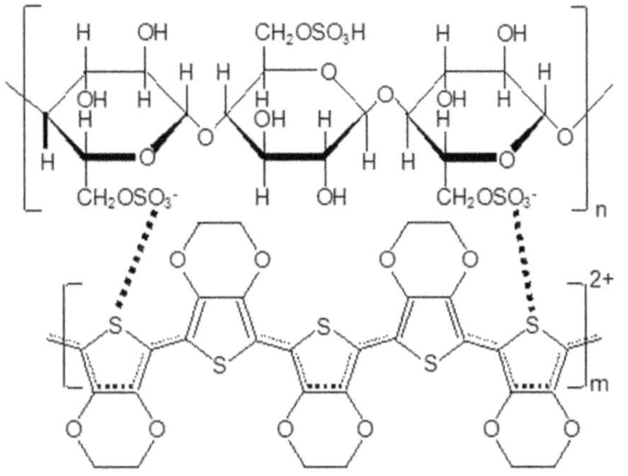

Preparation of PEDOT/CS and their Characterization

CS was prepared by reacting bulk cellulose (pulp) with chlorosulfonic acid. CS-1 to CS-5 with the different degree of substitution of sulfonate group (DS) per glucose unit were prepared. The DS values were calculated by elemental analysis. Consequently, the DS values for CS-1, CS-2, CS-3, CS-4, and CS-5 were as follows: 0.9, 1.03, 1.28, 1.58, and 2.06, respectively. 3,4-ethylenedioxythiophene (EDOT) was added to the aqueous CS dispersion, and the EDOT was oxidatively polymerized using potassium persulfate and iron sulfate as initiators. A PEDOT/CS dispersion was obtained by dialyzing and purification. The zeta potential of PEDOT/PSS particles was measured at -73.8 mV, whereas for all variations of PEDOT/CS particles, it averaged around -40 mV. These results indicate that the surfaces of the PEDOT/PSS particles were rich in sulfonyl groups. On the other hand, Z average particle diameters and PDIs for the PEDOT/PSS and PEDOT/CS particles in H_2O, determined by Dynamic Light Scattering (DLS). The Z average particle diameter of the PEDOT/CS particles ranged from about 250 to 350 nm (Horikawa, 2015, 2017, 2018, 2020).

Crystallinity of Cellulose Affecting Conductivity of PEDOT/CS Thin *Films*

Figure 9. XRD patterns of cellulose, PEDOT/PSS, and PEDOT/CS films

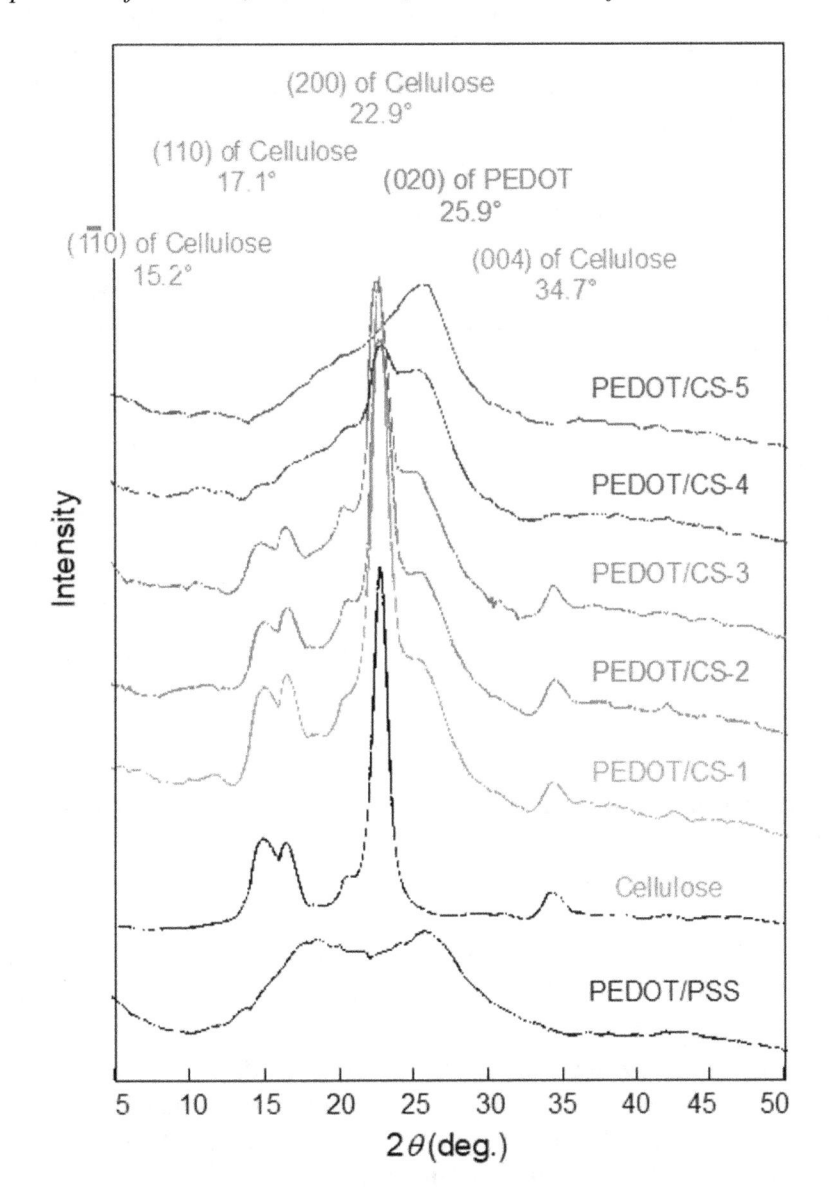

Figure 9 shows the structural evaluation of PEDOT/PSS and PEDOT/CS via XRD. The peaks at $2\theta =$ 15.2°, 17.1°, 22.9°, and 34.7° corresponded to the (11(−)0), (110), (200), and (004) planes of crystalline cellulose, respectively. The conformation of cellulose in CS-1 and CS-2 almost retained the characteristic rigid structure of cellulose chains despite the formation of a complex with PEDOT in water. The PEDOT molecules lined up around the rigid cellulose chains with linear β(1,4)-linked glucopyranose residues.

PEDOT/CS films were fabricated by spin coating PEDOT/CS on a glass substrate via the film formation method by Crispin et al. and heating at 120 °C for 20 min (Crispin et al., 2006). A PEDOT/PSS film was also fabricated for comparison. The volume resistivities of PEDOT/CS and PEDOT/PSS films were calculated by measuring the surface resistance using the four-probe method and obtaining the film thickness using the stylus method.

Figure 10. Electrical conductivity for the degree of substitution of sulfuric groups (a) and cellulose crystallinity (b)

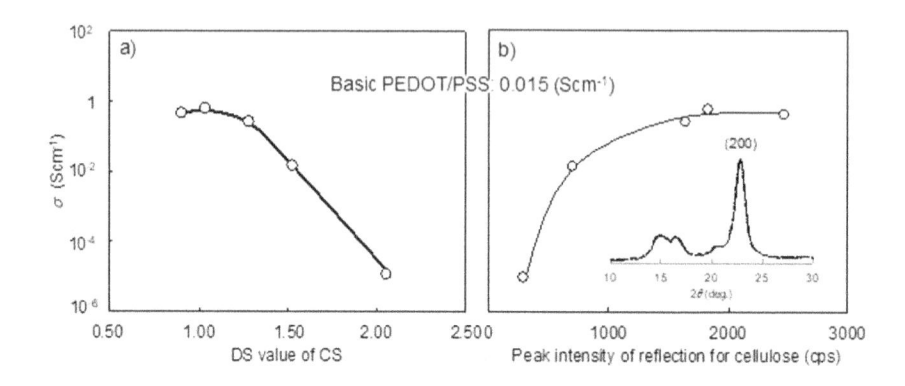

a) Relationship between electrical conductivity and the degree of substitution of sulfuric groups.
b) Peak intensity of 200 reflection for cellulose dependence of electrical conductivity of PEDOT/CS.
   Note. Inset: XRD pattern of PEDOT/CS

Figure 10(a) shows the relationship between DS and conductivity. Conductivity was the highest when the DS value was nearly 1, whereas it decreased when DS > 1. The sulfation of cellulose proceeded regioselectively in the order: C6 > C2 > C3. However, the structure became amorphous as DS increased similar to reports by Zang et al. (Zang et al., 2011) and Qin et al. (Qin et al., 2014). PEDOT/CS-4 had higher conductivity than PEDOT/CS-5. PEDOT was less likely to be arranged around CS-5 in PEDOT/CS-5 with a DS value of 2.06, indicating a lower conductivity than CS-4. As a result, PEDOT/CS-2 have highest conductivity of all kinds of PEDOT thin films.

Figure 10(b) shows the relationship between the conductivity of the PEDOT/CS series and cellulose crystallinity, as determined from the height of the (200) peak of cellulose at $2\theta = 22.9°$. The electrical conductivity of PEDOT/CS increased with cellulose crystallinity.

Figure 11. Raman spectra of PEDOT/PSS and PEDOT/CS-2 films (a) PEDOT/PSS, (b) PEDOT/CS-2

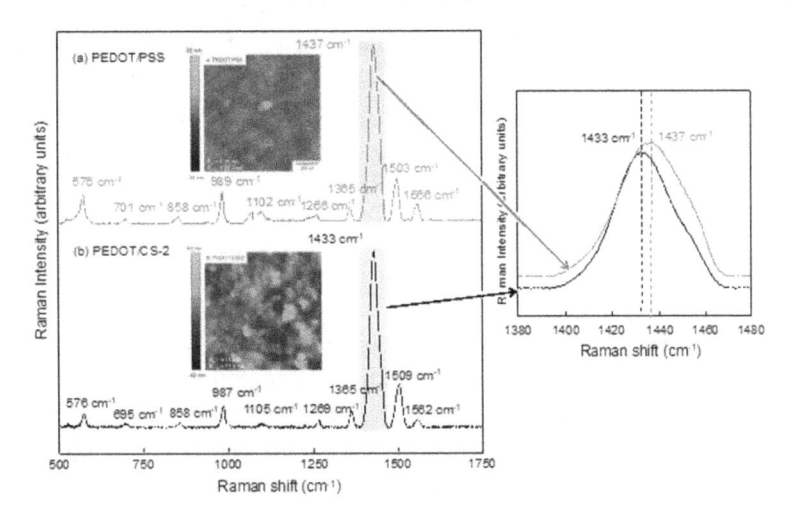

Figure 11(a) and (b) shows raman spectra of PEDOT/CS-2 and basic PEDOT/PSS. The $C_\alpha = C_\beta$ symmetric stretching vibration peaks of PEDOT/CS shifted from 1566 cm^{-1} to 1562 cm^{-1} and from 1437 cm^{-1} to 1433 cm^{-1}. The peak width decreased with the addition of PEDOT/CS, as shown in Figure 11(b). As reported by Ouyang et al., this suggests that PEDOT changed from benzoid to a quinoid structure that possibly delocalized the electrons (Ouyang et al., 2004). We also investigated the transparency of the sample. PEDOT/PSS was absorbed due to the phenyl group, while PEDOT/CS-2 had no phenyl group. Therefore, it was highly transparent in the UV region of 200–400 nm. Transmittance in the UV region also increased, which is an advantage of devices with UV light. Atomic force microscopy (AFM) results in Figure 11 revealed that R_a and R_{max} of PEDOT/PSS were 1.26 and 20.3 nm, respectively, while those of PEDOT/CS-2 were 8.51 and 71.6 nm, respectively. Thus, PEDOT/CS-2 was 38 times more conductive than basic PEDOT/PSS, despite its larger surface roughness.

In addition, the observation of transmission electron microscopy (TEM) results confirmed the presence of a nanocrystalline morphology in the PEDOT/CS material (Horikawa et al., 2017). Thus, PEDOT was oriented on cellulose fibers and suppressed the disturbance of the conduction path due to disorder within the PEDOT crystal lattice, leading to an improvement in conductivity. The PEDOT molecules were oriented along the cellulose structure and induced the formation of a quinoid structure, resulting in high electrical conductivity. Furthermore, cellulose was nanocrystallized via the sulfation of pulp. PEDOT/CS conductivity was affected by the crystallinity of cellulose. PEDOT/cellulose sulfate nanofiber (abbreviated as *s*-CNF) can be used as a conductive fiber, in the future.

Totally-organic NIR Shielding Materials using Conductive Cellulose Nanofibers

We studied the composites formed by cellulose nanofibers and PEDOT. PEDOT is a conductive polymer, which is lightweight and flexible. Hence, it is applied to devices such as touch panels. Conventionally, inorganic materials such as antimony tin oxide, indium tin oxide, and cesium tungstate have been used for heat shield glass. However, the use of these materials has been restricted due to their harmful effects on the human body or problems with the supply of rare metals. We envisaged that PEDOT/*s*-CNF with enhanced conductivity than that of PEDOT due to the presence of CNF might be capable of absorbing NIR rays. Thus, we investigated its application as a heat shield material (Yoshida et al., 2020).

In this section, we introduce the results of optimizing the polymerization conditions for conductivity and NIR absorption of the aforementioned conductive PEDOT/ s-CNF composite. In addition, we outline the results of incorporating it into laminated glass as an all-organic heat shield interlayer and confirm the resistance to temperature rise. Finally, we applied it to window sashes and used the results to confirm the energy-saving effect.

NIR Shielding Materials

NIR light in the range 700–2000 nm accounts for half of solar radiation energy (Avlasevich & Müllen, 2006). NIR irradiation increases the temperature inside cars, buildings, etc. Hence, more energy is consumed during air conditioning. The increase in energy consumption has contributed to global climate change. Thus, to suppress such temperature increases, researchers must develop materials and techniques that shield structures from NIR light. Various materials have been reported to show NIR-shielding properties, including 1) metal oxides (NIR absorbing) such as tin-doped indium oxide (ITO)(Dalapati et al., 2018; Ray et al., 2007; Katagiri et al., 2013), antimony-doped tin oxide (ATO) (Dalapati et al., 2018; Abendroth et al., 2017), and cesium tungsten oxide (Cs_2WO_4) (Li et al., 2020; Xu et al., 2018) and 2) metals (NIR reflective) such as Ni, Ag, and Au (Dalapati et al., 2018; Bennett & Ashley, 1965; Ahmad et al., 2012; Pratesi et al., 2014). The thin films of ATO, ITO, and Cs_2WO_4 have high transmittance in the visible-light range and high NIR optical absorption. However, the use of ATO is limited because antimony is harmful to humans. Indium and cesium are rare because their supply is limited. Moreover, they are in high demand as raw materials for the production of transparent electrodes and terminals for liquid-crystal displays, solar panels, organic light-emitting diodes, and touch panels. The thin films of these inorganic materials are rigid and inflexible, which limits their applications. Researchers have recently used conductive polymers for NIR shielding (Im et al., 2019; Lai et al., 2019; Chen et al., 2015). Among these polymers, poly(3,4-ethylenedioxythiophene) (PEDOT) is a totally organic material that absorbs NIR light (Lai et al., 2019; Chen et al., 2015), and has been most attractive for use in solar cells, LEDs, and thermoelectric devices (Sun et al., 2015; Seo et al., 2019; Yoon et al., 2018; Park et al., 2019; Fan & Ouyang, 2019). PEDOT thin films are simpler and less expensive to form than inorganic thin films because they can be prepared under atmospheric pressure. Additionally, the polarons and bipolarons of PEDOT exhibit NIR optical absorption bands, enhancing their conductivity (Zozoulenko et al., 2019; Mochizuki et al., 2012). Thus, the NIR absorption of PEDOT can be increased by increasing its conductivity (R.-Reyes et al., 2010; Zhao et al., 2014). It is generally doped with PSS. Recently, the conductivity of PEDOT/PSS films was increased by adding polar organic compounds to a PEDOT/PSS aqueous dispersion (Shi et al., 2015; Ouyang et al., 2004) and exposing the film to water vapor (Jikei et al., 2014). A transformation from benzoid to the quinoid structure is induced by the polar molecule, promoting the crystallization of PEDOT.

In the present study, we used sulfated cellulose as a dopant of PEDOT based on multiple factors: 1) presence of hydroxyl groups in glucopyranose residues; 2) presence of linear rigid chains composed of β (1→4)-linked glucopyranose residues; 3) alignment of PEDOT molecules around the rigid linear chains of cellulose, and 4) one can readily incorporated sulfated groups into cellulose can be easily controlled by using chlorosulfonic acid. Furthermore, the crystallinity of the cellulose structure influences the conductivity of PEDOT. PEDOT doped with microcrystalline cellulose sulfate had higher conductivity than PEDOT doped with amorphous cellulose sulfate (Horikawa et al., 2015).

Optimization of PEDOT Polymerization Conditions using s-CNF

s-CNF were obtained by dispersing CNF powder (Chuetsu Pulp Co., Ltd., Nanoforest, high fibrillation degree (C)) in DMF and reacting it with chlorosulfonic acid at room temperature. In this study, we used *s*-CNF with a DS value of 1.73 obtained under the specified conditions. The redox polymerization of EDOT with potassium persulfate ($K_2S_2O_8$)–iron(II) sulfate *n*-hydrate [$Fe_2(SO_4)_3$] as an initiator was conducted on the surface of the obtained *s*-CNF, according to a previous report (Horikawa et al., 2015).

Figure 12. Conductivity of PEDOT/s-CNF film prepared in various conditions a) EDOT/s-CNF ratio condition, b) Concentration of $K_2S_2O_8$, c) pH condition

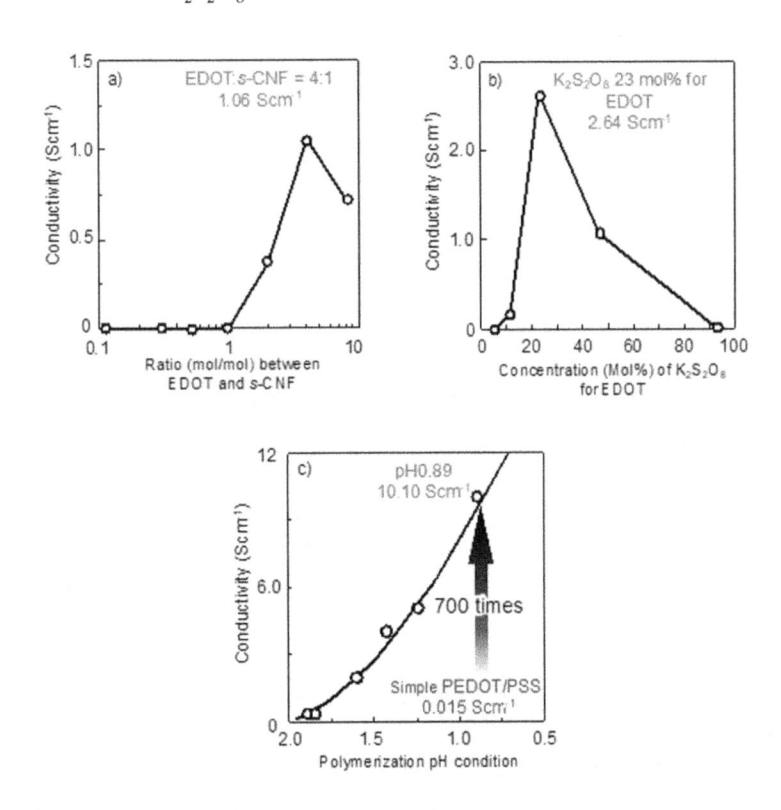

Table 2. Preparation condition of PEDOT/s-CNF

PEDOT Sample No.	Molar ratio		$K_2S_2O_8$ (mol x 10^{-4})	$Fe_2(SO_4)_3$ [1] (mL)	pH
	s-CNF	EDOT			
1	1.0	0.125	3.5	250	1.0
2	1.0	0.250	3.5	250	1.0
3	1.0	0.500	3.5	250	1.0
4	1.0	1.0	3.5	250	1.0
5	1.0	2.0	3.5	250	1.0
6	1.0	4.0	3.5	250	1.0
7	1.0	8.0	3.5	250	1.0
8	1.0	4.0	7.0	500	1.0
9	1.0	4.0	3.5	250	1.0
10	1.0	4.0	1.8	125	1.0
11	1.0	4.0	0.9	62.5	1.0
12	1.0	4.0	0.4	31.3	1.0
13	1.0	4.0	1.8	125	0.89
14	1.0	4.0	1.8	125	1.24
15	1.0	4.0	1.8	125	1.43
16	1.0	4.0	1.8	125	1.60
17	1.0	4.0	1.8	125	1.84
18	1.0	4.0	1.8	125	1.86

Note. 1) $Fe_2(SO_4)_3$ aqueous solution (1.4 mgmL^{-1})

Table 2 summarized 1) charge ratio (molar ratio) of EDOT to *s*-CNF, 2) concentration of $K_2S_2O_8$-$Fe_2(SO_4)_3$ in the solution, and 3) pH conditions to optimize the polymerization conditions. The polymerization reaction was performed for 24 h, and the physical properties were evaluated. Each product was purified via dialysis for 72 h to obtain PEDOT/*s*-CNF. 0.6 wt% PEDOT/*s*-CNF dispersion was used to fabricate a PEDOT/*s*-CNF thin film by the spin coating method. Figure 12(a) shows the relationship between the charging ratio of s-CNF to EDOT and the conductivity during polymerization. Polymerization was performed by varying the charging ratio of EDOT and s-CNF from 1:4 to 4:1. Several types of thin films were prepared from the obtained polymers using a spin coater, and their conductivities were analyzed. The highest conductivity was 1.06 S cm^{-1} at a charge ratio of 4:1. However, conductivity decreased when the charge ratio of EDOT > 4:1. The optimal ratio of KPS to EDOT during the polymerization was determined by keeping the ratio of EDOT to s-CNFs at 4:1. Figure 12(b) shows the relationship between the charging ratio of KPS to EDOT and conductivity during polymerization. The electrical conductivity was 2.64 S·cm^{-1} at a KPS ratio (mol%) of 23 mol%, and the electrical conductivity gradually decreased the ratio > 23 mol%. Furthermore, the polymerization was attempted by changing the pH conditions during the polymerization.

Figure 12(c) indicates that conductivity increased with the decrease of pH. The results of X-ray photoelectron spectroscopy (XPS) indicated low conductivity in the presence of a high amount of residual Na$^+$ ions, suggesting that the ions are responsible for the generation of polarons and bipolarons. Consequently, the highest conductivity of 10.1 S cm^{-1} was obtained, which was 27 times higher than that reported previously (Horikawa et al., 2015). The conductivity of the sample was 673 times higher than that of a thin film fabricated under the same conditions using basic PEDOT/PSS devoid of a secondary dopant.

Raman spectra showed that the $C_\alpha = C_\beta$ symmetric stretching vibration of PEDOT shifted from 1566 cm^{-1} to 1562 cm^{-1} and from 1437 cm^{-1} to 1433 cm^{-1}, indicating the change of PEDOT from a benzoid to a quinoid structure and were confirmed to have metastasized.

Figure 13. Optical properties of the PEDOT/s-CNF (PEDOT No.13-17) with various conductivities (a) and TEM image of PEDOT No.13 (b)

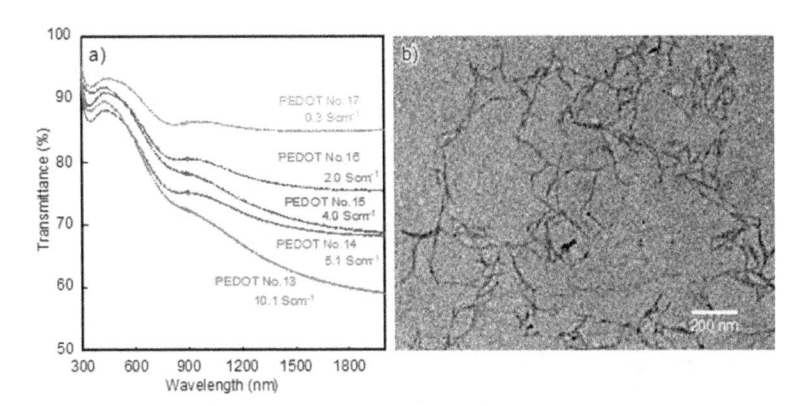

Figure 13(a) and (b) show and the relationship between the optical properties and conductivity of the PEDOT/*s*-CNF thin films prepared by varying pH during polymerization (a) and the TEM image of PEDOT/*s*-CNF No.13 (b), respectively. PEDOT/*s*-CNF has a microcrystalline structure with a size of approximately 200–400 nm. Light absorption in the NIR and IR wavelength regions from 700 to 2000 nm increased with increasing conductivity (decrease in pH during polymerization) of the PEDOT/*s*-CNF thin film. This is due to the extended conjugation resulting from the transition of benzoid to the quinoid structure of PEDOT.

We evaluated the transmittance in the visible ($\tau_{380-700}$) and the NIR light region ($\tau_{700-2000}$). The weight factor ($E\lambda \cdot \Delta\lambda$) can be obtained from the CIE daylight D65 spectrum and the wavelength distribution of CIE photopic luminosity (JIS R 3106:1998, Japanese Standards Association). The weighting coefficients ($E_\lambda \cdot \Delta\lambda$) are determined from the solar relative spectral distribution and wavelength interval. The visible light (380–700 nm) and NIR (700–2000 nm) transmittances were denoted as $\tau_{vis(380-700)}$ and $\tau_{NIR(700-2000)}$. We obtained $\tau_{vis(380-700)}$ by measuring the spectral transmittance $\tau(\lambda)$ from 380 to 700 nm using Eq. (1), which considers the weighting coefficients ($E_\lambda \cdot \Delta\lambda$) to calculate the weighted average. Similarly, we obtained the $\tau_{NIR(700-2000)}$ using Eq. (2).

$$\tau_{380-700} = \frac{\sum_{380}^{700} E_\lambda \cdot \Delta\lambda \cdot \tau(\lambda)}{\sum_{380}^{700} E_\lambda \cdot \Delta\lambda}$$

$$\tau_{700-2000} = \frac{\sum_{700}^{2000} E_\lambda \cdot \Delta\lambda \cdot \tau(\lambda)}{\sum_{700}^{2000} E_\lambda \cdot \Delta\lambda}$$

E_λ: Solar relative spectral distribution
$\Delta\lambda$: Wavelength interval
the weighting coefficients (JIS R 3106:1998, Japanese Standards Association).

Table 3. Vis and NIR transmittance ($\tau_{380-700}$ and $\tau_{700-2000}$) and conductivity of PEDOT/s-CNF thin films

PEDOT No.	Conductivity (Scm^{-1})	Transmittance (%) at 550 nm	$\tau_{380-700}$ (%)	$\tau_{700-2000}$ (%)
13	10.1	92.4	85.0	69.8
14	5.1	89.3	88.3	76.4
15	4.0	86.0	84.9	73.7
16	2.0	89.3	88.3	79.4
17	0.3	86.2	91.6	86.0

Table 3 summarizes the $\tau_{vis(380-700)}$ and $\tau_{NIR(700-2000)}$ values against the conductivity values of the obtained PEDOT/s-CNF thin films. An increase in conductivity gradually decreased $\tau_{NIR(700-2000)}$, maintaining a constant value of $\tau_{vis(380-700)}$. The $\tau_{vis(380-700)}$ value of the PEDOT/s-CNF thin film (PEDOT No. 13) with the highest conductivity was 85.0%, whereas the $\tau_{NIR(700-2000)}$ value was 69.8%.

Preparation and Strength of the PEDOT/s-CNF-PVA Film

A given amount of 1.2 wt% PEDOT/s-CNF was dispersed in a 10 wt% polyvinyl alcohol (PVA) aqueous solution containing a surfactant. The aqueous dispersion of PEDOT/s-CNF-PVA was coated onto the surface of a glass substrate and cleaned with a UV ozone cleaner via the doctor blade method to form a film.

Figure 14. PEDOT/s-CNF-PVA composite film

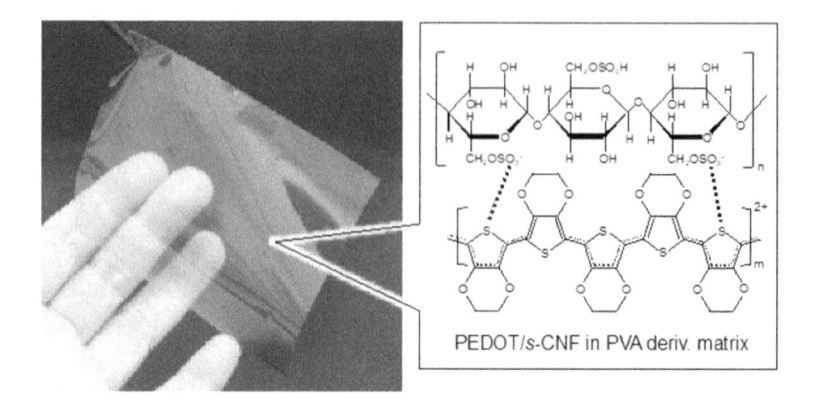

Figure 14 indicates a transparent PEDOT/*s*-CNF-PVA film with a thickness of approximately 11 µm. The tensile strengths of PVA and PEDOT/PSS-PVA films were also investigated for comparison. PVA and PEDOT/PSS-PVA films have a tensile strength of approximately 22–23 MPa, whereas the PEDOT/*s*-CNF-PVA film has a tensile strength of 31.8 MPa, which is approximately 1.4 times higher than the former (Yoshida et al., 2020). This is attributed to the reinforcing effect of *s*-CNF as reinforcing fibers. A PEDOT/*s*-CNF-PVA composite film glass was installed. A visible light cut filter (400–690 nm cut) was placed on top of it and irradiated with simulated sunlight for 30 min. The temperature inside the box was monitored (Figure 15). A glass plate and PVA film glass were used as blanks.

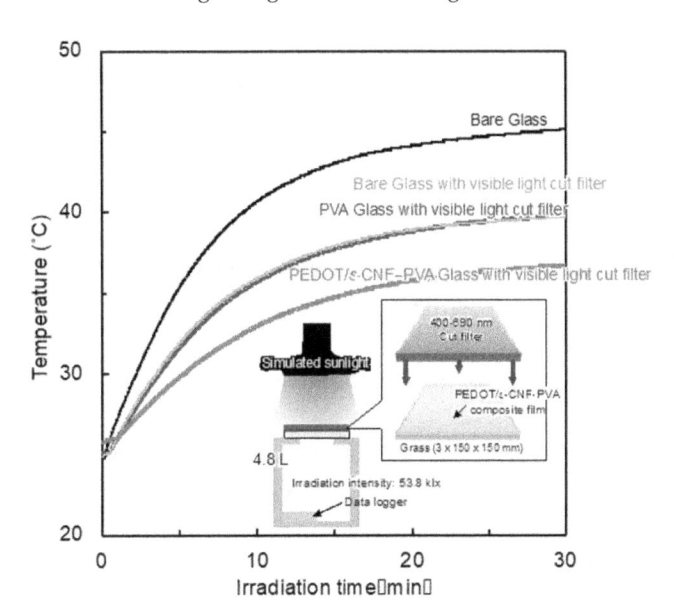

Figure 15. Temperature rise monitoring using simulate sunlight

Figure 15 shows that the internal temperature of the glass plate and the PVA film glass increased to 39.7 °C, while that of the PEDOT/*s*-CNF-PVA film glass increased to 36.7 °C. This indicated that the PEDOT/*s*-CNF-PVA film suppressed the temperature rise by 3 °C, owing to light absorption in the NIR region by PEDOT. We investigated the scale-up of a PEDOT/*s*-CNF-PVA dispersion for film deposition using roll-to-roll gravure printing.

Scaled-up Preparation of PEDOT/s-CNF-PVA Dispersion

We investigated the scale-up of a PEDOT/*s*-CNF-PVA dispersion for film deposition using roll-to-roll gravure printing (Yoshida et al., 2020). We used a 100 L glass-lined production batch of actual scale for the scaled-up production with the cooperation of KJ Chemicals Co., Ltd., which has a production base in Yatsushiro City, Kumamoto Prefecture, Japan.

A given amount of surfactant was dissolved in a 10 wt% PVA aqueous solution and 175 g of PEDOT/*s*-CNF was dispersed in it. 88 kg of the PEDOT/*s*-CNF-PVA dispersion-based coating solution was successfully prepared. The obtained water-based coating fluid was a viscous liquid with a viscosity of 285 mPa·s and a pH of 4.83. The obtained viscous water-based coating liquid was optimal for roll-to-roll gravure printing. The PEDOT/s-CNF-PVA composite layer was coated onto a 50 μm thickness of PET film using roll-to-roll gravure printing. The process comprises the following steps: 1) a PET film with a thickness of 50 μm, width of 900 mm, and length of 500 m was ejected at a speed of 5 m min^{-1}; 2) corona was treated; 3) micro-gravure was coated; 4) the PEDOT/s-CNF-PVA composite PET film was completely dried in a 22 m drying chamber at 110 °C and 5) a layer was coated on a PET film. The obtained PEDOT/s-CNF-PVA layer on the PET film was transparent with a thickness of approximately 11 μm.

Fabrication of Laminated Glass Incorporating the PEDOT/s-CNF-PVA Composite PET Film

We investigated the incorporation of the PEDOT/s-CNF-PVA composite film into a laminated glass plate (standard size) to fabricate glass for sashes. The EVA film (the second layer) was placed on a glass plate (the first layer) measuring 3 mm in thickness, 703 mm in width, and 1.565 mm in height. The PEDOT/s-CNF-PVA composite film (the third layer) was then placed, and finally, another layer of EVA film was added as the fourth layer. A glass plate as the fifth layer was placed on the top and bonded via vacuum thermocompressed bonding after tapping. The PEDOT/s-CNF-PVA composite film was incorporated as a light shielding interlayer between two glass plates.

Mounting as a Sash and Heat Shielding Effect

Based on these results, the performance of CNF-based products was evaluated from 2017 to 2020 (the fabrication of building materials using bamboo CNF, demonstration of the CO_2 reduction in existing housing complexes, 12 participating institutions, representative: Nikken Housing Systems Co., Ltd., Akiyoshi Furuyama). The PEDOT/s-CNF-PVA light-shielding laminated glass was installed in an existing housing complex as a window sash. The surface temperature of the house was confirmed using thermography on a sunny day in August. The saved energy of the three types of CNF building materials was verified by analyzing the cumulative amount of electricity consumed by the installed units. The cumulative power consumption of air conditioners for the entire dwelling unit from June to October was reduced by 6.5% (26.7 kWh) with CNF building materials compared to the conventional building.

SUMMARY OF LATTER HALF OF THIS SECTION

We synthesized conductive CNF (PEDOT/s-CNF) and evaluated its physical properties. The synthesis was scaled up successfully for the actual application. We developed a heat-shielding intermediate film formed by atmospheric-pressure roll-to-roll and incorporated it into the sash. Excellent energy-saving results were confirmed by the system installed in an apartment complex.

Although CNF are expensive, the amount of PEDOT/s-CNF-PVA required per m^2 of glass was approximately 200 mg, which is the same as that of commercially available light-shielding glass using antimony or ITO. Trial calculations have been performed to reduce CO_2 emissions by 78,000 L per dwelling unit (equivalent to four panes of glass) over the past 30 years. Thus, a small amount of material can provide significant light-shielding and heat-shielding effects. We anticipate our technology to be used as a power-saving support material to reduce the energy consumption of air conditioning and heating.

CONCLUSION

This chapter introduced the design, preparation, and application of cellulose particulates and the development of functional materials using cellulose nanofibers and their applications. Cellulose is sustainable as it can be produced by planned agroforestry. In Japan, forests occupy nearly 70% of the national land area, approximately two-thirds of which are artificial. The produced cedar and cypress are used as building and raw materials for pulp. Additionally, Japan's pulping technology is the top class in the world and has the potential to be converted into other forms, such as cellulose nanofibers. Thus, we believe that further utilization of cellulose is necessary. The use of petroleum-derived microplastics is restricted worldwide. Attention has been focused on fine cellulose materials, and their demand and expectations have increased. It is important to improve the processability of cellulose and develop its applications for environmental conservation. In the second half of this section, we introduce examples of the development of functional materials using cellulose nanofibers and their applications. In the future, it is expected that various research institutes, including the authors, will compete to accelerate cellulose research for future expansion and development of applications for cellulosic materials.

Author Note

The data related to the research in this chapter is the result of joint research with Kumamoto University. Several of these research results are also published in the cited references. We have no conflicts of interest to disclose. Correspondence concerning this article should be addressed to Dr. Shoji Nagaoka, Kumamoto Industrial Research Institute Materials Development Department, Higashi-ku, Higashi-machi, Kumamoto, Japan. Email: nagaoka@kumamoto-iri.jp

REFERENCES

Abendroth, T., Schumm, B., Alajlan, S. A., Almogbel, A. M., Mäder, G., Härtel, P., Althues, H., & Kaskel, S. (2017). Optical and thermal properties of transparent infrared blocking antimony doped tin oxide thin films. *Thin Solid Films*, 624(28), 152–159. DOI: 10.1016/j.tsf.2017.01.028

Adachi, T., Ida, J., Wakita, M., Hashimoto, M., Ihara, H., & Hirayama, C. (1999). Preparation of Spherical and Porous Chitosan Particles by Suspension Evaporation with O/W/O Multiple Emulsions. *Polymer Journal*, 31(4), 319–323. DOI: 10.1295/polymj.31.319

Ahmad, N., Stokes, J., Fox, N. A., Teng, M., & Cryan, M. J. (2012). Ultra-thin metal films for enhanced solar absorption. *Nano Energy*, 1(6), 777–782. DOI: 10.1016/j.nanoen.2012.08.004

Avlasevich, Y., & Müllen, K. (2006). Dibenzopentarylenebis(dicarboximide)s: Novel near-infrared absorbing dyes. *Chemical Communications*, 4440-4442(42), 4440–4442. Advance online publication. DOI: 10.1039/B610318A PMID: 17057870

Bai, Z., Wang, N., & Wang, M. (2021). Effects of microplastics on marine copepods. *Ecotoxicology and Environmental Safety*, 217(1), 112243–112254. DOI: 10.1016/j.ecoenv.2021.112243 PMID: 33915449

Bennett, J. M., & Ashley, E. J. (1965). Infrared Reflectance and Emittance of Silver and Gold Evaporated in Ultrahigh Vacuum. *Applied Optics*, 4(2), 221–224. DOI: 10.1364/AO.4.000221

CO_2 and Greenhouse Gas Emissions website, *Our World in Data*, accessed 10 August 2023, <https://ourworldindata.org/co2-and-greenhouse-gas-emissions>

Carbon Neutral Coalition website, accessed 10 August 2023, <https://carbon-neutrality.global/members/>

Chen, X., Yu, N., Zhang, L., Liu, Z., Wang, Z., & Chen, Z. (2015). Synthesis of polypyrrole nanoparticles for constructing full-polymer UV/NIR-shielding film. *RSC Advances*, 5(117), 96888–96895. DOI: 10.1039/C5RA20164K

Cooper, R. G., & Harrison, A. P. (2009). The exposure to and health effects of antimony. *Indian Journal of Occupational and Environmental Medicine*, 13(1), 3–10. DOI: 10.4103/0019-5278.50716 PMID: 20165605

Crispin, X., Jakobsson, F. L. E., Crispin, A., Grim, P. P. C. M., Andersson, P., Volodin, A., van Haesendonck, C., Van der Auweraer, M., Salaneck, W. R., & Berggren, M. (2006). The Origin of the High Conductivity of Poly(3,4-ethylenedioxythiophene)–Poly(styrenesulfonate) (PEDOT–PSS) Plastic Electrodes. *Chemistry of Materials*, 18(18), 4354–4360. DOI: 10.1021/cm061032+

Dalapati, G. K., Kushwaha, A. K., Sharma, M., Suresh, V., Shannigrahi, S., Zhuk, S., & Masudy-Panah, S. (2018). Transparent heat regulating (THR) materials and coatings for energy saving window applications: Impact of materials design, micro-structural, and interface quality on the THR performance. *Progress in Materials Science*, 95, 42–131. DOI: 10.1016/j.pmatsci.2018.02.007

Determann, V. H., Rehner, H., & Wieland, T. H. (1968). Ein perlförmiges Cellulosegel für die Chromatographie. *Die Makromolekulare Chemie*, 114(1), 263–274. DOI: 10.1002/macp.1968.021140122

Fan, Z., & Ouyang, J. (2019). Thermoelectric Properties of PEDOT:PSS. *Advanced Electronic Materials*, 5(1), 1800769–1800792. DOI: 10.1002/aelm.201800769

Fuji Chimera Research Institute. Inc, *2020Current status and future prospects of the fine powder market, Market research report*.

Hauser, R. A., Keith, J. M., King, J. A., & Holdren, J. L. (2008). Thermal conductivity models for single and multiple filler carbon/liquid crystal polymer composites. *Journal of Applied Polymer Science*, 110(5), 2914–2923. DOI: 10.1002/app.28869

He, H., Fu, R., Hun, Y., Shen, Y., & Song, X. (2007). Thermal conductivity of ceramic particle filled polymer composites and theoretical predictions. *Journal of Materials Science*, 42(16), 6749–6754. DOI: 10.1007/s10853-006-1480-y

Hermes, M. E., (1996). *Enough for One Lifetime: Wallace Carothers, Inventor of Nylon (History of Modern Chemical Sciences)*.

Hirayama, C., Ihara, H., Nagaoka, S., Furusawa, H., & Tsuruta, S. (1990). Regulation of Pore-size distribution of poly(γ-methyl L-glutamate) spheres as a gel permeation chromatography packings. *Polymer Journal*, 22(7), 614–619. DOI: 10.1295/polymj.22.614

Hirayama, C., Ihara, H., Shiba, M., Nakamura, M., Motozato, Y., & Kunitake, T. (1987). Macroporous glucomannan spheres for the size-exclusion separation of aqueous macromolecules. *Journal of Chromatography. A*, 409, 175–181. DOI: 10.1016/S0021-9673(01)86793-5

Hirayama, C., Nagaoka, S., Ihara, H., Honbo, J., Kurisaki, H., & Ikegami, S. (1995). Liquid chromatographic separation of geometrical isomers using spherical carbon packings prepared from spherical cellulose particles. *Journal of Liquid Chromatography*, 18(8), 1509–1520. DOI: 10.1080/10826079508009291

Ho, T., Zimmermann, T., Hauert, R., & Caseri, W. (2011). Preparation and characterization of cationic nanofibrillated cellulose from etherification and high-shear disintegration processes. *Cellulose (London, England)*, 18(6), 1391–1406. DOI: 10.1007/s10570-011-9591-2

Horikawa, M., Fujiki, T., Shirosaki, T., Ryu, N., Sakurai, H., Nagaoka, S., & Ihara, H. (2015). The development of a highly conductive PEDOT system by doping with partially crystalline sulfated cellulose and its electric conductivity. *Journal of Materials Chemistry. C, Materials for Optical and Electronic Devices*, 3(34), 8881–8887. DOI: 10.1039/C5TC02074C

Horikawa M., Nagaoka S., Shirosaki T., Ryu N., Takafuji M., Ihara H., Tanaka H., Sumi R., (2018). Electroconductive Material. PCT JP 2018-26300.

Horikawa M., Nagaoka S., Yoshida K., Ihara H., Noguchi H., & Tanaka H., (2020). Heat ray absorption material and method for producing the same, and heat ray absorption film. Jp. patent publication number, 2020111747.

Horikawa, M., Shirosaki, T., Ryu, N., Ohgi, Y., Sakurai, H., Nagaoka, S., & Ihara, H. (2017). PEDOT-sulfate nanocrystalline cellulose composites and their Characterization. *Kobunshi Ronbunshu*, 74(6), 565–571. DOI: 10.1295/koron.2017-0048

Im, S., Park, C., Cho, W., Kim, J., Jeong, M., & Kim, J. H. (2019). Synthesis of Solution-Stable PEDOT-Coated Sulfonated Polystyrene Copolymer PEDOT:P(SS-co-St) Particles for All-Organic NIR-Shielding Films. *Coatings*, 9(3), 151–161. DOI: 10.3390/coatings9030151

International Antimony Association, (2010). Proposed revision of RoHS directive concerns flame retardant industry. *Additives for Polymers, 2010*(2), 4 Page. DOI: 10.1016/S0306-3747(10)70023-3

Isa, A., Minamino, J., Kojima, Y., Suzuki, S., Ito, H., Makise, R., Okamoto, M., & Endo, T. (2016). The influence of dry-milled wood flour on the physical properties of wood flour/polypropylene composites. *Journal of Wood Chemistry and Technology*, 36(2), 105–113. DOI: 10.1080/02773813.2015.1083583

Isogai, A. (2013). Wood nanocelluloses: Fundamentals and applications as new bio-based nanomaterials. *Journal of Wood Science*, 59(6), 449–459. DOI: 10.1007/s10086-013-1365-z

Isogai, A., & Kato, Y. (1998). Preparation of Polyuronic Acid from Cellulose by TEMPO-mediated Oxidation. *Cellulose (London, England)*, 5(3), 153–164. DOI: 10.1023/A:1009208603673

Jambeck, J., Geyer, R., Wilcox, C., Siegler, T. R., Perryman, M., Andrady, A., Narayan, R., & Law, K. L. (2015). Plastic waste inputs from land into the ocean. *Science*, 347(6223), 768–771. DOI: 10.1126/science.1260352 PMID: 25678662

Jikei, M., Yamaya, T., Uramoto, S., & Matsumoto, K. (2014). Conductivity Enhancement of PEDOT/PSS Films by Solvent Vapor Treatment. *International Journal of Society Materials Engineering Resources*, 20(2), 158–162. DOI: 10.5188/ijsmer.20.158

JIS R 3106:1998, "Testing method on transmittance and emittance of flat glasses and evaluation of solar heat gain coefficient", Japanese Standards Association, 4-14-24, Akasaka, Minato-ku, Tokyo, 107-844 Japan (Reaffirmed 2012-10-22); http://www.webstore.jsa.or.jp/webstore/Top/index En.jsp?lang=en John W., & Tennant R. D. J., (1945). *Polymeric linear terephthalic esters*. US2465319A.

Jonas F., Heywang G., Schmidtberg W., (1988). Feststoff-elektrolyte und diese enthaltende elektrolytkondensatoren. DE 3814730 A1.

Katagiri, K., Takabatake, R., & Inumaru, K. (2013). Robust infrared-shielding coating films prepared using perhydropolysilazane and hydrophobized Indium Tin oxide Nanoparticles with Tuned Surface Plasmon Resonance. *ACS Applied Materials & Interfaces*, 5(20), 10240–10245. DOI: 10.1021/am403011t PMID: 24025399

Kato, Y., Habu, N., Yamaguchi, J., Kobayashi, Y., Shibata, I., Isogai, A., & Samejima, M. (2002). Biodegradation of β-1,4-linked polyglucuronic acid (cellouronic acid). *Cellulose (London, England)*, 9(1), 75–81. DOI: 10.1023/A:1015877416414

Kondo, T., Kose, R., Naito, H., & Kasai, W. (2014). Aqueous counter collision using paired water jets as a novel means of preparing bio-nanofibers. *Carbohydrate Polymers*, 112, 284–290. DOI: 10.1016/j.carbpol.2014.05.064 PMID: 25129746

Kuga, S. (1980). New cellulose gel for chromatography. *Journal of Chromatography. A*, 195(2), 221–230. DOI: 10.1016/S0021-9673(00)96813-4

Lai M.-K., Wu J.-R., Yeh J.-M., & Chang S. H., (2019). Unraveling the modified PEDOT:PSS thin films based near-infrared solar-heat shields by using broadband transmittance and Raman scattering spectrometers. *Physics Status Solidi A, 216*(11), Article ID 1900025, 5 page. DOI: 10.1002/pssa.201900025

Li, C.-P., Kang, C.-Y., Huang, S.-L., Lee, P.-T., Kuo, H.-C., & Hsu, F.-C. (2020). Near infrared radiation shielding using Cs_xWO_3 nanoparticles for infrared mini light-emitting diodes. *Materials Letters*, 260(1), 126961–126962. DOI: 10.1016/j.matlet.2019.126961

Lim, X. (2021). Microplastics are everywhere - but are they harmful? *Nature*, 593(7857), 22–25. DOI: 10.1038/d41586-021-01143-3 PMID: 33947993

Looney B., (2021). *Statistical Review of World Energy 2021*. BP

Maddah, H. A. (2016). Polypropylene as a promising plastic: A Review. *American Journal of Political Science*, 6(1), 1–11. DOI: 10.5923/j.ajps.20160601.01

Ministry of the Environmental website, Japanese government, accessed 10 August 2023, https://www.env.go.jp/content/900543574.pdf

Mochizuki, Y., Horii, T., & Okuzaki, H. (2012). Effect of pH on Structure and Conductivity of PEDOT/PSS. *Transactions of the Materials Research Society of Japan*, 37(2), 307–310. DOI: 10.14723/tmrsj.37.307

Moore, C. J. (2008). Synthetic polymers in the marine environment: A rapidly increasing, long-term threat. *Environmental Research*, 108(2), 131–139. DOI: 10.1016/j.envres.2008.07.025 PMID: 18949831

Motozato, Y., Ihara, H., Tomoda, T., & Hirayama, C. (1986). Preparation of Pullulan for High-speed Gel Chromatography Spheres. *Nippon Kagaku Kaishi*, 1986(9), 1187–1191. DOI: 10.1246/nikkashi.1986.1187

Motozato, Y., Matsumoto, K., & Hirayama, C. (1981). Preparations of bead-shaped cellulose gels and their utilities as gel chromatographic packings. *Nippon Kagaku Kaishi*, 1981(12), 1883–1889. DOI: 10.1246/nikkashi.1981.1883

Motozato, Y., Matsumoto, K., & Hirayama, C. (1984). Preparation and properties of porous cellulose beads using diluents. *Nippon Kagaku Kaishi*, 1984(5), 722–727. DOI: 10.1246/nikkashi.1984.722

Nagaoka, S., Arinaga, K., Kubo, H., Hamaoka, S., Takafuji, M., & Ihara, H. (2005). Cellulose/TiO_2 hybrid spherical microbeads prepared by a viscose phase separation method: Control of the distribution of TiO_2 particles in a sphering system. *Polymer Journal*, 37(3), 186–191. DOI: 10.1295/polymj.37.186

Nagaoka, S., Hamasaki, Y., Ishihara, S., Nagata, M., Iio, K., Nagasawa, C., & Ihara, H. (2001). Preparation of carbon/TiO_2 microsphere composites from cellulose/TiO_2 microsphere composites and their evaluation. *Journal of Molecular Catalysis A Chemical*, 177(2), 255–263. DOI: 10.1016/S1381-1169(01)00271-0

Nagaoka, S., Hirakawa, K., Kobayashi, K., Satoh, K., Nagata, M., Takafuji, M., & Ihara, H. (2008). Cellulose spherical microbeads interfacially-controlled by hard inorganic nano-particles using viscose phase separation method and their applications to abrasive material. *Kobunshi Ronbunshu*, 65(1), 80–89. DOI: 10.1295/koron.65.80

Nagaoka, S., Ihara, H., Honbo, J., Hirayama, C., Kurisaki, H., & Ikegami, S. (1994). Spherical carbon packings prepared from spherical cellulose particles for high-performance liquid Chromatography. *Analytical Sciences*, 10(4), 543–551. DOI: 10.2116/analsci.10.543

Nagaoka, S., Jodai, T., Kameyama, Y., Horikawa, M., Shirosaki, T., Ryu, N., Takafuji, M., Sakurai, H., & Ihara, H. (2016). Cellulose/boron nitride core–shell microbeads providing high thermal conductivity for thermally conductive composite sheets. *RSC Advances*, 6(39), 33036–30342. DOI: 10.1039/C6RA02950G

Nagaoka, S., Nagata, M., Arinaga, K., Shigemori, K., Takafuji, M., & Ihara, H. (2007). Environmentally friendly coloured materials: Cellulose/titanium dioxide/inorganic pigment composite spherical microbeads prepared by viscose phase-separation method. *Coloration Technology*, 123(6), 344–350. DOI: 10.1111/j.1478-4408.2007.00107.x

Nagaoka, S., Satoh, T., Sakamoto, K., & Ihara, H. (2005). Preparation and characterization of spherical polymer packings from polybutadiene for size-exclusion chromatography. *Journal of Chromatography. A*, 1082(2), 185–191. DOI: 10.1016/j.chroma.2005.05.063 PMID: 16035360

Nagaoka, S., Tobata, H., Takiguchi, Y., Satoh, T., Sakurai, T., Takafuji, M., & Ihara, H. (2005). Characterization of cellulose microbeads prepared by a viscose-phase-separation method and their chemical modification with acid anhydride. *Journal of Applied Polymer Science*, 97(1), 149–157. DOI: 10.1002/app.21539

Nakagaito, A. N., & Yano, H. (2005). Novel high-strength biocomposites based on microfibrillated cellulose having nano-order-unit web-like network structure. *Applied Physics. A, Materials Science & Processing*, 80(1), 155–159. DOI: 10.1007/s00339-003-2225-2

Noguchi, Y., Homma, I., & Matsubara, Y. (2017). Complete nanofibrillation of cellulose prepared by phosphorylation. *Cellulose (London, England)*, 24(3), 1295–1305. DOI: 10.1007/s10570-017-1191-3

Ohkuma S., Yamagishi K., Hara M., Suzuki K., & Yamamoto T., (1986). Fine cellulose particle and production thereof. Jpn publication number 61241337.

Ouyang, J., Xu, Q., Chu, C., Yang, Y., Lib, G., & Shinar, J. (2004). On the mechanism of conductivity enhancement in poly(3,4-ethylenedioxythiophene):poly(styrene sulfonate) film through solvent treatment. *Polymer*, 45(25), 8443–8450. DOI: 10.1016/j.polymer.2004.10.001

Park, J., Yoon, H., Kim, G., Lee, B., Lee, S., Jeong, S., Kim, T., Seo, J., Chung, S., & Hong, Y. (2019). Highly Customizable All Solution–Processed Polymer Light Emitting Diodes with Inkjet Printed Ag and Transfer Printed Conductive Polymer Electrodes. *Advanced Functional Materials*, 29(3), 1902412–1902421. DOI: 10.1002/adfm.201902412

Park, J.-W., Kim, G., Lee, S.-H., Kim, E.-H., & Lee, G.-H. (2010). The effect of film microstructures on cracking of transparent conductive oxide (TCO) coatings on polymer substrates. *Surface and Coatings Technology*, 205(3), 915–921. DOI: 10.1016/j.surfcoat.2010.08.055

Peterson, E. A., & Sorber, H. A. (1956). Chromatography of proteins. I. Cellulose ion-exchange adsorbents. *Journal of the American Chemical Society*, 78(4), 751–755. DOI: 10.1021/ja01585a016

Pratesi S., Sani E., & Lucia M. D., (2014). Optical and Structural Characterization of Nickel Coatings for Solar Collector Receivers. *International Journal of Photoenergy, Article ID 834128*, 7 Page. DOI: 10.1155/2014/834128

Qin, Z., Yin, X., Zhu, L., Lin, Q., & Qin, J. (2014). Synthesis and characterization of bacterial cellulose sulfates using a SO_3/pyridine complex in DMAc/LiCl. *Carbohydrate Polymers*, 101, 947–953. DOI: 10.1016/j.carbpol.2013.09.068 PMID: 24299860

Ray, S., Dutta, U., Das, R., & Chatterjee, P. (2007). Modelling of experimentally measured optical characteristics of ITO/TiO_2 transparent multi-layer heat shields. *Journal of Physics. D, Applied Physics*, 40(8), 2445–2451. DOI: 10.1088/0022-3727/40/8/006

Report, B. (1987). *Our Common Future, Report of the World Commission on Environment and Development*. Oxford University Press.

Reyes-Reyes, M., Cruz-Cruz, I., & López-Sandoval, R.R.-Reyes M. (2010). Enhancement of the Electrical Conductivity in PEDOT:PSS Films by the Addition of Dimethyl Sulfate. *The Journal of Physical Chemistry. C, Nanomaterials and Interfaces*, 114(47), 20220–20224. DOI: 10.1021/jp107386x

Saeedi, M. (2023). How microplastics interact with food chain: A short overview of fate and impacts. *Journal of Food Science and Technology*. Advance online publication. DOI: 10.1007/s13197-023-05720-4 PMID: 37360257

Sakata, M., Todokoro, M., & Kunitake, M. (2007). Pore-size Controlled and Polycation-immobilized Cellulose Spherical Particles for Removal of Endotoxin. *Kobunshi Ronbunshu*, 64(12), 821–829. DOI: 10.1295/koron.64.821

Sekimoto, E., Ryu, N., Shirosaki, T., Horikawa, M., Nagaoka, S., & Ihara, H. (2016). Nanoparticulation of acetyl cellulose by poor solvent induced phase separation method. *Preprints of 2016 Cellulose R&D The 23rd Annual Meeting of Cellulose Society of Japan*, 125.

Seo, K.-W., Lee, J., Jo, J., Cho, C., & Lee, J.-Y. (2019). Highly Efficient (>10%) Flexible Organic Solar Cells on PEDOT-Free and ITO-Free Transparent Electrodes. *Advanced Materials*, 31(36), 1902447–1902453. DOI: 10.1002/adma.201902447 PMID: 31304650

Shi, H., Liu, C., Jiang, Q., & Xu, J. (2015). Effective Approaches to Improve the Electrical Conductivity of PEDOT:PSS: A Review. *Advanced Electronic Materials*, 1(4), 1500017–1500033. DOI: 10.1002/aelm.201500017

Shi, W., Ching, Y. C., & Chuah, C. H. (2021). Preparation of aerogel beads and microspheres based on chitosan and cellulose for drug delivery: A review. *International Journal of Biological Macromolecules*, 170, 751–767. DOI: 10.1016/j.ijbiomac.2020.12.214 PMID: 33412201

Sun, K., Zhang, S., Li, P., Xia, Y., Zhang, X., Du, D., Isikgor, F. H., & Ouyang, J. (2015). Review on application of PEDOTs and PEDOT:PSS in energy conversion and storage devices. *Journal of Materials Science Materials in Electronics*, 26(7), 4438–4462. https://doi-org-443.webvpn.fjmu.edu.cn/10.1007/s10854-015-2895-5. DOI: 10.1007/s10854-015-2895-5

Sustainable Development Goals website, United Nations, accessed 10 August 2023, https://www.un.org/sustainabledevelopment/

Takano, T., Masunaga, H., Fujiwara, A., Okuzaki, H., & Sasaki, T. (2012). PEDOT Nanocrystal in Highly Conductive PEDOT:PSS Polymer Films. *Macromolecules*, 45(9), 3859–3865. DOI: 10.1021/ma300120g

Walz, Y., Janzen, S., Narvaez, L., Ortiz-Vargas, A., Woelki, J., Doswal, N., & Sebesvari, Z. (2021). Disaster-related losses of ecosystems and their services. Why and how do losses matter for disaster risk reduction? *International Journal of Disaster Risk Reduction*, 63, 102425–102440. DOI: 10.1016/j.ijdrr.2021.102425

Xanthos, D., & Walker, T. (2017). International policies to reduce plastic marine pollution from single-use plastics (plastic bags and microbeads): A review. *Marine Pollution Bulletin*, 118(1-2), 17–26. DOI: 10.1016/j.marpolbul.2017.02.048 PMID: 28238328

Xu, Q., Xiao, L., Ran, J., Tursun, R., Zhou, G., Deng, L., Tang, D., Shu, Q., Qin, J., Lu, G., & Peng, P. (2018). $Cs_{0.33}WO_3$ as a high-performance transparent solar radiation shielding material for windows. *Journal of Applied Physics*, 124(19), 193102. DOI: 10.1063/1.5050041

Yoon, D. G., Kang, M. G., Kim, J. B., & Kang, K.-T. (2018). Nozzle printed-PEDOT:PSS for organic light emitting diodes with various dilution rates of ethanol. *Applied Sciences (Basel, Switzerland)*, 8(2), 203–212. DOI: 10.3390/app8020203

Yorifuji, D., & Ando, S. (2011). Enhanced thermal conductivity over percolation threshold in polyimide blend films containing ZnO nano-pyramidal particles: Advantage of vertical double percolation structure. *Journal of Materials Chemistry*, 21(12), 4402–4407. DOI: 10.1039/c0jm04243a

Yoshida, K., Nagaoka, S., Horikawa, M., Noguchi, H., & Ihara, H. (2020). Totally-organic near-infrared shielding materials by conductive cellulose nanofibers. *Thin Solid Films*, 709, 138221–138221. DOI: 10.1016/j.tsf.2020.138221

Zhang, K., Brendler, E., Geissler, A., & Fisher, S. (2011). Synthesis and spectroscopic analysis of cellulose sulfates with regulable total degrees of substitution and sulfation patterns via 13C NMR and FT Raman spectroscopy. *Polymer*, 52(1), 26–32. DOI: 10.1016/j.polymer.2010.11.017

Zhao, Q., Jamal, R., Zhang, L., Wang, M., & Abdiryim, T. (2014). The structure and properties of PEDOT synthesized by template-free solution method. *Nanoscale Research Letters*, 9(1), 557–566. DOI: 10.1186/1556-276X-9-557 PMID: 25317104

Ziegler, K. (1964). Folgen und Werdegang einer Erfindung Nobel-Vortrag am 12. Dezember 1963. *Angewandte Chemie*, 76(13), 545–553. DOI: 10.1002/ange.19640761302

Zozoulenko, I., Singh, A., Singh, S. K., Gueskine, V., Crispin, X., & Berggren, M. (2019). Polarons, bipolarons, and absorption spectroscopy of PEDOT. *ACS Applied Polymer Materials*, 1(1), 83–94. DOI: 10.1021/acsapm.8b00061

Chapter 4
Cellulose Nanofibrils Composite Films

Huixin Jiang
Oak Ridge National Laboratory, USA

Hannah Snider
https://orcid.org/0000-0003-4806-5463
Oak Ridge National Laboratory, USA

Xianhui Zhao
Oak Ridge National Laboratory, USA

Saurabh Prakash Pethe
University of Tennessee, USA

Shuvodeep De
Oak Ridge National Laboratory, USA

Tolga Aytug
https://orcid.org/0000-0001-7971-5508
Oak Ridge National Laboratory, USA

Soydan Ozcan
Oak Ridge National Laboratory, USA

Kashif Nawaz
Oak Ridge National Laboratory, USA

Kai Li
https://orcid.org/0000-0003-1445-3206
Oak Ridge National Laboratory, USA

ABSTRACT

Nanocellulose derived from biomass is a sustainable, lightweight, and mechanically strong material. Extensive research has been directed towards applying nanocellulose for advanced applications. Nanocellulose-based films are novel and unique for designing functional materials considering their combination of sustainability with new properties. Nanocellulose films can be fabricated through simple strategies and make it easy to access various applications, such as barriers, sensors, energy storage, and so on. In this chapter, we summarized the preparation methods of nanocellulose-based films while focusing on nanocellulose as the film matrix and highlighting some representative applications. Given the sustainability of the nanocellulose and ease of introducing functional groups, we believe that nanocellulose-based films promise great potential for future advanced applications.

DOI: 10.4018/979-8-3693-0003-9.ch004

INTRODUCTION

Nanocellulose has received significant attention among the research community owing to its unique properties (Isogai, 2020; Lamm et al., 2021; Li, Clarkson, et al., 2021), such as sustainability, low density, good mechanical strength, and biodegradability (Copenhaver et al., 2022; Lamm et al., 2021; Mohammed et al., 2018; Sinquefield et al., 2020; Yanamala et al., 2014). Nanocellulose can be separated from various plants, algae, and other biological materials, which makes them a green and sustainable material (Abdul Khalil et al., 2012). Nanocellulose is comprised of both individual and bundles of nanoscale fibrils of polymeric repeating β-(1-4 linked), D-glucose linkages, held together tightly through hydrogen-bonding of the surface hydroxyl groups. There are mainly three kinds of nanocellulose, including cellulose nanocrystals (CNCs), cellulose nanofibrils (CNFs), and bacterial cellulose (BC). CNCs are rigid, rod-shaped particles with dimensions of 100–200 nm long and 5–20 nm wide (Thomas et al., 2018), and they are also referred to as "cellulose nanowhiskers" or "nanocrystalline cellulose". CNFs contain fibril structures with a length of >1 μm and a width of 5–200 nm (Copenhaver et al., 2021; Thomas et al., 2018). CNCs and CNFs are mainly isolated from plants, such as wood and bamboo, but can also be obtained from some animal tissues, such as tunicates (Thomas et al., 2018; L. Wang et al., 2021). BC is generated by bacteria (e.g., *Acetobacter xylinum*) through biosynthesis, and it has a length up to several micrometers and a width around 25–86 nm (Zhu et al., 2016).

CNFs can be isolated through mechanical, chemical, and biological approaches, and mechanical treatments (L. Wang et al., 2021), such as high-pressure homogenization, cryocrushing, and grinding, are the most commonly used. Depend on the raw materials source, chemical treatment is used to remove the lignin and hemicellulose. CNFs have a long aspect ratio and strong mechanical properties, which make them a good candidate for fiber reinforced composites. Compared to CNFs, which consist of significant amorphous regions, CNCs have a higher crystallinity of 54 - 88%. CNCs are produced through enzymatic and strong acid hydrolysis. CNCs have high axial stiffness (105−168 GPa), high Young's modulus (20−50 GPa), high tensile strength (~9 GPa), low coefficient of thermal expansion (~0.1 ppm/K), low density (1.5−1.6 g/cm^3), lyotropic liquid crystalline behavior, and shear thinning rheology (Iwamoto et al., 2011; Iwamoto et al., 2009; Lahiji et al., 2010; Thomas et al., 2018). BC are produced through cultivating bacteria for a few days in an aqueous culture media containing nutrition. BCs are composed of ribbon-shaped cellulose fibrils. The fibrils are fairly straight and continuous and have low size distribution. BCs have high crystallinity of 84−89%) and their elastic modulus can reach 78 GPa.

Nanocellulose has been processed into films, fibers, aerogels, and hydrogels (K. Li et al., 2022; Zhao et al., 2023) based on the type of the applications, ranging from nanocomposites (Hamedi et al., 2014), gas separation (Matsumoto & Kitaoka, 2016), flexible electronics (Hoeng et al., 2016), and fuel cells (Lu et al., 2015). Nanocellulose based film, where individual CNFs are stacked together to form lightweight and strong materials, is one important material platform for utilizing nanocellulose. To produce nanocellulose based films and to expand their application base, generally simple and versatile processes processing strategies have been adopted, such as solution casting and vacuum filtration. Nanocellulose based films with desired functional materials can be fabricated by introduction of functional additives in the process, or through surface modification of the nanocellulose with desired functional groups. These nanocellulose based films have unique properties, such as flexible, strong mechanical properties, good barrier properties, low thermal conductivities, and others.

In this chapter, research and development activities on nanocellulose-based films were summarized. The preparation methods of nanocellulose-based films were described with a focus on the nanocellulose as the film matrix. Some applications of the nanocellulose based films were highlighted. There are various other applications for nanocellulose-based films; here we focus on the packaging, functional nanocomposites, energy storage, thermal management, and sensors.

Preparation methods of nanocellulose film

Different strategies have been developed to fabricate nanocellulose-based films for various applications. These methods include vacuum filtration, solution casting, hydraulic press, layer-by-layer assembly, and others. Depending on the needs and the starting materials, researchers can choose suitable production methods. In this section, vacuum filtration, solution casting, hydraulic press, and layer-by-layer assembly are introduced. These methods can be used individually or in a combined fashion.

Vacuum filtration is a process used to separate solid materials from a liquid through a filter under vacuum. The vacuum creates a driving force that pulls the liquid through the filter while the solids cake on the outside of the filter. After filtration, a free-standing film can be peeled off from the filter. It is a simple and versatile strategy. Since nanocellulose are usually produced in water, vacuum filtration is an important method for making nanocellulose-based films, and many of papers have been published using this strategy. Li et al.(Li et al., 2019) developed a strong and tough CNF film using vacuum filtration (Figure 1) by utilizing the synergetic effect of hydrogen bonds and ionic interactions in the films. With the addition of chitosan (CS) and copper ion (Cu^{2+}), the tensile strength and Young's modulus of the newly developed film (CNF-CS- Cu) increased by 104% and 75%, respectively. During the drying of the film, pressure (e.g., hydraulic press) is usually applied on the film to ensure a smooth surface and reduce the porosity. A solvent exchange process can be added after the filtration to make porous films (Henriksson et al., 2008). For example, Henriksson et al (Henriksson et al., 2008) prepared micro fibrillated cellulose (MFC) wet films using vacuum filtration and followed by immersing the wet films in methanol, ethanol, and acetone to obtain films with different porosities.

Figure 1. Preparation of the CNF films using a vacuum filtration method (Adapted with permission from ref Li et al., 2019)

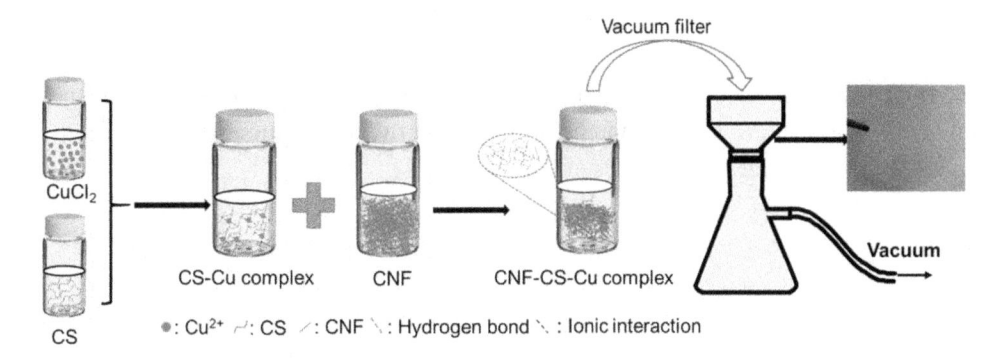

Solution casting is another easy and versatile process of producing thin films/sheets at the laboratory scale (Li et al., 2018). In the solution casting process, the nanocellulose is dispersed either in water or a non-aqueous volatile solvent, and the resulting dispersion is casted on a flat surface. During the process, additives can be added to the respective aqueous or non-aqueous-nanocellulose dispersions. The solvent phase is removed by evaporation and thereafter the dried film is released from the substrate. For instance, Arantes et al. fabricated CNF/chitosan biobased films using solution casting and evaluated the effect of the addition of the magnetite and glycerol on the properties (Arantes et al., 2019). The mixture solution in water was poured onto acrylic casting plates and held at 50 °C for 24 h to produce thin films. Solution casting enables researchers to make films with different functionalities by simply including functional additives.

Hydraulic press (Algar et al., 2016) has been used to prepare nanocellulose films at elevated temperature under pressure. This is a common process for processing polymer samples into desired shapes in a mold, including films. To make nanocellulose films using hydraulic press, two steps are usually involved: firstly, samples are pressurized at room temperature for dewatering, and then the temperature is increased to fully dry the sample and form a film. For instance, Algar et al (Algar et al., 2016) prepared BC/ Montmorillonite composite films by pressing BC/MMT membranes at 60 °C and 10 MPa for 5 min to remove the water from the composites. Similarly, wood chips were delignified by alkaline sulfite anthraquinone methanol pulping, and the resulting samples were dehydrated by pressing in a hydraulic press at 8 MPa for 3 h, and finally pressed at 90 °C for 3 min under 8 MPa to obtain nanocellulose films (Fang et al., 2020).

Layer by layer (LBL) deposition is also a very useful method for making film samples. The films are formed by depositing alternating layers with oppositely charged materials. Aulin et al. (Aulin et al., 2010) produced cellulose-based films through LBL assemblies using cationic/anionic MFC and MFC/PEI. Yang et al. (Yang et al., 2020) constructed thermal conductive CNF films by incorporating two-dimensional (2D) reduced graphene oxide (rGO) decorated with zero-dimension (0D) silver nanoparticles (AgNPs) using a LBL approach. The in-plane thermal conductivity of Ag-rGO/NFC showed a 995% increase over that of pure NFC. Wågberg and Erlandson (Wagberg & Erlandsson, 2021) published an extensive review on the nanocellulose based functional materials made from LBL self-assembly. The detailed principle and research progress are reviewed in their paper.

APPLICATION OF THE NANOCELLULOSE FILM

Owing to the sustainable nature and the unique properties of the nanocellulose films, they have been studied for various application. In this chapter, we focus on the packaging, functional nanocomposites, energy storage, thermal management, and sensors.

Package

Nanocellulose based package materials find important applications in packaging industries owing to their low cost and environmentally friendly nature. In general, the gas barrier properties of package materials are controlled by the diffusivity of the gas molecules into the materials. The incorporation of the high aspect ratio filler in the matrix can create a tortuous path for the penetrating gas molecules, thus improving the gas barrier properties of the package materials. Nanocellulose contains a high proportion of the crystalline regions (CNCs: 72%–80%, CNF: 50-60%), that are impermeable to gas molecules. The alternating crystalline and amorphous structure of nanocellulose increases the effective length of the gas flow path, creating unique barrier properties. Therefore, a dense, nonporous nanocellulose film can be used for gas barrier applications. Nanocellulose films have shown excellent mechanical properties, and superior oxygen barrier capabilities at low relative humidity. Additionally, transparent films with tunable transparency can be made from nanocellulose to meet different packaging needs. Because of this, nanocellulose has been widely use in packaging, including gas barrier additives or a main matrix (Wang et al., 2023). Herein, mainly the nanocellulose film as a main matrix for package related applications was focused. Nanocellulose has good barrier property at low relative humidity, however, due to its hydrophilic nature and high water absorption, the gas barrier properties, such as oxygen permittivity, at high relative humidity are poor. Therefore, different strategies including surface modification, hybrid fillers, application of hydrophobic coating, etc. have been developed to improve the barrier properties at high relative humidity.

For a hybrid approach, different additives, such as clay (Wu et al., 2012), lignin (Trifol & Moriana, 2022), oil (Valencia et al., 2019), layered double hydroxide (Wang et al., 2018), and nanoplates (Wu et al., 2022) have been used. Trifol et al. (Trifol & Moriana, 2022) fabricated CNF and lignocellulose nanofibrils (LCNF) films using a solution casting method, and found that by the incorporation of 25% LCNF to the CNF, films experienced enhanced mechanical properties (94% increase in tensile stress and a 414% increase in strain at break). The water vapor transmission rate (WVTR) and oxygen transmission rate (OTR) decreased by 16% and 53%, respectively. LCNFs improved the interfacial adhesion between CNFs due to the presence of lignin and created more tortuous paths for gas molecules, suggesting a promising pathway for sustainable package. Valencia et al. (Valencia et al., 2019) fabricated an CNF-oil composite film by casting CNFs stabilized using a Pickering emulsion solution. The embedded oil improved the mechanical properties and hydrophobicity, and the thermal stability also increased. They further evaluated the encapsulation of curcumin in the films and found increased antioxidant (up to 30% radical scavenging activity) and antimicrobial activity, inhibiting the growth of foodborne bacteria such as *Escherichia coli*.

Surface modification of the nanocellulose has proven to be an effective strategy to improve the barrier properties at high relative humidity (Vartiainen et al., 2020). Nanocellulose have abundant surface OH groups and are readily reacted through multiple reaction paths (Lamm, Li, Copenhaver, et al., 2022; Li, McGrady, et al., 2021). Visanko et al. (Visanko et al., 2015) studied butylamino modified CNC for package applications and achieved high resistance oxygen permeability even at high relative humidity (RH 80%), and values as low as 5.9 ± 0.2 cm^3 μm per m^2 per day per kPa. Tomé et al. (Tomé et al., 2010) an esterified BC film with hexanoyl chloride and found that introduction of the hydrophobicity decreased water vapor permeance by about 50%. Moreover, the permeabilities to O_2, CO_2, and N_2 at 100% RH were also decreased. Except covalent functionalization of the nanocellulose, utilizing interaction with nanocellulose with polymers, metal ions, surfactants can also be used for improving the barrier properties at high relative humidity environment (Hubbe et al., 2017).

Incorporate hydrophobic coating is another strategy to improve the hydrophobicity of the nanocellulose film and thus the barrier properties (Cherpinski et al., 2018; Vartiainen et al., 2020). Vartiainen et al. (Vartiainen et al., 2020) coated CNF film with inorganic–organic copolymers (ORMOCER®s) using a blade coating. The water contact angle of the film increased from 24° to the range between 54° and 102°, suggesting the improvement in the hydrophobicity. The WVTR decreased by 77% as compared to uncoated film. The oxygen transmission rates decreased from 107 cc/m^2/day to 51-86 cc/m^2/day at 80%

RH. Electrospinning has been used to coat polymers on nanocellulose films (Cherpinski et al., 2018), since it can produce uniform nanomats on substrates (Li et al., 2020). Cherpinski et al. (Cherpinski et al., 2018) electrospun polyhydroxyalkanoates on CNFs and LCNF films, and the results showed the improvement of water contact resistance, more balanced mechanical properties, and higher barrier performance against water vapor and oxygen in comparison to the neat nanocellulose films. However, the limonene permeance increased due to the intrinsic affinity of polyhydroxyalkanoates for limonene uptake.

Nanocomposites

Nanocomposites consist of two or more materials, of which at least one of the phases is a nanomaterial. The formation of nanocomposites usually involves embedding a reinforcing phase into another material (e.g., a polymer matrix) (Zhao et al., 2020). This often results in an improvement in material properties (e.g., tensile strength, Young' modulus, permeability, shape memory) for the reinforced composite. Cellulose nanofibrils (CNFs) have been used as a reinforcement phase for nanocomposites due to their excellent material properties, low density, abundance, and biodegradability. CNF-polymer nanocomposites have automotive, construction, and biomedical applications (Dagnon et al., 2012; Lamm, Li, Ker, et al., 2022; Wang et al., 2020; Zhao et al., 2023). Multiple processing methods have been developed to produce CNF-based nanocomposites, including extrusion, solvent casting, and hot pressing (Dufresne, 2018). Solvent casting and hot-pressing (hydraulic press) are used to fabricate film-based composites.

To decrease storage volume and transportation costs, nanocellulose is often dried prior to use for composite formation. Cellulose has strong interacting hydroxyl groups that result in the formation of intramolecular and intermolecular hydrogen bonds, particularly during drying. This causes irreversible aggregation of CNFs which decreases the dispersion ability during composite formation, and thereby compromises the material properties. Multiple methods have been developed to prevent aggregation and loss of nanostructure during drying. For example, Velasquez-Cock et al. (Velasquez-Cock et al., 2018) reported the surface modification of CNFs using poly(vinyl alcohol) (PVA) to improve the dispersibility in water. Fu et al. (Fu et al., 2021) studied the redistribution of lignin-containing CNFs under multiple different drying methods (e.g., freeze drying, oven drying, evaporation followed by vacuum drying, and centrifugal followed by vacuum drying) and found that aggregation was alleviated by freeze drying.

Table 1. Tensile properties of nanocellulose-polymer nanocomposites

Polymer Matrix	Fiber Treatment Method	Tensile Strength (MPa)	Modulus of Elasticity (GPa)	Ref.
Shape memory polyurethane (SMPU)	N/A	12.05	N/A	(Zhou et al., 2022)
Polyvinylpyrrolidone (PVP)	Hydrogen bond induced surface modification	162	N/A	(K. Li et al., 2022)
Poly(lactic acid) (PLA)	Lactic acid grafting	116.1	16	(Lafia-Araga et al., 2021)
Poly(vinyl alcohol) (PVA)	Lignin-containing cellulose nanofibrils	38.4	N/A	(Fu et al., 2021)

Table 1 exhibits some representative mechanical properties of CNF-based composites. The hydrophilic nature of CNFs compared to hydrophobic (non-polar) polymer matrices can compromise the material properties of CNF-polymer composites. Because of this, CNFs are often treated to adapt the surface chemistry and form a more compatible reinforcing material. Chemical treatment via molecule/fiber grafting, aqueous phase surface modification, esterification, and oxidation can be conducted to improve the adhesion of CNFs to a polymer matrix (Lafia-Araga et al., 2021; K. Li et al., 2022). Li et al. (K. Li et al., 2022) describe how the hydrogen bond-induced surface modification of CNFs using tannic acid and polyvinylpyrrolidone increased the tensile strength of the resulting nanocomposites by 76% when compared to the unmodified CNF composites (Figure 2). It is important to note that the tensile strength of a biocomposite is not always improved by the addition of CNFs; however, other aspects of the properties can be influenced as a result of initial chemical treatment of the fibers. Lafia-Agara et al. (Lafia-Araga et al., 2021) describe the esterification of fibers to form lactic-acid grafted CNFs. While neither the ungrafted nor the grafted composites demonstrated an improvement in tensile strength compared to neat PLA, the chemical treatment resulted in a significant improvement in barrier properties (decrease in water and oxygen permeability). Adapting material properties provides a route to forming nanocomposites that can be used in more sustainable applications, such as packaging.

Figure 2. Mechanical properties of the properties of the CNF and CNF-TA-PVP composites (Adapted with permission from ref K. Li et al., 2022)

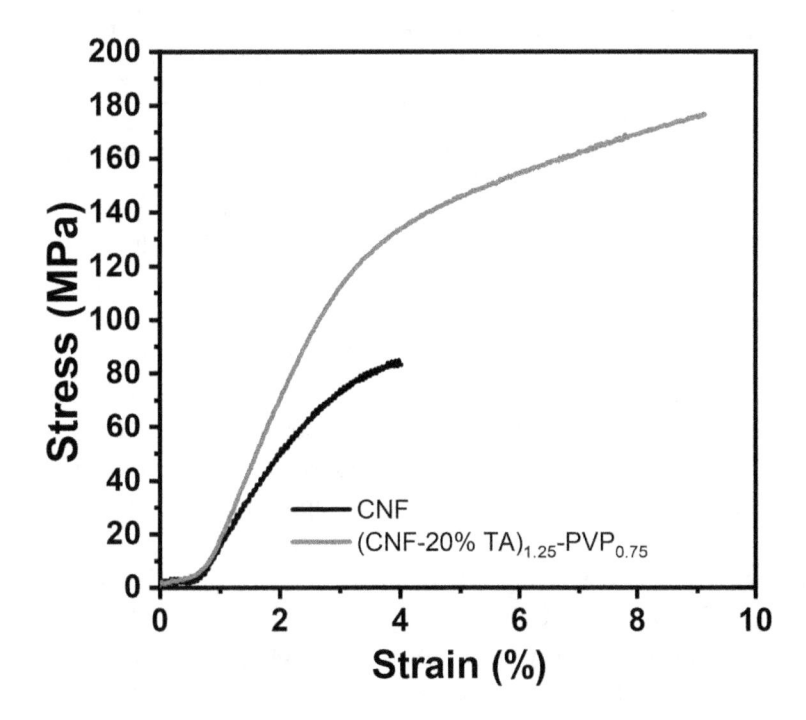

Beyond the above CNF-based composites, CNF has also been used for fabricating other types of composites. For example, Bian et al. (Bian et al., 2022) mixed lignin nanoparticles with CNFs to fabricate composite films, which were biodegradable and exhibited ultraviolet-blocking performance (95%). The neat CNF films exhibited a smooth and uniform surface because of the strong bonding between nanofibrils. The neat CNF films had a tensile strength of 137 MPa and Young's modulus of 6.1 GPa. Compared to neat CNF films, the composite films had higher tensile strength (164 MPa) and Young's modulus (8.0 GPa), shown in Table 2. In addition, the composite films exhibited a higher maximal weight loss temperature than neat CNF films at 310 °C and showed excellent biodegradability in the natural environment. After burying the composite films in soil for 18 days, their mass was reduced by approximately 45% (the mass of neat CNF films was reduced by approximately 30%). The mass loss of the films was mainly because of the existence of ions and microorganisms. However, the presence of more hydrogen bonds between nanofibrils in the neat CNF films probably resulted in them not being easily broken (Bian et al., 2022).

Table 2. The representative properties of CNF-based composites

Composite	Filler other than CNF	Composite fabrication	Representative property of composites	Ref.
CNF/lignin	Lignin nanoparticle	Mix CNF suspension and lignin nanoparticle suspension -> vacuum filtrate -> press and dry	Tensile strength: 164 MPa; Young's modulus: 8.0 GPa; Maximal weight loss temperature: 310 °C;	(Bian et al., 2022)
MoS_2/CNF	MoS_2	Mix MoS_2 and CNF suspension -> sonicate -> cast under vacuum -> dry	CNF addition reduced MoS_2 flake's anisotropy	(Camilo, de Menezes, Pereira-da-Silva, Guimarães, & Longaresi, 2019)
CNF/rubber	Rubber	Mix CNF suspension and aqueous rubber latex -> cast -> dry	Young's modulus: ~45 MPa; Coefficient of thermal expansion: ~20 ppm/K	(Fukui, Ito, Saito, Noguchi, & Isogai, 2018)
CNF/Ag	Silver nanoparticle	Mix CNF suspension and silver ammonia solution -> centrifuge -> vacuum dry	Maximum rejection of Na_2SO_4: 97%; Maximum rejection of NaCl: 33%	(M. He et al., 2022)
CNF/cellulose ether	Cellulose ether	Mix CNF suspension and cellulose ether/water solution -> degas -> cast -> dry	Tensile strength: ~44 MPa; Young's modulus: ~0.6 GPa;	(Okahashi et al., 2021)

Camilo et al. (Camilo et al., 2019) synthesized MoS_2/CNF nanocomposites using a liquid exfoliation process. A MoS_2 (molybdenum disulphide) monolayer is commonly used for sensors, transistors, and photovoltaic devices. The addition of CNFs reduced the anisotropy of the MoS_2 flakes. The addition of CNFs also promoted an increase in the absorption in the visible spectrum region (Camilo et al., 2019). Fukui et al. (Fukui et al., 2018) fabricated CNF/rubber composite films in heterogeneous and homogeneous systems, followed by casting and drying. The resulting composites had a low coefficient of thermal expansion of ~20 ppm/K and a high Young's modulus of ~45 MPa (Fukui et al., 2018).

He et al. (He et al., 2022) fabricated CNF/Ag composites by in situ immobilizing silver nanoparticles (Ag) on CNF during the interfacial polymerization. The resulting composites had a maximum rejection of Na_2SO_4 of 97%. In addition, the composites maintained a high water permeability and effective inhibition against Escherichia coli (He et al., 2022). Okahashi et al. (Okahashi et al., 2021) fabricated CNF/cellulose ether composites by casting and drying the aqueous mixtures. The resulting composites had a high light transparency. The composites' tensile strength could reach ~44 MPa and the Young's modulus could reach ~0.6 GPa (Okahashi et al., 2021).

Despite material and economic advantages, the incorporation of CNFs into a polymer phase is often limited by dispersibility and CNF-matrix interactions. Chemical treatment (e.g., grafting, esterification, oxidation) can be conducted to improve the adhesion between CNFs and the matrix. Other nanoparticles (e.g., lignin, silver) can be incorporated into CNFs to form composites with excellent mechanical or separation properties. In the future, the cost reduction of CNF-based composites needs to be explored for commercialization.

Energy storage

Energy storage can help transition from economy powered by fossil to one powered by sustainable renewable energy. One of the challenges in achieving this goal is that the current methods to store energy have high embodied energy cost in manufacturing them. Methods to reduce the embodied energy by introducing sustainable materials is the topic of ongoing research. CNFs play an important role in energy storage due to their low cost, tunable properties, naturally derived and inherent low embodied energy. In this section, we discuss research works on utilizing nanocellulose in energy storage systems, including batteries, supercapacitors, dielectric capacitors, and thermal energy storage.

Lithium-Ion Battery: Lithium-ion battery (LIBs) have emerged as the dominant from of energy storage for consumer electronics and electric vehicles (EV's). LIBs are constructed with a cathode and anode immersed in an electrolyte solution. The cathode and anode have an electronically insulating but ionically conducting separator between them. The cathode is made from transition metal oxide or phosphate, whereas the anode is usually made from graphite. The electrolyte can be inorganic solid or organic liquid state, with the solid state receiving increasing attention in the scientific community due to its inherent advantages (Famprikis et al., 2019). Such batteries store and release energy via the redox reaction that occurs when the lithium ion is shuttled between the cathode and anode. Nanocellulose-based films have hierarchical microstructures, strong mechanical performance, and good electrochemical stability, and showed promising applications as electrode components, solid electrolytes, and separator in batteries (Xu et al., 2021).

Figure 3. Lithium-ion battery separators based on carboxylated CNF from wood (Reproduced from ref (Kim et al., 2019) under terms of the CC-BY license)

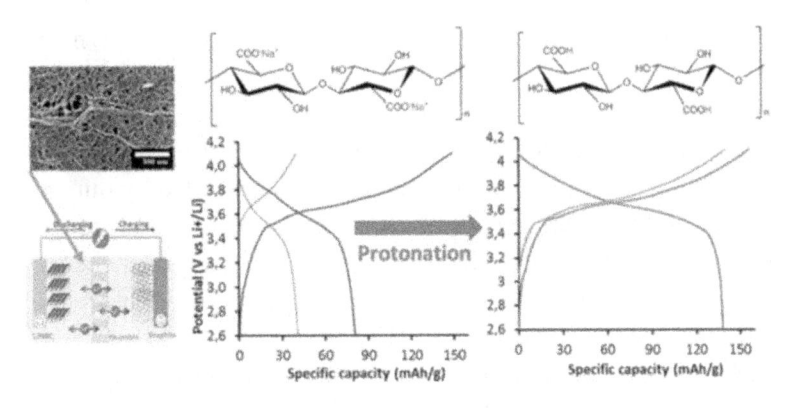

Kim et al. (Kim et al., 2019) fabricated a TEMPO oxidized CNF (T-CNF) asymmetric mesoporous membrane using the wet laid (paper making) approach (Figure 3). Protonated T-CNF-COOH membranes showed 94.5% discharge capacity at 1C rate for 100 cycles. Willgert et al. (Willgert et al., 2014) used propionyl chloride and acryloyl chloride modified CNF paper as a reinforcing agent in a two-phase electrolyte system. The two-phase electrolyte system consisted of polyethylene glycol (PEG)-methacrylate that is infiltrated with lithium hexa-fluorophosphate ($LiPF_6$) in ethylene carbonate to di-ethylene carbonate (1:1). The author reported an ionic conductivity of $5 * 10^{-5}$ S. cm^{-1} and elastic modulus of around 400 MPa at 25°C.

Nanocellulose films have also been used as the cathodes. For instance, Cao et al. (Cao et al., 2015) fabricated a freestanding lithium titanate (LTO)/CNT/CNF film, and reported a good high-rate cycling performance (142 mAh g^{-1} even at 10 °C). Nanocellulose based films have also been used in other type of the batteries, such as lithium-sulfur battery(Yu et al., 2017), Sodium-ion Batteries (Li et al., 2015), Zn-ion Batteries (Qu et al., 2018).

Supercapacitors: Supercapacitors (SCs) are one types of electrochemical energy storage devices to store energy at the interface of the surface of the conductor and electrolytic solution. SCs are made by stacking two electrodes in an electrolyte solution with an ion permeable but electron insulating separator. SCs have longer operational working lives, fast charging and discharging capabilities, and high-power density (Miller & Simon, 2008). Flexible SCs have all the inherent advantages of SCs, while being flexible, making them very attractive for portable electronics. Flexible SCs achieve flexibility by combining the electrode and current collector into one flexible layer (Shi et al., 2013). Nanocellulose have been used as nano spacers to build the desired hierarchical nanostructure. The hydrophilic nature of CNF improves the wettability of the electrode material allowing for better electrolyte ion diffusion. Furthermore, CNFs are low cost, low density and naturally derived (Gao et al., 2013). He et al. (He et al., 2020) combined CNFs and reduced graphene oxide (rGO) to form a self-supported flexible electrode material. The film tensile strength and conductivity were reported as 83 MPa and 202.94 S/m, respectively, for a CNFs/RGO ratio of 1:2. The areal capacitance under current density of 5 mAcm^{-2} was reported to be 146 mFcm^{-2}. The specific capacity retention rate after 2000 cycles was 83.7%. Yuan et al. (Yuan & Ma, 2021) reported a composite film made via vacuum filtration and in-situ polymerization of CNFs, molybdenum disulfide (MoS_2), and polypyrrole. The electrode had a tensile strength of 21.3 MPa, electrical conductivity of 9.70 S.cm^{-1}, specific capacitance of 0.734 F cm^{-2}, and high-capacity retention rate of 84.50% after 2000 charge/discharge cycles. The supercapacitor showed a high specific capacitance (1.39 F.cm^{-2} at 1 mA.cm^{-2} power density of 16.35 mW.cm^{-3}) and a volumetric energy density (6.36 mWh.cm^{-3}). The addition of hydrophilic CNFs increases the diffusion path of electrons and ions leading to superior electrical and mechanical properties.

Dielectric capacitors: Dielectric capacitors store energy in the form of an electrostatic field via dielectric polarization. A capacitor is made of conducting plates separated by thin insulating layers. The opposing plates are charged by a voltage source, and the charge that builds up is stored between the polarized insulating medium and the electrodes. Capacitors store electrical energy over a long period of time and release it in short periods as desired (Sarjeant et al., 2001). Nanocellulose based films are mainly used as the insulating medium. For instance, Goodman et al. (Goodman et al., 2022) synthesized TEMPO-oxidized CNF mixed with colloidal poly(vinylidene fluoride) nanoparticles in water to make composite films. The CNF was sandwiched between two thin polyvinyl alcohol layers. The sandwich was able to achieve an energy density of 7.22 J•cm^{-3} and a breakdown strength of 388 MV•m^{-1}, while also obtaining a stable power density of ~3 MW•cm^{-3} under an applied field of 300 MW•m^{-1} over 1000 charge/discharge cycles.

Thermal energy storage: Thermal energy storage (TES) technology stores heat in the form of sensible, latent, or thermochemical reactions (Akamo et al., 2023; Y. Z. Li et al., 2022). TES is best suited for processes that have a mismatch between supply and demand of thermal energy. This thermal energy can be stored and used later when there is demand. TES technologies can be classified into sensible heat storage which compromise air, water, and underground thermal energy storage, and latent heat such as phase change materials, and thermochemical methods such as chemical reactions and sorption systems (Alva et al., 2018). Shi et al. (Shi et al., 2019) reports casting composite films by aqueous dispersion of phase change material, poly(ethylene oxide) (PEO), and TEMPO oxidized cellulose nanofibrils (CNFs). The T-CNF/PEO composite films had clear melting point of 64 °C and crystallization point of 41 °C in the heating and cooling processes. T-CNF/PEO composite films showed high dimensional stability above melting point with a low coefficient of thermal expansion (CTE) of −53 ppm K^{-1} at 20% T-CNF loading. The T-CNFs restricted the formation of PEO crystal and formed well dispersed network structures in PEO leading to transparency, high thermal stability, and mechanical properties.

Thermal management

CNFs have low density (1.6 g/cm^3) and highly reactive surfaces due to the presence of -OH groups that allow grafting desirable chemicals to alter properties. Having low cost, naturally derived, and biodegradable characteristics make nanocellulose a perfect candidate to prepare conductive materials for thermal management. High thermal conductive fillers are generally used to enhance the thermal conductivity.

Zhang et al. (Zhang, Lu, et al., 2019) manufactured novel flexible composite films by means of vacuum filtration of silane treated aluminum nitride nanosheets (TAlN) blended with CNFs. Silane treatment led to formation of covalent bonds between AlN and CNF substrate, and hence better dispersibility. The silane treatment improves thermal conductivity by reducing the phonon scattering between the AlN and CNF substrate interfaces. When the proportion of TA1N is 25 wt%, the in-plane thermal conductivity of the composite film was reported to be 5.11 W/mK. The same author (Zhang, Tao, et al., 2019) in his prior work achieved an in-plane thermal conductivity of 4.20 W/mK without silane treatment.

Wang et al. (X. Wang et al., 2021) reports crosslinking hexagonal boron nitride (BN)-OH with lignin nanoparticle (LNP) with borax and built onto CNFs. The mixture is then freeze-dried and pressed into a composite film. Thermally conductive pathways were formed between BN-LNP where LNP was the point of cross-linking. A through-plane thermal conductivity of 2.577 W/mK was reported for a 50 wt% filler loaded BN-LNP/CNF composite film.

Zhang et al. (Zhang et al., 2022) used ultrasound treatment to exfoliate boron nitride nanosheets (BNNSs) with holocellulose nanofibrils (HCNFs), further combined with gallium liquid metal to fabricate nanocomposite films. The gallium metal flowed into the interspace gaps improving the interface between the structure. In-plane thermal conductivity of 11.78 W/mK was reported. Hu et al. (Hu et al., 2021) simultaneously improved thermally conductive and flame resistant of CNF/BNNS. BNNS was functionalized with ammonium polyphosphate before being vacuum assisted filtered with CNFs to form a compote film. A high flame resistance and in-plane thermal conductivity of 9.1 W/mK was reported.

Except enhancing thermal conductivity for thermal management, controlling the assembling of the nanocellulose to realize the control of the light to thermal management has also been studied (Smalyukh, 2021). However, the research is mainly focused on the CNCs owing to their high crystalline structure.

Sensors

Nanocellulose films have emerged as a highly promising option for sensor applications. They are created from sustainable and renewable materials and possesses tunable mechanical and optical properties. From biosensors to environmental monitoring, nanocellulose films offer unparalleled advantages that have the potential to revolutionize the sensor industry (S. Li et al., 2022; Nguyen et al., 2019). Nanocellulose films possess remarkable mechanical properties, including high tensile strength, stiffness, and flexibility. The intrinsic strength of cellulose, combined with the alignment of nanocellulose entities, contributes to their outstanding mechanical performance (Abitbol et al., 2016). These properties make nanocellulose films suitable for applications where mechanical stability is crucial, such as in flexible electronics and wearable sensors (Horta-Velázquez & Morales-Narváez, 2022). Additionally, nanocellulose films exhibit intriguing optical properties, including transparency, light scattering, and birefringence. The transparency of nanocellulose films can be attributed to their highly ordered structure and the absence of light-absorbing impurities (Dufresne, 2013). These properties make them suitable for optical sensor applications (ex. as a film-matrix), such as surface plasmon resonance sensors and optical waveguides (Misenan et al., 2022). Except some intrinsic properties of nanocellulose films, the ease of the incorporating functional additives to nanocellulose films is promising for other sensor applications, such as electronic and electrochemical sensors.

Biosensors: Biosensors are analytical tools that combine biological recognition components with a transducer in order to identify and measure particular biomolecules. The utilization of nanocellulose films in biosensor applications presents distinct benefits such as biocompatibility, a substantially high surface area, and efficient immobilization of biomolecules (Subhedar et al., 2021). Nanocellulose films can be used in glucose sensors, facilitating the non-invasive monitoring of blood glucose levels in individuals with diabetes. These films enable the immobilization of glucose oxidase enzymes and exhibit exceptional sensitivity, selectivity, and stability (Soriano & Dueñas-Mas, 2019). They can be functionalized with various enzymes to detect specific analytes in biological and environmental samples. These films serve as an ideal platform for immobilizing enzymes, ensuring high enzyme activity, prolonged stability, and enhanced sensor performance (Wei et al., 2014). Moreover, through modifying nanocellulose films with DNA or protein probes, biosensors for genetic analysis or protein detection can be developed. These films possess a high surface density, leading to improved capture efficiency and enhanced signal transduction, thereby enabling the sensitive and specific detection of target molecules (Zha et al., 2022).

Environmental Sensors: Nanocellulose films have also been used in environmental sensing, allowing real-time monitoring of various parameters for pollution control and environmental assessment. For instance, nanocellulose films can be functionalized with specific receptors or nanoparticles to detect gases and vapors. The films' high surface area and porosity facilitate efficient gas adsorption, enhancing sensor response and enabling the detection of volatile organic compounds and toxic gases (Fisher et al., 2023; Norrrahim et al., 2022). Owing to its hydrophilic nature, nanocellulose films exhibit high sensitivity to changes in humidity and moisture levels. By incorporating these films into humidity sensors, precise monitoring of humidity in diverse environments can be achieved (Kumar et al., 2022). Nanocellulose films-based pH sensors have also been investigated. These films undergo structural changes or exhibit alterations in electrical properties in response to variations in pH, making them suitable for pH sensing applications (Nguyen et al., 2019).

Optical Sensors: Nanocellulose films possess exceptional optical properties, making them well-suited for the development of optical sensors. Transparent nanocellulose films have been extensively studied (Zhu et al., 2014). Nanocellulose films have be utilized as sensing layers in surface plasmon resonance (SPR) sensors. These films provide a stable and biocompatible matrix for immobilizing biomolecules, enabling label-free and real-time monitoring of molecular interactions (Misenan et al., 2022). They can be integrated into fiber optic platforms to create optical sensors. By coating or embedding nanocellulose films on fiber optic substrates, the interaction of light with the films can be exploited for sensing various parameters such as temperature, strain, refractive index, and chemical species (Lv et al., 2021). They have also been engineered into photonic crystal structures, which exhibit unique interactions between light and matter. These structures enable the development of highly sensitive sensors based on changes in optical properties resulting from variations in environmental conditions or analyte concentrations (Thiruganasambanthan et al., 2022).

Mechanical and Pressure Sensors: Nanocellulose films possess exceptional mechanical properties that make them suitable for mechanical and pressure sensing applications. These films can be used as sensing elements in devices that detect mechanical deformation, pressure, or strain, allowing for applications in touch sensors, wearable devices, and structural health monitoring (Ji et al., 2022).

Nanocellulose films have emerged as a groundbreaking material for sensor technology, offering distinctive properties and versatile fabrication approaches. With their tunable properties, nanocellulose films hold tremendous potential for diverse sensor applications, ranging from biosensors to environmental monitoring and optical sensing. Successfully overcoming fabrication challenges, ensuring compatibility, and addressing scale-up requirements will pave the way for the commercialization and widespread adoption of nanocellulose-based sensors. This transformative development will revolutionize various industries and enable new technological advancements.

CONCLUSION AND OUTLOOK

There is an increasing need to develop sustainable materials for various applications. As a nature derived material, nanocellulose have shown a great potential and been widely used in producing high performance and environmentally friendly materials for different needs. As one of the materials platforms, nanocellulose based films are a promising for making sustainable functional materials. This chapter summarized the commonly used strategy to fabricate nanocellulose films and some representative applications of the nanocellulose films. Nanocellulose fibril film can be fabricated using scalable approaches, such as vacuum filtration, solution casting, hydraulic press, layer-by-layer assembly, and others, and been applied in different field, such as packaging, functional nanocomposites, energy storage, thermal management, and sensors.

Moreover, the properties of the nanocellulose films are tunable. During the fabrication of films, their proproteins can be through functional additives or surface modification to achieve desired properties, such as barrier, mechanical, optical, chemical, magnetic, or electrical properties. Therefore, the application of the nanocellulose film is not limited to the ones discussed in the chapter. Many other applications are also developing rapidly. Owing to the sustainable nature and tunable properties of the nanocellulose, nanocellulose based films are promising materials for replacing petroleum based plastic films for various applications, and will play an important role in building a sustainable society.

Author Note

This research is supported by the US Department of Energy (DOE), Building Technologies Office, under Contract DE-AC05-00OR22725 with UT-Battelle LLC. This manuscript has been authored by UT-Battelle LLC under Contract DE-AC05-00OR22725 with DOE. The US government retains and the publisher, by accepting the article for publication, acknowledges that the US government retains a nonexclusive, paid-up, irrevocable, worldwide license to publish or reproduce the published form of this manuscript, or allow others to do so, for US government purposes. DOE will provide public access to these results of federally sponsored research in accordance with the DOE Public Access Plan (http://energy.gov/downloads/ doe-public-access-plan).

Correspondence concerning this article should be addressed to Kai Li, Buildings and Transportation Science Division, Oak Ridge National Laboratory, Oak Ridge, TN 37830, USA. Email: lik1@ornl.gov.

REFERENCES

Abdul Khalil, H. P. S., Bhat, A. H., & Ireana Yusra, A. F. (2012). Green composites from sustainable cellulose nanofibrils: A review. *Carbohydrate Polymers*, 87(2), 963–979. DOI: 10.1016/j.carbpol.2011.08.078

Abitbol, T., Rivkin, A., Cao, Y., Nevo, Y., Abraham, E., Ben-Shalom, T., Lapidot, S., & Shoseyov, O. (2016). Nanocellulose, a tiny fiber with huge applications. *Current Opinion in Biotechnology*, 39, 76–88. DOI: 10.1016/j.copbio.2016.01.002 PMID: 26930621

Akamo, D. O., Kumar, N., Li, Y., Pekol, C., Li, K., Goswami, M., Hirschey, J., LaClair, T. J., Keffer, D. J., Rios, O., & Gluesenkamp, K. R. (2023). Stabilization of low-cost phase change materials for thermal energy storage applications. *iScience*, 26(7), 107175. DOI: 10.1016/j.isci.2023.107175 PMID: 37426345

Algar, I., Garcia-Astrain, C., Gonzalez, A., Loli, M., Gabilondo, N., Retegi, A., & Eceiza, A. (2016). Improved permeability properties for bacterial cellulose/montmorillonite hybrid bionanocomposite membranes by in-situ assembling. *Journal of Renewable Materials*, 4(1), 57–65. DOI: 10.7569/JRM.2015.634124

Alva, G., Lin, Y., & Fang, G. (2018). An overview of thermal energy storage systems. *Energy*, 144, 341–378. DOI: 10.1016/j.energy.2017.12.037

Arantes, A. C. C., Silva, L. E., Wood, D. F., Almeida, C. D., Tonoli, G. H. D., de Oliveira, J. E., da Silva, J. P., Williams, T. G., Orts, W. J., & Bianchi, M. L. (2019). Bio-based thin films of cellulose nanofibrils and magnetite for potential application in green electronics. *Carbohydrate Polymers*, 207, 100–107. DOI: 10.1016/j.carbpol.2018.11.081 PMID: 30599989

Aulin, C., Johansson, E., Wågberg, L., & Lindström, T. (2010). Self-organized films from cellulose I nanofibrils using the layer-by-layer technique. *Biomacromolecules*, 11(4), 872–882. DOI: 10.1021/bm100075e PMID: 20196583

Bian, H., Shu, X., Su, W., Luo, D., Dong, M., Liu, X., Ji, X., & Dai, H. (2022). Biodegradable, flexible and ultraviolet blocking nanocellulose composite film incorporated with lignin nanoparticles. *International Journal of Molecular Sciences*, 23(23), 14863. DOI: 10.3390/ijms232314863 PMID: 36499190

Camilo, A. C. E., de Menezes, A. J., Pereira-da-Silva, M. A., Guimarães, F. E. G., & Longaresi, R. H. (2019). Optical properties of the nanocomposite of molybdenum disulphide monolayers/cellulose nanofibrils. *Cellulose (London, England)*, 27(2), 713–728. DOI: 10.1007/s10570-019-02854-7

Cao, S., Feng, X., Song, Y., Xue, X., Liu, H., Miao, M., Fang, J., & Shi, L. (2015). Integrated fast assembly of free-standing lithium titanate/carbon nanotube/cellulose nanofiber hybrid network film as flexible paper-electrode for lithium-ion batteries. *ACS Applied Materials & Interfaces*, 7(20), 10695–10701. DOI: 10.1021/acsami.5b02693 PMID: 25938940

Cherpinski, A., Torres-Giner, S., Vartiainen, J., Peresin, M. S., Lahtinen, P., & Lagaron, J. M. (2018). Improving the water resistance of nanocellulose-based films with polyhydroxyalkanoates processed by the electrospinning coating technique. *Cellulose (London, England)*, 25(2), 1291–1307. DOI: 10.1007/s10570-018-1648-z

Copenhaver, K., Li, K., Lamm, M. E., Walker, C., Johnson, D., Han, Y., Wang, L., Zhao, X., Pu, Y., Hinton, H., Tekinalp, H., Bhagia, S., Ragauskas, A. J., Gardner, D. J., & Ozcan, S. (2021). Recycled cardboard containers as a low energy source for cellulose nanofibrils and their use in poly(l-lactide) nanocomposites. *ACS Sustainable Chemistry & Engineering*, 9(40), 13460–13470. DOI: 10.1021/acssuschemeng.1c03890

Copenhaver, K., Li, K., Wang, L., Lamm, M., Zhao, X. H., Korey, M., Neivandt, D., Dixon, B., Sultana, S., Kelly, P., Gramlich, W. M., Tekinalp, H., Gardner, D. J., MacKay, S., Nawaz, K., & Ozcan, S. (2022). Pretreatment of lignocellulosic feedstocks for cellulose nanofibril production. *Cellulose (London, England)*, 29(9), 4835–4876. DOI: 10.1007/s10570-022-04580-z

Dagnon, K. L., Shanmuganathan, K., Weder, C., & Rowan, S. J. (2012). Water-triggered modulus changes of cellulose nanofiber nanocomposites with hydrophobic polymer matrices. *Macromolecules*, 45(11), 4707–4715. DOI: 10.1021/ma300463y

Dufresne, A. (2013). Nanocellulose: A new ageless bionanomaterial. *Materials Today*, 16(6), 220–227. DOI: 10.1016/j.mattod.2013.06.004

Dufresne, A. (2018). Cellulose nanomaterials as green nanoreinforcements for polymer nanocomposites. *Philosophical Transactions. Series A, Mathematical, Physical, and Engineering Sciences*, 376(2112), 20170040. DOI: 10.1098/rsta.2017.0040 PMID: 29277738

Famprikis, T., Canepa, P., Dawson, J. A., Islam, M. S., & Masquelier, C. (2019). Fundamentals of inorganic solid-state electrolytes for batteries. *Nature Materials*, 18(12), 1278–1291. DOI: 10.1038/s41563-019-0431-3 PMID: 31427742

Fang, Z., Li, B., Liu, Y., Zhu, J., Li, G., Hou, G., Zhou, J., & Qiu, X. (2020). Critical role of degree of polymerization of cellulose in super-strong nanocellulose films. *Matter*, 2(4), 1000–1014. DOI: 10.1016/j.matt.2020.01.016

Fisher, C., Skolrood, L. N., Li, K., Joshi, P. C., & Aytug, T. (2023). Aerosol-Jet printed sensors for environmental, safety, and health monitoring: A review. *Advanced Materials Technologies*, 2300030(15), 2300030. Advance online publication. DOI: 10.1002/admt.202300030

Fu, H., Li, Y., Wang, B., Li, J., Zeng, J., Li, J., & Chen, K. (2021). Structural change and redispersion characteristic of dried lignin-containing cellulose nanofibril and its reinforcement in PVA nanocomposite film. *Cellulose (London, England)*, 28(12), 7749–7764. DOI: 10.1007/s10570-021-04041-z

Fukui, S., Ito, T., Saito, T., Noguchi, T., & Isogai, A. (2018). Surface-hydrophobized TEMPO-nanocellulose/rubber composite films prepared in heterogeneous and homogeneous systems. *Cellulose (London, England)*, 26(1), 463–473. DOI: 10.1007/s10570-018-2107-6

Gao, K., Shao, Z., Li, J., Wang, X., Peng, X., Wang, W., & Wang, F. (2013). Cellulose nanofiber–graphene all solid-state flexible supercapacitors. *Journal of Materials Chemistry. A, Materials for Energy and Sustainability*, 1(1), 63–67. DOI: 10.1039/C2TA00386D

Goodman, S. M., Che, J. J., Neri, W., Yuan, J. K., & Dichiara, A. B. (2022). Water-processable cellulosic nanocomposites as green dielectric films for high-energy storage. *Energy Storage Materials*, 48, 497–506. DOI: 10.1016/j.ensm.2022.03.047

Hamedi, M. M., Hajian, A., Fall, A. B., Håkansson, K., Salajkova, M., Lundell, F., Wågberg, L., & Berglund, L. A. (2014). Highly conducting, strong nanocomposites based on nanocellulose-assisted aqueous dispersions of single-wall carbon nanotubes. *ACS Nano*, 8(3), 2467–2476. https://pubs.acs.org/doi/pdfplus/10.1021/nn4060368. DOI: 10.1021/nn4060368 PMID: 24512093

He, M., Li, W.-D., Chen, J.-C., Zhang, Z.-G., Wang, X.-F., & Yang, G.-H. (2022). Immobilization of silver nanoparticles on cellulose nanofibrils incorporated into nanofiltration membrane for enhanced desalination performance. *npj. NPJ Clean Water*, 5(1), 64. DOI: 10.1038/s41545-022-00217-7

He, W., Wu, B., Lu, M. T., Li, Z., & Qiang, H. (2020). Fabrication and performance of self-supported flexible cellulose nanofibrils/reduced graphene oxide supercapacitor electrode materials. *Molecules (Basel, Switzerland)*, 25(12), 2793. DOI: 10.3390/molecules25122793 PMID: 32560428

Henriksson, M., Berglund, L. A., Isaksson, P., Lindström, T., & Nishino, T. (2008). Cellulose nanopaper structures of high toughness. *Biomacromolecules*, 9(6), 1579–1585. DOI: 10.1021/bm800038n PMID: 18498189

Hoeng, F., Denneulin, A., & Bras, J. (2016). Use of nanocellulose in printed electronics: A review. *Nanoscale*, 8(27), 13131–13154. DOI: 10.1039/C6NR03054H PMID: 27346635

Horta-Velázquez, A., & Morales-Narváez, E. (2022). Nanocellulose in wearable sensors. *Green Analytical Chemistry*, 1, 100009. DOI: 10.1016/j.greeac.2022.100009

Hu, D., Liu, H., Ding, Y., & Ma, W. (2021). Synergetic integration of thermal conductivity and flame resistance in nacre-like nanocellulose composites. *Carbohydrate Polymers, 264*, 118058, 118058. https://doi.org/DOI: 10.1016/j.carbpol.2021.118058

Hubbe, M. A., Ferrer, A., Tyagi, P., Yin, Y., Salas, C., Pal, L., & Rojas, O. J. (2017). Nanocellulose in thin films, coatings, and plies for packaging applications: A review. *BioResources*, 12(1), 2143–2233. DOI: 10.15376/biores.12.1.Hubbe

Isogai, A. (2020). Emerging nanocellulose technologies: Recent developments. *Advanced Materials*, 2000630. Advance online publication. DOI: 10.1002/adma.202000630 PMID: 32686197

Iwamoto, S., Isogai, A., & Iwata, T. (2011). Structure and Mechanical Properties of Wet-Spun Fibers Made from Natural Cellulose Nanofibers. *Biomacromolecules*, 12(3), 831–836. DOI: 10.1021/bm101510r PMID: 21302950

Iwamoto, S., Kai, W., Isogai, A., & Iwata, T. (2009). Elastic Modulus of Single Cellulose Microfibrils from Tunicate Measured by Atomic Force Microscopy. *Biomacromolecules*, 10(9), 2571–2576. DOI: 10.1021/bm900520n PMID: 19645441

Ji, F., Sun, Z., Hang, T., Zheng, J., Li, X., Duan, G., Zhang, C., & Chen, Y. (2022). Flexible piezoresistive pressure sensors based on nanocellulose aerogels for human motion monitoring: A review. *Composites Communications*, 101351, 101351. Advance online publication. DOI: 10.1016/j.coco.2022.101351

Kim, H., Guccini, V., Lu, H., Salazar-Alvarez, G., Lindbergh, G., & Cornell, A. (2019). Lithium ion battery separators based on carboxylated cellulose nanofibers from wood. *ACS Applied Energy Materials*, 2(2), 1241–1250. DOI: 10.1021/acsaem.8b01797

Kumar, S., Ngasainao, M. R., Sharma, D., Sengar, M., Gahlot, A. P. S., Shukla, S., & Kumari, P. (2022). Contemporary nanocellulose-composites: A new paradigm for sensing applications. *Carbohydrate Polymers*, 120052, 120052. Advance online publication. DOI: 10.1016/j.carbpol.2022.120052 PMID: 36241259

Lafia-Araga, R. A., Sabo, R., Nabinejad, O., Matuana, L., & Stark, N. (2021). Influence of lactic acid surface modification of cellulose nanofibrils on the properties of cellulose nanofibril films and cellulose nanofibril-poly(lactic acid) composites. *Biomolecules*, 11(9), 1346. DOI: 10.3390/biom11091346 PMID: 34572560

Lahiji, R. R., Xu, X., Reifenberger, R., Raman, A., Rudie, A., & Moon, R. J. (2010). Atomic Force Microscopy Characterization of Cellulose Nanocrystals. *Langmuir*, 26(6), 4480–4488. DOI: 10.1021/la903111j PMID: 20055370

Lamm, M., Li, K., Copenhaver, K., Kelly, P. V., Senkum, H., Tekinalp, H., Gramlich, W. M., & Ozcan, S. (2022). Aqueous-based polyimine functionalization of cellulose nanofibrils for effective drying and polymer composite reinforcement. *ACS Applied Polymer Materials*, 4(10), 7674–7684. DOI: 10.1021/acsapm.2c01286

Lamm, M., Li, K., Ker, D., Zhao, X., Hinton, H., Copenhaver, K., Tekinalp, H., & Ozcan, S. (2022). Exploiting chitosan to improve the interface of nanocellulose reinforced polymer composites. *Cellulose (London, England)*, 29(7), 3859–3870. DOI: 10.1007/s10570-021-04327-2

Lamm, M. E., Li, K., Qian, J., Wang, L., Lavoine, N., Newman, R., Gardner, D. J., Li, T., Hu, L., Ragauskas, A. J., Tekinalp, H., Kunc, V., & Ozcan, S. (2021). Recent Advances in Functional Materials through Cellulose Nanofiber Templating. *Advanced Materials*, 33(12), e2005538. DOI: 10.1002/adma.202005538 PMID: 33565173

Li, K., Berton, P., Kelley, S. P., & Rogers, R. D. (2018). Singlet oxygen production and tunable optical properties of deacetylated chitin-porphyrin crosslinked films. *Biomacromolecules*, 19(8), 3291–3300. DOI: 10.1021/acs.biomac.8b00605 PMID: 29901993

Li, K., Clarkson, C. M., Wang, L., Liu, Y., Lamm, M., Pang, Z., Zhou, Y., Qian, J., Tajvidi, M., Gardner, D. J., Tekinalp, H., Hu, L., Li, T., Ragauskas, A. J., Youngblood, J. P., & Ozcan, S. (2021). Alignment of cellulose nanofibers: Harnessing nanoscale properties to macroscale benefits. *ACS Nano*, 15(3), 3646–3673. DOI: 10.1021/acsnano.0c07613 PMID: 33599500

Li, K., Li, Y., Tekinalp, H., Kumar, V., Zhao, X., Pu, Y., Ragauskas, A. J., Nawaz, K., Aytug, T., & Ozcan, S. (2022). Hydrogen bond-induced aqueous-phase surface modification of nanocellulose and its mechanically strong composites. *Journal of Materials Science*, 57(17), 8127–8138. DOI: 10.1007/s10853-022-07161-4

Li, K., McGrady, D., Zhao, X., Ker, D., Tekinalp, H., He, X., Qu, J., Aytug, T., Cakmak, E., Phipps, J., Ireland, S., Kunc, V., & Ozcan, S. (2021). Surface-modified and oven-dried microfibrillated cellulose reinforced biocomposites: Cellulose network enabled high performance. *Carbohydrate Polymers*, 256, 117525. https://doi.org/https://doi.org/10.1016/j.carbpol.2020.117525. DOI: 10.1016/j.carbpol.2020.117525 PMID: 33483046

Li, K., McGuire, M., Lupini, A., Skolrood, L., List, F., Ozpineci, B., Ozcan, S., & Aytug, T. (2020). Copper–carbon nanotube composites enabled by electrospinning for advanced conductors. *ACS Applied Nano Materials*, 3(7), 6863–6875. DOI: 10.1021/acsanm.0c01236

Li, K., Skolrood, L., Aytug, T., Tekinalp, H., & Ozcan, S. (2019). Strong and tough cellulose nanofibrils composite films: Mechanism of synergetic effect of hydrogen bonds and ionic interactions. *ACS Sustainable Chemistry & Engineering*, 7(17), 14341–14346. DOI: 10.1021/acssuschemeng.9b03442

Li, S., Chen, H., Liu, X., Li, P., & Wu, W. (2022). Nanocellulose as a promising substrate for advanced sensors and their applications. *International Journal of Biological Macromolecules*, 218, 473–487. DOI: 10.1016/j.ijbiomac.2022.07.124 PMID: 35870627

Li, Y., Zhu, H., Shen, F., Wan, J., Lacey, S., Fang, Z., Dai, H., & Hu, L. (2015). Nanocellulose as green dispersant for two-dimensional energy materials. *Nano Energy*, 13, 346–354. DOI: 10.1016/j.nanoen.2015.02.015

Li, Y. Z., Kumar, N., Hirschey, J., Akamo, D. O., Li, K., Tugba, T., Goswami, M., Orlando, R., LaClair, T. J., Graham, S., & Gluesenkamp, K. R. (2022). Stable salt hydrate-based thermal energy storage materials. *Composites. Part B, Engineering*, 233, 109621. Advance online publication. DOI: 10.1016/j.compositesb.2022.109621

Lu, Y., Armentrout, A. A., Li, J., Tekinalp, H. L., Nanda, J., & Ozcan, S. (2015). A cellulose nanocrystal-based composite electrolyte with superior dimensional stability for alkaline fuel cell membranes. *Journal of Materials Chemistry. A, Materials for Energy and Sustainability*, 3(25), 13350–13356. DOI: 10.1039/C5TA02304A

Lv, P., Lu, X., Wang, L., & Feng, W. (2021). Nanocellulose-based functional materials: From chiral photonics to soft actuator and energy storage. *Advanced Functional Materials*, 31(45), 2104991. DOI: 10.1002/adfm.202104991

Matsumoto, M., & Kitaoka, T. (2016). Ultraselective gas separation by nanoporous metal– organic frameworks embedded in gas-barrier nanocellulose films. *Advanced Materials*, 28(9), 1765–1769. DOI: 10.1002/adma.201504784 PMID: 26669724

Miller, J. R., & Simon, P. (2008). Electrochemical capacitors for energy management. *Science*, 321(5889), 651–652. DOI: 10.1126/science.1158736 PMID: 18669852

Misenan, M. S. M., Akhlisah, Z., Shaffie, A., Saad, M. A. M., & Norrrahim, M. (2022). Nanocellulose in sensors. *Industrial applications of nanocellulose and its nanocomposites*, 213-243. https://doi.org/ DOI: 10.1016/B978-0-323-89909-3.00005-5

Mohammed, N., Grishkewich, N., & Tam, K. C. (2018). Cellulose nanomaterials: Promising sustainable nanomaterials for application in water/wastewater treatment processes. *Environmental Science. Nano*, 5(3), 623–658. DOI: 10.1039/C7EN01029J

Nguyen, L. H., Naficy, S., Chandrawati, R., & Dehghani, F. (2019). Nanocellulose for sensing applications. *Advanced Materials Interfaces*, 6(18), 1900424. DOI: 10.1002/admi.201900424

Norrrahim, M. N. F., Knight, V. F., Nurazzi, N. M., Jenol, M. A., Misenan, M. S. M., Janudin, N., Kasim, N. A. M., Shukor, M. F. A., Ilyas, R. A., Asyraf, M. R. M., & Naveen, J. (2022). The frontiers of functionalized nanocellulose-based composites and their application as chemical sensors. *Polymers*, 14(20), 4461. DOI: 10.3390/polym14204461 PMID: 36298039

Okahashi, K., Takeuchi, M., Zhou, Y., Ono, Y., Fujisawa, S., Saito, T., & Isogai, A. (2021). Nanocellulose-containing cellulose ether composite films prepared from aqueous mixtures by casting and drying method. *Cellulose (London, England)*, 28(10), 6373–6387. DOI: 10.1007/s10570-021-03897-5

Qu, R., Zhang, W., Liu, N., Zhang, Q., Liu, Y., Li, X., Wei, Y., & Feng, L. (2018). Antioil Ag_3PO_4 nanoparticle/polydopamine/Al_2O_3 sandwich structure for complex wastewater treatment: Dynamic catalysis under natural light. *ACS Sustainable Chemistry & Engineering*, 6(6), 8019–8028. DOI: 10.1021/acssuschemeng.8b01469

Sarjeant, W., Clelland, I. W., & Price, R. A. (2001). Capacitive components for power electronics. *Proceedings of the IEEE*, 89(6), 846–855. DOI: 10.1109/5.931475

Shi, S., Xu, C., Yang, C., Li, J., Du, H., Li, B., & Kang, F. (2013). Flexible supercapacitors. *Particuology*, 11(4), 371–377. DOI: 10.1016/j.partic.2012.12.004 PMID: 24008931

Shi, Z., Xu, H., Yang, Q., Xiong, C., Zhao, M., Kobayashi, K., Saito, T., & Isogai, A. (2019). Carboxylated nanocellulose/poly(ethylene oxide) composite films as solid-solid phase-change materials for thermal energy storage. *Carbohydrate Polymers*, 225(115215), 115215. Advance online publication. DOI: 10.1016/j.carbpol.2019.115215 PMID: 31521315

Sinquefield, S., Ciesielski, P. N., Li, K., Gardner, D. J., & Ozcan, S. (2020). Nanocellulose dewatering and drying: Current state and future perspectives. *ACS Sustainable Chemistry & Engineering*, 8(26), 9601–9615. DOI: 10.1021/acssuschemeng.0c01797

Smalyukh, I. I. (2021). Thermal management by engineering the alignment of nanocellulose. *Advanced Materials*, 33(28), e2001228. DOI: 10.1002/adma.202001228 PMID: 32519371

Soriano, M. L., & Dueñas-Mas, M. J. (2019). Promising sensing platforms based on nanocellulose. *Carbon-based Nanosensor Technology*, 273-301. https://doi.org/DOI: 10.1007/5346_2018_26

Subhedar, A., Bhadauria, S., Ahankari, S., & Kargarzadeh, H. (2021). Nanocellulose in biomedical and biosensing applications: A review. *International Journal of Biological Macromolecules*, 166, 587–600. DOI: 10.1016/j.ijbiomac.2020.10.217 PMID: 33130267

Thiruganasambanthan, T., Ilyas, R. A., Norrrahim, M. N. F., Kumar, T. S. M., Siengchin, S., Misenan, M. S. M., Farid, M. A. A., Nurazzi, N. M., Asyraf, M. R. M., Zakaria, S. Z. S., & Razman, M. (2022). Emerging developments on nanocellulose as liquid crystals: A biomimetic approach. *Polymers*, 14(8), 1546. DOI: 10.3390/polym14081546 PMID: 35458295

Thomas, B., Raj, M. C., Joy, J., Moores, A., Drisko, G. L., & Sanchez, C. (2018). Nanocellulose, a versatile green platform: From biosources to materials and their applications. *Chemical Reviews*, 118(24), 11575–11625.

Tomé, L. C., Brandão, L., Mendes, A. M., Silvestre, A. J. D., Neto, C. P., Gandini, A., Freire, C. S. R., & Marrucho, I. M. (2010). Preparation and characterization of bacterial cellulose membranes with tailored surface and barrier properties. *Cellulose (London, England)*, 17(6), 1203–1211. DOI: 10.1007/s10570-010-9457-z

Trifol, J., & Moriana, R. (2022). Barrier packaging solutions from residual biomass: Synergetic properties of CNF and LCNF in films. *Industrial Crops and Products*, 177, 114493. DOI: 10.1016/j.indcrop.2021.114493

Valencia, L., Nomena, E. M., Mathew, A. P., & Velikov, K. P. (2019). Biobased cellulose nanofibril-oil composite films for active edible barriers. *ACS Applied Materials & Interfaces*, 11(17), 16040–16047. DOI: 10.1021/acsami.9b02649 PMID: 30977999

Vartiainen, J., Rose, K., Kusano, Y., Mannila, J., & Wikstrom, L. (2020). Hydrophobization, smoothing, and barrier improvements of cellulose nanofibril films by sol-gel coatings. *Journal of Coatings Technology and Research*, 17(1), 305–314. DOI: 10.1007/s11998-019-00292-5

Velasquez-Cock, J., Gomez, H. B., Posada, P., Serpa, G. A., Gomez, H. C., Castro, C., Ganan, P., & Zuluaga, R. (2018). Poly (vinyl alcohol) as a capping agent in oven dried cellulose nanofibrils. *Carbohydrate Polymers*, 179, 118–125. DOI: 10.1016/j.carbpol.2017.09.089 PMID: 29111034

Visanko, M., Liimatainen, H., Sirviö, J. A., Mikkonen, K. S., Tenkanen, M., Sliz, R., Hormi, O., & Niinimäki, J. (2015). Butylamino-functionalized cellulose nanocrystal films: Barrier properties and mechanical strength. *RSC Advances*, 5(20), 15140–15146. DOI: 10.1039/C4RA15445B

Wagberg, L., & Erlandsson, J. (2021). The use of layer-by-layer self-assembly and nanocellulose to prepare advanced functional materials. *Advanced Materials*, 33(28), e2001474. DOI: 10.1002/adma.202001474 PMID: 32767441

Wang, L., Gardner, D. J., Wang, J., Yang, Y., Tekinalp, H. L., Tajvidi, M., Li, K., Zhao, X., Neivandt, D. J., Han, Y., Ozcan, S., & Anderson, J. (2020). Towards industrial-scale production of cellulose nanocomposites using melt processing: A critical review on structure-processing-property relationships. *Composites. Part B, Engineering*, 201, 108297. https://doi.org/https://doi.org/10.1016/j.compositesb.2020.108297. DOI: 10.1016/j.compositesb.2020.108297

Wang, L., Kelly, P. V., Ozveren, N., Zhang, X., Korey, M., Chen, C., Li, K., Bhandari, S., Tekinalp, H., Zhao, X., Wang, J., Seydibeyoğlu, M. Ö., Alyamac-Seydibeyoglu, E., Gramlich, W. M., Tajvidi, M., Webb, E., Ozcan, S., & Gardner, D. J. (2023). Multifunctional polymer composite coatings and adhesives by incorporating cellulose nanomaterials. *Matter*, 6(2), 344–372. DOI: 10.1016/j.matt.2022.11.024

Wang, L., Li, K., Copenhaver, K., Mackay, S., Lamm, M. E., Zhao, X., Dixon, B., Wang, J., Han, Y., Neivandt, D., Johnson, D. A., Walker, C. C., Ozcan, S., & Gardner, D. J. (2021). Review on nonconventional fibrillation methods of producing cellulose nanofibrils and their applications. *Biomacromolecules*, 22(10), 4037–4059. DOI: 10.1021/acs.biomac.1c00640 PMID: 34506126

Wang, M. M., Li, H. L., Du, C., Liang, Y., & Liu, M. R. (2018). Preparation and barrier properties of nanocellulose/layered double hydroxide composite film. *BioResources*, 13(1), 1055–1064. DOI: 10.15376/biores.13.1.1055-1064

Wang, X., Qu, Y. F., Jiao, L., Bian, H. Y., Wang, R. B., Wu, W. B., Fang, G. G., & Dai, H. Q. (2021). Boosting the thermal conductivity of CNF-based composites by cross-linked lignin nanoparticle and BN-OH: Dual construction of 3D thermally conductive pathways. *Composites Science and Technology*, 204, 108641. Advance online publication. DOI: 10.1016/j.compscitech.2020.108641

Wei, H., Rodriguez, K., Renneckar, S., & Vikesland, P. J. (2014). Environmental science and engineering applications of nanocellulose-based nanocomposites. *Environmental Science. Nano*, 1(4), 302–316. DOI: 10.1039/C4EN00059E

Willgert, M., Leijonmarck, S., Lindbergh, G., Malmström, E., & Johansson, M. (2014). Cellulose nanofibril reinforced composite electrolytes for lithium ion battery applications. *Journal of Materials Chemistry. A, Materials for Energy and Sustainability*, 2(33), 13556–13564. DOI: 10.1039/C4TA01139B

Wu, C. N., Saito, T., Fujisawa, S., Fukuzumi, H., & Isogai, A. (2012). Ultrastrong and high gas-barrier nanocellulose/clay-layered composites. *Biomacromolecules*, 13(6), 1927–1932. DOI: 10.1021/bm300465d PMID: 22568705

Wu, Y., Liang, Y., Mei, C., Cai, L., Nadda, A., Le, Q. V., Peng, Y., Lam, S. S., Sonne, C., & Xia, C. (2022). Advanced nanocellulose-based gas barrier materials: Present status and prospects. *Chemosphere*, 286, 131891. DOI: 10.1016/j.chemosphere.2021.131891 PMID: 34416587

Xu, T., Du, H., Liu, H., Liu, W., Zhang, X., Si, C., Liu, P., & Zhang, K. (2021). Advanced nanocellulose-nased composites for flexible functional energy storage devices. *Advanced Materials*, 33(48), 2101368. https://doi.org/https://doi.org/10.1002/adma.202101368. DOI: 10.1002/adma.202101368 PMID: 34561914

Yanamala, N., Farcas, M. T., Hatfield, M. K., Kisin, E. R., Kagan, V. E., Geraci, C. L., & Shvedova, A. A. (2014). *In Vivo* evaluation of the Pulmonary toxicity of cellulose nanocrystals: A renewable and sustainable nanomaterial of the future. *ACS Sustainable Chemistry & Engineering*, 2(7), 1691–1698. DOI: 10.1021/sc500153k PMID: 26753107

Yang, S., Xue, B., Li, Y., Li, X., Xie, L., Qin, S., Xu, K., & Zheng, Q. (2020). Controllable Ag-rGO heterostructure for highly thermal conductivity in layer-by-layer nanocellulose hybrid films. *Chemical Engineering Journal*, 383, 123072. Advance online publication. DOI: 10.1016/j.cej.2019.123072

Yu, M., Ma, J., Xie, M., Song, H., Tian, F., Xu, S., Zhou, Y., Li, B., Wu, D., Qiu, H., & Wang, R. (2017). Freestanding and sandwich-structured electrode material with high areal mass loading for long-life lithium–sulfur batteries. *Advanced Energy Materials*, 7(11), 1602347. DOI: 10.1002/aenm.201602347

Yuan, Q., & Ma, M. G. (2021). Conductive polypyrrole incorporated nanocellulose/MoS2 film for preparing flexible supercapacitor electrodes. *Frontiers of Materials Science*, 15(2), 227–240. DOI: 10.1007/s11706-021-0549-5

Zha, R., Shi, T., He, L., & Zhang, M. (2022). Nanoengineering and green chemistry-oriented strategies toward nanocelluloses for protein sensing. *Advances in Colloid and Interface Science*, 102758, 102758. Advance online publication. DOI: 10.1016/j.cis.2022.102758 PMID: 36037672

Zhang, C., Wang, M., Lin, X., Tao, S., Wang, X., Chen, Y., Liu, H., Wang, Y., & Qi, H. (2022). Holocellulose nanofibrils assisted exfoliation of boron nitride nanosheets for thermal management nanocomposite films. *Carbohydrate Polymers, 291*, 119578, 119578. https://doi.org/DOI: 10.1016/j.carbpol.2022.119578

Zhang, K., Lu, Y. X., Hao, N. K., & Nie, S. X. (2019). Enhanced thermal conductivity of cellulose nanofibril/aluminum nitride hybrid films by surface modification of aluminum nitride. *Cellulose (London, England)*, 26(16), 8669–8683. DOI: 10.1007/s10570-019-02694-5

Zhang, K., Tao, P., Zhang, Y., Liao, X., & Nie, S. (2019). Highly thermal conductivity of CNF/AlN hybrid films for thermal management of flexible energy storage devices. *Carbohydrate Polymers*, 213, 228–235. DOI: 10.1016/j.carbpol.2019.02.087 PMID: 30879664

Zhao, X., Bhagia, S., Gomez-Maldonado, D., Tang, X., Wasti, S., Lu, S., Zhang, S., Parit, M., Rencheck, M. L., Korey, M., Jiang, H., Zhu, J., Meng, X., Lamm, M. E., Copenhaver, K., Peresin, M. S., Wang, L., Tekinalp, H., Yang, G., & Ozcan, S. (2023). Bioinspired design toward nanocellulose-based materials. *Materials Today*, 66, 409–430. DOI: 10.1016/j.mattod.2023.04.010

Zhao, X., Li, K., Wang, Y., Tekinalp, H., Richard, A., Webb, E., & Ozcan, S. (2020). Bio-treatment of poplar via amino acid for interface control in biocomposites. *Composites. Part B, Engineering*, 199, 108276. DOI: 10.1016/j.compositesb.2020.108276

Zhu, H., Fang, Z., Preston, C., Li, Y., & Hu, L. (2014). Transparent paper: Fabrications, properties, and device applications. *Energy & Environmental Science*, 7(1), 269–287. DOI: 10.1039/C3EE43024C

Zhu, H., Luo, W., Ciesielski, P. N., Fang, Z., Zhu, J. Y., Henriksson, G., Himmel, M. E., & Hu, L. (2016). Wood-derived materials for green electronics, biological devices, and energy applications. *Chemical Reviews*, 116(16), 9305–9374. DOI: 10.1021/acs.chemrev.6b00225 PMID: 27459699

Chapter 5
Environmentally Sustainable Production of Bacterial Nanocellulose in Waste-Based Cell Culture Media and Applications

Takaomi Kobayashi
Nagaoka University of Technology, Japan

ABSTRACT

Bacterial cellulose has attracted great attention due to the demand for eco-friendly materials and sustainable products. It possesses properties superior to those of plant cellulose and has potential uses in various applications. The wider application of bacterial cellulose depends on the practical considerations such as the scale-up capability and production costs. The high cost of bacterial cellulose production is the main drawback that hinders industrial implementation. The cost-competitiveness can be improved, and bacterial cellulose production can be maximized by the utilization of agricultural and industrial waste and by-products as bacterial cell culture media. This chapter provides an overview of cost-effective culture media for bacterial cellulose production by using agricultural waste as a primary nutrient source. The applications related to nanocellulose, and the future challenges are also described.

INTRODUCTION

A low-carbon society to reduce the emissions of greenhouse gases has become a worldwide trend to achieve climate protection goals. The greenhouse gases such as carbon dioxide are not only caused by the energy or fuel consumption in the industrial production and transportation, but also by plastics value chains (Bauer, Nielsen et al. 2022). Plastics have a large amount of carbon, which is released into the atmosphere at their end of life depending on polymer type and production technique. Therefore, recycling the plastics can keep the embedded carbon in the carbon loop (Bauer, Nielsen et al. 2022). On the other hands, producing plastics from biopolymer avoids adding more carbon footprint to the loop,

DOI: 10.4018/979-8-3693-0003-9.ch005

but maintain carbon neutrality to drive the carbon cycle (Schirmeister and Mülhaupt 2022). The use of bio-based alternatives to replace petroleum–based plastics is a key of the UN Sustainable Development Goals (SDGs) (Walker 2021). These sustainable alternatives can mitigate the single use plastics built up in the environment due to non-biodegradability and inadequate waste disposal (RameshKumar, Shaiju et al. 2020). Among different biopolymers, cellulose has been widely considered as good candidates to be applied in a variety of everyday applications for plastic replacement (Yun, Tao et al. 2023).

Cellulose, the most abundant biopolymer, has emerged as a multifunctional material widely used in the field of functional papers (Zhu, Zhu et al. 2015, Zhang, Zhang et al. 2018), packaging (Li, Wang et al. 2020, Tardy, Mattos et al. 2021), and other paper–based materials such as papers in flexible electronics (Guan, Song et al. 2020, Yang, Lu et al. 2021), and construction (Sandolo, Matricardi et al. 2009, Yang, Bai et al. 2018) due to its physical and mechanical properties. It is noteworthy, engineering with high industrialization and automation enables various cellulose–related products to be cost–efficient and available, thus demonstrating the high potential for plastic replacement (Liu, Du et al. 2021, Sun, Liu et al. 2021). Cellulose is a plant structural component bound to other biopolymers such as hemicellulose, lignin, and pectin (Brethauer, Shahab et al. 2020). These impurities are eliminated by acid/alkali pre-treatments and bleaching yielding the pure plant microcellulose. Nanocellulose from plants can also be obtained using a high-pressure homogenizer, grinders, and other mechanical processes whereby the high shear fibrillation process converts microcellulose into nanocellulose fibers (Abdul Khalil, Davoudpour et al. 2014).

Cellulose can also be synthesized by bacteria, so called bacterial cellulose, bacterial nanocellulose, or biocellulose (BC). A rod–shaped, aerobic gram–negative bacterium such as *Gluconacetobacter xylinus* is mainly used to produce a BC pellicle, a gelatinous cellulose sheet, with a denser surface on one side near the air and a looser layer on the other side (Jozala, Pértile et al. 2015). This biological process is more eco–friendly and does not use chemical treatment that generates wastewater effluent containing harmful chemicals like the process used for plant cellulose (Kongklieng, Kobayashi et al. 2023). Summative comparison between plant-derived nanocellulose and BC is shown in Figure 1. Because BC does not contain any hemicellulose, lignin, or other natural components, it is a high purity source of cellulose (Avcioglu 2022). The chemical structure of BC is identical to that of plant cellulose, but their physicochemical properties are different (Coseri 2021). One of the superior properties of BC over the plant cellulose is ultrafine nanoscale mat structure (Figure 2). This nanofibrous network structure of BC with high degree of polymerization (between 2000 and 6000) results in a high degree of crystallinity (> 70%), mechanical strength, water holding capacity, porosity (Keshk and Sameshima 2006, Mohite and Patil 2014), and high surface area (Ul-Islam, Khan et al. 2012). The properties of BC can be adjusted by choosing appropriate fermentation strategies, microbial strains, and modification (Raval, Raval et al. 2020).

The nanoporous network and the simple modification make BC a very attractive material for biotechnological applications (Avcioglu 2022). Modifications of BC by environmentally benign techniques have been developed, which fall into the categories of *in–situ* and *ex–situ* modifications to obtain the novel BC structure for each application (Taokaew, Phisalaphong et al. 2016). The *in–situ* modification is the addition of other materials to the culture medium before the biosynthesis (Taokaew, Seetabhawang et al. 2013). These exogenous materials are embedded in BC matrices resulting in new entanglements of nanofibrils (Taokaew, Phisalaphong et al. 2016). The addition of additives to the culture medium, such as ethanol (Park, Jung et al. 2003), agar (Bae, Sugano et al. 2004), sodium alginate (Zhou, Sun et al. 2007), not only have certain changes in BC crystallinity, mechanical strength, but also increase BC production. The *ex–situ* modification implies the modification of BC matrix after biosynthesis by the addition of other materials over the BC surface and fiber layers creating the new physicochemical (Suratago, Taokaew et al. 2015) and biological properties (Taokaew, Nunkaew et al. 2014, Taokaew, Phisalaphong et al. 2014, Taokaew, Phisalaphong et al. 2015). The versatility and range if properties provided by typical and the modified BC make it an interesting raw material for applications such as nutritional enhancement, thickening/stabilizing agent in food industry (Shi, Zhang et al. 2014), wound healing materials (Pang, Huang et al. 2020), composite reinforcement (Nogi and Yano 2008), and opto-electronics materials (Fortunato, Gaspar et al. 2016). However, due to the high cost of nutrient sources for large–scale industrial production, the use of waste as a nutrient source has been considered to overcome this limitation. Regarding sustainability and economic viability, which are important aspects, industrial waste utilization in BC production is a promising solution that may satisfy both issues.

Figure 1. Summative comparison between top-down approach to prepare nanocellulose from plant biomass and bottom-up approach to prepare bacterial nanocellulose

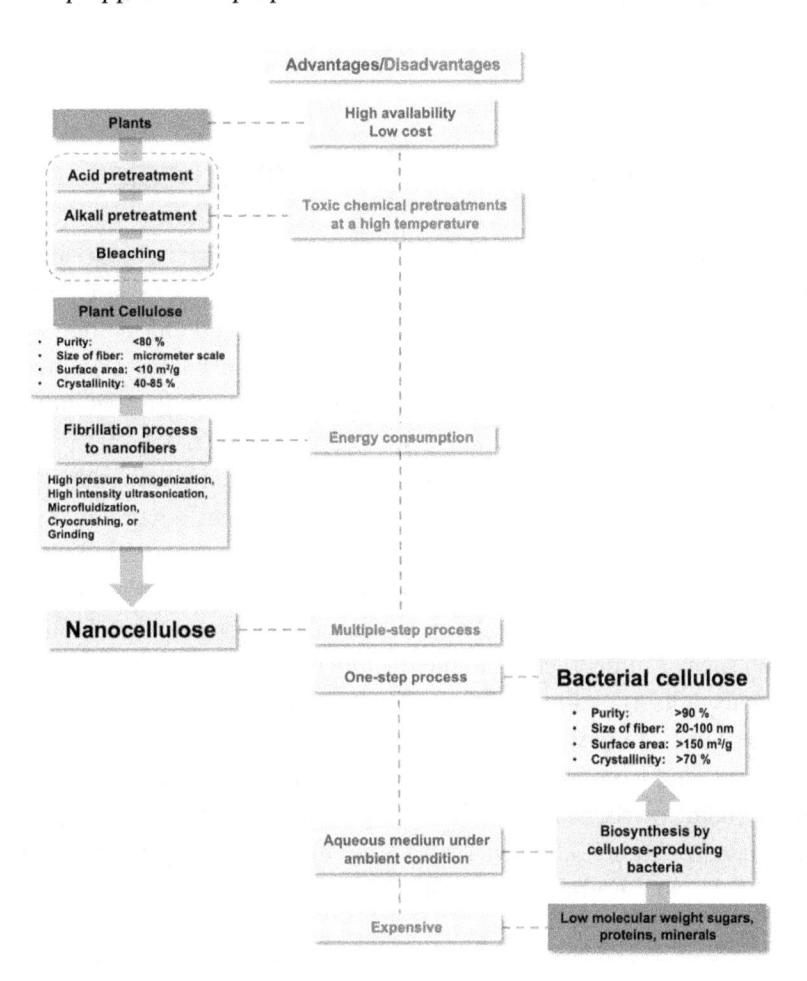

Figure 2. Microscopic images of freeze–dried cellulose extracting from sugarcane (left) and freeze–dried bacterial cellulose (right) visualized by using Scanning electron microscopy (TM3030 Plus, Hitachi, Japan)

CONVENTIONAL BACTERIAL CELLULOSE BIOSYNTHESIS

The effective sources for the production of BC are *Acetobacter xylinus*, *Acetobacter hansenii*, and *Acetobacter pasteurianus*. Among these, *Acetobacter xylinus*, also called *Gluconacetobacter xylinus* (*G. xylinus*), has been used for producing commercially available BC due to its high productivity (Wang, Tavakoli et al. 2019). These aerobic gram–negative bacteria are capable of extracellularly producing BC by moving towards the oxygen–rich interface surface of the culture medium at temperatures of 25–30 °C and pH of 3–7 for 1–14 days (Wang, Tavakoli et al. 2019, Taokaew, Thienchaimongkol et al. 2022). Carbon sources used in BC production are generally monosaccharides such as glucose and fructose, disaccharides such as sucrose, and polysaccharides such as dextran and starch (Islam, Ullah et al. 2017, Suwanmajo and Taokaew 2022, Taokaew, Thienchaimongkol et al. 2022). Peptone, yeast extract, $(NH_4)_2SO_4$, and KNO_3 can be used as nitrogen sources in BC synthesis (Santos, Carbajo et al. 2013, Suwanmajo and Taokaew 2022). The BC is produced as a primary metabolite to protect the cells from external stimuli such as ultraviolet light and to retain moisture (González-García, Meza-Contreras et al. 2022). In the biosynthesis, cellulose subfibrils (1.5 nm) extracellularly produced by bacteria assemble into nanofibrils of 2–4 nm width, which are composed of 10–250 single polymeric chains and are 1–9 nm length equivalent from 18,000–20,000 glucose units organized into of 40–60 nm or 20–100 nm–wide ribbon depending on the biosynthesis conditions (Chawla, Bajaj et al. 2009, Cacicedo, Castro et al. 2016) as illustrated in Figure 3. A matrix of these interwoven ribbons constitutes the BC pellicle (Iguchi, Yamanaka et al. 2000, Klemm, Schumann et al. 2001).

Figure 3. Schematic illustration of cellulose fibril formation (left) (modified from (Chawla, Bajaj et al. 2009) and microscopic image of freeze–dried cellulose producing bacteria (Gluconacetobacter xylinus) and the produced bacterial cellulose nanofibers (right) visualized by using Scanning electron microscopy (TM3030 Plus, Hitachi, Japan)

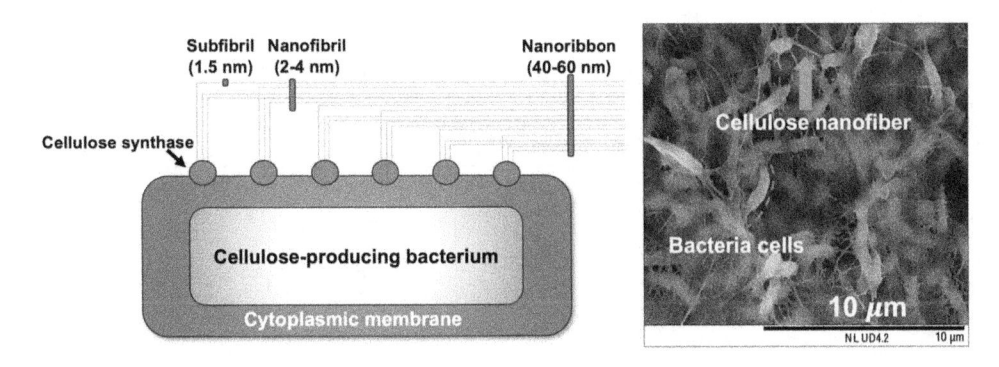

In conventional method, simple static cultivation has been used for the BC production, by simply culturing the bacteria in shallow bottles or trays containing the liquid culture to form dense, fibrous pellicles at the interface of the culture medium and air. Owing to its limited air contact area, the BC productivity is limited (Mohite, Salunke et al. 2013). Hence, cultures with agitation or shaking have been developed to overcome the oxygen transport limitation in the medium. The formation of the pellicle is modified by the aeration during biosynthesis (Mohite, Salunke et al. 2013). This affects the supramolecular organization of microfibrils by disrupting the formation of hydrogen bonds between cellulose chains. The BC produced by agitation culture has a lower degree of polymerization, crystallinity and mechanical strength compared to those produced by static culture (Czaja, Romanovicz et al. 2004). In some case, the aeration condition induces formation of non–cellulose producing strains due to the applied shear force during agitation or shaking, resulting in a decreased BC yield (Jozala, de Lencastre-Novaes et al. 2016). The BC pellicles from agitated culture conditions are obtained in irregular masses and shapes suspended in the culture medium depending on the rotating/agitating speed and time (Algar, Fernandes et al. 2015, Wang, Tavakoli et al. 2019, Fukuzaki, Thienchaimongkol et al. 2022). Mixing static and agitated culture conditions is more efficient than the static culture alone (Taokaew, Thienchaimongkol et al. 2022). To increase the productivity of BC, fed–batch operations (Bae and Shoda 2004) and bioreactors (Wang, Tavakoli et al. 2019) can be employed. The efficiency of these modes is higher than batch or shaking.

Traditionally, BC is produced from expensive culture media, containing glucose, fructose, or mannitol as a carbon source and other nutrient sources such as proteins and minerals resulting in high production costs. The typical Hestrin–Schramm (HS) medium is the most widely used for culturing BC, which accounts for 30–65% of the total cost of the process (Vazquez, Foresti et al. 2013, Revin, Liyaskina et al. 2018, Raval, Raval et al. 2020). This culture medium formulation containing glucose and pure proteins i.e., yeast extract and peptone provides much higher BC production than the minimal culture medium like Yeast extract Glucose Chloramphenicol Medium (YGC medium) (Taokaew, Thienchaimongkol et al. 2022). The formulation of bacterial cell culture media in water (%w/v) is listed in Table 1. However, the amount of BC production is higher in Zhou (Z) medium containing corn steep liquor/powder rather than HS and Yamanaka (Y) media, when glucose, sucrose, and especially mannitol are used as a carbon source (Mohammadkazemi, Azin et al. 2015). This suggests that the corn steep liquor/powder, an agricultural by–product, shows a positive effect on cellulose production (Ruka, Simon et al. 2012). Among sugars, mannitol and fructose are better carbon sources, but they are expensive. Therefore, some or all these commercial nitrogenous and carbonaceous components should be replaced in the original culture medium to minimize the cost of BC production.

Table 1. Composition of Yeast extract Glucose Chloramphenicol (YGC), Hestrin–Schramm (HS), Yamanaka (Y), and Zhou (Z) media (%w/v)

Components	YGC	HS	Y	Z
Glucose	5	2	5	4
Yeast extract	0.5	0.5	0.5	–
Peptone	–	0.5	–	–
Na_2HPO_4	–	0.27	–	–
Corn steep liquor	–	–	–	2
Citric acid·H_2O	–	0.115	–	–
$(NH_4)_2SO_4$	–	–	0.5	0.4
KH_2PO_4	–	–	0.3	0.2
$MgSO_4·7H_2O$	–	–	0.005	0.04
$CaCO_3$	1.25	–	–	–

ALTERNATIVE NUTRIENT SOURCES FOR ECONOMIC PRODUCTION OF BACTERIAL CELLULOSE

Cost of the culture medium plays a critical role in BC production (Heydorn, Lammers et al. 2023). Thus, one of the important aspects in the cellulose biosynthesis process is finding a new cost–effective culture medium and obtaining a high yield of BC. In order to decrease the costs of BC production, culture media based on waste sugars like those from fruits and vegetables has been gaining notable attention (Rodrigues, Fontão et al. 2019) or even hydrocarbon wastes like naphthalene (Marín, Martirani-Von Abercron et al. 2019), cellulose–based waste textiles (Kuo, Lin et al. 2010, Hong, Guo et al. 2012), and waste fiber sludge (Cavka, Guo et al. 2013). Organic waste is obtained from different stages of agro–industrial productions that, in many cases, cannot be marketed due to poor quality and end up by being discarded. However, they are rich with sugars such as glucose, fructose, and sucrose, as well as nitrogen and vitamins that can be used as feedstock in cellulose biosynthesis (Carreira, Mendes et al. 2011). While some waste such as fruit juices can be used directly in the culture medium, macromolecular waste such as cellulose are required to undergo hydrolysis to obtain smaller molecules such as mono and disaccharides prior to being fed into the culture medium. One of the eco–friendly methods is direct hydrolysis via enzymatic digestion with cellulase (Hong, Guo et al. 2012) and amyloglucosidase (Voon, Muhialdin et al. 2019), or fermentation with enzyme–producing fungi such as *Aspergillus niger* (Kuo, Lin et al. 2010), *Trichoderma reesei* (Cavka, Guo et al. 2013, Guo, Cavka et al. 2013), *Aspergillus oryzae*, and *Aspergillus awamori* (Tsouko, Kourmentza et al. 2015). Routes for preparing BC by supplementation with agricultural waste and monosaccharides prepared from enzymatic hydrolysis in the cell culture media is summarized in Figure 4. To quantify the components in nutritional waste, chemical characterization of the nutritional waste sources can be employed. Such chemical analysis must be understood in order to control the composition of the culture medium. The methods are summarized in Table 2.

Figure 4. Overview schematic preparation of BC by an addition of agricultural waste with and without pretreatment from various approaches in the cell culture medium

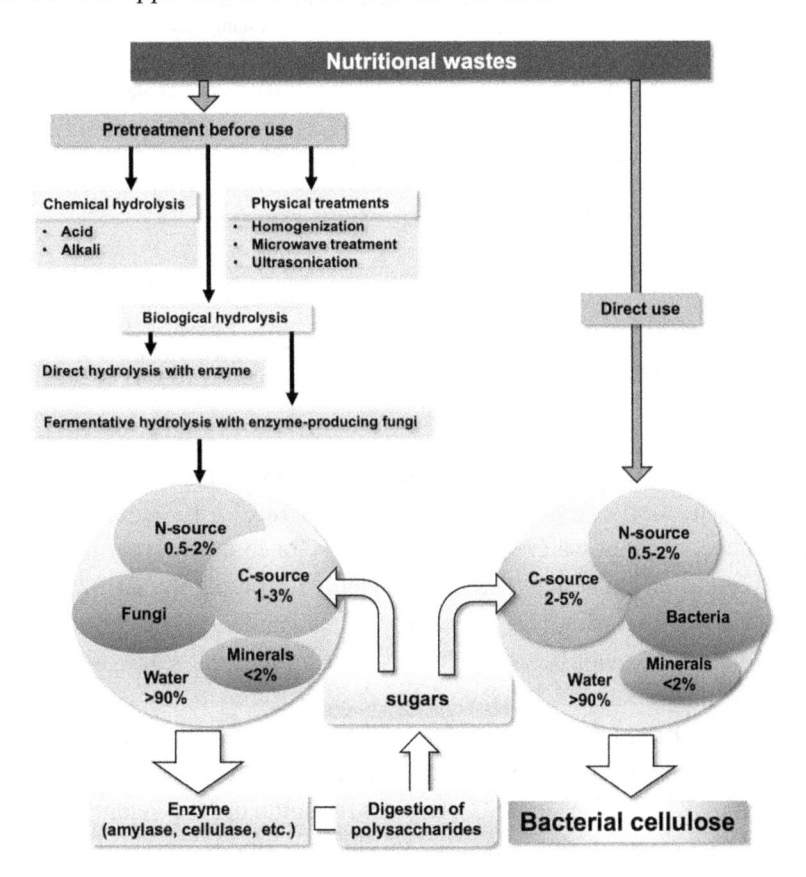

Table 2. Chemical characterization methods used for analysis of compositions in nutritional wastes before BC production

Components	Methods	Ref.
Reducing sugars	3, 5 dinitrosalicylic acid (DNS) assay	(Miller 1959, Lin, Lopez-Sanchez et al. 2014, Dórame-Miranda, Gámez-Meza et al. 2019)
	Liquid chromatography–mass spectrometry	(Ciecholewska-Juśko, Broda et al. 2021)
Starch	Iodine starch method	(Ciecholewska-Juśko, Broda et al. 2021)
Protein	AOAC[a]984.13[b]	(Dórame-Miranda, Gámez-Meza et al. 2019)
	Bradford protein assay	(Bradford 1976, Ciecholewska-Juśko, Broda et al. 2021)
	DUMAS method using Nitrogen/Protein analyzer	(Guimarães, de Oliveira Barros et al. 2023)
Amino acids	High–Performance Liquid Chromatography	(Taokaew, Zhang et al. 2020, Taokaew, Nakson et al. 2023)

continued on following page

Table 2. Continued

Components	Methods	Ref.
Fat/lipids	AOAC[a]920.39	(Cunniff and Washington 1995, Dórame-Miranda, Gámez-Meza et al. 2019)
	Official Method Am 5–04 using Fat analyzer	(Am 2005, Guimarães, de Oliveira Barros et al. 2023)
Moisture Ash Fiber	AOAC[a] 930.15 AOAC[a]942.05 AOAC[a]962.09	(Cunniff and Washington 1995, Dórame-Miranda, Gámez-Meza et al. 2019)
Vitamins	High–Performance Liquid Chromatography	(Guimarães, de Oliveira Barros et al. 2023)
Organic acids	High–Performance Liquid Chromatography	(Huang, Yang et al. 2015)
Minerals	X–ray fluorescence analysis	(Taokaew, Nakson et al. 2022)

[a] standard test of the Association of Official Agricultural Chemists
[b] Protein = content of nitrogen (N) × 6.25 according to Kjeldahl method

To mitigate problems associated with food waste and effectively utilize it as a resource, these by–product streams, primarily from the food industry, can be used as substrate for BC production (Carreira, Mendes et al. 2011). For instance, coconut water and molasses have been extensively used for BC production. Coconut water is known as an abundant source of nutrients for producing BC, known as *nata de coco*. It is employed to replace distilled water for preparing BC in Thailand (Phisalaphong, Tran et al. 2016). Molasses derived from sugarcane or beets is a by–product in the final stage of crystallization of sugar production processes in many countries. Replacement of glucose with molasses can reduce BC production costs by about 20% (Machado, Meneguin et al. 2018). Thus, molasses is frequently reported as an alternative nutrient source, especially as a carbon source, after it is pretreated or diluted sufficiently (Çakar, Özer et al. 2014, Heydorn, Lammers et al. 2023). Besides the study in the conventional cultivation system, molasses has been studied in semi–continuous BC production modes. Due to a high sucrose content of raw sugar beet molasses (60% sucrose) (Sjölin, Thuvander et al. 2020) or raw sugarcane molasse (34% sucrose) with low nitrogen content (Sen, Hussin et al. 2019), the sucrose is hydrolyzed to simple sugars (monosaccharides) such as glucose and fructose, facilitating metabolization by bacteria into gluconic and/or acetic acids. These acids produced by glucose digestion decrease the pH of the culture medium from about 5 to 3 (Gullo, La China et al. 2019), and 40% of the glucose is enzymatically converted into gluconic acid by *G. xylinus* (Zhong, Zhang et al. 2013). Thus, the decrease of the culture medium's pH affects the final BC yield, especially when the hydrolyzed molasses is used in the medium of *G. xylinus* as compared to unhydrolyzed molasses in the conventional cultivation system (Huang, Chen et al. 2022). Cakar et. al. (Çakar, Özer et al. 2014) studied that the highest cellulose yield (1.6 g/L) was obtained after 7 days by static semi–continuous operation mode. The series static culture system (four stages) was studied by Oz et. al. (Öz and Kalender 2023). In stage 1, *G. xylinus* was inoculated under aseptic conditions in a laminar air flow cabinet at 30 °C. Every 10 days, the culture medium under BC produced in the stage 1 was transferred to stages 2 and 3, respectively using a valve and peristaltic pump. The BC production in stage 4 was only carried out in molasses medium containing 100 g/L sugar from beet molasses. The BC yield at this sugar concentration was the highest among 20–100 g/L sugars, and it was about a 20% increase from the conventional cultivation system (Wang, Tavakoli et al. 2019).

Other agricultural waste by–products and rotten fruits from the food industry have also been investigated in the past decade as alternative sources for the production of BC (Table 3), such as, residues of dry olive mill, sago, potatoes, and fruit juices. Dry olive mill residue, which accounts for ~35% of massive high water content solid residue (40 kilotons) is produced every year from olive oil production in Portugal and the Mediterranean basin. Normally, it is burned for cogeneration of electric power, used as organic fertilizers or animal feed supplements (Gomes, Silva et al. 2013). Gomes et. al (Gomes, Silva et al. 2013) used this residue from the olive oil production as nutrient and carbon source for the BC production by *Gluconacetobacter sacchari*. The residue was dried at 40 and 100 °C and hydrolyzed with 1M H_2SO_4 to obtain sugar rich aqueous extracts prior to bacterial cultivation. After 96 h, BC yields were 0.81 and 0.85 g/L, respectively, which is lower than that produced in HS medium (2.5 g/L). After being supplemented with a combination of nitrogen and phosphate sources to overcome possible nutritional limitations, the BC yield was increased to 21–43% compared to cultivation with no supplementation (Gomes, Silva et al. 2013). Sago starch by–products are found in massive quantities (183,405 tons in 2014) in Malaysia (Voon, Muhialdin et al. 2019). Due to 60–70% dry weight of starch granules contained in the by–product, it has been used as a low–cost substrate to produce fermentable sugars, bioethanol, and BC (Voon, Muhialdin et al. 2019). Prior to adding to the culture medium, sago by–product was enzymatically hydrolyzed by amyloglucosidase enzyme (6 U/mL). After 14 days under static culture, the yield of BC after culturing with sago hydrolysate in the presence of 2% glucose, 0.5% peptone, 0.5% yeast extract, and 0.27% disodium phosphate in the culture medium, is about 1.5 g/L similar to the yield after culturing in HS medium. Pecan nutshell can be used as a carbon source in the production of BC due a high content of cellulose, lignin, and other carbohydrates (Flores-Córdova, Sánchez Chávez et al. 2016). Around 40–50% of the total pecan production, which is more than 460 kilotons worldwide in 2016, is waste (Dórame-Miranda, Gámez-Meza et al. 2019). For BC production, nutshell of pecan with a content of reducing sugars of 40 g/L was used as a sole carbon source and supplemented with peptone (10 g/L), yeast extract (10 g/L), glacial acetic acid and absolute ethanol (10 g/L), dibasic ammonium phosphate (2.4 g/L), ammonium sulphate (0.3 g/L), and magnesium sulphate (0.1 g/L). After 14 and 28 days, yields of BC are about 1 and 2.8 g/L, respectively (Dórame-Miranda, Gámez-Meza et al. 2019).

Table 3. Waste nutrient sources used in bacterial nanocellulose production from the recent findings

Nutrient Sources	Bacteria name	Conditions to produce BC	Cellulose Yield (g/L)	Ref.
Olive residue	*Gluconacetobacter sacchari*	Static culture, pH 4.5, 4 days	0.81–1.5 HS[a]:2.5	(Gomes, Silva et al. 2013)
Sago residue	*Gluconacetobacter xylinus*	Static culture, pH 6.0, 14 days, 30 °C	1.55 HS[a]:1.57	(Voon, Muhialdin et al. 2019)
Pecan nutshell	*Gluconacetobacter entanii*	Static culture, pH 3.5, 28 days, 30 °C	2.8	(Dórame-Miranda, Gámez-Meza et al. 2019)
Potato Peel	*Gluconacetobacter xylinus*	Static culture, pH 6.0, 4 days, 30 °C	1.27–2.61 HS[a]:1.21	(Abdelraof, Hasanin et al. 2019)
juice	*Komagataeibacter xylinus*	Static culture, pH 6.0, 7 days, 28 °C	2–4 HS[a]:3–5	(Ciecholewska-Juśko, Broda et al. 2021)
Residue of cashew apple juice processing	*Komagataeibacter xylinus*	Static culture, pH 4.3, 12 days, 30 °C	1–3 HS[a]:0.5–3	(Guimarães, de Oliveira Barros et al. 2023)

continued on following page

Table 3. Continued

Nutrient Sources	Bacteria name	Conditions to produce BC	Cellulose Yield (g/L)	Ref.
Dairy and Soy–related wastes • cheese whey	*Gluconacetobacter xylinus*	Static culture, pH 5.5, 14 days, 28 °C	1–3.55 HS[a]:3.26	(Salari, Sowti Khiabani et al. 2019)
	Gluconacetobacter sucrofermentans	Shaking culture at 250 rpm, pH 5.0–6.0, 3 days, 28 °C	5.4 HS[a]: 2.14	(Revin, Liyaskina et al. 2018)
• soybean whey	*Komagataeibacter xylinus*	Static culture, pH 6.2, 7 days, 28 °C	4.1 HS[a]:1.15	(Suwanposri, Yukphan et al. 2014)
• tofu wastewater	*Komagataeibacter xylinus*	Static culture, pH 4.5, 15 days, 30 °C	10.6	(Apriyana, Andriani et al. 2020)
• Soybean residue	*Gluconacetobacter xylinus*	Static culture, pH 4.5, 15 days, 30 °C	1.5–2.8 HS[a]: 5.1	(Taokaew, Nakson et al. 2023)
Wastewater of candied jujube–processing industry	*Gluconacetobacter xylinus*	Static culture, 6 days, 30 °C	0.5–2	(Li, Wang et al. 2015)
Wine fermentation waste broth (Stillage)	*Gluconacetobacter xylinus*	Static culture, pH 6.0, 7 days, 30 °C	3–10.3 HS[a]: 4.2	(Wu and Liu 2012, Wu and Liu 2013)
	Gluconacetobacter sucrofermentans	Shaking culture at 250 rpm, pH 5.0–6.0, 3 days, 28 °C	6.2 HS[a]: 2.14	(Revin, Liyaskina et al. 2018)
Microbial ethanol fermentation wastewater (Vinasse)	*Gluconacetobacter hansenii*	Static culture, pH 5.0, 7–14 days, 30 °C	1.2–7.0	(Lin, Lopez-Sanchez et al. 2014)
Acetone–butanol–ethanol fermentation wastewater	*Gluconacetobacter xylinus*	Static culture, pH 6.0, 7 days, 28 °C	1.7 HS[a]: 0.9	(Huang, Yang et al. 2015)
Waste and by–product streams from biodiesel and confectionery industries	*Komagataeibacter sucrofermentans*	Shaking (100 rpm)–Static cultures, pH 6.0, 15 days, 30 °C	13.3	(Tsouko, Kourmentza et al. 2015)
	Gluconacetobacter xylinus	Static culture, pH 5.0, 7 days, 30 °C	1–7.32	(Ho Jin, Lee et al. 2019)
Spruce hydrolysate	*Gluconacetobacter xylinus*	Static culture, pH 5.0, 14 days, 30 °C	5–8.2	(Guo, Cavka et al. 2013)
Wood hot water extracts	*Gluconacetobacter xylinus*	Static culture, 14 days, 26–30 °C	0.04–0.15	(Erbas Kiziltas, Kiziltas et al. 2015)
Tobacco waste extract	*Gluconacetobacter xylinus*	Static–shaking (150 rpm) cultures, pH 6.5 7–15 days, 30 °C	0.5–5.2 HS[a]: 3.26	(Ye, Zheng et al. 2019)
Rice–washed water	*Komagataeibacter xylinus*	Static culture, pH 4.5, 15 days, 30 °C	6.57	(Apriyana, Andriani et al. 2020)

[a]Yield of BC in HS medium used as a reference.

Potatoes are extensively cultivated in Eastern Europe, China, and India due to high availability and low price throughout the year cycle. In developed countries, around 69.5% (in 2012) of total produced potatoes are processed. Large amount of peel waste remains after industrial potato processing, which ranges from 15 to 40% of initial product mass. In addition, over 10 million tons of starch and starch derivatives are produced, and approximately 3.5 tons of potato tuber waste is produced during the manufacture of a single ton of potato starch (Ciecholewska-Juśko, Broda et al. 2021). The potato peel hydrolysate medium has the buffering capacity, which maintains BC production in the optimal pH range during the cultivation similar to corn steep liquor. The medium containing potato peel hydrolyzed by HNO_3 promoted the relatively high production of BC (2.61 g/L) after 4 days of cultivation when compared with those containing the peels hydrolyzed by HCl, H_2SO_4, and H_3PO_4. The BC yield can reach 3.5 and 4 g/L after 7 days of cultivations of *Komagataeibacter xylinus* (*K. xylinus*) ATCC 53582 and 53524, respectively, in potato juice medium (Ciecholewska-Juśko, Broda et al. 2021). Similarly, the yield reaches 3 and 1 g/L after 12 days of cultivations of *K. xylinus* ATCC 53582 and ARS B42, respectively in the medium containing cashew apple juice residue (Guimarães, de Oliveira Barros et al. 2023). The cashew apple juice residue is waste effluent discarded from the industrial production of cashew apple juice. In Brazil, about 10% of the processed juice retained in the ultrafiltration membrane is discarded annually, equivalent to 730,000 L (Guimarães, de Oliveira Barros et al. 2023). The °Brix ranging from 5 to 15 and 100–500 mg/100 mL of vitamin C are found in the effluents (Das and Arora 2017). Thus, it can be a supplement source for BC production. Pictures of BC products from various cost–effective media are shown in Figure 5.

Figure 5. Photos of BC obtained from Hestrin–Schramm (a) and natural ingredients–based media prepared from potato juice (b), potato peels (c), orange peels (d), beetroots (e), and apples (f) (Reprinted from ref. (Ciecholewska-Juśko, Broda et al. 2021) open access, Copyright © 2021 by the authors. Licensee MDPI)

Waste disposal from the dairy industry is particularly concerning because factories use a lot of water. The effluents containing lipids, lactose, detergents, and sanitizing agents must be treated before being released into the environment (Ganta, Bashir et al. 2022, El-Bestawy, Eltaweil et al. 2023). However, dairy wastewater contains 14–830 mg/L of total nitrogen and 9–280 mg/L of total phosphorus. Therefore, dairy waste utilization has the dual benefits of purification of dairy wastewater and feasible BC production (Jin Chung, Shim et al. 2022, El-Bestawy, Eltaweil et al. 2023). Whey is a by–product of cheese or soybean curd manufacturing. It contains about 55% of nutrients in the original milk including soluble protein, lactose, minerals and B group vitamins (Salari, Sowti Khiabani et al. 2019). Cultivation using raw cheese whey that is untreated and pretreated to remove lactose due to the lack of β–galactosidase of *G. xylinus* yields BC of 1 and 3 g/L, respectively after 14 days of cultivation under static conditions (Salari, Sowti Khiabani et al. 2019). For soybean whey medium containing sucrose and $(NH_4)_2SO_4$, at the optimum condition of pH 6.2 and 1.6%v/v ethanol, BC production of 4.1 g/L is obtained after 7 days of cultivating 10% inoculum of *K. xylinus* under static conditions (Suwanposri, Yukphan et al. 2014). The yield is increased to 10.6 g/L, when the tofu wastewater is used as the sole nitrogen source in the medium with the supplement of sucrose, and cultivation of 20% inoculum of *K. xylinus* is conducted at 30 °C for 15 days. Soybean residue with various types of amino acids comparable to commercial yeast extract (Taokaew, Zhang et al. 2020) can be used as a sole nitrogen source with sucrose added in the culture medium (Taokaew, Nakson et al. 2023) as shown in Figure 6. After 7 days of cultivating 10% inoculum of *G. xylinus*, 1.5–2.8 g/L of BC is produced depending on the pretreatment to increase water solubility of the soybean residue (Taokaew, Nakson et al. 2023). Without any nitrogen source i.e., the soybean residue in the HS–based culture medium, the production of BC drops to about 1 g/L (Taokaew, Nakson et al. 2023). Besides, the large amount of waste from soybean industries in Asian countries such as China and Japan. The massive amount of wastewater from candied jujube manufacturing is generated during the production of 5.88 million ton/year of jujube at Cangzhou city in China. This wastewater contains various saccharides, amino acids, vitamins, and other nutrients suitable for growing BC–producing bacteria (Li, Wang et al. 2015). After cultivation of *G. xylinus* in the jujube waste with and without pretreatment and supplement with ammonium citrate for 6 days under static conditions, BC yields 1.5 and 2 g/L, respectively. Without ammonium citrate, the production of BC drastically drops to lower than 0.5 g/L (Li, Wang et al. 2015).

Figure 6. Photos of soybean residue (a), BC pellicle produced in soybean residue based medium (b) and harvested BC (c). Yield of dry nanocellulose and pH of the cell culture medium measured at seventh day after harvesting the BC pellicles (d), cellulose density in a pellicle and thickness of the pellicle obtained from cultivation of Gluconacetobacter xylinus in culture media containing homogenized okara (BCH), okara extracted protein (BCD), and the modified protein by 1 M acetic acid at 65 °C (BCM) at the concentrations of 0–3%w/v (Reprinted with permission from ref. (Taokaew, Nakson et al. 2023) Copyright © 2022 The Society for Biotechnology, Japan, Elsevier, B.V.)

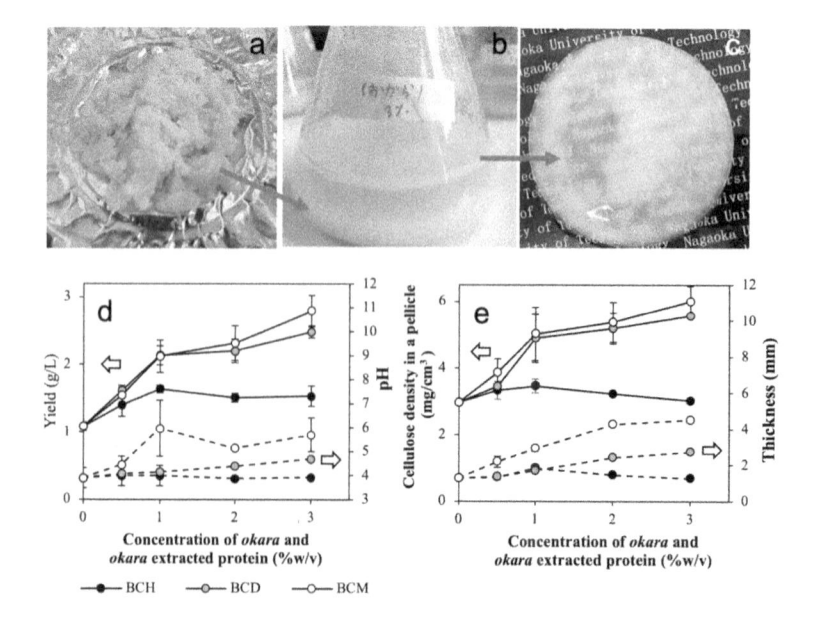

Thin stillage or distillery wastewater is a liquid by–product of microbial ethanol fermentation of grain–based feedstock by yeast and subsequent distillation of the fermented mash. Carbohydrates such as rice, wheat, and barley are the preferred raw material to product alcohol (Mustafa, McKinnon et al. 2000). Rice thin stillage is a wastewater from rice wine distillery for preparation of Chinese food. Due to being rich in reducing sugars and organic acids such as succinic acid, it is a supplement in HS medium. At 50% of the rice thin stillage in HS medium, a BC yield of 6.2 g/L is obtained after 7 days of static cultivation (Wu and Liu 2013). Wheat thin stillage is a by–product of ethanol production generated in large amounts in Russia (Revin, Liyaskina et al. 2018). The thin stillage contains salts, carbohydrates, organic compounds, and nitrogen nutrient such as betaine and glycerophosphocholine (Ratanapariyanuch, Shen et al. 2011). Therefore, it is considered a nutrient for the production of value–added products including BC (Wu and Liu 2012). After the thin stillage is used to replaced distilled water in the preparation of HS medium, BC production is enhanced by about 2.5–fold to a concentration of 10.3 g/L after 7 days of cultivation (Wu and Liu 2012). For waste beer or ethanol fermentation broth (vinasse), intense pretreatment is needed to obtain enough reducing sugars, unless the carbon source is compensated by glucose addition (Lin, Lopez-Sanchez et al. 2014, Heydorn, Lammers et al. 2023). Those treatments include 0.1 M NaOH treatment, high speed homogenizer, ultrasonication, and microwave treatment followed by hydrolysis at 121 °C for 20 min under acidic conditions (pH 2) as a second step (Lin, Lopez-Sanchez et al. 2014). BC production using waste beer broth treated by ultrasonication with a reducing sugar yield of 29.1%w/w provide the highest yield of 3.8 g/L. Since high sugar concentrations inhibit BC production, waste beer broth is diluted to a sugar concentration of 3%. This dilution increases BC yield to 7.0 g/L after 10 days. However, a sugar concentration of 5% led to a maximum of 5.7 g/L after 8 days and the maximum yield for 7% sugar concentration is 2.8 g/L after 7 days (Lin, Lopez-Sanchez et al. 2014). Without broth pretreatment, the BC production yield is low. For instance, the fermentation broth from acetone, butanol, and ethanol production, the yield of BC is 1.4 g/L on day 7 and drop to 1.2 g/L on day 8 due to the insufficient reducing sugars in the feed (0.45 g/L xylose and 0.25 g/L glucose) (Huang, Yang et al. 2015).

In BC production by *G. xylinus*, glycerol is one of the carbon sources found in a considerable number of studies conducted over the past few decades, besides glucose, sucrose, and fructose (Vazquez, Foresti et al. 2013, Zhong, Zhang et al. 2013). Crude glycerol is generated for about 10%w/w of the total product in the biodiesel production process during transesterification of triglycerides with an alcohol, most frequently methanol (Tsouko, Kourmentza et al. 2015, Ho Jin, Lee et al. 2019). The worldwide biodiesel production is increasing annually, and approximately 40% of the worldwide production capacity is in Europe. In 2012, production was globally more than 431 thousand barrels per day, increasing 39% over the last five years. After biodiesel production from oilseeds, by–product streams such as crude glycerol and oilseed meals are obtained (Tsouko, Kourmentza et al. 2015). The oilseed meals are the solid residues after the oil extraction. For sunflower meal used for biodiesel production in southern European countries, its worldwide production is 16.04 million tons in 2014/2015 with 25% of the total production in the European Union (Tsouko, Kourmentza et al. 2015). In the BC production using *Komagataeibacter sucrofermentans* under mixed dynamic and static conditions, the highest BC yield is 13.3 g/L after 15 days when the combination of crude glycerol and sunflower meal hydrolysates as the sole fermentation media is used, which is higher than using the crude glycerol (3.2 g/L) and commercial sucrose (4.9 g/L) in the culture (Tsouko, Kourmentza et al. 2015). However, under static culture conditions for 7 days using *G. xylinus*, BC yield is 6.9 and 7.3 g/L, when the crude glycerol and pure glycerol are used, respectively (Ho Jin, Lee et al. 2019).

Residues from forestry such as branches, tops, sawdust have been used as agricultural feedstocks and waste. In the northern hemisphere such as Sweden, softwood (conifer) forests mainly Norway spruce and Scots pine are a major source of lignocellulosic feedstock. Residues of the softwood are rich in carbohydrates such as glucan, mannan, and galactan, which can be hydrolyzed to hexose sugars and then used as substrates in BC production (Guo, Cavka et al. 2013). However, BC production inhibitor needs to be removed including aliphatic acids, furan aldehydes, and especially phenolic compounds to obtain the BC yield of 8.2 g/L after cultivation with *G. xylinus* under the static condition for 14 days (Guo, Cavka et al. 2013). Similarly, for the use of tobacco waste extract, nicotine needs to be removed to obtain a BC of yield 2.27 g/L after cultivation with *G. xylinus* under mixed static and shaking condition for 7 days (Ye, Zheng et al. 2019). Without pretreatment, the yield of BC is about 0.5 g/L for 2.4 g/L nicotine contained in the culture medium (Ye, Zheng et al. 2019) or 0.15 g/L for wood hot water extraction medium (Erbas Kiziltas, Kiziltas et al. 2015). The BC yield can reach 5.2 and 6.5 g/L after second stage cultivation of *G. xylinus* by adjusting the pH of the culture medium to 6.5 (Ye, Zheng et al. 2019) and single static cultivation of *K. xylinus* for 15 days in rice–washed water before rice cooking medium supplemented with sucrose is added (Apriyana, Andriani et al. 2020).

BACTERIAL CELLULOSE QUANTIFICATION AND CHARACTERIZATION

Although BC can be efficiently produced in alternative culture media, different nutrient sources and cultivation modes also determine structural features and properties related to the microstructure (Campano, Balea et al. 2016). This can be characterized by scanning electron microscopy by using wet or dry BC. The wet BC pellicle without drying or gold coating can be directly visualized using Cryo–scanning electron microscopy (Figure 7) (Taokaew, Chiaoprakobkij et al. 2021). Characterization of other properties derived from different nutrient resources is influential and should be performed by different approaches such as kinetic studies, degree of polymerization/molecular weight, and crystallinity. The BC sample preparation and operating conditions of the analytical equipment for BC characterization are listed in Table 4.

Figure 7. Microscopic image of wet bacterial cellulose pellicle without drying visualized by using Cryo–Scanning electron microscopy (JEOL JSM–IT200, Japan) (Reprinted with permission from supplementary data of ref. (Taokaew, Chiaoprakobkij et al. 2021) Copyright © 2020 Elsevier, B.V.).

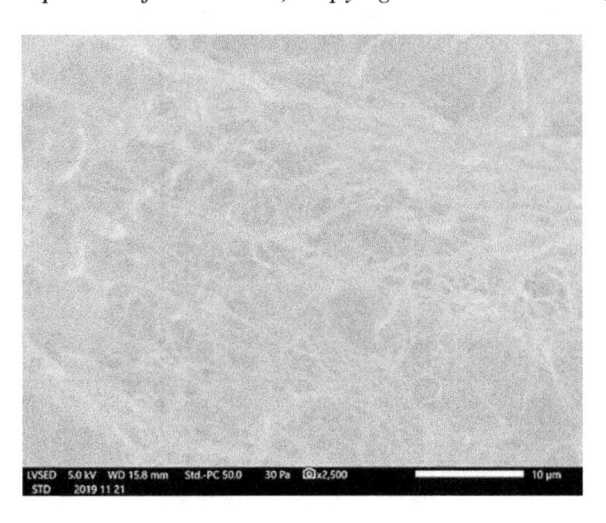

Table 4. Characterization techniques of bacterial cellulose, specimen preparation and operating conditions to analyze physical and chemical properties

	Characterization Techniques	Specimen preparation	Operating condition	Ref.
Produced cellulose content	Kinetic studies	Dry by air drying/freeze drying	Calculation based on Equation (1)–(5)	(Jozala, Pértile et al. 2015, Abdelraof, Hasanin et al. 2019, Taokaew, Nakson et al. 2022, Taokaew, Nakson et al. 2023)
Microstructure of nanofibers	Scanning electron microscopy	• Wet specimen[a]	5 kV voltage, 2500× magnification in frozen chamber	(Taokaew, Chiaoprakobkij et al. 2021)
		• Dry by freeze drying or liquid CO_2 critical point drying and then gold coating	5–15 kV voltage, 5000–10000× magnification	(Guimarães, de Oliveira Barros et al. 2023, Taokaew, Nakson et al. 2023)
Unique functional groups	Fourier transform infrared spectroscopy	5%w/w dried cellulose in KBr	32–100 times scanning at 4 cm^{-1} resolution between 4000 and 400 cm^{-1} wavenumber	(Guimarães, de Oliveira Barros et al. 2023, Taokaew, Nakson et al. 2023)
Degree of polymerization/ molecular weight	• Gel Permeation Chromatography • Viscosity–based technique	• 0.1%w/w of dried cellulose in 1%w/w of LiCl/ DMAc • 0.02%w/v of dried cellulose in 0.5 mol/L copper–ethylenediamine/water	1 ml/min flow rate at 40 °C column temperature Calculation based on Equation (6)–(9)	(Taokaew, Nakson et al. 2023) (Guimarães, de Oliveira Barros et al. 2023)

continued on following page

Table 4. Continued

	Characterization Techniques	Specimen preparation	Operating condition	Ref.
Produced cellulose content	Kinetic studies	Dry by air drying/freeze drying	Calculation based on Equation (1)–(5)	(Jozala, Pértile et al. 2015, Abdelraof, Hasanin et al. 2019, Taokaew, Nakson et al. 2022, Taokaew, Nakson et al. 2023)
Crystallinity	X–Ray diffraction analysis	Dry by air drying/freeze drying	40 kV voltage and 30–40 mA current scanning at 0.01° steps, 0.5–3.0°/min sweep speed, and 5–40° angular range (2 θ), calculation based on Equation (10)–(12)	(Guimarães, de Oliveira Barros et al. 2023, Taokaew, Nakson et al. 2023)
Tensile strength and Young's modulus	Universal Testing machine, mechanical tester	Cut into rectangular specimens	ASTM [b] D882–12	(Guimarães, de Oliveira Barros et al. 2023, Öz and Kalender 2023)
Thermal degradation behavior	Thermogravimetry	Dry by air drying/freeze drying	25 to 950 °C with a heating rate of 10 °C/min heating rate under 20–40 mL/min N_2 flow rate	(Guimarães, de Oliveira Barros et al. 2023, Öz and Kalender 2023)
Elemental composition and purity	X–ray fluorescence analysis	Dry by air drying/freeze drying, 1–3 cm specimen diameter	30–keV voltage and 100–mA current x–ray tube, 2θ of 63°	(Taokaew, Nakson et al. 2022)

[a] characterization by Cryo–Scanning electron microscope or cool–stage equipped in normal scanning electron microscope
[b] standard test of the American Society for Testing and Materials

Kinetic Studies

To identify the efficiency of using a particular substrate, the kinetics during BC production is measured. The efficiency of BC production is evaluated after cultivation in terms of BC production yield (Y_{BC}) that is defined in equation (1) (Jozala, Pértile et al. 2015, Taokaew, Nakson et al. 2022, Guimarães, de Oliveira Barros et al. 2023, Taokaew, Nakson et al. 2023):

$$Y\ (g/L) = \frac{m_{BC}}{V} \tag{1}$$

$$Y\ (\%) = \frac{\frac{m_{BC}}{V}}{S_i - S_f} \times 100 \tag{2}$$

where m_{BC} and V denote the dried weights of BC and volume of cell culture medium. Also, S_i and S_f are the initial and final concentrations of substrate (g/L), respectively.

Productivity of BC (P_{BC}) or BC production rate in g/L/d or g/L/h (Jozala, Pértile et al. 2015)

$$P_{BC}\ (g/L/d) = \frac{m_{BC}}{V \cdot t} \tag{3}$$

BC yield coefficient ($Y_{P/S}$) (Jozala, Pértile et al. 2015)

$$Y_{P/S} = \frac{dS/dT}{dP_{BC}/dt} \tag{4}$$

Conversion of carbon source (sugars) into cellulose and overall mass balance expressed in terms of cellulose yield and carbon source loss in the process are calculated by substrate conversion ratio (S) (Abdelraof, Hasanin et al. 2019, Taokaew, Nakson et al. 2022)

$$S (\%) = \frac{S_i - S_f}{S_i} \times 100 \tag{5}$$

Sugar consumption can be calculated by the DNS method or High–Performance Liquid Chromatography (Guimarães, de Oliveira Barros et al. 2023)

Degree Of Polymerization and Molecular Weight Distribution

Differences in the degree of polymerization (DP) are quantified by Gel Permeation Chromatography (GPC) measurement, which elucidates details of the molecular weight distribution (Heydorn, Lammers et al. 2023). The BC solutions are prepared in dimethylacetamide (DMAc) containing concentrated dehydrated LiCl. The solution is further diluted with pure DMAc and filtered through a syringe filter prior to injection (Heydorn, Lammers et al. 2023, Taokaew, Nakson et al. 2023). Alternatively, the DP can be calculated based on viscosity by the Mark–Houwink–Sakurada equation for DPs greater than 950 (equation 6) measured using a Canon–Fenske viscometer according to TAPPI method T 230–08 ($a = 2.28$) (Guimarães, de Oliveira Barros et al. 2023) and using an Ubbelholde capillary viscosimeter described in American Society for Testing and Materials (ASTM) D1795 standard method ($a = 2.42$) (Algar, Fernandes et al. 2015):

$$\eta = a \cdot DP^{0.76} \tag{6}$$

$$\eta = \sqrt{(2(\eta_{sp} - \ln(\eta_r)))}/C \tag{7}$$

$$\eta = \eta_r - 1 \tag{8}$$

$$\eta_r = t/t_0 \tag{9}$$

In equations (6)–(9), η is intrinsic viscosity, η_r is relative viscosity, η_{sp} is specific viscosity, C is the concentration of the BC solution, t and t_0 are the flow time of the copper–ethylenediamine solution (0.5 mol/L) with and without BC (Guimarães, de Oliveira Barros et al. 2023).

Crystallinity

Crystallinity determined by X–Ray diffraction analysis (XRD) has often been regarded as a BC quality indicator (Öz and Kalender 2023, Taokaew, Nakson et al. 2023). The Segal and peak deconvolution method using computer software are applied for crystallinity determination. For Segal method (Segal, Creely et al. 1959), the degree of crystallinity or crystallinity index (CI) can be simply calculated from the diffracted intensity values using the method described in equation (10).

$$CI (\%) = 100 \tag{10}$$

where I_{200} is the maximum intensity of the diffraction peak for the crystalline region referring to the crystallographic plane 200 (cellulose I) at 2θ of ~22.7° and I_{am} is the intensity scattered by the amorphous part of BC located at 2 of ~16–18° (Algar, Fernandes et al. 2015). The crystallite size (D, equation 11) and d spacing between the planes in the atomic lattice (d, equation 12) are calculated based on Scherrer equation (Cheng, Catchmark et al. 2009) by using the Bragg method or computer softwares.

$$D = (K \cdot \lambda\_1)/(FWHM \cdot \cos\theta) \tag{11}$$

$$d = (n \cdot \lambda\_2)/(2 \sin\theta) \tag{12}$$

K is the form factor (0.94), λ_1 is the x–ray wavelength (1.788965 Å), $FWHM$ is the full width at half the height of the peak I_{200}, is the Bragg angle concerning peak I_{200}, n is an integer, λ_2 is the wavelength of the X–ray beam (0.154 nm).

APPLICATIONS OF NANOCELLULOSE

Nanofibers for Reinforcement

The concept of using nanocellulose as reinforcing agent by filling in polymer matrices and by coating is of great practical significance (Balea, Fuente et al. 2020). As a kind of typical cellulose, BC exhibits large aspect ratio (length >20 μm and diameter <100 nm), which can construct a strong network (Zhu, Zhang et al. 2021). Besides its large surface area, biodegradability, and biocompatibility, high stiffness of cellulose crystals approximately 100–160 GPa is desirable in reinforcement application (Eichhorn and Davies 2006, Hsieh, Yano et al. 2008). The addition of BC leads to property improvement of paper made from low-grade fiber or waste resources; mechanical properties of the paper sheets prepared from waste office paper pulp are improved by BC loading (Kalyoncu and Peşman 2020). As compared to the non-reinforced paper, the tensile index and burst index values are nearly identical but the tear index values decrease. This is due to the enhancement of fiber to fiber bonding properties, whereas $CaCO_3$ filler in the waste office papers has negative effect on the formation of hydrogen bonds between the fibers (Kalyoncu and Peşman 2020). In a paper reinforced epoxy resin with BC loading of 70%w/w, a tensile modulus and strength of up to 21 GPa and 325 MPa are obtained (Yano, Sugiyama et al. 2005). Nonetheless, BC and nanofibrillated cellulose (NFC) from wood have the similar ability to act as reinforcement to produce high strength materials (Lee, Tammelin et al. 2012). In a bioplastic film reinforced with 5%w/w BC or nanocellulose loading, tensile strength of 3.0 and 1.7 MPa, and Young's modulus of 74 and 46 MPa are obtained, respectively (Efthymiou, Tsouko et al. 2022). At an equivalent fiber volume fraction of 60%v/v, epoxy nanocomposites reinforced by nanocellulose from both sources have a similar stiffness and strength of approximately ~8 GPa and ~100 MPa (Lee, Tammelin et al. 2012).

In the field of coating, electro-conductive cellulosic paper can be produced via surface coating of multiwalled carbon nanotube (MWCNT)/graphene oxide (GO) nanocomposites using nanocellulose as a binder. Zeta potential, sedimentation stability, and mechanical properties of MWCNT/GO cellulosic paper are improved by an addition of nanocellulose in surface coating (Tang, He et al. 2014). Nanocellulose can also increase the barrier properties of paper (Vaezi, Asadpour et al. 2019, Balea, Fuente et al.

2020). For example, increased nanocellulose loading in nanocellulose/cationic starch nanocomposite coating improved the tensile strength, oil resistance, and air resistance of the coated kraft paper. Due to the incorporation of 5%w/w nanocellulose, water absorption of the coated paper decreases 50% (Vaezi, Asadpour et al. 2019). Hence, nanocellulose can be used as an effective coating agent for developing high-performance coated paper (Jin, Tang et al. 2021) with improved printing quality (Lourenço, Martins et al. 2023).

Hydrophobic Papers and Bioplastics

Cellulose has been used as fibers, textiles, and paper-based materials (Balea, Fuente et al. 2020, Felgueiras, Azoia et al. 2021). Paper materials are the main alternative to fossil plastics (Nilsen-Nygaard, Fernández et al. 2021). Nanocellulose paper or films have a high oxygen barrier property (oxygen transmission rates of 17 ± 1 ml/m^2/day for a film with a thickness of 21 μm), which is competitive with synthetic polymers such as ethylene vinyl alcohol (3-5 ml/m^2/day) and polyvinylidene chloride coated polyester films (9-15 ml/m^2/day) at approximately same thickness (Nair, Zhu et al. 2014). In particular, BC and NFC papers have a unique property such as high tensile strength and durability (Santos, Matos et al. 2023). Papers prepared from BC and NFC from wood possess similar tensile moduli and strengths of 12 GPa and 110 MPa, respectively (Lee, Tammelin et al. 2012). However, thermal degradation temperatures in both nitrogen and air atmosphere of BC are higher than that of NFC due to the higher crystallinity of BC (71% for BC and 41% for NFC) (Lee, Tammelin et al. 2012). Compared with various plastic materials, the cellulose paper-based materials have the hydrophilicity due to the presence of numerous hydroxyl groups on cellulose macromolecule, which can easily absorb water from surroundings (Li, Hu et al. 2019). Their mechanical property is then decreased. Therefore, the development of hydrophobic cellulosic paper materials has been developed to displace plastic products by simply coating with oil, waxes, or other hydrophobic coatings such as palm tree wax, cationic starch (Santos, Matos et al. 2023), fluoroalkyl functionalized cellulose nanofibers (Baidya, Ganayee et al. 2017), tetraethylorthosilicate (TEOS)/trimethylethoxysilane (TMES) (Wang, Xiong et al. 2016), micro/nano SiO_2 (Teng, Shi et al. 2020), nano TiO_2 modified by γ-aminopropyltriethoxysilane (Teng, Wang et al. 2020), and epoxy resin (Zhang, Zhang et al. 2021). In addition, a mixture of nanocellulose, montmorillonite, and soy protein as a coating agent synergistically enables oil and grease resistance to the base paper comparable to the use of fluorochemicals widely used in nonstick, stain-resistant, and waterproof food packaging because of their hydrophobic and lipophobic properties (Tyagi, Lucia et al. 2019). After coating with nanocellulose mixture, oxygen transmission rate is reduced by a factor of about 260 compared to uncoated paper due to the decrease in porosity (Tyagi, Lucia et al. 2019).

Biofoams

Packaging materials provide sufficient barriers during distribution and handling, which are important in preserving and protecting quality and safety of products to reduce waste (Reichert, Bugnicourt et al. 2020). The production of BC paper in the form of thin sheets having characteristics of nanofibers, which are high mechanical strength and good physical properties have been described (Balea, Fuente et al. 2020, Kalyoncu and Peşman 2020). Besides, foams are very popular as packaging because of lightweight, high strength, thermal insulation, good cushioning performance, which readily absorb impact to protect packaged goods from damage (Nechita and Năstac 2022), and flame retardancy (He, Liu et al. 2019).

The popular plastic foams are made of polystyrene and polyurethane which are non-biodegradable (Li, Du et al. 2017, Sun, Yuan et al. 2020). Cellulose-based foams have been developed in the form of aerogel materials in replacing traditional petrochemical-based foam products (Liao, Luan et al. 2022). The aerogels/foams are obtained by sol-gel method (Li, Du et al. 2017), crosslinking, and 3D printing (Françon, Wang et al. 2020, Lee, Kim et al. 2021). The crosslinkers added in cellulose fiber suspension to improve the water barrier of the cellulose foam include N,N′-methylene bisacrylamide (Geng 2018), butane tetracarboxylic acid (Wang, Aliheidari et al. 2019), borate (He, Liu et al. 2019), citric acid (Hassan, Tucker et al. 2020), γ-glycidoxypropyltrimethoxysilane (Lu, Lin et al. 2021), and polyamide epichlorohydrin (Liao, Luan et al. 2022). In the sol-gelmethod, cellulose is dissolved in solvents such as NaOH/Urea in the presence of surfactant such as sodium dodecyl sulfate (SDS) as a foaming agent to prepare bubble solution and then regenerating into a gel (Li, Du et al. 2017). To preserve the microstructure and produce nano- or microporous foam materials, the wet foams are dried by freeze drying, supercritical drying (Li, Du et al. 2017), ambient-air drying (Françon, Wang et al. 2020), hot-air drying (Liao, Luan et al. 2022), and microwave radiation (Hafez and Tajvidi 2021). Properties of the regenerated cellulose prepared by dissolution of cellulose and then coagulation processes into hydrogel depends on the change of the regeneration parameters such as type of solvents/coagulants and time of coagulation (Taokaew 2023). The coagulant type strongly influences the dimensional accuracy of regenerated cellulose. Shrinkage of water-saturated individual cellulose fibers in longitudinal and transverse directions are ~1% and 20–30%, respectively (Baranov, Sommerhoff et al. 2023). The incorporation of cellulose fibers yields the stability of the wet fiber foam and thermal insulation of the resulting dry foam comparable to dry polystyrene foam (Lohtander, Herrala et al. 2022). In particular, nanocellulose serves as a binder with pulp fibers to enhance fiber-fiber adhesion via hydrogen bond network, enabling the foam with high compressive strength of 3.72 MPa at low apparent density of 0.1 g/cm^3 (Hafez and Tajvidi 2021). By regulating the content and size of nanocellulose, the cellulose foams have stable structure with porosity of 91.7–94.5%), low apparent density (0.08–0.12 g/cm^3), and high compression modulus (0.49–2.96 MPa) (Sun 2023).

Without a solvent to dissolve cellulose, liquid cellulose nanofiber suspension as the wet foam is produced by simple stirring of cellulose nanofiber suspension with SDS as a foaming agent (Lee, Kim et al. 2021). The foam volume is increased with increasing SDS content and stirring rate, but the pore structure is destroyed at high stirring rates due to reduced wall thickness (Lee, Kim et al. 2021). The average wall thickness of the nanocellulose foam is increased with an increase in nanocellulose content, whereas the average pore area decreases. For freeze drying, a decrease in freezing temperature leads to the decreased wall thickness (with no wall structure at −196 °C) (Li, Cheng et al. 2018). The liquid cellulose nanofiber foam with shear thinning behavior enables printing through a needle using computer-aided design (Lee, Kim et al. 2021). Tunable densities (23–38 kg/m^3) and compressive moduli (97–275 kPa) can be prepared by using different nanocellulose concentrations (Françon, Wang et al. 2020).

Containers and 3D-Shaped Composites

Nanocellulose with other polymers can be molded into complex geometric 3D features by molding (Glenn, Orts et al. 2023) and 3D printing (Ma, Xue et al. 2023). The dimensional accuracy of the final 3D forms can be analyzed by 3D laser scanning (Baranov, Sommerhoff et al. 2023). Compression molding is a method to form cellulose slurry or wet foam into finished products such as water bottle by using platens and low clamping force. The mold is removed from the foam before drying in the oven (Glenn, Orts et al. 2023). The 3D printing with direct-ink-writing technique assembles 3D structures layer by layer through extrusion of ink filaments to shape to hydrogels and aerogels containing various sources of raw materials (Ma, Xue et al. 2023). By combining BC with polyimide solution, PAA/BC ink containing low solid contents (PAA 50 mg/ml and BC 21.6 mg/ml) has shear-thinning behavior suitable for smooth and continuous extrusion to different 3D shapes (Ma, Xue et al. 2023).

The well-known 3D packages are made of molded papers from waste pulps, that have been used as industrial packaging for electronic devices, auto parts, food trays, and egg cartons (Didone, Saxena et al. 2017). During the practical condition for a regional supply chain e.g., cold storage followed by ambient storage, material and size of the trays have an impact on quality changes and shelf life of fresh fruits such as cherry tomatoes (Korte, Albrecht et al. 2023) and strawberries (Efthymiou, Tsouko et al. 2022). Groundwood pulp and sugarcane trays shows higher preservation efficiency in comparison with the references such as corrugated cardboard and recycled polyethylene terephthalate (Korte, Albrecht et al. 2023). To increase the barrier properties such as resistance to oil and grease, their surface can be coated using nanocellulose such as nanocellulose coating on molded pulp lunch boxes (Han 2022) and cellulose nanofiber/carboxymethyl cellulose paste coating on molded pulp food trays (Zhang and Youngblood 2023). By using lignin-containing cellulose nanofibers coated onto bowl-like containers, commercial cooking oils can be held for over five months (H. Tayeb, Tajvidi et al. 2020). Such good oil and grease barrier property is obtained by the cellulose nanofiber coated layers of the container. Because of extensive hydrogen bonding among the fibrils, the cohesive energy density between the nanofiber coating layers is high and thus the diffusion rate of oil and grease is low through these layers (Hossain, Tajvidi et al. 2021). Antimicrobial feature can also be added for the cellulose-based food storage containers by coating or mounting on the bottom of the container with essential oil (Bardenstein, Storgaard et al. 2020), herbal extracts (Moradian, Almasi et al. 2018), and synthetic antimicrobial agents (Viscusi, Bugatti et al. 2021). By coating with cellulose acetate as a carrier of paraben-family preservative, cellulose food tray can prolong the shelf-life of fresh-cut iceberg lettuce to 16 days (Viscusi, Bugatti et al. 2021) and ready-to-eat pasta to 2 months (Bugatti, Viscusi et al. 2020) . By mounting BC membrane containing herbal extracts from pomegranate peel, green tea, and rosemary on the bottom of the food tray, shelf life of button mushrooms can be extended to over 15 days under storage at 4 °C (Moradian, Almasi et al. 2018).

CONCLUSION AND FUTURE CHALLENGES

The use of petroleum-based plastics, which emits huge amounts of greenhouse gases has a strong impact on the climate change. In comparison with the conventional plastics, the high cost of bioplastics prepared from biopolymer such as cellulose for using in many applications is still an obstacle to overcome. Therefore, environmental friendliness of production and cost control of the cellulose-based bioplastic products is important. One of the eco-friendly and simple methods to produce cellulose is biosynthesis by bacteria. Cost–effectiveness of cell culture medium is considered an important factor in the economic feasibility of bacterial cellulose production for large scale industrial applications. In order to meet the industrial sector's requirements, an inexpensive culture medium that is rich in nutrient, necessary for bacteria to produce cellulose, is very desirable. The utilization of agro– and food industrial residues to produce bacterial cellulose represents an important pathway for the implementation of the zero–waste concept of sustainability. This approach also provides the dual benefits of decreasing the environmental problems associated with the disposal of the waste and creating a method to efficiently produce bacterial cellulose.

Recently, successful results have been reported about different applications of nanocellulose including bacterial cellulose. It is a potential solution for many challenges regarding the plastic replacement. However, there are knowledge gaps that need to be addressed and research database should be continually collected. For instance, a comprehensive understanding based on the interaction between nanocellulose filler and polymer matrix for enhancing the physico-chemical properties is prerequisite in the design of industrial products. For the bioplastic film/paper with enhanced hydrophobic surface, more studies need to be conducted about effects of declining properties of hydrophobicity of the coatings, since hydrophobic surfaces tend to be destroyed by the adsorbed impurities or harsh surroundings such as high temperatures, UV irradiation, and corrosive media. The stability of these coatings and durability of the final products may be one of the limitations in the large-scale applications. For the preparation of cellulose foams as a cushion packaging, two initial approaches can be performed, that are through cellulose slurry and cellulose solution. For the later approach, the hydrogel is firstly prepared using dissolved cellulose in solvent prior to mixing with foaming agents and then coagulation. The challenge of this method involves solvent recovery issue or expensive production process such as drying by freeze-drying or supercritical carbon dioxide drying, which are high processing cost, time consuming, and are not simple to operate in an industrial scale. Therefore, more study should base on solvent-free and simple technique under ambient condition that will provide an extensive potential of cellulose-based materials.

The current state of knowledge in bacterial cellulose production using agro–waste with and without pretreatments as described in this chapter, provides innovative ideas for using the agro–waste and other wastes in bacterial cellulose production to overcome the challenges related to bioplastic development.

Author Note

The author declares no conflict of interest. Correspondence concerning this article is addressed to Siriporn Taokaew, Department of Materials Science and Bioengineering, Nagaoka University of Technology, 1603-1 Kamitomioka, Nagaoka, Niigata 940-2188 Japan. Email: t.siriporn@mst.nagaokaut.ac.jp and siriporntaokaew@gmail.com

REFERENCES

Abdelraof, M., Hasanin, M. S., & El -Saied, H. (2019). Ecofriendly green conversion of potato peel wastes to high productivity bacterial cellulose. *Carbohydrate Polymers*, 211(1), 75–83. DOI: 10.1016/j.carbpol.2019.01.095 PMID: 30824106

Abdul Khalil, H. P. S., Davoudpour, Y., Islam, M. N., Mustapha, A., Sudesh, K., Dungani, R., & Jawaid, M. (2014). Production and modification of nanofibrillated cellulose using various mechanical processes: A review. *Carbohydrate Polymers*, 99, 99649–99665. DOI: 10.1016/j.carbpol.2013.08.069 PMID: 24274556

Algar, I., Fernandes, S. C. M., Mondragon, G., Castro, C., Garcia-Astrain, C., Gabilondo, N., Retegi, A., & Eceiza, A. (2015). Pineapple agroindustrial residues for the production of high value bacterial cellulose with different morphologies. *Journal of Applied Polymer Science*, 132(1), 41237. DOI: 10.1002/app.41237

Am, A. A. P. (2005). Rapid Determination of Oil/Fat Utilizing High Temperature Solvent Extraction, American Oil Chemists Society: Urbana, IL, USA. **1:** 1-2.

Apriyana, A. Y., Andriani, D., & Karina, M. (2020). Production of bacterial cellulose from tofu liquid waste and rice-washed water: Morphological property and its functional groups analysis. *IOP Conference Series. Earth and Environmental Science*, 483(1), 012005. DOI: 10.1088/1755-1315/483/1/012005

Avcioglu, N. H. (2022). Bacterial cellulose: Recent progress in production and industrial applications. *World Journal of Microbiology & Biotechnology*, 38(5), 1–13. DOI: 10.1007/s11274-022-03271-y PMID: 35397756

Bae, S., & Shoda, M. (2004). Bacterial Cellulose Production by Fed-Batch Fermentation in Molasses Medium. *Biotechnology Progress*, 20(5), 1366–1371. DOI: 10.1021/bp0498490 PMID: 15458319

Bae, S., Sugano, Y., & Shoda, M. (2004). Improvement of bacterial cellulose production by addition of agar in a jar fermentor. *Journal of Bioscience and Bioengineering*, 97(1), 33–38. DOI: 10.1016/S1389-1723(04)70162-0 PMID: 16233586

Baidya, A., Ganayee, M. A., Jakka Ravindran, S., Tam, K. C., Das, S. K., Ras, R. H. A., & Pradeep, T. (2017). Organic Solvent-Free Fabrication of Durable and Multifunctional Superhydrophobic Paper from Waterborne Fluorinated Cellulose Nanofiber Building Blocks. *ACS Nano*, 11(11), 11091–11099. DOI: 10.1021/acsnano.7b05170 PMID: 29059514

Balea, A., Fuente, E., Monte, M. C., Merayo, N., Campano, C., Negro, C., & Blanco, A. (2020). Industrial Application of Nanocelluloses in Papermaking: A Review of Challenges, Technical Solutions, and Market Perspectives. *Molecules (Basel, Switzerland)*, 25(3), 526. Advance online publication. DOI: 10.3390/molecules25030526 PMID: 31991802

Baranov, A., Sommerhoff, F., Curnow, O. J., & Staiger, M. P. (2023). Prospects for manufacturing of complex 3D-shaped all-cellulose composites. *Composites. Part A, Applied Science and Manufacturing*, 173107627, 107627. Advance online publication. DOI: 10.1016/j.compositesa.2023.107627

Bardenstein, A., Storgaard, B. G., Landa, S., Olsen, M. H., Kirilov, K. A., Allermann, H., Kisbye, K., ... (2020). Implementation of Cellulose-Based Antimicrobial Packaging with Enhanced Barrier Performance. 22nd World Conference on Packaging "Future of Packaging".

Bauer, F., Nielsen, T. D., Nilsson, L. J., Palm, E., Ericsson, K., Fråne, A., & Cullen, J. (2022). Plastics and climate change—Breaking carbon lock-ins through three mitigation pathways. *One Earth*, 5(4), 361–376. DOI: 10.1016/j.oneear.2022.03.007

Bradford, M. M. (1976). A rapid and sensitive method for the quantitation of microgram quantities of protein utilizing the principle of protein-dye binding. *Analytical Biochemistry*, 72(1), 248–254. DOI: 10.1016/0003-2697(76)90527-3 PMID: 942051

Brethauer, S., Shahab, R. L., & Studer, M. H. (2020). Impacts of biofilms on the conversion of cellulose. *Applied Microbiology and Biotechnology*, 104(12), 5201–5212. DOI: 10.1007/s00253-020-10595-y PMID: 32337627

Bugatti, V., Viscusi, G., & Gorrasi, G. (2020). Formulation of a Bio-Packaging Based on Pure Cellulose Coupled with Cellulose Acetate Treated with Active Coating: Evaluation of Shelf Life of Pasta Ready to Eat. *Foods*, 9(10), 1414. Advance online publication. DOI: 10.3390/foods9101414 PMID: 33036319

Cacicedo, M. L., Castro, M. C., Servetas, I., Bosnea, L., Boura, K., Tsafrakidou, P., Dima, A., Terpou, A., Koutinas, A., & Castro, G. R. (2016). Progress in bacterial cellulose matrices for biotechnological applications. *Bioresource Technology*, 213, 213172–213180. DOI: 10.1016/j.biortech.2016.02.071 PMID: 26927233

Çakar, F., Özer, I., Aytekin, A. Ö., & Şahin, F. (2014). Improvement production of bacterial cellulose by semi-continuous process in molasses medium. *Carbohydrate Polymers*, 106(1), 7–13. DOI: 10.1016/j.carbpol.2014.01.103 PMID: 24721044

Campano, C., Balea, A., Blanco, A., & Negro, C. (2016). Enhancement of the fermentation process and properties of bacterial cellulose: A review. *Cellulose (London, England)*, 23(1), 57–91. DOI: 10.1007/s10570-015-0802-0

Carreira, P., Mendes, J. A. S., Trovatti, E., Serafim, L. S., Freire, C. S. R., Silvestre, A. J. D., & Neto, C. P. (2011). Utilization of residues from agro-forest industries in the production of high value bacterial cellulose. *Bioresource Technology*, 102(15), 7354–7360. DOI: 10.1016/j.biortech.2011.04.081 PMID: 21601445

Cavka, A., Guo, X., Tang, S.-J., Winestrand, S., Jönsson, L. J., & Hong, F. (2013). Production of bacterial cellulose and enzyme from waste fiber sludge. *Biotechnology for Biofuels*, 6(1), 1–10. DOI: 10.1186/1754-6834-6-25 PMID: 23414733

Chawla, P. R., Bajaj, I. B., Survase, S. A., & Singhal, R. S. (2009). Microbial cellulose: Fermentative production and applications. *Food Technology and Biotechnology*, 47(2), 107–124.

Cheng, K.-C., Catchmark, J. M., & Demirci, A. (2009). Enhanced production of bacterial cellulose by using a biofilm reactor and its material property analysis. *Journal of Biological Engineering*, 3(1), 1–10. DOI: 10.1186/1754-1611-3-12 PMID: 19630969

Ciecholewska-Juśko, D., Broda, M., Żywicka, A., Styburski, D., Sobolewski, P., Gorący, K., Migdał, P., Junka, A., & Fijałkowski, K. (2021). Potato Juice, a Starch Industry Waste, as a Cost-Effective Medium for the Biosynthesis of Bacterial Cellulose. *International Journal of Molecular Sciences*, 22(19), 10807. DOI: 10.3390/ijms221910807 PMID: 34639147

Coseri, S. (2021). Insights on Cellulose Research in the Last Two Decades in Romania. *Polymers*, 13(5), 689. Advance online publication. DOI: 10.3390/polym13050689 PMID: 33668896

Cunniff, P., & Washington, D. (1995). *AOAC, Official methods of analysis of AOAC international*. AOAC.

Czaja, W., Romanovicz, D., & Brown, R. (2004). Structural investigations of microbial cellulose produced in stationary and agitated culture. *Cellulose (London, England)*, 11(3), 403–411. DOI: 10.1023/B:CELL.0000046412.11983.61

Das, I., & Arora, A. (2017). Post-harvest processing technology for cashew apple – A review. *Journal of Food Engineering*, 194(1), 87–98. DOI: 10.1016/j.jfoodeng.2016.09.011

Didone, M., Saxena, P., Brilhuis-Meijer, E., Tosello, G., Bissacco, G., McAloone, T. C., Pigosso, D. C. A., & Howard, T. J. (2017). Moulded Pulp Manufacturing: Overview and Prospects for the Process Technology. *Packaging Technology & Science*, 30(6), 231–249. DOI: 10.1002/pts.2289

Dórame-Miranda, R. F., Gámez-Meza, N., Medina-Juárez, L. Á., Ezquerra-Brauer, J. M., Ovando-Martínez, M., & Lizardi-Mendoza, J. (2019). Bacterial cellulose production by Gluconacetobacter entanii using pecan nutshell as carbon source and its chemical functionalization. *Carbohydrate Polymers*, 207(1), 91–99. DOI: 10.1016/j.carbpol.2018.11.067 PMID: 30600072

Efthymiou, M.-N., Tsouko, E., Papagiannopoulos, A., Athanasoulia, I.-G., Georgiadou, M., Pispas, S., Briassoulis, D., Tsironi, T., & Koutinas, A. (2022). Development of biodegradable films using sunflower protein isolates and bacterial nanocellulose as innovative food packaging materials for fresh fruit preservation. *Scientific Reports*, 12(1), 6935. DOI: 10.1038/s41598-022-10913-6 PMID: 35484184

Eichhorn, S. J., & Davies, G. R. (2006). Modelling the crystalline deformation of native and regenerated cellulose. *Cellulose (London, England)*, 13(3), 291–307. DOI: 10.1007/s10570-006-9046-3

El-Bestawy, E., Eltaweil, A. S., & Khallaf, N. S. (2023). Effective production of bacterial cellulose using acidic dairy industry by-products and agro wastes. *Sustainable Chemistry and Pharmacy*, 33(1), 101064. DOI: 10.1016/j.scp.2023.101064

Erbas Kiziltas, E., Kiziltas, A., & Gardner, D. J. (2015). Synthesis of bacterial cellulose using hot water extracted wood sugars. *Carbohydrate Polymers*, 124(1), 131–138. DOI: 10.1016/j.carbpol.2015.01.036 PMID: 25839803

Felgueiras, C., Azoia, N. G., Gonçalves, C., Gama, M., & Dourado, F. (2021). Trends on the Cellulose-Based Textiles: Raw Materials and Technologies. *Frontiers in Bioengineering and Biotechnology*, 9, 608826. Advance online publication. DOI: 10.3389/fbioe.2021.608826 PMID: 33869148

Flores-Córdova, M. A., Sánchez Chávez, E., Chávez-Mendoza, C., García-Hernández, J. L., & Preciado-Rangel, P. (2016). Bioactive compounds and phytonutrients in edible part and nutshell of pecan (Carya illinoinensis). *Cogent Food & Agriculture*, 2(1), 1262936. DOI: 10.1080/23311932.2016.1262936

Fortunato, E., Gaspar, D., Duarte, P., Pereira, L., Águas, H., Vicente, A., Dourado, F., ... (2016). Chapter 11 - Optoelectronic Devices from Bacterial NanoCellulose. Bacterial Nanocellulose M. Gama, F. Dourado and S. Bielecki. Amsterdam, Elsevier, *1*, 179-197.

Françon, H., Wang, Z., Marais, A., Mystek, K., Piper, A., Granberg, H., Malti, A., Gatenholm, P., Larsson, P. A., & Wågberg, L. (2020). Ambient-Dried, 3D-Printable and Electrically Conducting Cellulose Nanofiber Aerogels by Inclusion of Functional Polymers. *Advanced Functional Materials*, 30(12), 1909383. DOI: 10.1002/adfm.201909383

Fukuzaki, S., Thienchaimongkol, J., Taokaew, S., & Kobayashi, T. (2022). Biopolymer Based Hydrogels; Hydrogels Bio-Synthesized by Bacteria and Their Industrial Applications. Encyclopedia of Materials: Plastics and Polymers M. S. J. Hashmi. Oxford, Elsevier, *1*, 487-496.

Ganta, A., Bashir, Y., & Das, S. (2022). Dairy Wastewater as a Potential Feedstock for Valuable Production with Concurrent Wastewater Treatment through Microbial Electrochemical Technologies. *Energies*, 15(23), 9084. DOI: 10.3390/en15239084

Geng, H. (2018). A facile approach to light weight, high porosity cellulose aerogels. *International Journal of Biological Macromolecules*, 118, 118921–118931. DOI: 10.1016/j.ijbiomac.2018.06.167 PMID: 29964109

Glenn, G., Orts, W., Klamczynski, A., Shogren, R., Hart-Cooper, W., Wood, D., Lee, C., & Chiou, B.-S. (2023). Compression molded cellulose fiber foams. *Cellulose (London, England)*, 30(6), 3489–3503. DOI: 10.1007/s10570-023-05111-0

Gomes, F. P., Silva, N. H. C. S., Trovatti, E., Serafim, L. S., Duarte, M. F., Silvestre, A. J. D., Neto, C. P., & Freire, C. S. R. (2013). Production of bacterial cellulose by Gluconacetobacter sacchari using dry olive mill residue. *Biomass and Bioenergy*, 55(1), 205–211. DOI: 10.1016/j.biombioe.2013.02.004

González-García, Y., Meza-Contreras, J. C., Gutiérrez-Ortega, J. A., & Manríquez-González, R. (2022). In Vivo Modification of Microporous Structure in Bacterial Cellulose by Exposing Komagataeibacter xylinus Culture to Physical and Chemical Stimuli. *Polymers*, 14(20), 4388. DOI: 10.3390/polym14204388 PMID: 36297965

Guan, F., Song, Z., Xin, F., Wang, H., Yu, D., Li, G., & Liu, W. (2020). Preparation of hydrophobic transparent paper via using polydimethylsiloxane as transparent agent. *Journal of Bioresources and Bioproducts*, 5(1), 37–43. DOI: 10.1016/j.jobab.2020.03.004

Guimarães, D. T., de Oliveira Barros, M., de Araújo e Silva, R., Silva, S. M. F., de Almeida, J. S., de Freitas Rosa, M., Gonçalves, L. R. B., & Brígida, A. I. S. (2023). Superabsorbent bacterial cellulose film produced from industrial residue of cashew apple juice processing. *International Journal of Biological Macromolecules*, 242(1), 124405. DOI: 10.1016/j.ijbiomac.2023.124405 PMID: 37100327

Gullo, M., La China, S., Petroni, G., Di Gregorio, S., & Giudici, P. (2019). Exploring K2G30 Genome: A High Bacterial Cellulose Producing Strain in Glucose and Mannitol Based Media. *Frontiers in Microbiology*, 10(1), 00058. DOI: 10.3389/fmicb.2019.00058 PMID: 30761107

Guo, X., Cavka, A., Jönsson, L. J., & Hong, F. (2013). Comparison of methods for detoxification of spruce hydrolysate for bacterial cellulose production. *Microbial Cell Factories*, 12(1), 12–93. DOI: 10.1186/1475-2859-12-93 PMID: 24119691

Hafez, I., & Tajvidi, M. (2021). Comprehensive Insight into Foams Made of Thermomechanical Pulp Fibers and Cellulose Nanofibrils via Microwave Radiation. *ACS Sustainable Chemistry & Engineering*, 9(30), 10113–10122. DOI: 10.1021/acssuschemeng.1c01816

Han, G. (2022). Characterization and properties of nanocellulose-enhanced pulp-molded lunch boxes. *Journal of Physics: Conference Series*, 2393(1), 012006. DOI: 10.1088/1742-6596/2393/1/012006

Hassan, M. M., Tucker, N., & Le Guen, M. J. (2020). Thermal, mechanical and viscoelastic properties of citric acid-crosslinked starch/cellulose composite foams. *Carbohydrate Polymers*, 230115675, 115675. Advance online publication. DOI: 10.1016/j.carbpol.2019.115675 PMID: 31887917

He, S., Liu, C., Chi, X., Zhang, Y., Yu, G., Wang, H., Li, B., & Peng, H. (2019). Bio-inspired lightweight pulp foams with improved mechanical property and flame retardancy via borate cross-linking. *Chemical Engineering Journal*, 371, 37134–37142. DOI: 10.1016/j.cej.2019.04.018

Heydorn, R. L., Lammers, D., Gottschling, M., & Dohnt, K. (2023). Effect of food industry by-products on bacterial cellulose production and its structural properties. *Cellulose (London, England)*, 30(7), 4159–4179. DOI: 10.1007/s10570-023-05097-9

Ho Jin, Y., Lee, T., Kim, J. R., Choi, Y.-E., & Park, C. (2019). Improved production of bacterial cellulose from waste glycerol through investigation of inhibitory effects of crude glycerol-derived compounds by Gluconacetobacter xylinus. *Journal of Industrial and Engineering Chemistry*, 75(1), 158–163. DOI: 10.1016/j.jiec.2019.03.017

Hong, F., Guo, X., Zhang, S., Han, S., Yang, G., & Jönsson, L. J. (2012). Bacterial cellulose production from cotton-based waste textiles: Enzymatic saccharification enhanced by ionic liquid pretreatment. *Bioresource Technology*, 104(1), 503–508. DOI: 10.1016/j.biortech.2011.11.028 PMID: 22154745

Hossain, R., Tajvidi, M., Bousfield, D., & Gardner, D. J. (2021). Multi-layer oil-resistant food serving containers made using cellulose nanofiber coated wood flour composites. *Carbohydrate Polymers*, 267118221, 118221. Advance online publication. DOI: 10.1016/j.carbpol.2021.118221 PMID: 34119175

Hsieh, Y. C., Yano, H., Nogi, M., & Eichhorn, S. J. (2008). An estimation of the Young's modulus of bacterial cellulose filaments. *Cellulose (London, England)*, 15(4), 507–513. DOI: 10.1007/s10570-008-9206-8

Huang, C., Yang, X. Y., Xiong, L., Guo, H. J., Luo, J., Wang, B., Zhang, H. R., Lin, X.-Q., & Chen, X.-D. (2015). Evaluating the possibility of using acetone-butanol-ethanol (ABE) fermentation wastewater for bacterial cellulose production by Gluconacetobacter xylinus. *Letters in Applied Microbiology*, 60(5), 491–496. DOI: 10.1111/lam.12396 PMID: 25615895

Huang, W.-M., Chen, J.-H., Nagarajan, D., Lee, C.-K., Varjani, S., Lee, D.-J., & Chang, J.-S. (2022). Immobilization of Chlorella sorokiniana AK-1 in bacterial cellulose by co-culture and its application in wastewater treatment. *Journal of the Taiwan Institute of Chemical Engineers*, 137(1), 104286. DOI: 10.1016/j.jtice.2022.104286

Iguchi, M., Yamanaka, S., & Budhiono, A. (2000). Bacterial cellulose - a masterpiece of nature's arts. *Journal of Materials Science*, 35(2), 261–270. DOI: 10.1023/A:1004775229149

Islam, M. U., Ullah, M. W., Khan, S., Shah, N., & Park, J. K. (2017). Strategies for cost-effective and enhanced production of bacterial cellulose. *International Journal of Biological Macromolecules*, 102(1), 1166–1173. DOI: 10.1016/j.ijbiomac.2017.04.110 PMID: 28487196

Jin, K., Tang, Y., Liu, J., Wang, J., & Ye, C. (2021). Nanofibrillated cellulose as coating agent for food packaging paper. *International Journal of Biological Macromolecules*, 168, 168331–168338. DOI: 10.1016/j.ijbiomac.2020.12.066 PMID: 33310098

Jin Chung, W., Shim, J., & Ravindran, B. (2022). Characterization of cheese processed wastewater and treatment using calcium nanoparticles synthesised by Senna auriculata L flower extract. *Journal of King Saud University. Science*, 34(2), 101793. DOI: 10.1016/j.jksus.2021.101793

Jozala, A. F., de Lencastre-Novaes, L. C., Lopes, A. M., de Carvalho Santos-Ebinuma, V., Mazzola, P. G., Pessoa-Jr, A., Grotto, D., Gerenutti, M., & Chaud, M. V. (2016). Bacterial nanocellulose production and application: A 10-year overview. *Applied Microbiology and Biotechnology*, 100(5), 2063–2072. DOI: 10.1007/s00253-015-7243-4 PMID: 26743657

Jozala, A. F., Pértile, R. A. N., dos Santos, C. A., de Carvalho Santos-Ebinuma, V., Seckler, M. M., Gama, F. M., & Pessoa, A.Jr. (2015). Bacterial cellulose production by Gluconacetobacter xylinus by employing alternative culture media. *Applied Microbiology and Biotechnology*, 99(3), 1181–1190. DOI: 10.1007/s00253-014-6232-3 PMID: 25472434

Kalyoncu, E. E., & Peşman, E. (2020). Bacterial cellulose as reinforcement in paper made from recycled office waste pulp. *BioResources*, 15(4), 8496–8514. DOI: 10.15376/biores.15.4.8496-8514

Keshk, S., & Sameshima, K. (2006). Influence of lignosulfonate on crystal structure and productivity of bacterial cellulose in a static culture. *Enzyme and Microbial Technology*, 40(1), 4–8. DOI: 10.1016/j.enzmictec.2006.07.037

Klemm, D., Schumann, D., Udhardt, U., & Marsch, S. (2001). Bacterial synthesized cellulose — Artificial blood vessels for microsurgery. *Progress in Polymer Science*, 26(9), 1561–1603. DOI: 10.1016/S0079-6700(01)00021-1

Kongklieng, P., Kobayashi, T., & Taokaew, S. (2023). Elastic transparent lignocellulose bioplastics from corncob waste: Positive effects of alkali treatment without bleaching process. *Biomass Conversion and Biorefinery*. Advance online publication. DOI: 10.1007/s13399-023-04772-y

Korte, I., Albrecht, A., Mittler, M., Waldhans, C., & Kreyenschmidt, J. (2023). Influence of different bio-based and conventional packaging trays on the quality loss of fresh cherry tomatoes during distribution and storage. *Packaging Technology & Science*, 36(7), 569–583. DOI: 10.1002/pts.2728

Kuo, C.-H., Lin, P.-J., & Lee, C.-K. (2010). Enzymatic saccharification of dissolution pretreated waste cellulosic fabrics for bacterial cellulose production by Gluconacetobacter xylinus. *Journal of Chemical Technology and Biotechnology*, 85(10), 1346–1352. DOI: 10.1002/jctb.2439

Lee, H., Kim, S., Shin, S., & Hyun, J. (2021). 3D structure of lightweight, conductive cellulose nanofiber foam. *Carbohydrate Polymers*, 253117238, 117238. Advance online publication. DOI: 10.1016/j.carbpol.2020.117238 PMID: 33278994

Lee, K.-Y., Tammelin, T., Schulfter, K., Kiiskinen, H., Samela, J., & Bismarck, A. (2012). High Performance Cellulose Nanocomposites: Comparing the Reinforcing Ability of Bacterial Cellulose and Nanofibrillated Cellulose. *ACS Applied Materials & Interfaces*, 4(8), 4078–4086. DOI: 10.1021/am300852a PMID: 22839594

Li, J., Cheng, R., Xiu, H., Zhang, M., Liu, Q., Song, T., Dong, H., Yao, B., Zhang, X., Kozliak, E., & Ji, Y. (2018). Pore structure and pertinent physical properties of nanofibrillated cellulose (NFC)-based foam materials. *Carbohydrate Polymers*, 201, 201141–201150. DOI: 10.1016/j.carbpol.2018.08.008 PMID: 30241805

Li, K., Wang, S., Chen, H., Yang, X., Berglund, L. A., & Zhou, Q. (2020). Self-Densification of Highly Mesoporous Wood Structure into a Strong and Transparent Film. *Advanced Materials*, 32(42), 2003653. DOI: 10.1002/adma.202003653 PMID: 32881202

Li, R., Du, J., Zheng, Y., Wen, Y., Zhang, X., Yang, W., Lue, A., & Zhang, L. (2017). Ultra-lightweight cellulose foam material: Preparation and properties. *Cellulose (London, England)*, 24(3), 1417–1426. DOI: 10.1007/s10570-017-1196-y

Li, W., Hu, J., Cheng, L., Chen, L., Zhou, L., Zhang, J., & Yuan, Y. (2019). Study on thermal behavior of regenerated micro-crystalline cellulose containing slight amount of water induced by hydrogen-bonds transformation. *Polymer*, 185121989, 121989. Advance online publication. DOI: 10.1016/j.polymer.2019.121989

Li, Z., Wang, L., Hua, J., Jia, S., Zhang, J., & Liu, H. (2015). Production of nano bacterial cellulose from waste water of candied jujube-processing industry using Acetobacter xylinum. *Carbohydrate Polymers*, 120(1), 115–119. DOI: 10.1016/j.carbpol.2014.11.061 PMID: 25662694

Liao, J., Luan, P., Zhang, Y., Chen, L., Huang, L., Mo, L., Li, J., & Xiong, Q. (2022). A lightweight, biodegradable, and recyclable cellulose-based bio-foam with good mechanical strength and water stability. *Journal of Environmental Chemical Engineering*, 10(3), 107788. DOI: 10.1016/j.jece.2022.107788

Lin, D., Lopez-Sanchez, P., Li, R., & Li, Z. (2014). Production of bacterial cellulose by Gluconacetobacter hansenii CGMCC 3917 using only waste beer yeast as nutrient source. *Bioresource Technology*, 151(1), 113–119. DOI: 10.1016/j.biortech.2013.10.052 PMID: 24212131

Liu, H., Du, H., Zheng, T., Liu, K., Ji, X., Xu, T., Zhang, X., & Si, C. (2021). Cellulose based composite foams and aerogels for advanced energy storage devices. *Chemical Engineering Journal*, 426(1), 130817. DOI: 10.1016/j.cej.2021.130817

Lohtander, T., Herrala, R., Laaksonen, P., Franssila, S., & Österberg, M. (2022). Lightweight lignocellulosic foams for thermal insulation. *Cellulose (London, England)*, 29(3), 1855–1871. DOI: 10.1007/s10570-021-04385-6

Lourenço, A. F., Martins, D., Dourado, F., Sarmento, P., Ferreira, P. J. T., & Gamelas, J. A. F. (2023). Impact of bacterial cellulose on the physical properties and printing quality of fine papers. *Carbohydrate Polymers*, 314120915, 120915. Advance online publication. DOI: 10.1016/j.carbpol.2023.120915 PMID: 37173044

Lu, B., Lin, Q., Yin, Z., Lin, F., Chen, X., & Huang, B. (2021). Robust and lightweight biofoam based on cellulose nanofibrils for high-efficient methylene blue adsorption. *Cellulose (London, England)*, 28(1), 273–288. DOI: 10.1007/s10570-020-03553-4

Ma, Z., Xue, T., Wali, Q., Miao, Y.-E., Fan, W., & Liu, T. (2023). Direct ink writing of polyimide/bacterial cellulose composite aerogel for thermal insulation. *Composites Communications*, 39101528, 101528. Advance online publication. DOI: 10.1016/j.coco.2023.101528

Machado, R. T. A., Meneguin, A. B., Sábio, R. M., Franco, D. F., Antonio, S. G., Gutierrez, J., Tercjak, A., Berretta, A. A., Ribeiro, S. J. L., Lazarini, S. C., Lustri, W. R., & Barud, H. S. (2018). Komagataeibacter rhaeticus grown in sugarcane molasses-supplemented culture medium as a strategy for enhancing bacterial cellulose production. *Industrial Crops and Products*, 122(1), 637–646. DOI: 10.1016/j.indcrop.2018.06.048

Marín, P., Martirani-Von Abercron, S. M., Urbina, L., Pacheco-Sánchez, D., Castañeda-Cataña, M. A., Retegi, A., Eceiza, A., & Marqués, S. (2019). Bacterial nanocellulose production from naphthalene. *Microbial Biotechnology*, 12(4), 662–676. DOI: 10.1111/1751-7915.13399 PMID: 31087504

Miller, G. L. (1959). Use of dinitrosalicylic acid reagent for determination of reducing sugar. *Analytical Chemistry*, 31(3), 426–428. DOI: 10.1021/ac60147a030

Mohammadkazemi, F., Azin, M., & Ashori, A. (2015). Production of bacterial cellulose using different carbon sources and culture media. *Carbohydrate Polymers*, 117(1), 518–523. DOI: 10.1016/j.carbpol.2014.10.008 PMID: 25498666

Mohite, B. V., & Patil, S. V. (2014). A novel biomaterial: Bacterial cellulose and its new era applications. *Biotechnology and Applied Biochemistry*, 61(2), 101–110. DOI: 10.1002/bab.1148 PMID: 24033726

Mohite, B. V., Salunke, B. K., & Patil, S. V. (2013). Enhanced Production of Bacterial Cellulose by Using Gluconacetobacter hansenii NCIM 2529 Strain Under Shaking Conditions. *Applied Biochemistry and Biotechnology*, 169(5), 1497–1511. DOI: 10.1007/s12010-013-0092-7 PMID: 23319186

Moradian, S., Almasi, H., & Moini, S. (2018). Development of bacterial cellulose-based active membranes containing herbal extracts for shelf life extension of button mushrooms (Agaricus bisporus). *Journal of Food Processing and Preservation*, 42(3), e13537. DOI: 10.1111/jfpp.13537

Mustafa, A. F., McKinnon, J. J., Ingledew, M. W., & Christensen, D. A. (2000). The nutritive value for ruminants of thin stillage and distillers' grains derived from wheat, rye, triticale and barley. *Journal of the Science of Food and Agriculture*, 80(5), 607–613. DOI: 10.1002/(SICI)1097-0010(200004)80:5<607::AID-JSFA582>3.0.CO;2-F

Nair, S. S., Zhu, J. Y., Deng, Y., & Ragauskas, A. J. (2014). High performance green barriers based on nanocellulose. *Sustainable Chemical Processes*, 2(1), 23. DOI: 10.1186/s40508-014-0023-0

Nechita, P., & Năstac, S. M. (2022). Overview on Foam Forming Cellulose Materials for Cushioning Packaging Applications. *Polymers*, 14(10), 1963. Advance online publication. DOI: 10.3390/polym14101963 PMID: 35631844

Nilsen-Nygaard, J., Fernández, E. N., Radusin, T., Rotabakk, B. T., Sarfraz, J., Sharmin, N., Sivertsvik, M., Sone, I., & Pettersen, M. K. (2021). Current status of biobased and biodegradable food packaging materials: Impact on food quality and effect of innovative processing technologies. *Comprehensive Reviews in Food Science and Food Safety*, 20(2), 1333–1380. DOI: 10.1111/1541-4337.12715 PMID: 33547765

Nogi, M., & Yano, H. (2008). Transparent Nanocomposites Based on Cellulose Produced by Bacteria Offer Potential Innovation in the Electronics Device Industry. *Advanced Materials*, 20(10), 1849–1852. DOI: 10.1002/adma.200702559

Öz, Y. E., & Kalender, M. (2023). A novel static cultivation of bacterial cellulose production from sugar beet molasses: Series static culture (SSC) system. *International Journal of Biological Macromolecules*, 225(1), 1306–1314. DOI: 10.1016/j.ijbiomac.2022.11.190 PMID: 36435464

Pang, M., Huang, Y., Meng, F., Zhuang, Y., Liu, H., Du, M., Ma, Q., Wang, Q., Chen, Z., Chen, L., Cai, T., & Cai, Y. (2020). Application of bacterial cellulose in skin and bone tissue engineering. *European Polymer Journal*, 122(1), 109365. DOI: 10.1016/j.eurpolymj.2019.109365

Park, J. K., Jung, J. Y., & Park, Y. H. (2003). Cellulose production by Gluconacetobacter hansenii in a medium containing ethanol. *Biotechnology Letters*, 25(24), 2055–2059. DOI: 10.1023/B:BILE.0000007065.63682.18 PMID: 14969408

Phisalaphong, M., Tran, T. K., Taokaew, S., Budiraharjo, R., Febriana, G. G., Nguyen, D. N., Chu-Ky, S., et al. (2016). Nata de coco Industry in Vietnam, Thailand, and Indonesia. Bacterial Nanocellulose: From Biotechnology to Bio-Economy, *1*, 231-236.

RameshKumar, S., Shaiju, P., & O'Connor, K. E. (2020). Bio-based and biodegradable polymers-State-of-the-art, challenges and emerging trends. *Current Opinion in Green and Sustainable Chemistry*, 21, 75–81.

Ratanapariyanuch, K., Shen, J., Jia, Y., Tyler, R. T., Shim, Y. Y., & Reaney, M. J. T. (2011). Rapid NMR Method for the Quantification of Organic Compounds in Thin Stillage. *Journal of Agricultural and Food Chemistry*, 59(19), 10454–10460. DOI: 10.1021/jf2026007 PMID: 21875138

Raval, A. A., Raval, U. G., & Sayyed, R. Z. (2020). Utilization of industrial waste for the sustainable production of bacterial cellulose. *Environmental Sustainability*, 3(4), 427–435. DOI: 10.1007/s42398-020-00126-w

Reichert, C. L., Bugnicourt, E., Coltelli, M.-B., Cinelli, P., Lazzeri, A., Canesi, I., Braca, F., Martínez, B. M., Alonso, R., Agostinis, L., Verstichel, S., Six, L., Mets, S. D., Gómez, E. C., Ißbrücker, C., Geerinck, R., Nettleton, D. F., Campos, I., Sauter, E., & Schmid, M. (2020). Bio-Based Packaging: Materials, Modifications, Industrial Applications and Sustainability. *Polymers*, 12(7), 1558. Advance online publication. DOI: 10.3390/polym12071558 PMID: 32674366

Revin, V., Liyaskina, E., Nazarkina, M., Bogatyreva, A., & Shchankin, M. (2018). Cost-effective production of bacterial cellulose using acidic food industry by-products. *Brazilian Journal of Microbiology*, 49(1), 151–159. DOI: 10.1016/j.bjm.2017.12.012 PMID: 29703527

Rodrigues, A. C., Fontão, A. I., Coelho, A., Leal, M., Soares da Silva, F. A. G., Wan, Y., Dourado, F., & Gama, M. (2019). Response surface statistical optimization of bacterial nanocellulose fermentation in static culture using a low-cost medium. *New Biotechnology*, 49(1), 19–27. DOI: 10.1016/j.nbt.2018.12.002 PMID: 30529474

Ruka, D. R., Simon, G. P., & Dean, K. M. (2012). Altering the growth conditions of Gluconacetobacter xylinus to maximize the yield of bacterial cellulose. *Carbohydrate Polymers*, 89(2), 613–622. DOI: 10.1016/j.carbpol.2012.03.059 PMID: 24750766

Salari, M., Sowti Khiabani, M., Rezaei Mokarram, R., Ghanbarzadeh, B., & Samadi Kafil, H. (2019). Preparation and characterization of cellulose nanocrystals from bacterial cellulose produced in sugar beet molasses and cheese whey media. *International Journal of Biological Macromolecules*, 122(1), 280–288. DOI: 10.1016/j.ijbiomac.2018.10.136 PMID: 30342939

Sandolo, C., Matricardi, P., Alhaique, F., & Coviello, T. (2009). Effect of temperature and cross-linking density on rheology of chemical cross-linked guar gum at the gel point. *Food Hydrocolloids*, 23(1), 210–220. DOI: 10.1016/j.foodhyd.2008.01.001

Santos, A. A., Matos, L. C., Mendonça, M. C., Lago, R. C., Muguet, M. C. S., Damásio, R. A. P., Ponzecchi, A., Soares, J. R., Sanadi, A. R., & Tonoli, G. H. D. (2023). Evaluation of paper coated with cationic starch and carnauba wax mixtures regarding barrier properties. *Industrial Crops and Products*, 203117177, 117177. Advance online publication. DOI: 10.1016/j.indcrop.2023.117177

Santos, S. M., Carbajo, J. M., & Villar, J. C. (2013). The effect of carbon and nitrogen sources on bacterial cellulose production and properties from gluconacetobacter sucrofermentans CECT 7291 focused on its use in degraded paper restoration. *BioResources*, 8(3), 3630–3645. DOI: 10.15376/biores.8.3.3630-3645

Schirmeister, C. G., & Mülhaupt, R. (2022). Closing the Carbon Loop in the Circular Plastics Economy. *Macromolecular Rapid Communications*, 43(13), 2200247. DOI: 10.1002/marc.202200247 PMID: 35635841

Segal, L., Creely, J. J., Martin, A. E.Jr, & Conrad, C. M. (1959). An Empirical Method for Estimating the Degree of Crystallinity of Native Cellulose Using the X-Ray Diffractometer. *Textile Research Journal*, 29(10), 786–794. DOI: 10.1177/004051755902901003

Sen, K. Y., Hussin, M. H., & Baidurah, S. (2019). Biosynthesis of poly(3-hydroxybutyrate) (PHB) by Cupriavidus necator from various pretreated molasses as carbon source. *Biocatalysis and Agricultural Biotechnology*, 17, 1751–1759. DOI: 10.1016/j.bcab.2018.11.006

Shi, Z., Zhang, Y., Phillips, G. O., & Yang, G. (2014). Utilization of bacterial cellulose in food. *Food Hydrocolloids*, 35(1), 539–545. DOI: 10.1016/j.foodhyd.2013.07.012

Sjölin, M., Thuvander, J., Wallberg, O., & Lipnizki, F. (2020). Purification of Sucrose in Sugar Beet Molasses by Utilizing Ceramic Nanofiltration and Ultrafiltration Membranes. *Membranes (Basel)*, 10(1), 5. Advance online publication. DOI: 10.3390/membranes10010005 PMID: 31892103

Sun, H., Liu, Y., Guo, X., Zeng, K., Kanti Mondal, A., Li, J., Yao, Y., & Chen, L. (2021). Strong, robust cellulose composite film for efficient light management in energy efficient building. *Chemical Engineering Journal*, 425(1), 131469. DOI: 10.1016/j.cej.2021.131469

Sun, L., Liu, L., Wu, M., Wang, D., Shen, R., Zhao, H., Lu, J., & Yao, J. (2023). Nanocellulose interface enhanced all-cellulose foam with controllable strength via a facile liquid phase exchange route. *Carbohydrate Polymers*, 299120192, 120192. Advance online publication. DOI: 10.1016/j.carbpol.2022.120192 PMID: 36876806

Sun, X.-D., Yuan, X.-Z., Jia, Y., Feng, L.-J., Zhu, F.-P., Dong, S.-S., Liu, J., Kong, X., Tian, H., Duan, J.-L., Ding, Z., Wang, S.-G., & Xing, B. (2020). Differentially charged nanoplastics demonstrate distinct accumulation in Arabidopsis thaliana. *Nature Nanotechnology*, 15(9), 755–760. DOI: 10.1038/s41565-020-0707-4 PMID: 32572228

Suratago, T., Taokaew, S., Kanjanamosit, N., Kanjanaprapakul, K., Burapatana, V., & Phisalaphong, M. (2015). Development of bacterial cellulose/alginate nanocomposite membrane for separation of ethanol-water mixtures. *Journal of Industrial and Engineering Chemistry*, 32(1), 305–312. DOI: 10.1016/j.jiec.2015.09.004

Suwanmajo, T., & Taokaew, S. (2022). Biosynthesized Nanocomposite of Bacterial Nanocellulose—Tacca leontopetaloides Starch for Sustainable UV Resistant Film. *Stärke*, 74(11-12), 2200065. DOI: 10.1002/star.202200065

Suwanposri, A., Yukphan, P., Yamada, Y., & Ochaikul, D. (2014). Statistical optimisation of culture conditions for biocellulose production by Komagataeibacter sp. PAP1 using soya bean whey. *Maejo International Journal of Science and Technology*, 8(1), 1–14.

Tang, Y., He, Z., Mosseler, J. A., & Ni, Y. (2014). Production of highly electro-conductive cellulosic paper via surface coating of carbon nanotube/graphene oxide nanocomposites using nanocrystalline cellulose as a binder. *Cellulose (London, England)*, 21(6), 4569–4581. DOI: 10.1007/s10570-014-0418-9

Taokaew, S. (2023). Recent Advances in Cellulose-Based Hydrogels Prepared by Ionic Liquid-Based Processes. *Gels (Basel, Switzerland)*, 9(7), 546. Advance online publication. DOI: 10.3390/gels9070546 PMID: 37504425

Taokaew, S., Chiaoprakobkij, N., Siripong, P., Sanchavanakit, N., Pavasant, P., & Phisalaphong, M. (2021). Multifunctional cellulosic nanofiber film with enhanced antimicrobial and anticancer properties by incorporation of ethanolic extract of Garcinia mangostana peel. *Materials Science and Engineering C*, 120(1), 111783. DOI: 10.1016/j.msec.2020.111783 PMID: 33545910

Taokaew, S., Nakson, N., Thienchaimongkol, J., & Kobayashi, T. (2023). Enhanced production of fibrous bacterial cellulose in Gluconacetobacter xylinus culture medium containing modified protein of okara waste. *Journal of Bioscience and Bioengineering*, 135(1), 71–78. DOI: 10.1016/j.jbiosc.2022.10.007 PMID: 36437213

Taokaew, S., Nakson, N., Zhang, X., Kongklieng, P., & Kobayashi, T. (2022). Biotransformation of okara extracted protein to nanocellulose and chitin by Gluconacetobacter xylinus and Bacillus pumilus. *Bioresource Technology Reports*, 17(1), 100904. DOI: 10.1016/j.biteb.2021.100904

Taokaew, S., Nunkaew, N., Siripong, P., & Phisalaphong, M. (2014). Characteristics and anticancer properties of bacterial cellulose films containing ethanolic extract of mangosteen peel. *Journal of Biomaterials Science. Polymer Edition*, 25(9), 907–922. DOI: 10.1080/09205063.2014.913464 PMID: 24802115

Taokaew, S., Phisalaphong, M., & Newby, B. M. Z. (2015). Modification of bacterial cellulose with organosilanes to improve attachment and spreading of human fibroblasts. *Cellulose (London, England)*, 22(4), 2311–2324. DOI: 10.1007/s10570-015-0651-x PMID: 26478661

Taokaew, S., Phisalaphong, M., & Newby, B. M. Z. (2016). Bacterial cellulose: Biosyntheses, modifications, and applications. *Applied Environmental Materials Science for Sustainability*, 1, 255–283.

Taokaew, S., Phisalaphong, M., & Zhang Newby, B. M. (2014). In vitro behaviors of rat mesenchymal stem cells on bacterial celluloses with different moduli. *Materials Science and Engineering C*, 38(1), 263–271. DOI: 10.1016/j.msec.2014.02.005 PMID: 24656377

Taokaew, S., Seetabhawang, S., Siripong, P., & Phisalaphong, M. (2013). Biosynthesis and characterization of nanocellulose-gelatin films. *Materials (Basel)*, 6(3), 782–794. DOI: 10.3390/ma6030782 PMID: 28809339

Taokaew, S., Thienchaimongkol, J., Nakson, N., & Kobayashi, T. (2022). Valorisation of okara Waste as an Alternative Nitrogen Source in the Biosynthesis of Nanocellulose. *Chemical Engineering Transactions*, 92(1), 649–654. DOI: 10.3303/CET2292109

Taokaew, S., Zhang, X., Chuenkaek, T., & Kobayashi, T. (2020). Chitin from fermentative extraction of crab shells using okara as a nutrient source and comparative analysis of structural differences from chemically extracted chitin. *Biochemical Engineering Journal*, 159(1), 107588. DOI: 10.1016/j.bej.2020.107588

Tardy, B. L., Mattos, B. D., Otoni, C. G., Beaumont, M., Majoinen, J., Kämäräinen, T., & Rojas, O. J. (2021). Deconstruction and Reassembly of Renewable Polymers and Biocolloids into Next Generation Structured Materials. *Chemical Reviews*, 121(22), 14088–14188. DOI: 10.1021/acs.chemrev.0c01333 PMID: 34415732

Tayeb, H., A., Tajvidi, M.&Bousfield, D. (. (2020). Paper-Based Oil Barrier Packaging using Lignin-Containing Cellulose Nanofibrils. *Journal*, 25(6). Advance online publication. DOI: 10.3390/molecules25061344

Teng, Y., Shi, B., Zhang, J., Chen, Y., & Wang, Y. (2020). Preparation of Robust Superhydrophobic Paper by Roll Coating with Modified Micro/Nano SiO2. *Chemistry Letters*, 49(9), 1095–1098. DOI: 10.1246/cl.200381

Teng, Y., Wang, Y., Shi, B., & Chen, Y. (2020). Facile preparation of economical, eco-friendly superhydrophobic surface on paper substrate with excellent mechanical durability. *Progress in Organic Coatings*, 147105877, 105877. Advance online publication. DOI: 10.1016/j.porgcoat.2020.105877

Tsouko, E., Kourmentza, C., Ladakis, D., Kopsahelis, N., Mandala, I., Papanikolaou, S., Paloukis, F., Alves, V., & Koutinas, A. (2015). Bacterial Cellulose Production from Industrial Waste and by-Product Streams. *International Journal of Molecular Sciences*, 16(7), 14832–14849. DOI: 10.3390/ijms160714832 PMID: 26140376

Tyagi, P., Lucia, L. A., Hubbe, M. A., & Pal, L. (2019). Nanocellulose-based multilayer barrier coatings for gas, oil, and grease resistance. *Carbohydrate Polymers*, 206, 206281–206288. DOI: 10.1016/j.carbpol.2018.10.114 PMID: 30553323

Ul-Islam, M., Khan, T., & Park, J. K. (2012). Water holding and release properties of bacterial cellulose obtained by in situ and ex situ modification. *Carbohydrate Polymers*, 88(2), 596–603. DOI: 10.1016/j.carbpol.2012.01.006

Vaezi, K., Asadpour, G., & Sharifi, S. H. (2019). Effect of coating with novel bio nanocomposites of cationic starch/cellulose nanocrystals on the fundamental properties of the packaging paper. *Polymer Testing*, 80106080, 106080. Advance online publication. DOI: 10.1016/j.polymertesting.2019.106080

Vazquez, A., Foresti, M. L., Cerrutti, P., & Galvagno, M. (2013). Bacterial Cellulose from Simple and Low Cost Production Media by Gluconacetobacter xylinus. *Journal of Polymers and the Environment*, 21(2), 545–554. DOI: 10.1007/s10924-012-0541-3

Viscusi, G., Bugatti, V., & Gorrasi, G. (2021). Active packaging based on cellulose trays coated with layered double hydroxide as nano-carrier of parahydroxybenzoate: Application to fresh-cut iceberg lettuce. *Packaging Technology & Science*, 34(6), 353–360. DOI: 10.1002/pts.2565

Voon, W. W. Y., Muhialdin, B. J., Yusof, N. L., Rukayadi, Y., & Meor Hussin, A. S. (2019). Bio-cellulose Production by Beijerinckia fluminensis WAUPM53 and Gluconacetobacter xylinus 0416 in Sago By-product Medium. *Applied Biochemistry and Biotechnology*, 187(1), 211–220. DOI: 10.1007/s12010-018-2807-2 PMID: 29915916

Walker, T. R. (2021). (Micro)plastics and the UN Sustainable Development Goals. *Current Opinion in Green and Sustainable Chemistry*, 30100497, 100497. Advance online publication. DOI: 10.1016/j.cogsc.2021.100497

Wang, J., Tavakoli, J., & Tang, Y. (2019). Bacterial cellulose production, properties and applications with different culture methods – A review. *Carbohydrate Polymers*, 219(1), 63–76. DOI: 10.1016/j.carbpol.2019.05.008 PMID: 31151547

Wang, N., Xiong, D., Pan, S., Deng, Y., Shi, Y., & Wang, K. (2016). Superhydrophobic paper with superior stability against deformations and humidity. *Applied Surface Science*, 389, 389354–389360. DOI: 10.1016/j.apsusc.2016.07.110

Wang, P., Aliheidari, N., Zhang, X., & Ameli, A. (2019). Strong ultralight foams based on nanocrystalline cellulose for high-performance insulation. *Carbohydrate Polymers*, 218, 218103–218111. DOI: 10.1016/j.carbpol.2019.04.059 PMID: 31221311

Wu, J.-M., & Liu, R.-H. (2012). Thin stillage supplementation greatly enhances bacterial cellulose production by Gluconacetobacter xylinus. *Carbohydrate Polymers*, 90(1), 116–121. DOI: 10.1016/j.carbpol.2012.05.003 PMID: 24751018

Wu, J.-M., & Liu, R.-H. (2013). Cost-effective production of bacterial cellulose in static cultures using distillery wastewater. *Journal of Bioscience and Bioengineering*, 115(3), 284–290. DOI: 10.1016/j.jbiosc.2012.09.014 PMID: 23102658

Yang, S., Bai, S., & Wang, Q. (2018). Sustainable packaging biocomposites from polylactic acid and wheat straw: Enhanced physical performance by solid state shear milling process. *Composites Science and Technology*, 158(1), 34–42. DOI: 10.1016/j.compscitech.2017.12.026

Yang, Y., Lu, Y.-T., Zeng, K., Heinze, T., Groth, T., & Zhang, K. (2021). Recent Progress on Cellulose-Based Ionic Compounds for Biomaterials. *Advanced Materials*, 33(28), 2000717. DOI: 10.1002/adma.202000717 PMID: 32270900

Yano, H., Sugiyama, J., Nakagaito, A. N., Nogi, M., Matsuura, T., Hikita, M., & Handa, K. (2005). Optically Transparent Composites Reinforced with Networks of Bacterial Nanofibers. *Advanced Materials*, 17(2), 153–155. DOI: 10.1002/adma.200400597

Ye, J., Zheng, S., Zhang, Z., Yang, F., Ma, K., Feng, Y., Zheng, J., Mao, D., & Yang, X. (2019). Bacterial cellulose production by Acetobacter xylinum ATCC 23767 using tobacco waste extract as culture medium. *Bioresource Technology*, 274(1), 518–524. DOI: 10.1016/j.biortech.2018.12.028 PMID: 30553964

Yun, T., Tao, Y., Li, Q., Cheng, Y., Lu, J., Lv, Y., Du, J., & Wang, H. (2023). Superhydrophobic modification of cellulosic paper-based materials: Fabrication, properties, and versatile applications. *Carbohydrate Polymers*, 305120570, 120570. Advance online publication. DOI: 10.1016/j.carbpol.2023.120570 PMID: 36737208

Zhang, J., & Youngblood, J. P. (2023). Cellulose Nanofibril (CNF)-Coated PFAS-Free, Grease-Resistant All-Bio-Based Molded Pulp Containers for Food Packaging. *ACS Applied Polymer Materials*, 5(7), 5696–5706. DOI: 10.1021/acsapm.3c00979 PMID: 37469883

Zhang, J., Zhang, L., & Gong, X. (2021). Large-Scale Spraying Fabrication of Robust Fluorine-Free Superhydrophobic Coatings Based on Dual-Sized Silica Particles for Effective Antipollution and Strong Buoyancy. *Langmuir*, 37(19), 6042–6051. DOI: 10.1021/acs.langmuir.1c00706 PMID: 33939432

Zhang, Y., Zhang, L., Cui, K., Ge, S., Cheng, X., Yan, M., Yu, J., & Liu, H. (2018). Flexible Electronics Based on Micro/Nanostructured Paper. *Advanced Materials*, 30(51), 1801588. DOI: 10.1002/adma.201801588 PMID: 30066444

Zhong, C., Zhang, G.-C., Liu, M., Zheng, X.-T., Han, P.-P., & Jia, S.-R. (2013). Metabolic flux analysis of Gluconacetobacter xylinus for bacterial cellulose production. *Applied Microbiology and Biotechnology*, 97(14), 6189–6199. DOI: 10.1007/s00253-013-4908-8 PMID: 23640364

Zhou, L. L., Sun, D. P., Hu, L. Y., Li, Y. W., & Yang, J. Z. (2007). Effect of addition of sodium alginate on bacterial cellulose production by Acetobacter xylinum. *Journal of Industrial Microbiology & Biotechnology*, 34(7), 483–483. DOI: 10.1007/s10295-007-0218-4 PMID: 17440758

Zhu, C., Zhang, J., Qiu, S., Jia, Y., Wang, L., & Wang, H. (2021). Tailoring the pore size of polyphenylene sulfide nonwoven with bacterial cellulose (BC) for heat-resistant and high-wettability separator in lithium-ion battery. *Composites Communications*, 24100659, 100659. Advance online publication. DOI: 10.1016/j.coco.2021.100659

Zhu, H., Zhu, S., Jia, Z., Parvinian, S., Li, Y., Vaaland, O., Hu, L., & Li, T. (2015). Anomalous scaling law of strength and toughness of cellulose nanopaper. *Proceedings of the National Academy of Sciences of the United States of America*, 112(29), 8971–8976. DOI: 10.1073/pnas.1502870112 PMID: 26150482

Chapter 6
Production of Sustainable Bioplastics Through Biomass Wastes Valorization to Mitigate Carbon Footprint Emissions

Takaomi Kobayashi
Nagaoka University of Technology, Japan

Debbie Dominic
Universiti Sains Malaysia, Malaysia

Nurul Alia Syufina Abu Bakar
https://orcid.org/0009-0009-9157-8986
School of Industrial Technology, Universiti Sains Malaysia, Malaysia

Siti Baidurah
https://orcid.org/0000-0003-3210-7470
School of Industrial Technology, Universiti Sains Malaysia, Malaysia

ABSTRACT

This chapter focuses on the production of bioplastics such as polyhydroxyalkanoates (PHAs) via biomass wastes valorization. The high production costs of PHAs, particularly substrate of the base material, has limited its application. The utilization of biomass wastes from agro-industrial such as molasses, banana trunk juice, palm oil waste effluent, and animal derived chitin from crustacean as an alternative carbon feedstock is explored by many researchers, due to the abundancy, inexpensive, contains high sugar and oil contents. These industrial biomass wastes can be further exploited for lowering the production costs, simultaneously reduce biomass waste accumulation, and accelerate the application of the bioplastics at large scale. This chapter also covers the challenges of utilizing industrial waste as feedstocks for PHA production and its impact on carbon footprint mitigation. These initiatives are in parallel with various Sustainable Development Goals such as number 12, 13, 14, and 15.

DOI: 10.4018/979-8-3693-0003-9.ch006

INTRODUCTION

Bioplastics such as polyhydroxyalkanoates (PHAs) offered many benefits such as complete biodegradable, biocompatible, nontoxic, antioxidant, immunotolerant, and ability to be produced from various waste streams. The physical properties of PHAs are very similar to those of commercial conventional plastics packaging (Alper et al., 1963; Gowda & Shivakumar, 2019). Although PHAs offers many benefits over the conventional plastics, the production costs of PHAs are financially infeasible as compared to the conventional petrochemical-derived plastics such as polyethylene (PE). The production cost of PHAs is approximately 4-10th fold higher as compared to PE per kg. The high price of carbon source utilize during fermentation stage is the main factors contribute to the high production cost (Nonato et al., 2001; Roland-Holst et al., 2013). Approximately 50% of the production cost is allotted to the supply of raw materials, especially in terms of the carbon source (Koller et al., 2005). The carbon sources used for PHA production are expensive because of purified sugars, fatty acids, or edible plant oils as those are relevant to human nutrition (Sen & Baidurah, 2021). Therefore, the final production cost of PHAs depends on the cost of the carbon source. The utilization of low-cost carbon source from industrial biomass wastes touted as the most viable strategy to reduce the production cost of PHA (Koller et al., 2012; Koller & Braunegg, 2018).

The initiative to utilize the biomass wastes from industry sector are in parallel with Sustainable Development Goals 12, whereby to ensure the sustainable consumption and production patterns. There are various attributes that must be met in order to determine an economical carbon source for fermentation. The economic feasibility of preferred carbon sources is characterised to have an adequate supply, uniform substrate composition, same quality from batch to batch, the ease for transportation, storage, resistance to spoilage, and no competition with food and feed. Additionally, other fermentation factors should be considered, including substrate fermentability by specific bacteria, carbon supply concentration, and the presence of PHAs accumulation inhibitors (Koller & Braunegg, 2018).

Biomass waste can be defined as the organic matter derived from plants, animals, and microorganisms that can be utilized as a source of energy, materials, and chemicals. The industry sector generates a significant amount of biomass waste, which can be utilized for the production of various biobased products, including PHAs (Ooi et al., 2023). The example of biomass wastes from the industry sector that can be utilized for PHAs production includes crude glycerol, wastewater and waste cooking oil from the food processing industry, seafood processing industry, as well as lignocellulosic waste from the pulp and paper industry.

Among of the most noteworthy PHAs makers include *Cupriavidus necator*, *Bacillus megaterium*, and *Pseudomonas* sp. Bacteria produce PHAs biopolymers as carbon and energy storage materials under nutrient limiting conditions such as low nitrogen, oxygen, sulphur, phosphate, magnesium, and potassium in conjunction with an abundant carbon source. PHA accumulates intracellularly as cytosolic granules and is found in protein- and lipid-bound subcellular organelles. The quantity and size of the granules differed between bacterium species and growth conditions (Alper et al., 1963).

Various analytical methods are available and practical to elucidate the complex composition of PHAs and its degradation mechanisms such as physical methods (SEM, TEM, weighing analytical balance, etc.), chromatographic methods (GC, THM-GC, SEC/GPC), spectroscopic methods (NMR, FTIR, XRD, XRF), respirometric methods, and thermal methods (DSC, DTA, TGA) [10].

Bacteria, eukaryotic species such as yeast, and mammalian cells can accumulate modest levels of polyhydroxybutyrate (PHB), the simplest form of PHAs. With the advancement of genetic engineering, living organisms such as plants, bacteria, diatoms, and insect cells are genetically altered to excessively produce PHAs. Plants were once assumed to be a low-cost option for PHB synthesis, however genetically altered plants only gathered a limited amount of PHB. This was attributable to the fact that increased PHB production had a negative impact on plant growth and development.

This review will concentrate on the utilization of industrial biomass wastes for the production of PHAs through bacteria fermentation. It is expected that with the utilization of industrial biomass waste will mitigate carbon footprint, eliminate greenhouse gases (GHG) emission such as methane and carbon dioxide, and simultaneously reduce climate change.

Types of Industrial biomass wastes suitable for feedstocks of PHAs production

Agro-industry: Molasses and banana trunk juice

Figure 1. Various suitable industrial biomass wastes as feedstocks together with two routes of treatments: (1) conventional landfill treatment and (2) fermentation to produce biodegradable plastics

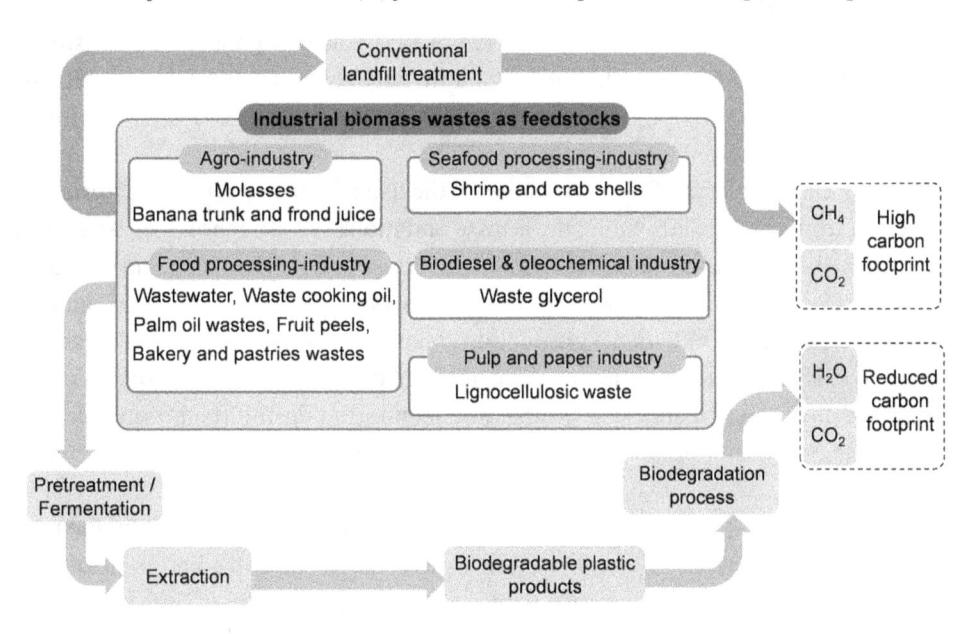

Figure 1 delineates the schematic diagram of various suitable industrial biomass wastes as feedstocks together with two routes of treatments: (1) conventional landfill treatment and (2) fermentation to produce biodegradable plastics. The conventional landfill treatment will lead to high carbon footprint. Whilst, through fermentation, the biomass wastes serve as feedstocks to produce biodegradable plastics and simultaneously able to reduce the carbon footprint.

The production of PHAs from renewable resources such as molasses has been of great interest due to their low cost, availability, and sustainable alternative *in lieu* of traditional petroleum-derived plastic production. Molasses is a by-product of the sugar refining process and is commonly used in the production of various fermented products such as alcohol, vinegar, and lactic acid. It is a rich source of carbohydrates, mainly sucrose, glucose, and fructose, making it an ideal carbon source for the production of PHAs (Sen et al., 2019). The process of biovalorization of molasses for PHAs production involves several steps. First, the molasses must be pre-treated to remove impurities and pH should be adjusted to the optimal range for the specific microbial growth. The molasses is then inoculated with a selected strain of bacteria or archaea that are capable of producing PHAs. The microorganisms utilize the sugar components in the molasses as a carbon source to synthesize PHAs, which can then be harvested and purified for various applications.

Many studies have reported successful production of PHAs from molasses using different bacterial strains. For example, Sen et al. (2019) reported the production of PHAs using *Cupriavidus necator* and pre-treated molasses as the sole carbon source. The molasses consists of 34.49 ± 0.13% sucrose, 5.93 ± 0.07% glucose and 5.00 ± 0.04% fructose respectively. The study found that the cell dry weight (CDW) and PHAs produced by *C. necator* using hydrothermal acid pretreated molasses is the highest which are 2.86 ± 0.82 g and 27% respectively.

Albuquerque et al. (2010) evaluated the production of PHAs utilizing molasses feedstock in various concentration ranging from 30-60 Cmmol VFA/L (volatile fatty acid/L) using a mixed microbial culture. The molasses substrate concentration of 45 Cmmol VFA/L showed the best PHA-storing capacity of 88% in the cells, with maximum PHAs content of 74.6%.

Another study by Dalsasso et al. (2019) investigated the use of a pure culture of *Cupriavidus necator* for PHB production from pre-treated sugarcane vinasse and molasses. The molasses was initially hydrolyzed by acid (HCl or H_2SO_4) or enzymatic procedures (invertase or baker's yeast), and the microorganism growth kinetics were compared, whereby enzymatic hydrolysis was chosen for subsequent tests based upon maximum specific growth rate. In the production stage, the molasses was added, reaching 11.7 g/L of PHB and 56% of PHB content accumulation. The nitrogen content in molasses approximately 1.8 g/L was accountable for an atypical increase in biomass concentration during the production phase, which renders this substrate beneficial, based on the cultivation condition utilized in the reported experiment condition.

The production of PHAs using molasses as a carbon source is an attractive option due to its low cost and widespread availability. Molasses is a by-product of the sugar industry and is readily available in large quantities. The low cost of molasses compared to other carbon sources such as pure glucose and sucrose makes it an attractive option for large-scale PHAs production. Furthermore, the use of molasses in PHAs production contributes to the reduction of waste generated by the sugar industry, making it an environmentally friendly option.

In addition to molasses, other sugar-based feedstocks such as banana trunk and frond extract (BTFE) have also been investigated for PHAs production through biovalorization. BTFE is a waste material generated during banana cultivation and processing that can be utilized as a low-cost substrate due to its abundancy and high sugar content. BTFE contains high amount of glucose, fructose, sucrose, and other simple fermentable sugars that can be utilized by microorganisms to produce PHAs. Several studies have reported the use of BTFE as a substrate for PHAs production using various microorganisms, including *Bacillus* sp., *Pseudomonas* sp., and *Cupriavidus* sp. (Low et al., 2021).

Low et al. (2021) reported that the total amount of sugar in the banana frond extract (BFE) was 42.7 g/L, which in part consist of 10.5 g/L fructose. The effect of various BFE concentration on the PHB production was determined by conducting the fermentation utilizing BFE ranging from 10-50% (v/v) at 30 oC, 200 rpm for 72 hours of incubation. The authors reported that fermentation with 40% (v/v) BFE resulted in the highest PHB content in biomass of 32.1%) and PHB concentration of 0.9 g/L. Furthermore, the production of PHB was also examined by comparing the fermentation with 40% (v/v) raw and sucrase enzyme pre-treated BFE, respectively. The fructose content in the enzyme pre-treated BFE after enzymatic hydrolysis was increased to 14.6 g/L. The fermentation with 40% (v/v) pre-treated BFE showed higher PHB content in biomass (37.4%) and PHB concentration (1.3 g/L) as compared to the 40% (v/v) original BFE. The result indicates that 40% (v/v) of enzyme pre-treated BFE can be utilized as an alternative renewable and sustainable carbon feedstocks for PHB production.

Vijay and Tarika (2018) conducted an experiment using locally isolated bacteria from soil sample at landfill area in Maharashtra, India and added banana peel as carbon source with yeast extract in the fermentation medium. The authors prepare various ratio of carbon and nitrogen source ranging from 1:1, 2:1, 3:1, and 4:1, and fermentation conducted for 96 hours. *Geobacillus stearothermophilus* R-35646 accumulated the highest PHAs with 84.6%, 71.58%, and 62.68% of PHAs for 4:1, 3:1, and 2:1 C:N ratios respectively. *Bacillus subtilis* JCM 1465 accumulated 71.78% and 62.23% of PHAs for 4:1 and 3:1 C:N ratios respectively at 24 hours fermentation. *Bacillus siamensis* PD-A10 accumulated 77.55% at 24 hours, 69.70% at 72 hours, and 65.75% at 72 hours of incubation with 3:1, 3:1, and 4:1 C:N ratios respectively. However, the utilization of BTFJ as a substrate for PHAs production also presents some challenges. One of the main challenges is the presence of inhibitors such as phenolic compounds, which can affect microbial growth and PHAs production. Overall, the authors highlighted that the banana peels proved to be a suitable substitute for the carbon source which is available easily at free of cost and able to supports PHAs production up to considerable amount of levels by almost all selected bacterial isolates. Therefore, the problem of waste disposal is addressed by producing a new value-added product of PHAs.

Seafood processing industry: crustacean waste such as exoskeleton of shrimp and crab

Sustainable Development Goals (SDG) 14 highlights the importance of conserving and sustainably use the oceans, seas, and marine resources. The ocean absorbs approximately one quarter of the world's annual carbon dioxide (CO_2) emissions, thereby mitigating climate change and alleviating its impacts. Various initiatives are taken to further enhance the conservation and sustainable use of oceans and their resources. Among the initiatives is to utilize the wastes generated in seafood processing industry.

Recently, there has been growing interest in utilizing crustacean waste, such as shrimp, prawns, and crab exoskeleton as a source of chitin, which can be exploited as feedstocks for chitosan biofilm production. Among the application of chitin and chitosan are as biomaterials and devices applicable to repair, replace, or strengthen the living tissues and organs of the human body. Generally, chitin is extracted using either biological or chemical extraction. Both extraction methods involve the process of isolation, demineralization, and deproteinization of chitin. However, the chemical extraction method produce chitin with higher degree of purity (Fernando et al., 2016). Subsequently after chitin extraction, polymer film formation is conducted by mixing chitin with dissolution solvent and moulded into film via cold-pressing (Fernando et al., 2016). Akin to chitin, chitosan biofilms are also biodegradable, non-toxic, edible, and renewable (Fernandez & Ingber, 2014). Taking advantage of its soluble properties, chitosan is used in various application forms such as solutions, gels, films, and fibres. Chitosan biofilm production involves purification steps whereby chitin is dissolved in acid and filter through membranes, followed by chemical adjustment of altering the pH to approximately 7.5, and final steps includes rinsing and drying processes (Rinaudo, 2006).

For example, chitin from crab, crayfish, and shrimp shells was isolated with sodium hypochlorite (NaClO) for 10 minutes, prior demineralization and deproteinization procedures. This treatment brings the yield of chitin approximately 13.4% for crab, 15.3% for crayfish, and 14.8% for shrimp shells (Kaya et al., 2015).

In another approach, Chakravarty et al. utilized protease producing bacteria such as *Bacillus megaterium* NH21 and *Serratia marcescens* Db11, together with organic acid-producing bacterium of *Lactobacillus plantarum* to ferment lobster shell wastes. A combination of *Serratia marcescens* Db11 and *Lactobacillus plantarum* resulted high chitin yield of approximately 82.56% from lobster biomass with high efficiency of total deproteinization of 87.19% and demineralization of 89.59% (Chakravarty et al., 2018).

Generally, the fermentation process of chitin is considerably high as compared to chemical and enzymatic techniques, however, the extraction cost can be lowered by exploiting agro-wastes as the carbon source for microbial fermentation (Gadgey & Bahekar, 2017).

Food processing industry: wastewater, palm oil wastes, fruit peels, bakery and pastries waste, etc.

The wastewater from the food processing industry is another example of waste that can be utilized for PHAs production. Most industrial wastewater from the food industry is abundant source of organic carbon that can be recovered in the form of PHAs. However, the wastewater composition varies considerably among the different industries. For example, the wastewater from the yeast industry, contains high concentration in organic carbon, but may contains a high ammonium concentration which can lead to the decrease of PHAs production. Bhalerao et al. (2020) investigated the PHAs production using yeast industry wastewater through an enriched mixed microbial culture (MMC). The authors conducted continuous cultivation for PHAs production and compared to the widely used batch cultivation. PHAs accumulating MMC was enriched using a sequencing batch reactor (SBR) operated under aerobic dynamic feeding. An MMC dominated by the *Thauera* species was successfully enriched in the SBR. Their experimental results showed that ammonium is required for PHAs accumulation, and the complete absence of ammonium negatively affects the accumulation process. Using wastewater, batch, and continuous feeding strategies, respectively, yielded the PHAs accumulation of 72% and 65% per CDW. Despite the slightly lower PHAs accumulation with continuous cultivation, four-fold increment of biomass growth accumulated. Thus, higher theoretical PHAs production (270 ton/year) can be expected using continuous cultivation in half of the reactor volume (45 m^3). Bhalerao et al. claims that the viability of continuous cultivation as a feasible investigatory tool and PHAs production approach.

In more detail an extensive study, Tamang et al. (2021) explored the PHAs production from several industrial wastewaters using a MMC enriched in bacteria of the genus *Thauera* sp. The authors classified the wastewaters based on the volatile fatty acid (VFA) concentration and the presence of ethanol; namely W1 for the wastewater with VFA <12 mM (anaerobically treated dairy, yeast and paper mill wastewaters), W2 refers to the wastewater with VFA >12 mM (acidified sugar, paper mill and brewery wastewaters), and W3 refers to the wastewater with VFA >12 mM and contains ethanol (acidified starch, distillery and dairy wastewaters). The authors reported that no PHAs accumulation was obtained with wastewaters in the category W1 due to the VFA concentration was too low and only utilized for biomass growth and maintenance. The maximum PHAs accumulation of 46.5% CDW was achieved from acidified paper mill wastewater (W2). The PHAs accumulation was in the range of 37.0–45.6% CDW for the remaining type of wastewaters in the categories W2 and W3. The PHAs accumulation was inhibit due to the VFA concentration in the W2 category, whilst inhibition in the W3 category triggered by the wastewater matrix itself. The authors suggested to overcome this inhibition by conducting wastewater dilution. The presence of ethanol in wastewaters category W3 of approximately more than 27 mM led to the decrease of PHAs accumulation circa 20%. The thermal properties of the copolymer poly(hydroxybutyrate-*co*-hydroxyvalerate) (P(HB-HV)) obtained in the present study are similar to those of commercial PHAs. Furthermore, the produced copolymer was successfully utilized as a carbon source for nitrate removal (0.72 mg NO/mg PHAs). Tamang et al. (2021) emphasize that resource recovery in the form of biopolymers is viable option for several industrial wastewaters.

The waste cooking oil from food industry is a low-cost carbon source that can be utilized for the production of various biobased products, including PHAs. In a study by Ruiz et al. (2019), the authors developed a high cell density bioreactor-based process for the production of medium chain length PHAs (mcl-PHAs) utilizing waste cooking oil as the sole carbon and energy source. *Pseudomonas chlororaphis* 555 recorded high biomass of 73 g/l medium and considerably high biomass yield (including intracellular PHAs) of 0.52 g/g substrate. *P. chlororaphis* 555 accumulated 13.9 g mcl-PHA/L and achieved polymer productivity of 0.29 g mcl-PHA/L/h. The mcl-PHA contained predominantly (*R*)-3-hydroxyoctanoic acid and (*R*)-3-hydroxydecanoic acid monomers, with a high fraction of (*R*)-3-hydroxydodecanoic acid monomers.

Palm oil waste is among the most promising biofeedstocks for PHAs production, which is generated in large quantities by the palm oil industry, especially in Indonesia and Malaysia. Palm oil is one of the most widely produced and consumed vegetable oils in the world, with an estimated global production of 77.5 million tonnes in 2020. However, the production of palm oil generates voluminous amount of waste, including empty fruit bunches, palm kernel shells, and palm oil mill effluent (POME). POME is a particularly problematic waste stream, as it contains high levels of organic pollutants and nutrients that can cause environmental problems if not properly treated.

One approach to mitigating the environmental impact of POME is to convert it into a valuable product, such as PHAs. Several studies have investigated the production of PHAs using POME as a carbon source. For example, Mozejko and Ciesielski (2013) successfully synthesize mclPHAs by *Pseudomonas* sp. Gl01 using saponified waste palm oil (SWPO) as the sole carbon source. More than 43% of mcl-PHAs were produced at 17 hours fermentation time, when Pseudomonas sp. Gl01 was grown in a biofermentor containing 15 g/l of SWPO. The authors reported that lower carbon source supplementation will decrease the mcl-PHAs production. Moreover, the authors also explore the mcl-PHAs biosynthesis at the molecular level. The expression of PHA synthase genes (*phaC1* and *phaC2*) and PHA depolymerase gene (*phaZ*) was analyzed by means of RT real-time PCR method. The obtained data indicate that the *phaZ* gene could be transcribed together with the *phaC1* or *phaC2* gene, which means that PHAs synthesis and degradation followed concurrently. The extracted and purified polymers comprised of monomers ranging from C6 to C16 with beneficial physical and chemical properties.

A study by Lee et al. (2015) explores the sustainable utilization of wastewater by resource recovery approach and transform wastes into valuable resources. The authors examine the biovalorization of fermented POME into PHAs through fermentation of PHAs accumulating organisms in activated sludge and the subsequent production of PHAs by the cultivated sludge. The enrichment of PHA-accumulating organisms via aerobic feeding process was reported efficient and had significantly enhanced the PHAs storage capacity of the sludge from 4 wt% (dry weight of seed sludge) to 40–64 wt% (dry weight of sludge cultivated for more than 50 days). By using fluorescent in situ hybridization, the cultivated sludge is identified to be consisted of 42±12% *Betaproteobacteria*, 35±7% *Alphaproteobacteria*, and 13±4% *Gammaproteobacteria*. The authors also investigated the influence of pH on the production of PHAs by the cultivated sludge. Neutral pH was the most favourable for PHAs production, resulting in a PHAs content of 64 wt% in 8 hours. The PHAs produced was made up of 77 mol% 3-hydroxybutyrate and 23 mol% 3-hydroxyvalerate. These findings imply that the combination of fermented POME and activated sludge offers an alternative to the palm oil and the plastics industries for a more sustainable POME management and create an economical PHAs production pathway.

The PHAs obtained from POME varies depending on the type of microorganisms used, the fermentation conditions, and the composition of the POME. However, the studies discussed above demonstrate that POME can be a promising substrate for PHAs production. In addition to mitigating the environmental impact of POME, the production of PHAs from this waste stream could also provide a valuable product that can be used in various applications, including packaging, agriculture, and medicine. Overall, PHAs are biodegradable polymers that have the potential to replace traditional petroleum-based plastics. Palm oil waste, including POME, is a promising biofeedstocks of PHAs, and several studies have proved the high potential production of PHAs using this waste stream. Further research is necessary to optimize the production of PHAs from POME and to develop cost-effective and scalable processes for their production.

Fruit peels are waste material that has been studied for PHAs production due to their abundance and high content of fermentable sugars. Fruit peels are rich in pectin, cellulose, and hemicellulose, which can be hydrolyzed into simple sugars and used as a carbon source for PHAs synthesis. Several studies have been conducted to investigate the potential of fruit peels for PHAs production. Orange peels are an abundant food waste stream that can be converted into useful products, such as PHAs. However, limonene, the major component in the oil of citrus fruitpeels, creates a significant hurdle to developing a successful biopolymer synthesis from orange peels due to its characteristics of hindering microbial development. Davaritouchaee et al. (2023) successfully created a one-pot oxidation system that extracts sugars from orange peels while removing limonene via superoxide ($O_2^{\bullet-}$) produced by potassium superoxide (KO_2). The ideal conditions were determined to be a one-hour treatment with 0.05 M KO_2, which recovered 55% of the sugars contained in orange peels. The orange peel sugars were then readily utilized as a carbon source by genetically modified *Escherichia coli* to produce PHB. The inclusion of orange peel liquor with 3 w/v% boosted cell proliferation and exhibiting 90-100% cell viability. Over a 24 to 96-hour fermentation period, the bacterial generation of PHB utilising orange peel liquor resulted in 1.7-3.0 g/L cdw and 136-393 mg (8-13 w/w%) ultra-high molecular weight PHB content (Mw = 1900 kDa). The extracted PHB thermal characterization indicated polymeric characteristics similar to PHBs derived from pure glucose or fructose. The authors reported that the one-pot oxidation process for separating sugars and removing inhibitory compounds is an efficient and simple method for separating sugars from orange peels and removing limonene, or residual limonene after limonene extraction, and it holds great potential for extracting sugars from other complex biomass feedstocks.

Liu et al. (2021) compiled a comprehensive review of the PHAs production from apple industrial waste residues. Apple pomace is the waste left over after apple processing, serves as a possible carbon source for the synthesis of PHAs. It includes an abundance of carbs, fibre, and polyphenols. Biovalorizing these waste resources offers two social benefits: trash management and waste conversion to an eco-friendly biopolymer. This reduces the process's overall profitability. The high cost of biopolymer in compared to petroleum-derived polymer is a key barrier to commercialization. Liu et al. provides an overview of the valorization of apple pomace for the synthesis of biopolymer, including the various methodologies used, limitations, and future prospects.

Dharani et al. (2022) concentrated on developing a sustainable method for producing PHAs by *Pseduomonas aeruginosa* and *Bacillus substilis* with pomegranate wastes as a carbon source. Pomegranate wastes was processed with water and acid hydrolysis to effectively produce PHAs and degrade lignin. When compared to water hydrolysis, acid hydrolysis results in better lignin breakdown. Under optimal temperature, pH, and substrate concentration, the authors reported that PHAs production by *Pseudomonas aeruginosa* (4.9 g/l) outperforms *Bacillus subtilis* (4.4 g/l). The produced PHAs was qualitatively examined using the Sudan Black Test and the Nile Blue Test before being validated using the FTIR spectroscopic technique. The PHAs yield from fruit peel waste varies depending on various factors, such as the type of microorganism used, the carbon source concentration, and the fermentation conditions. However, the literature review discussed above suggest that fruit peel waste can be a potential carbon source for PHAs production.

The bakery and pastries industry are among the significant contributor to the global food industry, generating a large amount of inedible food waste in the form of breadcrumbs and spoiled cakes, among others. This waste can be used as a potential substrate for PHAs production, as it contains significant amounts of fermentable sugars and other organic compounds that can serve as precursors for PHAs synthesis. Several studies have investigated the use of bakery and pastries waste for PHAs production.

In a study conducted by Pleissner et al. (2014), the authors reported a promising PHB production method by *Halomonas boliviensis* using bakery waste hydrolysate and seawater in batch and fed-batch fermentation [30]. Three methods for bakery waste hydrolysis were explored for feedstock preparation, which include: (1) use of crude enzyme extracts from *Aspergillus awamori*, (2) *Aspergillus awamori* solid mashes, and (3) commercial glucoamylase. In the first method, the resultant free amino nitrogen (FAN) concentration in hydrolysates was 150 and 250 mg/l after 20 hours at enzyme-to-solid ratios of 6.9 and 13.1U/g, respectively. In both cases, the final glucose concentration was around 130–150 g/l. In the second method, the resultant FAN and glucose concentrations were 250 mg/l and 150 g/l, respectively. In the third method, highest glucose and lowest FAN concentrations of 170–200 g/l and 100 mg/l, respectively, were obtained in hydrolysates after only 5 hours.

The study by Pleissner et al. (2014) demonstrates the potential of bakery and pastries waste as a substrate for PHAs production. However, there are several challenges associated with utilizing this waste, including its high moisture content, low pH, and high salt content. To overcome these challenges, pretreatment techniques such as drying, grinding, and hydrolysis may be necessary to enhance the availability of fermentable sugars and other organic compounds for PHAs synthesis. Moreover, additional research is required to optimize PHAs production from bakery waste and to develop cost-effective as well as scalable processes for industrial-scale production.

Biodiesel and oleochemical industry

One example of biomass waste that can be utilized for PHAs production is glycerol, a by-product of the biodiesel and oleochemistry industry. Glycerol is considered as a cheap carbon source for PHAs production (Kumar et al., 2015). *Bacillus thuringiensis* EGU45 was found to produce 1.5–3.5 g/l PHA from feed containing 1–10% crude glycerol (v/v) and 125 ml nutrient broth (NB). *B. thuringiensis* EGU45 could produce PHAs at the rate of 1.54–1.83 g/L, from 1% CG (v/v) on media with high nitrogen contents. Kumar et al. conducted three different fermentation conditions: (i) NB, (ii) NB + 0.5% NH_4Cl (w/v), and (iii) peptone + yeast extract + 0.5% NH_4Cl (w/v). *B.thuringiensis* EGU45 reported was able to produce co-polymer of P(3HB-*co*-3HV) with 13.4% 3HV content on high nitrogen containing feed supplemented with propionic acid. The study demonstrated that the *B. thuringiensis* able to convert crude glycerol into PHAs co-polymer under non-limiting nitrogen conditions.

Pulp and paper industry

The lignocellulosic waste from the pulp and paper industry is also a potential source of biomass waste for PHAs production. Examples of biomass containing lignocellulosic materials are coir pith (a by-product of coconut fibers), rice straw, wheat straw, and empty fruit bunch. These wastes contain cellulose, hemicellulose, and lignin, which can be utilized as a carbon source for PHAs production. Nevertheless, the utilization of lignocellulose material as biofeedstocks during fermentation processes represent a challenge due to the requirements of hydrolysis and detoxification. The hydrolysis is required due to yield fermentable sugars and detoxification is necessary to remove abundance antimicrobial agents in the lignocellulosic materials (Obruca et al., 2015).

Concept of carbon footprint mitigation by utilizing wastes as feedstocks for PHAs productions

Carbon footprint mitigation can be defined as the process of reducing or offsetting the amount of greenhouse gas (GHG) emissions, such as carbon dioxide and methane that are produced by an individual, organization, or activity, either directly or indirectly. The concept of carbon footprint mitigation has gained considerable attention in recent years due to the increasing global concern over climate change. According to the Intergovernmental Panel on Climate Change (IPCC), global carbon dioxide emissions need to be reduced by 45% prior 2030 and reach net-zero by 2050 to avoid the worst effects of climate change (*Global Warming of 1.5 °C*, 2018).

One of the major contributors to carbon emissions is the production and disposal of conventional petroleum derived plastics, which not only have adverse effects on the environment but also pose a threat to human health. PHAs are biodegradable, eco-friendly polymers that can replace conventional plastics. PHAs can be produced using waste materials as feedstocks, which not only mitigates carbon footprint but also reduces waste and pollution.

PHAs can decompose naturally in the environment and break down into water, carbon dioxide, and other natural substances. The utilization of biodegradable plastics such as PHAs can help to reduce carbon footprint in a few ways:

1. Reduced production emissions: The production of biodegradable plastics typically generates fewer GHG emissions compared to conventional petroleum-derived plastics. This is because plant-based materials used in biodegradable plastics exhibit low carbon footprint as compared to petroleum-based materials.

2. Reduced landfill emissions and wastes accumulation: When conventional plastics are disposed in landfills, they can take hundreds of years to decompose and release GHG such as methane. Biodegradable plastics, on the other hand, can break down rapidly and release fewer emissions, simultaneously reducing the amount of plastic waste in the environment.

In conclusion, the utilization of biodegradable plastics can help to reduce carbon footprint by reducing emissions during production and disposal, and by reducing the amount of plastic waste in the environment. However, it is important to note that biodegradable plastics are not a complete solution to the problem of plastic waste and should be used in conjunction with other waste reduction initiatives and recycling efforts.

ACKNOWLEDGEMENT

The authors gratefully acknowledge the financial support from the Malaysian Ministry of Higher Education to the Universiti Sains Malaysia (USM), through the Fundamental Research Grant Scheme (FRGS), with the project code of FRGS/1/2021/STG01/USM/02/12.

REFERENCES

Albuquerque, M. G. E., Torres, C. A. V., & Reis, M. A. M. (2010). Polyhydroxyalkanoate (PHA) production by a mixed microbial culture using sugar molasses: Effect of the influent substrate concentration on culture selection. *Water Research*, 44(11), 3419–3433. DOI: 10.1016/j.watres.2010.03.021 PMID: 20427069

Alper, R., Lundgren, D. G., Marchessault, R. H., & Cote, W. A. (1963). Properties of poly-β-hydroxybutyrate. I. General considerations concerning the naturally occurring polymer. *Biopolymers*, 1(6), 545–556. DOI: 10.1002/bip.360010605

Bhalerao, A., Banerjee, R., & Nogueira, R. (2020). Continuous cultivation strategy for yeast industrial wastewater-based polyhydroxyalkanoate production. *Journal of Bioscience and Bioengineering*, 129(5), 595–602. DOI: 10.1016/j.jbiosc.2019.11.006 PMID: 31836378

Chakravarty, J., Yang, C. L., Palmer, J., & Brigham, C. J. (2018). Chitin extraction from lobster shell waste using microbial culture-based methods. *Applied Food Biotechnology*, 5(3), 141–154. DOI: 10.22037/afb.v%vi%i.20787

Dalsasso, R. R., Pavan, F. A., Bordignon, S. E., de Aragão, G. M. F., & Poletto, P. (2019). Polyhydroxybutyrate (PHB) production by Cupriavidus necator from sugarcane vinasse and molasses as mixed substrate. *Process Biochemistry (Barking, London, England)*, 85, 12–18. DOI: 10.1016/j.procbio.2019.07.007

Davaritouchaee, M., Mosleh, I., Dadmohammadi, Y., & Abbaspourrad, A. (2023). One-Step Oxidation of Orange Peel Waste to Carbon Feedstock for Bacterial Production of Polyhydroxybutyrate. *Polymers*, 15(3), 697. Advance online publication. DOI: 10.3390/polym15030697 PMID: 36771998

Dharani, D., Jayashree, M., Shanmugapriya, K., Sathya, R., Gayathri, D. M., & Ruthra, S. (2022). Utilization of Pomegranate Waste for The Production of Biopolymer. *International Journal of Progressive Research In Engineering Management and Science*, 2(5), 432–439.

Fernandez, J. G., & Ingber, D. E. (2014). Manufacturing of large-scale functional objects using biodegradable chitosan bioplastic. *Macromolecular Materials and Engineering*, 299(8), 932–938. DOI: 10.1002/mame.201300426

Fernando, L. A. T., Poblete, M. R. S., Ongkiko, A. G. M., & Diaz, L. J. L. (2016). Chitin Extraction and Synthesis of Chitin-Based Polymer Films from Philippine Blue Swimming Crab (Portunus pelagicus) Shells. *Procedia Chemistry*, 19, 462–468. DOI: 10.1016/j.proche.2016.03.039

Gadgey, K. K., & Bahekar, D. A. (2017). Studies on Extraction Methods of Chitin From Crab Shell and Investigation of Its Mechanical Properties. *International Journal of Mechanical Engineering and Technology*, 8(2), 220–231.

Global Warming of 1.5 ºC. (2018). Intergovernmental Panel On Climate Change (IPCC), United Nations. https://www.ipcc.ch/sr15/

Gowda, V., & Shivakumar, S. (2019). Novel biocontrol agents: Short chain fatty acids and more recently, polyhydroxyalkanoates. In *Biotechnological Applications of Polyhydroxyalkanoates* (pp. 323–345). Springer Singapore., DOI: 10.1007/978-981-13-3759-8_12

Kaya, M., Baran, T., & Karaarslan, M. (2015). A new method for fast chitin extraction from shells of crab, crayfish and shrimp. *Natural Product Research*, 29(15), 1477–1480. DOI: 10.1080/14786419.2015.1026341 PMID: 25835041

Koller, M., Bona, R., Braunegg, G., Hermann, C., Horvat, P., Kroutil, M., Martinz, J., Neto, J., Pereira, L., & Varila, P. (2005). Production of polyhydroxyalkanoates from agricultural waste and surplus materials. *Biomacromolecules*, 6(2), 561–565. DOI: 10.1021/bm049478b PMID: 15762613

Koller, M., & Braunegg, G. (2018). Advanced approaches to produce polyhydroxyalkanoate (PHA) biopolyesters in a sustainable and economic fashion. *The EuroBiotech Journal*, 2(2), 89–103. DOI: 10.2478/ebtj-2018-0013

Koller, M., Salerno, A., Reiterer, A., Malli, H., Malli, K., Kettl, K. H., Narodoslawsky, M., Schnitzer, H., Chiellini, E., & Braunegg, G. (2012). Sugarcane as feedstock for biomediated polymer production. In *Sugarcane* (pp. 105–136). Production, Cultivation and Uses.

Kumar, P., Ray, S., Patel, S. K. S., Lee, J. K., & Kalia, V. C. (2015). Bioconversion of crude glycerol to polyhydroxyalkanoate by Bacillus thuringiensis under non-limiting nitrogen conditions. *International Journal of Biological Macromolecules*, 78, 9–16. DOI: 10.1016/j.ijbiomac.2015.03.046 PMID: 25840150

Lee, W. S., Chua, A. S. M., Yeoh, H. K., Nittami, T., & Ngoh, G. C. (2015). Strategy for the biotransformation of fermented palm oil mill effluent into biodegradable polyhydroxyalkanoates by activated sludge. *Chemical Engineering Journal*, 269, 288–297. DOI: 10.1016/j.cej.2015.01.103

Liu, H., Kumar, V., Jia, L., Sarsaiya, S., Kumar, D., Juneja, A., Zhang, Z., Sindhu, R., Binod, P., Bhatia, S. K., & Awasthi, M. K. (2021). Biopolymer poly-hydroxyalkanoates (PHA) production from apple industrial waste residues: A review. *Chemosphere*, 284, 131427. DOI: 10.1016/j.chemosphere.2021.131427 PMID: 34323796

Low, T. J., Mohammad, S., Sudesh, K., & Baidurah, S. (2021). Utilization of banana (Musa sp.) fronds extract as an alternative carbon source for poly(3-hydroxybutyrate) production by Cupriavidus necator H16. *Biocatalysis and Agricultural Biotechnology*, 34, 102048. DOI: 10.1016/j.bcab.2021.102048

Mozejko, J., & Ciesielski, S. (2013). Saponified waste palm oil as an attractive renewable resource for mcl-polyhydroxyalkanoate synthesis. *Journal of Bioscience and Bioengineering*, 116(4), 485–492. DOI: 10.1016/j.jbiosc.2013.04.014 PMID: 23706994

Nonato, R. V., Mantelatto, P. E., & Rossell, C. E. V. (2001). Integrated production of biodegradable plastic, sugar and ethanol. *Applied Microbiology and Biotechnology*, 57(1–2), 1–5. DOI: 10.1007/s002530100732 PMID: 11693904

Obruca, S., Benesova, P., Marsalek, L., & Marova, I. (2015). Use of lignocellulosic materials for PHA production. *Chemical and Biochemical Engineering Quarterly*, 29(2), 135–144. DOI: 10.15255/CABEQ.2014.2253

Ooi, W. C., Dominic, D., Kassim, M. A., & Baidurah, S. (2023). Biomass Fuel Production through Cultivation of Microalgae Coccomyxa dispar and Scenedesmus parvus in Palm Oil Mill Effluent and Simultaneous Phycoremediation. *Agriculture*, 13(2), 336. Advance online publication. DOI: 10.3390/agriculture13020336

Pleissner, D., Lam, W. C., Han, W., Lau, K. Y., Cheung, L. C., Lee, M. W., Lei, H. M., Lo, K. Y., Ng, W. Y., Sun, Z., Melikoglu, M., & Lin, C. S. K. (2014). Fermentative polyhydroxybutyrate production from a novel feedstock derived from bakery waste. *BioMed Research International*, 2014, 1–8. Advance online publication. DOI: 10.1155/2014/819474 PMID: 25136626

Rinaudo, M. (2006). Chitin and chitosan: Properties and applications. *Progress in Polymer Science*, 31(7), 603–632. DOI: 10.1016/j.progpolymsci.2006.06.001

Roland-Holst, D., Triolo, R., Heft-Neal, S., & Bayrami, B. (2013). *Bioplastics in California: Economic Assessment of Market Conditions for PHA/PHB Bioplastics Produced from Waste Methane.* https://bearecon.com/portfolio-item/calepa-bioplastics/

Ruiz, C., Kenny, S. T., Narancic, T., Babu, R., & Connor, K. O. (2019). Conversion of waste cooking oil into medium chain polyhydroxyalkanoates in a high cell density fermentation. *Journal of Biotechnology*, 306, 9–15. DOI: 10.1016/j.jbiotec.2019.08.020 PMID: 31476332

Sen, K. Y., & Baidurah, S. (2021). Renewable biomass feedstocks for production of sustainable biodegradable polymer. In *Current Opinion in Green and Sustainable Chemistry* (Vol. 27). DOI: 10.1016/j.cogsc.2020.100412

Sen, K. Y., Hussin, M. H., & Baidurah, S. (2019). Biosynthesis of poly(3-hydroxybutyrate) (PHB) by Cupriavidus necator from various pretreated molasses as carbon source. *Biocatalysis and Agricultural Biotechnology*, 17, 51–59. DOI: 10.1016/j.bcab.2018.11.006

Tamang, P., Arndt, C., Bruns-Hellberg, J., & Nogueira, R. (2021). Polyhydroxyalkanoates production from industrial wastewaters using a mixed culture enriched with Thauera sp.: Inhibitory effect of the wastewater matrix. *Environmental Technology & Innovation*, 21, 101328. Advance online publication. DOI: 10.1016/j.eti.2020.101328

Vijay, R., & Tarika, K. (2018). Banana peel as an inexpensive carbon source for microbial polyhydroxyalkanoate (PHA) production. *International Research Journal of Environmental Aciences*, 7(1), 28–36.

Chapter 7
Chitosan and Its Biomass Composites in Applications

Truong Thi Cam Trang
Vietnam National University, Vietnam

Khoa Dang Nguyen
https://orcid.org/0000-0001-8831-989X
Van Lang University, Vietnam

ABSTRACT

Biomass-based materials have been received a lot of attractive attentions due to their renewablity and low-carbon emmsion during processings. Chitosan is a deactylated product from chitin, which is the second most found bio-polymer in nature. Moreover, with the transformation of science and technology as well as the demand of society, chitosan-based composites have been extensively studied to adopt the changes towards sustainable development. Therefore, this chapter summarizes the existing researches of chitosan-based composites in different fields including biomedical applications, degradable food-packaging material and envrionmental remediation.

INTRODUCTION

In recent years, environmental issues such as climate change are one of the serious concerns caused by the increase in green-house-gas emission (Noohian & Mahmoudi, 2023). In particular, carbon dioxide, which is a significant contributor to global warming, has shown a dramatic increase over the past 20 years due to the demands of human consumption and the use of various materials (Kabir et al., 2023) Therefore, resilient efforts to create environmentally friendly materials based on green technologies and renewable sources are crucial for building sustainable development (Mikunda et al., 2021).

Regarding waste-based raw materials, the global consumption of seafood has been steadily increasing in response to the demands of each country (Boenish et al., 2022). It was reported that nearly 10 miliion tons of crustaceans, mostly shrimpt, prawns, crabs and crayfish are sold annually (Research, 2016). Consequently, the waste and by-products generated by this substantial industry can be considered as a significant source for chitin extraction, such as the shells of crustaceans. In the consumer market, the global consumption of chitin is growing at an average rate of 6.8%. Additionally, the worldwide chitin market is projected to increase from 45 million to 59 million US dollars between 2019 and 2024 (More, 2019).

Chitin, which is the second most abundant polysaccharide in nature, is found in crustacean shells, insects, and fungal cell walls (see Figure 1). As a biopolymer material, chitin possesses a wide range of functional properties (Hajji et al., 2015). However, its limited solubility in common solvents hampers its applications due to high crystallization and strong intra-/intermolecular hydrogen bonds (Gong et al., 2016), (Jardine & Sayed, 2018). Therefore, a conversion method called deacetylation has been employed to prepare chitosan from chitosan through chemical (Younes & Rinaudo, 2015) or biological processes (Schmitz et al., 2019). Chitosan is a cationic linear polysaccharide with multiple reactive amino and hydroxyl groups and consists of randomly distributed N-acetyl-D-glucosamine and β-(1–4)-linked D-glucosamine units (Sadiq et al., 2023). Due to its excellent properties such as biocompatibility, biodegradability, non-toxicity, and adsorption capabilities, chitosan has gained considerable attention for various applications, including food packaging, medical and pharmaceutical fields, drug delivery, and environmental remediation (Antonino et al., 2017) (Figure 1). Recently, the combination of polymer hydrogel networks with nanoparticles (metals, non-metals, metal oxides, and polymeric moieties) has shown promise in providing enhanced functionality to composite materials (Thoniyot et al., 2015). This chapter provide a summary of the preparation of chitosan composites and their utilization in various fields.

Figure 1. Flowing chitosan production process to its potential application from biomass sources through extraction process to chitin and extractionto of chitosan with deacetylation to chitosan

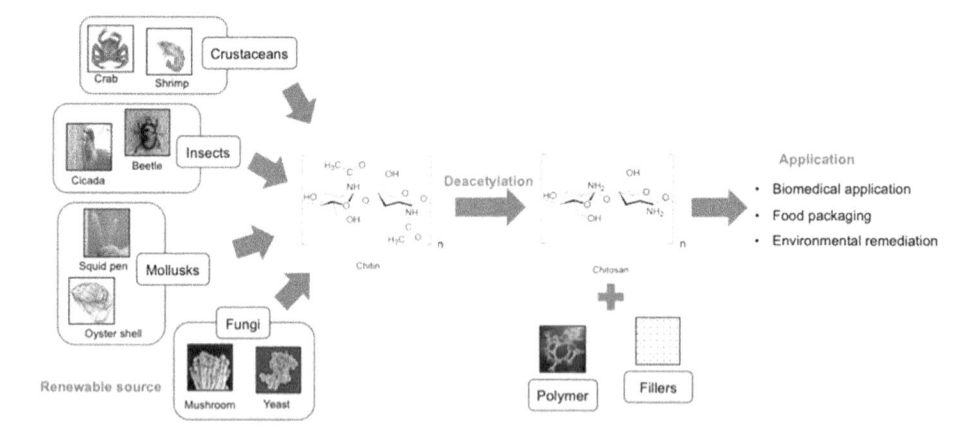

CHITOSAN COMPOSITES FOR THE BIOMEDICAL APPLICATIONS

Biopolymers have been globally recognized as promising materials for tissue engineering development (Thambiliyagodage et al., 2023). Chitosan is known as a naturally benign material for tissue engineering due to its outstanding properties, such as flexible modification, non-toxicity, biocompatibility, biodegradability, and regulation of drug release (Kim et al., 2023). Therefore, there is no doubts that chitosan-based materials have been an interesting candidate for the drug delivery system based on the instinctly excellent chracteristics. Hybrid hydrogels were developed to precisely control drug release, utilizing gamma irradiation as an environmentally friendly method to create high-molecular-weight chitosan and polyvinyl alcohol (PVA) hydrogels without the use of toxic chemicals. The resulting composite hydrogel exhibited significant drug release capabilities of amoxicillin drug due to its porous structure. *In vitro* drug release studies highlighted the influence of solution pH and release media on antibiotic drug release, following a Non-Fickian diffusion mechanism for swelling and drug release. These gamma-irradiated chitosan/PVA hydrogels are suggested as potential carriers for medication delivery applications (Tran Vo et al., 2022). Chen et al. reported that chitosan-based selenium composites showed great potential as carriers for inorganic, organic, and nanoselenium (Chen, Li, et al., 2022). Several studies indicated that selenium loading could be effectively controlled to improve selenium bioavailability (Zhang et al., 2011) as well as antioxidant and anticancer activity (L. Liu et al., 2019). Through chemical reduction at different temperatures and durations, chitosan-selenium nanocomposites have been found with sizes ranging from 43.2 to 83 nm (Chen, Cheng, et al., 2022; Chen et al., 2018). However, when citrate was added to the chitosan polymer matrix, selenium nanoparticle-loaded CTS/citrate complexes showed an increase in size from 1 to 30 µm, while retaining antioxidant activity (Bai et al., 2017). It is known that pH responce of chitosan-based composite in aqueous solution could be an important indicator. The pH response of chitosan-based composites in aqueous solutions is an important indicator. A pH-responsive, magnetic, biocompatible chitosan-based nanocomposite carrier for the release of ciprofloxacin was prepared by Panrinaz Jabbari et al. (Jabbari et al., 2023). Through the co-precipitation of Fe^{3+}/Fe^{2+} ion, the nano clay Montmorillonite was magnetized and incorporated into a chitosan matrix crosslinked with citric acid. This prepared nanocomposite carrier demonstrated high encapsulation efficiency for ciprofloxacin, up to 98%. The release rate in a neutral environment was significantly higher than in acidic media. Other pH-responsive hydrogel loaded with magnetic nanocomposite in based on κ-carrageenan/chitosan/silver nanoparticles were found to be effective in antibacterial carriers for cancer treatment. The loading capacity of the anticancer drug Doxorubicin in the nanocomposite hydrogels was approximately 98% (Fathi & Mohammadi, 2023). As reported, the release amount was significantly controlled on the dependent of pH which had the higher drug release at pH 5. In another study, novel carboxymethylcellulose/ZnO/chitosan hydrogel microbeads loaded with curcumin were designed to enhance the encapsulation efficiency of curcumin for targeted drug delivery to the small intestine (Y. Li et al., 2023). The swelling ratio of the composite hydrogel microbeads was strongly influenced by solution pH, being about 25% at pH 1.2 and sharply increasing to 320% - 370% at pH 6.8 and 7.4. The addition of ZnO nanoparticles to the composite hydrogel microbeads aimed to improve mechanical properties, antibacterial properties, stability, and modulate the drug release rate, achieving a 91.7% release of curcumin. It is worth noting that chitosan is soluble in dilute acidic solutions below pH 6.0 (Goñi et al., 2017). To address this issue, chitosan derivatives were functionalized into carboxymethyl chitosan, enabling water solubility. Ionic-crosslinking carboxymethyl chitosan -gelatin composite hydrogels with varying loads of graphene oxide (GO) exhibited excellent swelling properties, high biocompatibility, and drug delivery efficiency (C. Li et al., 2023). Increasing the GO content up to 10% in the composite hydrogel composition resulted in the highest cumulative release rate of Ibuprofen, at 74.6% and 89.1% at pH 6.5 and 7.4 of PBS, respectively. The drug release kinetics suggested that GO played a vital role in the controlled drug release from the composite hydrogel. As chitosan-based reduced GO-CeO_2 nanocomposite was prepared, the presence of GO improved drug release efficiency (Sanmugam et al., 2023). The results showed that the percentage of drug release in the case of the chitosan-rGO-CeO_2 sample was higher than those of the other samples as the function of time. After 72 h to archieve the release plateaus, the regulated drug release ratios were measured as 218.7%, 214.9%, and 198.4% for chitosan-rGO-CeO_2, chitosan-rGO, and chitosan composites, respectively. Chitosan-based nanocomposites have been strongly considered as excellent candidates for potential anticancer activities through diverse mechanisms, encompassing enzymatic, anti-angiogenic, immune-enhancement, and apoptotic pathways (Adhikari & Yadav, 2018). These bioactive compounds targeted cancer cells while sparing healthy ones, ensuring focused and sustained release. Moreover, magnetic properties facilitate systemic distribution when administered intravenously, thereby reducing the risk of toxicity throughout the body, making magnetic thermotherapy, for example, a potentially useful method of cancer treatment (Yusefi M, 2023). Coating magnetic nanoparticles with a chitosan layer enhanced the targeting of cancer cells for hyperthermia treatment (Dhavale et al., 2021) and chitosan-templated magnetic nanoparticles exhibited promising uses in cancer cell treatments (Arias et al., 2012).

Antimicrobial Activity of Chitosan Materials

Many materials incorporating chitosan are known to exhibit excellent antimicrobial properties. The following are some examples. Chitosan-based composites exhibited antimicrobial activity against a wide range of microorganisms, including yeasts, filamentous fungi, as well as both gram-negative and gram-positive bacteria (Kumar et al., 2020; Yilmaz Atay, 2019). Notably, degree of deacetylation (DD) emerged as the most influential factor enhancing or decreasing the antibacterial performance of chitosan-based composites (Bhardwaj et al., 2020). Furthermore, the findings of Rodrigues and his team highlighted that these biosynthetic materials exhibited strong antimicrobial potential due to the conversion from proton to ammonium under acidic conditions (Rodrigues et al. 2020). The mucoadhesion of chitosan can be improved by increasing the degree of deacetylation (DD), resulting in a higher number of positive charges of the amino group backbone, directly relating to its hemostatic activity. Thus, chitosan can interact with the negatively charged parts of cell membranes and then lead to the reorganization, opening tight junction proteins and enhancing the permeation properties of this polysaccharide (Kołodziejska et al., 2021). After completing experiments with the 125I-fibronectin protein, it was hypothesized that protonated amino groups derived from glucosamine units of chitosan for modulate cell adhesion and ehnaces the adsorption of adhesion proteins to the cell membrane, such as adsorption of an adhesive protein, fibronectin (Fn). Therefore, higher DD in chitosan enhanced the Fn surface concentration in both single protein solution and diluted serum. The rat bone marrow stromal cells cultured in the presence of 96%-DD of chitosan showed higher level of osteogenic differentiation than on the control of lower DD samples (Amaral et al., 2005), indicating processes of cell adhesion, migration, and proliferation on chitosan matrices.

Additionally, chitosan-zinc composite exhibited antibacterial efficacy compared to the pristine chitosan (Khan et al. 2013) and the combination of chitosan with ZnO nanoparticles presented excellent anti-bacterial and anticancer properties. These displayed effectiveness against both Gram-positive and

Gram-negative bacterial pathogens, along with human cancer cell lines such as HeLa and MCF-7 (Ahmad et al. 2019). Furthermore, in composites of metals and chitosan antibacterial application was exhibited (Kumar-Krishnan et al. 2015; Karthikeyan, Nithya, and Jothivenkatachalam 2017; Pinto et al. 2012). In other cases, the following studies were conducted in chomposites with chitosann materials and inorganic materials; chitosan-silver nanocomposite films against S. aureus and *E. coli* (Raghavendra et al. 2016*)*, chitosan films combined with Ag nanoparticles (Dai et al. 2022), cellulose/chitosan and copper oxide (CuO) nanoparticles for bacteria and fungi while maintaining compatibility with human fibroblasts (Tran et al. 2017), chitosan-Cu@Ag nanocomposite gauged against *E. coli* and Bacillus cereus (Mallick et al. 2015), nanocomposite composed of chitosan and $AgIO_3$ for several pathogenic microorganisms of *P. aeruginosa, K. pneumoniae, S. saprophyticus, E. Coli*, and *S. aureus* (Ahghari et al. 2022) .

In the other hand, many photocatalysts showed antibacterial candidates (Qian, Su, and Tan 2011; Karthikeyan, Nithya, and Jothivenkatachalam 2017; Zhang et al. 2017). As used as various types of membranes to assess their effectiveness in preventing the growth of bacteria and fungi, a blend of chitosan-1% TiO_2 demonstrated resistance against a range of bacteria (*S. aureus, E. coli, P. aeruginosa, and S. Typhimurium*) and fungi (*Aspergillus and Penicillium*) (Sripatrawan et al., 2018). Conversely, samples without UV-light exposure exhibited the continued bacterial growth. When subjected to UV light, both the chitosan membrane and the chitosan-TiO_2 composite displayed a remarkable inhibitory effect on bacteria and fungi. This outcome stemmed from the fact that these membranes generated various forms of reactive oxygen species upon light exposure, including O_2^- and OH radicals, which loweried the functionality of cell membranes of bacteria and played a pivotal role in impeding their growth (Siripatrawan and Kaewklin, 2018).

Figure 2. Illustration of enhancing cell-proliferation on chitosan-based composite scaffold for processes of cell adhesion, tissue integration and appearances of biocompatibility, antimicrobial and cell proliferation

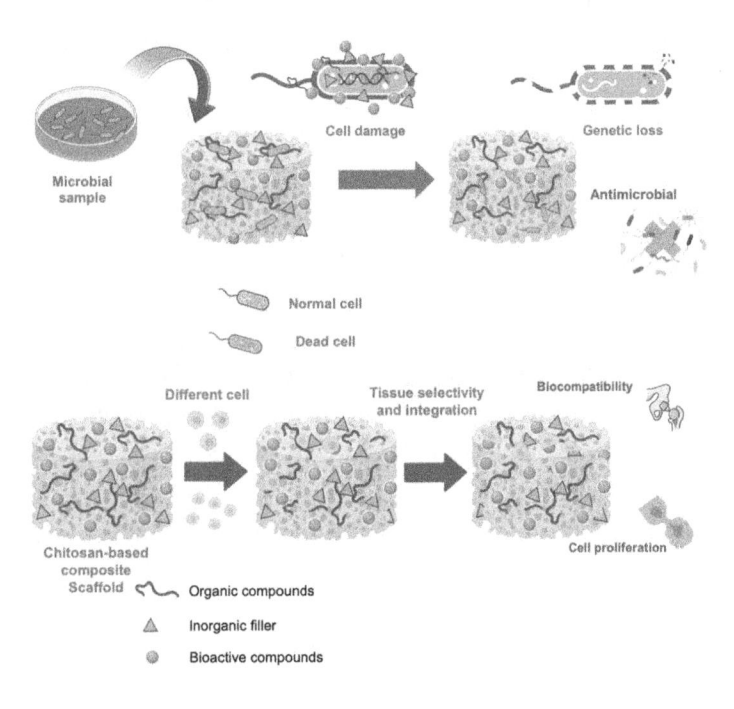

Effect of Chitosan on Cell Growing

Besides bacteria, effects on cells have also been reported. Different cell types have successfully grown on scaffolds made from chitosan-based composites. Figure 2 presents the enhancement in different cell-proliferation of chitosan-based composite.

A bioactive scaffold with preferred properties for bone tissue engineering has been reported as polycaprolactone-based polyurethane urea/chitosan composites used for the investigation of human bone marrow mesenchymal stem cells (hBMSCs) into osteoblasts in term of growth, adhesion, and differentiation (Amiryaghoubi et al., 2022). The composites promoted calcium level formation and alkaline phosphatase activity of hBMSCs, when cultured on the polymer scaffold in both the growth and osteogenic medium. When observed with the presence of osteocalcin genes, the addition of chitosan into the modified polyurethane urea could promote the hydrophilicity for enhancing the cell viability. In addition, alizarin red staining indicated the enhancement of extracellular calcium deposition of the hBMSCs cultured on the polymer scaffold.

Collagen-blended chitosan membranes provided a good attachment surface for rat hepatocytes survived in *in-vitro* experiments for at least 14 days (Murat et al., 1998). Optimal attachment was $3 - 4 \times 10^4$ cells/cm^2 after 3 h of incubation when seeding media was prepared from medium-molecular-weight chitosan films. The biomimetic core-shell nanofibers composed of hydroxyapatite/gelatin-chitosan was used as composite scaffolds and investigated for biocompatibility and osteogenesis for bone tissue engineering (Chen et al., 2019). The chitosan core-shell nanofibers formed like RGD(Arginine-Glycine-Aspartic acid) structure to mimic the organic component of natural bone extracellular matrix, resulting the Human osteoblast (MG-63) cell proliferation was enhanced compared to other samples. It indicated that the materials exhibited better biocompatibility, and were suitable for cells growth. Also, in 3D-hydroxyapatite–chitosan using a natural crosslinking with genipin, the scaffolds for bone tissue engineering applications were developed (Zafeiris et al., 2021). The cytocompatibility showed that viability of the MG63 human osteosarcoma cells seeding on surface of the prepared materials increased from 24 h to 7 days in comparison with the corresponding of the tissue culture in polystyrene surfaces. The produced 3D printed hybrid scaffolds are promising candidate for the regeneration of damaged and/or defected tissue, hence allowing for the rapid patient's healing.

Carbon-based materials such as Graphene oxide/ reduced Graphene oxide (GO/RGO) have been known as effective fillers for enhancing not only the mechanical strength but also the osteoinductive potential for biomedical application. The preparation of Xylan/chitosan/ hydroxyapatite matrix in the presence of GO/RGO for bone tissue application was proposed. The results of protein adsoprtion indicated that the addition of GO in the X/CTS/HAP was more effective than those of RGO and attributed on the dose-dependence of these inorganic fillers in the mixture. The MG-63 cell attachment was observed in all the scaffold samples cultured for 72 h, followed by the increase in Alkaline phosphatase activity over a period of 7 and 14 days which was observed in the samples with GO and RGO in comparison to the composite.

CHITOSAN COMPOSITES FOR FILM APPLICATIONS

Alternative Film to Petroleum-Derived Plastics

Plastic is a widely used material for food and beverage packaging, which accounted for the majority of the global market's revenue at 52% in 2022 (Research, 2022). The packaging and processed food manufacturing industries have expanded due to shifts in consumer lifestyles and preferences, likely increasing the popularity of plastic packaging. Plastic offers good performance for single-serve consumer packaging due to its convenience.

It has been known that synthetic plastics were derived from fossil fuels including non-degradable polymers like polyethylene (PE), polypropylene (PP), nylon, polyester (PS), polytetrafluoroethylene (PTFE), and epoxy (Nagalakshmaiah et al., 2019). During the plastic packaging production process, considerable resources such as energy, water, chemicals, and petroleum are utilized, often leading to air emissions including greenhouse gases, heavy metals, and particulates, as well as wastewater and/or sludge containing toxic contaminants. Unfortunately, most packaging is designed for single-use and is typically discarded rather than being reused or recycled. As a result, a significant amount of plastic packaging materials is discarded in the same year they are produced. This, combined with the large volume of packaging material and the limited recyclability of plastics in many regions worldwide, has created a significant solid waste issue, particularly when plastics are not properly collected by waste management systems and end up in aquatic ecosystems (Kan & Miller, 2022; L. Wang et al., 2022). As given for the finite nature of petroleum reserves and their depletion, the development of alternative materials that serve the same purpose as conventional plastics is highly necessary. Thus, degradable polymer-based plastics are excellent candidates as eco-friendly materials that can help improve the environment by reducing waste, consuming less energy, and lowering greenhouse gas emissions. Renewable sources such as plants or animals have been used to extract biopolymers and its derivates (Babaremu et al., 2023).

Figure 3. Food-packaging application of chitosan-based composite with different components

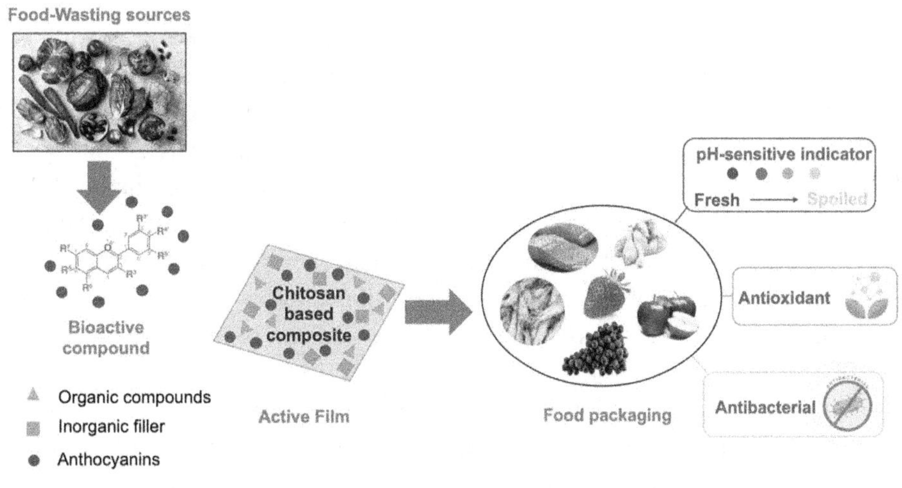

Chitosan Composite Films for Application of Food Packaging

Figure 3 shows the food-packaging applications of chitosan-based composite with different components. As seen, chitosan composite films holds activities for storing foods, when bioactive species is composed with chitosan. Fully biodegradable composites were developed using plasticized chitosan blended with varying amounts of hydrolyzed orange peel waste to produce films for active packaging of dry foods. The maximum elastic modulus value of about 400 MPa was found for the blend with 70% orange peel, more than twice the elastic modulus of plasticized chitosan (150 MPa). Notably, the plasticized chitosan /70% orange peel film exhibited an elastic modulus similar to that of LDPE (55 – 380 MPa) and HDPE (413 – 1490 MPa) and the tensile strength of the resulting sample was 20 MPa, comparable to that of HDPE (14.5–38 MPa). Strong interaction between chitosan and pectin molecules improved the mechanical and barrier properties of the composite materials. The water vapor permeability of the plasticized chitosan-citrus peel composite was 1.5×10^{-9} g/s.m.Pa, lower than that of quinoa starch films having 4.5×10^{-9} g/s.m.Pa (Galus et al., 2020) or pectin films of 2.2×10^{-9} g/s.m.Pa (Merino & Athanassiou, 2022). The composites with different orange peel contents had an excellent oxygen barrier properties, with O_2 permeability values, but did lower than those of potato peel-based films of 561.2×10^3 $cm^3.\mu m^2/$ $m^2.day.atm$ (Merino et al., 2021), carrot pomace-based flim of 928.5×10^3 $cm^3.\mu m^2/$ $m^2.day.atm$ (Perotto et al., 2018) and even PE materials as $50 – 200 \times 10^3$ $cm^3.\mu m^2/$ $m^2.day.atm$ (Michiels et al., 2017). Additionally, the developed composite provided high transparency, antioxidant capacity, and rapid biodegradability in soil, with a 62.9% weight loss after 26 days.

Biomass-based fillers showed potential as eco-friendly bioactive packaging candidates for food preservation. Chitosan/polyvinyl alcohol composite films incorporated with different loading amounts of biowaste orange peel in the range of 0.25–1.25 wt/wt% were introduced for food packaging applications (Terzioğlu et al., 2021). The results demonstrated that the physical properties of the prepared chitosan/PVA composite films improved with the addition of orange peel, enhancing flexibility, thermal stability, and water vapor permeability, while reducing transparency, hydrophobicity, and oxygen transmission rate. Furthermore, the addition of orange peel notably enhanced the antioxidant activity of the films compared to butylated hydroxyanisole and α-tocopherol, used as antioxidant standards. Chitosan-based films were reinforced with different micron-sized plant fibers, including wheat straw fiber, maize straw fiber, rice straw fiber, and sisal fiber, at 85 wt% (Ji et al., 2023), indicating that the tensile strength increased from 41.74 MPa for chitosan-based film to over 58 MPa for the fiber-containing composite samples. Also, the addition of 20 wt% micro ramie fiber in chitosan/glycerol-based film enhanced the tensile strength by 29.6%, while the same amount of lignin increased antioxidant activity by 288% and reduced water absorption by 41.2%. Both ramie fiber and lignin had a similar effect on thermal stability, raising the T_{max} of the composite film by more than 5°C. In food preservation experiments, the composite films showed better preservation meats such as chicken breast than cherry tomatoes. After 7 days of storage, the pH of chicken breast packaged using chitosan-based composite films was still within the appropriate range, and the number of growing microorganisms was significantly lower than that packaged in PE film (Ji et al., 2022).

Nanoparticles are often introduced into biodegradable films to improve its desired characteristics like antimicrobial, mechanical and gas barrier properties for food packaging, providing new substitutes to conventional packaging materials (das Neves et al., 2023; C. Wang et al., 2022). Chitosan-based films reinforced with 1.25 wt% zinc oxide nanoparticles and different concentrations of bamboo leaves extract (BLE) were applied for active food packaging (Liu et al., 2021). The addition of ZnO-nanoparticle significantly elevated the antimicrobial activity of chitosan film, resulting in higher antimicrobial activity of the films against *E. coli* and *S. aureus*. This effect was attributed to the interaction between Zn^{2+} ions released from ZnO nanoparticles and intracellular contents flowing out of destroyed cell membranes. Also, hydroxyethyl cellulose/carboxymethyl chitosan and zinc oxide nanoparticles were applied as a promising food-packaging material. This composite reduced water solubility by 94.3%, enhanced UV shielding ability by up to 45.73%, improved elasticity by 494.34%, and increased maximum load capacity by 142.24%, in addition with potent inhibition against *Listeria monocytogenes* and *Pseudomonas aeruginosa* due to the synergistic effect of carboxymethyl chitosan and ZnO-nanoparticles (Cen et al., 2023). ZnO-nanoparticls were also shown to be a promising filler for increasing the potential application of food packaging when it was added to the mixture of chitosan/guar gum with roselle calyx extract (Dehankar et al., 2023). Although the incorporation of the extract and 0.5 wt% ZnO-nanoparticles decreased film transparency, it enhanced UV–Vis light barrier properties and tensile strength (69.42 MPa), while decreasing elongation at break (9.69%). The addition of ZnO into the composite films also reduced water vapor permeability by 13%. The chitosan composite films having Zn nanoparticles exhibited the highest radical scavenging activity. In carboxymethyl chitosan/fish skin gelatin based ZnO-nanocomposite film showed promising antibacterial properties for food packaging applications. In contrast, carboxymethyl chitosan/gelatinl/nano ZnO composite film displayed better solvent resistance, higher elasticity, and ductility and exhibited strong antibacterial activity against *E. coli* (99.20%) and *S. aureus* (84.70%) for food packages. The presence of pomegranate peel extract in carboxymethyl chitosan/gelatin composite film reduced solubility by more than 5%, increased mechanical resistance with tensile strength of 15 MPa, and improved light barrier properties by 30% (Bertolo et al., 2022). The extract also imparted antioxidant activity to the composite film, which was able to inhibit 20% of 2,2-diphenyl-1-picrylhydrazyl radical at 38 mg/mL. An editable food packaging film composed of carboxymethyl chitosan and carboxymethyl starch composite crosslinked with date palm kernel extract was prepared to inhibit foodborne pathogens during fruit preservation. As the amount of the extract increased from 0 to 4 mL, the tensile strength of the carboxymethyl chitosan/carboxymethyl starch film increased from 18.7 to 21.9 MPa, respectively, while the opposite trend was observed in the case of elongation at break due to phenolic compounds containing several –OH that formed hydrogen bonds with both components. In the case of the highest loaded extract, the composite film exhibited antibacterial activities with a zone of inhibition of 31±0.37, 35±0.76, 27±0.24, 25±0.52, 23±0.25, and 22±0.49 mm, respectively, for *E.coli O157*, *S. typhimuruimm*, *L. monocytogenes*, *S. aureus*, *R.oryzae*, and *A. niger*, respectively, promoting the safety and freshness of fruits and vegetables (Zidan et al., 2023). Natural kaolinite clay and Ficus leaf extract were combined with chitosan to create a novel antioxidant and biodegradable composite film for food packaging (El Mouzahim et al., 2023). Despite the lower opacity observed, the addition of 10 wt% leave extract and 10 wt% kaolinite clay significantly improved moisture barrier properties. The water vapor permeability of the composite decreased to 6.26×10^{-11}/ s.m.Pa, while this value was measured as 9.84×10^{-11}/ s.m.Pa for the neat carboxymethyl chitosan film. However, the increase in tensile strength and elongation at break was enhanced from 14.07 MPa, 29.81% to 21.08 MPa, 33.35% for the carboxymethyl chitosan film and the ternary-mixed composite, respectively. Despite its high mechanical properties, the biodegradation of the carboxymethyl chitosan/kaolinite clay/Ficus leaf extract film was observed at 54% after 1 week of soil burial. Moreover, the composite film presented the strongest DPPH radical scavenging activity, suggesting its potential usefulness in protecting foods against free radical oxidation.

The combination between chitosan and tannic acid to synthesize the composite under different neutralizing conditions in Tris buffer at pH 8.5 and PBS at pH 7.4 was studied for the active food packaging (Lee et al., 2023). The findings revealed that the higher crosslinking density, as seen in chitosan-tannic acid composite films exhibited lower water vapor and oxygen transmission rates and were significantly better than the pristine chitosan film and improved in the mechanical properties. A higher content of tannic acid in the chitosan matrix also increased the antioxidant activity, showing superior activity in scavenging radicals. Additionally, the antibacterial property of the composite film against *S. aureus* was higher than against *E. coli*, which could be attributed to the loaded amount of tannic acid in the composite film. As the loaded amount increased, the chitosan matrix was filled with tannic acid molecules in the polymeric matrix, reducing water permeation and leading to a significant reduction in enzymatic browning and weight loss during extended storage time for bananas package. The chitosan/hexagonal boron nitride composite membrane film wrapped around fresh-cut potatoes revealed better appearance quality, which clearly verified the excellent oxygen barrier property, as well as the antibacterial activities of the chitosan component (Wang et al., 2023). In another example, citric acid cross-linked chitosan/poly(vinyl alcohol) composite films for cherry packaging were proposed (Jiang et al., 2023). Compared to samples prepared without adding citric acid, the obtained chitosan/PVA composite showed higher water vapor permeability and water vapor adsorption capacity, contributing to the improved preservation of cherries after 10 days of storage.

Electrospun fibrous composite nano-layers prepared from chitosan blended with PVA in the presence of silver nanoparticle showed effective antimicrobial activity against *Listeria monocytogenes* and *E. coli*, inhibiting microbial degradation of packaged food and extending meat shelf-life by one week, thus maintaining quality in an eco-friendly manner (Pandey et al., 2020). The chitosan/nitrogen-doped carbon dots composite was hydrothermally prepared using the solvent casting method for the preservation of pork and storage of blueberries (Chen et al., 2023). The tensile strength and elongation of the composite film increased, when the content of nitrogen doped carbon was 3 wt% in chitosan, the value of water vapor permeability significantly increased up to 7 wt% as 2.59 g/Pa.m.s as compared to the neat chitosan film of 1.41 g/Pa.m.s. Under ligh irradiation conditions, the composite having 7 wt% nitrogen doped carbon showed remarkable photodynamic antibacterial properties at 91.2% for *E. coli* and 99.9% for *S. aureus*. N-CDs. The photo-sensitively activated by visible light created reactive oxygen radicals, inducing the oxidation of intracellular substances, damaging and dysfunction of internal organs, and ultimately leading to bacterial death (Juan et al., 2021). In the food preservation experiment using the composite film with 3 wt% nitrogen doped carbon extended the shelf life of pork and effectively reduced pH, volatile base nitrogen, and total viable count. In blueberry storage, the composite film stabilized anthocyanin content and delayed blueberry spoilage.

Intelligent food packaging with pH-indicator-based anthocyanins, as extracted from different sources, has received so many attractive attentions. Depending on the pH of the surrounding environment, the color of anthocyanin-loaded films is relatively obtained suggesting the change in chemical compositions of the tesing materials. A chitosan/PVA was composited with anthocyanin extracted from *Brassica oleracea var. capitata* (Red Cabbage) was studied (Pereira et al., 2015). After coming into contact with the milk, a visible colour change detected at pH 6.7 which was unspoiled milk showing a dark grey color. As the pH decreased from 6.7 to 5.0, gradually the color became clearer. At pH 4.6 the color of the the chitosan composite flim changed to a dark pink color, clearly indicating milk spoilage. Colorimetric indicator based on chitosan/ gelatin reinforced with nano-ZnO and black peanut seed coat anthocyanins was investigated for intelligent packaging as shrimp-perservation materials (Lu et al., 2022). The distinctive color change of thecomposite film was highly correlated with the change of the total volatile basic nitrogen, pH and total viable count of shrimp. pH-responsive chitosan-based film incorporated with alizarin for intelligent packaging applications showed great antibacterial activity against *E. coli and L. monocytogenes* (Ezati & Rhim, 2020). The color of the composite film changed vividly from slightly yellow to purple in response at pH 4–10, which was strongly sensitive to ammonia vapor, indicating the onset spoilage of fish, when showing color change was from khaki to light brown.

CHITOSAN COMPOSITES FOR ENVIRONMENTAL REMEDIATIONS

Metal Ion Elimination by Chitosan Composites

Contamination in water is an important problem causing serious impacts on the environment. Among the convensional methods towards water-treatment, adsorption is highly consideres as a effective process due to its simplicity, lowcost, opreational flexibility in various adsorbents (De Gisi et al., 2016; Momina & Ahmad, 2023). As a regenerated polymer from biomass, chitosan is one of bioadsorbent owing to the presence of amino (–NH, –NH$_2$) and hydroxyl (–OH) in the chemical structure, enhancing the adorption capacity (Benettayeb et al., 2022). To increase adsorption capacity of the removal of heavy metal ions, a combination with chitosan and dexan showed the adsorption capacities for Cu (II), Co (II), Ni (II), Pb (II) and Cd (II) were 342 mg/g, 232 mg/g, 184 mg/g, 395 mg/g, and 269 mg/g, respectively at 20°C, pH 7. The chitosan/dexan film adsorbent had a good reusability, retaining about 75% of its adsorption capacity after five cycles (Yang Liu et al., 2019).Also, maximum adsorption amounts of Pb (II) and Cd (II) by the polyethyleneimine-grafted freeze-dried chitosan/epichlorohydrin composite beads was 315 mg/g at pH 5 and 278 mg/g at pH 6, respectively (Igberase et al., 2023). An eco-friendly chitosan-diethylamino-ethyl cellulose composite demonstrated adsorption capacity of 218.71 mg/g for Pb (II) at pH 6 (Majdoubi et al., 2023). The crosslinked glucan/chitosan composite found to be an attractive material for adsorption capacities for Cu (II), Co (II), Ni (II), Pb (II), and Cd (II) of 342 mg/g, 232 mg/g, 184 mg/g, 395 mg/g, and 269 mg/g, respectively, at pH 7 (Jiang et al., 2019). I other cases of metal adsorbents, following examples were reported for chitosan/ hydroxyapatite composite for Pb (II) up-taking with 570 mg/g at pH 2 (Liaw et al., 2020), chitosan loading wt% tannic acid composite membrane immobilized with *Stenotrophomonas maltophilia* fo Mg (II) (Song et al., 2022), glutaraldehyde crosslinked coconut fiber/chitosan for the elimination of heavy metal ions (Trang et al., 2018), showing adsorption in the order of Cr > As > Cd > Pb at pH 5.5, and also a thiol-grafted composite of chitosan/rice straw biochar with a maximum adsorption capacity of cadmium (II) of 261.47 mg/g at pH 5.5 (Hamid et al., 2023).

Graphene-based chitosan adsorbents were considered as potential materials for the removal of heavy metals such as Pb (II), Cd (II), and Hg (II) (Purnendu & Satapathi, 2017). GO was simply modified with sulfydryls (–SH) to increase adsorption efficiency, leading to uptake amounts for Pb (II) (447 mg/g), Cu (II) (425 mg/g), and Cd (II) (177 mg/g) at pH 5 and 20°C (Li et al., 2015). The combination of Fe (III) and chitosan-polymeric scaffold with FeOOH also increased the removal percentage of Pb (II) at pH 5.5 (Sheshmani et al., 2015). In the adsorption of various heavy metal ions, magnetic thiolated/quaternized-chitosan composites exhibited for binding Pb (II)with the capacity of 235.63 mg/g at pH 7 (Song et al., 2018). Another conductive material, polypyrrole-chitosan composited with nickel-ferrite nanoparticle was introduced to detecting the amount of different metal ion (Sadrolhosseini et al., 2017). A metal-organic framework modified zeolitic imidazolate framework-67 (ZIF-67) on bacterial cellulose/chitosan composite showed an adsorption capacity of Cu (II) and Cr (VI) of 200.6 mg/g and 152.1 mg/g at pH 6, respectively (Li et al., 2020). For toxic and carcinogenic hexavalent chromium or Cr (VI), there were problems of the metal discharged by several industries in chrome plating, textile dyeing, and leather tanning (Sultan & Hasnain, 2005; Thacker et al., 2006). The maximum adsorption capacity of Cr (VI) from aqueous solution by chitosan-GO composite was reported as 104.16 mg/g at pH 2 (Samuel et al., 2019). Another concept involves the nitrogen-doped method, which theoretically increases the electrostatic interactions between the adsorbent and adsorbate through the positive charges of nitrogen-doped graphene oxide (Gong et al., 2009). Further examples of heavy metal removal are listed below; Ferrite magnetic (Fe_3O_4) nanoparticles with GO moiety in chitosan composite for removal of Cr (VI) from wastewater (Samuel et al., 2018), a tetraoctyl ammonium bromide-impregnated magnetic CTS-GO composite used for the adsorption of Cr (VI) having maximum adsorption capacity of 145.35 mg/g at pH 3.5 (Li et al., 2014), and nanocomposite materials based on magnetic glutaraldehyde cross-linked chitosan for Cr (VI) ions (Zeraatkar et al., 2019).

In addition, there were studies of magnetic nanocomposite based on maghemite (γ-Fe_2O_3), chitosan and polypyrrole for the elimination of Cr (IV) (Reis et al., 2021), chitosan/metal oxyhydroxides like chitosan engraved iron-lanthanum mixed oxyhydroxide (Preethi et al., 2017; Preethi eta l., 2019) for Cr (VI)/g with 106 mg/g and nanocellulose/chitosan composite for Cr (VI) removal via vacuum filtration (Goswami et al., 2021). In PVA composite of chitosan enhanced mechanical properties by forming -OH bonds with the amino group of the chitosan molecule, whereas additional active sites for adsorption were introduced by Zeolite to uptake Cr(VI) having 450 mg/g (Habiba et al., 2017). This exhibited good reusability at higher than 90% after five cycles.

Since Arsenic (As) contamination in drinking water is a global concern (Sharma & Bhattacharya, 2017), chitosan-based composite was also designed for the treatment of As-contaminated water. Using CuO-ZnO-doped crosslinked chitosan adosrbent was effective for rapidlly uptaking As ion in water, and recycle for the application of low cost As removal filters (Purohit et al., 2020). Another binary metal-oxide impregnated chitosan composite was reported for the elimination of As (Qi et al., 2015). The Fe-Mg oxide-engraved chitosan composite showed superiorly removal performance for As (III) at pH 7. After the fourth regeneration, over 85% and 83% of the original adsorption efficiency was well maintained for As(III) and As(V), respectively. Chitosan beads having rod-shaped FeOOH nanoparticles used arsenic-contamination for drinking water, where the maximum adsorption capacity was found to be 5.4 mg/g for As(V) and 7.2 mg/g for As(III) (Hasan et al., 2014) and also SiO_2/chitosan composite adsorbed As(V) at 198.6 mg/g within 2 minutes (J. Liu et al., 2019).

Moreover, zeolite-mixed chitosan membrane demonstrated as a potential candidate anion elimination for zirconium (Zr (IV)) (Salehi et al., 2020). In the modified clinoptilolite/chitosan nanocomposite exhibited a high adsorption capacity for vanadium (V) through several mechanisms, including electrostatic and hydrogen bonding.

Elimination of Anionic Ions Using Chitosan Composites

On the other hands, In contrast to removal of inorganic ions from water using chitosan-based composites, providing simple and effective (Yazdi et al., 2023), the presence of other anions such as nitrate, phosphate and fluoride becomes also influenced quality of water affecting human life (Narsimha & Sudarshan, 2018; Rotiroti et al., 2023; Saka et al., 2023; Smyntek et al., 2022). For nitrate ions, the chitosan polymer matrix loaded with various fillers such as bentonite, titanium oxide, and alumina were prepared and utilized for the removal of nitrate from water through batch biosorption (Golie & Upadhyayula, 2017). The adsorption amounts of nitrate were found in the order of chitosan composite with / bentonite (35.68 mg/g) > titanium oxide (43.62 mg/g) > alumina oxide (45.38 mg/g). In chitosan-clinoptilolite nanocomposites, effective adsorbents was observed for nitrate (Yazdi et al., 2019). The absorptivity was compared for nitrate removal efficiency of different adsorbents of nano-chitosan/clinoptilolite composite, acidic treated nano-chitosan/clino acid and pentaethylene hexamine modified one. In the case of the pentaethylene hexamine-modified composite exhibited an enhanced nitrate adsorption capacity of 277.77 mg/g, while the acidic treated and original composite showed 227.27 mg/g and 185.18 mg/g, respectively. Notably, under competitive conditions, the selective removal of nitrate ions was better than that of chloride, carbonate, and sulfate in water.

Excessive phosphate concentration in the aquatic environment due to anthropogenic activities can damage the ecological system and harm human health (Zhou et al., 2022). Phosphate-containing water was treated using a self-crosslinked chitosan/cationic guar gum composite developed in good adsorption capacity of 10.72 mg/g (Li et al., 2020). An increase in the phosphate uptake amount to 19.24 mg/g at pH 4 was reported when using Fe(III)-loaded chitosan composites with biochar derived from paper mill sludge under CO_2 conditions at 60°C (Palansooriya et al., 2021). The maximum adsorption amount of the lanthanum-loaded chitosan/attapulgite composite toward phosphate ions was obtained at 298 K and was 102.9 mg/g. This was reduced to 75.6 mg/g at 318 K, when measured at a neutral pH (Kong et al., 2023). The adsorption of phosphate from water within a wide pH range by the La/Al bimetallic organic frameworks loaded on chitosan composite showed a great uptake amount of up to 264.48 mg/g (Liu et al., 2023).

Flouride contamination in drinking water becomes an urgent issue (Ahmad et al., 2022; Yadav et al., 2019). As crosslinked sawdust biochar-chitosan composite was applied, the composite bead behaved biosorbent for defluoridation in groundwater (Rupasinghe et al., 2022). For the maximum uptake of 4.4 mg/g at pH 7 within 60 min, this was due to the formation of hydrogen bonding between flouride ion and protonated amino groups of chitosan. A La (III)-ectrapped silica gel/chitosan composite was also used to enhance fluoride removal, having defluoridation capacity as 4.9 mg/g (Viswanathan et al., 2014) and in addition, Zr (IV) incorporated in chitosan/gelatin composite showed facilitatively having an adosorption amount of 12.12 mg/g at 323 K and pH 6 (Preethi et al., 2021). In other cases of fluoride adsorption, Zr-chitosan/GO membrane for 29.06 mg/g capacity binding (Zhang et al., 2017) and nanoZr-chitosan composite with 96.58 mg/g at pH 7 (Prasad et al., 2014) were observed.

Figure 4. Illustration of adsorption processes of chitosan-based composite for environmental remediation

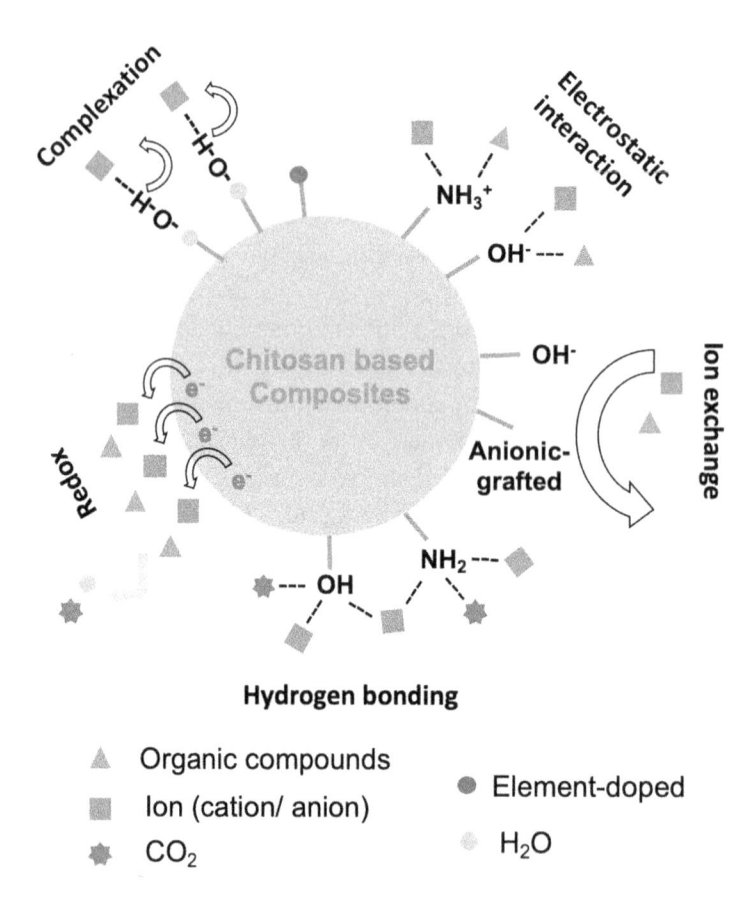

▲ Organic compounds ● Element-doped
■ Ion (cation/ anion) H₂O
✦ CO$_2$

ORGANIC MOLECULE POLLUTANTS FOR PHARMACEUTICAL AND DYE WASTES

As mentioned in heavy metals and anionic counter ions, chitosan-based composites have proven to be useful for the treatments. Figure 4 shows illustration of adsorption processes of chitosan-based composite for environmental remediation. Because chitosan based composites have several types of binding sites for pollutants by aiding with complexation, electrostatic interaction, ion exchange, redox reaction and hydrogen bonding.

Pharmaceutical Wastes

Among them, organic molecules in water becomes possible by using chitosan based composites for removal of pharmaceutical pollutants. A glutaraldehyde-crosslinked chitosan-biochar composite was reported for removing sulfamethoxazole (SMX) antibiotics, when derived the biochar from orange peel or used coffee grounds (Son Tran et al., 2023). The highest adsorption capacity was 7.24 mg/g for chitosan-orange peeled biochar and 14.73 mg/g for chitosan-caffee beans biochar, respectively, attributing to interactions of ion exchange, π bonding and hydrogen bonding. Waste coffee grounds were proven effective when combined with a chitosan/PVA composite to enhance the elimination of metamizol, acetylsalicylic acid, acetaminophen and caffeine at pH 6. The removal followed the order of acetylsalicylic acid > caffeine > acetaminophen > metamizol (Lessa et al., 2018). In another way, chitosan composite nanofiltration membrane was applied by furosemide-modified chitosanas assisted with pectin extracted from citrus peel and then functionalized with basement polymer of polyethersulfone (PES). This membrane showed a high removal percentage for elimination of Ceftriaxone and Cefixime in wastewater (Moradi et al., 2023). The introduction of magnetite into bio-polymer matrix with chemical modifications was able to remarkably improve pollutant-removal efficiency in water (Ling et al., 2019; Zhang et al., 2024). Preparation of a zinc-ferrite-magnetized chitosan/curdlan composite for efficient adsorption of tetracycline was carried out in water medium at pH 6 with uptaken tetracycline of 326.8 mg/g at 298 K and 371 mg/g at 328 K (Valizadeh et al., 2023). A ternary composite of hydroxyapatite, chitosan, and magnetite was fabricated for high removal performance of Ibuprofen, deflazacort, methylprednisolone, and norfloxacin in aqueous phase. The Langmuir adsorption capacity was reported as 535.6 mg/g for norfloxacin, 833 mg/g for Ibuprofen, 625 mg/g for methylprednisolone, and 714 mg/g for deflazacort (Nayak et al., 2023). Moreover, core-brush-shaped chitosan/Fe_3O_4 composite particles attracted practical applications in pharmaceutical removal due to the numerous functional groups on branches and their flexibility towards various contaminants (Zhou et al., 2018). For instance, the composites were used for the adsorption of typical pharmaceuticals (norfloxacin, tylosin, and diclofenac sodium). The polyanions (poly(sodium p-styrenesulfonate))-modified chitosan composite exhibited a high equilibrium uptake of 165 mg/g for norfloxacin and 134 mg/g for tylosin. For diclofenac sodium, chitosan composite modified with polycations (poly(p-vinylbenzyl trimethylammonium chloride)) branches reached 151 mg/g. Grafting co-polymerization with three types of grafting branches, including polycations (poly(2-methyl acryloyloxyethyl trimethyl ammonium chloride)), polyanions (polyacrylic acid), and neutral polymer (polymethylmethacrylate) provided chitosan-composite particles with high performance in treating waste pharmaceutical water (Zhang et al., 2016). By using excellent adsorption capacity of GO, the removal of pharmaceuticals by a magnetic genipin-crosslinked chitosan/graphene oxide-SO_3H composite eliminated mixture of pharmaceutical compounds in wastewater. Also, bio-based chitosan/polyvinyl alcohol/graphene oxide derivatives were prepared for the removal of pharmaceutical pollutants from aqueous mixtures (Malesic-Eleftheriadou et al.,2023), investigating three different categories of pharmaceuticals of non-steroidal anti-inflammatory drugs (diclofenac, ibuprofen, ketoprofen, paracetamol), anti-epileptics (carbamazepine), and antihypertensives (valsartan and irbesartan). The highest adsorption capacity was achieved after 3 hours at pH 10 and 55°C, showing strong selectivity for the removal of valsartan and diclofenac.

Organic Dye Waste

Not only are chitosan-based composites effective in treating pharmaceutical-contaminated water, but they have also shown potential for enhancing the adsorption capacity of methylene blue from aqueous solutions. For instance, the adsorption capacity of methylene blue from alkali pH solutions using a GO-crosslinked chitosan composite was reported as 259.5 mg/g at room temperature (Tran et al., 2023). Similarly, the removal of anionic dyes such as methyl orange and Acid red 1 at pH 3 was achieved using chitosan/GO nanoplates composite spheres, reaching 230.91 mg/g for methyl orange and 132.94 mg/g for Acid red 1. Another study employed a GO-impregnated chitosan/PVA nanocomposite for the treatment of anionic Congo red dye in wastewater, when the initial concentration was 20 mg/L of Congo red and a dose of 6 g/L of the composite, the adsorption efficiency was found to be 88.17% at pH 2, 81% at neutral pH, and decreased to 78.2% at higher pH (Das et al., 2020). For micro-nano metal-organic framework MIL-68(Fe)/chitosan had composited sponge structure and proved effective for the treatment of congo red, with a maximum adsorption capacity of 1184 mg/g at 318 K when adsorption was performed in a neutral environment (Jin et al., 2023). For reactive orange adsorption by the cross-linked chitosan-epichlorohydrin/bentonite composite, variation with environmental temperature was seen when increasing from 55.27 mg/g at 30°C to 70.80 mg/g at 50°C (Benhouria et al., 2023). Similarly, the crosslinked chitosan/sepiolite composite exhibited the highest adsorption amount of methylene blue at pH 9 (41 mg/g) and reactive orange at pH 3 (191 mg/g) (Marrakchi et al., 2016).

The application of chitosan-treated cotton composite for the removal of reactive Remazol Brilliant Red F3B dye at pH 7 resulted in a maximum adsorption of 169.33 mg/g (Salman et al., 2023). The adsorption of methyl orange by the TEMPO-oxidized pineapple leaf pulp and chitosan aerogel composite was reported as 136.64 mg/g within only 6 minutes (Do et al., 2022). Utilizing cellulose, a sustainable cellulose nanofiber/chitosan composite film designed through a facile procedure exhibited efficient selective adsorption of methyl orange, with an adsorption capacity of 655.23 mg/g, as proposed by Hou et al. (Huo et al., 2021). Furthermore, the maximum uptake amount of methyl orange was increased to 742.68 mg/g by using the regenerated cellulose and chitosan aerogel composite in the presence of glutaraldehyde as a cross-linker (He et al., 2023).

Functionalized carboxymethyl chitosan/phytic acid hydrogels were utilized for the adsorption of methyl orange and Congo red at pH 7, achieving higher maximum uptake capacities of 13.62 mg/g and 8.49 mg/g, respectively (Han et al., 2021). For other cationic dyes like methylene blue, a superabsorbent prepared from alginate/carboxymethyl chitosan composite hydrogel beads exhibited a significant adsorption capacity of 1010 mg/g (Zhang et al., 2023). In contrast, Crystal violet, a recalcitrant dye molecule that persists in the environment for a prolonged period and poses toxic effects, but it was removed by cross-linked chitosan-coated bentonite beads at pH 5, achieving an uptake amount of 169.5 mg/g (Vithalkar & Jugade, 2020). Additionally, the maximum adsorption capacity of Crystal violet performed by the chitosan-modified L-cysteine/Bentonite nanocomposite gradually increased from 219 mg/g at 303 K to 240 mg/g at 323 K, when the pH of the solution was adjusted to 8 (Ahmad & Ejaz, 2023). Enhanced adsorption capacity for Crystal violet was achieved using a chitosan/hydroxyapatite hybrid composite with 294.13 mg/g at pH 6.5 (Jebli et al., 2023).

Chitosan having poly-$\beta1\rightarrow4$-*N*-acetylglucosamine segments, is a hydrophilic and expresses biodegradable, and biocompatible. In addition, it possesses good adhesive and film-forming properties, along with a high affinity for CO_2 due to the presence of primary amine and hydroxyl functional groups in chitosan. Therefore, chitosan composite also applied for gas separation capabilities. Chitosan-based composite membranes effectively behaved as gas barriers for mixed CO_2/CH_4 gas mixture (Torre-Celeizabal et al., 2023). The results indicated that the increase in permeance of both CO_2 and CH_4 gas components depended on the loading of fillers. Notably, membranes filled with 5 wt% of zeolite 4A demonstrated high CO_2 permeance with separation factors of up to 30.

Another nanosheet filler of hexagonal boron nitride nanosheets (hBNNSs), incorporated into the chitosan membrane, was found to enhance gas adsorption and UV shielding properties (Wang et al., 2023). The hHBNs/chitosan composite films exhibited exceptional UV resistance, with rates reaching up to 98.5% for UV-A and 96.6% for UV-B. The gas barrier properties for oxygen (O_2) and carbon dioxide (CO_2) permeability were attributed to the dosage of hHBNs within the chitosan membrane. At a relative humidity of 0% and a temperature of 23°C, the O_2 permeability of the chitosan film was 4.93 cm^3.mm/(m^2.24h.atm). In contrast, the hHBNs/chitosan composite membrane, particularly with 2 wt% and 5 wt% hBNNSs, exhibited significantly reduced O_2 permeabilities of 0.32 and 0.18 cm^3.mm/(m^2.24h.atm) for the 2 and 5 wt% hBNNSs, repectively. Similarly, the permeability values of CO_2 showed gas barrier improvements with the hHBNs/chitosan composite membrane. Notably, the 5 wt% hHBNs content led to a reduced CO_2 permeability of 0.50 cm^3.mm/(m^2.24h.atm) compared to 0.64 cm^3.mm/(m^2.24h.atm) for 2 wt% and 8.42 cm^3.mm/(m^2.24h.atm) for 0 wt%.

CONCLUSION

Chitosan has been extensively studied due to its excellent properties in many applications. Moreover, the combination of chitosan with many different materials from shape, composition, properties and size has been studied in order to improve the physical and chemical properties as well as increase the practical applications of chitosan-based composite. This chapter has summarized the recent utilizations of chitosan-based composite materials applied in biomedical, packaging and environmental treatment to raise awareness about the use of natural-origin materials obtained from nature or regenerating from waste in order to reduce environmental impact and less consumption in intial materials during the production processes. Furthermore, it would contribute to reducing emissions and towards sustainable development in the future.

REFERENCES

Adhikari, H. S., & Yadav, P. N. (2018). Anticancer Activity of Chitosan, Chitosan Derivatives, and Their Mechanism of Action. *International Journal of Biomaterials*, 2952085, 1–29. Advance online publication. DOI: 10.1155/2018/2952085 PMID: 30693034

Ahmad, R., & Ejaz, M. O. (2023). Efficient adsorption of crystal violet (CV) dye onto benign chitosan-modified l-cysteine/bentonite (CS-Cys/Bent) bionanocomposite: Synthesis, characterization and experimental studies. *Dyes and Pigments*, 216, 111305. https://doi.org/https://doi.org/10.1016/j.dyepig.2023.111305. DOI: 10.1016/j.dyepig.2023.111305

Ahmad, S., Singh, R., Arfin, T., & Neeti, K. (2022). Fluoride contamination, consequences and removal techniques in water: a review []. *Environmental Science: Advances*, 1(5), 620-661. https://doi.org/DOI: 10.1039/D1VA00039J

Amaral, I. F., Lamghari, M., Sousa, S. R., Sampaio, P., & Barbosa, M. A. (2005). Rat bone marrow stromal cell osteogenic differentiation and fibronectin adsorption on chitosan membranes: The effect of the degree of acetylation. *Journal of Biomedical Materials Research. Part A*, 75A(2), 387–397. https://doi.org/https://doi.org/10.1002/jbm.a.30436. DOI: 10.1002/jbm.a.30436 PMID: 16092111

Amiryaghoubi, N., Noroozi Pesyan, N., Fathi, M., & Omidi, Y. (2022). The design of polycaprolactone-polyurethane/chitosan composite for bone tissue engineering. *Colloids and Surfaces. A, Physicochemical and Engineering Aspects*, 634, 127895. https://doi.org/https://doi.org/10.1016/j.colsurfa.2021.127895. DOI: 10.1016/j.colsurfa.2021.127895

Babaremu, K., Oladijo, O. P., & Akinlabi, E. (2023). Biopolymers: A suitable replacement for plastics in product packaging. *Advanced Industrial and Engineering Polymer Research*. https://doi.org/https://doi.org/10.1016/j.aiepr.2023.01.001

Bai, K., Hong, B., Hong, Z., Sun, J., & Wang, C. (2017). Selenium nanoparticles-loaded chitosan/citrate complex and its protection against oxidative stress in d-galactose-induced aging mice. *Journal of Nanobiotechnology*, 15(1), 92. DOI: 10.1186/s12951-017-0324-z PMID: 29262862

Benettayeb, A., Ghosh, S., Usman, M., Seihoub, F. Z., Sohoo, I., Chia, C. H., & Sillanpää, M. (2022). Some Well-Known Alginate and Chitosan Modifications Used in Adsorption: A Review. *Water (Basel)*, 14(9), 1353. DOI: 10.3390/w14091353

Benhouria, A., Zaghouane-Boudiaf, H., Bourzami, R., Djerboua, F., Hameed, B. H., & Boutahala, M. (2023). Cross-linked chitosan-epichlorohydrin/bentonite composite for reactive orange 16 dye removal: Experimental study and molecular dynamic simulation. *International Journal of Biological Macromolecules*, 242, 124786. https://doi.org/https://doi.org/10.1016/j.ijbiomac.2023.124786. DOI: 10.1016/j.ijbiomac.2023.124786 PMID: 37169046

Bertolo, M. R. V., Dias, L. D., Oliveira Filho, J. G., Alves, F., Marangon, C. A., Amaro Martins, V. C., Ferreira, M. D., Bagnato, V. S., Guzzi Plepis, A. M., & Bogusz, S. (2022). Central composite design optimization of active and physical properties of food packaging films based on chitosan/gelatin/pomegranate peel extract. *Food Packaging and Shelf Life*, 34, 100986. https://doi.org/https://doi.org/10.1016/j.fpsl.2022.100986. DOI: 10.1016/j.fpsl.2022.100986

Bhardwaj, S., Bhardwaj, N. K., & Negi, Y. S. (2020). Effect of degree of deacetylation of chitosan on its performance as surface application chemical for paper-based packaging. *Cellulose (London, England)*, 27(9), 5337–5352. DOI: 10.1007/s10570-020-03134-5

Boenish, R., Kritzer, J. P., Kleisner, K., Steneck, R. S., Werner, K. M., Zhu, W., Schram, F., Rader, D., Cheung, W., Ingles, J., Tian, Y., & Mimikakis, J. (2022). The global rise of crustacean fisheries. *Frontiers in Ecology and the Environment*, 20(2), 102–110. https://doi.org/https://doi.org/10.1002/fee.2431. DOI: 10.1002/fee.2431

Cen, C., Wang, F., Wang, Y., Li, H., Fu, L., Li, Y., Chen, J., & Wang, Y. (2023). Design and characterization of an antibacterial film composited by hydroxyethyl cellulose (HEC), carboxymethyl chitosan (CMCS), and nano ZnO for food packaging. *International Journal of Biological Macromolecules*, 231, 123203. https://doi.org/https://doi.org/10.1016/j.ijbiomac.2023.123203. DOI: 10.1016/j.ijbiomac.2023.123203 PMID: 36623619

Chen, P., Liu, L., Pan, J., Mei, J., Li, C., & Zheng, Y. (2019). Biomimetic composite scaffold of hydroxyapatite/gelatin-chitosan core-shell nanofibers for bone tissue engineering. *Materials Science and Engineering C*, 97, 325–335. https://doi.org/https://doi.org/10.1016/j.msec.2018.12.027. DOI: 10.1016/j.msec.2018.12.027 PMID: 30678918

Chen, S., Zeng, Q., Tan, X., Ye, M., Zhang, Y., Zou, L., Liu, S., Yang, Y., Liu, A., He, L., & Hu, K. (2023). Photodynamic antibacterial chitosan/nitrogen-doped carbon dots composite packaging film for food preservation applications. *Carbohydrate Polymers*, 314, 120938. https://doi.org/https://doi.org/10.1016/j.carbpol.2023.120938. DOI: 10.1016/j.carbpol.2023.120938 PMID: 37173034

Chen, W., Cheng, H., Chen, L., Zhan, X., & Xia, W. (2022). Synthesis, characterization, and biological evaluation of novel selenium-containing chitosan derivatives. *Carbohydrate Polymers*, 284, 119185. https://doi.org/https://doi.org/10.1016/j.carbpol.2022.119185. DOI: 10.1016/j.carbpol.2022.119185 PMID: 35287904

Chen, W., Li, X., Cheng, H., & Xia, W. (2022). Chitosan-based selenium composites as potent Se supplements: Synthesis, beneficial health effects, and applications in food and agriculture. *Trends in Food Science & Technology*, 129, 339–352. https://doi.org/https://doi.org/10.1016/j.tifs.2022.10.008. DOI: 10.1016/j.tifs.2022.10.008

Chen, W., Yue, L., Jiang, Q., Liu, X., & Xia, W. (2018). Synthesis of varisized chitosan-selenium nanocomposites through heating treatment and evaluation of their antioxidant properties. *International Journal of Biological Macromolecules*, 114, 751–758. https://doi.org/https://doi.org/10.1016/j.ijbiomac.2018.03.108. DOI: 10.1016/j.ijbiomac.2018.03.108 PMID: 29588203

Das, L., Das, P., Bhowal, A., & Bhattachariee, C. (2020). Synthesis of hybrid hydrogel nano-polymer composite using Graphene oxide, Chitosan and PVA and its application in waste water treatment. *Environmental Technology & Innovation*, 18, 100664. https://doi.org/https://doi.org/10.1016/j.eti.2020.100664. DOI: 10.1016/j.eti.2020.100664 das Neves, M. S., Scandorieiro, S., Pereira, G. N., Ribeiro, J. M., Seabra, A. B., Dias, A. P., Yamashita, F., Martinez, C. B. R., Kobayashi, R. K. T., & Nakazato, G.das Neves. (2023). Antibacterial Activity of Biodegradable Films Incorporated with Biologically-Synthesized Silver Nanoparticles and the Evaluation of Their Migration to Chicken Meat. *Antibiotics (Basel, Switzerland)*, 12(1), 178. Advance online publication. DOI: 10.3390/antibiotics12010178 PMID: 36671379

De Gisi, S., Lofrano, G., Grassi, M., & Notarnicola, M. (2016). Characteristics and adsorption capacities of low-cost sorbents for wastewater treatment: A review. *Sustainable Materials and Technologies, 9*, 10-40. https://doi.org/https://doi.org/10.1016/j.susmat.2016.06.002

De Queiroz Antonino, R. S., Lia Fook, B. R., De Oliveira Lima, V. A., De Farias Rached, R. Í., Lima, E. P., Da Silva Lima, R. J., Peniche Covas, C. A., & Lia Fook, M. V. (2017). Preparation and Characterization of Chitosan Obtained from Shells of Shrimp (Litopenaeus vannamei Boone). *Marine Drugs*, 15(5), 141. DOI: 10.3390/md15050141 PMID: 28505132

Dehankar, H. B., Mali, P. S., & Kumar, P. (2023). Edible composite films based on chitosan/guar gum with ZnONPs and roselle calyx extract for active food packaging. *Applied Food Research*, 3(1), 100276. https://doi.org/https://doi.org/10.1016/j.afres.2023.100276. DOI: 10.1016/j.afres.2023.100276

Dhavale, R. P., Dhavale, R. P., Sahoo, S. C., Kollu, P., Jadhav, S. U., Patil, P. S., Dongale, T. D., Chougale, A. D., & Patil, P. B. (2021). Chitosan coated magnetic nanoparticles as carriers of anticancer drug Telmisartan: pH-responsive controlled drug release and cytotoxicity studies. *Journal of Physics and Chemistry of Solids*, 148, 109749. https://doi.org/https://doi.org/10.1016/j.jpcs.2020.109749. DOI: 10.1016/j.jpcs.2020.109749

Do, N. H. N., Truong, B. Y., Nguyen, P. T. X., Le, K. A., Duong, H. M., & Le, P. K. (2022). Composite aerogels of TEMPO-oxidized pineapple leaf pulp and chitosan for dyes removal. *Separation and Purification Technology*, 283, 120200. https://doi.org/https://doi.org/10.1016/j.seppur.2021.120200. DOI: 10.1016/j.seppur.2021.120200

El Mouzahim, M., Eddarai, E. M., Eladaoui, S., Guenbour, A., Bellaouchou, A., Zarrouk, A., & Boussen, R. (2023). Food packaging composite film based on chitosan, natural kaolinite clay, and Ficus. carica leaves extract for fresh-cut apple slices preservation. *International Journal of Biological Macromolecules*, 233, 123430. https://doi.org/https://doi.org/10.1016/j.ijbiomac.2023.123430. DOI: 10.1016/j.ijbiomac.2023.123430 PMID: 36716844

Ezati, P., & Rhim, J.-W. (2020). pH-responsive chitosan-based film incorporated with alizarin for intelligent packaging applications. *Food Hydrocolloids*, 102, 105629. https://doi.org/https://doi.org/10.1016/j.foodhyd.2019.105629. DOI: 10.1016/j.foodhyd.2019.105629

Fathi, R., & Mohammadi, R. (2023). Preparation of pH-responsive magnetic nanocomposite hydrogels based on k-carrageenan/chitosan/silver nanoparticles: Antibacterial carrier for potential targeted anticancer drug delivery. *International Journal of Biological Macromolecules*, 246, 125546. https://doi.org/https://doi.org/10.1016/j.ijbiomac.2023.125546. DOI: 10.1016/j.ijbiomac.2023.125546 PMID: 37355059

Galus, S., Arik Kibar, E. A., Gniewosz, M., & Kraśniewska, K. (2020). Novel Materials in the Preparation of Edible Films and Coatings—A Review. *Coatings*, 10(7), 674. DOI: 10.3390/coatings10070674

Golie, W. M., & Upadhyayula, S. (2017). An investigation on biosorption of nitrate from water by chitosan based organic-inorganic hybrid biocomposites. *International Journal of Biological Macromolecules*, 97, 489–502. https://doi.org/https://doi.org/10.1016/j.ijbiomac.2017.01.066. DOI: 10.1016/j.ijbiomac.2017.01.066 PMID: 28099890

Gong, K., Du, F., Xia, Z., Durstock, M., & Dai, L. (2009). Nitrogen-Doped Carbon Nanotube Arrays with High Electrocatalytic Activity for Oxygen Reduction. *Science*, 323(5915), 760–764. DOI: 10.1126/science.1168049 PMID: 19197058

Gong, P., Wang, J., Liu, B., Ru, G., & Feng, J. (2016). Dissolution of chitin in aqueous KOH. *Cellulose (London, England)*, 23(3), 1705–1711. DOI: 10.1007/s10570-016-0932-z

Goñi, M. G., Tomadoni, B., Roura, S. I., & Moreira, M. R. (2017). Lactic acid as potential substitute of acetic acid for dissolution of chitosan: Preharvest application to Butterhead lettuce. *Journal of Food Science and Technology*, 54(3), 620–626. DOI: 10.1007/s13197-016-2484-5 PMID: 28298675

Goswami, R., Mishra, A., Bhatt, N., Mishra, A., & Naithani, P. (2021). Potential of chitosan/nanocellulose based composite membrane for the removal of heavy metal (chromium ion). *Materials Today: Proceedings*, 46, 10954–10959. https://doi.org/https://doi.org/10.1016/j.matpr.2021.02.036. DOI: 10.1016/j.matpr.2021.02.036

Habiba, U., Siddique, T. A., Joo, T. C., Salleh, A., Ang, B. C., & Afifi, A. M. (2017). Synthesis of chitosan/polyvinyl alcohol/zeolite composite for removal of methyl orange, Congo red and chromium(VI) by flocculation/adsorption. *Carbohydrate Polymers*, 157, 1568–1576. https://doi.org/https://doi.org/10.1016/j.carbpol.2016.11.037. DOI: 10.1016/j.carbpol.2016.11.037 PMID: 27987870

Hajji, S., Ghorbel-Bellaaj, O., Younes, I., Jellouli, K., & Nasri, M. (2015). Chitin extraction from crab shells by Bacillus bacteria. Biological activities of fermented crab supernatants. *International Journal of Biological Macromolecules*, 79, 167–173. DOI: 10.1016/j.ijbiomac.2015.04.027 PMID: 25910648

Hamid, Y., Liu, L., Haris, M., Usman, M., Lin, Q., Chen, Y., Rashid, M. S., Ulhassan, Z., Hussain, M. I., & Yang, X. (2023). Novel thiol-grafted composite of chitosan and rice straw biochar (TH@CT-BC): A two-step fabrication for highly selective adsorption of cadmium from contaminated water. *Journal of Environmental Chemical Engineering*, 11(5), 110527. https://doi.org/https://doi.org/10.1016/j.jece.2023.110527. DOI: 10.1016/j.jece.2023.110527

Han, D., Zhao, H., Gao, L., Qin, Z., Ma, J., Han, Y., & Jiao, T. (2021). Preparation of carboxymethyl chitosan/phytic acid composite hydrogels for rapid dye adsorption in wastewater treatment. *Colloids and Surfaces. A, Physicochemical and Engineering Aspects*, 628, 127355. https://doi.org/https://doi.org/10.1016/j.colsurfa.2021.127355. DOI: 10.1016/j.colsurfa.2021.127355

Hannah Ritchie, M. R. (2019). Saefood Production. *Our World in Data*. https://ourworldindata.org/seafood-production#global-seafood-production

Hasan, S., Ghosh, A., Race, K., Schreiber, R.Jr, & Prelas, M. (2014). Dispersion of FeOOH on Chitosan Matrix for Simultaneous Removal of As(III) and As(V) from Drinking Water. *Separation Science and Technology*, 49(18), 2863–2877. DOI: 10.1080/01496395.2014.949774

He, S., Li, J., Cao, X., Xie, F., Yang, H., Wang, C., Bittencourt, C., & Li, W. (2023). Regenerated cellulose/chitosan composite aerogel with highly efficient adsorption for anionic dyes. *International Journal of Biological Macromolecules*, 244, 125067. https://doi.org/https://doi.org/10.1016/j.ijbiomac.2023.125067. DOI: 10.1016/j.ijbiomac.2023.125067 PMID: 37245747

Huo, M.-X., Jin, Y.-L., Sun, Z.-F., Ren, F., Pei, L., & Ren, P.-G. (2021). Facile synthesis of chitosan-based acid-resistant composite films for efficient selective adsorption properties towards anionic dyes. *Carbohydrate Polymers*, 254, 117473. https://doi.org/https://doi.org/10.1016/j.carbpol.2020.117473. DOI: 10.1016/j.carbpol.2020.117473 PMID: 33357927

Igberase, E., Sithole, N. T., & Mashifana, T. (2023). Synergic binding of rhodamine-B and heavy metal ions onto polyethyleneimine grafted freeze-dried chitosan/epichlorohydrin composite beads and possible mechanism. *International Journal of Biological Macromolecules*, 249, 125983. https://doi.org/https://doi.org/10.1016/j.ijbiomac.2023.125983. DOI: 10.1016/j.ijbiomac.2023.125983 PMID: 37494996

Jabbari, P., Mahdavinia, G. R., Rezaei, P. F., Heragh, B. K., Labib, P., Jafari, H., & Javanshir, S. (2023). pH-responsive magnetic biocompatible chitosan-based nanocomposite carrier for ciprofloxacin release. *International Journal of Biological Macromolecules*, 250, 126228. https://doi.org/https://doi.org/10.1016/j.ijbiomac.2023.126228. DOI: 10.1016/j.ijbiomac.2023.126228 PMID: 37558030

Jardine, A., & Sayed, S. (2018). Valorisation of chitinous biomass for antimicrobial applications. *90*(2), 293-304. https://doi.org/doi:10.1515/pac-2017-0707 (Pure and Applied Chemistry)

Jebli, A., Amri, A. E., Hsissou, R., Lebkiri, A., Zarrik, B., Bouhassane, F. Z., Hbaiz, E., Rifi, E. H., & Lebkiri, A. (2023). Synthesis of a chitosan@hydroxyapatite composite hybrid using a new approach for high-performance removal of crystal violet dye in aqueous solution, equilibrium isotherms and process optimization. *Journal of the Taiwan Institute of Chemical Engineers*, 149, 105006. https://doi.org/https://doi.org/10.1016/j.jtice.2023.105006. DOI: 10.1016/j.jtice.2023.105006

Ji, M., Li, J., Li, F., Wang, X., Man, J., Li, J., Zhang, C., & Peng, S. (2022). A biodegradable chitosan-based composite film reinforced by ramie fibre and lignin for food packaging. *Carbohydrate Polymers*, 281, 119078. https://doi.org/https://doi.org/10.1016/j.carbpol.2021.119078. DOI: 10.1016/j.carbpol.2021.119078 PMID: 35074129

Ji, M., Liu, X., Li, J., Li, F., Man, J., Li, J., Zhang, C., Sun, K., & Qiu, Y. (2023). Comparison of biodegradable chitosan-based composite films reinforced by different micron-sized plant fibers for food packaging. *Materials Letters*, 348, 134669. https://doi.org/https://doi.org/10.1016/j.matlet.2023.134669. DOI: 10.1016/j.matlet.2023.134669

Jiang, C., Wang, X., Wang, G., Hao, C., Li, X., & Li, T. (2019). Adsorption performance of a polysaccharide composite hydrogel based on crosslinked glucan/chitosan for heavy metal ions. *Composites. Part B, Engineering*, 169, 45–54. https://doi.org/https://doi.org/10.1016/j.compositesb.2019.03.082. DOI: 10.1016/j.compositesb.2019.03.082

Jiang, S., Qiao, C., Liu, R., Liu, Q., Xu, J., & Yao, J. (2023). Structure and properties of citric acid cross-linked chitosan/poly(vinyl alcohol) composite films for food packaging applications. *Carbohydrate Polymers*, 312, 120842. https://doi.org/https://doi.org/10.1016/j.carbpol.2023.120842. DOI: 10.1016/j.carbpol.2023.120842 PMID: 37059567

Jin, Y., Liu, F., Li, Y., Du, Q., Song, F., Chen, B., Chen, K., Zhang, Y., Wang, M., Sun, Y., Zhao, S., Jing, Z., Pi, X., Wang, Y., & Wang, D. (2023). Efficient adsorption of azo anionic dye Congo Red by micro-nano metal-organic framework MIL-68(Fe) and MIL-68(Fe)/chitosan composite sponge: Preparation, characterization and adsorption performance. *International Journal of Biological Macromolecules*, 126198, 126198. https://doi.org/https://doi.org/10.1016/j.ijbiomac.2023.126198. DOI: 10.1016/j.ijbiomac.2023.126198 PMID: 37586626

Juan, C. A., Pérez de la Lastra, J. M., Plou, F. J., & Pérez-Lebeña, E. (2021). The Chemistry of Reactive Oxygen Species (ROS) Revisited: Outlining Their Role in Biological Macromolecules (DNA, Lipids and Proteins) and Induced Pathologies. *International Journal of Molecular Sciences*, 22(9), 4642. Advance online publication. DOI: 10.3390/ijms22094642 PMID: 33924958

Kabir, M., Habiba, U. E., Khan, W., Shah, A., Rahim, S., Rios-Escalante, P. R. D., Farooqi, Z.-U.-R., Ali, L., & Shafiq, M. (2023). Climate change due to increasing concentration of carbon dioxide and its impacts on environment in 21st century; a mini review. *Journal of King Saud University. Science*, 35(5), 102693. https://doi.org/https://doi.org/10.1016/j.jksus.2023.102693. DOI: 10.1016/j.jksus.2023.102693

Kan, M., & Miller, S. A. (2022). Environmental impacts of plastic packaging of food products. *Resources, Conservation and Recycling*, 180, 106156. https://doi.org/https://doi.org/10.1016/j.resconrec.2022.106156. DOI: 10.1016/j.resconrec.2022.106156

Kim, Y., Zharkinbekov, Z., Raziyeva, K., Tabyldiyeva, L., Berikova, K., Zhumagul, D., Temirkhanova, K., & Saparov, A. (2023). Chitosan-Based Biomaterials for Tissue Regeneration. *Pharmaceutics*, 15(3), 807. DOI: 10.3390/pharmaceutics15030807 PMID: 36986668

Kołodziejska, M., Jankowska, K., Klak, M., & Wszoła, M. (2021). Chitosan as an Underrated Polymer in Modern Tissue Engineering. *Nanomaterials (Basel, Switzerland)*, 11(11), 3019. DOI: 10.3390/nano11113019 PMID: 34835782

Kong, H., Li, Q., Zheng, X., Chen, P., Zhang, G., & Huang, Z. (2023). Lanthanum modified chitosan-attapulgite composite for phosphate removal from water: Performance, mechanisms and applicability. *International Journal of Biological Macromolecules*, 224, 984–997. https://doi.org/https://doi.org/10.1016/j.ijbiomac.2022.10.183. DOI: 10.1016/j.ijbiomac.2022.10.183 PMID: 36306917

Kumar, S., Mukherjee, A., & Dutta, J. (2020). Chitosan based nanocomposite films and coatings: Emerging antimicrobial food packaging alternatives. *Trends in Food Science & Technology*, 97, 196–209. https://doi.org/https://doi.org/10.1016/j.tifs.2020.01.002. DOI: 10.1016/j.tifs.2020.01.002

Lee, S. J., Gwak, M. A., Chathuranga, K., Lee, J. S., Koo, J., & Park, W. H. (2023). Multifunctional chitosan/tannic acid composite films with improved anti-UV, antioxidant, and antimicrobial properties for active food packaging. *Food Hydrocolloids*, 136, 108249. https://doi.org/https://doi.org/10.1016/j.foodhyd.2022.108249. DOI: 10.1016/j.foodhyd.2022.108249

Lessa, E. F., Nunes, M. L., & Fajardo, A. R. (2018). Chitosan/waste coffee-grounds composite: An efficient and eco-friendly adsorbent for removal of pharmaceutical contaminants from water. *Carbohydrate Polymers*, 189, 257–266. https://doi.org/https://doi.org/10.1016/j.carbpol.2018.02.018. DOI: 10.1016/j.carbpol.2018.02.018 PMID: 29580407

Li, C., Li, F., Wang, K., Wang, Q., Liu, H., Sun, X., & Xie, D. (2023). Synthesis, characterizations, and release mechanisms of carboxymethyl chitosan-graphene oxide-gelatin composite hydrogel for controlled delivery of drug. *Inorganic Chemistry Communications*, 155, 110965. https://doi.org/https://doi.org/10.1016/j.inoche.2023.110965. DOI: 10.1016/j.inoche.2023.110965

Li, D., Tian, X., Wang, Z., Guan, Z., Li, X., Qiao, H., Ke, H., Luo, L., & Wei, Q. (2020). Multifunctional adsorbent based on metal-organic framework modified bacterial cellulose/chitosan composite aerogel for high efficient removal of heavy metal ion and organic pollutant. *Chemical Engineering Journal*, 383, 123127. https://doi.org/https://doi.org/10.1016/j.cej.2019.123127. DOI: 10.1016/j.cej.2019.123127

Li, L., Luo, C., Li, X., Duan, H., & Wang, X. (2014). Preparation of magnetic ionic liquid/chitosan/graphene oxide composite and application for water treatment. *International Journal of Biological Macromolecules*, 66, 172–178. https://doi.org/https://doi.org/10.1016/j.ijbiomac.2014.02.031. DOI: 10.1016/j.ijbiomac.2014.02.031 PMID: 24560948

Li, X., Zhou, H., Wu, W., Wei, S., Xu, Y., & Kuang, Y. (2015). Studies of heavy metal ion adsorption on Chitosan/Sulfydryl-functionalized graphene oxide composites. *Journal of Colloid and Interface Science*, 448, 389–397. https://doi.org/https://doi.org/10.1016/j.jcis.2015.02.039. DOI: 10.1016/j.jcis.2015.02.039 PMID: 25746192

Li, Y., Luo, X.-E., Tan, M.-J., Yue, F.-H., Yao, R.-Y., Zeng, X.-A., Woo, M.-W., Wen, Q.-H., & Han, Z. (2023). Preparation of carboxymethylcellulose / ZnO / chitosan composite hydrogel microbeads and its drug release behaviour. *International Journal of Biological Macromolecules*, 247, 125716. https://doi.org/https://doi.org/10.1016/j.ijbiomac.2023.125716. DOI: 10.1016/j.ijbiomac.2023.125716 PMID: 37419258

Liaw, B.-S., Chang, T.-T., Chang, H.-K., Liu, W.-K., & Chen, P.-Y. (2020). Fish scale-extracted hydroxyapatite/chitosan composite scaffolds fabricated by freeze casting—An innovative strategy for water treatment. *Journal of Hazardous Materials*, 382, 121082. https://doi.org/https://doi.org/10.1016/j.jhazmat.2019.121082. DOI: 10.1016/j.jhazmat.2019.121082 PMID: 31472467

Ling, C., Zhao, Y., Ren, Z., Han, J., Zhu, C., & Liu, F.-Q. (2019). Synergistic co-removal of zinc(II) and cefazolin by a Fe/amine-modified chitosan composite. *Chinese Chemical Letters*, 30(12), 2196–2200. https://doi.org/https://doi.org/10.1016/j.cclet.2019.09.035. DOI: 10.1016/j.cclet.2019.09.035

Liu, J., Chen, Y., Han, T., Cheng, M., Zhang, W., Long, J., & Fu, X. (2019). A biomimetic SiO_2@chitosan composite as highly-efficient adsorbent for removing heavy metal ions in drinking water. *Chemosphere*, 214, 738–742. https://doi.org/https://doi.org/10.1016/j.chemosphere.2018.09.172. DOI: 10.1016/j.chemosphere.2018.09.172 PMID: 30293027

Liu, J., Huang, J., Hu, Z., Li, G., Hu, L., Chen, X., & Hu, Y. (2021). Chitosan-based films with antioxidant of bamboo leaves and ZnO nanoparticles for application in active food packaging. *International Journal of Biological Macromolecules*, 189, 363–369. https://doi.org/https://doi.org/10.1016/j.ijbiomac.2021.08.136. DOI: 10.1016/j.ijbiomac.2021.08.136 PMID: 34450140

Liu, L., Xiao, Z., Niu, S., He, Y., Wang, G., Pei, X., Tao, W., & Wang, M. (2019). Preparation, characteristics and feeble induced-apoptosis performance of non-dialysis requiring selenium nanoparticles@chitosan. *Materials & Design*, 182, 108024. https://doi.org/https://doi.org/10.1016/j.matdes.2019.108024. DOI: 10.1016/j.matdes.2019.108024

Liu, Y., Hu, L., Tan, B., Li, J., Gao, X., He, Y., Du, X., Zhang, W., & Wang, W. (2019). Adsorption behavior of heavy metal ions from aqueous solution onto composite dextran-chitosan macromolecule resin adsorbent. *International Journal of Biological Macromolecules*, 141, 738–746. https://doi.org/https://doi.org/10.1016/j.ijbiomac.2019.09.044. DOI: 10.1016/j.ijbiomac.2019.09.044 PMID: 31499105

Liu, Y., Liu, R., Li, M., Yu, F., & He, C. (2019). Removal of pharmaceuticals by novel magnetic genipin-crosslinked chitosan/graphene oxide-SO3H composite. *Carbohydrate Polymers*, 220, 141–148. https://doi.org/https://doi.org/10.1016/j.carbpol.2019.05.060. DOI: 10.1016/j.carbpol.2019.05.060 PMID: 31196533

Liu, Y., Zhong, D., Xu, Y., Chang, H., Dong, L., Han, Z., Li, J., & Zhong, N. (2023). Adsorption of phosphate in water by La/Al bimetallic-organic frameworks-chitosan composite with wide adaptable pH range. *Journal of Environmental Chemical Engineering*, 11(5), 110309. https://doi.org/https://doi.org/10.1016/j.jece.2023.110309. DOI: 10.1016/j.jece.2023.110309

Lu, M., Zhou, Q., Yu, H., Chen, X., & Yuan, G. (2022). Colorimetric indicator based on chitosan/gelatin with nano-ZnO and black peanut seed coat anthocyanins for application in intelligent packaging. *Food Research International*, 160, 111664. https://doi.org/https://doi.org/10.1016/j.foodres.2022.111664. DOI: 10.1016/j.foodres.2022.111664 PMID: 36076380

Majdoubi, H., El Kaim Billah, R., Aminul Islam, M., Nazal, M. K., Shekhawat, A., Alrashdi, A. A., Alberto Lopez-Maldonado, E., Soulaimani, A., Tamraoui, Y., Jugade, R., & Lgaz, H. (2023). An eco-friendly chitosan-diethylaminoethyl cellulose composite: In-depth analysis of lead (II) and arsenic(V) decontamination from water with molecular perspectives. *Journal of Molecular Liquids*, 387, 122680. https://doi.org/https://doi.org/10.1016/j.molliq.2023.122680. DOI: 10.1016/j.molliq.2023.122680

Malesic-Eleftheriadou, N., Trikkaliotis, D. G., Evgenidou, E., Kyzas, G. Z., Bikiaris, D. N., & Lambropoulou, D. A. (2023). New biobased chitosan/polyvinyl alcohol/graphene oxide derivatives for the removal of pharmaceutical compounds from aqueous mixtures. *Journal of Molecular Liquids*, 387, 122673. https://doi.org/https://doi.org/10.1016/j.molliq.2023.122673. DOI: 10.1016/j.molliq.2023.122673

Marrakchi, F., Khanday, W. A., Asif, M., & Hameed, B. H. (2016). Cross-linked chitosan/sepiolite composite for the adsorption of methylene blue and reactive orange 16. *International Journal of Biological Macromolecules*, 93, 1231–1239. https://doi.org/https://doi.org/10.1016/j.ijbiomac.2016.09.069. DOI: 10.1016/j.ijbiomac.2016.09.069 PMID: 27663552

Merino, D., & Athanassiou, A. (2022). Biodegradable and Active Mulch Films: Hydrolyzed Lemon Peel Waste and Low Methoxyl Pectin Blends with Incorporated Biochar and Neem Essential Oil. *ACS Sustainable Chemistry & Engineering*, 10(33), 10789–10802. DOI: 10.1021/acssuschemeng.2c01539

Merino, D., Simonutti, R., Perotto, G., & Athanassiou, A. (2021). Direct transformation of industrial vegetable waste into bioplastic composites intended for agricultural mulch films []. *Green Chemistry*, 23(16), 5956-5971. https://doi.org/DOI: 10.1039/D1GC01316E

Michiels, Y., Puyvelde, P. V., & Sels, B. (2017). Barriers and Chemistry in a Bottle: Mechanisms in Today's Oxygen Barriers for Tomorrow's Materials. *Applied Sciences (Basel, Switzerland)*, 7(7), 665. DOI: 10.3390/app7070665

Mikunda, T., Brunner, L., Skylogianni, E., Monteiro, J., Rycroft, L., & Kemper, J. (2021). Carbon capture and storage and the sustainable development goals. *International Journal of Greenhouse Gas Control*, 108, 103318. https://doi.org/https://doi.org/10.1016/j.ijggc.2021.103318. DOI: 10.1016/j.ijggc.2021.103318

Momina, & Ahmad, K. (2023). Feasibility of the adsorption as a process for its large scale adoption across industries for the treatment of wastewater: Research gaps and economic assessment. *Journal of Cleaner Production, 388*, 136014. https://doi.org/https://doi.org/10.1016/j.jclepro.2023.136014

Moradi, G., Heydari, R., Zinadini, S., & Rahimi, M. (2023). Chitosan-furosemide/pectin surface functionalized thin film nanofiltration membrane with improved antifouling behavior for pharmaceutical wastewater treatment. *Journal of Industrial and Engineering Chemistry*, 124, 368–380. https://doi.org/https://doi.org/10.1016/j.jiec.2023.04.031. DOI: 10.1016/j.jiec.2023.04.031

More, A. (2019). Chitin Market 2019: Global Industry Trends, Future Growth, Regional Overview, Market Share, Size, Revenue, and Forecast Outlook till 2024. https://www.marketwatch.com/press-release/chitin-market-2019-global-industry-trends-future-growth-regional-overview-market-share-size-revenue-and-forecast-outlook-till-2024-2019-11-07 (The Express Wire)

Murat, E., Dixit, V., & Gitnic, G. (1998). Hepatocyte Attachment on Biodegradable Modified Chitosan Membranes: In Vitro Evaluation for the Development of Liver Organoids. *Artificial Organs*, 22(10), 837–846. https://doi.org/https://doi.org/10.1046/j.1525-1594.1998.06182.x. DOI: 10.1046/j.1525-1594.1998.06182.x PMID: 9790081

Nagalakshmaiah, M., Afrin, S., Malladi, R. P., Elkoun, S., Robert, M., Ansari, M. A., Svedberg, A., & Karim, Z. (2019). Biocomposites: Present trends and challenges for the future. In Koronis, G., & Silva, A. (Eds.), *Green Composites for Automotive Applications* (pp. 197–215). Woodhead Publishing., https://doi.org/https://doi.org/10.1016/B978-0-08-102177-4.00009-4 DOI: 10.1016/B978-0-08-102177-4.00009-4

Narsimha, A., & Sudarshan, V. (2018). Drinking water pollution with respective of fluoride in the semi-arid region of Basara, Nirmal district, Telangana State, India. *Data in Brief*, 16, 752–757. https://doi.org/https://doi.org/10.1016/j.dib.2017.11.087. DOI: 10.1016/j.dib.2017.11.087 PMID: 29270457

Nayak, A., Bhushan, B., & Kotnala, S. (2023). Evaluation of hydroxyapatite-chitosan-magnetite nanocomposites for separation of pharmaceuticals from water: A mechanistic and comparative approach. *Journal of Hazardous Materials Advances*, 10, 100308. https://doi.org/https://doi.org/10.1016/j.hazadv.2023.100308. DOI: 10.1016/j.hazadv.2023.100308

Noohian, M., & Mahmoudi, J. (2023). Energy simulation on how to go green buildings in an earth's dry climate. *International Journal of Thermofluids*, 20, 100405. https://doi.org/https://doi.org/10.1016/j.ijft.2023.100405. DOI: 10.1016/j.ijft.2023.100405

Palansooriya, K. N., Kim, S., Igalavithana, A. D., Hashimoto, Y., Choi, Y.-E., Mukhopadhyay, R., Sarkar, B., & Ok, Y. S. (2021). Fe(III) loaded chitosan-biochar composite fibers for the removal of phosphate from water. *Journal of Hazardous Materials*, 415, 125464. https://doi.org/https://doi.org/10.1016/j.jhazmat.2021.125464. DOI: 10.1016/j.jhazmat.2021.125464 PMID: 33730647

Pandey, V. K., Upadhyay, S. N., Niranjan, K., & Mishra, P. K. (2020). Antimicrobial biodegradable chitosan-based composite Nano-layers for food packaging. *International Journal of Biological Macromolecules*, 157, 212–219. https://doi.org/https://doi.org/10.1016/j.ijbiomac.2020.04.149. DOI: 10.1016/j.ijbiomac.2020.04.149 PMID: 32339572

Pereira, V. A.Jr, de Arruda, I. N. Q., & Stefani, R. (2015). Active chitosan/PVA films with anthocyanins from Brassica oleraceae (Red Cabbage) as Time–Temperature Indicators for application in intelligent food packaging. *Food Hydrocolloids*, 43, 180–188. https://doi.org/https://doi.org/10.1016/j.foodhyd.2014.05.014. DOI: 10.1016/j.foodhyd.2014.05.014

Perotto, G., Ceseracciu, L., Simonutti, R., Paul, U. C., Guzman-Puyol, S., Tran, T.-N., Bayer, I. S., & Athanassiou, A. (2018). Bioplastics from vegetable waste via an eco-friendly water-based process []. *Green Chemistry, 20*(4), 894-902. https://doi.org/DOI: 10.1039/C7GC03368K

Prasad, K. S., Amin, Y., & Selvaraj, K. (2014). Defluoridation using biomimetically synthesized nano zirconium chitosan composite: Kinetic and equilibrium studies. *Journal of Hazardous Materials*, 276, 232–240. https://doi.org/https://doi.org/10.1016/j.jhazmat.2014.05.038. DOI: 10.1016/j.jhazmat.2014.05.038 PMID: 24887125

Preethi, J., Karthikeyan, P., Vigneshwaran, S., & Meenakshi, S. (2021). Facile synthesis of Zr4+ incorporated chitosan/gelatin composite for the sequestration of Chromium(VI) and fluoride from water. *Chemosphere*, 262, 128317. https://doi.org/https://doi.org/10.1016/j.chemosphere.2020.128317. DOI: 10.1016/j.chemosphere.2020.128317 PMID: 33182083

Preethi, J., Prabhu, S. M., & Meenakshi, S. (2017). Effective adsorption of hexavalent chromium using biopolymer assisted oxyhydroxide materials from aqueous solution. *Reactive & Functional Polymers*, 117, 16–24. https://doi.org/https://doi.org/10.1016/j.reactfunctpolym.2017.05.006. DOI: 10.1016/j.reactfunctpolym.2017.05.006

Preethi, J., Vigneshwaran, S., & Meenakshi, S. (2019). Performance of chitosan engraved iron and lanthanum mixed oxyhydroxide for the detoxification of hexavalent chromium. *International Journal of Biological Macromolecules*, 130, 491–498. https://doi.org/https://doi.org/10.1016/j.ijbiomac.2019.02.101. DOI: 10.1016/j.ijbiomac.2019.02.101 PMID: 30794904

Purohit, S., Chini, M. K., Chakraborty, T., Yadav, K. L., & Satapathi, S. (2020). Rapid removal of arsenic from water using metal oxide doped recyclable cross-linked chitosan cryogel. *SN Applied Sciences*, 2(4), 768. DOI: 10.1007/s42452-020-2525-6

Qi, J., Zhang, G., & Li, H. (2015). Efficient removal of arsenic from water using a granular adsorbent: Fe–Mn binary oxide impregnated chitosan bead. *Bioresource Technology*, 193, 243–249. https://doi.org/https://doi.org/10.1016/j.biortech.2015.06.102. DOI: 10.1016/j.biortech.2015.06.102 PMID: 26141284

Reis, E. S., Gorza, F. D. S., Pedro, G. C., Maciel, B. G., da Silva, R. J., Ratkovski, G. P., & de Melo, C. P. (2021). (Maghemite/Chitosan/Polypyrrole) nanocomposites for the efficient removal of Cr (VI) from aqueous media. *Journal of Environmental Chemical Engineering*, 9(1), 104893. https://doi.org/https://doi.org/10.1016/j.jece.2020.104893. DOI: 10.1016/j.jece.2020.104893

Research, P. (2022). *Plastic Packaging Market.* https://www.precedenceresearch.com/plastic-packaging-market

Research, T. M. (2016). *Crustacean Materket - Global Industry Analysis, Size, Share, Growth, Trends, and Forecast 2016 - 2026*https://www.transparencymarketresearch.com/crustacean-market.html

Rotiroti, M., Sacchi, E., Caschetto, M., Zanotti, C., Fumagalli, L., Biasibetti, M., Bonomi, T., & Leoni, B. (2023). Groundwater and surface water nitrate pollution in an intensively irrigated system: Sources, dynamics and adaptation to climate change. *Journal of Hydrology (Amsterdam)*, 623, 129868. https://doi.org/https://doi.org/10.1016/j.jhydrol.2023.129868. DOI: 10.1016/j.jhydrol.2023.129868

Rupasinghe, N. K. L. C., Senanayake, S. M. A. E., & Nanayakkara, K. G. N. (2022). Development, characterization and mechanisms study of protonated sawdust biochar-chitosan composite bead biosorbent for defluoridation of contaminated groundwater. *Bioresource Technology Reports*, 17, 100946. https://doi.org/https://doi.org/10.1016/j.biteb.2022.100946. DOI: 10.1016/j.biteb.2022.100946

Sadiq, A. C., Olasupo, A., Rahim, N. Y., Ngah, W. S. W., Hanafiah, M. A. K. M., & Suah, F. B. M. (2023). Fabrication and characterisation of novel chitosan-based polymer inclusion membranes and their application in environmental remediation. *International Journal of Biological Macromolecules*, 244, 125400. https://doi.org/https://doi.org/10.1016/j.ijbiomac.2023.125400. DOI: 10.1016/j.ijbiomac.2023.125400 PMID: 37330084

Sadrolhosseini, A. R., Naseri, M., & Rashid, S. A. (2017). Polypyrrole-chitosan/nickel-ferrite nanoparticle composite layer for detecting heavy metal ions using surface plasmon resonance technique. *Optics & Laser Technology*, 93, 216–223. https://doi.org/https://doi.org/10.1016/j.optlastec.2017.03.008. DOI: 10.1016/j.optlastec.2017.03.008

Saka, D., Adu-Gyamfi, J., Skrzypek, G., Antwi, E. O., Heng, L., & Torres-Martínez, J. A. (2023). Disentangling nitrate pollution sources and apportionment in a tropical agricultural ecosystem using a multi-stable isotope model. *Environmental Pollution*, 328, 121589.

Salehi, S., Alijani, S., & Anbia, M. (2020). Enhanced adsorption properties of zirconium modified chitosan-zeolite nanocomposites for vanadium ion removal. *International Journal of Biological Macromolecules*, 164, 105–120. https://doi.org/https://doi.org/10.1016/j.ijbiomac.2020.07.055. DOI: 10.1016/j.ijbiomac.2020.07.055 PMID: 32652153

Salman, M. S., Sheikh, M. C., Hasan, M. M., Hasan, M. N., Kubra, K. T., Rehan, A. I., Awual, M. E., Rasee, A. I., Waliullah, R. M., Hossain, M. S., Khaleque, M. A., Alsukaibi, A. K. D., Alshammari, H. M., & Awual, M. R. (2023). Chitosan-coated cotton fiber composite for efficient toxic dye encapsulation from aqueous media. *Applied Surface Science*, 622, 157008. https://doi.org/https://doi.org/10.1016/j.apsusc.2023.157008. DOI: 10.1016/j.apsusc.2023.157008

Samuel, M. S., Bhattacharya, J., Raj, S., Santhanam, N., Singh, H., & Pradeep Singh, N. D. (2019). Efficient removal of Chromium(VI) from aqueous solution using chitosan grafted graphene oxide (CS-GO) nanocomposite. *International Journal of Biological Macromolecules*, 121, 285–292. https://doi.org/https://doi.org/10.1016/j.ijbiomac.2018.09.170. DOI: 10.1016/j.ijbiomac.2018.09.170 PMID: 30267821

Samuel, M. S., Shah, S. S., Subramaniyan, V., Qureshi, T., Bhattacharya, J., & Pradeep Singh, N. D. (2018). Preparation of graphene oxide/chitosan/ferrite nanocomposite for Chromium(VI) removal from aqueous solution. *International Journal of Biological Macromolecules*, 119, 540–547. https://doi.org/ https://doi.org/10.1016/j.ijbiomac.2018.07.052. DOI: 10.1016/j.ijbiomac.2018.07.052 PMID: 30009902

Sanmugam, A., Abbishek, S., Kumar, S. L., Sairam, A. B., Palem, V. V., Kumar, R. S., Almansour, A. I., Arumugam, N., & Vikraman, D. (2023). Synthesis of chitosan based reduced graphene oxide-CeO2 nanocomposites for drug delivery and antibacterial applications. *Journal of the Mechanical Behavior of Biomedical Materials*, 145, 106033. https://doi.org/https://doi.org/10.1016/j.jmbbm.2023.106033. DOI: 10.1016/j.jmbbm.2023.106033 PMID: 37478544

Satapathi, S. (2017). Graphene-based 3D xerogel as adsorbent for removal of heavy metal ions from industrial wastewater. *Journal of Renewable Materials*, 5(2), 96–102.

Schmitz, C., González Auza, L., Koberidze, D., Rasche, S., Fischer, R., & Bortesi, L. (2019). Conversion of Chitin to Defined Chitosan Oligomers: Current Status and Future Prospects. *Marine Drugs*, 17(8), 452. DOI: 10.3390/md17080452 PMID: 31374920

Sharma, S., & Bhattacharya, A. (2017). Drinking water contamination and treatment techniques. *Applied Water Science*, 7(3), 1043–1067. DOI: 10.1007/s13201-016-0455-7

Sheshmani, S., Akhundi Nematzadeh, M., Shokrollahzadeh, S., & Ashori, A. (2015). Preparation of graphene oxide/chitosan/FeOOH nanocomposite for the removal of Pb(II) from aqueous solution. *International Journal of Biological Macromolecules*, 80, 475–480. https://doi.org/https://doi.org/10.1016/j.ijbiomac.2015.07.009. DOI: 10.1016/j.ijbiomac.2015.07.009 PMID: 26187194

Smyntek, P. M., Lamagna, N., Cravotta, C. A. III, & Strosnider, W. H. J. (2022). Mine drainage precipitates attenuate and conceal wastewater-derived phosphate pollution in stream water. *The Science of the Total Environment*, 815, 152672. https://doi.org/https://doi.org/10.1016/j.scitotenv.2021.152672. DOI: 10.1016/j.scitotenv.2021.152672 PMID: 34968601

Son Tran, V., Hao Ngo, H., Guo, W., Ha Nguyen, T., Mai Ly Luong, T., Huan Nguyen, X., Lan Anh Phan, T., Trong Le, V., Phuong Nguyen, M., & Khai Nguyen, M. (2023). New chitosan-biochar composite derived from agricultural waste for removing sulfamethoxazole antibiotics in water. *Bioresource Technology*, 385, 129384. https://doi.org/https://doi.org/10.1016/j.biortech.2023.129384. DOI: 10.1016/j.biortech.2023.129384 PMID: 37355142

Song, W., Huang, T., Zuo, H., Deng, D., & Tang, C. (2022). Application of microbial immobilization on chitosan composite membrane for manganese removal in water treatment. *Polymer*, 243, 124531. https://doi.org/https://doi.org/10.1016/j.polymer.2022.124531. DOI: 10.1016/j.polymer.2022.124531

Song, X., Li, L., Zhou, L., & Chen, P. (2018). Magnetic thiolated/quaternized-chitosan composites design and application for various heavy metal ions removal, including cation and anion. *Chemical Engineering Research & Design*, 136, 581–592. https://doi.org/https://doi.org/10.1016/j.cherd.2018.06.025. DOI: 10.1016/j.cherd.2018.06.025

Sultan, S., & Hasnain, S. (2005). Chromate Reduction Capability of a Gram Positive Bacterium Isolated from Effluent of Dying Industry. *Bulletin of Environmental Contamination and Toxicology*, 75(4), 699–706. DOI: 10.1007/s00128-005-0808-7 PMID: 16400550

Terzioğlu, P., Güney, F., Parın, F. N., Şen, İ., & Tuna, S. (2021). Biowaste orange peel incorporated chitosan/polyvinyl alcohol composite films for food packaging applications. *Food Packaging and Shelf Life*, 30, 100742. https://doi.org/https://doi.org/10.1016/j.fpsl.2021.100742. DOI: 10.1016/j.fpsl.2021.100742

Thacker, U., Parikh, R., Shouche, Y., & Madamwar, D. (2006). Hexavalent chromium reduction by Providencia sp. *Process Biochemistry (Barking, London, England)*, 41(6), 1332–1337. https://doi.org/https://doi.org/10.1016/j.procbio.2006.01.006. DOI: 10.1016/j.procbio.2006.01.006

Thambiliyagodage, C., Jayanetti, M., Mendis, A., Ekanayake, G., Liyanaarachchi, H., & Vigneswaran, S. (2023). Recent Advances in Chitosan-Based Applications—A Review. *Materials (Basel)*, 16(5), 2073. DOI: 10.3390/ma16052073 PMID: 36903188

Thoniyot, P., Tan, M. J., Karim, A. A., Young, D. J., & Loh, X. J. (2015). Nanoparticle–Hydrogel Composites: Concept, Design, and Applications of These Promising, Multi-Functional Materials. *Advancement of Science*, 2(1-2), 1400010. https://doi.org/https://doi.org/10.1002/advs.201400010. DOI: 10.1002/advs.201400010 PMID: 27980900

Torre-Celeizabal, A., Casado-Coterillo, C., Gomis-Berenguer, A., Iniesta, J., & Garea, A. (2023). Chitosan-based mixed matrix composite membranes for CO2/CH4 mixed gas separation. Experimental characterization and performance validation. *Separation and Purification Technology*, 325, 124535. https://doi.org/https://doi.org/10.1016/j.seppur.2023.124535. DOI: 10.1016/j.seppur.2023.124535

Tran, M. L., Tran, T. T. V., Juang, R.-S., & Nguyen, C. H. (2023). Graphene oxide crosslinked chitosan composites for enhanced adsorption of cationic dye from aqueous solutions. *Journal of the Taiwan Institute of Chemical Engineers*, 142, 104678. https://doi.org/https://doi.org/10.1016/j.jtice.2023.104678. DOI: 10.1016/j.jtice.2023.104678

Tran Vo, T. M., Piroonpan, T., Preuksarattanawut, C., Kobayashi, T., & Potiyaraj, P. (2022). Characterization of pH-responsive high molecular-weight chitosan/poly (vinyl alcohol) hydrogel prepared by gamma irradiation for localizing drug release. *Bioresources and Bioprocessing*, 9(1), 89. DOI: 10.1186/s40643-022-00576-6 PMID: 38647766

Trang, T., Takaomi, K., & Bui, H. (2018). Chitosan/zeolite composite membranes for eliminating trace metal ions in the evacuation permeability process. *Journal of the Serbian Chemical Society*, 84(1), 85–85. DOI: 10.2298/JSC180606085T

Valizadeh, K., Bateni, A., Sojoodi, N., Rafiei, R., Behroozi, A. H., & Maleki, A. (2023). Preparation and characterization of chitosan-curdlan composite magnetized by zinc ferrite for efficient adsorption of tetracycline antibiotics in water. *International Journal of Biological Macromolecules*, 235, 123826. https://doi.org/https://doi.org/10.1016/j.ijbiomac.2023.123826. DOI: 10.1016/j.ijbiomac.2023.123826 PMID: 36828094

Viswanathan, N., Pandi, K., & Meenakshi, S. (2014). Synthesis of metal ion entrapped silica gel/chitosan biocomposite for defluoridation studies. *International Journal of Biological Macromolecules*, 70, 347–353. https://doi.org/https://doi.org/10.1016/j.ijbiomac.2014.06.010. DOI: 10.1016/j.ijbiomac.2014.06.010 PMID: 25008134

Vithalkar, S. H., & Jugade, R. M. (2020). Adsorptive removal of crystal violet from aqueous solution by cross-linked chitosan coated bentonite. *Materials Today: Proceedings*, 29, 1025–1032. https://doi.org/https://doi.org/10.1016/j.matpr.2020.04.705. DOI: 10.1016/j.matpr.2020.04.705

Wang, C., Gong, C., Qin, Y., Hu, Y., Jiao, A., Jin, Z., Qiu, C., & Wang, J. (2022). Bioactive and functional biodegradable packaging films reinforced with nanoparticles. *Journal of Food Engineering*, 312, 110752. https://doi.org/https://doi.org/10.1016/j.jfoodeng.2021.110752. DOI: 10.1016/j.jfoodeng.2021.110752

Wang, K., Li, F., Sun, X., Wang, F., Xie, D., & Wei, Y. (2023). Transparent chitosan/hexagonal boron nitride nanosheets composite films with enhanced UV shielding and gas barrier properties. *International Journal of Biological Macromolecules*, 251, 126308. https://doi.org/https://doi.org/10.1016/j.ijbiomac.2023.126308. DOI: 10.1016/j.ijbiomac.2023.126308 PMID: 37573919

Wang, L., Elahi, E., Zhou, Y., Wang, L., & Zhang, S. (2022). A Review of Packaging Materials’ Consumption Regulation and Pollution Control. *Sustainability*, 14(23).

Yadav, K. K., Kumar, S., Pham, Q. B., Gupta, N., Rezania, S., Kamyab, H., Yadav, S., Vymazal, J., Kumar, V., Tri, D. Q., Talaiekhozani, A., Prasad, S., Reece, L. M., Singh, N., Maurya, P. K., & Cho, J. (2019). Fluoride contamination, health problems and remediation methods in Asian groundwater: A comprehensive review. *Ecotoxicology and Environmental Safety*, 182, 109362. https://doi.org/https://doi.org/10.1016/j.ecoenv.2019.06.045. DOI: 10.1016/j.ecoenv.2019.06.045 PMID: 31254856

Yazdi, F., Anbia, M., & Salehi, S. (2019). Characterization of functionalized chitosan-clinoptilolite nanocomposites for nitrate removal from aqueous media. *International Journal of Biological Macromolecules*, 130, 545–555. https://doi.org/https://doi.org/10.1016/j.ijbiomac.2019.02.127. DOI: 10.1016/j.ijbiomac.2019.02.127 PMID: 30807801

Yazdi, F., Anbia, M., & Sepehrian, M. (2023). Recent advances in removal of inorganic anions from water by chitosan-based composites: A comprehensive review. *Carbohydrate Polymers*, 320, 121230. https://doi.org/https://doi.org/10.1016/j.carbpol.2023.121230. DOI: 10.1016/j.carbpol.2023.121230 PMID: 37659817

Yilmaz Atay, H. (2019). Antibacterial Activity of Chitosan-Based Systems. In Jana, S., & Jana, S. (Eds.), *Functional Chitosan: Drug Delivery and Biomedical Applications* (pp. 457–489). Springer Singapore., DOI: 10.1007/978-981-15-0263-7_15

Younes, I., & Rinaudo, M. (2015). Chitin and Chitosan Preparation from Marine Sources. Structure, Properties and Applications. *Marine Drugs*, 13(3), 1133–1174. DOI: 10.3390/md13031133 PMID: 25738328

Yusefi, M. S. K., Jahangirian H, Teow SY, Afsah-Hejri L, Mohamad Sukri SNA, Kuča K. (2023). How Magnetic Composites are Effective Anticancer Therapeutics? A Comprehensive Review of the Literature. *International Journal of Nanomedicine, 18*, 3535-3575. https://doi.org/https://doi.org/10.2147/IJN.S375964

Zafeiris, K., Brasinika, D., Karatza, A., Koumoulos, E., Karoussis, I. K., Kyriakidou, K., & Charitidis, C. A. (2021). Additive manufacturing of hydroxyapatite–chitosan–genipin composite scaffolds for bone tissue engineering applications. *Materials Science and Engineering C*, 119, 111639. https://doi.org/https://doi.org/10.1016/j.msec.2020.111639. DOI: 10.1016/j.msec.2020.111639 PMID: 33321677

Zeraatkar Moghaddam, A., Esmaeilkhanian, E., & Shakourian-Fard, M. (2019). Immobilizing magnetic glutaraldehyde cross-linked chitosan on graphene oxide and nitrogen-doped graphene oxide as well-dispersible adsorbents for chromate removal from aqueous solutions. *International Journal of Biological Macromolecules*, 128, 61–73. https://doi.org/https://doi.org/10.1016/j.ijbiomac.2019.01.086. DOI: 10.1016/j.ijbiomac.2019.01.086 PMID: 30682473

Zhang, B., Zhao, Z., Ma, R., Chen, N., Kong, Z., Lei, Z., & Zhang, Z. (2024). Unveiling the mechanisms of Fe(III)-loaded chitosan composite (CTS-Fe) in enhancing anaerobic digestion of waste activated sludge. *Journal of Environmental Sciences (China)*, 138, 200–211. https://doi.org/https://doi.org/10.1016/j.jes.2023.04.001. DOI: 10.1016/j.jes.2023.04.001 PMID: 38135389

Zhang, J., Chen, N., Su, P., Li, M., & Feng, C. (2017). Fluoride removal from aqueous solution by Zirconium-Chitosan/Graphene Oxide Membrane. *Reactive & Functional Polymers*, 114, 127–135. https://doi.org/https://doi.org/10.1016/j.reactfunctpolym.2017.03.008. DOI: 10.1016/j.reactfunctpolym.2017.03.008

Zhang, S., Dong, Y., Yang, Z., Yang, W., Wu, J., & Dong, C. (2016). Adsorption of pharmaceuticals on chitosan-based magnetic composite particles with core-brush topology. *Chemical Engineering Journal*, 304, 325–334. https://doi.org/https://doi.org/10.1016/j.cej.2016.06.087. DOI: 10.1016/j.cej.2016.06.087

Zhang, S., Luo, Y., Zeng, H., Wang, Q., Tian, F., Song, J., & Cheng, W.-H. (2011). Encapsulation of selenium in chitosan nanoparticles improves selenium availability and protects cells from selenium-induced DNA damage response. *The Journal of Nutritional Biochemistry*, 22(12), 1137–1142. https://doi.org/https://doi.org/10.1016/j.jnutbio.2010.09.014. DOI: 10.1016/j.jnutbio.2010.09.014 PMID: 21292467

Zhang, Z., Abidi, N., & Lucia, L. (2023). Smart superabsorbent alginate/carboxymethyl chitosan composite hydrogel beads as efficient biosorbents for methylene blue dye removal. *Journal of Materials Science and Technology*, 159, 81–90. https://doi.org/https://doi.org/10.1016/j.jmst.2023.02.045. DOI: 10.1016/j.jmst.2023.02.045

Zhou, J., Leavitt, P. R., Zhang, Y., & Qin, B. (2022). Anthropogenic eutrophication of shallow lakes: Is it occasional? *Water Research*, 221, 118728. https://doi.org/https://doi.org/10.1016/j.watres.2022.118728. DOI: 10.1016/j.watres.2022.118728 PMID: 35717711

Zhou, X., Dong, C., Yang, Z., Tian, Z., Lu, L., Yang, W., Wang, Y., Zhang, L., Li, A., & Chen, J. (2018). Enhanced adsorption of pharmaceuticals onto core-brush shaped aromatic rings-functionalized chitosan magnetic composite particles: Effects of structural characteristics of both pharmaceuticals and brushes. *Journal of Cleaner Production*, 172, 1025–1034. https://doi.org/https://doi.org/10.1016/j.jclepro.2017.10.207. DOI: 10.1016/j.jclepro.2017.10.207

Zidan, N., Alalawy, A. I., Al-Duais, M. A., Alzahrani, S., Kasem, M., Tayel, A. A., & Nagib, R. M. (2023). Active and smart antimicrobial food packaging film composed of date palm kernels extract loaded carboxymethyl chitosan and carboxymethyl starch composite for prohibiting foodborne pathogens during fruits preservation. *European Polymer Journal*, 197, 112353.

Chapter 8
Chitosan–Based Hydrogels:
Current Strategic Fabrication and Practical Application Perspectives

Tu Minh Tran Vo
https://orcid.org/0000-0002-6785-4188
Chulalongkorn University, Thailand

Takaomi Kobayashi
Nagaoka University of Technology, Japan

ABSTRACT

Chitosan holds appeal as an inventing sustainable green material because of the sources of crab and shrimp shell wastes and encompasses applications in the medical, agriculture, and water treatment fields, and the biomass polymer consisting glycoconjugate backbone having amino groups makes it water soluble and can be gelatinized. This chapter highlights the properties of chitosan hydrogels used as several applications. Notably, providing a better understanding of hydrogels and utilization of chitosan-based hydrogels, the preparation methods of chitosan hydrogels involving numerous "green" techniques. The promising properties of chitosan hydrogels are extensively explored in this chapter. Also, the unique properties of chitosan making these hydrogels a powerful bio-friendly material are described in agriculture purpose, applied for slow control release of fertilizer, and in water purification

INTRODUCTION

Background of Chitosan Hydrogel

Biomass hydrogels represent an innovative and emerging class of sustainable materials that are gaining prominence in various fields of research and development since the availability of abundant natural polysaccharide sources and the regenerative potential of these natural polymers from biomass resources and waste materials are primary factors driving the creation of eco-friendly materials. Because of the huge amounts of water retention in the hydrogels, closing human body environment, effective delivery can be accomplished with polymeric systems designed to serve as drug carriers or fertilizer transporters that control release rates (Sung & Kim, 2020). Among these polymer materials, biomass chitosan has a linear polysaccharide composed of repeating units of glucosamine and *N*-acetylglucosamine, as shown in Figure 1, and exhibits impressive adsorption characteristics because of the presence of highly functional groups, namely, amines and hydroxyls, within its chemical structure, which serve as active sites for the adsorption of metal ions (Upadhyay et al., 2021). Chitosan-based hydrogel typically involves combining a polymer with an active cross-linkage agent to form a heterogeneous solid matrix retaining a large amount of water. Hydrogels have garnered considerable attention from all polymers due to their specific properties, such as biocompatibility and hydrophilicity. These properties enable hydrogels to absorb water and swell while maintaining their structural integrity, making them an attractive option for developing drug delivery systems (Kesharwani et al., 2021).

Figure 1. Chemical structure of chitosan

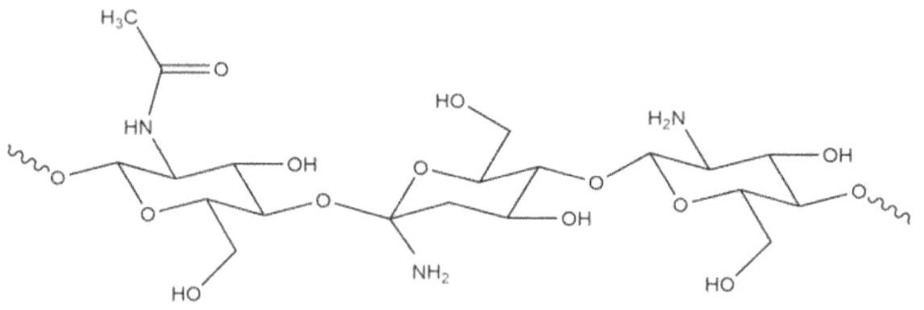

Furthermore, smart chitosan hydrogels can respond to external stimuli, such as changes in temperature (Don et al., 2008), pH (Hrubý et al., 2005), light (Shamay et al., 2011), or ultrasound (Wei et al., 2021), by undergoing reversible changes in their swelling behavior, structure, or permeability. These unique properties make them promising candidates for drug delivery systems (DDS), as they can provide a controllably targeted release of drugs in response to specific stimuli. The development of smart hydrogels for DDS is challenging, as they require careful selection and design of the hydrogel composition, crosslinking density, and functional groups that can respond to the desired stimuli. It is known that chitosan is a versatile biomaterial that is produced by the deacetylation of chitin, resulting in a polymer with a cationic charge due to the presence of amino groups in its structure (Casadidio et al., 2019). Thus, chitosan possesses antimicrobial, antioxidant, and immunomodulatory properties (Xia et al., 2022) which can provide a biocompatible and biodegradable platform for drug delivery, minimizing the risk of adverse effects and enabling safe and efficient drug delivery. However, chitosan hydrogels have lower mechanical strength compared to synthetic hydrogels (Ye et al., 2023) limiting their use in practical applications such as DDSs, remediation process, or fertilizer carrier that require high mechanical stability. Pure chitosan hydrogels have a tendency to swell rapidly and can reach their maximum swelling capacity quickly leading to burst release in a short period of time (Martino et al., 2017). To overcome this obstacle, crosslinking chitosan hydrogel gained the attention of scientists through chemical reactions with crosslinker agents or reinforced by synthetic polymers.

Hydrogelation Processes of Chitosan

There are two pathways to synthesizing hydrogel: physical and chemical crosslinking. The physical method involves the formation of a hydrogel network through non-covalent interactions between polymer chains. The most common physical method for preparing chitosan hydrogels is through the process of heating or cooling due to the helix formation, association of the helices, or creating junction zones. Additionally, physical cross-linking hydrogels are formed via ionic interaction, hydrogen bonding, and freeze-thawing (Pita-López et al., 2021). On the other hand, chemically cross-linked hydrogels can be obtained by (i) radical polymerization of low molecular weight monomers in the presence of crosslinking agents, (ii) chemical reaction of complementary groups: the presence of some hydrophilic functional groups in their structures, mainly hydroxyl (OH), amide ($CONH_2$), and amine (NH_2) can be crosslinked by aldehyde or epoxy, (iii) high energy radiation such as gamma, electron beams, or X-rays (Qin et al., 2018). The advantage of chemical crosslinks compared to physical crosslinks is to provide enhanced mechanical properties which are necessary for various applications (Dodero et al., 2021). Unfortunately, the most commonly used crosslinking agents for chitosan hydrogels, such as glutaraldehyde and formaldehyde, have some disadvantages for drug delivery applications. These agents can be toxic and their use can result in the formation of unexpected by-products, which can affect the properties of the hydrogel and potentially cause harm to the patient (McGregor et al., 2006). Therefore, there is a need to develop alternative crosslinking methods that are safer and more effective for use in drug delivery application and other approaches to reach the standard of Sustainable Development Goals (SDGs).

For this reason, the main aim of these current studies is to investigate in the point of view for a biosafety system externally employing chitosan hydrogels as carriers for medication and fertilizers. Moreover, irradiation technologies in the gelation process (M. A. Raza et al., 2021) by chemically crosslinkable substance are considered as aspects of green methods, when crosslinker fabricates the hydrogel matrix which plays a crucial factor for smart DDSs in the medical fields. In this current chapter, several crucial points are considered stating in following points of the properties of chitosan hydrogels, fabrication of chitosan-based hydrogels via "green" approaches, utilization of chitosan-based hydrogels in water treatment, agriculture, and DDSs.

Properties of Chitosan Hydrogels for Sustainable Development

The physicochemical properties of chitosan hydrogels (Figure 2) have a significant impact on the applications, making them important considerations in material engineering. This section is dedicated to deles into the properties of hydrogel, which are primarily responsible for its behavior and its utilization.

Figure 2. Illustration of promising properties of chitosan (CS)-based hydrogels

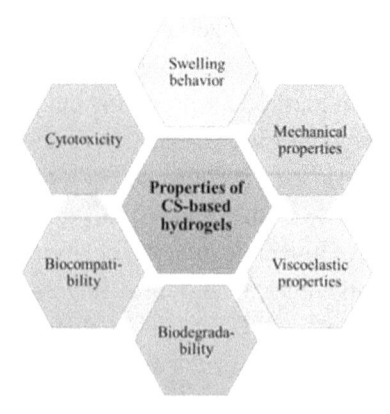

Swelling Behavior

It is known that hydrogels are characterized by their ability to hold large amounts of water and thus retain both solid and liquid properties due to the following reason that the presence of the hydrophilic group in the network give hydrogel the ability to absorb, swell, and retain a high amount of water. Sometimes, as characteristic of hydrogels, this becomes up to several thousand times of their own dried weight. To determine the swelling behavior of hydrogels, the sample is immersed in distilled water at room temperature and monitored by frequent weighing. To ensure accuracy, the swollen polymer is removed from the water and any surface moisture is wiped with tissue paper before weighing (Ahmed, 2015). The swelling degree (SD) is used to express the ability of the hydrogel to swell, and is calculated using the equation (1). Where W_d is the weight of the dried hydrogel (g), W_t is the weight of the swollen hydrogels at time t (g).

$$SD = \frac{(W_t - W_d) \times 100}{W_d} \% \tag{1}$$

The swelling behavior of chitosan hydrogels was pH-dependent, influenced by the dissociation of $-NH_3^+$ ions, which is characterized by a dissociation rate constant known as pKa (Spinks et al., 2006), which was reported as pKa = 6.3. The swelling properties of chitosan-based porous hydrogels are significantly impacted by the methods of drying and the density of crosslinking (Park et al., 2006). Furthermore, new pH-sensitive chitosan hydrogels were developed and extensively studied to comprehend their swelling behavior and water states (Qu et al., 2000). These investigations emphasize the importance of comprehending the swelling characteristics of chitosan hydrogels under diverse conditions, as they hold potential for applications in drug delivery, tissue engineering, agricultural issues, and water treatment. Qu et al also demonstrated significant swelling behavior under acidic conditions, while their swelling remains relatively stable within the pH range of 6 to 10 (Qu et al., 2000). The high swelling observed at low pH is attributed to the electrostatic repulsion between ionized amino groups, causing the positive charge-charge repulsion of NH_3^+ in the hydrogels and thus the volumatic hydrogel expands. This expansion facilitates the penetration of water and acidic substances into the hydrogel sample, contributing to its swelling behavior.

Chitosan and its derivative-based hydrogel have garnered attention due to their capability of absorbent materials. Its amino groups readily form chelates or ionic bonds with metal ions, so that such hydrogels can be applied for metal adsorbents, for example, Francis et al. conducted a review on gel adsorbents composed of chitosan, examining and deliberating upon the adsorption mechanisms of these gels concerning Cr(IV), Cd(II), phenol, and methylene blue (Francis et al., 2023). In addition, controlled-release fertilizer characteristics, were applied in agriculture field. A new type of superabsorbent called chitosan-g-poly(acrylic acid)/sodium humate (CTS-g-PAA/sodium humate) was developed, and the swelling behaviors of this superabsorbent were explored, highlighting its biodegradable and environmentally friendly attributes. The inclusion of sodium humate led to an enhancement in the water absorption capacity of the CTS-g-PAA superabsorbent. This innovation not only brings about a significant reduction in production costs but also introduces novel functional properties. The most notable absorption performance was observed, reaching 183 grams per gram of material in distilled water and 41 grams per gram of material in a 0.9 wt% NaCl solution (Liu et al., 2007).

Such swelling hydrogel can more potentially hold a greater quantity of drugs, thus influencing the overall drug loading capacity (Wan et al., 1993). A limitation inherent to swelling-controlled systems lies in their comparatively sluggish response, particularly in the case of larger hydrogels, where the gradual diffusion of water hampers the speed of change. To accelerate this process, one approach involves reducing the size of hydrogel, thereby shortening the diffusion path. Another method is to create a network of interconnected macropores within the hydrogel structure (Ankareddi & Brazel, 2007). Meanwhile, the swelling characteristics of hydrogels can impact their long-term stability. If there is excessive swelling followed by rapid deswelling, it can result in the degradation of the hydrogel (Bajpai, 2001). Achieving a well-balanced swelling behavior is essential to uphold the integrity and durability of hydrogel-based drug delivery systems.

Mechanical and Viscoelastic Properties

Visually, hydrogel appears as a soft, rubbery, and transparent material. When mentioned about mechanical properties of hydrogels, many factors like shear module (G' and G"), viscoelasticity and tensile strength are appointed (Lee et al., 2018). The mechanical properties of hydrogels from chitosan were studied, and the results confirmed that hydrogels behave as elastic materials (Duan et al., 2015). One the important factors impact on the mechanical property of the hydrogel is the degree of crosslinking (Schoenmakers et al., 2018). Usually the denser the cross-linking gel requires a higher external force to deflect, leading to the fact that higher density crosslinking gel have better strength, hardness, or stiffness. Besides the density of crosslinking, polymer sources and type of crosslink also affected the mechanical properties of the hydrogel (Zhang et al., 2018).

Nayan et al. conducted the creation of PVA/Chitosan hydrogels crosslinked with GA using a superficial freeze-thawing technique (Nayan et al., 2018). The hydrogels exhibited their optimal mechanical properties and the highest swelling ratio when loaded with 6 wt.% chitosan, demonstrating a moderate water absorbency. In this case, specifically 0.58 g/g in a phosphate buffered saline (PBS) solution exhibited significant water retention in soil, retaining 48% of their moisture content after 15 days. Furthermore, the incorporation of these hydrogels into the soil had a positive impact on the growth of okra plants, leading to increased length and weight (Nayan et al., 2018).

The viscoelastic property of hydrogels plays a crucial role as it provides insights into their specific flexible or tight nature, which is influenced for application (Roy et al., 2009). Hydrogels possess inherent viscoelasticity, meaning they exhibit both viscous and elastic behavior. However, due to their porosity and the presence of water, their viscoelastic behavior includes both characters of solid and liquid. To assess and understand the viscoelastic properties of hydrogels, dynamic (oscillatory) rheology has been employed (Stojkov et al., 2021). This technique allows for the characterization of the frequency-dependent mechanical properties of materials, enabling a comprehensive understanding of the viscoelastic behavior of hydrogels (Kocen et al., 2017). In particular, the viscoelastic properties of chitosan-based hydrogel hold a comprehensive influence on its applicability to the desired site. In medical applications, whether by injection, oral, transdermal, or other route of administration, the interaction between the flow and gelation behavior of the hydrogel profoundly shapes its ability to adhere to the intended target site (Herrada-Manchón et al., 2023). Therefore, such tailored viscoelastic properties became important. Chitosan hydrogels exhibit the dual ability to retain their structural integrity post-application or seamlessly flow into intricate tissue crevices and irregularities. This intrinsic characteristic is valuable in enabling precise drug delivery to specific anatomical locales, offering enhanced therapeutic outcomes. Chitosan/xanthan gum-based hydrogels possessed a notable level of viscoelasticity, a quality that held significance for their potential applications in drug delivery (Martínez-Ruvalcaba et al., 2007).

Biodegradable Properties

Biodegradation refers to the capability of a material to decompose through enzymatic or chemical reactions caused by microorganisms when implanted with *in vivo* in a living organism. Under physiological conditions, labile bonds was present in the polymer backbone or cross-links became to be hydrolyzed, leading to the breakdown of the material (Hennink & Nostrum, 2002, Feksa et al., 2018). In such way for biodegradation, there are two methods that are commonly used to assess the rate of biodegradation in hydrogels by enzymatic degradation and microbial degradation.

For enzyme degradation, in chitosan-based semi-IPN hydrogels (Ullah et al., 2018), the specific enzyme lysozyme showed greater degradation compared to the non-specific collagenase. The hydrogels underwent degradation both in the presence and absence of lysozyme and collagenase. When enzymes were absent, but the weight loss was minimal, reaching a maximum of 4.95% by the end of the fourth week. Conversely, lysozyme exhibited significant activity, leading to a weight loss of 17.44%. This effect stemmed from lysozyme's ability to cleave β (1-4)-linked GlcNAc and Glc subunits, resulting in chain breakage and subsequent weight loss. In contrast, collagenase, while nonspecific to chitosan, still caused a total weight loss of 9.89%. This was due to collagenase's random degradation of macro-chains within the polymer network, leading to an overall reduction in weight. These biodegradibility property holds promise for tumor-targeted administration (Craciun et al., 2019).

Secondly, microbial degradation is applied to evaluate the rate of biodegradation in hydrogels. Din et al. made a groundbreaking by introducing a cell-based approach (Din et al., 2016). Their investigation revealed that even after oral administration, the surviving bacteria managed to regenerate their population, resulting in a distinctive pulsatile delivery mode. In the realm of mucosal drug delivery, lactococci and lactobacilli have been established as reliable vectors for years. Wang et al. achieved a remarkable feat by effectively transporting beneficial bacteria, specifically Bacillus subtilis, into the gut microbiome through encapsulation within its own biofilm (Wang et al., 2023). This strategy was pivotal in amplifying intestinal colonization by an impressive factor of 17 and boosting bioavailability by an astounding 125-fold when compared to non-coated bacteria.

The study involved chitosan xerogel beads loaded with dicyandiamide (DCD), aiming to assess whether the longevity of DCD could be prolonged in soil through gradual release from a chitosan hydrogel (Minet et al., 2013). The findings indicated that a smaller proportion of the encapsulated DCD was promptly released from these chitosan xerogel beads. Specifically, only 19% of the encapsulated DCD was released after 9 hours when placed in water, and after 7 days in soil under conditions of heavy rainfall, a mere 33% of the DCD was released suggesting availability over an extended period, potentially enhancing its effectiveness in soil applications.

Biocompatibility and Cytotoxicity

Biocompatibility and cytotoxicity are important characteristics for biomedical applications. Here, biocompatibility consists of two main elements: biosafety and bio-functionality. In vitro tests for biocompatibility are generally performed in two different ways. The first method involves positioning the hydrogels in direct contact with host environmental cells and incubating them for a specific period at 37°C. In the other method, the material is placed in a suitable physiological solution and incubated for a specified period of time at 37°C to allow for any leaching from the material. The obtained leachates are then used to conduct biocompatibility tests in the presence of cells (Pal et al., 2009). In contrast, cytotoxicity tests usually involve assessing cell viability and cell proliferation. Cell proliferation is visualized by microscopy and by carrying out the MTT assay, a colorimetric method that allows for the quantification of cell growth and proliferation (Khadem et al., 2023). A new covalently cross-linked hydrogel, composed of N,O-carboxymethyl chitosan and oxidized alginate, was developed for drug delivery applications. The hydrogel's cytocompatibility and biocompatibility were tested *in vitro* and in vivo. The hydrogel showed good cytocompatibility with NH3T3 cells over 3 days in vitro. Acute toxicity tests revealed no harm to major organs during a 21-day intraperitoneal administration. Skin reactions were absent after subcutaneous injection for 72 hours. The hydrogel degraded slowly over time. Hemolysis tests demonstrated excellent compatibility with blood cells (Eivazzadeh-Keihan et al., 2021). In another research, injectable blends containing chitosan, phospholipid, and lauric aldehyde (PoLigel-LA) or lauric chloride (PoLigel-LCl) were studied for biocompatibility (Souza et al., 2009). In vitro tests on L929 and HeLa cell lines indicated acceptable biocompatibility for both blends, with PoLigel-LA showing better cell viability. In vivo tests on healthy CD-1 mice revealed that subcutaneous injection of PoLigel-LA caused no local or systemic toxic effects over four weeks, while PoLigel-LCl caused immediate local toxicity. Intraperitoneal injection of PoLigel-LA in mice did not lead to physical or behavioral changes, and body weight changes were similar to control animals.

In addition, chitosan hydrogels are highly biocompatible due to their hydrophilic surface and low interfacial free energy, resulting in low protein and cell adhesion, which could affect the stability and performance of the carried drug. Chitosan serves as a mucoadhesive polymer and has been widely employed in creating micro and nanoparticles. These particles have been harnessed as carriers for diverse therapeutic agents, including peptides, proteins, vaccines, DNA, and drugs. This carrier system proves effective for both parenteral and non-parenteral administration methods. Notably, chitosan micro- or nanoparticles loaded with therapeutic agents exhibit enhanced stability, permeability, and bioactivity (Ahmed and Aljaeid, 2016). A polyelectrolyte complex (PEC)hydrogel was developed using chitosan as the cationic polyelectrolyte and γ-poly(glutamic acid) (γ-PGA) as the anionic polyelectrolyte. The chitosan-γ-PGA PEC hydrogels demonstrated favorable biocompatibility, and specifically, displayed a strong affinity for NIH 3T3 fibroblast cells, due to positive charge dominance of chitosan. This indicated the potential of the chitosan-γ-PGA PEC hydrogel for promising biomedical applications (Tsao et al., 2010).

"Green" Approaches to Fabricated Chitosan Hydrogels

Green crosslinkers for chitosan gelation

Green methods for creating and utilizing chitosan hydrogels are centered on eco-friendly and sustainable techniques that aim to reduce the reliance on harmful chemicals and lessen the environmental footprint. These strategies are in harmony with the tenets of green chemistry and the principles of sustainability. In case of material design, these perspectives are necessary for building a sustainable society, and this concept should not be forgotten when creating chitosan hydrogels. The properties of hydrogels including their porosity, swelling capacity, mechanical strength, and biodegradability, can be determined by the method of crosslinking (Pellá et al., 2018). Therefore, the selection of preparation techniques has significant impact in the produced hydrogels, which can be gelled by physically or chemically based on crosslinker the primary forming the polymer network. Physically crosslinked hydrogels could be created by secondary interactions which involved electrostatic attraction, hydrogen bonding, and hydrophobic interaction (Mittal et al., 2018). Meanwhile, the chemical crosslinked hydrogels created through covalent bonds occured between the functional groups of crosslinkers and chitosan (Fu et al., 2018). In the latter case, chemical reagents might be used to cross-linked chitosan, however, some cross-linking agents were environmentally hazardous and toxic and require caution.

Physically Crosslinked Hydrogels

The polymeric chain in the hydrogels is mainly connected by entanglement or physical interaction including electrostatic interaction and hydrogen bonding. Chitosan, being a positively charged polysaccharide, is capable of eliciting electrostatic interactions with molecules that possess a negative charge anionic ions like sulphates (Cui et al., 2008) and citrates (Shu et al., 2001). The strength of these ionic interactions is influenced by the degree of deacetylation and M_w of used chitosan, and concentration of chitosan (Shu & Zhu, 2002). In addition, as a polymeric crosslinker of chitosan anionic polymers can form polyelectrolyte complexes (PECs) through electrostatic interactions with polyanions with opposite charges to chitosan. Various anionic polysaccharides like alginate, pectin, dextran, and gellan gum, proteins such as silk fibroin protein and collagen, as well as synthetic polymers like poly(acrylic acid), poly(L-lactide), and polyphosphate, could all create PECs when combined with chitosan (Fu et al., 2018). A versatile method was developed to create dynamic hydrogels through the controlled interactions between positively charged biopolymers and polyoxometalate (POM) anions. By immersing primary networks in aqueous solutions of POMs with different nuclearity and charges, chitosan-POM hydrogels were formed. The integration of predispersed chitosan in this process resulted in the formation of double network hydrogels, which exhibited significantly higher toughness, surpassing the toughness of previous composite hydrogels by 2-3 orders of magnitude (Huang et al., 2020).

On the other hand, chtosan has aminogroups and oh groups in the framework, there are hydrogen bonds, which involve the attraction between a hydrogen atom attached to a molecule and a highly electronegative atom on another molecule (Bi et al., 2020). The most common hydrogen bonding sites within hydrogels are amine, amide, alcohol, and carboxylic acid groups (Szalewicz, 2003). A self-healing hydrogel based on chitosan was successfully created using noncovalent interactions, specifically hydrogen bonding between acrylamide segments and chitosan backbone (Zhang et al., 2019). This autonomous hydrogel had the ability to spontaneously repair itself due to the reversible nature of the hydrogen bonds formed between the polymer segments and chitosan.

In addition with hydrogen bonds and electrostatic force for chiosan gelation, entanglement is also important gelation force. A tangle of long polymer chains exists, which facilitates gelation. Under conditions of higher temperature and concentration, the regenerated chitosan polymer chains possess the ability to entangle and create cross-links via hydrogen bonds, resulting in the formation of hydrogels. The impact of chitosan alkaline solution concentrations on G' and G'' values was studied. The temperature where G' and G'' intersect was identified as the gelation temperature. Notably, the gelation temperature decreased from 49 °C to 37 °C as chitosan concentration increased from 2 to 4 wt % indicating increased aggregation and entanglement among chitosan chains in concentrated solutions due to a higher number of hydrogen-bonded junction zones (Duan et al., 2015). Lin et al. found that, during the creation of calcium–alginate–N, O-carboxymethyl chitosan (NOCC) beads, the alginate became entangled within the NOCC network, leading to the development of an interpenetrating polymeric network. This was because some protein molecules might be entangled within the hydrogel network, and those cannot be released unless polymer matrixes are degraded (Lin et al., 2005). The nature of reversible and unstable network produced the limitation of physical crosslinking.

Chemically Crosslinked Hydrogels

Primarily, hydrogels that create the chemical bonds network of crosslinker and chitosan are caused by covalent bond reaction of various reactive amino and hydroxyl groups of chitosan with the functional crosslinker used. The chemical crosslinkers were added during the gelation process generally in chitosan aqueous solution, and their concentration and reactivity were adjusted to control for changing the degree of crosslinking. Chemical cross-linking involves the formation of permanent bonds through following reactions, for example, radical crosslinking (Salmawi, 2007), photo-crosslinking (Maiz-Fernández et al., 2022), enzymatic reactions (Tran et al., 2020), and covalent cross-linking using chemical reactions like aldehydes (Yeh et al., 2022), epoxy (Li et al., 2022), genipin (Vo et al., 2021), and carbodiimide (Lin et al., 2017).

Some of these cross-linkers are toxic or contaminating reagents in favor of gelation. In order to meet the requirements of sustainable development goals, the synthesized materials should possess non-toxicity and biocompatibility. Among the chemical crosslinking methods, utilizing high-energy radiation and epoxy crosslinkers has a significant influence on producing hydrogels with elevated mechanical strength and enhanced safety for human-being.

Table 1. Common crosslinker agents for crosslinking chitosan hydrogel

Crosslinkers	Active groups	Mechanism	Advantage	Disadvantage	References
Aldehyde (Glutaraldehyde)	Aldehydic groups (R–CH=O)	Covalent bonds are established through Schiff base or imine linkages with the amino groups present in chitosan.	- Highly reactive - Good mechanical strength for crosslinked hydrogel	- Potential toxicity of residual aldehyde groups	(Kocemba & Mucha, 2017)
Epoxy	Epoxy functional groups $R-\overset{O}{\overset{\diagup\diagdown}{CH-CH_2}}$	By interacting with the hydroxyl or amine groups on chitosan molecules, stable covalent bonds are formed.	-Highly reactive -Improved mechanical properties for hydrogel. - Lower toxicity than aldehyde	- Causing skin or respiratory irritation upon direct exposure in their uncrosslinked form.	(Jabeen et al., 2015)
Carbodiimide (1-ethyl-3-(3-dimethylaminopropyl) carbodiimide)	Carbodiimide groups -N=C=N-	Facilitating the creation of amide bonds between carboxyl and amine groups present in chitosan.	- Improved stability and mechanical properties	Require careful pH control to avoid side reactions	(Rafat et al., 2008)
Genipin	(chemical structure)	Creates covalent bonds with amino groups found in proteins and polymers.	-Its biocompatibility -Less toxic than glutaraldehyde	-Cytotoxic effects at higher concentrations. -Lower mechanical strength.	(Gao et al., 2014)

For chitosan based-hydrogels, typical crosslinkers have included epoxy (Jawad et al., 2019), carbodiimides (Rafat et al., 2008), aldehydes (Lou et al., 2020), and genipin (Gao et al., 2014) as summarized in Table 1. Epoxy and carbodiimide crosslinkers are commonly chosen in biomedical contexts due to their reduced toxicity, which is essential for maintaining biocompatibility. These crosslinkers facilitate the enhancement of mechanical strength in hydrogels through the establishment of covalent bonds. In contrast, Genipin isolated from gardenia fruit becomes a natural alternative crosslinker having lower toxicity as compared to aldehyde-based crosslinkers like glutaraldehyde. Due to its plant biomass origin, its biocompatibility has attracted the interest of researchers seeking non-toxic crosslinking agents for applications in the biomedical field. However, a drawback of using natural-derived crosslinkers like genipin is the potential limitation in achieving high mechanical strength in the resulting hydrogels (Roberts & Taylor, 1989).

Among chitosan, hydrogels crosslinking chitosan with epoxy groups via a process called epoxy-amine crosslinking (Bratskaya et al., 2019).The process involves linkage reaction of amine groups of chitosan with the epoxy crosslinker, which is generally a bifunctional epoxy compound, in the presence of catalyst like acidic or basic. The linkage reaction between the amino groups and epoxy groups, which are highly reactive, can form covalent bonds with both ends epoxy groups, including amine and hydroxyl groups on polysaccharides, allowing for efficient reaction to form stable hydrogels (Abdulhameed et al., 2019). Particularly, di-glycidyl ethers exhibit reactivity in alkaline environments and have consequently been employed to produce hydrogels with the pH rising from 7.5 to 9.5, since the primary reaction occurred between the amino groups and ethylene glycol di-glycidyl ether (Lu et al., 2006). Against alkalinity, in acidic conditions, the interaction between di-glycidyl ethers of glycols and chitosan primarily occurred through the hydroxyl groups located at the C6 position of the glucosamine unit (Bratskaya et al., 2019). The degree of crosslinking was controlled by adjusting the concentration of the epoxy compounds. This allowed the products swelling hydrogel to be tailored to biomedical application (Liu et al., 2014).

Some common epoxy compounds used for crosslinking polysaccharides including glycidyl methacrylate (GMA), di-glycidyl ether of bisphenol A (DGEBA), epichlorohydrin (ECH), ethylene glycol di-glycidyl ether (EGDE), and polyethylene glycol di-glycidyl ether (PEGDGE) that were summarized in Table 2.

Table 2. Prevalent epoxy crosslinkers

Crosslinker	Composition	Chemical structure	Advantage	Disadvantage	References
Glycidyl methacrylate (GMA)	An epoxy group and a methacrylate group		-Highly reactive	-Cytotoxicity	(M. He & Chu, 2013) (Yin et al., 2021)
Diglycidyl ether of bisphenol A (DGEBA)	Two epoxy groups		-Highly reactive -Excellent mechanical properties for hydrogel	-Endocrine disruption and toxicity -High viscosity	(Jabeen et al., 2015)
Epichlorohydrin (ECH)	An epoxy group and a chlorohydrin group		-Highly reactive	-Potential health concerns: carcinogenicity and mutagenicity	(Garnica-Palafox et al., 2014)
Ethylene glycol diglycidyl ether (EGDE)	Bifunctional epoxy		-Highly reactive -EGDE-crosslinked hydrogels have good swelling properties -Biocompatibility at certain concentration	-Flammability	(Abdulhameed et al., 2019)
Polyethylene glycol diglycidyl ether (PEGDGE)	Bifunctional epoxy		-Low toxicity	-High viscosity -Longer reaction times	(Tripodo et al., 2018)

Additionally, EGDE is a less toxic compound with bifunctional epoxy groups that are reactive towards hydroxyl, carboxyl, amino, and sulfhydryl groups (Zalipsky, 1995). This property made it a popular choice for crosslinking biopolymers like DNA, proteins, and polysaccharides (He et al., 2020). Compared to other dialdehydes, EGDE has a lower toxicity level and has been widely used for crosslinking. The reaction of EGDE occurs under alkaline conditions, i.e., pH above 7 (He et al., 2020).

The effectiveness of the reaction between chitosan and di-glycidyl ethers of glycols is significantly influenced by factors such as pH and the type of acid used to dissolve chitosan. When hydrochloric acid was employed as catalyst, crosslinking with di-glycidyl ethers of ethylene glycol (EGDGE) and polyethylene glycol (PEGDGE) at both room and subzero temperatures formed mechanically robusted chitosan gels and cryogels (Bratskaya et al., 2019). However, in acetic acid solutions, only weak chitosan gels were formed under the same conditions. Research involving chitosan cryogels demonstrated their biocompatibility in a mouse model. An observed moderate inflammatory reaction around the implanted cryogels was accompanied by the formation of normal granulation tissue. Importantly, no toxic, immunosuppressive, or sensitizing effects were noted in the recipient's tissues, underscoring the favorable biocompatibility of these materials. Another effective strategy to fabricate a novel composite hydrogel comprising chitosan and polyvinyl alcohol (PVA) was demonstrated. The uniqueness of this hydrogel improving mechanical strength and antibacterial properties was achieved by incorporating polyethylene glycol di-glycidyl ether (PEGDE) as a crosslinking agent and PVA as a reactive toughening agent (Hu et al., 2023). The stability of all prepared hydrogels was maintained in a PBS buffer for a duration of at least 6 days. These hydrogels exhibited a pH-responsive swelling behavior, with the maximum swelling ratio being observed under acidic conditions. Remarkably, the composite hydrogel demonstrated exceptional antibacterial activity against both *Escherichia coli* (E. coli) and Staphylococcus aureus, as demonstrated through inhibition zone measurements. This research suggests that the developed the chitosan/PVA composite hydrogel held promise for scalability and could serve as a potent antibacterial material with potential applications in the field of wound healing.

A straightforward method to create a chemical crosslinking of glycol-chitosan through a one-step process took place in water at 37 °C. Here, glycol-chitosan is a derivative of chitosan that is water-soluble in physiological conditions and has advantageous biological properties, which make it potentially useful for tissue engineering applications (Tripodo et al., 2018). As known for non-toxic and stimulation to chondrocyte growth at low concentrations, glycol-chitosan can create chemical crosslinking in efficient and practical approach to leverage the desirable. Recently, there has been considerable interest in using ethylene glycol di-glycidyl ether (EGDE) as a successful alternative for modifying chitosan due to its desirable features, including non-toxicity, water solubility, and bifunctional diepoxy groups. EGDE contains two epoxide functional groups located at both ends of each molecule, which are highly reactive compared to other ethers. The reaction between the epoxide groups of EGDE and the amino groups ($-NH_2$) of D-glucosamine units in the polymeric matrix of chitosan resulted in successful cross-linking (Kalidason et al., 2022).

Radio-active method of chitosan hydrogels without chemical agent

A covalent bonding method that does not use a chemical reagent cross-linker is a gelation method that uses radiation or electron beams. Radiation chemistry is a field of chemistry that studies the chemical changes that occur due to the absorption of high-energy ionizing radiation. This type of radiation can be generated in the form of electromagnetic radiation with a short wavelength between 10-100 nm and an energy of approximately 10-100 keV, such as X-rays, γ-rays, or particulate radiation like electrons (E-beam) and protons (Collins & Ottinger, 2003). The most commonly utilized source of γ-rays for chemical studies in industry, medicine, and research is cobalt-60. Gamma radiation, also referred to as gamma ray, was first identified by French scientist Paul Villard in 1900 and named by Ernest Rutherford in 1903. This type of radiation is ionizing and originates from the nucleus of a radioactive atom. The wavelength of gamma radiation is incredibly small, measuring less than 0.01 nm, which makes it the shortest wavelength in the electromagnetic spectrum. Due to this short wavelength, gamma radiation has the highest frequency, which is approximately 10^{19} Hz. Furthermore, gamma radiation has the greatest energy of all electromagnetic radiation types (Clarke & Valentin, 2009).

The radionuclide cobalt-60 can be created in a nuclear power reactor by irradiation of cobalt-59 with neutrons as displayed in below equation (Eden, 1956).

$$^{59}_{27}Co + ^{1}_{0}n \rightarrow ^{60}_{27}Co$$

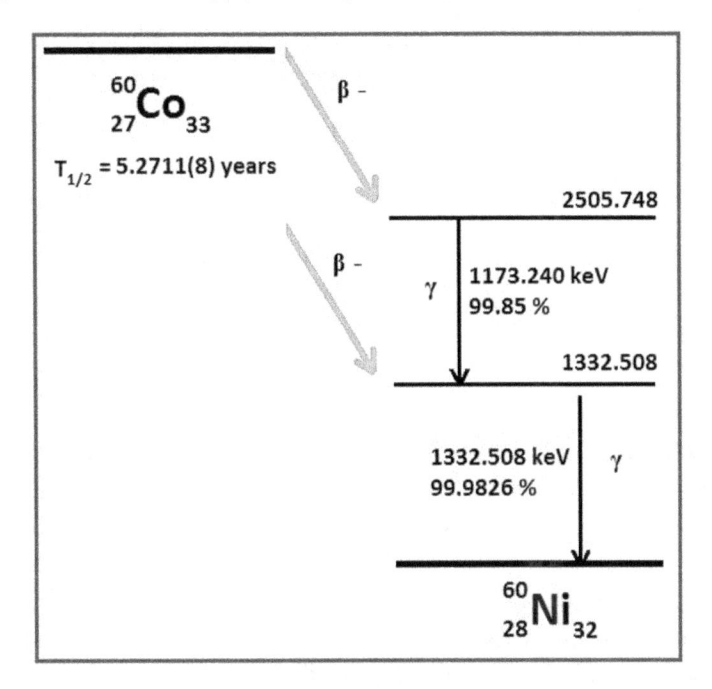

Figure 3. Diagram of radioactive decay by disintegration of ^{60}Co

Cobalt-60 is an unstable atom due to the excess neutron it contains (L'Annunziata, 2007). As illustrated in Figure 3, it undergoes radioactive decay and transforms into a stable nickel-60 by emitting photons with energies of 1.17 and 1.33 MeV (Ioan et al., 2017). Radiation effects on materials exposed to radiation can be classified into two categories: direct and indirect effects referring to the decomposition of chemical bonds via radiation energy (Boujelbane et al., 2022), causing molecules to become ionization or excitation to a higher energy state. Indirect effects, on the other hand, are also important in the overall action of radiation and occur through the continuous reactions of radiolytic products, particularly hydroxyl radicals (OH•), e^-_{aq}, hydrogen radicals (H•), hydrogen peroxide (H_2O_2), and hydroperoxyl radicals ($HO_2°$).

The initial steps of radiolysis involve the generation of reactive species when aqueous solutions are exposed to radiation (Lousada et al., 2016). Solvated electrons, which react slowly with carbohydrates, and -OH radicals and atoms, which react more easily, are examples of these reactive species. Polymers having their carbon-bound hydrogen atoms formed both OH• and H• in the radiolysis of water and the incorporation of free radicals (Jiménez-Becerril et al., 2016), respectively.

$$H_2O \rightarrow H_2O^+ + e^-$$

$$e^- + H_2O \rightarrow H_2O^-$$

$$H_2O^+ \rightarrow H^+ + OH^o$$

$$H_2O^- \rightarrow H^o + OH^-$$

$$H^o + OH^o \rightarrow H_2O$$

$$e^- + OH^o \rightarrow OH^-$$

$$e^- + H_3O^+ \rightarrow H_2O + H^o$$

$$OH^o + OH^o \rightarrow H_2O_2$$

$$H^o + H^o \rightarrow H_2$$

$$H_2O_2 + e^- \rightarrow OH + OH^-$$

$$H_2 + OH^o \rightarrow H_2O + H^o$$

In secondary stages of polymer crosslinking reactions (Naikwadi et al., 2022), hydroxyl and hydrogen radicals are highly reactive species that can interact with a variety of molecules, including polymers like chitosan. When these radicals contacted with the polymer chain, they can abstract a hydrogen atom from the carbon backbone, leaving behind a carbon-centered radical. If this part If this site encounters other radicals in chitosan, the both sites recombine, and form new covalent bonds, leading crosslinking between both polymer chains. To counteract the dominant chain scission observed in pure chitosan under gamma radiation, it becomes imperative to create hybrid hydrogels by combining chitosan with other synthetic polymers like PVA (Zhao et al., 2014), Polyvinylpyrrolidone (Dergunov & Mun, 2009), and Polyacrylic acid (Chen et al., 2014). Interestingly, gamma irradiation combines crosslinking and sterilized process in one step to eliminates all forms of life, especially microorganisms such as viruses, bacteria, and fungi (Parsons, 2012) and its dose of 25 kGy sterilized devices and materials recommending for medical purpose (Beerlage et al., 2021). On the hydrogel fabrication with chitosan and another polymer, their composites using gamma irradiation were applied for biomedical uses (Nisar et al., 2021). Chitosan-based hydrogel beads were prepared using L-glutamic acid as a monomer and gamma-irradiated graft copolymerization technique. These hydrogel beads were loaded with the anti-cancer drug Doxorubicin and showed potential as an anti-cancer drug delivery system with high swelling ratio (426%) and drug release (81.33% in 144 hours). The system demonstrated effective MCF-7 cancer cell toxicity and exhibited effective cancer cell toxicity of about 78.18%, thus it can be used as a potent anti-cancer drug carrier.

Chitosan/Au nanocomposites, consisting of gold nanoparticles and chitosan, were fabricated by irradiating chitosan/Au solutions with varying doses (Sokary et al., 2021). The spherical-shaped gold nanoparticles formed were free of aggregation. These nanocomposites were tested as an anticancer agent and found to be more effective in inhibiting the proliferation of HepG-2 and CACO-2 cells compared to neat chitosan. The concentration required to inhibit 50% of the cells was lower for the nanocomposites than for chitosan on both cell lines. Physically mixing chitosan and P407 were cross-linking it with gamma irradiation and the products were evaluated for its wound-healing properties. The gelatinous products maintained its antimicrobial/antifungal properties through acidic pH, demonstrating the highest wound-closure rate in the first week of testing on mice (Leyva-Gómez et al., 2017).

In other instances, hydrogels composed of chitosan and PVA were prepared under gamma irradiation and used as wound dressings (Fan et al., 2016). The swelling ratios of all hydrogels decreased when the pH values of PBS exceeded 6.0. Despite this, these hydrogels had a great potential in wound dressings. Also, PVA-based chitosan composited hydrogels could have chemical and physical linkages in the presence of tannic acid (TA), when irradiated under gamma ray. The hydrogel was obtained first using gamma irradiation and then immersed in TA solution to form a second physical network and the double linkages caused improved mechanical properties. Furthermore, the subsequent inclusion of TA created a strong physical crosslinking network and gave the composite hydrogels antibacterial properties, making to withstand medical specifications in biomedical applications and soft devices (Guo et al., 2021). In the composite of PVA and chitosan using gamma radiation, the chemical structure of chitosan underwent chain scission reactions and altered it hydrophilicity, but molecular weight was reduced by degradation to sharply be 5.53×10^4 g/mol, when irradiated at 100 kGy. The radiation-degraded chitosan was then utilized in the dissolution casting method to produce the mixture (Islam et al., 2014).

Apart from the use of PVA polymers, examples of chitosan composite with vinyl monomer under radiation included radiation grafting for 2-acrylamido-2-methyl propane sulfonic acid (AMPS) (Gad, 2008). The hydrogel formed from chitosan-AMPS had the highest swelling in the equilibrium, when the AMPS content was at 40% and the absorbed dose was 10 kGy. The hydrogel containing polyelectrolyte parts acted as sorbents and ion exchangers. For treating wastewaters, the chitosan-AMPS hydrogel was used in the sorption of dyes or metal ions (Rattanawongwiboon et al., 2020). In contrast, the method of creating chitosan-poly(ethylene glycol) diacrylate (PEGDA) beads through radiation-induced crosslinking could provide porous chitosan-PEGDA beads with maximum porosity of $97.50 \pm 0.73\%$ and enhanced swelling degree ($48.32 \pm 4.90\%$) of the beads, when irradiated with gamma radiation at 25 kGy. The loaded PEGDA and radiation dose were important factors in controlling the porosity and swelling degree, respectively. To sorb peppermint oil into the beads, a concentration of 0.25 mg/ml was used, the loaded efficiency was $93 \pm 0.51\%$. In addition, the release of peppermint oil from the beads was controlled for approximately ten days (230 h).

In another radiation method for gelation of chitosan, electron beams were available beside of gamma radiation technique. The electron irradiation can emerge as a highly appealing method due to its simplicity, speed, and energy adjustability for sample treatment. The energy level of these electrons dictates their depth of penetration within the sample. Notably, this technique can facilitate crosslinking reactions without the need for supplementary chemicals, solvents, or initiator, making electron irradiation method promising avenue for engineering advanced biomedical materials (Wisotzki et al., 2014). In the same way, ionizing radiation served convenient tool for synthesizing hydrogels based on chitosan and its derivative, carboxymethyl chitosan. Consequently, the transient products of water radiolysis reacted with the polysaccharide, forming hydroxyl radical (•OH) being particularly reactive towards carbohydrates (Czechowska-Biskup et al., 2016). This radical rapidly abstracts hydrogen atoms from glucosamine rings, generating macroradicals and subsequent reactions and makes following crosslinking reaction. Due to chitosan radicals' instability and propensity for linkage bond formation, the crosslinking reactions involving chitosan might be less efficient and polysaccharide, but, when PGDA was present with chitosan, the gelation was enhanced. This was because that recombination of PEGDA and chitosan radicals contributed to the formation of a network structure as shown in Figure 4. This innovative approach not only manipulates radiation energy, but also sheds light on the intricate interplay of reactions in the formation of hydrogel networks.

Figure 4. Possible network architecture of polysaccharide chains crosslinked with PEGDA. Numbers represent crosslinking points of (1) PEGDA-PEGDA; (2) & (4) PEGDA-polysaccharide; (3) CMCS-CMCS; and (5) free or entangled polysaccharide chain (Czechowska-Biskup et al., 2016)

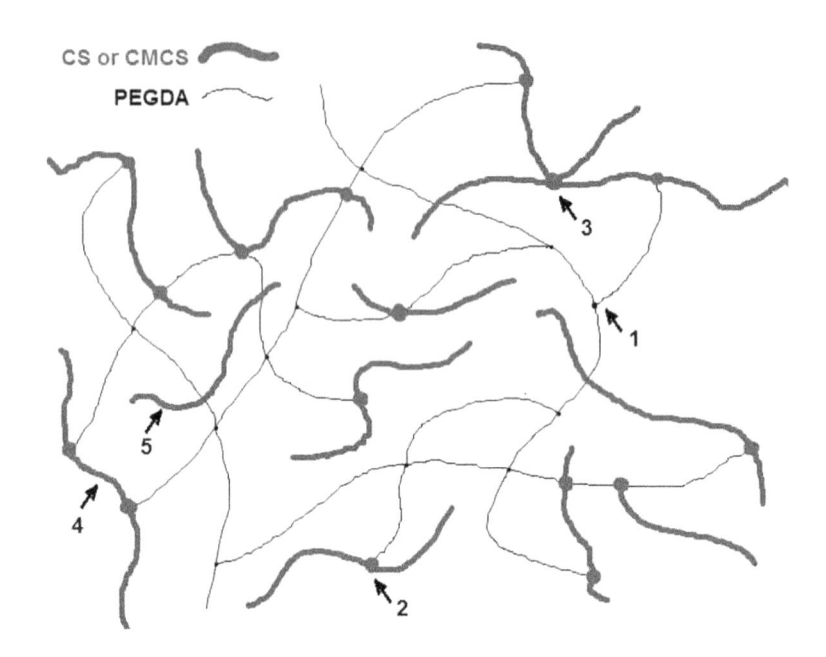

Certainly, the degree of crosslinking of chitosan hydrogels plays a crucial role in determining the release rate of medicine and fertilizer, because it affect the gel content (%) and swelling behavior (g/g) of irradiated hydrogel. In the cases of chitosan and carboxymethyl chitosan hydrogels using for gelation by gamma irradiation (Ghobashy et al., 2021) and electron beam irradiation (Czechowska-Biskup et al., 2016), the degree of crosslinking exerted an influence on the porosity of resultant chitosan hydrogels. In particular, chitosan solution was subjected to irradiation at varying doses, resulting in the creation of porous structure in the hydrogel films (Baroudi et al., 2018).

Slow-Release Fertilizer of Chitosan Hydrogels Applied in Agriculture

For optimizing fertilizer use and minimizing environmental consequences in agriculture traditional farming practices have tendency to apply excessive amounts of fertilizers in pursuit of higher crop yields. This results in more than half of the applied nitrogen (N) and a staggering 85% of the applied phosphorus (P) not being utilized by crops, contributing to issues of nutrient runoff and leaching that have adverse ecological impacts (Fertahi et al., 2021). Presently, chitosan hydrogels are recognized as substances that can enhance soil quality and increase crop yields by effectively retaining water and nutrients, gradually releasing them as needed due to high capacity holding water. The pace at which nutrients are released during the degradation of hydrogels can be adjusted to match the specific nutritional needs of plants. For eco-friendly fertilizers by using gelous entrapment, chitosan hydrogels enhance the soil's ability to retain moisture. Practically, the controlled fertilizer released from chitosan hydrogel for slow-releasing (Jamnongkan & Kaewpirom, 2010). The agricultural materials merged the properties of superabsorbent hydrogels with those of fertilizers. These hydrogels served to enhance soil quality and caused effectiveness of fertilizers (Azeem et al., 2023). When using gamma radiation for chitosan hydrogels, a water retention capacity was 28–36% of chitosan hydrogel even after 72 hours, and this slow retention continued for up to 9 days (Elbarbary & Ghobashy, 2017). Notably, when immersed in fertilizer solutions, these hydrogels displayed lower swelling compared to their swelling in water and the ability to adsorb and release fertilizers, which contributed to their slow-release property. Specifically, the release rate of urea was approximately 10 times higher than that of phosphate. Within 3 days, around 60% of the urea was released, whereas only 10–12% of the phosphate was released. When applied in agricultural fields, hydrogels had a positive impact on the growth of plants, since the hydrogel-loaded fertilizers showed slow-release fertilization in soil. The high swelling capacity, and gradual water retention behavior in hydrogels made them a promising choice for safer fertilizer release systems and soil conditioning in agricultural applications.

Moreover, in one approach, chitosan served as the inner coating material, while acrylamide (AM) and acrylic acid (AA) (poly(AA-co-AM)) used for hydrogels loaded with nitrogen, phosphorous, and potassium (NPK). The fertilizer hydrogel acted as the outer coating layer to absorb water and regulate the release of essential N, P, and K nutrients. The release of these nutrients into the soil followed a non-Fickian diffusion mechanism. The extent of swelling in the resulting hydrogels was influenced by several factors, including the overall concentration of reactants (monomers and initiator), the composition of the liquid fertilizer, the presence of a cross-linking agent, and initiator concentrations. Among the nutrients, N was released at the highest rate, while K exhibited the lowest release rate from the NPK fertilizer granules during 30 days of water immersion. The release of both N and P nutrients from the trilayer-coated NPK fertilizer hydrogel was controlled (Noppakundilograt et al., 2015).

A highly biodegradable composited hydrogel was developed by incorporating 2-dimensional (2D) Montmorillonite (MMT) nanoflakes into chitosan hydrogel, rendering it both elastic and flexible (Dou et al., 2022). The release of phosphorus (P) and potassium (K) increased significantly, going from 22.0% to 94.9% for P and from 9.6% to 31.4% for K. In soil, the chitosan-MMT nanocomposite hydrogel effectively controlled the overall release of fertilizer. Over 15 days period, 55.3% of the loaded fertilizer was released at a daily rate of 2.8%. In contrast, with traditional fertilizer powder, a substantial 89.2% of the fertilizer was washed away during the initial irrigation using the same conditions. Therefore, the chitosan-MMT biopolymer scaffold, serving as a controlled-release fertilizer carrier, presented a promising opportunity for advancing sustainable agriculture.

Pollutant Removal of Chitosan-based Hydrogels in Water Treatment

The fast growth of industries has resulted in a large increase in the release of heavy metal pollutants, which has caused a number of environmental problems. These hazardous heavy metal ions remain in water bodies for protracted periods of time and can build up through the food chain. Heavy metals in wastewater are toxic and extremely dangerous to both human health and the ecology, even in extremely low concentrations. Presently, various methods are employed to remove heavy metal ions from wastewater and pollutant water. These methods have included chemical precipitation (Abdel-Raouf et al., 2023), coagulation (Perumal et al., 2019), membrane filtration (Trikkaliotis et al., 2022), electrodialysis, ion exchange, and adsorption (Fu and Wang, 2011). In these technologies, hydrogels have emerged as a favored choice for heavy metal pollutant removal due to their rich surface functional groups and outstanding water-absorption capabilities. These hydrogels boast well-defined three-dimensional porous structures and exhibit hydrophilic properties. Moreover, they feature various functional groups like -NH_2, -OH, -SO_3H, and -COOH, which enable them to effectively capture dyes and metal ions from wastewater (Jang et al., 2008). Consequently, chitosan and its derivative based hydrogels have been gaining increasing attention in this context because of the presence of -NH_2 groups in the polymeric framework. Specifically, depending on the pH, chitosan having -NH_2 group changes the body charges in negative or positive, allowing it to effectively retain dyes or metal ions that carry opposite charges (Zhu et al., 2023).

For example, the removal multiple heavy metal ions from water using hydrogel particles made from a mix of chitosan and gelatin was investigated (Perumal et al., 2019). These hydrogel particles were prepared in different ratios of chitosan to gelatin and then dried using two methods. The results showed that these hydrogel particles were most effective at adsorbing Hg(II), as compared to other heavy metal ions like Pb(II), Cd(II), and Cr(II). This efficiency in Hg removal was attributed to the higher binding energy of the Hg-N bond. Interestingly, all compositions of the hydrogel were effective at removing all metal ions when they were present in higher concentrations (54–95%). This was different from their individual removal, where only Hg was removed efficiently (>50%), while the removal of other metal ions was very low (<10%). However, even in the mixture of metal ions, the removal of Hg remained more prominent than that of the other metal ions, consistent with the trend observed when removing metal ions individually.

Chitosan/sodium alginate/calcium ion double-network hydrogel showed effective removing to metal ions including Pb^{2+}, Cu^{2+}, and Cd^{2+} with adsorption capacities of 176.50 mg/g, 70.83 mg/g and 81.25 mg/g, respectively (Tang et al., 2020). Using pseudo-first-order kinetics and the Langmuir isotherm model, diethylene triamine penta-acetic acid modified chitosan/polyethylene oxide nanofibers and efficiently adsorbed Cu^{2+}, Cd^{2+} and Ni^{2+} was observed in the following order: $Cu^{2+} > Pb^{2+} > Ni^{2+}$ and adsorption capacities were 177 mg/g, 142 mg/g, and 56 mg/g, respectively. In another topic, chitosan hydrogel beads were subjected to an etherization reaction using chloroacetic acid as the etherification agent for modification of carboxymethylated chitosan beads surface (Yan et al., 2011). This significantly improved the selective adsorption of Cu^{2+} ions as compared to Pb^2 and Mg^{2+} ions in aqueous solutions depending upon factors like pH and temperature. Consequently, the adsorbent could be effortlessly regenerated at lower pH levels and reused with almost no loss of adsorption capacity. Additionally, the loaded with Cu^{2+} ion exhibited stability at higher pH and could be effectively utilized as a novel adsorbent for phosphate removal. Under these conditions, a maximum phosphate uptake was occurred in the capacity of approximately 58.0 mg/ g at around pH 5.

A double network hydrogel was successfully synthesized by cross-linking ethylene diamine tetraacetic acid with chitosan and N,N-methylene bis(acrylamide) with polyacrylamide through a two-step method and then employed for the adsorption of heavy metal ions (Ma et al., 2017). Various adsorption conditions, including pH, ionic strength, adsorbent dosage, and contact time, were thoroughly investigated under the experimental conditions. A theoretical maximum sorption capacity was 86 mg/g for Cd^{2+}, 99 mg/g for Cu^{2+}, or 138 mg/g for Pb^{2+}. The adsorption process on heavy metal ions was found to be endothermic and followed an ionexchange reaction proving highly effective in treating practical wastewater containing various heavy metals. Specifically, at an adsorbent dosage of 8 g/L, the concentration of total metal ions in the wastewater decreased from 448.5 mg/L to 5.0 mg/L.

Controlled-Drug Release Technology of Chitosan Hydrogel Medicine

A drug delivery system (DDS) refers to a product or tool that facilitates the administration of a medicinal substance to the body. It enhances the effectiveness and safety of the drug by regulating the timing, location, and pace of its release within the body (Bruschi, 2015). DDS has been evolving over time, starting from traditional large-scale systems to more advanced controlled, targeted, and responsive nanoscale systems. The modern era of drug delivery technology began with the development of Spansule® sustained-release capsules in 1952, which could provide drug release for up to 12 hours after oral ingestion (Haesun Park et al., 2022). Oral and transdermal formulations were the primary DDSs used until the 1980s. In the period between 1980 and 2010, also known as the second-generation era of DDSs, significant advancements were made, but their clinical applications were not particularly impressive (Reza Rezaie et al., 2018). Currently, smart drug delivery system (SDDS) has focused on the preparation of drug carrier from natural polymers, which enable the introduction of a therapeutic substance into the body that improves its efficacy and safety by controlling the rate, time, and place of release of drugs into the body. SDDs are also known as stimuli-sensitive delivery systems, and they can be triggered by chemical or physical stimuli. The stimulants can be categorized as exogenous and endogenous stimulants (Figure 5). The term "exogenous" refers to external stimuli such as temperature, electric fields, light, ultrasound, and magnetic fields that can trigger drug release in drug delivery systems. In contrast, "endogenous" stimuli refer to internal chemical processes such as pH changes, enzyme activity, and redox changes that can also trigger drug release in targeted areas (Kundu et al., 2020). Endogenous stimuli-based drug delivery systems have the advantage of being more specific and targeted to the site of action, as they respond to changes in the body's natural environment.

Figure 5. Exogenous and endogenous stimulants in DDSs of chitosan hydrogel medicine

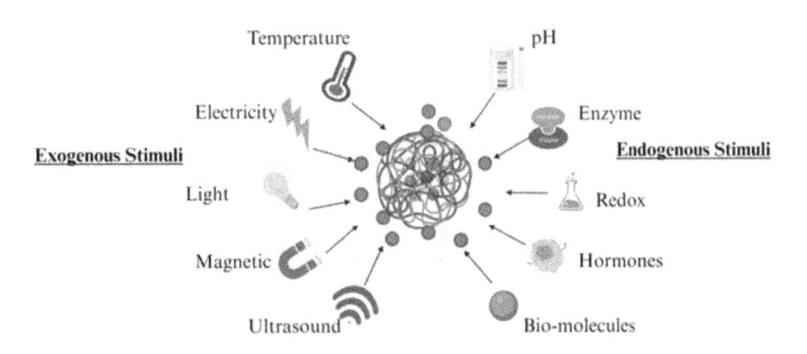

On the other hand, exogenous stimuli-based drug delivery systems, can be more easily controlled and manipulated, as the external stimuli can be applied and removed as needed (Raza et al., 2019). The two types of drug delivery systems, endogenous and exogenous stimuli-based, have undergone significant research and development, and their efficacy is determined by the particular application and the intended result. Ultimately, the decision to utilize endogenous or exogenous stimuli-based drug delivery systems will depend on the individual requirements of the patient and the therapeutic goal.

pH-Simulated Drugs Release from CS Hydrogels

Since chitosan has -NH_2 group, unique pH-responsive response is possessed in addition with its biocompatibility, and nontoxicity, meaning convenience in targeted drug delivery. Such chitosan is of pronounced interest owing to the change of swelling capacity or polymer degradation which promote the medicine release faster from the polymer matrix in the specific conditions. These variations in pH levels are significant because they can be exploited to design drug delivery systems that selectively release drugs in response to changes in pH (Figure 6). For example, cationic hydrogels like chitosan swells in acidic media at low pH, due to the protonation of amino/imine groups having weakly based groups, which dissociate in a pH-dependent manner. Conversely, the positively charged moieties on the chitosan chains repel each other, leading to swelling, because of similar charge repulsion in the gel matrix, as shown in Figure 6, indicating that the expanded volume of the hydrogel enables to retain much water. Such hydrogels can be used for drug delivery to the stomach for an injectable drug delivery system. In particular, the majority of human tissues have a physiological pH range of approximately 7 to 7.4, with a few exceptions. Notably, the pH in the stomach is between pH 1 and 3, while the pH in the small intestine falls is between 4 and 7 and blood range is from 7.3 to 7.45 (Gaohua et al., 2021).

Figure 6. Conceptual drawing of pH responsive swelling for cationic and anionic hydrogels

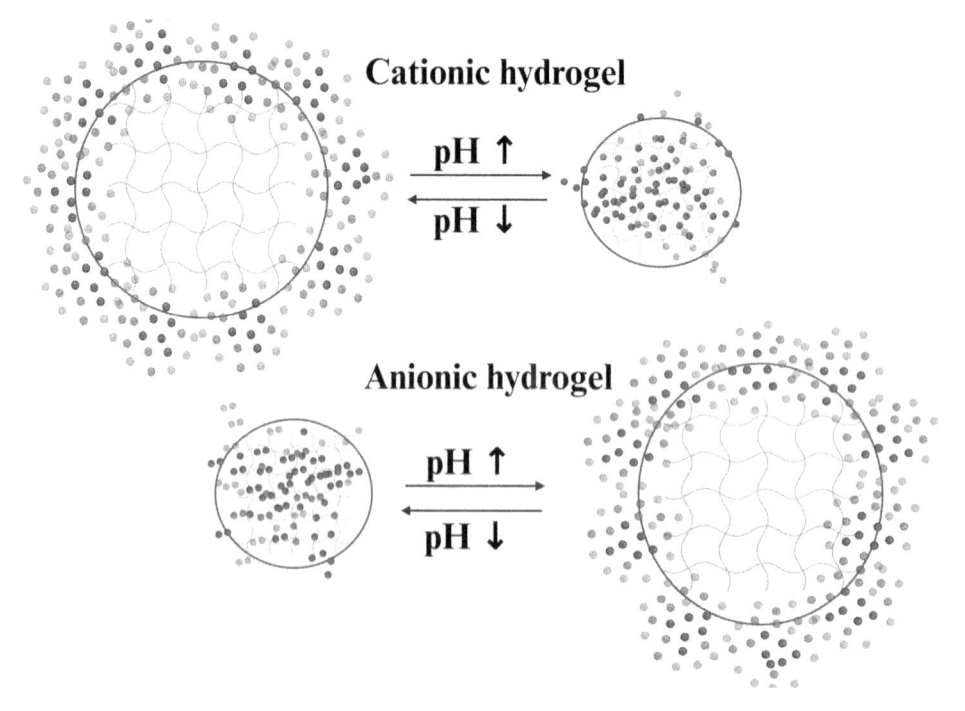

By releasing drugs in response to changes in pH, these systems can target specific tissues or organs while avoiding non-target areas. The presence of ionic pendant groups, and the impact of their pKa values affect ionization, swelling, and targeted drug release and allows enhancing the therapeutic efficacy. In the drug release mechanism, the intestine involves both diffusion and microbial degradation, offering a dual release approach. This strategy ensured safe passage through the stomach's acidity for peptides and proteins, releasing them in the intestine's less proteolytic environment for better absorption and caused optimizing drug delivery and improving therapeutic outcomes (Rizwan et al., 2017). The synthesized pH-responsive compounds using chitosan and hyaluronic acid was applied to observe changes in various properties of a polymer-based drug delivery system for traditional Chinese medicine Cortex Moutan (CM) (Chatterjee et al., 2021). The highest cumulative drug release (86.5%) after 5 days under mild acidic conditions (pH 6.4), leading to more effective treatments with reduced side effects. As a result, chitosan hydrogel responded to the environmental pH of a tissue, which, when existing within a certain acidic range, behaved structural and chemical changes so they could deliver drugs in a targeted and controlled manner and provided suitable for overcoming the shortcomings of conventional drug formulations.

Figure 7. pH-Responsive γ-Irradiated Chitosan/Poly (Vinyl Alcohol) hydrogel for localizing amoxicillin release

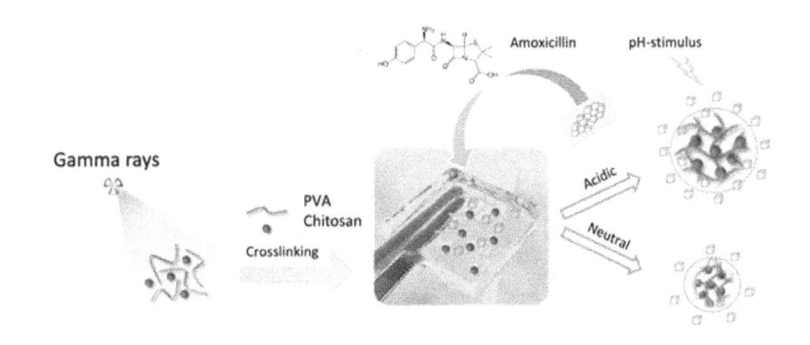

A novel approach taken to create pH-responsive hybrid hydrogels (Figure 7) for precise control of drug release, gamma irradiation was employed as an eco-friendly technique to craft high-molecular-weight chitosan and PVA hydrogels, avoiding toxic chemicals. The resultant composite hydrogel exhibited remarkable drug release capabilities due to its porous structure. In vitro studies uncovered the impact of solution pH and release media on antibiotic drug release. The release of amoxicillin from chitosan/PVA hydrogel in the physiological environment of PBS became 85%, and 50% at pH 2.1 and 7.4 in PBS media, respectively; but 34% was obtained at pH 5.5 in DI water. A comprehensive analysis involving five kinetic models, including Higuchi and Korsmeyer–Peppas, was used to understand the release profile. The Korsmeyer–Peppas model indicated Non-Fickian diffusion, as shown by the diffusion exponent (n) values for swelling and drug release. These findings suggest that the gamma-irradiated CS/PVA hydrogels hold promise as effective carriers for medication delivery (Tran Vo et al., 2022).

Thermal-Responsive Drugs Release from CS-Based Hydrogels

Temperature-control drug release is one of the primary stimulants to tailor the amount of drug uptake or release from the chitosan polymer matrix (Rizzo & Kehr, 2021). The process of temperature-controlled drug release from chitosan polymer matrices was influenced by the responsiveness of polymer to temperature changes. The concept of control-release mechanism of a porous grafted polymer hydrogels involved integrating hydrogel molecular chains into a porous material's matrix, shown in Figure 8. When the temperature was below the lower critical solution temperature (LCST), the grafted hydrogel chains extended freely, covering the micropores in the matrix. Conversely, when the temperature surpassed the LCST, the grafted hydrogel's macromolecular chains contract, revealing micropores on the surface of the material. In this state, drug diffusion occurred outward, enabling controlled drug release (Ankareddi & Brazel, 2007).

Figure 8. Thermal-stimuli drug release mechanism

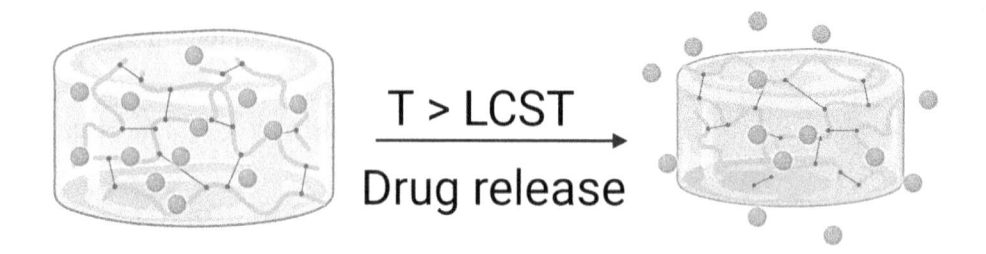

An injectable hydrogel composed of chitosan-co-poly(*N*-isopropyl-acrylamide) (CSN) with thermo-responsive properties was explored as an option for gastro-protective and controlled delivery of loxoprofen sodium, a model drug for combined effects of temperature and pH on the drug release (Ahmad et al., 2019). At pH 1.2 and 25°C, the drug release rate for CSN reached approximately 60% in 5 hours, escalating to 80% in 10 hours. Under pH 7.4 and 25°C conditions, the drug release rate was around 40% in 6 hours, gradually reaching 50-55% in 10 hours. Similarly, at pH 1.2 and 37°C, the CSN 4 hydrogel exhibited a drug release rate of approximately 50% in 5 hours, increasing to 75% in 10 hours. When exposed to pH 7.4 and 37°C, the drug release rate was approximately 30% in 5 hours, reaching 40-45% in 10 hours. Notably, the hydrogel showed controlled drug release in the extended range of pH 7.4 and temperature 37°C, suggesting its potential for tailored drug delivery applications.

Thermo-sensitive poly(*N*-isopropyl acrylamide-co-vinyl pyrrolidone)/chitosan [P(NIPAM-co-NVP)/CS] semi-IPN hydrogels were developed, showcasing heightened drug loading capacity and sustained release properties. Incorporating chitosan into the semi-interpenetrating network (semi-IPN) provided a robust affinity for anionic drugs, contributing to drug stability and efficient diffusion within the hydrogel matrix. The release profile of NAP from the hydrogels exhibited a significant burst release phenomenon within the initial 8 hours, accounting for 85-90% of the total NAP release. Notably, the release rate of NAP at 37°C became comparatively slower than that at 25°C. During the first 8 hours, only approximately 50% of NAP was released when the temperature was 37°C (Li et al., 2012). Following the initial rapid release, a gradual and sustained drug release profile persisted for an extended 36-hour duration, underscoring the sustained release nature of the hydrogel system. This phenomenon can be attributed to the temperature-triggered phase transition of only *N*-isopropyle acrylamide at 37°C, leading to hydrophobic interactions that impede the free diffusion of the drug molecules.

US-Triggered Drugs Release from CS Hydrogels

Ultrasound (US) has emerged as a powerful and non-invasive technology for DDS, with a wide range of applications (Kobayashi, 2023). It can be used to achieve controlled drug release and activation in DDS, and is an attractive option due to its precision, low toxicity, and non-invasive nature. Until now, US has potential in biomedical imaging for implantable drug delivery systems, enabling sustained drug release. In addition, it can be utilized to target and modulate drug release in advanced drug delivery systems. Smart US-responsive drug delivery systems have been developed that can release drugs at the tumor site upon demand. Although the use of US in drug delivery systems is a relatively new concept, recent studies have demonstrated promising results, highlighting its potential as an innovative and effective approach (Feril & Tachibana, 2012).

Figure 9. Illustration of ultrasound effect of chitosan hydrogel loaded-drug release. US provides both thermal and mechanical effects on the hydrogels in addition to soften and self-healing gelation for the medicine hydrogel

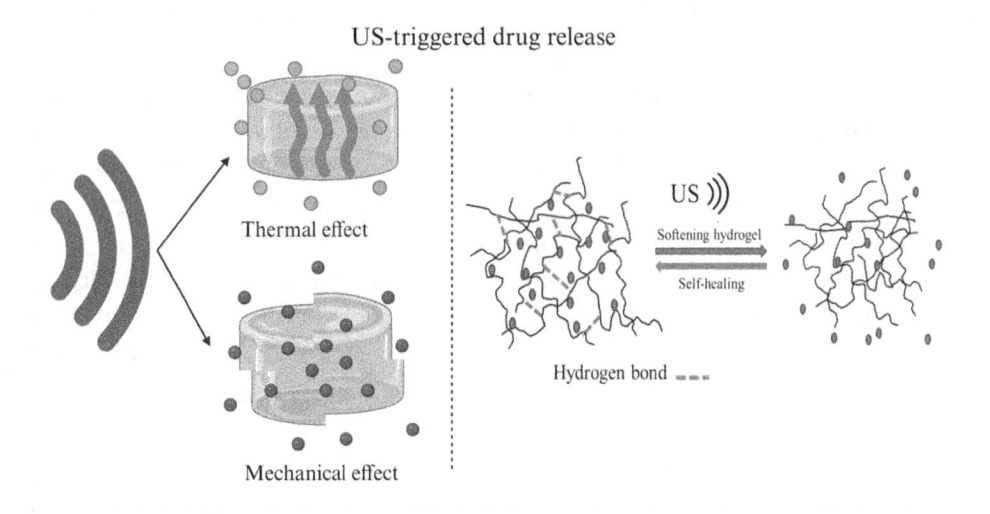

US, which can be classified as low frequency (<100 kHz) or high frequency (>100 kHz and MHz range), is a widely used as physical factor in disease diagnosis and therapy. Research has shown that US can enhance the permeability of agents into living cells since the mid-1990s. The delivery efficiency of drugs is improved through thermal and non-thermal effects induced under sonication, as summarized in Figure 9. Thermal effects are produced by the absorption of acoustic energy in biological tissues. US can increase the temperature of the target area, causing the drug to release faster due to increased molecular mobility and solubility because of increase the permeability of the cell membrane, allowing for faster drug uptake by the cells. Meanwhile, non-thermal effects are mainly generated by acoustic pressure and streaming, shockwaves, liquid microjet, and ultrasound-induced oscillation or cavitation. The non-thermal effects of US can bring about alterations in the structure of the drug delivery system, leading to improved drug release kinetics. Such characteristic actions are caused by the ability to create cavitation, which results in the formation and collapse of gas bubbles within the drug delivery system, thus leading to the temporary creation of pores or holes. This allows for a faster drug release. These effects were widely studied and are considered promising for various applications in drug delivery systems (Cai et al., 2020). Furthermore, US promoted the release of drug from the hydrogel by breaking the hydrogen bonds present in the hydrogel matrix through chemical effect of US (Figure 9). When exposing the gallic acid (GA)-chitin hydrogel to 43 kHz US the release of GA was significantly enhanced, with a release rate of 0.74 µg/mL·min. The release rate of GA from the hydrogel was approximately nine times higher, when exposed to US. As compared to the sample without US, US was acted as effective in disrupting the hydrogen bonds within the GA-chitin hydrogel matrix. These findings demonstrate the potential of US as a stimulus to facilitate the delivery of gallic acid from hydrogels of chitin, which is precursor of chitosan (Jiang and Kobayashi, 2017). The US responsiveness of the physical properties of chitin and cellulose hydrogels was also reported that US irradiation softened the gels of these polysaccharides (Iresha and Kobayashi, 2021), finding that such hydrogels softened by US and had self-healing properties that returned to their original state when US irradiation was stopped (Figure 9).

The chitosan-coated nanobubbles displayed the capability to form complexes with DNA and offer protection to it (Cavalli et al., 2012). Under US stimulation, these nanobubbles exhibited the ability to transfect DNA in vitro. During in vitro release tests, no release of plasmid DNA (pDNA) occurred from the nanobubbles without US, indicating robust interaction between the chitosan shell and the phosphate groups of pDNA. Upon subjecting them to 30 seconds of US, a moderate level of transfection was achieved. These findings underscore the potential of chitosan nanobubbles as a valuable tool for developing US-responsive formulations designed for targeted DNA delivery. The innovative system of O-carboxymethyl chitosan nanodroplets showed undergo pH-dependent charge conversion and responded to US, providing a platform for precise drug delivery in cancer therapy (Meng et al., 2019). Noteworthy features encompass a negative charge at pH 7.4, extending blood circulation duration; charge reversal within acidic tumor environments, facilitating interaction with tumor cells; robust stability in serum and impressive US-enhanced imaging capabilities. The system enabled drug release triggered by both pH and US, resulting in enhanced eradication of PC-3 cells when loaded with doxorubicin and exposed to US. Thus, it was considerable potential for advancing effective cancer treatment strategies.

Challenges in Chitosan Hydrogel in Clinical Practicals

In overview of the current state of chitosan-based hydrogels, encompassing their properties, these fabrication methods, and potential formulations of chitosan-based hydrogels in DDSs demonstrating promising outcomes, the clinical application of drug delivery systems would be based on chitosan hydrogels as drug carrier encounters for various obstacles.

- While chitosan hydrogel has demonstrated minimal to no toxicity in animal models, there is a limit of comprehensive clinical data to support their integration into medical practice.
- Achieving controlled and consistent fabrication of chitosan hydrogels is essential for their clinical applications. However, this can be challenging due to the complexity of the hydrogel fabrication process. The reproducibility of chitosan hydrogels is essential for their clinical applications. Achieving consistent properties and performance across different batches is challenging.
- Numerous cross-linkers employed during the creation of chitosan hydrogels could potentially raise worries about toxicity and patient safety.

- Swift degradation of chitosan-based hydrogel might result in insufficient drug release or compromised mechanical integrity.
- Sustaining the stability of chitosan hydrogels across a prolonged duration is a critical factor in their applicability for chronic conditions. The prevention of hydrogel deterioration or leakage of drugs over time presents a notable challenge.

Scale-up and large-scale production of drug-loaded chitosan hydrogel also pose challenges. For instance, a wide size distribution may lead to inconsistencies in therapeutic results between batches.

CONCLUSION

In this chapter, we described that chitosan hydrogels have established themselves as advanced and valuable materials for a wide range of uses and other applications such as water treatment, agriculture, pharmaceuticals and medical suppies. The creation of chitosan hydrogels involves numerous techniques that have been extensively investigated in prior research. Nonetheless, employing chitosan hydrogel necessitates a comprehensive evaluation of cross-linkers, pivotal in gelation of chitosan hydrogels. Additionally, enhanced performance of chitosan hydrogels often stems from the development of intricate systems where two or more polymers are integrated, utilizing various crosslinkers or other polymer composites. While diverse cross-linkers with varying reaction mechanisms can bond chitosan to polymers, only a limited number are deemed suitable for multiple applications related to sustainable development due to concerns surrounding toxicity, biocompatibility, and degradability. Composited chitosan hydrogel systems reveals their potential benefits in multi-component chitosan hydrogel arrangements, boasting superior properties such as heightened mechanical stability, biocompatibility, and targeted release capabilities. Beside that, further application of chitosan-based hydrogels should take into account a variety of other factors, such as toxicity of cross-linking, low mechanical resistance, and the inability to control hydrogel pore size in the scale-up process.

REFERENCES

Abdel-Raouf, M. E.-S., Farag, R. K., Farag, A. A., Keshawy, M., Abdel-Aziz, A., & Hasan, A. (2023). Chitosan-Based Architectures as an Effective Approach for the Removal of Some Toxic Species from Aqueous Media. *ACS Omega*, 8(11), 10086–10099. DOI: 10.1021/acsomega.2c07264 PMID: 36969416

Abdulhameed, A. S., Jawad, A. H., & Mohammad, A. T. (2019). Synthesis of chitosan-ethylene glycol diglycidyl ether/TiO(2) nanoparticles for adsorption of reactive orange 16 dye using a response surface methodology approach. *Bioresource Technology*, 293, 122071. DOI: 10.1016/j.biortech.2019.122071 PMID: 31491651

Ahmad, U., Sohail, M., Ahmad, M., Minhas, M. U., Khan, S., Hussain, Z., Kousar, M., Mohsin, S., Abbasi, M., Shah, S. A., & Rashid, H. (2019). Chitosan based thermosensitive injectable hydrogels for controlled delivery of loxoprofen: Development, characterization and in-vivo evaluation. *International Journal of Biological Macromolecules*, 129, 233–245. DOI: 10.1016/j.ijbiomac.2019.02.031 PMID: 30738157

Ahmed, E. M. (2015). Hydrogel: Preparation, characterization, and applications: A review. *Journal of Advanced Research*, 6(2), 105–121. DOI: 10.1016/j.jare.2013.07.006 PMID: 25750745

Ahmed, T. A., & Aljaeid, B. M. (2016). Preparation, characterization, and potential application of chitosan, chitosan derivatives, and chitosan metal nanoparticles in pharmaceutical drug delivery. *Drug Design, Development and Therapy*, 10, 483–507. DOI: 10.2147/DDDT.S99651 PMID: 26869768

Ankareddi, I., & Brazel, C. S. (2007). Synthesis and characterization of grafted thermosensitive hydrogels for heating activated controlled release. *International Journal of Pharmaceutics*, 336(2), 241–247. DOI: 10.1016/j.ijpharm.2006.11.065 PMID: 17234371

Azeem, M. K., Islam, A., Khan, R. U., Rasool, A., Qureshi, M. A., Rizwan, M., Sher, F., & Rasheed, T. (2023). Eco-friendly three-dimensional hydrogels for sustainable agricultural applications: Current and future scenarios. *Polymers for Advanced Technologies*, 34(9), 3046–3062. DOI: 10.1002/pat.6122

Bajpai, S. K. (2001). Swelling–deswelling behavior of poly(acrylamide-co-maleic acid) hydrogels. *Journal of Applied Polymer Science*, 80(14), 2782–2789. DOI: 10.1002/app.1394

Baroudi, A., García-Payo, C., & Khayet, M. (2018). Structural, Mechanical, and Transport Properties of Electron Beam-Irradiated Chitosan Membranes at Different Doses. In *Polymers* (Vol. 10, Issue 2). https://doi.org/DOI: 10.3390/polym10020117

Beerlage, C., Wiese, B., Kausch Annemie, R., & Arsenijevic, M. (2021). Change in Radiation Sterilization Process from Gamma Ray to X-ray. *Biomedical Instrumentation &amp. Technology*, 55(s3), 78–84. DOI: 10.2345/0899-8205-55.s3.78 PMID: 34153997

Bi, S., Pang, J., Huang, L., Sun, M., Cheng, X., & Chen, X. (2020). The toughness chitosan-PVA double network hydrogel based on alkali solution system and hydrogen bonding for tissue engineering applications. *International Journal of Biological Macromolecules*, 146, 99–109. DOI: 10.1016/j.ijbiomac.2019.12.186 PMID: 31874265

Boujelbane, F., Nasr, K., Sadaoui, H., Bui, H. M., Gantri, F., & Mzoughi, N. (2022). Decomposition mechanism of hydroxychloroquine in aqueous solution by gamma irradiation. *Chemicke Zvesti*, 76(3), 1777–1787. DOI: 10.1007/s11696-021-01969-1 PMID: 35106020

Bratskaya, S., Privar, Y., Nesterov, D., Modin, E., Kodess, M., Slobodyuk, A., Marinin, D., & Pestov, A. (2019). Chitosan Gels and Cryogels Cross-Linked with Diglycidyl Ethers of Ethylene Glycol and Polyethylene Glycol in Acidic Media. *Biomacromolecules*, 20(4), 1635–1643. DOI: 10.1021/acs.biomac.8b01817 PMID: 30726063

Bruschi, M. L. (2015). *Strategies to modify the drug release from pharmaceutical systems*. Woodhead Publishing.

Cai, X., Jiang, Y., Lin, M., Zhang, J., Guo, H., Yang, F., Leung, W., & Xu, C. (2020). Ultrasound-Responsive Materials for Drug/Gene Delivery. *Frontiers in Pharmacology*, 10, 1650. Advance online publication. DOI: 10.3389/fphar.2019.01650 PMID: 32082157

Casadidio, C., Peregrina, D. V., Gigliobianco, M. R., Deng, S., Censi, R., & Di Martino, P. (2019). Chitin and chitosans: Characteristics, eco-friendly processes, and applications in cosmetic science. *Marine Drugs*, 17(6), 369. Advance online publication. DOI: 10.3390/md17060369 PMID: 31234361

Cavalli, R., Bisazza, A., Trotta, M., Argenziano, M., Civra, A., Donalisio, M., & Lembo, D. (2012). New chitosan nanobubbles for ultrasound-mediated gene delivery: Preparation and in vitro characterization. *International Journal of Nanomedicine*, 7, 3309–3318. DOI: 10.2147/IJN.S30912 PMID: 22802689

Chatterjee, S., Hui, P. C., Siu, W. S., Kan, C., Leung, P.-C., Wanxue, C., & Chiou, J.-C. (2021). Influence of pH-responsive compounds synthesized from chitosan and hyaluronic acid on dual-responsive (pH/temperature) hydrogel drug delivery systems of Cortex Moutan. *International Journal of Biological Macromolecules*, 168, 163–174. DOI: 10.1016/j.ijbiomac.2020.12.035 PMID: 33309656

Chen, Y., He, F., Ren, Y., Peng, H., & Huang, K. (2014). Fabrication of chitosan/PAA multilayer onto magnetic microspheres by LbL method for removal of dyes. *Chemical Engineering Journal*, 249, 79–92. DOI: 10.1016/j.cej.2014.03.093

Clarke, R. H., & Valentin, J. (2009). The History of ICRP and the Evolution of its Policies: Invited by the Commission in October 2008. *Annals of the ICRP*, 39(1), 75–110. DOI: 10.1016/j.icrp.2009.07.009 PMID: 19853177

Collins, E. D., & Ottinger, C. L. (2003). Isotopes, Separation and Application. In Meyers, R. A. (Ed.), *Encyclopedia of Physical Science and Technology* (3rd ed., pp. 109–126). Academic Press., DOI: 10.1016/B0-12-227410-5/00355-0

Craciun, A. M., Mititelu Tartau, L., Pinteala, M., & Marin, L. (2019). Nitrosalicyl-imine-chitosan hydrogels based drug delivery systems for long term sustained release in local therapy. *Journal of Colloid and Interface Science*, 536, 196–207. DOI: 10.1016/j.jcis.2018.10.048 PMID: 30368091

Cui, Z., Xiang, Y., Si, J., Yang, M., Zhang, Q., & Zhang, T. (2008). Ionic interactions between sulfuric acid and chitosan membranes. *Carbohydrate Polymers*, 73(1), 111–116. DOI: 10.1016/j.carbpol.2007.11.009

Czechowska-Biskup, R., Wach, R. A., Stojek, P., Kamińska, M., Rosiak, J. M., & Ulański, P. (2016). Synthesis of chitosan and carboxymethyl chitosan hydrogels by electron beam irradiation. *Progress on Chemistry and Application of Chitin and Its Derivatives*, 21(October), 27–45. DOI: 10.15259/PCACD.21.03

De Souza, R., Zahedi, P., Allen, C. J., & Piquette-Miller, M. (2009). Biocompatibility of injectable chitosan–phospholipid implant systems. *Biomaterials*, 30(23), 3818–3824. DOI: 10.1016/j.biomaterials.2009.04.003 PMID: 19394688

Dergunov, S. A., & Mun, G. A. (2009). γ-irradiated chitosan-polyvinyl pyrrolidone hydrogels as pH-sensitive protein delivery system. *Radiation Physics and Chemistry*, 78(1), 65–68. DOI: 10.1016/j.radphyschem.2008.07.003

Di Martino, A., Kucharczyk, P., Capakova, Z., Humpolicek, P., & Sedlarik, V. (2017). Chitosan-based nanocomplexes for simultaneous loading, burst reduction and controlled release of doxorubicin and 5-fluorouracil. *International Journal of Biological Macromolecules*, 102, 613–624. DOI: 10.1016/j.ijbiomac.2017.04.004 PMID: 28431942

Din, M. O., Danino, T., Prindle, A., Skalak, M., Selimkhanov, J., Allen, K., Julio, E., Atolia, E., Tsimring, L. S., Bhatia, S. N., & Hasty, J. (2016). Synchronized cycles of bacterial lysis for in vivo delivery. *Nature*, 536(7614), 81–85. DOI: 10.1038/nature18930 PMID: 27437587

Dodero, A., Scarfi, S., Mirata, S., Sionkowska, A., Vicini, S., Alloisio, M., & Castellano, M. (2021). Effect of Crosslinking Type on the Physical-Chemical Properties and Biocompatibility of Chitosan-Based Electrospun Membranes. In *Polymers* (Vol. 13, Issue 5). https://doi.org/DOI: 10.3390/polym13050831

Don, T.-M., Huang, M.-L., Chiu, A.-C., Kuo, K.-H., Chiu, W.-Y., & Chiu, L.-H. (2008). Preparation of thermo-responsive acrylic hydrogels useful for the application in transdermal drug delivery systems. *Materials Chemistry and Physics*, 107(2), 266–273. DOI: 10.1016/j.matchemphys.2007.07.009

Dou, Z., Bini Farias, M. V., Chen, W., He, D., Hu, Y., & Xie, X. (2022). Highly degradable chitosan-montmorillonite (MMT) nano-composite hydrogel for controlled fertilizer release. *Frontiers of Environmental Science & Engineering*, 17(5), 53. DOI: 10.1007/s11783-023-1653-9

Duan, J., Liang, X., Cao, Y., Wang, S., & Zhang, L. (2015). High Strength Chitosan Hydrogels with Biocompatibility via New Avenue Based on Constructing Nanofibrous Architecture. *Macromolecules*, 48(8), 2706–2714. DOI: 10.1021/acs.macromol.5b00117

Eden, R. J. (1956). No Title. *Nuclear Physics*, 1(8), 311. DOI: 10.1016/S0029-5582(56)80039-4

Eivazzadeh-Keihan, R., Radinekiyan, F., Aliabadi, H. A. M., Sukhtezari, S., Tahmasebi, B., Maleki, A., & Madanchi, H. (2021). Chitosan hydrogel/silk fibroin/Mg(OH)2 nanobiocomposite as a novel scaffold with antimicrobial activity and improved mechanical properties. *Scientific Reports*, 11(1), 650. DOI: 10.1038/s41598-020-80133-3 PMID: 33436831

El Salmawi, K. M. (2007). Gamma radiation-induced crosslinked PVA/chitosan blends for wound dressing. *Journal of Macromolecular Science, Part A: Pure and Applied Chemistry*, 44(5), 541–545. DOI: 10.1080/10601320701235891

Elbarbary, A. M., & Ghobashy, M. M. (2017). *Controlled release fertilizers using superabsorbent hydrogel prepared by gamma radiation. 105*(10), 865–876. https://doi.org/doi:10.1515/ract-2016-2679

Fan, L., Yang, H., Yang, J., Peng, M., & Hu, J. (2016). Preparation and characterization of chitosan/gelatin/PVA hydrogel for wound dressings. *Carbohydrate Polymers*, 146, 427–434. DOI: 10.1016/j.carbpol.2016.03.002 PMID: 27112893

Feksa, L. R., Troian, E. A., Muller, C. D., Viegas, F., Machado, A. B., & Rech, V. C. (2018). Hydrogels for biomedical applications. In Grumezescu, A. M. (Ed.), *Nanostructures for the Engineering of Cells, Tissues and Organs* (pp. 403–438). William Andrew Publishing., DOI: 10.1016/B978-0-12-813665-2.00011-9

Feril, L. B.Jr, & Tachibana, K. (2012). Use of ultrasound in drug delivery systems: Emphasis on experimental methodology and mechanisms. *International Journal of Hyperthermia*, 28(4), 282–289. DOI: 10.3109/02656736.2012.668640 PMID: 22621730

Fertahi, S., Ilsouk, M., Zeroual, Y., Oukarroum, A., & Barakat, A. (2021). Recent trends in organic coating based on biopolymers and biomass for controlled and slow release fertilizers. *Journal of Controlled Release*, 330, 341–361. DOI: 10.1016/j.jconrel.2020.12.026 PMID: 33352245

Francis, A. O., Zaini, M. A. A., Muhammad, I. M., Abdulsalam, S., & El-Nafaty, U. A. (2023). Physicochemical modification of chitosan adsorbent: A perspective. *Biomass Conversion and Biorefinery*, 13(7), 5557–5575. DOI: 10.1007/s13399-021-01599-3

Fu, F., & Wang, Q. (2011). Removal of heavy metal ions from wastewaters: A review. *Journal of Environmental Management*, 92(3), 407–418. DOI: 10.1016/j.jenvman.2010.11.011 PMID: 21138785

Fu, J., Yang, F., & Guo, Z. (2018). The chitosan hydrogels: From structure to function. *New Journal of Chemistry*, 42(21), 17162–17180. DOI: 10.1039/C8NJ03482F

Gad, Y. H. (2008). Preparation and characterization of poly(2-acrylamido-2-methylpropane-sulfonic acid)/Chitosan hydrogel using gamma irradiation and its application in wastewater treatment. *Radiation Physics and Chemistry*, 77(9), 1101–1107. DOI: 10.1016/j.radphyschem.2008.05.002

Gao, L., Gan, H., Meng, Z., Gu, R., Wu, Z., Zhang, L., Zhu, X., Sun, W., Li, J., Zheng, Y., & Dou, G. (2014). Effects of genipin cross-linking of chitosan hydrogels on cellular adhesion and viability. *Colloids and Surfaces. B, Biointerfaces*, 117, 398–405. DOI: 10.1016/j.colsurfb.2014.03.002 PMID: 24675278

Gaohua, L., Miao, X., & Dou, L. (2021). Crosstalk of physiological pH and chemical pKa under the umbrella of physiologically based pharmacokinetic modeling of drug absorption, distribution, metabolism, excretion, and toxicity. *Expert Opinion on Drug Metabolism & Toxicology*, 17(9), 1103–1124. DOI: 10.1080/17425255.2021.1951223 PMID: 34253134

Garnica-Palafox, I. M., Sanchez-Arevalo, F. M., Velasquillo, C., Garcia-Carvajal, Z. Y., Garcia-Lopez, J., Ortega-Sanchez, C., Ibarra, C., Luna-Barcenas, G., & Solis-Arrieta, L. (2014). Mechanical and structural response of a hybrid hydrogel based on chitosan and poly(vinyl alcohol) cross-linked with epichlorohydrin for potential use in tissue engineering. *Journal of Biomaterials Science. Polymer Edition*, 25(1), 32–50. DOI: 10.1080/09205063.2013.833441 PMID: 24007370

Ghobashy, M. M., Elbarbary, A. M., & Hegazy, D. E. (2021). Gamma radiation synthesis of a novel amphiphilic terpolymer hydrogel pH-responsive based chitosan for colon cancer drug delivery. *Carbohydrate Polymers*, 263(March), 117975. DOI: 10.1016/j.carbpol.2021.117975 PMID: 33858572

Guo, W., Yang, M., Liu, S., Zhang, X., Zhang, B., & Chen, Y. (2021). Chitosan/polyvinyl alcohol/tannic acid multiple network composite hydrogel: Preparation and characterization. *Iranian Polymer Journal*, 30(11), 1159–1168. DOI: 10.1007/s13726-021-00966-1

He, M., & Chu, C.-C. (2013). Dual stimuli responsive glycidyl methacrylate chitosan-quaternary ammonium hybrid hydrogel and its bovine serum albumin release. *Journal of Applied Polymer Science*, 130(5), 3736–3745. DOI: 10.1002/app.39635

He, Q., Kobayashi, K., Kusumi, R., Kimura, S., Enomoto, Y., Yoshida, M., Kim, U.-J., & Wada, M. (2020). In Vitro Synthesis of Branchless Linear (1 → 6)-α-d-Glucan by Glucosyltransferase K: Mechanical and Swelling Properties of Its Hydrogels Crosslinked with Diglycidyl Ethers. *ACS Omega*, 5(48), 31272–31280. DOI: 10.1021/acsomega.0c04699 PMID: 33324837

He, Q., Kusumi, R., Kimura, S., Kim, U.-J., Deguchi, K., Ohki, S., Goto, A., Shimizu, T., & Wada, M. (2020). Highly swellable hydrogel of regioselectively aminated (1→3)-α-d-glucan crosslinked with ethylene glycol diglycidyl ether. *Carbohydrate Polymers*, 237, 116189. DOI: 10.1016/j.carbpol.2020.116189 PMID: 32241412

Hennink, W. E., & van Nostrum, C. F. (2002). Novel crosslinking methods to design hydrogels. *Advanced Drug Delivery Reviews*, 54(1), 13–36. DOI: 10.1016/S0169-409X(01)00240-X PMID: 11755704

Herrada-Manchón, H., Fernández, M. A., & Aguilar, E. (2023). Essential Guide to Hydrogel Rheology in Extrusion 3D Printing: How to Measure It and Why It Matters? In *Gels* (Vol. 9, Issue 7). https://doi.org/DOI: 10.3390/gels9070517

Hrubý, M., Koňák, Č., & Ulbrich, K. (2005). Polymeric micellar pH-sensitive drug delivery system for doxorubicin. *Journal of Controlled Release*, 103(1), 137–148. DOI: 10.1016/j.jconrel.2004.11.017 PMID: 15710507

Hu, J.-L., Zhang, Y.-X., Song, C.-H., Wei, H.-G., Zhong, C., & Chu, L.-Q. (2023). Preparation and characterization of carboxymethyl chitosan/polyvinyl alcohol composite hydrogel with improved mechanical and antibacterial properties. *Reactive & Functional Polymers*, 187, 105592. DOI: 10.1016/j.reactfunctpolym.2023.105592

Huang, S.-C., Xia, X.-X., Fan, R.-X., & Qian, Z.-G. (2020). Programmable Electrostatic Interactions Expand the Landscape of Dynamic Functional Hydrogels. *Chemistry of Materials*, 32(5), 1937–1945. DOI: 10.1021/acs.chemmater.9b04726

Ioan, M. R., Postolache, C., Fugaru, V., Bercea, S., Celarel, A. & Cenuă, C. (2017). *COMPUTATIONAL METHOD FOR THE DETERMINATION OF INTENSE GAMMA-RAYS SOURCES ACTIVITY BY USING GEANT4.*

Iresha, H., & Kobayashi, T. (2021). In Situ Viscoelasticity Behavior of Cellulose-Chitin Composite Hydrogels during Ultrasound Irradiation. *Gels (Basel, Switzerland)*, 7(3), 81. Advance online publication. DOI: 10.3390/gels7030081 PMID: 34209349

Islam, A., Yasin, T., & Rehman, I. U. (2014). Synthesis of hybrid polymer networks of irradiated chitosan/poly(vinyl alcohol) for biomedical applications. *Radiation Physics and Chemistry*, 96, 115–119. DOI: 10.1016/j.radphyschem.2013.09.009

Jabeen, S., Saeed, S., Kausar, A., Muhammad, B., Gul, S., & Farooq, M. (2015). Influence of chitosan and epoxy cross-linking on physical properties of binary blends. *IJPAC. International Journal of Polymer Analysis and Characterization*, 21(2), 163–174. DOI: 10.1080/1023666X.2016.1131404

Jamnongkan, T., & Kaewpirom, S. (2010). *Controlled-Release Fertilizer Based on Chitosan Hydrogel: Phosphorus Release Kinetics*. https://api.semanticscholar.org/CorpusID:98695516

Jang, S. H., Jeong, Y. G., Min, B. G., Lyoo, W. S., & Lee, S. C. (2008). Preparation and lead ion removal property of hydroxyapatite/polyacrylamide composite hydrogels. *Journal of Hazardous Materials*, 159(2), 294–299. DOI: 10.1016/j.jhazmat.2008.02.018 PMID: 18430514

Jawad, A. H., Mamat, N. F. H., Hameed, B. H., & Ismail, K. (2019). Biofilm of cross-linked Chitosan-Ethylene Glycol Diglycidyl Ether for removal of Reactive Red 120 and Methyl Orange: Adsorption and mechanism studies. *Journal of Environmental Chemical Engineering*, 7(2), 102965. DOI: 10.1016/j.jece.2019.102965

Jiang, H., & Kobayashi, T. (2017). Ultrasound stimulated release of gallic acid from chitin hydrogel matrix. *Materials Science and Engineering C*, 75, 478–486. DOI: 10.1016/j.msec.2017.02.082 PMID: 28415488

Jiménez-Becerril, J., Moreno-López, A., & Jiménez-Reyes, M. (2016). Radiocatalytic degradation of dissolved organic compounds in wastewater. *Nukleonika*, 61(4), 473–476. https://doi.org/doi:10.1515/nuka-2016-0077. DOI: 10.1515/nuka-2016-0077

Kalidason, A., Saito, K., Nanbu, Y., Sasaki, H., Ohsumi, R., Kanazawa, A., & Kuroiwa, T. (2022). Biodegradable Crosslinked Chitosan Gel Microbeads with Controlled Size, Prepared by Membrane Emulsification-External Gelation and Their Application as Reusable Adsorption Materials. *Journal of Chemical Engineering of Japan*, 55(1), 61–70. DOI: 10.1252/jcej.21we061

Kesharwani, P., Bisht, A., Alexander, A., Dave, V., & Sharma, S. (2021). Biomedical applications of hydrogels in drug delivery system: An update. *Journal of Drug Delivery Science and Technology*, 66, 102914. DOI: 10.1016/j.jddst.2021.102914

Khadem, E., Kharaziha, M., & Salehi, S. (2023). Colorimetric pH-responsive and hemostatic hydrogel-based bioadhesives containing functionalized silver nanoparticles. *Materials Today. Bio*, 20, 100650. DOI: 10.1016/j.mtbio.2023.100650 PMID: 37206880

Kobayashi, T. (2023). *Chapter 20 - Ultrasound-triggered drug delivery* (A. K. Nayak, M. S. Hasnain, B. Laha, & S. B. T.-A. and M. A. for D. D. Maiti (Eds.); pp. 577–591). Academic Press. https://doi.org/ DOI: 10.1016/B978-0-323-91668-4.00025-3

Kocemba, A., & Mucha, M. (2017). Chitosan/poly(vinyl alcohol) hydrogels as controlled drug delivery systems. *Progress on Chemistry and Application of Chitin and Its Derivatives*, 22, 97–105. DOI: 10.15259/PCACD.22.09

Kocen, R., Gasik, M., Gantar, A., & Novak, S. (2017). Viscoelastic behaviour of hydrogel-based composites for tissue engineering under mechanical load. *Biomedical Materials (Bristol, England)*, 12(2), 25004. DOI: 10.1088/1748-605X/aa5b00 PMID: 28106535

Kundu, P., Das, S., & Chattopadhyay, N. (2020). Switching from endogenous to exogenous delivery of a model drug to DNA through micellar engineering. *Journal of Photochemistry and Photobiology. B, Biology*, 203, 111765. DOI: 10.1016/j.jphotobiol.2019.111765 PMID: 31923806

L'Annunziata, M. F. (2007). 3 - Gamma- and X-Radiation — Photons. In L'Annunziata, M. F. (Ed.), *Radioactivity* (pp. 187–215). Elsevier Science B.V., DOI: 10.1016/B978-044452715-8.50006-2

Lee, D., Zhang, H., & Ryu, S. (2018). Elastic Modulus Measurement of Hydrogels. In Mondal, M. I. H. (Ed.), *Cellulose-Based Superabsorbent Hydrogels* (pp. 1–21). Springer International Publishing., DOI: 10.1007/978-3-319-76573-0_60-1

Leyva-Gómez, G., Santillan-Reyes, E., Lima, E., Madrid-Martínez, A., Krötzsch, E., Quintanar-Guerrero, D., Garciadiego-Cázares, D., Martínez-Jiménez, A., Hernández Morales, M., Ortega-Peña, S., Contreras-Figueroa, M. E., Cortina-Ramírez, G. E., & Abarca-Buis, R. F. (2017). A novel hydrogel of poloxamer 407 and chitosan obtained by gamma irradiation exhibits physicochemical properties for wound management. *Materials Science and Engineering C*, 74, 36–46. DOI: 10.1016/j.msec.2016.12.127 PMID: 28254305

Li, D., Zhang, J., Li, L., Cai, G., Zuo, W., Zhan, W., Wang, P., & Tian, Y. (2022). Di-cationic epoxy resin one-step crosslinked chitosan/cellulose hydrogel for efficient dye and oily wastewater remediation. *Journal of Cleaner Production*, 371, 133650. Advance online publication. DOI: 10.1016/j.jclepro.2022.133650

Li, G., Guo, L., Chang, X., & Yang, M. (2012). Thermo-sensitive chitosan based semi-IPN hydrogels for high loading and sustained release of anionic drugs. *International Journal of Biological Macromolecules*, 50(4), 899–904. DOI: 10.1016/j.ijbiomac.2012.02.013 PMID: 22679630

Lin, L.-X., Luo, J.-W., Yuan, F., Zhang, H.-H., Ye, C.-Q., Zhang, P., & Sun, Y.-L. (2017). In situ crosslinking carbodiimide-modified chitosan hydrogel for postoperative adhesion prevention in a rat model. *Materials Science and Engineering C*, 81, 380–385. DOI: 10.1016/j.msec.2017.07.024 PMID: 28887987

Lin, Y.-H., Liang, H.-F., Chung, C.-K., Chen, M.-C., & Sung, H.-W. (2005). Physically crosslinked alginate/N,O-carboxymethyl chitosan hydrogels with calcium for oral delivery of protein drugs. *Biomaterials*, 26(14), 2105–2113. DOI: 10.1016/j.biomaterials.2004.06.011 PMID: 15576185

Liu, C., Thormann, E., Claesson, P. M., & Tyrode, E. (2014). Surface grafted chitosan gels. Part II. Gel formation and characterization. *Langmuir*, 30(29), 8878–8888. DOI: 10.1021/la501319r PMID: 25006685

Liu, J., Wang, Q., & Wang, A. (2007). Synthesis and characterization of chitosan-g-poly(acrylic acid)/sodium humate superabsorbent. *Carbohydrate Polymers*, 70(2), 166–173. DOI: 10.1016/j.carbpol.2007.03.015

Lou, C., Tian, X., Deng, H., Wang, Y., & Jiang, X. (2020). Dialdehyde-beta-cyclodextrin-crosslinked carboxymethyl chitosan hydrogel for drug release. *Carbohydrate Polymers*, 231, 115678. DOI: 10.1016/j.carbpol.2019.115678 PMID: 31888806

Lousada, C. M., Soroka, I. L., Yagodzinskyy, Y., Tarakina, N. V., Todoshchenko, O., Hänninen, H., Korzhavyi, P. A., & Jonsson, M. (2016). Gamma radiation induces hydrogen absorption by copper in water. *Scientific Reports*, 6(1), 24234. DOI: 10.1038/srep24234 PMID: 27086752

Lu, X., Xu, Y., Zheng, C., Zhang, G., & Su, Z. (2006). Ethylene glycol diglycidyl ether as a protein cross-linker: A case study for cross-linking of hemoglobin. *Journal of Chemical Technology and Biotechnology*, 81(5), 767–775. DOI: 10.1002/jctb.1441

Ma, J., Zhou, G., Chu, L., Liu, Y., Liu, C., Luo, S., & Wei, Y. (2017). Efficient Removal of Heavy Metal Ions with An EDTA Functionalized Chitosan/Polyacrylamide Double Network Hydrogel. *ACS Sustainable Chemistry & Engineering*, 5(1), 843–851. DOI: 10.1021/acssuschemeng.6b02181

Maiz-Fernández, S., Pérez-Álvarez, L., Silván, U., Vilas-Vilela, J. L., & Lanceros-Mendez, S. (2022). Photocrosslinkable and self-healable hydrogels of chitosan and hyaluronic acid. *International Journal of Biological Macromolecules*, 216, 291–302. DOI: 10.1016/j.ijbiomac.2022.07.004 PMID: 35798076

Martínez-Ruvalcaba, A., Chornet, E., & Rodrigue, D. (2007). Viscoelastic properties of dispersed chitosan/xanthan hydrogels. *Carbohydrate Polymers*, 67(4), 586–595. DOI: 10.1016/j.carbpol.2006.06.033

Mat Nayan, N. H., Hamzah, M. S. A., Mohd Tahir, A. A., Rajali, A. A. A., Muslih, E. F., & Mazlan, R. (2018). Development of Polyvinyl Alcohol/Chitosan Hydrogel Loaded with Fertilizer Compound: Preparation, Properties and Effect on Seed Germination. *Journal of Science and Technology, 10*(4 SE-Articles). https://publisher.uthm.edu.my/ojs/index.php/JST/article/view/3649

McGregor, D., Bolt, H., Cogliano, V., & Richter-Reichhelm, H.-B. (2006). Formaldehyde and Glutaraldehyde and Nasal Cytotoxicity: Case Study Within the Context of the 2006 IPCS Human Framework for the Analysis of a Cancer Mode of Action for Humans. *Critical Reviews in Toxicology*, 36(10), 821–835. DOI: 10.1080/10408440600977669 PMID: 17118731

Meng, D., Guo, L., Shi, D., Sun, X., Shang, M., Zhou, X., & Li, J. (2019). Charge-conversion and ultrasound-responsive O-carboxymethyl chitosan nanodroplets for controlled drug delivery. *Nanomedicine (London)*, 14(19), 2549–2565. DOI: 10.2217/nnm-2019-0217 PMID: 31271101

Minet, E. P., O'Carroll, C., Rooney, D., Breslin, C., McCarthy, C. P., Gallagher, L., & Richards, K. G. (2013). Slow delivery of a nitrification inhibitor (dicyandiamide) to soil using a biodegradable hydrogel of chitosan. *Chemosphere*, 93(11), 2854–2858. DOI: 10.1016/j.chemosphere.2013.08.043 PMID: 24035690

Mittal, H., Ray, S. S., Kaith, B. S., Bhatia, J. K., Sukriti, , Sharma, J., & Alhassan, S. M. (2018). Recent progress in the structural modification of chitosan for applications in diversified biomedical fields. *European Polymer Journal*, 109, 402–434. DOI: 10.1016/j.eurpolymj.2018.10.013

Naikwadi, A. T., Sharma, B. K., Bhatt, K. D., & Mahanwar, P. A. (2022). Gamma Radiation Processed Polymeric Materials for High Performance Applications: A Review. *Frontiers in Chemistry*, 10, 837111. Advance online publication. DOI: 10.3389/fchem.2022.837111 PMID: 35360545

Nisar, S., Pandit, A. H., Nadeem, M., Pandit, A. H., Rizvi, M. M. A., & Rattan, S. (2021). γ-Radiation induced L-glutamic acid grafted highly porous, pH-responsive chitosan hydrogel beads: A smart and biocompatible vehicle for controlled anti-cancer drug delivery. In *International Journal of Biological Macromolecules* (Vol. 182, pp. 37–50). https://doi.org/DOI: 10.1016/j.ijbiomac.2021.03.134

Noppakundilograt, S., Pheatcharat, N., & Kiatkamjornwong, S. (2015). Multilayer-coated NPK compound fertilizer hydrogel with controlled nutrient release and water absorbency. *Journal of Applied Polymer Science*, 132(2), 1–11. DOI: 10.1002/app.41249

Pal, K., Banthia, A. K., & Majumdar, D. K. (2009). Polymeric Hydrogels: Characterization and Biomedical Applications. *Designed Monomers and Polymers*, 12(3), 197–220. DOI: 10.1163/156855509X436030

Park, H., Otte, A., & Park, K. (2022). Evolution of drug delivery systems: From 1950 to 2020 and beyond. *Journal of Controlled Release*, 342, 53–65. DOI: 10.1016/j.jconrel.2021.12.030 PMID: 34971694

Park, H., Park, K., & Kim, D. (2006). Preparation and swelling behavior of chitosan-based superporous hydrogels for gastric retention application. *Journal of Biomedical Materials Research. Part A*, 76A(1), 144–150. DOI: 10.1002/jbm.a.30533 PMID: 16258961

Parsons, B. J. (2012). 8 - Sterilisation of healthcare products by ionising radiation: sterilisation of drug-device products and tissue allografts. In Lerouge, S., & Simmons, A. (Eds.), *Sterilisation of Biomaterials and Medical Devices* (pp. 212–239). Woodhead Publishing., DOI: 10.1533/9780857096265.212

Pellá, M. C. G., Lima-Tenório, M. K., Tenório-Neto, E. T., Guilherme, M. R., Muniz, E. C., & Rubira, A. F. (2018). Chitosan-based hydrogels: From preparation to biomedical applications. *Carbohydrate Polymers*, 196(March), 233–245. DOI: 10.1016/j.carbpol.2018.05.033 PMID: 29891292

Perumal, S., Atchudan, R., Yoon, D. H., Joo, J., & Cheong, I. W. (2019). Spherical Chitosan/Gelatin Hydrogel Particles for Removal of Multiple Heavy Metal Ions from Wastewater. *Industrial & Engineering Chemistry Research*, 58(23), 9900–9907. DOI: 10.1021/acs.iecr.9b01298

Pita-López, M. L., Fletes-Vargas, G., Espinosa-Andrews, H., & Rodríguez-Rodríguez, R. (2021). Physically cross-linked chitosan-based hydrogels for tissue engineering applications: A state-of-the-art review. *European Polymer Journal*, 145, 110176. DOI: 10.1016/j.eurpolymj.2020.110176

Qin, H., Wang, J., Wang, T., Gao, X., Wan, Q., & Pei, X. (2018). Preparation and Characterization of Chitosan/β-Glycerophosphate Thermal-Sensitive Hydrogel Reinforced by Graphene Oxide. *Frontiers in Chemistry*, 6, 565. Advance online publication. DOI: 10.3389/fchem.2018.00565 PMID: 30555817

Qu, X., Wirsén, A., & Albertsson, A. C. (2000). Novel pH-sensitive chitosan hydrogels: Swelling behavior and states of water. *Polymer*, 41(12), 4589–4598. DOI: 10.1016/S0032-3861(99)00685-0

Rafat, M., Li, F., Fagerholm, P., Lagali, N. S., Watsky, M. A., Munger, R., Matsuura, T., & Griffith, M. (2008). PEG-stabilized carbodiimide crosslinked collagen-chitosan hydrogels for corneal tissue engineering. *Biomaterials*, 29(29), 3960–3972. DOI: 10.1016/j.biomaterials.2008.06.017 PMID: 18639928

Rattanawongwiboon, T., Hemvichian, K., Lertsarawut, P., & Suwanmala, P. (2020). Chitosan-poly(ethylene glycol) diacrylate beads prepared by radiation-induced crosslinking and their promising applications derived from encapsulation of essential oils. *Radiation Physics and Chemistry*, 170, 108656. Advance online publication. DOI: 10.1016/j.radphyschem.2019.108656

Raza, A., Rasheed, T., Nabeel, F., Hayat, U., Bilal, M., & Iqbal, H. M. N. (2019). Endogenous and Exogenous Stimuli-Responsive Drug Delivery Systems for Programmed Site-Specific Release. In *Molecules* (Vol. 24, Issue 6). https://doi.org/DOI: 10.3390/molecules24061117

Raza, M. A., Jeong, J.-O., & Park, S. H. (2021). State-of-the-Art Irradiation Technology for Polymeric Hydrogel Fabrication and Application in Drug Release System. *Frontiers in Materials*, 8, 769436. Advance online publication. DOI: 10.3389/fmats.2021.769436

Reza Rezaie, H., & Esnaashary, M. Aref arjmand, A. & Öchsner, A. (2018). The History of Drug Delivery Systems. In H. Reza Rezaie, M. Esnaashary, A. Aref arjmand & A. Öchsner (Eds.), *A Review of Biomaterials and Their Applications in Drug Delivery* (pp. 1–8). Springer Singapore. https://doi.org/ DOI: 10.1007/978-981-10-0503-9_1

Rizwan, M., Yahya, R., Hassan, A., Yar, M., Azzahari, A. D., Selvanathan, V., Sonsudin, F., & Abouloula, C. N. (2017). pH Sensitive Hydrogels in Drug Delivery: Brief History, Properties, Swelling, and Release Mechanism, Material Selection and Applications. In *Polymers* (Vol. 9, Issue 4). https://doi.org/ DOI: 10.3390/polym9040137

Rizzo, F., & Kehr, N. S. (2021). Recent Advances in Injectable Hydrogels for Controlled and Local Drug Delivery. *Advanced Healthcare Materials*, 10(1), 2001341. DOI: 10.1002/adhm.202001341 PMID: 33073515

Roberts, G. A. F., & Taylor, K. E. (1989). Chitosan gels, 3. The formation of gels by reaction of chitosan with glutaraldehyde. *Die Makromolekulare Chemie*, 190(5), 951–960. DOI: 10.1002/macp.1989.021900504

Roy, N., Saha, N., Kitano, T., Saha, P., & Zatloukal, M. (2009). Importance of Viscoelastic Property Measurement of a New Hydrogel for Health Care. *AIP Conference Proceedings*, 1152(1), 210–216. DOI: 10.1063/1.3203272

Schoenmakers, D. C., Rowan, A. E., & Kouwer, P. H. J. (2018). Crosslinking of fibrous hydrogels. *Nature Communications*, 9(1), 2172. DOI: 10.1038/s41467-018-04508-x PMID: 29867185

Shamay, Y., Adar, L., Ashkenasy, G., & David, A. (2011). Light induced drug delivery into cancer cells. *Biomaterials*, 32(5), 1377–1386. DOI: 10.1016/j.biomaterials.2010.10.029 PMID: 21074848

Shu, X. Z., & Zhu, K. J. (2002). The influence of multivalent phosphate structure on the properties of ionically cross-linked chitosan films for controlled drug release. *European Journal of Pharmaceutics and Biopharmaceutics*, 54(2), 235–243. DOI: 10.1016/S0939-6411(02)00052-8 PMID: 12191697

Shu, X. Z., Zhu, K. J., & Song, W. (2001). Novel pH-sensitive citrate cross-linked chitosan film for drug controlled release. *International Journal of Pharmaceutics*, 212(1), 19–28. DOI: 10.1016/S0378-5173(00)00582-2 PMID: 11165817

Sokary, R., Abu el-naga, M. N., Bekhit, M., & Atta, S. (2021). A potential antibiofilm, antimicrobial and anticancer activities of chitosan capped gold nanoparticles prepared by γ–irradiation. *Materials Technology*, 37(7), 493–502. DOI: 10.1080/10667857.2020.1863555

Spinks, G. M., Lee, C. K., Wallace, G. G., Kim, S. I., & Kim, S. J. (2006). Swelling Behavior of Chitosan Hydrogels in Ionic Liquid−Water Binary Systems. *Langmuir*, 22(22), 9375–9379. DOI: 10.1021/la061586r PMID: 17042556

Stojkov, G., Niyazov, Z., Picchioni, F., & Bose, R. K. (2021). Relationship between Structure and Rheology of Hydrogels for Various Applications. *Gels (Basel, Switzerland)*, 7(4), 255. Advance online publication. DOI: 10.3390/gels7040255 PMID: 34940315

Sung, Y. K., & Kim, S. W. (2020). Recent advances in polymeric drug delivery systems. *Biomaterials Research*, 24(1), 12. DOI: 10.1186/s40824-020-00190-7 PMID: 32537239

Szalewicz, K. (2003). Hydrogen Bond. In Meyers, R. A. (Ed.), *Encyclopedia of Physical Science and Technology* (3rd ed., pp. 505–538). Academic Press., DOI: 10.1016/B0-12-227410-5/00322-7

Tang, S., Yang, J., Lin, L., Peng, K., Chen, Y., Jin, S., & Yao, W. (2020). Construction of physically crosslinked chitosan/sodium alginate/calcium ion double-network hydrogel and its application to heavy metal ions removal. *Chemical Engineering Journal*, 393, 124728. DOI: 10.1016/j.cej.2020.124728

Tran, D. L., Le Thi, P., Hoang Thi, T. T., & Park, K. D. (2020). Novel enzymatically crosslinked chitosan hydrogels with free-radical-scavenging property and promoted cellular behaviors under hyperglycemia. *Progress in Natural Science*, 30(5), 661–668. DOI: 10.1016/j.pnsc.2020.08.004

Tran Vo, T. M., Piroonpan, T., Preuksarattanawut, C., Kobayashi, T., & Potiyaraj, P. (2022). Characterization of pH-responsive high molecular-weight chitosan/poly (vinyl alcohol) hydrogel prepared by gamma irradiation for localizing drug release. *Bioresources and Bioprocessing*, 9(1), 89. DOI: 10.1186/s40643-022-00576-6 PMID: 38647766

Trikkaliotis, D. G., Ainali, N. M., Tolkou, A. K., Mitropoulos, A. C., Lambropoulou, D. A., Bikiaris, D. N., & Kyzas, G. Z. (2022). Removal of Heavy Metal Ions from Wastewaters by Using Chitosan/Poly(Vinyl Alcohol) Adsorbents: A Review. In *Macromol* (Vol. 2, Issue 3, pp. 403–425). https://doi.org/ DOI: 10.3390/macromol2030026

Tripodo, G., Trapani, A., Rosato, A., Di Franco, C., Tamma, R., Trapani, G., Ribatti, D., & Mandracchia, D. (2018). Hydrogels for biomedical applications from glycol chitosan and PEG diglycidyl ether exhibit pro-angiogenic and antibacterial activity. *Carbohydrate Polymers*, 198, 124–130. DOI: 10.1016/j.carbpol.2018.06.061 PMID: 30092982

Tsao, C. T., Chang, C. H., Lin, Y. Y., Wu, M. F., Wang, J.-L., Han, J. L., & Hsieh, K. H. (2010). Antibacterial activity and biocompatibility of a chitosan–γ-poly(glutamic acid) polyelectrolyte complex hydrogel. *Carbohydrate Research*, 345(12), 1774–1780. DOI: 10.1016/j.carres.2010.06.002 PMID: 20598293

Ullah, K., Sohail, M., Mannan, A., Rashid, H., Shah, A., Murtaza, G., & Khan, S. A. (2018). Facile Synthesis of Chitosan Based-(AMPS-co-AA) Semi-IPNs as a Potential Drug Carrier: Enzymatic Degradation, Cytotoxicity, and Preliminary Safety Evaluation. *Current Drug Delivery*, 16(3), 242–253. DOI: 10.2174/1567201815666181024152101 PMID: 30360742

Upadhyay, U., Sreedhar, I., Singh, S. A., Patel, C. M., & Anitha, K. L. (2021). Recent advances in heavy metal removal by chitosan based adsorbents. *Carbohydrate Polymers*, 251, 117000. DOI: 10.1016/j.carbpol.2020.117000 PMID: 33142569

Vo, N. T. N., Huang, L., Lemos, H., Mellor, A. L., & Novakovic, K. (2021). Genipin-crosslinked chitosan hydrogels: Preliminary evaluation of the in vitro biocompatibility and biodegradation. *Journal of Applied Polymer Science*, 138(34), 50848. DOI: 10.1002/app.50848

Wan, L. S. C., Heng, P. W. S., & Wong, L. F. (1993). Relationship Between Swelling and Drug Release in a Hydrophilic Matrix. *Drug Development and Industrial Pharmacy*, 19(10), 1201–1210. DOI: 10.3109/03639049309063012

Wang, X., Cao, Z., Zhang, M., Meng, L., Ming, Z. & Liu, J. (2023). Bioinspired oral delivery of gut microbiota by self-coating with biofilms. *Science Advances,* 6(26), eabb1952. https://doi.org/DOI: 10.1126/sciadv.abb1952

Wei, P., Jan, E. & Jianzhong, C. (2021). *Ultrasound - responsive polymer - based drug delivery systems.* 1323–1339.

Wisotzki, E. I., Hennes, M., Schuldt, C., Engert, F., Knolle, W., Decker, U., Käs, J. A., Zink, M., & Mayr, S. G. (2014). Tailoring the material properties of gelatin hydrogels by high energy electron irradiation. *Journal of Materials Chemistry. B*, 2(27), 4297–4309. DOI: 10.1039/C4TB00429A PMID: 32261568

Xia, Y., Wang, D., Liu, D., Su, J., Jin, Y., Wang, D., Han, B., Jiang, Z., & Liu, B. (2022). Applications of Chitosan and its Derivatives in Skin and Soft Tissue Diseases. *Frontiers in Bioengineering and Biotechnology*, 10, 894667. Advance online publication. DOI: 10.3389/fbioe.2022.894667 PMID: 35586556

Yan, H., Dai, J., Yang, Z., Yang, H., & Cheng, R. (2011). Enhanced and selective adsorption of copper(II) ions on surface carboxymethylated chitosan hydrogel beads. *Chemical Engineering Journal*, 174(2), 586–594. DOI: 10.1016/j.cej.2011.09.064

Ye, R., Liu, S., Zhu, W., Li, Y., Huang, L., Zhang, G., & Zhang, Y. (2023). Synthesis, Characterization, Properties, and Biomedical Application of Chitosan-Based Hydrogels. In *Polymers* (Vol. 15, Issue 11). https://doi.org/DOI: 10.3390/polym15112482

Yeh, Y.-Y., Tsai, Y.-T., Wu, C.-Y., Tu, L.-H., Bai, M.-Y., & Yeh, Y.-C. (2022). The Role of Aldehyde-Functionalized Crosslinkers on the Property of Chitosan Hydrogels. *Macromolecular Bioscience*, 22(5), 2100477. DOI: 10.1002/mabi.202100477 PMID: 35103401

Yin, M., Li, X., Liu, Y., & Ren, X. (2021). Functional chitosan/glycidyl methacrylate-based cryogels for efficient removal of cationic and anionic dyes and antibacterial applications. *Carbohydrate Polymers*, 266, 118129. DOI: 10.1016/j.carbpol.2021.118129 PMID: 34044945

Zalipsky*, S. (1995). Functionalized Poly(ethylene glycol) for Preparation of Biologically Relevant Conjugates. *Bioconjugate Chem, 6*(2), 150–165. https://doi.org/DOI: 1043-1802/95/2906-0150$09.00/0

Zhang, F., Xiong, L., Ai, Y., Liang, Z., & Liang, Q. (2018). Stretchable Multiresponsive Hydrogel with Actuatable, Shape Memory, and Self-Healing Properties. *Advancement of Science*, 5(8), 1800450. Advance online publication. DOI: 10.1002/advs.201800450 PMID: 30128253

Zhang, Z.-X., Liow, S. S., Xue, K., Zhang, X., Li, Z., & Loh, X. J. (2019). Autonomous Chitosan-Based Self-Healing Hydrogel Formed through Noncovalent Interactions. *ACS Applied Polymer Materials*, 1(7), 1769–1777. DOI: 10.1021/acsapm.9b00317

Zhao, L., Gwon, H. J., Lim, Y. M., Nho, Y. C., & Kim, S. Y. (2014). Hyaluronic acid/chondroitin sulfate-based hydrogel prepared by gamma irradiation technique. *Carbohydrate Polymers*, 102(1), 598–605. DOI: 10.1016/j.carbpol.2013.11.048 PMID: 24507324

Zhu, H., Chen, S., & Luo, Y. (2023). Adsorption mechanisms of hydrogels for heavy metal and organic dyes removal: A short review. *Journal of Agriculture and Food Research*, 12, 100552. DOI: 10.1016/j.jafr.2023.100552

Chapter 9
Insect Resources for Chitin Biomass

Guillermo Ignacio Guangorena Zarzosa
Nagaoka University of Technology, Japan

Takaomi Kobayashi
Nagaoka University of Technology, Japan

ABSTRACT

Among all chitin sources, insects have recently gained the attention of researchers and entrepreneurs. In addition to being supplied primarily from unutilized products of the marine food industry, chitin sources from insect sources are expected to increase in the future as the insect food industry grows. Insect sources of chitin, in particular, are currently the focus of much attention and research, as it is important to know the properties and uses of chitin and how it relates to marine food products. So, this chapter highlights current extraction methods and some characteristics of insect chitin. In addition, some information about insects will be provided to have a holistic view when thinking about insect industrialization and its impact on chitin.

INTRODUCTION TO ARTHROPODS AND HEXOPODA.

Within the Animalia kingdom, there is a phylum denominated Arthropoda representing about 84% of all known species in the world. Arthropoda phylum is subdivided in Crustacea (Marine arthropods such as prawn, crabs, etc.), Onychophora, Chelicerata, Myriapoda, and Hexapoda (Gironza & Obando, 2022). Generally, Arthropoda share the following characteristics: Their bodies usually are bilaterally symmetrical, these segments are grouped into specialized region; they possess jointed appendages which can serve for feeding, locomotion and sensing, and an exoskeleton that contains chitin (Gironza & Obando, 2022). Nowadays, most of the current food industry and chitin industry rely in crustaceans due to the acceptance rate of costumers around the world. Crustaceans are mostly used as food and their shells, compose of chitin and inorganic material, become a byproduct that is treated as waste. However, insect resources are more abundant than crustaceans as source of protein and chitin biomass. Around 30,000 species of crustaceans have been described, but entomologist estimate the actual number of living hexopoda is around 5 million to 10 million. Among these insects, according to Food and Agriculture

DOI: 10.4018/979-8-3693-0003-9.ch009

Organization of the United Nation (FAO), more than 2000 species are considered as edible species (Errico et al., 2022). As seen for Figure 1, the distribution of some edible species around the world can be seen. The reported species are prominent in Latin America, Africa and Asia.

Chitin market outlook indicates that the global chitin market size is valued at around US$1801.3 million in 2023 and expect to be US$5746.2 million by 2033 (Future Market Insights, 2023). Since the increase consumption of chitin in various end-use industries of food & beverages, agrochemicals, and healthcare. Therefore, the development of the insect-derived chitin industry holds great promise for the future.

Hexopada can be further subdivided as the following subclases: apterygote, palaeoptera (dragonflies), polyneoptera (grasshoppers, stick insects, etc), paraneopteran (lice), and endopterygota (beetles, moths, flies, wasps)(Amat-García & Andrade-C, 2007). The following classification can be seen in Figure 2. In their classification when insects exhibit wing marks, they are referred as Pterygota. Usually, wings are located only in the mesothorax and metathorax. Pterygote insects have hemimetabulus development which consist on its immature stages going through metamorphosis to reach their adult winged form. On the other hand, the more primitive hexapods are ametabulus, which means that nymph and adult forms are not that different, so they have an incomplete metamorphosis.

Insects are monophyletic group with three major anatomical structures as the head, thorax, and abdomen that share single pair of locomotor appendages within the thoracic segments (3 pairs)(Gironza & Obando, 2022). Independently of the insect type, they periodically shed their exoskeletons when they grow (J. Zhang et al., 2014). This process is called molting and it continues until the insect reaches adulthood, generating in the process exuviae which can come from larval/pupal stages or nymph stages depending in the organism.

In Figure 2, the difference in surface morphology in the cuticle and wings form various insects can be seen. These differences have been describing various subclasses of insect (Kukalova-Peck, 1978). From the SEM pictures of Figure 2, insects can show a plain surface such as the one seen in dragonflies, corrugated with some arrangements as seen in the stink bug, or possessing some microstructures as seen in the rhinoceros beetle. The same difference can be observed in the wings. All these structures contain chitin and may contribute to different characteristics of the biopolymer.

It is well known that insects such as beetles, and butterflies, have different colors within their cuticles, chitin layers on insects cause a structural coloration trough chitin fiber arrangement that scatter light exhibiting birefringence (Fernández Del Río et al., 2014; Hou et al., 2021). Light reflection is not the only factor that contributes to chitins color on insects which will be discuss later in synthesis of insect chitin.

Figure 1. Edible insects around the world and percentages in eaten insects and insect food companies iaround world

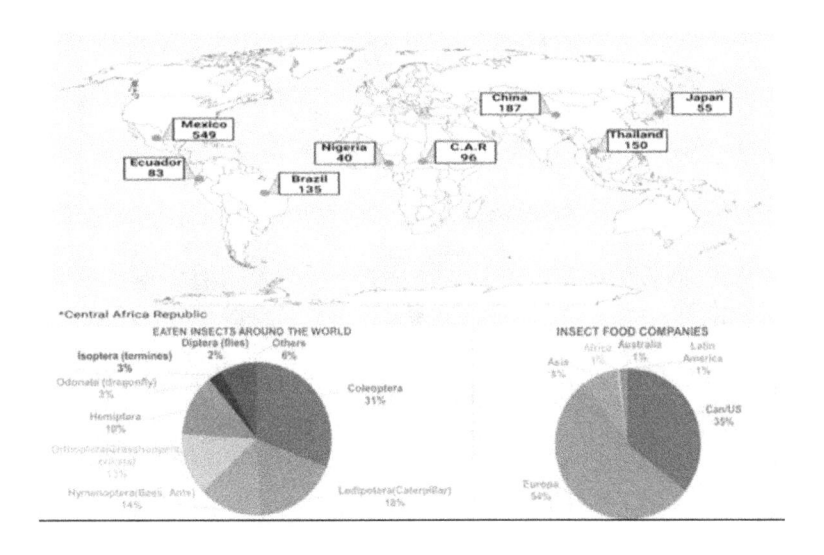

INSECT INDUSTRY

Entomophagy

By 2050, as the human population continues to grow, the demand for protein production is projected to double while food supplies are limited (Gahukar, 2016). In addition, 30% of the Earth's land surface is used for raising livestock and humans consume 40% of the biomass of the land and oceans. For these reasons, insect-based industries have attracted the attention of entrepreneurs and scientists in recent years because of their high protein content, micronutrients, and sustainable agriculture. This situation has prompted research on insects as an alternative protein source.

Figure 2. Arthropods classification and SEM pictures of each insect parts

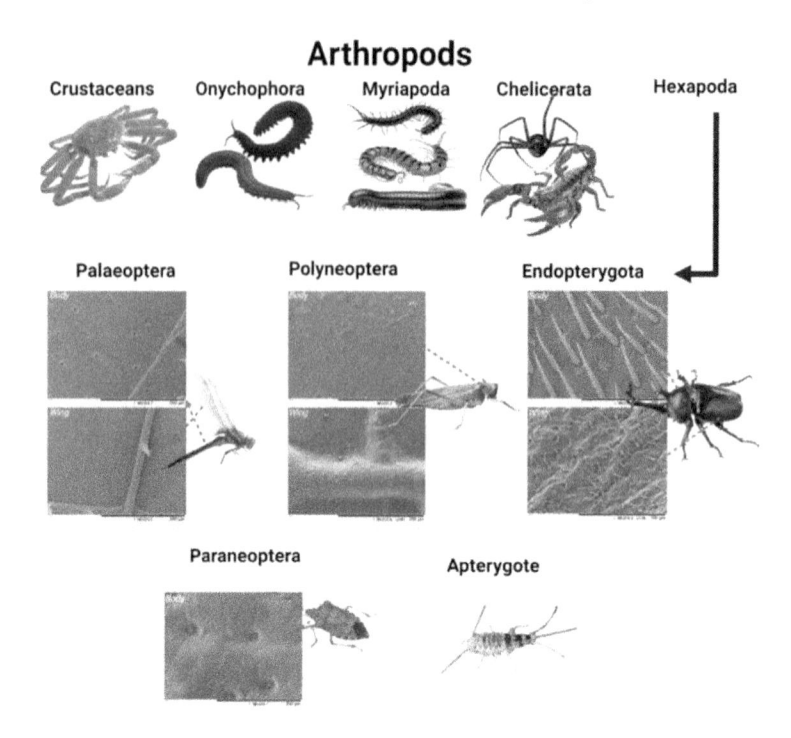

By controlling the food that insects eat, protein mass conversion can be performed more effectively. In addition, insects are thermostatic animals that regulate their body temperature according to environmental temperature, and unlike mammals, which are thermostatic animals, their growth can also be controlled. The life cycle of insects is shorter than that of mammal protein sources and reproduce in large numbers. On the other hand, from the perspective of greenhouse gas emissions, livestock as a source of protein emits CO^2, methane, and ammonia into the environment, but insect breeding does not, and in addition has the advantage of requiring less water for breeding (Gahukar, 2016; Nikkhah et al., 2021). For these reasons, the insect industry has been expanded globally in recent year, as seen in Figure 1 (a) for the percentage of companies around the world. Most of the companies related to insect industry are located within Europe and America (Dossey et al., 2016). For instance, in the case of chitin production, most of the chitin production worldwide is made from shrimp, crab, and lobster waste, but currently insects as a source for chitin has grown in Europe. Namely as insect chitin, which is found in the byproduct of the rearing and production of feed from insects. Within the prospects for industrial production the most prominent candidates are silkworm, house crickets, tenebrio, weaver ants, grasshoppers, and black solider fly(Ayieko et al., 2021; Dagevos, 2021; Gahukar, 2016; Nikkhah et al., 2021). In Figure 1 (b), it is noted that the most eaten insects according to the subclass, coleoptera is account for the 30% of the total. Currently the edible insect market is around 0.75 billion dollars but is expected to grow to 4.63 billion by 2027(Errico et al., 2022). As extraction techniques and production methods for insect proteins advance, insect-derived chitin, like crustaceans, will become a byproduct of the food industry.

Increasing Development of Insect Chitin and Chitin Biomass Sources

In recent years, insects have been used as an alternative recycling source of petroleum-derived plastics by microbiome action. Within them the family of Tenebrionidae (Tenebrio Molitor, Zophobas Morio, etc.) is known to eat and degrade plastics like polystyrene (PS), polyvinyl (PVC), polyethylene (PE), and polypropylene (PP) by their depolymerization due to an intermediate oxidation, or an intermediated mineralization to CO_2, and H_2O (Kuan et al., 2022; Peng, Chen et al., 2020; Peng, Li et al., 2020; S. S. Yang et al., 2020). Also, the Indian mealmoths, greater waxworms can degrade plastic (W. M. Wu & Criddle, 2021). These plastics, in granular or powdered form, were served to the insects, who ingested the plastics. This fact could lead to chitin conversion using waste as a food source, and thus secure chitin biomass, unlike wastes such as shrimp and crab.

On the other hand, nowadays, it is known that chitin is present in different living organism, but the first discovery was in 1800, when a substance resistant to sulfuric acid was reported in mushrooms, which was called fanzinene. The same substance was subsequently found in insects in 1823 and in plants in 1984.Then, nitrogen species within chitin structure were found (Khoushab & Yamabhai, 2010).

Figure 3. Chemical structures of chitin and N-acetylglucosamine

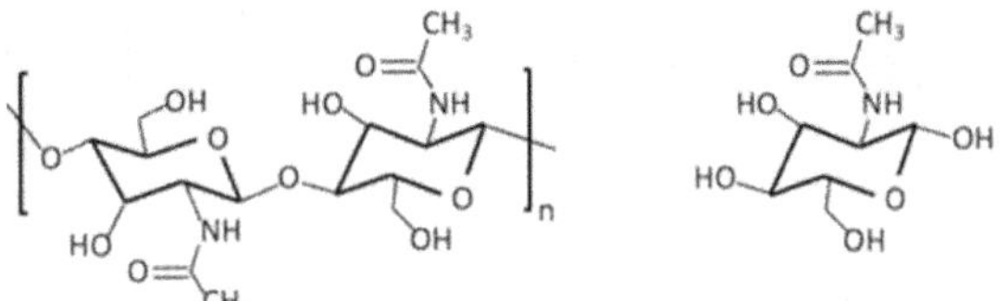

Chitin is a linear polysaccharide linked with 2-acetamido-2-deoxy-D-glucopyranose group through its β (1 4) and Chitin is said to be a biomass that is structurally and physiologically intermediate between cellulose (plants) and collagen (mammals) (Hou et al.) All of these biopolymers mainly serve as structural components supporting cell and body surfaces. Namely, cellulose strengthens the cell wall of plant cells, whereas chitin contributes to the mechanical strength of fungal cell and exoskeletons of arthropods, as collagen functions in a similar way as part of the connective tissue (Merzendorfer & Zimoch, 2003). The ideal chemical structure of chitin is rarely found in nature. Usually the degree of N-acetylation is limited, so that the percentage of acetylated units in chitin is almost 90% or more, and partially aminated hydrolysis occurs. This phenomenon may be due to the action of chitin deacetylases in chitin-containing organisms, but it may also be the result of extraction processes during chitin purification (Merzendorfer & Zimoch, 2003).

Insect chitin is not limited to be present in its cuticle, but is also in other tissues such peritrophic matrix, ducts (salivary glands, trachea), muscle attachment points, and wings, etc.(Barbosa et al., 2014; Moreira et al., 2017; Sarwade & Bhawane, 2013). In the cuticle, the epidermal cells cover acellular and solid components to prevent water evaporation and provide protection from microorganisms and physical damage. The cuticle is divided by 2 layers: 1) epicuticle having water-resistant wax and then 2) procuticle composed mostly of chitin fibers associated with structural proteins and pigment (Noh et al., 2016; Sun et al., 2012). Before ecdysis, the new epicuticle and procuticle are formed and then chitinases and peptidases in the molting fluid aids the breakdown of the old cuticle in epidermal cells. As once the molting has occurred, the tanning process takes places giving its color and hardness which provide the mechanical properties of the cuticle. There are specific forms of chitin in different life stages as well, since each produces chitin fibers in this process in insects (Merzendorfer & Zimoch, 2003; Noh et al., 2016). This is well illustrated in Figure 4, which shows the life cycle of Zophobus morio and the chitin after chemical extraction. The physicochemical properties of each chitin were different (Zarzosa & Kobayashi, 2023).

Figure 4. Life cycle of Zophobus morio from exuviae to larva, pupa and adult beetle and pictures of resultant chitins

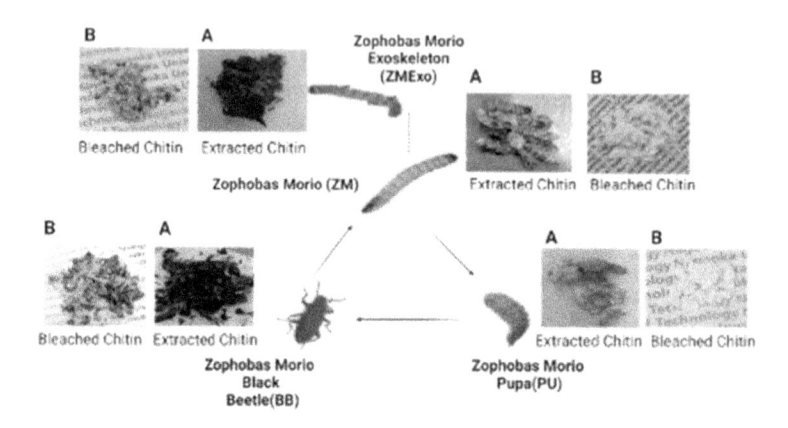

On the other hand, during the biosynthesis of chitin, *N*-acetylglucosamine is monomer (Figure 3) and joined in a reaction catalyzed by the membrane-integral enzyme chitin synthase, a member of the family 2 of glycosyltransferases. The polymerization requires UDP–*N*-acetylglucosamine as a substrate and divalent cations as co-factors. Chitin formation is divided into three distinct steps. In the first step, the catalytic domain of chitin synthase formed the polymer by the cytoplasmic sites. The second step involves the translocation of the polymer form the intracellular space to extracellular space. The final step contains a chemical change of chitin by its deacetylation and crosslinking with structural proteins and pigment.(Merzendorfer & Zimoch, 2003). Helically, the arranged globular proteins are bound to the periphery of these parallel chitin crystals. The chitin-protein units are grouped into fibrils of various diameters where the chitin crystals appear as rods segmented with proteins. The chitin-protein fibrils build up sheets of successive planes of horizontal and parallel fibrils having directions changing from one plane to the other by continuous rotation (Hou et al., 2021), creating horizontal laminae whose thickness varies from a tenth of a micron to several micrometers. The chitin fibers form together with lamina, leading that the formation of layers contributes to the resistance to physical harm.

Contrary to crustacean chitin, insects possess a lower calcium carbonate concentration. For example, crustaceans possessed around 10 to 20%, while insects had around 3% (Huet et al., 2020; Rong et al., 2019). When combined with proteins and chitin in the cuticle, this promotes nucleation and growth of calcium carbonate in the form of calcite crystals, appearing and growing within the fibrillar organic network and ultimately forming into a rigid acellular composite material in the exoskeleton. Hance, chitins in the exoskeleton coexistence also in the embedded calcium carbonate matrix and in the cross-linking of proteins and pigments. The pigment in insects has different chemical structures but the synthesis starts with tyrosine, then its hydroxylated into 3,4-dihydroxyphenylalanine, turning into dopamine, which is transformed into different quinones. Depending on the quinone molecule it is known that the color of exoskeleton changes during the tanning-sclerotization process (Noh et al., 2016; Sugumaran, 2009). In addition, there is chitin nanofiber of mushrooms consisting of a network of chitin embedded mainly in β- 1,3 glucans. The morphology of the cell wall of mushroom is different from that of arthropods, where the highly branched β-1,3/1,6-glucans are connected in the cells together. Also, the α-1,3-glucan serves as an amorphous matrix, and interdigitates with the inner side of the β-1,3-glucan layer, as embedded form in crystalline of chitin nanofibers (Ifuku et al., 2011).

Chitin Extraction Process from Insect Sources

In order to use the chitin contained in insects as a raw material, the first step for chitin is extraction process. However, the extraction method is still being explored. For example, for chitin extraction from insects, there are several factors to take into consideration. One is the duration of the lifecycle. As related to the life cycle of the insects, for example, zophobas morio to be the darkling beetle has a cycle period around 6 months, but dynastes hercules is around 2 years (Iannacone-Oliver & Soras-Vega, 2010), meaning shorter rearing periods and higher yields in a better efficiency for in chitin extraction. Thus, species such as hermetia illucens, crickets, and tenebrionidae are prospective for industrial production of protein and chitin.

As reported in zophobas morio (Zarzosa & Kobayashi, 2023), adapting the same extraction process to samples from different LIFE cycles did not yield the same results. As shown in Figure 4, in both life stages, acid-alkali treatment resulted in chitin containing brown solid material, but after the bleaching process, exoskeleton and adult beetles had components that could not be bleached. In addition, within life cycle of same species, chitin properties influence the extraction processes. In the case of Zophobas Morio or hermetia illucens of larvae, pupa and adult stages became different, in chitin yield, deacetylation degree, crystallinity and molecular weight. For larva and pupa, Zophobas Morio showed similar yields around 4%, but when up to 16% in the adult beetle, maintaining its deacetylation degree in similar levels, while *Hermetia illucens* larva yielded with 3% and pupal state was in 6%, and its deacetylation degree went from 80 to 92% respectively (Smets et al., 2020).

Since insects contain protein, fats and pigments and the amount of these components varies with life stage, the use of insect source to chitin leads to different factors. Also, exuviae of insects also present such components, but its chitin contents are higher than the whole insect. Hence, persistence in separating these impurities in the process of producing chitin is necessary in the case of insect sources for defatting, demineralization, deproteinization and bleaching in the extraction (Figure 5).

Figure 5. Chitin extraction process es via chemical and biological processes

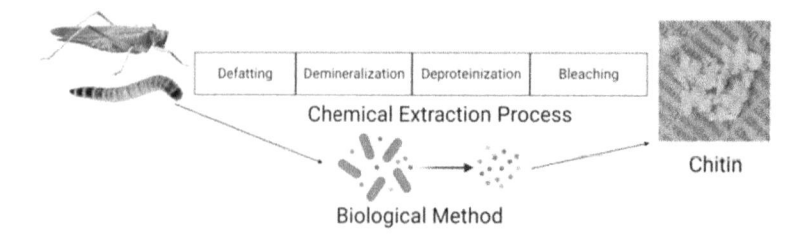

Defatting

Because lipids are present in the insects within an organism and its own life cycle, on insects that have a complete metamorphosis cycle, larval stages contain a higher tendency to store lipids than the adult form (Smets et al., 2020). Exuvia also exists lipids in the form of wax in the cuticles to allowed them a better control of internal moisture (Sun et al., 2012). About 80% of fats are in the triacylglycerol form, along with cholesterol (Mohan et al., 2020) and the amount of lipid influences, when the chitin is extracted, and sometimes fats are not able to be fully removed by the acid or alkali treatment. For this purpose, preliminary washing out with ethanol and acetone, or n-hexane was used (Draczynski, 2008; Huet et al., 2020; Son et al., 2021), or also, soxhlet extraction of n-hexane and petroleum ether (Smets et al., 2020). Therefore, selecting an appropriate defatting agent influenced subsequent results for the quantity yield of chitin. In another thing, the use of the extracted fats was focused for its use as biodiesel (Zheng et al., 2013) . The efforts of the biotechnology industries are pushing to single step process. Table 1 lists the use of different insect for chitin extraction.

Demineralization by Acid Treatment

In the case of insect resources, although insects have fewer inorganic components in their matrix than marine arthropods (Rong et al., 2019; Smets et al., 2020), they are characterized by the presence of calcium carbonate, which also affects extraction. In the step consisting in the removal of calcium carbonate, the acidic process is commonly achieved by using sulfuric acid, hydrochloric acid, nitric acid, acetic acid, oxalic acid, or formic acid(Hahn et al., 2022), as seen in Table 1. Calcium carbonate is an inverse solubility salt, whose solubility decreases with increasing temperature, and the generated salts after the acid hydrolysis (Berce et al., 2021; Hahn et al., 2022). For example, sulfuric acid and phosphoric acid was effective for elimination of calcium carbonate salts, even though it is low solubility in water (Hahn et al., 2022). Also, nitric acid, hydrochloric acid, acetic acid, and formic acid showed a better efficiency in mineral removal. Regardless, HCl treatment was found to be superior to all of these acids, thus being the most common used agent (Abidin et al., 2020; Mohan et al., 2020; Rehman et al., 2023). Also, currently more sustainable methods such as lactic acid fermentation are being considered as options for mineral removal in the chitin extrication process.

Deproteinization by Alkaline Treatment

Insect chitin binds pigments and structural proteins in the cuticle. The deproteinization step cleavage the chemical bonds between the chitin and proteins. In the chemical acid treatment elutes inorganic substances, whereas there were reports on alkali treatment for protein elution like some quinones and proteins (Hahn et al., 2022; Xu et al., 1997; M. Zhang et al., 2000). Usually sodium hydroxide (NaOH) was used as the preferred agent, leading to partial or total hydrolysis of the peptide bonds (Hahn et al., 2022) . In the chemical extraction, NaOH varied in concentration from 0.124 to 5M at different temperatures up to 175°C and different times (Abidin et al., 2020; Mohan et al., 2020; Rehman et al., 2023). In addition, deproteinization at alkaline conditions and elevated temperatures caused partial deacetylation of chitin (Pires et al., 2014). Whereas the drawbacks of such chemical treatment on demineralization process, alternative methos involved the use of enzyme such as alkaline proteases for protein removal (Taokaew et al., 2020). However, the amount of protein remaining was higher and required longer reaction time than following chemical treatment(Mohan et al., 2020). At the moment the use enzymatic methods for protein removal at an industrial level seems to become challenging, nonetheless, alkaline hydrolysis is conveniently used for the general process for chitosan extraction, when the acetylation degree reached 50% of lower, thus enhancing its biological properties (Wattjes et al., 2020).

Bleaching Using Chemical Agents

After defatting, demineralization and deproteinization, still the extracts contains tough impurities. However, In arthropod shells having a faint red coloration attributed to non-covalently bound carotenes, thus the bleaching process make it easier by using ethanol (Vilasoa-Martínez et al., 2008). In contrast, because insects are covalently crosslinked to the matrix, generally several chemical bleaching agents were used as seen in Table 1 (Abidin et al., 2020; Mohan et al., 2020; Rehman et al., 2023). Among them sodium hypochlorite (NaOCl) and hydrogen peroxide in NaOH were effective when used (Hahn et al., 2022). As seen in Figure 4, the effect of bleach is remarkable for Zophobas Morio (Zarzosa & Kobayashi, 2023).

Biological Process With Green Extraction Methods

One of the major challenges of chitin extraction using the chemical extraction is that properties such as molecular weight, crystallinity and deacetylation degree change substantially in comparison of its native state. In addition, the release of toxic effluent wastewater in to the environment is another demerit of the chemical method. Therefore, green extraction methods are gaining popularity due to their cleaner and more ecofriendly approaches. Eco-friendly methods such as biological extraction of chitin have gained interest recently. Bacteria were used for demineralization, deproteinization, and deacetylation (Rehman et al., 2023). These biological methods refer to the fermentation in microbial suspensions(Arbia et al., 2013). The fermentation products such as protease secreted from microorganisms were utilized to mainly remove proteins from crustacean shells under milder condition comparing to the chemical extraction method(Taokaew et al., 2020). Microbial protease is produced extracellularly into the fermentation culture medium. Several microbes produce protease including fungi and bacteria such as *L plantarum* (Rao et al., 2000), *L paracasaei* (Cremades et al., 2001), *Pesudomonas aeruginosa* (Wang & Chio, 1998), *bacillus pumilus, bacillus subtilis, bacillus licheniformis, bacillus cerus, bacillus majovensis, bacillus amyloliquefaciens*, and *serratia marcescens* (Hajji et al., 2015). Generally, these biological extractions were carried in the range from 3 to 7 days, and the demineralization yields were from 73 to 97%, and deproteinization yields from 43 to 94%, depending on the microorganism used. Microbial methods, despite the simplicity of the chitin production procedure, appear to be more effective for desalting, as the amount needed to obtain the same level of deproteinization with chemical method. Because of this, ways to enhance microbial efficacy were investigated with using the addition of microwaves (Hongkulsup et al., 2016), ultrasound (Valdez-Peña et al., 2010) and redox reaction (Gopal et al., 2019).

Table 1. Lists of extraction methods for different insect species

Hexopoda	Defatting	Mineral removal	Protein Removal	Bleaching	Reference
Apterygota					
S. fonscolombii (dragonfly)	NA	1 M HCl room temperature 1 hour	1 M NaOH 50°C for 15 hours	chloroform: methanoland distilled water (1: 2: 4, v/v)	(Kaya, Sargin, Al-jaf, et al., 2016)
Paraneoptera					
Palomena prasine	NA	2 M HCl 100°C for 2 hours	2 M NaOH 140°C for 20 hours	water, methanol, and chloroform at a mixing ratio of 4:2:1 for 2 hours	(Kaya, Baublys, et al., 2015)
Cicada orni Sloughs	NA	1 M HCl 30°C for 2 hours	1 M NaOH 90°C 2 hours	1%(v/v) Sodium Hyphochlorite 25° 30 minutes	(Poerio et al., 2021)
Cicada sloughs	NA	1 N HCl 100°C for 20 minutes	1 N NaOH 80°C for 36 hours	6%(v/v) Sodium Hyphochlorite	(Sajomsang & Gonil, 2010)
Polyneoptera					
Ruspolia differens	NA	1 M HCl 25°C 1 hour	20% NaOH at 25°C for 1-2 weeks	6% hydrogen peroxide in 4% NaOH for 6 hours	(Q. Wu et al., 2020)
B. portentosus	NA	Oxalic Acid solution 25°C for 3 hours	1 M NAOH 95°C for 6 hours	1%(w/v) sodium hypochlorite 25°C for 3 hours	(Ibitoye et al., 2018)
D. maroccanus					
Adult	NA	2 M HCl 55°C for 1 hour	2 M NaOH 50°C for 18 hours	methanol, chloroform and distilled water (in the ratio of 2:1:4)	(Erdogan & Kaya, 2016)
Nymph					
Ailopus simulatrix	NA	4 M HCl 75°C for 1 hour	2 M NaOH 175°C for 18 hours	chloroform: methanoland distilled water (1: 2: 4, v/v)	(Kaya, Erdogan, et al., 2015)
Ailopus strepens					
Duroniella fracta					
Duroniella laticornis					
Oedipoda miniate					
Oedipoda caerulescens					
Pyrgomorpha cognata					
Celes variabilis	NA	4 M HCl 75°C for 2 hours	4M NaOH 150°C for 20 hours	NA	(Kaya, Lelešius, et al., 2015)
Decticus verrucivorus					
Melanogryllus desertus					
Paracyptera labiate					

continued on following page

Table 1. Continued

Hexopoda	Defatting	Mineral removal	Protein Removal	Bleaching	Reference
Apterygota					
Periplaneta americana	NA	4 M HCl 75°C for 2 hours	4M NaOH 150°C for 20 hours	chloroform: methanoland distilled water (1: 2: 4, v/v)	(Kaya & Baran, 2015)
Endopterygota					
Argynnis pandora	NA	2 M HCl 50°C for 24 hours	2 M NaOH 50°C for 24 hours	chloroform: methanoland distilled water (1: 2: 4, v/v)	(Kaya, Bitim, et al., 2015)
Vespa crabro	NA	2 M HCl 75°C for 2 hours	4M NaOH 150°C for 18 hours	chloroform: methanoland distilled water (1: 2: 4, v/v)	(Kaya, Bağriaçik, et al., 2015)
Vespa orientalis					
Vespa germanica					
Bombyx eri	ethanol and acetone	1 M HCl 80°C for 35 minutes	1M NaOH 80°C for 24 hours	NA	(Huet et al., 2020)
M. domestica	NA	2 N HCl 21°C for 3 hours	1.25 N NaOH 95°C for 3 hours	NA	(Kim et al., 2016)
Hermetia illucens					
Adult	NA	1M HCl 100°C for 30 minutes	1M NaOH 80°C for 24 hours	1% potassium permanganete solution for 60 minutes	(Purkayastha & Sarkar, 2020)
Pupa Exuviae					
Hermetia illucens					
Larva	NA	0.5M Formic Acid 25°C for 1 hour	1M NaOH 80°C for 2 hours	5%(v/v) hydrogen peroxide	(Triunfo et al., 2022)
Pupa					
Adult					
Hermetia illucens					
Larva	Soxhlet extraction with petroleum ether for 18 hours	1M HCl 100°C for 30 minutes	1 M NaOH 80°C for 24 hours	NA	(Smets et al., 2020)
Prepupae					
Pupa					
Tenebrio Molitor	n-hexane 1:5 w/w for 6 hours	1.5 M HCl 20°C for 6 hours	1.25 M NaOH 80°C for 24 hours	NA	(Son et al., 2021)
Zophobas Morio					
Larva	NA	1 M HCl 20°C for 24 hours	1M NaOH 90°C for 5 hours	2.5%(v/v) sodium hypochlorite 20°C for 2 hours	(Guangorena Zarzosa & Kobayashi, 2023)
Exuviae					
Pupa					
Adult					
Holotrichia parallela Motschulsky	NA	1 M HCl 100°C for 35 minutes	1 M NaOH 80°C 24 hours	1% potassium permanganete solution for 60 minutes	(Liu et al., 2012)

Table 2. Yield, degree deacetylation and crystallinity for different insects' species

Hexopoda	Yield (%)	Degree of Deacetylation (%)	Crystallinity (%)	Reference
Apterygota				
S. fonscolombii (dragonfly)	20.3	102	96.40	(Kaya, Sargin, Al-jaf, et al., 2016)
Paraneoptera				
Palomena prasina	10.8	83.70	84.90	(Kaya, Baublys, et al., 2015)
Cicada orni Sloughs	NA	74-102.7%	61.5-72.1%	(Poerio et al., 2021)
Cicada sloughs	36.6	90.8-102.3%	82.2-89.7%	(Sajomsang & Gonil, 2010)
Polyneoptera				
Ruspolia differens	NA	99%	84-86%	(Q. Wu et al., 2020)
B. portentosus	4.3-7.1	108.10%	88.02%	(Ibitoye et al., 2018)
D. maroccanus				
Adult	14	232	71	(Erdogan & Kaya, 2016)
Nymph	12	187	74	
Ailopus simulatrix	5.3	NA	76	(Kaya, Lelešius, et al., 2015)
Ailopus strepens	7.4		75	
Duroniella fracta	5.7		72	
Duroniella laticornis	6.5		71	
Oedipoda miniata	8.1		74	
Oedipoda caerulescens	8.9		74	
Pyrgomorpha cognata	6.6		63	
Celes variabilis	6.65-9.93	125-180	76-80	(Kaya, Lelešius, et al., 2015)
Decticus verrucivorus	10.03-11.84	108-115	80	
Melanogryllus desertus	4.71-7.35	131-168	77-78	
Paracyptera labiate	6.80-7.60	130-162	75-79	
Periplaneta americana	13-18%	98.67	86.7	(Kaya & Baran, 2015)
Endopterygota				
Argynnis pandora	8-22%	95-104.2	64-66	(Kaya, Bitim, et al., 2015)
Vespa crabro	8.3	96.85	69.9	(Kaya, Bağriaçik, et al., 2015)
Vespa orientalis	6.4	99.82	53.9	
Vespa germanica	11.9	79.83	50	
Bombyx eri	3.3	93	63	(Huet et al., 2020)
M. domestica	7.71-8.50	89.76-92.39	NA	(Kim et al., 2016)
Hermetia illucens				
Adult	23	71.56-86.2	49.4	(Purkayastha & Sarkar, 2020)
Pupa Exuviae	9	111-115	25.5	
Hermetia illucens				

continued on following page

Table 2. Continued

Hexopoda	Yield (%)	Degree of Deacetylation (%)	Crystallinity (%)	Reference
Apterygota				
Larva	10±0.7	92-94	80-90	(Triunfo et al., 2022)
Pupa	23±1.9	89-91	67-84	
Adult	6±0.1	96-98	93-896	
Hermetia illucens				
Larva	3.85	89		(Smets et al., 2020)
Prepupae	4.72	90		
Pupa	6.34	92		
Tenebrio Molitor	4.72	95	57	(Son et al., 2021)
Zophobas Morio				
Larva	3.7-4.1	63.8-74.3	74.2--85.2	(Guangorena Zarzosa & Kobayashi, 2023)
Exuviae	10.2-14.9	66-78.7	73.3-79.5	
Pupa	3.1-4.5	68.74-71.4	58.2-82.5	
Adult	14.4-16.5	71.6-73.3	60.6-73.8	
Holotrichia parallela Motschulsky	15	93.1-94.3	89.05	(Liu et al., 2012)

INSECT CHITIN CHARACTERIZATION

In Table 2, different insect chitin's yields, and properties such as deacetylation and crystallinity are listed. As well as other polysaccharides, characterization of chitin was investigated by its three basic parameters, degree of acetylation, crystallinity, and molecular weight. In normal basis, chitin has a high molecular weight, a high degree of acetylation and a high crystallinity, thus these locates chitin in the insolubility range. However, once chitin is turn into chitosan as shown in Figure 6, gaining a quite broad biological activity for chitosan. Then different characterization measurements can be done depending of the main objective of the research work on deacetylation degree, crystallinity, and other physical chemical properties.

Figure 6. Transformation of chitin to chitosan

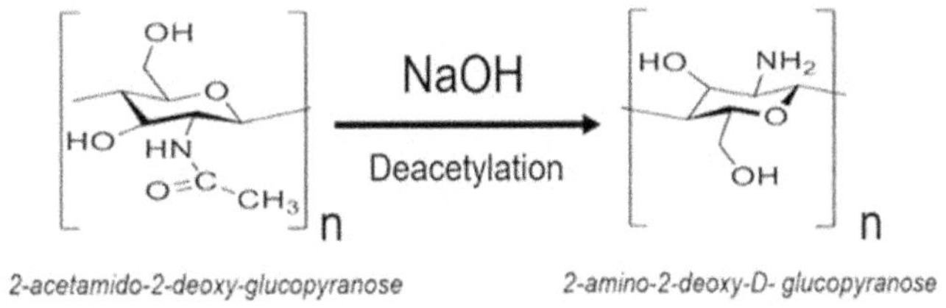

2-acetamido-2-deoxy-glucopyranose →(NaOH, Deacetylation)→ 2-amino-2-deoxy-D-glucopyranose

Deacetylation Degree and FTIR

According to Figure 6 scheme, the acetoamide group of chitin is known to dissociate to amino group -NH_2 and CH_3COOH, when generally the hydrolysis reaction occurs in the presence of acid or base. In the cases of purified chitin, the degree of acetylation (DA) is typically in the range of 0.85 – 0.95, expressing with % ratio of the 2-acetamido-2-deoxy-glucopyranose to the 2-amino-2-deoxy-D-glucopyranose. When the degree was changed in chitin a striking effect was seen in solubility and solution properties (Pillai et al, 2009). In the definition of chitin, if the value of the DA is higher than 0.5, it is called as chitin and that is lower than 0.5 for chitosan. To determine DA, the following experimental measurements were performed with ^{1}H-NMR (Dahmane et al., 2014), ^{13}C-NMR(Dahmane et al., 2014; Kaya, Mujtaba, et al., 2017; Van De Velde & Kiekens, 2004), IR spectroscopy (Adeosun, et al., 2017; Guangorena Zarzosa & Kobayashi, 2023; Velde & Kiekens, 2004), titration method (Liu et al., 2012), elemental analysis (Erdogan, 2016; Bağriaçik, et al., 2015; Bitim, et al., 2015; Erdogan, et al., 2015; Kaya, et al., 2016; Kaya & Baran, 2015; Mol et al., 2018).

Among these methods, FTIR spectroscopy becomes conventional method to analyze chitin's structure. On a chitin spectrum as shown in Figure 7, there are several characteristic peaks for OH and NH groups peaks attributed to the stretching vibration of Oh and NH groups at 3700 and 3000 cm^1. Also observing the CH symmetric stretching bands at 2920 cm^{-1}, 2850cm^{-1} and CH asymmetric stretching at 1380 cm^{-1}. One of the main peaks of chitin is recognized as the ones caused by amide I, II, and III which appear around 1660, 1550, and 1310 cm^1, the extend of the absorption of these substances depends on the degree of hydrolysis of chitin and chitosan. The C-O-C asymmetric stretch in phase ring and CH rind stretch, also appeared in 1070 cm^{-1}, and 890 cm^{-1}. The presence of two sharp bands at 1654-1652 cm^1 and 1619-1618 cm^{-1} for amide and carboxyl bands were also reported in different insect, and marine arthropod chitin, in specific α-chitin(Jang et al., 2004; Sajomsang & Gonil, 2010). DA can be calculated using different equations but one of the most used ones is the following:

$$DA(\%) = \frac{1}{1.33}\left(\frac{A1655}{A3450}\right) X 100$$

where A1625 and A3450 are absorbances measured at 1625 and 3450 cm^{-1}, respectively.

Figure 7. FTIR spectra and XRD patterns of chitin and chitosan

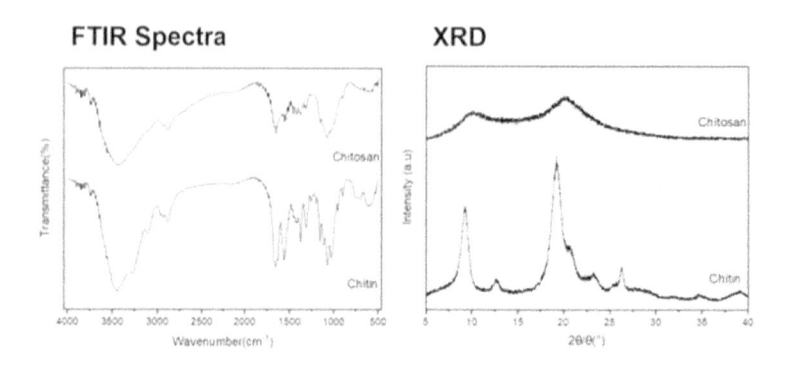

Crystallinity Measurements

When the state of chitin's intra and inter molecular forces changes the crystallinity of the biopolymer, the driving force of hydrogen-bonded chitin-chitin interaction linked by C=O and H-N-groups provide the inter-molecular forces resulting from hydrogen bonds between the neighboring chitin rings. So, these manners influence on crystallinity of chitin. Also, the carbonyl group and OH-group of chitosan rings participate to the hydrogen bonding as well as the case of cellulose (Haleem & Parker, 1976; Minke & Blackwell, 1978). In other words, the interactions of the inter-intra molecular forces affect chitin/chitosan crystallinity which provides the fiber mechanical properties of the biopolymer. It is known that chitin is arranged in three crystalline structures for its polymorphic forms as α-, β- and γ-chitin depending on the distribution of the chains within the crystalline regions (Hackman & Goldberg, 1965; Jang et al., 2004). The difference in polymer chains arrangement influences the difference in their hydration degree and polymeric unit size and in the numbers.

For example, in α-chitin, the chains are anti-parallel, meanwhile, β-chitin chains are parallel, and the γ-chitin has two of three chains involving in parallel and the third is anti-parallel. In those arrangements, the α- form is the most abundant allomorph and is found in hard structures, whereas the β- and γ-forms are flexible in the structures. Similar with marine arthropods, insects also possess α-chitin. Even though the inter-intra molecular forces of chitin provide good mechanical properties, these hydrogen-bonding linkages play a role in chitin insolubility in water (Handbook of Chitin and Chitosan, 2020). In the α-chitin chemical structure the interaction of inter-intramolecular bonds is followed: intramolecular C(3')OH---OC(5) and intermolecular C(21)NH---O=C(73) hydrogen bonding and C(6'1)OH---O=C(71)/ C(6'1)O---HOC(62) intra-/intermolecular hydrogen bonds(Kumirska et al., 2010; Pillai, 2009). While β-chitin is distinguished by its loose packing of molecules caused by their parallel arrangement, thus more soluble in and reactive towards solvent. In γ-chitin, the chains are a mixture of α- and β-chitin(Jang et al., 2004).

For measuring the crystalline structure of chitin X-ray diffraction (XRD) was mainly used for distinguishing there forms of chitin. This method uses the diffraction phenomenon, in which strong X rays are observed in a specific direction, when X rays are incident on a crystal with regularly arranged atom, information about the crystal structure can be obtain in form of diffraction patterns. In the pattern, the strongest peaks at 2θ appeared around 9.6° and 19.6° as represented for the (020) and (110) planes in the α-chitin(Akpan et al., 2018; Gbenebor, et al., 2017; Kaya, et al., 2016; Mohan et al., 2019; Suenaga et al., 2017) . In case of β-chitin, the crystal-like structural peaks appeared at 9.1° and 20.3° (Gbenebor, Akpan, et al., 2017; Song et al., 2019; Suenaga et al., 2017). For γ-chitin, due to the chemical structure was composed of both form of α- and β-, the reflection peaks are at 9.6 and 19.8° (Kaya, Mujtaba, et al., 2017).

In addition to the diffractograms, the crystallinity index (CI) can be calculated to evaluate the crystalline percentage (Gbenebor, Adeosun, et al., 2017; Kaya, Mujtaba, et al., 2017; Kumirska et al., 2010; Van De Velde & Kiekens, 2004) with the following equation:

$$CrI(\%) = \left(\frac{I110 - Iam}{I110}\right) \times 100$$

where Icr is the maximum intensity at 2θ around the reflection peak, for example $I110$ or $I020$ and Iam is the amorphous peak around $2\theta = 16°$.

Physical Chemistry Properties of Chitin

By using mass loss, thermal stability of chitin can be obtained by TGA and DSC. Chitin usually has two peaks (Nam et al., 2010). The first one is due to the water bound within chitin which evaporates. This affinity to water is changed by the N-acetylation degree(Nam et al., 2010; Ofem et al., 2022; Taokaew et al., 2020). The second one represents the chitin degradation which is influenced by molecular weight, crystallinity, and *N*-acetylation degree of chitin molecule. The maximum degradation temperature of insect chitins ranged from 307°C to 412°C (Abidin et al., 2020; Mohan et al., 2020; Rehman et al., 2023) which varied with the species and life cycle of this ones.

Since chitin has repetition units of *N*-acetylglucosamine groups like polysaccharide, the molecular weight is also responsible for the biological activity of chitin and is evaluated with viscometry methods and high-performance liquid chromatography, but for these purposes, chitin needs to be dissolved in solvents. However, there is a difficulty for chitin to be dissolve in solvents, with some exception as DMAc/LiCl (Jiang & Kobayashi, 2017). Another approached was used for deacetylated part to dissolve in acetic acid and water. However, in this case, acid cleavage of the polysaccharide segments, determining to be a lower molecular weight. For example, insect chitin molecular weight changes within the life cycle for the case of *zophobas morio* from larva to adult resulting in a lower to higher molecular weight, respectively (Zarzosa & Kobayashi, 2023) and also the molecular weights in the range of 2.7 to 426 KDa(Mohan et al., 2020).

On the other hand, several solvents are available for chitin. When chitin was applied for NaOH/Urea aqueous solutions at 8wt% NaOH/4 wt% urea at -20°C (Zhou & Zhang, 2000), but even if this alternative was treated as a green solvent, gelation occurred as temperature raised (Hu et al., 2007). Another used was $CaCl_2$/methanol solvent which showed effectiveness, but the solubility was influenced by the degree of *N*-acetylation and the molecular weight of chitin (Tamura et al., 2006). In this case calcium ion reacted with the amide bond leading to a decrease of crystalline structure of chitin.

In case of *N,N*-dimethylacetamide/lithium chloride (DMAc/LiCl), chitin dissolubles under moderate conditions up to 15 wt% concentration ("Handbook of Chitin and Chitosan," 2020) . The solubility mechanism was studied by NMR revealing that there was a strong interaction of LiCl with chitin OH group and acetamide group. Then concluded the formation of macro-cation of (DMAc-Li) $^+$. In the ion cluster with chitin, the Cl- ion dissociated from the Li^+ cation intercalates with one or more DMAc molecules. Also, the Li^+ cation associates with the carbonyl oxygen of DMAc. The interacted solvent-LiCl can solvate to labile proton groups of –OH and –$NHCOCH_3$ of chitin, disrupting the extensive hydrogen bonds (De Vasconcelos et al., 2011; Poirier & Charlet, 2002) . Interestingly, the DMAc/LiCl solvent was considered as a non-degradative for dissolving chitin (Muthu, 2014; Nguyen, 2021)allow the chitin fibers to swell and dissolve. But any remanent of protein or pigment decreased the solubility of insect chitin in the cosolvent(Zarzosa & Kobayashi, 2023).

As well as cellulose, ionic liquids (ILs) are promising volatile organic solvents for chitin with favorable advantages for the environment such as near zero vapor pressure, high thermal stability, wide electrochemical window, and low flammability(Mohan et al., 2020) but some specific ILs have some disadvantages, like high cost and toxicity, which making them unsuitable for biological applications(Q. Yang et al., 2018). Ionic liquids as deep eutectic solvents (DES) were recently reported for production of chitin from H illiucenss using acid detergent fiber and acid detergent lignin methods(P. Zhou et al., 2019).

Figure 8. Relation of deacetylation degree, molecular weight and chitin and its effects

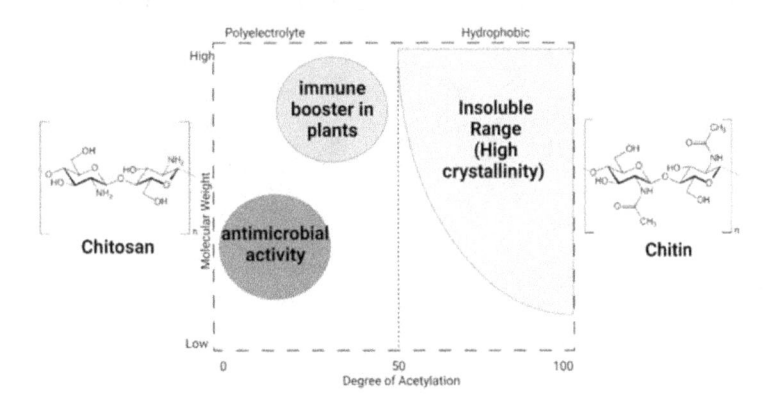

CURRENT APPLICATIONS OF INSECT CHITIN

As shown in Figure 8, chitosan and chitin have a large and small DA relationship, and their properties and functions will be different. Chitosan with more amino groups is water soluble and has properties such as antibacterial and immunogenic expression. On the other hand, chitin is insoluble in water, often crystallized, and well-known biocompatible material (Yang, 2011) and most of the biological activity of this biopolymer is present. Once most of chitin are made from *cicada cryptotympana atrata*, and *clanis bilineata*, showing antibacterial activity against *B subtilis*, *S. aureus*, *E. Coli* (Wu, 2012; Wu et al., 2013) . Also, chitin films sourced from *B.giganteus* cockroach wings inhibited the biofilm formation of a. *baumanni* and *s. sonnei* bacteria, as well as a reduction in the spores of *A. Niger* (Kaya, et al., 2017) . Furthermore, insect chitin-hyaluronan nanoparticles proved effectiveness to increase fibroblast and lowering inflammation reactants (Morganti et al., 2013) and chito oligo saccharides from *clanis bilineata* had hypolipidemic activity (Xia et al., 2013). However, although bioplastic films were sources from grasshopper (*Ruspolia differens*) and exhibited high tensile strengthen (Wu et al., 2020), insect sourced chitins are limited in the examples of applications. Among them, currently insect chitin hydrogels were fabricated by using chitin extracted from zophobas morio in lava, exoskeleton, pupa and beetle and identified its viscoelastic properties (Figure 9). Namely, chitin hydrogel contains hydrophilic structure having ability to retain excess amount of water inside their three-dimensional polymeric network (Ahmed, 2015; Boonmahitthisud et al., 2017; Tovar-Carrillo et al., 2013). As result for insect chitin hydrogel films, the modulus became high in larva (LAR) and pupa (PU) chitin, showing good as hydrogel material, but exuviae (EX) and adult black beetle (BB) were in looser gel condition, because of less solubility to DMAc/LiCl solution (Zarzosa & Kobayashi, 2023). The latter two chitins were due to retained linkage of chitin-pigment, causing solubility decline, especially for BB having brown colored hydrogel body (inserted pictures in Figure).

Figure 9. Viscoelastic storage (G') and loss (G") moduli at each mechanical strain% to the chitin hydrogel deformation. Inserted pictures are chitin hydrogels from each parts of life cycle of Zophobras morio.

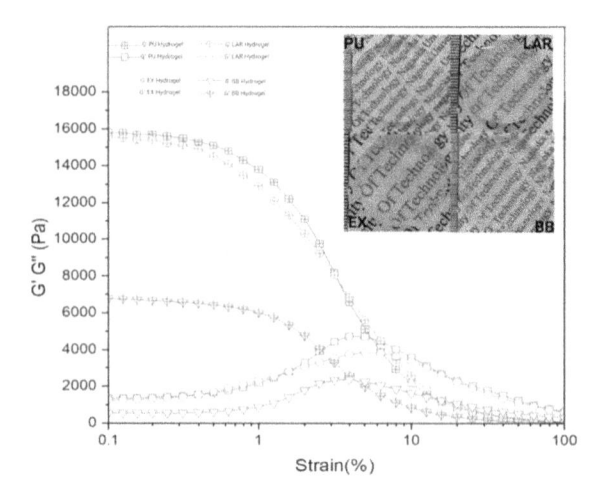

FUTURE PERSPECTIVES

Insects are sources of different kind of bioproducts such as proteins, fats, and chitin. Thus, processes should develop to extract multiple components. Being why chitin industry should develop along food industry, but proteins are seemed to be used as the future protein foods, while lipids also will be transform to substitute for biodiesel. There are insects that have comparable characteristics to those of marine sources. Also, biological methods such as lactic fermentation can be further explored on insect chitin due to its lower content of inorganic material within the cuticle structure. In addition, some insects have the ability to degradable plastic so a further evaluation of how a plastic diet affects. For future research since chitin has a wide range of uses in the cosmetics, textile, culinary, and biotechnological industries, as well as in the pharmaceutical, medical, and agricultural fields. Accordingly, the development of insect-derived chitin holds great promise.

CONCLUSION

The current trend toward creating a sustainable society will strongly promote the insect business, and research on insect-derived chitin become focusing much attention in the future. The growth of food and medical fields industry is pushing the development of insect industry and its chitin as an alternative sustainable source. As this chapter introduced for insect sources chitin, its significance and the current state of industry and research, the importance of insects for the economy, the differences of insect chitin within other sources, and aspects needs to take into consideration more. Against this background, we expect that the industrialization of insect farming and chitin extraction will develop more actively and that its application fields will be broadly expanded.

REFERENCES

Abidin, N. A. Z., Kormin, F., Abidin, N. A. Z., Anuar, N. A. F. M., & Bakar, M. F. A. (2020). The potential of insects as alternative sources of chitin: An overview on the chemical method of extraction from various sources. In *International Journal of Molecular Sciences* (Vol. 21, Issue 14). DOI: 10.3390/ijms21144978

Ahmed, E. M. (2015). Hydrogel: Preparation, characterization, and applications: A review. In *Journal of Advanced Research* (Vol. 6, Issue 2). DOI: 10.1016/j.jare.2013.07.006

Akpan, E. I., Gbenebor, O. P., & Adeosun, S. O. (2018). Synthesis and characterisation of chitin from periwinkle (Tympanotonus fusatus (L.)) and snail (Lissachatina fulica (Bowdich)) shells. *International Journal of Biological Macromolecules*, 106, 1080–1088. Advance online publication. DOI: 10.1016/j.ijbiomac.2017.08.106 PMID: 28842202

Amat-García, G., & Andrade-C, G. (2007). Libro rojo de los invertebrados terrestres de Colombia. In … *Nacional de Colombia*, ….

Arbia, W., Arbia, L., Adour, L., & Amrane, A. (2013). Chitin extraction from crustacean shells using biological methods -A review. *Food Technology and Biotechnology*, 51(1).

Ayieko, I. A., Onyango, M., Ngadze, R. T., & Ayieko, M. A. (2021). Edible Insects as New Food Frontier in the Hospitality Industry. In *Frontiers in Sustainable Food Systems* (Vol. 5). DOI: 10.3389/fsufs.2021.693990

Barbosa, P., Berry, D. L., & Kary, C. S. (2014). Insect Histology: Practical Laboratory Techniques. In *Insect Histology: Practical Laboratory Techniques* (Vol. 9781444336955). DOI: 10.1002/9781118876114

Berce, J., Zupančič, M., Može, M., & Golobič, I. (2021). A review of crystallization fouling in heat exchangers. In *Processes* (Vol. 9, Issue 8). DOI: 10.3390/pr9081356

Boonmahitthisud, A., Nakajima, L., Nguyen, K. D., & Kobayashi, T. (2017). Composite effect of silica nanoparticle on the mechanical properties of cellulose-based hydrogels derived from cottonseed hulls. *Journal of Applied Polymer Science*, 134(10), app.44557. Advance online publication. DOI: 10.1002/app.44557

Cremades, O., Ponce, E., Corpas, R., Gutiérrez, J. F., Jover, M., Alvarez-Ossorio, M. C., Parrado, J., & Bautista, J. (2001). Processing of crawfish (Procambarus clarkii) for the preparation of carotenoproteins and chitin. *Journal of Agricultural and Food Chemistry*, 49(11), 5468–5472. Advance online publication. DOI: 10.1021/jf0104174 PMID: 11714345

Dagevos, H. (2021). A Literature Review of Consumer Research on Edible Insects: Recent Evidence and New Vistas from 2019 Studies. *Journal of Insects as Food and Feed*, 7(3), 249–259. Advance online publication. DOI: 10.3920/JIFF2020.0052

Dahmane, E. M., Taourirte, M., Eladlani, N., & Rhazi, M. (2014). Extraction and Characterization of Chitin and Chitosan from Parapenaeus longirostris from Moroccan Local Sources. *IJPAC. International Journal of Polymer Analysis and Characterization*, 19(4), 342–351. Advance online publication. DOI: 10.1080/1023666X.2014.902577

De Vasconcelos, C. L., Bezerril, P. M., Pereira, M. R., Ginani, M. F., & Fonseca, J. L. C. (2011). Viscosity-temperature behavior of chitin solutions using lithium chloride/DMA as solvent. *Carbohydrate Research*, 346(5), 614–618. Advance online publication. DOI: 10.1016/j.carres.2010.12.016 PMID: 21295769

Dossey, A. T., Tatum, J. T., & McGill, W. L. (2016). Modern Insect-Based Food Industry: Current Status, Insect Processing Technology, and Recommendations Moving Forward. In *Insects as Sustainable Food Ingredients: Production, Processing and Food Applications*. DOI: 10.1016/B978-0-12-802856-8.00005-3

Draczynski, Z. (2008). Honeybee corpses as an available source of chitin. *Journal of Applied Polymer Science*, 109(3), 1974–1981. Advance online publication. DOI: 10.1002/app.28356

Erdogan, S., & Kaya, M. (2016). High similarity in physicochemical properties of chitin and chitosan from nymphs and adults of a grasshopper. *International Journal of Biological Macromolecules*, 89, 118–126. Advance online publication. DOI: 10.1016/j.ijbiomac.2016.04.059 PMID: 27112982

Errico, S., Spagnoletta, A., Verardi, A., Moliterni, S., Dimatteo, S., & Sangiorgio, P. (2022). Tenebrio molitor as a source of interesting natural compounds, their recovery processes, biological effects, and safety aspects. *Comprehensive Reviews in Food Science and Food Safety*, 21(1), 148–197. Advance online publication. DOI: 10.1111/1541-4337.12863 PMID: 34773434

Fernández Del Río, L., Arwin, H., & Järrendahl, K. (2014). Polarizing properties and structural characteristics of the cuticle of the scarab Beetle Chrysina gloriosa. *Thin Solid Films*, 571(P3), 410–415. Advance online publication. DOI: 10.1016/j.tsf.2013.11.149

Future Market Insights, inc. (2023, September 15). *Chin Market*. https://www.futuremarketinsights.com/reports/chitin-market

Gahukar, R. T. (2016). Edible Insects Farming: Efficiency and Impact on Family Livelihood, Food Security, and Environment Compared With Livestock and Crops. In *Insects as Sustainable Food Ingredients: Production, Processing and Food Applications*. DOI: 10.1016/B978-0-12-802856-8.00004-1

Gbenebor, O. P., Adeosun, S. O., Lawal, G. I., Jun, S., & Olaleye, S. A. (2017). Acetylation, crystalline and morphological properties of structural polysaccharide from shrimp exoskeleton. *Engineering Science and Technology, an International Journal, 20*(3). DOI: 10.1016/j.jestch.2017.05.002

Gbenebor, O. P., Akpan, E. I., & Adeosun, S. O. (2017). Thermal, structural and acetylation behavior of snail and periwinkle shells chitin. *Progress in Biomaterials*, 6(3), 97–111. Advance online publication. DOI: 10.1007/s40204-017-0070-1 PMID: 28726011

Gironza, N. S. C., & Obando, R. G. (2022). Zoología de artrópodos. In *Zoología de artrópodos*. DOI: 10.2307/j.ctv2gvdmkf

Gopal, J., Muthu, M., Dhakshanamurthy, T., Kim, K. J., Hasan, N., Kwon, S. J., & Chun, S. (2019). Sustainable ecofriendly phytoextract mediated one pot green recovery of chitosan. *Scientific Reports*, 9(1), 13832. Advance online publication. DOI: 10.1038/s41598-019-50133-z PMID: 31554844

Guangorena Zarzosa, G. I., & Kobayashi, T. (2023). Insect Chitins and Hydrogels Sourced from Zophobas Morio in Different Lifes Stage and Their Properties. *Chemistry Letters*, 52(8), 674–677. DOI: 10.1246/cl.230248

Hackman, R. H., & Goldberg, M. (1965). Studies on chitin. VI. The nature of alpha- and beta-chitins. *Australian Journal of Biological Sciences*, 18(4), 935. Advance online publication. DOI: 10.1071/BI9650935 PMID: 5861258

Hahn, T., Tafi, E., von Seggern, N., Falabella, P., Salvia, R., Thomä, J., Febel, E., Fijalkowska, M., Schmitt, E., Stegbauer, L., & Zibek, S. (2022). Purification of Chitin from Pupal Exuviae of the Black Soldier Fly. *Waste and Biomass Valorization*, 13(4), 1993–2008. Advance online publication. DOI: 10.1007/s12649-021-01645-1

Hajji, S., Ghorbel-Bellaaj, O., Younes, I., Jellouli, K., & Nasri, M. (2015). Chitin extraction from crab shells by Bacillus bacteria. Biological activities of fermented crab supernatants. *International Journal of Biological Macromolecules*, 79, 167–173. Advance online publication. DOI: 10.1016/j.ijbiomac.2015.04.027 PMID: 25910648

Haleem, M. A., & Parker, K. D. (1976). X-Ray Diffraction Studies on the Structure of α Chitin. *Zeitschrift für Naturforschung. C, A Journal of Biosciences*, 31(7–8), 383, b. Advance online publication. DOI: 10.1515/znc-1976-7-806 PMID: 134580

Handbook of Chitin and Chitosan. (2020). *Handbook of Chitin and Chitosan.*, DOI: 10.1016/C2018-0-03015-7

Hongkulsup, C., Khutoryanskiy, V. V., & Niranjan, K. (2016). Enzyme assisted extraction of chitin from shrimp shells (Litopenaeus vannamei). *Journal of Chemical Technology and Biotechnology*, 91(5), 1250–1256. Advance online publication. DOI: 10.1002/jctb.4714

Hou, J., Aydemir, B. E., & Dumanli, A. G. (2021). Understanding the structural diversity of chitins as a versatile biomaterial. In *Philosophical Transactions of the Royal Society A: Mathematical, Physical and Engineering Sciences* (Vol. 379, Issue 2206). DOI: 10.1098/rsta.2020.0331

Hu, X., Du, Y., Tang, Y., Wang, Q., Feng, T., Yang, J., & Kennedy, J. F. (2007). Solubility and property of chitin in NaOH/urea aqueous solution. *Carbohydrate Polymers*, 70(4), 451–458. Advance online publication. DOI: 10.1016/j.carbpol.2007.05.002

Huet, G., Hadad, C., Husson, E., Laclef, S., Lambertyn, V., Araya Farias, M., Jamali, A., Courty, M., Alayoubi, R., Gosselin, I., Sarazin, C., & Van Nhien, A. N. (2020). Straightforward extraction and selective bioconversion of high purity chitin from Bombyx eri larva: Toward an integrated insect biorefinery. *Carbohydrate Polymers*, 228, 115382. Advance online publication. DOI: 10.1016/j.carbpol.2019.115382 PMID: 31635752

Iannacone-Oliver, J., & Soras-Vega, A. (2010). Dynastes (Macleay, 1819) (Coleoptera: Scarabaeidae): Distribución, lista de especies para Sudamérica y crianza en cautiverio. *Scientia*, 12(12).

Ibitoye, E. B., Lokman, I. H., Hezmee, M. N. M., Goh, Y. M., Zuki, A. B. Z., & Jimoh, A. A. (2018). Extraction and physicochemical characterization of chitin and chitosan isolated from house cricket. *Biomedical Materials (Bristol, England)*, 13(2), 025009. Advance online publication. DOI: 10.1088/1748-605X/aa9dde PMID: 29182521

Ifuku, S., Nomura, R., Morimoto, M., & Saimoto, H. (2011). Preparation of chitin nanofibers from mushrooms. *Materials (Basel)*, 4(8), 1417–1425. Advance online publication. DOI: 10.3390/ma4081417 PMID: 28824151

Jang, M. K., Kong, B. G., Jeong, Y., Lee, C. H., & Nah, J.-W. (2004). Physicochemical characterization of α-chitin, β-chitin, and γ-chitin separated from natural resources. *Journal of Polymer Science. Part A, Polymer Chemistry*, 42(14), 3423–3432. Advance online publication. DOI: 10.1002/pola.20176

Jiang, H., & Kobayashi, T. (2017). Ultrasound stimulated release of gallic acid from chitin hydrogel matrix. *Materials Science and Engineering C*, 75, 478–486. Advance online publication. DOI: 10.1016/j.msec.2017.02.082 PMID: 28415488

Kaya, M., Bağriaçik, N., Seyyar, O., & Baran, T. (2015). Comparison of chitin structures derived from three common wasp species (vespa crabro linnaeus, 1758, vespa orientalis linnaeus, 1771 and vespula germanica (FABRICIUS, 1793)). *Archives of Insect Biochemistry and Physiology*, 89(4), 204–217. Advance online publication. DOI: 10.1002/arch.21237 PMID: 25850818

Kaya, M., & Baran, T. (2015). Description of a new surface morphology for chitin extracted from wings of cockroach (Periplaneta americana). *International Journal of Biological Macromolecules*, 75, 7–12. Advance online publication. DOI: 10.1016/j.ijbiomac.2015.01.015 PMID: 25597430

Kaya, M., Baublys, V., Šatkauskiene, I., Akyuz, B., Bulut, E., & Tubelyte, V. (2015). First chitin extraction from Plumatella repens (Bryozoa) with comparison to chitins of insect and fungal origin. *International Journal of Biological Macromolecules*, 79, 126–132. Advance online publication. DOI: 10.1016/j.ijbiomac.2015.04.066 PMID: 25940531

Kaya, M., Bitim, B., Mujtaba, M., & Koyuncu, T. (2015). Surface morphology of chitin highly related with the isolated body part of butterfly (Argynnis pandora). *International Journal of Biological Macromolecules*, 81, 443–449. Advance online publication. DOI: 10.1016/j.ijbiomac.2015.08.021 PMID: 26277749

Kaya, M., Erdogan, S., Mol, A., & Baran, T. (2015). Comparison of chitin structures isolated from seven Orthoptera species. *International Journal of Biological Macromolecules*, 72, 797–805. Advance online publication. DOI: 10.1016/j.ijbiomac.2014.09.034 PMID: 25290985

Kaya, M., Lelešius, E., Nagrockaite, R., Sargin, I., Arslan, G., Mol, A., Baran, T., Can, E., & Bitim, B. (2015). Differentiations of Chitin content and surface morphologies of chitins extracted from male and female grasshopper species. *PLoS One*, 10(1), e0115531. Advance online publication. DOI: 10.1371/journal.pone.0115531 PMID: 25635814

Kaya, M., Mujtaba, M., Ehrlich, H., Salaberria, A. M., Baran, T., Amemiya, C. T., Galli, R., Akyuz, L., Sargin, I., & Labidi, J. (2017). On chemistry of γ-chitin. *Carbohydrate Polymers*, 176, 177–186. Advance online publication. DOI: 10.1016/j.carbpol.2017.08.076 PMID: 28927596

Kaya, M., Sargin, I., Al-jaf, I., Erdogan, S., & Arslan, G. (2016). Characteristics of corneal lens chitin in dragonfly compound eyes. *International Journal of Biological Macromolecules*, 89, 54–61. Advance online publication. DOI: 10.1016/j.ijbiomac.2016.04.056 PMID: 27109757

Kaya, M., Sargin, I., Aylanc, V., Tomruk, M. N., Gevrek, S., Karatoprak, I., Colak, N., Sak, Y. G., & Bulut, E. (2016). Comparison of bovine serum albumin adsorption capacities of α-chitin isolated from an insect and β-chitin from cuttlebone. *Journal of Industrial and Engineering Chemistry*, 38, 146–156. Advance online publication. DOI: 10.1016/j.jiec.2016.04.015

Kaya, M., Sargin, I., Sabeckis, I., Noreikaite, D., Erdonmez, D., Salaberria, A. M., Labidi, J., Baublys, V., & Tubelytė, V. (2017). Biological, mechanical, optical and physicochemical properties of natural chitin films obtained from the dorsal pronotum and the wing of cockroach. *Carbohydrate Polymers*, 163, 162–169. Advance online publication. DOI: 10.1016/j.carbpol.2017.01.022 PMID: 28267493

Khoushab, F., & Yamabhai, M. (2010). Chitin research revisited. In *Marine Drugs* (Vol. 8, Issue 7). DOI: 10.3390/md8071988

Kim, M. W., Han, Y. S., Jo, Y. H., Choi, M. H., Kang, S. H., Kim, S. A., & Jung, W. J. (2016). Extraction of chitin and chitosan from housefly, Musca domestica, pupa shells. *Entomological Research*, 46(5), 324–328. Advance online publication. DOI: 10.1111/1748-5967.12175

Kobayashi, T. (2015). Fabrication of cellulose hydrogels and characterization of their biocompatible films. In *Studies in Natural Products Chemistry* (Vol. 45). DOI: 10.1016/B978-0-444-63473-3.00001-0

Kuan, Z. J., Chan, B. K. N., & Gan, S. K. E. (2022). Worming the Circular Economy for Biowaste and Plastics: Hermetia illucens, Tenebrio molitor, and Zophobas morio. In *Sustainability (Switzerland)* (Vol. 14, Issue 3). DOI: 10.3390/su14031594

Kukalova-Peck, J. (1978). Origin and evolution of insect wings and their relation to metamorphosis, as documented by the fossil record. *Journal of Morphology*, 156(1), 53–125. Advance online publication. DOI: 10.1002/jmor.1051560104 PMID: 30231597

Kumirska, J., Czerwicka, M., Kaczyński, Z., Bychowska, A., Brzozowski, K., Thöming, J., & Stepnowski, P. (2010). Application of spectroscopic methods for structural analysis of chitin and chitosan. In *Marine Drugs* (Vol. 8, Issue 5). DOI: 10.3390/md8051567

Liu, S., Sun, J., Yu, L., Zhang, C., Bi, J., Zhu, F., Qu, M., Jiang, C., & Yang, Q. (2012). Extraction and characterization of chitin from the beetle Holotrichia parallela motschulsky. *Molecules (Basel, Switzerland)*, 17(4), 4604–4611. Advance online publication. DOI: 10.3390/molecules17044604 PMID: 22510609

Merzendorfer, H., & Zimoch, L. (2003). Chitin metabolism in insects: Structure, function and regulation of chitin synthases and chitinases. In *Journal of Experimental Biology* (Vol. 206, Issue 24). DOI: 10.1242/jeb.00709

Minke, R., & Blackwell, J. (1978). The structure of α-chitin. *Journal of Molecular Biology*, 120(2), 167–181. Advance online publication. DOI: 10.1016/0022-2836(78)90063-3 PMID: 642008

Mohan, K., Ganesan, A. R., Muralisankar, T., Jayakumar, R., Sathishkumar, P., Uthayakumar, V., Chandirasekar, R., & Revathi, N. (2020). Recent insights into the extraction, characterization, and bioactivities of chitin and chitosan from insects. In *Trends in Food Science and Technology* (Vol. 105). DOI: 10.1016/j.tifs.2020.08.016

Mohan, K., Ravichandran, S., Muralisankar, T., Uthayakumar, V., Chandirasekar, R., Rajeevgandhi, C., Karthick Rajan, D., & Seedevi, P. (2019). Extraction and characterization of chitin from sea snail Conus inscriptus (Reeve, 1843). *International Journal of Biological Macromolecules*, 126, 555–560. Advance online publication. DOI: 10.1016/j.ijbiomac.2018.12.241 PMID: 30594627

Mol, A., Kaya, M., Mujtaba, M., & Akyuz, B. (2018). Extraction of high thermally stable and nanofibrous chitin from Cicada (Cicadoidea). *Entomological Research*, 48(6), 480–489. Advance online publication. DOI: 10.1111/1748-5967.12299

Moreira, N. R., Cardoso, C., Dias, R. O., Ferreira, C., & Terra, W. R. (2017). A physiologically-oriented transcriptomic analysis of the midgut of Tenebrio molitor. *Journal of Insect Physiology*, 99, 58–66. Advance online publication. DOI: 10.1016/j.jinsphys.2017.03.009 PMID: 28341416

Morganti, P., Palombo, M., Fabrizi, G., Guarneri, F., Svolacchia, F., Cardillo, A., Del Ciotto, P., Carezzi, F., & Morganti, G. (2013). New insights on anti-aging activity of chitin nanofibril-hyaluronan block copolymers entrapping active ingredients: In vitro and in vivo study. *Journal of Applied Cosmetology*, 31(1).

Muthu, S. S. (2014). Roadmap to Sustainable Textiles and Clothing. Eco-friendly Raw Materials, Technologies, and Processing Methods. In *Roadmap to Sustainable Textiles and Clothing*. Eco-friendly Raw Materials, Technologies, and Processing Methods.

Nam, Y. S., Park, W. H., Ihm, D., & Hudson, S. M. (2010). Effect of the degree of deacetylation on the thermal decomposition of chitin and chitosan nanofibers. *Carbohydrate Polymers*, 80(1), 291–295. Advance online publication. DOI: 10.1016/j.carbpol.2009.11.030

Nguyen, K. D. (2021). Temperature effect of water coagulation bath on chitin fiber prepared through wet-spinning process. *Polymers*, 13(12), 1909. Advance online publication. DOI: 10.3390/polym13121909 PMID: 34201247

Nikkhah, A., Van Haute, S., Jovanovic, V., Jung, H., Dewulf, J., Cirkovic Velickovic, T., & Ghnimi, S. (2021). Life cycle assessment of edible insects (Protaetia brevitarsis seulensis larvae) as a future protein and fat source. *Scientific Reports*, 11(1), 14030. Advance online publication. DOI: 10.1038/s41598-021-93284-8 PMID: 34234157

Noh, M. Y., Muthukrishnan, S., Kramer, K. J., & Arakane, Y. (2016). Cuticle formation and pigmentation in beetles. In *Current Opinion in Insect Science* (Vol. 17). DOI: 10.1016/j.cois.2016.05.004

Ofem, M. I., Ubi, P. A., & Christian, A. (2022). Modeling the effect of gauge length on the mechanical properties of Chitin Whiskers reinforced composites. *Nigerian Journal of Technology*, 40(5), 846–854. Advance online publication. DOI: 10.4314/njt.v40i5.10

Peng, B. Y., Chen, Z., Chen, J., Yu, H., Zhou, X., Criddle, C. S., Wu, W. M., & Zhang, Y. (2020). Biodegradation of Polyvinyl Chloride (PVC) in Tenebrio molitor (Coleoptera: Tenebrionidae) larvae. *Environment International*, 145, 106106. Advance online publication. DOI: 10.1016/j.envint.2020.106106 PMID: 32947161

Peng, B. Y., Li, Y., Fan, R., Chen, Z., Chen, J., Brandon, A. M., Criddle, C. S., Zhang, Y., & Wu, W. M. (2020). Biodegradation of low-density polyethylene and polystyrene in superworms, larvae of Zophobas atratus (Coleoptera: Tenebrionidae): Broad and limited extent depolymerization. *Environmental Pollution*, 266, 115206. Advance online publication. DOI: 10.1016/j.envpol.2020.115206 PMID: 32682160

Pillai, K. S. Paul, W and C. P., S. (2009). Chitin and chitosan polymers: Chemistry, solubility and fiber formation. Progress in Polymer Science 34, (7), 641–678. *Progress in Polymer Science*, •••, 34.

Pires, C. T. G. V. M. T., Vilela, J. A. P., & Airoldi, C. (2014). The Effect of Chitin Alkaline Deacetylation at Different Condition on Particle Properties. *Procedia Chemistry*, 9, 220–225. Advance online publication. DOI: 10.1016/j.proche.2014.05.026

Poerio, A., Girardet, T., Petit, C., Fleutot, S., Jehl, J. P., Arab-Tehrany, E., Mano, J. F., & Cleymand, F. (2021). Comparison of the physicochemical properties of chitin extracted from cicada orni sloughs harvested in three different years and characterization of the resulting chitosan. *Applied Sciences (Basel, Switzerland)*, 11(23), 11278. Advance online publication. DOI: 10.3390/app112311278

Poirier, M., & Charlet, G. (2002). Chitin fractionation and characterization in N,N-dimethylacetamide/lithium chloride solvent system. *Carbohydrate Polymers*, 50(4), 363–370. Advance online publication. DOI: 10.1016/S0144-8617(02)00040-1

Purkayastha, D., & Sarkar, S. (2020). Physicochemical Structure Analysis of Chitin Extracted from Pupa Exuviae and Dead Imago of Wild Black Soldier Fly (Hermetia illucens). *Journal of Polymers and the Environment*, 28(2), 445–457. Advance online publication. DOI: 10.1007/s10924-019-01620-x

Rao, M. S., Muñoz, J., & Stevens, W. F. (2000). Critical factors in chitin production by fermentation of shrimp biowaste. *Applied Microbiology and Biotechnology*, 54(6), 808–813. Advance online publication. DOI: 10.1007/s002530000449 PMID: 11152073

Rehman, K., Hollah, C., Wiesotzki, K., Heinz, V., Aganovic, K., Rehman, R., Petrusan, J.-I., Zheng, L., Zhang, J., Sohail, S., Mansoor, M. K., Rumbos, C. I., Athanassiou, C., & Cai, M. (2023). Insect-Derived Chitin and Chitosan: A Still Unexploited Resource for the Edible Insect Sector. *Sustainability (Basel)*, 15(6), 4864. Advance online publication. DOI: 10.3390/su15064864

Rong, J., Lin, Y., Sui, Z., Wang, S., Wei, X., Xiao, J., & Huang, D. (2019). Amorphous calcium phosphate in the pupal cuticle of Bactrocera dorsalis Hendel (Diptera: Tephritidae): A new discovery for reconsidering the mineralization of the insect cuticle. *Journal of Insect Physiology*, 119, 103964. Advance online publication. DOI: 10.1016/j.jinsphys.2019.103964 PMID: 31604063

Sajomsang, W., & Gonil, P. (2010). Preparation and characterization of α-chitin from cicada sloughs. *Materials Science and Engineering C*, 30(3), 357–363. Advance online publication. DOI: 10.1016/j.msec.2009.11.014

Sarwade, A. B., & Bhawane, G. P. (2013). Anatomical and histological structure of digestive tract of adult *Platynotus belli* F. (Coleoptera : Tenebrionidae). *Biological Forum : An International Journal*, 5(2).

Smets, R., Verbinnen, B., Van De Voorde, I., Aerts, G., Claes, J., & Van Der Borght, M. (2020). Sequential Extraction and Characterisation of Lipids, Proteins, and Chitin from Black Soldier Fly (Hermetia illucens) Larvae, Prepupae, and Pupae. *Waste and Biomass Valorization*, 11(12), 6455–6466. Advance online publication. DOI: 10.1007/s12649-019-00924-2

Son, Y. J., Hwang, I. K., Nho, C. W., Kim, S. M., & Kim, S. H. (2021). Determination of carbohydrate composition in mealworm (Tenebrio molitor l.) larvae and characterization of mealworm chitin and chitosan. *Foods*, 10(3), 640. Advance online publication. DOI: 10.3390/foods10030640 PMID: 33803569

Song, S., Zhao, Y., Yuan, X., & Zhang, J. (2019). β-Chitin nanofiber hydrogel as a scaffold to in situ fabricate monodispersed ultra-small silver nanoparticles. *Colloids and Surfaces. A, Physicochemical and Engineering Aspects*, 574, 36–43. Advance online publication. DOI: 10.1016/j.colsurfa.2019.04.047

Suenaga, S., Totani, K., Nomura, Y., Yamashita, K., Shimada, I., Fukunaga, H., Takahashi, N., & Osada, M. (2017). Effect of acidity on the physicochemical properties of α- and β-chitin nanofibers. *International Journal of Biological Macromolecules*, 102, 358–366. Advance online publication. DOI: 10.1016/j.ijbiomac.2017.04.011 PMID: 28410951

Sugumaran, M. (2009). Complexities of cuticular pigmentation in insects. In *Pigment Cell and Melanoma Research* (Vol. 22, Issue 5). DOI: 10.1111/j.1755-148X.2009.00608.x

Sun, M., Liang, A., Watson, G. S., Watson, J. A., Zheng, Y., & Jiang, L. (2012). Compound Microstructures and Wax Layer of Beetle Elytral Surfaces and Their Influence on Wetting Properties. *PLoS One*, 7(10), e46710. Advance online publication. DOI: 10.1371/journal.pone.0046710 PMID: 23056414

Tamura, H., Nagahama, H., & Tokura, S. (2006). Preparation of chitin hydrogel under mild conditions. *Cellulose (London, England)*, 13(4), 357–364. Advance online publication. DOI: 10.1007/s10570-006-9058-z

Taokaew, S., Zhang, X., Chuenkaek, T., & Kobayashi, T. (2020). Chitin from fermentative extraction of crab shells using okara as a nutrient source and comparative analysis of structural differences from chemically extracted chitin. *Biochemical Engineering Journal*, 159, 107588. Advance online publication. DOI: 10.1016/j.bej.2020.107588

Tovar-Carrillo, K. L., Tagaya, M., & Kobayashi, T. (2013). Bamboo Fibers Elaborating Cellulose Hydrogel Films for Medical Applications. *Journal of Materials Science and Chemical Engineering*, 01(07), 7–12. Advance online publication. DOI: 10.4236/msce.2013.17002

Triunfo, M., Tafi, E., Guarnieri, A., Salvia, R., Scieuzo, C., Hahn, T., Zibek, S., Gagliardini, A., Panariello, L., Coltelli, M. B., De Bonis, A., & Falabella, P. (2022). Characterization of chitin and chitosan derived from Hermetia illucens, a further step in a circular economy process. *Scientific Reports*, 12(1), 6613. Advance online publication. DOI: 10.1038/s41598-022-10423-5 PMID: 35459772

Valdez-Peña, A. U., Espinoza-Perez, J. D., Sandoval-Fabian, G. C., Balagurusamy, N., Hernandez-Rivera, A., de-la-Garza-Rodriguez, I. M., & Contreras-Esquivel, J. C. (2010). Screening of industrial enzymes for deproteinization of shrimp head for chitin recovery. In *Food Science and Biotechnology* (Vol. 19, Issue 2). DOI: 10.1007/s10068-010-0077-z

Van De Velde, K., & Kiekens, P. (2004). Structure analysis and degree of substitution of chitin, chitosan and dibutyrylchitin by FT-IR spectroscopy and solid state13C NMR. *Carbohydrate Polymers*, 58(4), 409–416. Advance online publication. DOI: 10.1016/j.carbpol.2004.08.004

Vilasoa-Martínez, M., Calaza-Ramos, C., López-Hernández, J., Lage-Yusty, M. A., Losada, P. P., & Rodríguez-Bernaldo de Quirós, A. (2008). Determination of vitamin E and carotenoid pigments by high performance liquid chromatography in shell of Chionoecetes opilio. *Analytica Chimica Acta*, 617(1–2), 225–229. Advance online publication. DOI: 10.1016/j.aca.2008.03.001 PMID: 18486662

Wang, S. L., & Chio, S. H. (1998). Deproteinization of shrimp and crab shell with the protease of Pseudomonas aeruginosa K-187. *Enzyme and Microbial Technology*, 22(7), 629–633. Advance online publication. DOI: 10.1016/S0141-0229(97)00264-0

Wattjes, J., Sreekumar, S., Richter, C., Cord-Landwehr, S., Singh, R., El Gueddari, N. E., & Moerschbacher, B. M. (2020). Patterns matter part 1: Chitosan polymers with non-random patterns of acetylation. In *Reactive and Functional Polymers* (Vol. 151). DOI: 10.1016/j.reactfunctpolym.2020.104583

Wu, Q., Mushi, N. E., & Berglund, L. A. (2020). High-Strength Nanostructured Films Based on Well-Preserved α-Chitin Nanofibrils Disintegrated from Insect Cuticles. *Biomacromolecules*, 21(2), 604–612. Advance online publication. DOI: 10.1021/acs.biomac.9b01342 PMID: 31742385

Wu, S. (2012). Preparation of chitooligosaccharides from clanis bilineata larvae skin and their antibacterial activity. *International Journal of Biological Macromolecules*, 51(5), 1147–1150. Advance online publication. DOI: 10.1016/j.ijbiomac.2012.08.035 PMID: 22981818

Wu, S. J., Pan, S. K., Wang, H., & Wu, J.-H. (2013). Preparation of chitooligosaccharides from cicada slough and their antibacterial activity. *International Journal of Biological Macromolecules*, 62, 348–351. Advance online publication. DOI: 10.1016/j.ijbiomac.2013.09.042 PMID: 24095661

Wu, W. M., & Criddle, C. S. (2021). Characterization of biodegradation of plastics in insect larvae. In *Methods in Enzymology* (Vol. 648). DOI: 10.1016/bs.mie.2020.12.029

Xia, Z., Chen, J., & Wu, S. (2013). Hypolipidemic activity of the chitooligosaccharides from Clanis bilineata (Lepidoptera), an edible insect. *International Journal of Biological Macromolecules*, 59, 96–98. Advance online publication. DOI: 10.1016/j.ijbiomac.2013.04.017 PMID: 23591472

Xu, R., Huang, X., Hopkins, T. L., & Kramer, K. J. (1997). Catecholamine and histidyl protein cross-linked structures in sclerotized insect cuticle. *Insect Biochemistry and Molecular Biology*, 27(2), 101–108. Advance online publication. DOI: 10.1016/S0965-1748(96)00083-5

Yang, Q., Zhang, Z., Sun, X. G., Hu, Y. S., Xing, H., & Dai, S. (2018). Ionic liquids and derived materials for lithium and sodium batteries. In *Chemical Society Reviews* (Vol. 47, Issue 6). DOI: 10.1039/C7CS00464H

Yang, S. S., Ding, M. Q., He, L., Zhang, C. H., Li, Q. X., Xing, D. F., Cao, G. L., Zhao, L., Ding, J., Ren, N. Q., & Wu, W. M. (2020). Biodegradation of polypropylene by yellow mealworms (Tenebrio molitor) and superworms (Zophobas atratus) via gut-microbe-dependent depolymerization. *The Science of the Total Environment*. Advance online publication. DOI: 10.1016/j.scitotenv.2020.144087 PMID: 33280873

Yang, T. L. (2011). Chitin-based materials in tissue engineering: Applications in soft tissue and epithelial organ. In *International Journal of Molecular Sciences* (Vol. 12, Issue 3). DOI: 10.3390/ijms12031936

Zhang, J., Lu, A., Kong, L., Zhang, Q., & Ling, E. (2014). Functional analysis of insect molting fluid proteins on the protection and regulation of ecdysis. *The Journal of Biological Chemistry*, 289(52), 35891–35906. Advance online publication. DOI: 10.1074/jbc.M114.599597 PMID: 25368323

Zhang, M., Haga, A., Sekiguchi, H., & Hirano, S. (2000). Structure of insect chitin isolated from beetle larva cuticle and silkworm (Bombyx mori) pupa exuvia. *International Journal of Biological Macromolecules*, 27(1), 99–105. Advance online publication. DOI: 10.1016/S0141-8130(99)00123-3 PMID: 10704991

Zheng, L., Hou, Y., Li, W., Yang, S., Li, Q., & Yu, Z. (2013). Exploring the potential of grease from yellow mealworm beetle (Tenebrio molitor) as a novel biodiesel feedstock. *Applied Energy*, 101, 618–621. Advance online publication. DOI: 10.1016/j.apenergy.2012.06.067

Zhou, J., & Zhang, L. (2000). Solubility of cellulose in NaOH/urea aqueous solution. *Polymer Journal*, 32(10), 866–870. Advance online publication. DOI: 10.1295/polymj.32.866

Zhou, P., Li, J., Yan, T., Wang, X., Huang, J., Kuang, Z., Ye, M., & Pan, M. (2019). Selectivity of deproteinization and demineralization using natural deep eutectic solvents for production of insect chitin (Hermetia illucens). *Carbohydrate Polymers*, 225, 115255. Advance online publication. DOI: 10.1016/j.carbpol.2019.115255 PMID: 31521314

KEY TERMS AND DEFINITIONS

Arthropod: Any of a phylum of invertebrate animals (as insects, arachnids, and crustaceans) having a segmented body, jointed limbs, and a shell of chitin that is shed periodically.

Biomass: Organic matter is derived from living and generally can be used as a source of energy. Commonly this refers to plants, but also is applicable to chitin.

Chitin: Chitin is a linear polymer of β (1 4) linked 2-acetamido-2-deoxy-D-glucopyranose.

Hydrogel: Is a structure network-type of polymer, cross-linkage fabricates water insoluble matrix showing highly hydrophilic nature to retain over 90% of water.

Insect: Any small air-breathing arthropod of the class Hexpoda, having a body divided into head, thorax, and abdomen, three pairs of legs, and (in most species) two pairs of wings. Insects comprise about five sixths of all known animal species, with a total of over one million named species.

Chapter 10
Bioconversion of Waste Materials for the Production of Polylactic Acid to Alleviate Carbon Footprint

Debbie Dominic
Universiti Sains Malaysia, Malaysia

Nurul Alia Syufina Abu Bakar
https://orcid.org/0009-0009-9157-8986
Universiti Sains Malaysia, Malaysia

Siti Baidurah
https://orcid.org/0000-0003-3210-7470
Universiti Sains Malaysia, Malaysia

ABSTRACT

This chapter presents a comprehensive study on the utilization of various waste sources as feedstocks for the production of polylactic acid (PLA), that is a biodegradable and biocompatible polymer with versatile applications. In the case of PLA, food products containing mainly starch are sources, so that the development of utilizing waste materials instead of relying on such plant resources contributes to securing food resources. This waste approach is discussed in depth in this chapter, including PLA chemical structure, thermal, mechanical, and rheological behaviours and biodegradability. In addition, processing methods of the PLA obtained by bioconversion, as well as its multifaced properties and its applications across various industries are highlighted. Since integration of sustainable waste-PLA strategies provides a deep understanding of PLA, sheds light on its strengths, and identifies its limitations and future prospects, this approach can contribute highly to both environmental protection and a circular economy.

INTRODUCTION

Polylactic acid (PLA) is a biodegradable and bio-based polymer that has emerged as a promising alternative to conventional petroleum-derived plastics, driven by environmental concerns and the need for sustainable materials. PLA has numerous applications across different industries. Some common applications of lactic acid include biodegradable plastic packaging, textile fabric, medical devices, cosmetics, and soil conditioners (Baidurah et al., 2022). As seen in Figure 1, the chemical structure of PLA composed of lactic acid units having a chiral segment, which exist in two distinct mirror-image forms as known as enantiomers. These enantiomers are labeled as L-lactic acid and D-lactic acid, based on their optical activity in levorotatory and dextrorotatory, respectively, for rotation of the plane-polarize light. In biological systems, L-lactic acid is the more prevalent form and is produced during anaerobic metabolism in various organisms, including humans. As considered the "natural" form of lactic acid, this form is commonly found in fermented foods, such as yogurt, sauerkraut, pickles and in the muscles during strenuous exercise. In contrast, PLA having D-lactic acid enantiomer is produced by certain microorganisms, particularly lactic acid bacteria, during the fermentation of carbohydrates. Whilst they are chemically identical in terms of their molecular formula, but their different stereochemistry or three-dimensional structures result in different optical activities, potential physiological effects, mechanical properties, degradation rate, biocompatibility, crystallinity and melting point (Djukić-Vuković et al., 2019; Pohanka, 2020).

Figure 1. Chemical structures of polylactic acid, monomers of lactic acid, and lactide

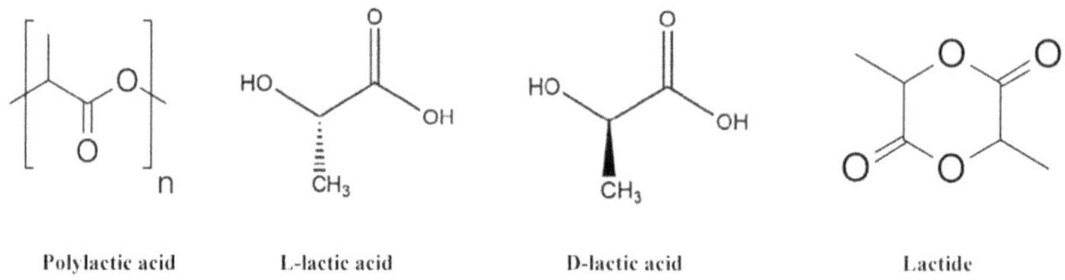

Lactic Acid Production, Purification, Polymerization, and Characterization Methods

Biological fermentation process to sugar and lactic acid from cellulosic wastes.

PLA is a polymer made of lactic acid connected via ester bonds and formed long-linked polymers, and is classified as a polyester. The monomer, lactic acid can be produced through both biological fermentation and chemical synthesis, however, biological fermentation is the more common and sustainable method. The two enantiomers of L and D formed lactic acids have different properties and their production can be achieved using different microorganisms. In fermentation process, lactic acid bacteria (LAB) such as *Lactobacillus* and *Streptococcus* species, are commonly used. These microorganisms ferment sugars like hexose or pentose and also typically glucose or other carbohydrates, through a series of intricate metabolic reactions via pyruvate into lactic acid. Basically, the fermentation pathway can be categorized into two processes, *viz.*, homolactic and heterolactic fermentation. In homolactic fermentation, only lactic acid is produced, whereas heterolactic fermentation produces three products namely ethanol, lactic acid, and acetic acid (Martinez et al., 2013; Yu et al., 2023). Relative to biological process, chemical synthesis involves hydrolysis of lactonitrile, which can be derived from acetaldehyde or acrylonitrile. This process, however, is less commonly used due to economic and negative impact to the environment (Gao et al., 2011)

Through the biological process of LAB, various renewable feedstocks are available for fermentation in a controlled bioreactor and the resultant lactic acid is separated and purified through distillation or extraction. Recently, various waste biomasses were utilized as carbon sources for production of value-added products (Sen et al., 2019; Sen and Baidurah, 2021; Ooi et al., 2023; Low et al., 2021; Mohammad et al., 2021a; Mohammad et al., 2021b). Such waste materials became feedstock to produce PLA and offered a significant opportunity to alleviate carbon footprint, contributing to environmental sustainability and carbon neutral. As renewable feedstocks, numerous types of waste materials can be bioconverted into PLA through various processes. Table 1 summarizes some examples of waste materials including agricultural residues, agro-industrial residues, aquaculture wastes, food wastes, municipal solid wastes, textile wastes, dairy wastes, wood, and paper wastes.

Table 1. Examples of waste materials

No.	Waste Materials	Remarks	Type of feedstocks	References
1	Agricultural residues	Crop residues such as wood bark, corncob, corn stover, wheat straw, rice husks, and sugarcane bagasse contains high cellulose and hemicellulose, which can be hydrolyzed into sugars and further ferment to produce lactic acid.	Sugars	Ozturk et al., 2017
2	Food wastes	Fruit peels, vegetable scraps, and expired food products residues contains high amount of fermentable carbohydrates for lactic acid production.	Fermentable carbohydrates	Peinemann et al., 2019; Carmona-Cabello et al., 2020; Yaradoddi et al., 2022
3	Wood and paper	Sawdust, wood chips, and pulping residues can be enzymatically hydrolyzed to release sugars and ferment to produce lactic acid.	Sugars	Castoldi et al., 2014
4	Textile wastes	Cotton waste and cuttings from garment production contain cellulose that can be broken down into fermentable sugars.	Sugars	El-Sheshtawy et al., 2022
5	Agro-industrial residues	Coffee, cocoa, and olive processing, as well as brewery and distillery waste, contain sugars and other organic compounds.	Sugars and other organic compounds	Singh et al., 2022
6	Aquaculture wastes	Fish, shellfish, and crab shell aquaculture wastes are chitin-rich materials which can be converted into chitosan, a precursor for PLA synthesis.	Chitin/chitosan	Palaniyappan et al., 2023
7	Dairy wastes	Whey, a byproduct of cheese and yogurt production, contains lactose that can be enzymatically converted into lactic acid.	Lactose	Singh et al., 2022

For example, Malaysia generated a total of 168 million tons of agricultural biomass waste annually (Ozturk et al., 2017). Agricultural residues or crop residues such as wood bark, corncob, corn stover, wheat straw, rice husks, and sugarcane bagasse were abundant agricultural wastes that can be utilized for PLA production (Ozturk et al., 2017). Also, wood bark contains high lignin, approximately 43.8%, followed by 29.8% hemicellulose, and 24.8% cellulose. In contrast, agricultural residues of corncob and corn stover contain high amount of 51.2-52.0% of cellulose, 30.7-32.0% of hemicellulose, and 14.4-15.0% of lignin (Ozturk et al., 2017). The cellulose and hemicellulose can be hydrolyzed into sugars and further ferment to produce lactic acid. Qiu et al. studied the D-lactic acid production using sugarcane bagasse waste by pre-fermentation of water-soluble carbohydrates prior acid pretreatment in the presence of *Pediococcus acidilactici XH11,* under optimized fermentation conditions (temperature 30°C, pH 5.0-5.5, agitation 150 rpm). *P. acidilactici* is a robust D-lactic acid-producing strain, which achieved 57.0 g/L of D-lactic acid produced via one-pot simultaneous saccharification and co-fermentation from the whole slurry of the undetoxified and pretreated sugarcane bagasse, with the overall yield of 0.58 g/g dry feedstock. This highlighted the strain's potential as a viable candidate for lactic acid production and deduced an efficient strategy for enhancing lactic acid production, especially in sugar industry (Qiu et al., 2023).

For another case, food waste including fruit peels, vegetable scraps, and expired food products can be valuable in source for lactic acid production. Carbohydrates have a significant portion in food waste, sometimes exceeding 50%. In the residues of food waste, starch is the most abundant carbohydrate with approximately 30% (w/w dry basis), lipids (12–37.7%), proteins (10–20%), and dietary fibers (2%) (Peinemann et al., 2019). Due to its high content of fermentable carbohydrates, such food waste is widely utilized for lactic acid production through fermentation (Carmona-Cabello et al., 2020; Yaradoddi et al., 2022).

Waste materials from wood processing and paper production, such as sawdust, wood chips, and pulping residues can be enzymatically hydrolyzed to release sugars. These sugars can then be fermented to produce lactic acid. Eucalyptus sawdust biologically pretreated in the presence of white-rot fungi of *Pleurotus ostreatus* and *Pleurotus pulmonarius* and increased yield of the reducing sugar became twenty-fold the total. In detail, the amount of reducing sugars released after 48 h of saccharification of sawdust was increased from 2.5 μmol/mL to 48.0 μmol/mL under the condition set in the experiment (Castoldi et al., 2014).

Waste textiles including cotton waste and cuttings from garment production contain cellulose that could be broken down into fermentable sugars (El-Sheshtawy et al., 2022) and then the products were converted into lactic acid through fermentation, successfully generating 8015.90 mg/L of reducing sugar from cotton via chemical hydrolysis using 5% hydrochloric acid and then isolated *Kosakonia cowanii* (B2) bacteria produced the highest lactic acid concentration of 24.97 g/L for 24 hours, 27.91 g/L for 48 hours and with the 28.14 g/L for 72 hours, reaching to maximum production. Also, agro-industrial residues such as coffee, cocoa, and olive processing, as well as brewery and distillery waste containing cellulose fibers and other organic compounds utilized to be harnessed for PLA production (Singh et al., 2022).

Moreover, aquaculture waste from fish, shellfish, and crab shell and processing byproducts can be used for PLA production (Palaniyappan et al., 2023). According the process, such chitin-rich waste was converted into chitosan, which is a biopolymer, serving as a precursor for PLA synthesis. Dairy waste such as whey, a by-product of cheese and yogurt production has lactose that can be enzymatically converted into lactic acid (Singh et al., 2022).

Dedenaro et al., investigates the potential of using agri-food waste as a feedstock for the production of L-lactic acid through fermentation. The study discusses the challenges with L-lactic acid production from food waste, including low yield due to slow hydrolysis rate. The study uses a mixture of food waste, including pear pomace and ricotta cheese whey as a low-cost source of nutrients for lactic acid fermentation of *Lactobacillus casei* and *Lactobacillus farciminis* in microaerophilic conditions and mild sterility. The yield factor for biomass was 8%, corresponding to 2.50 g/L of cells, and 2.5 g/L of l-lactic acid was produced from the waste mixture (Dedenaro et al., 2016). In this study, the authors conducted a batch fermentation of a 25:75 mixture (in 1 liter working volume reactor) of pear pomace and ricotta cheese whey, respectively, reached an overall yield factor of 90% and a volumetric productivity of 0.42 g/L·h.

This lactic acid was then used as a feedstock for PLA production. While these waste materials offer potential for lactic acid production, several technical and economic challenges ought to be addressed. When used the waste materials possessing rich in cellulose components and then microbiologically converted sugars or other fermentable components, the entire production process to lactic acid requires optimization for efficiency and cost-effectiveness.

After lactic acid is produced via fermentation, since the object is mixed with the culture medium, then the purification or extraction process of lactic acid is next step by following distillation, crystallization, and filtration. Fractional distillation, which is a separation process by using vaporized temperature of fraction part used was generally applied for the fermentation broth. Upon rapid heating and cooling process, the latter process condenses liquid and finally lactic acid causes it to crystallize in distilled part. The crystals can then be collected by filtration. A wide range of membranes can be used to separate lactic acid from any impurities based on size and charge, such as microfiltration, ultrafiltration, nanofiltration, pervaporation, reverse osmosis and electrodialysis (Swetha et al., 2023).

PLA polymerization process from lactic acid.

It is known that PLA synthesis from lactic acid by polymerization process is divided into two processes of polycondensation of lactic acid and ring-opening polymerization of lactide. The polymerization of lactic acid into PLA is initiated using a catalyst, often tin-based in the presence of tin (II) octanoate, lactic acid is heated and maintained at a specific temperature, generally in the range of 140-180°C, promoting the condensation reaction that forms PLA chains. However, the exact temperature and conditions for optimized condition varies depending on the type of catalyst used, either L- or D- lactic acids and also lactide (LT) as precursors. The reaction progress was monitored by taking samples at different time intervals, when used for the ring opening polymerization method (Widjaja et al., 2023). Due to the importance and economic value of the preparation and purification of LA and LT throughout the entire PLA industry, the main objectives were summarized as recent achievements in LA and LT production as well as their industrialization progress (Yu et al., 2023). In several reports of ring-opening polymerization of lactide, the ring-opening polymerization of cyclic diesters LA synthesized from α-hydroxycarboxylic acids provided high-molecular-weight polyester in high yields in the presence of tin (II) 2-ethylhexanoate catalyst having high reactivity and low toxicity. For the synthesis of PLA from LT precursor, purity of monomers and the amount of water and alcohol in the reaction system are significant factors for increasing molecular weight and conversion of polyesters (Dechy-Cabaret et al., 2004; Kaihara et al., 2007).

Applications of PLA.

Applications of PLA cover broad range in many industries such as plastics packaging, textiles, biomedical devices, agriculture, and automotive due to its unique properties such as biocompatibility, biodegradability, and versatility (Baidurah et al., 2022). In the pharmaceutical and medical industries,

PLA has been extensively used for drug delivery systems, tissue engineering, and medical implants due to its biocompatibility and ability to degrade into harmless lactic acid within the body. PLA-based materials have been employed in the production of bioresorbable suture threads, bone fixation screws, stent coatings, and other medical devices, showcasing their potential in advancing healthcare technologies. Furthermore, PLA's hydrophobic nature and ability to degrade through hydrolysis make it suitable for applications in nanomedicine, particularly in the synthesis of nanoparticles for targeted drug delivery systems. The use of PLA in nanocarriers offers promising opportunities for enhancing drug delivery efficiency and reducing side effects, highlighting its importance in the field of nanomedicine. Furthermore, PLA's properties have led to its application in the production of biodegradable packaging materials, disposable items, and 3-D printing filaments, contributing to the reduction of non-degradable plastic waste and environmental impact.

Characterization and Properties of PLA.

Many researchers successfully produce waste-derived PLA with the same characteristics with commercially available PLA (Peinemann et al., 2019; Carmona-Cabello et al., 2020; Yaradoddi et al., 2022). It is well known that PLA is a biodegradable plastic that undergoes hydrolysis by moisture in the environment to become low molecular weight and is ultimately degraded to carbon dioxide and water by microbes. Therefore, it is a carbon-neutral material because it can be synthesized from plant-derived starch raw materials via LA. However, coexistence with food raw materials is currently difficult, and processes for waste-derived PLA will become increasingly important.

On the other hand, as mentioned before, since PLA has a helix structurethe content of lactic acid enantiomers, the PLA chain affects the final characteristics of PLA such as the degree of crystallization rate, melting temperature, and glass transition temperature (Dana & Ebrahimi, 2023). PLA chains comprising either optically pure L-lactic acid or D-lactic acid, PLA homopolymers are semicrystalline polyesters with a melting temperature (T_m) of about 175°C. While due to disorders of polymer chains PLA heteropolymers (poly DL-lactic acid) are amorphous (Narancic et al., 2020). In addition, based on the processing grade, PLA glass transition temperature (T_g) varies from 50 to 59°C with a melting point of 159–178°C (Ejaz et al., 2020).

Thermal Properties.

The glass transition temperature (T_g) is the temperature at which PLA transitions from a glassy, brittle state to a rubbery, more flexible state. For PLA, the T_g typically ranges from approximately 55°C to 65°C, depending on the stereochemistry (L-, D-, or DL-lactic acid), molecular weight, and crystallinity. Above the T_g, PLA becomes more flexible and less rigid, which affects its processability and mechanical properties (Zhang et al., 2022). The melting temperature (T_m) of PLA, is the temperature at which the polymer chain transitions from a solid crystalline state to a liquid state. It is a critical thermodynamic parameter that characterizes the polymer's thermal behavior and stability. In the case of PLA, T_m refers to the temperature at which the polymer chains become mobile enough to break their organized crystalline structure and transition into a more disordered liquid phase. The T_m of PLA typically falls between 110°C and 180°C, influencing on the crystallinity and thermal properties of PLA. Higher crystallinity led to a higher T_m (Chuensangjun et al., 2013; Widjaja et al., 2023). The crystallization temperature (T_c) is the temperature at which amorphous PLA molecules begin to arrange into a more ordered, crystalline structure, of which PLA is generally slightly lower than its T_m. Accordingly, crystallization can affect mechanical properties, such as stiffness and strength.

In contrast, PLA starts to degrade under heat condition by depolymerization, leading to the release of volatile products. The thermal decomposition temperature (T_d) of PLA ranges between 200°C and 240°C, but the thermal degradation sensitively affects its processing conditions and limits the maximum temperature when exposed to without significant degradation. These parameters are especially important when processing PLA into a product. Namely, because PLA is a thermoplastic material, PLA can be processed using extrusion, injection molding, and 3D printing. The processing temperature range for PLA is generally selected between T_g and T_m. To prevent thermal decomposition, the process temperatures need to be carefully controlled to prevent thermal degradation while achieving desired material properties.

Mechanical Properties.

Tensile strength measures a material's ability to withstand tensile (pulling) forces without breaking. The tensile strength of PLA is influenced by factors such as crystallinity and molecular weight. Tensile strength for PLA generally ranges from 45 MPa to 80 MPa. The average molecular weight for PLA is approximately 81 kDa (Detyothin et al., 2013). Young's modulus, also known as the modulus of elasticity, represents a material's stiffness and its resistance to deformation under an applied load and the modulus of PLA is typically around 3 - 4 GPa. This is comparable to many common engineering plastics and elongation at break is usually in the range of 2% to 6%, indicating that it is relatively less stretch and brittle compared to some other polymers. Flexural strength is measure of a material's ability to withstand bending forces and PLA has usually around 60 MPa to 100 MPa measuring a material's stiffness under bending conditions. The observed value of flexural modulus of PLA is similar to Young's modulus, usually around 3 - 4 GPa. Also, impact strength measuring a material's ability to absorb energy during an impact without breaking for PLA is relatively lower compared to some other polymers, such as acrylonitrile butadiene styrene (ABS). However, certain PLA formulations and processing conditions can enhance its impact resistance (Li et al., 2019).

In their mechanical properties processing condition, some parameters can be modified by using additives and blending with other polymers. For example, additives like plasticizers, impact modifiers, and reinforced the incorporated PLA to ameliorate certain mechanical properties (Carrasco et al., 2021). Additionally, post-processing treatments such as annealing can affect the crystallinity and mechanical behaviour of PLA. When selecting PLA for a specific application, it is extremely important to consider its mechanical properties in relation to the intended use and environmental conditions. Although PLA has inferior mechanical properties compared to petroleum-derived engineering plastics, its combination of biodegradability and appropriate mechanical properties make it a valuable material for various sustainable applications.

Degradability Properties.

Figure 2. Schematic diagram of PLA cycle involving production and degradation, supporting the circular economy concept

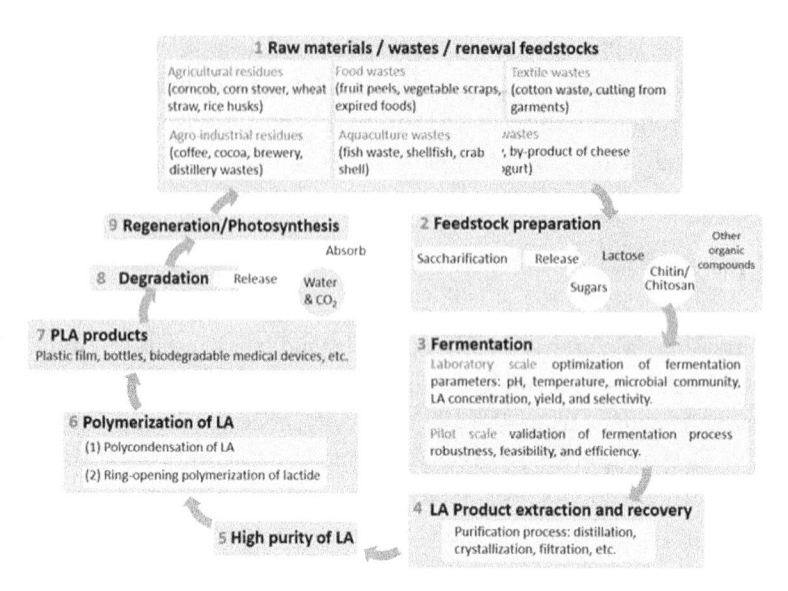

In circular economy, as seen in the schematic diagram of PLA cycle for Figure 2, involving production and degradation process, should align cycled relation through the cycle. The production of PLA from LA and its recycling after decomposition, rather than disposal after commercialization, would make a significant contribution to a sustainable society. When biodegraded in the PLA cycle, PLA decomposes into water and carbon dioxide, but lactic acid synthesis by carbon dioxide consumption through microbial fermentation is possible, and many biological processes, as described above, have achieved this as one of the distinctive features in the cycle. For biodegradability of PLA, degradation occurs through various mechanisms, depending on various parameters under exposed environment and conditions such as temperature, humidity, and microbial activity to release carbon dioxide and water. The primary mechanisms of PLA degradation were highlighted for hydrolytic degradation, biodegradation with microbial activity, photodegradation exposed to UV radiation, thermal degradation, and mechanical degradation (Boey et al., 2021). Hydrolytic degradation of PLA occurs in moist and humid conditions, including soil, compost, and aquatic environments. PLA undergoes hydrolysis, whereby water molecules break the ester bonds in the polymer chains, leading to the cleavage of the polymer into smaller fragments. This process was known to be accelerated by elevated temperatures, exposure to high moisture, and highly basic and highly acidic mediums (as compared to neutral conditions). Higher temperatures generally accelerate hydrolytic degradation due to increased water mobility and reaction rates (Zare and Rhee, 2019; Elsawy et al., 2017).

It is well-known that biodegradation with microbial activity of PLA can occur in compost, soil, and other environments with microbial activity (Butbunchu & Pathom-Aree, 2019; Janczak et al., 2020; Kamarudin et al., 2021). Microorganisms such as bacteria and fungi secrete enzymes that hydrolyze PLA, breaking the long polymer chain into simpler compounds of its lactic acid oligomers, water, carbon dioxide, methane (anaerobic degradation), and residual impurities. These simpler compounds can be further metabolized by microorganisms as a carbon source. Generally, the biodegradation rate and the microbial activity increased monotonically with temperature (Teixeira et al., 2021).

Photodegradation refers to the process by which PLA chemical structure undergoes deterioration due to exposure to light, particularly ultraviolet (UV) radiation from the sun or other light sources. This degradation process is initiated by the absorption of photons, which are particles of light energy, by the molecules in the PLA material. Normally, this degradation occurs in outdoor environments exposed to sunlight, whereby the UV radiation from sunlight causes the breakdown of PLA long polymer chain through photo-oxidation. The UV light initiates chain scission reactions, leading to the formation of free radicals, which can react with oxygen in the air, promoting further degradation. Virág et al., studied the effects of PLA long-term exposure to UV-C germicidal irradiation on its properties and reported the changes of amorphous PLA films (170 ± 20 μm) subjected up to 8 hours of exposure. The samples were irradiated in a custom-built chamber equipped with two light sources of 254 nm wavelength commonly used for sterilisation, resulting in formation of the number-average molecular weights when photo-degraded by 94%. The results indicated less-changed PLA film with 13 cycles of UV-C sterilisation in tensile strength and modulus. The PLA film used in their study was an extrusion grade PLA, namely Ingeo 4032D (supplied by Nature Works LLC, Plymouth, MN, USA), with a density of 1.24 g/cm^3 and 1.4–2.0% D-lactide. In the context of PLA polymers, photodegradation can lead to a variety of observable physical changes in the material's properties, such as color fading, reduced material properties, surface cracking and chalking, which simultaneously lead to a decrease in service life. In order to determine degradability properties of PLA, the synthesized PLAs were generally characterized using Fourier-transform infrared spectroscopy and nuclear magnetic resonance spectroscopy for its chemical structure and gel permeation chromatography for molecular weight and its distribution in addition of X-ray methods (Baidurah, 2022).

CHALLENGES TO PRODUCE PLA FROM WASTE MATERIALS

In the PLA cycle, producing PLA using waste materials presents several challenges related to sustainable process in successfully achieving the close-loop perspective. Some of the key challenges include variability in quantity, feedstock quality, composition inconsistency, pre-processing, contamination, purity, low reaction efficiency, low economic viability, and scaling-up difficulties. One of the primary issues is the unpredictable quantity of waste feedstock available. This variability can be seasonal, weekly or daily. For example, agricultural waste, such as crop residues, is seasonal, as concentrated in the harvest season, thus may not be available year-round, meaning shortage supply during peak demand periods and causing inefficiencies in production stages.

For biological fermentation to produce LA, the type of waste can vary and influenced greatly in terms of their composition and quality. Different sources of waste vary levels of contaminants, moisture, and other impurities. Inconsistent quality containing higher degree of cellulosic moieties can affect the efficiency of bioconversion processes. Therefore, in some cases, additional treatment steps are required to ensure that a raw material of the feedstock to meet the required specifications. For example, biomass feedstocks derived from plants involve lignin content, affecting their suitability for fermentable processes to occur. Inconsistent feedstock quality can lead to variations in PLA properties, affecting its usability, since waste materials often require extensive pre-processing to remove impurities, contaminants, and non-biodegradable components. In addition, ensuring a consistent supply of waste materials with suitable properties is essential and also the pre-processing step can be energy-intensive and increase the overall production costs.

Such impurities of waste sources for biological fermentation stage to LA synthesis ultimately affect the polymerization process and the properties of PLA leading to poor PLA quality (Dechy-Cabaret et al., 2004; Kaihara et al., 2007). Thus, developing effective purification methods to remove contaminants is crucial. Optimizing reaction conditions and catalysts for waste-based feedstocks is essential to achieve acceptable yields. Referring to Figure 2, commonly, the optimization of fermentation parameters at laboratory scale involves the pH, temperature, microbial community, LA concentration, yield and its selectivity. Then, validation of fermentation process is conducted at pilot scale.

In another point, PLA production using waste materials needs to be economically viable to compete with traditional production methods. The costs associated with feedstock collection, pre-processing, and conversion must be carefully balanced against the market value of PLA. Namely, while using waste materials are surely environmentally beneficial, the entire production process's life cycle impact should be considered, as included for evaluating energy consumption, greenhouse gas emissions, and waste generation during feedstock collection, conversion, and disposal. Furthermore, regulatory approvals and standards for using waste materials in PLA production becomes important to be establish compliance with health, safety, and environmental regulations.

Transitioning from lab-scale to industrial-scale production can be challenging. Scaling up waste-based PLA production requires addressing challenges related to equipment, process optimization, and maintaining consistent quality.

ADVANTAGES OF PRODUCING PLA FROM WASTE MATERIALS

Producing PLA from waste materials offers several significant advantages, contributing to both environmental sustainability and resource efficiency. Some key advantages include waste reduction, resource conservation, and alleviation of carbon footprint. Valorization of waste materials into PLA conversion reduces the burden on landfills and waste disposal facilities. This initiative transforms waste that would otherwise be discarded into a valuable resource, creating a circular economy approach. Moreover, using waste feedstocks for PLA production reduces the demand for virgin resources such as fossil fuels and agricultural crops. This initiative enables conservation of natural resources and mitigate environmental degradation associated with resource extraction. Therefore, in waste-derived PLA, one main merit is the reduction of carbon footprint in comparison to PLA produced from virgin resources, due to the concept of carbon neutrality associated with waste materials and the overall reduction in greenhouse gas emissions during the waste-to-PLA conversion process (Subbarao et al., 2023). Figure 3 shows the illustration of the waste as a resource incorporated into the PLA cycle and its positive impact in mitigating greenhouse gas emissions. So as advantages, waste-derived PLA results in fewer net greenhouse gas emissions through various ways such as:

(1) carbon neutrality of waste materials,
(2) reduced virgin resource consumption, and
(3) conversion efficiency.

Figure 3. Illustration of the wastes as a resource with incorporated into the PLA cycle and its positive impact in mitigating greenhouse gas emissions

Waste materials are out of the natural carbon cycle, if not utilized as a resource. During their growth, plants absorb carbon dioxide (CO_2) from the atmosphere through photosynthesis. When these plants decompose or convert into waste, the CO_2 they release upon decomposition is roughly equivalent to the CO_2 they absorb during their growth. In contrast, using waste materials as feedstock eliminates the need for these resource-intensive extraction and processing steps, resulting in lower emissions.

In terms of conversion efficiency, waste-to-PLA conversion processes are typically designed to be energy-efficient, especially, when used biological fermentation process to lactic acid source. The processes can harness the energy content of waste materials, particularly organic waste, through techniques such as anaerobic digestion or fermentation. These processes release biogas of mainly methane that can be used to generate energy. However, methane is often associated with the high global warming potential needs to be recovered and prevented from being released into the environment. Then arises the need to use it for fuel. The energy generated can be used in the conversion process, reducing the need for external energy sources and their associated emissions.

To quantitatively estimate the potential reduction in greenhouse gas emissions when producing PLA from waste materials, a life cycle assessment (LCA) approach is commonly used. This method involves evaluating the emissions associated with various stages of PLA production and comparing the waste-derived PLA scenario with a conventional PLA production scenario. To simplify, the calculation method can be divided into 4 steps.

Step 1-Emissions from Waste Material Decomposition: Calculate the CO_2 emissions that would occur if the waste materials were disposed of without conversion. This is the baseline scenario against which emissions from PLA production are compared.

Step 2-Emissions from PLA Production: Calculate the emissions associated with producing PLA from waste materials. This includes emissions from waste processing, fermentation, polymerization, and other relevant processes.

Step 3-Emissions from virgin PLA Production: Calculate the emissions associated with producing PLA from virgin resources. Consider emissions from resource extraction, transportation, processing, and manufacturing.

Step 4-Net Emissions Comparison: Compare the emissions calculated in steps 1, 2, and 3. Generally, the emissions associated with waste-derived PLA will be lower than those associated with virgin PLA, given the carbon-neutral nature of waste materials and the reduced reliance on energy-intensive extraction processes.

The LCA tools and software can facilitate these calculations, taking into account various factors such as energy consumption, emissions, waste generation, and transportation. Specific results can vary depending on factors such as waste composition, conversion processes, and energy sources used. The LCA analysis provides a comprehensive view of the environmental impact and helps in making informed decisions for sustainable PLA production.

FUTURE DIRECTIONS

Developing efficient processes to convert waste materials into PLA promotes innovation and technological advancements in the field of bioplastics. This can lead to discoveries that improve waste management practices and resource utilization. Incorporating waste materials into PLA production underscores the potential of bioplastics to contribute to a more circular and sustainable economy. It capitalizes on the value present in waste streams, transforming them into useful products and driving positive environmental outcomes. In other words, the upcycling process, which utilizes the value present in waste materials and converts them into useful products, not only contributes to improving people's lives through materials, but also has positive environmental results.

CONCLUSION

In conclusion, this chapter highlighted the multifaceted properties of PLA, emphasizing its potential as a sustainable option to traditional non-degradable plastics. By providing a comprehensive analysis of PLA's chemical, thermal, mechanical attributes, as well as its biodegradability and production utilizing wastes, this chapter contributes to a deeper understanding of PLA's role in achieving circular economy and carbon neutrality. The feasibility and efficiency of PLA production from waste sources depends on several factors, such as the type and quality of waste, the pretreatment process, and the conversion efficiency of the biomass to PLA. Additionally, the cost-effectiveness of PLA production from waste sources is the main bottleneck, as the conversion process may require additional energy and resources compared to traditional PLA production methods. Collaboration between researchers, industry stakeholders, and policymakers is crucial to developing sustainable and cost-effective solutions for PLA production from waste sources.

ACKNOWLEDGEMENT

The authors gratefully acknowledge the financial support from the Malaysian Ministry of Higher Education to the Universiti Sains Malaysia (USM), through the Fundamental Research Grant Scheme (FRGS), with the project code of FRGS/1/2021/STG01/USM/02/12.

REFERENCES

Baidurah, S. (2022). Methods of Analyses for Biodegradable Polymers: A Review. *Polymers*, 14(22), 4928. DOI: 10.3390/polym14224928 PMID: 36433054

Baidurah, S., Kobayashi, T., & Abd Aziz, A. (2022). PLA Based Plastics for Enhanced Sustainability of the Environment. In Hashmi, M. S. J. (Ed.), *Encyclopedia of Materials: Plastics and Polymers* (Vol. 2, pp. 511–519). Elsevier., DOI: 10.1016/B978-0-12-820352-1.00175-9

Boey, J. Y., Mohamad, L., Khok, Y. S., Tay, G. S., & Baidurah, S. (2021). A Review of the Applications and Biodegradation of Polyhydroxyalkanoates and Poly(lactic acid) and Its Composites. *Polymers*, 13(10), 1544. DOI: 10.3390/polym13101544 PMID: 34065779

Butbunchu, N., & Pathom-Aree, W. (2019). Actinobacteria as Promising Candidate for Polylactic Acid Type Bioplastic Degradation. *Frontiers in Microbiology*, 10, 2834. Advance online publication. DOI: 10.3389/fmicb.2019.02834 PMID: 31921021

Carmona-Cabello, M., García, I. L., Sáez-Bastante, J., Pinzi, S., Koutinas, A. A., & Dorado, M. P. (2020). Food waste from restaurant sector – Characterization for biorefinery approach. *Bioresource Technology*, 301, 122779. DOI: 10.1016/j.biortech.2020.122779 PMID: 31958693

Carrasco, F., Santana Pérez, O., & Maspoch, M. L. (2021). Kinetics of the Thermal Degradation of Poly(lactic acid) and Polyamide Bioblends. *Polymers*, 13(22), 3996. DOI: 10.3390/polym13223996 PMID: 34833295

Castoldi, R., Bracht, A., de Morais, G. R., Baesso, M. L., Correa, R. C. G., Peralta, R. A., Moreira, R. D. F. P. M., de Moraes, M. D. L. T., de Souza, C. G. M., & Peralta, R. M. (2014). Biological pretreatment of *Eucalyptus grandis* sawdust with white-rot fungi: Study of degradation patterns and saccharification kinetics. *Chemical Engineering Journal*, 258, 240–246. DOI: 10.1016/j.cej.2014.07.090

Chuensangjun, C., Pechyen, C., & Sirisansaneeyakul, S. (2013). Degradation Behaviors of Different Blends of Polylactic Acid Buried in Soil. *Energy Procedia*, 34, 73–82. DOI: 10.1016/j.egypro.2013.06.735

Dana, H. R., & Ebrahimi, F. (2023). Synthesis, properties, and applications of polylacticacid-based polymers. *Polymer Engineering and Science*, 63(1), 22–43. DOI: 10.1002/pen.26193

Dechy-Cabaret, O., Martin-Vaca, B., & Bourissou, D. (2004). Controlled Ring-Opening Polymerization of Lactide and Glycolide. *Chemical Reviews*, 104(12), 6147–6176. DOI: 10.1021/cr040002s PMID: 15584698

Dedenaro, G., Costa, S., Rugiero, I., Pedrini, P., & Tamburini, E. (2016). Valorization of Agri-Food Waste via Fermentation: Production of l-lactic Acid as a Building Block for the Synthesis of Biopolymers. *Applied Sciences (Basel, Switzerland)*, 6(12), 379. DOI: 10.3390/app6120379

Detyothin, S., Selke, S. E., Narayan, R., Rubino, M., & Auras, R. (2013). Reactive functionalization of poly(lactic acid), PLA: Effects of the reactive modifier, initiator and processing conditions on the final grafted maleic anhydride content and molecular weight of PLA. *Polymer Degradation & Stability*, 98(12), 2697–2708. DOI: 10.1016/j.polymdegradstab.2013.10.001

Djukić-Vuković, A., Mladenović, D., Ivanović, J., Pejin, J., & Mojović, L. (2019). Towards sustainability of lactic acid and poly-lactic acid polymers production. *Renewable & Sustainable Energy Reviews*, 108, 238–252. DOI: 10.1016/j.rser.2019.03.050

Ejaz, M., Azad, M., Shah, A. U. R., Afaq, S. K., & Song, J. (2020). Mechanical and Biodegradable Properties of Jute/Flax Reinforced PLA Composites. *Fibers and Polymers*, 21(11), 2635–2641. DOI: 10.1007/s12221-020-1370-y

El-Sheshtawy, H. S., Fahim, I., Hosny, M., & El-Badry, M. A. (2022). Optimization of lactic acid production from agro-industrial wastes produced by *Kosakonia cowanii*. *Current Research in Green and Sustainable Chemistry*, 5, 100228. DOI: 10.1016/j.crgsc.2021.100228

Elsawy, M. A., Kim, K. H., Park, J. W., & Deep, A. (2017). Hydrolytic degradation of polylactic acid (PLA) and its composites. *Renewable & Sustainable Energy Reviews*, 79, 1346–1352. DOI: 10.1016/j.rser.2017.05.143

Gao, C., Ma, C., & Xu, P. (2011). Biotechnological routes based on lactic acid production from biomass. *Biotechnology Advances*, 29(6), 930–939. DOI: 10.1016/j.biotechadv.2011.07.022 PMID: 21846500

Janczak, K., Dąbrowska, G. B., Raszkowska-Kaczor, A., Kaczor, D., Hrynkiewicz, K., & Richert, A. (2020). Biodegradation of the plastics PLA and PET in cultivated soil with the participation of microorganisms and plants. *International Biodeterioration & Biodegradation*, 155, 105087. DOI: 10.1016/j.ibiod.2020.105087

Kaihara, S., Matsumura, S., Mikos, A. G., & Fisher, J. P. (2007). Synthesis of poly L-lactide and polyglycolide by ring-opening polymerization. *Nature Protocols*, 2(11), 2767–2771. DOI: 10.1038/nprot.2007.391 PMID: 18007612

Kamarudin, N., H., A., Noor, N. D., M., Rahman, R., N., Z., R. (. (2021). Microbial degradation of polylactic acid bioplastic. *Journal of Sustainability Science and Management*, 16(7), 299–317. DOI: 10.46754/jssm.2021.10.021

Li, Y., Qiang, Z., Chen, X., & Ren, J. (2019). Understanding thermal decomposition kinetics of flame-retardant thermoset polylactic acid. *RSC Advances*, 9(6), 3128–3139. DOI: 10.1039/C8RA08770A PMID: 35518982

Low, T. J., Mohammad, S., Sudesh, K., & Baidurah, S. (2021). Utilization of banana (*Musa* sp.) fronds extract as an alternative carbon source for poly(3-hydroxybutyrate) production by *Cupriavidus necator* H16. *Biocatalysis and Agricultural Biotechnology*, 34, 102048. DOI: 10.1016/j.bcab.2021.102048

Martinez, F. A. C., Balciunas, E. M., Salgado, J. M., González, J. M. D., Converti, A., & de Souza Oliveira, R. P. (2013). Lactic acid properties, applications and production: A review. *Trends in Food Science & Technology*, 30(1), 70–83. DOI: 10.1016/j.tifs.2012.11.007

Mohammad, S., Baidurah, S., Kamimura, N., Matsuda, S., Bakar, N. A. S. A., Muhamad, N. N. I., Ahmad, A. H., Dominic, D., & Kobayashi, T. (2021b). Fermentation of Palm Oil Mill Effluent in the Presence of *Lysinibacillus* sp. LC 556247 to Produce Alternative Biomass Fuel. *Sustainability (Basel)*, 13(21), 11915. DOI: 10.3390/su132111915

Mohammad, S., Baidurah, S., Kobayashi, T., Ismail, N., & Leh, C. P. (2021a). Palm Oil Mill Effluent Treatment Processes—A Review. *Processes (Basel, Switzerland)*, 9(5), 739. DOI: 10.3390/pr9050739

Narancic, T., Cerrone, F., Beagan, N., & O'Conno, K. E. (2020). Recent Advances in Bioplastics: Application and Biodegradation. *Polymers*, 12(4), 920. DOI: 10.3390/polym12040920 PMID: 32326661

Ooi, W. C., Dominic, D., Kassim, M. A., & Baidurah, S. (2023). Biomass Fuel Production through Cultivation of Microalgae *Coccomyxa dispar* and *Scenedesmus parvus* in Palm Oil Mill Effluent and Simultaneous Phycoremediation. *Agriculture*, 13(2), 336. DOI: 10.3390/agriculture13020336

Ozturk, M., Saba, N., Altay, V., Iqbal, R., Hakeem, K. R., Jawaid, M., & Ibrahim, F. H. (2017). Biomass and bioenergy: An overview of the development potential in Turkey and Malaysia. *Renewable & Sustainable Energy Reviews*, 79, 1285–1302. DOI: 10.1016/j.rser.2017.05.111

Palaniyappan, S., & Sivakumar, N. K. (2023). Development of crab shell particle reinforced polylactic acid filaments for 3D printing application. *Materials Letters*, 341, 134257. DOI: 10.1016/j.matlet.2023.134257

Peinemann, J. C., Demichelis, F., Fiore, S., & Pleissner, D. (2019). Techno-economic assessment of non-sterile batch and continuous production of lactic acid from food waste. *Bioresource Technology*, 289, 121631. DOI: 10.1016/j.biortech.2019.121631 PMID: 31220764

Pohanka, M. (2020). D-Lactic Acid as a Metabolite: Toxicology, Diagnosis, and Detection. *BioMed Research International*, 2020, 1–9. Advance online publication. DOI: 10.1155/2020/3419034 PMID: 32685468

Qiu, Z., Han, X., Fu, A., Jiang, Y., Zhang, W., Jin, C., Li, D., Xia, J., He, J., Deng, Y., Xu, J., Liu, X., He, A., Gu, H., & Xu, J. (2023). Enhanced cellulosic d-lactic acid production from sugarcane bagasse by pre-fermentation of water-soluble carbohydrates before acid pretreatment. *Bioresource Technology*, 368, 128324. DOI: 10.1016/j.biortech.2022.128324 PMID: 36400276

Sen, K. Y., & Baidurah, S. (2021). Renewable biomass feedstocks for production of sustainable biodegradable polymer. *Current Opinion in Green and Sustainable Chemistry*, 27, 100412. DOI: 10.1016/j.cogsc.2020.100412

Sen, K. Y., Hussin, M. H., & Baidurah, S. (2019). Biosynthesis of poly(3-hydroxybutyrate) (PHB) by *Cupriavidus necator* from various pretreated molasses as carbon source. *Biocatalysis and Agricultural Biotechnology*, 17, 51–59. DOI: 10.1016/j.bcab.2018.11.006

Singh, T. A., Sharma, M., Sharma, M., Sharma, G. D., Passari, A. K., & Bhasin, S. (2022). Valorization of agro-industrial residues for production of commercial biorefinery products. *Fuel*, 322, 124284. DOI: 10.1016/j.fuel.2022.124284

Subbarao, P. M., D'Silva, T. C., Adlak, K., Kumar, S., Chandra, R., & Vijay, V. K. (2023). Anaerobic digestion as a sustainable technology for efficiently utilizing biomass in the context of carbon neutrality and circular economy. *Environmental Research*, 234, 116286. DOI: 10.1016/j.envres.2023.116286 PMID: 37263473

Swetha, T. A., Ananthi, V., Bora, A., Sengottuvelan, N., Ponnuchamy, K., Muthusamy, G., & Arun, A. (2023). A review on biodegradable polylactic acid (PLA) production from fermentative food waste - Its applications and degradation. *International Journal of Biological Macromolecules*, 234, 123703. DOI: 10.1016/j.ijbiomac.2023.123703 PMID: 36801291

Teixeira, S., Eblagon, K. M., Miranda, F. R., Pereira, M. F., & Figueiredo, J. L. (2021). Towards Controlled Degradation of Poly(lactic) Acid in Technical Applications. *C*, 7(2), 42. DOI: 10.3390/c7020042

Virág, Á. D., Tóth, C., & Molnár, K. (2023). Photodegradation of polylactic acid: Characterisation of glassy and melt behaviour as a function of molecular weight. *International Journal of Biological Macromolecules*, 126336, 126336. Advance online publication. DOI: 10.1016/j.ijbiomac.2023.126336 PMID: 37586636

Widjaja, T., Alifatul, A., Nurchamidah, S., Altway, A., Yusuf, B., Fakhrizal, F., & Pahlevi, A. (2023). Poly Lactic Acid Production Using the Ring Opening Polymerization (ROP) Method Using Lewis Acid Surfactant Combined Iron (Fe) Catalyst (Fe(DS)$_3$). *Heliyon*, 9(8), e17985. DOI: 10.1016/j.heliyon.2023.e17985 PMID: 37520956

Yaradoddi, J. S., Banapurmath, N. R., Ganachari, S. V., Soudagar, M. E. M., Sajjan, A. M., Kamat, S., Mujtaba, M. A., Shettar, A. S., Anqi, A. E., Safaei, M. R., & Ali, M. A. (2022). Bio-based material from fruit waste of orange peel for industrial applications. *Journal of Materials Research and Technology*, 17, 3186–3197. DOI: 10.1016/j.jmrt.2021.09.016

Yu, J., Xu, S., Liu, B., Wang, H., Qiao, F., Ren, X., & Wei, Q. (2023). PLA bioplastic production: From monomer to the polymer. *European Polymer Journal*, 193, 112076. DOI: 10.1016/j.eurpolymj.2023.112076

Zare, Y., & Rhee, K. Y. (2019). Following the morphological and thermal properties of PLA/PEO blends containing carbon nanotubes (CNTs) during hydrolytic degradation. *Composites. Part B, Engineering*, 175, 107132. DOI: 10.1016/j.compositesb.2019.107132

Zhang, R., Du, F., Jariyavidyanont, K., Zhuravlev, E., Schick, C., & Androsch, R. (2022). Glass transition temperature of poly(d,l-lactic acid) of different molar mass. *Thermochimica Acta*, 718, 179387. DOI: 10.1016/j.tca.2022.179387

Chapter 11
Pectin Materials Sourced From Agriculture Waste:
Extraction, Purification, Properties, and Applications

Tapanee Chuenkaek
https://orcid.org/0009-0004-4307-0516
Nagaoka University of Technology, Japan

Tu Minh Tran Vo
https://orcid.org/0000-0002-6785-4188
Chulalongkorn University, Thailand

Keita Nakajima
Nagaoka University of Technology, Japan

Takaomi Kobayashi
Nagaoka University of Technology, Japan

ABSTRACT

For abundant citrus fruits, the perspective involves effectively utilizing waste materials such as citrus peels and apple pomace. The pursuit of sustainable pectin sourcing has also driven innovation in extraction techniques. Conventional pectin extraction methods are gradually giving way to more environmentally friendly and green extraction methods. The appeal of pectin lies in its diverse range of applications across various industries. Apart from its traditional uses in the food industry as a gelling agent and stabilizer, pectin has ventured into new frontiers in fields such as packaging, moisturizers, drug delivery, wound healing, and tissue engineering. This chapter provides a comprehensive review of the chemical structure of pectin, its conventional sources, and pectin extraction, with a particular focus on its physical, chemical, and functional properties within the context of various industrial applications.

DOI: 10.4018/979-8-3693-0003-9.ch011

Copyright ©2025, IGI Global. Copying or distributing in print or electronic forms without written permission of IGI Global is prohibited.

INTRODUCTION

Pectin, a fundamental polysaccharide primarily sourced from plants, plays a central role in the structural composition of plant cell walls (Willats et al., 2006). It is found in various parts of a plant including the middle lamella, primary cell walls, and secondary walls, and is deposited during the early stages of cell growth (Figure 1) (Crombie et al., 2003). Pectin serves multiple essential functions within a plant. It is a key contributor to the formation of plant cell walls, providing strength and support to these dynamic structures (Crombie et al., 2003; Fry, 1988). Generally, the primary cell walls of dicotyledonous plants consist of approximately 35% pectin, 30% cellulose, 30% hemicellulose, and 5% protein (Fry, 1988). Beyond its role in plant biology, significance of pectin extends to impacting the physical characteristics and nutritional properties of various plant-derived products. In the realm of biopolymers, pectin has gained global recognition due to its unique functional attributes, which have extensive applications in diverse sectors, including food, pharmaceuticals, and biomedicine (Chandel et al.). Pectin is distinguished by its intricate molecular structure, characterized by its capacity to form flexible polymer chains that ultimately result in the creation of hydrogel-like structures (Freitas et al., 2021). The expression of these functional properties is closely linked to the conditions under which pectin is extracted, a factor significantly influenced by the choice of botanical source. This comprehensive overview delves into the multifaceted world of pectin, encompassing its botanical origins, structural attributes, and the myriad applications that underscore its importance in various industries.

Figure 1. Pectin distribution in plant cell walls

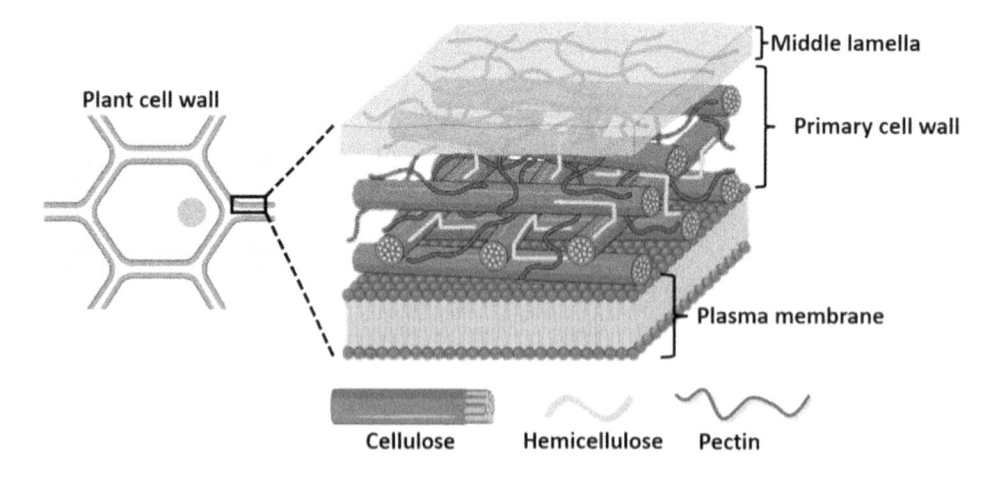

Figure 2. Flow illustration of pectin extraction process from raw materials through conventional chemical extraction and non-conventional extraction methods

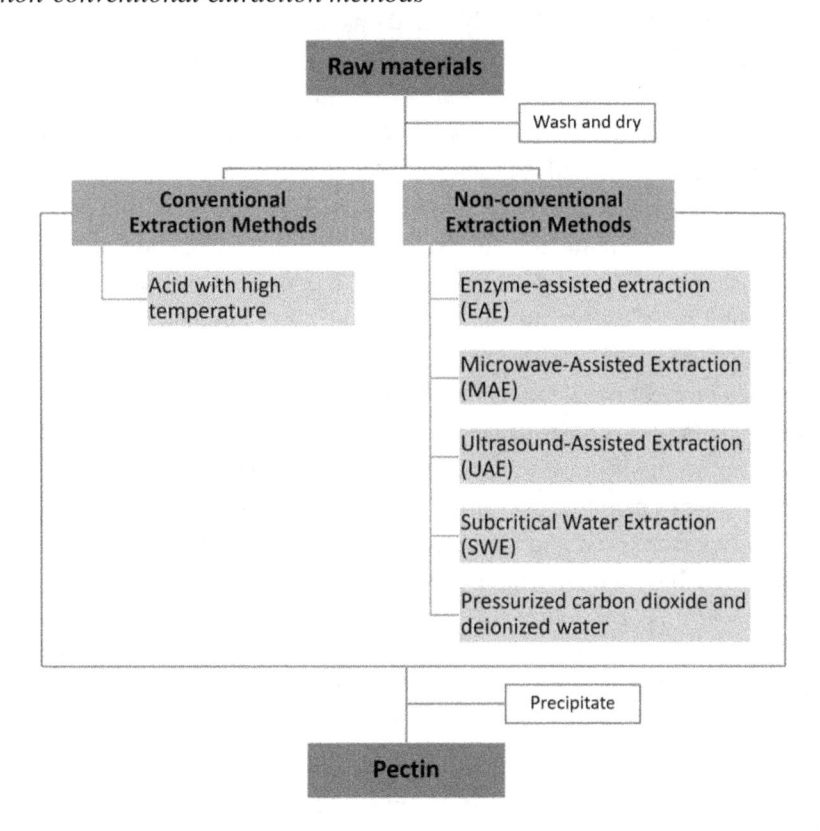

Figure 3. Chemical structure of High Methoxyl (HM) pectin and Low Methoxyl (LM) pectin

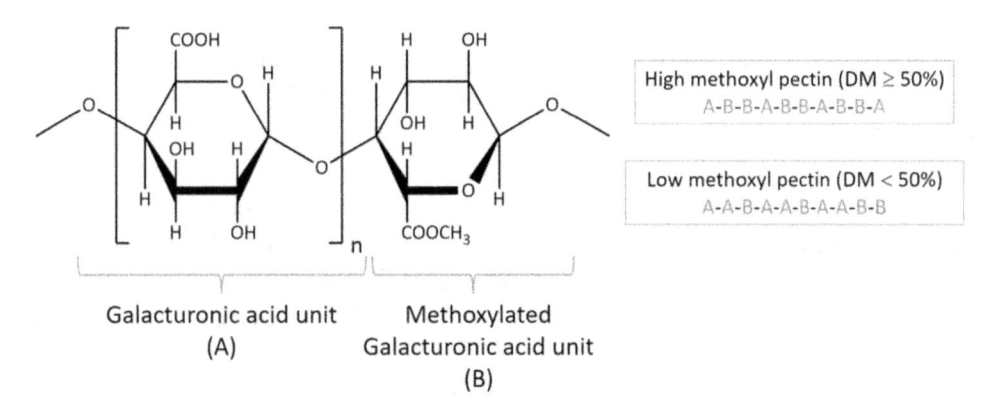

High methoxyl pectin (DM ≥ 50%)
A-B-B-A-B-B-A-B-B-A

Low methoxyl pectin (DM < 50%)
A-A-B-A-A-B-A-A-B-B

Galacturonic acid unit (A)

Methoxylated Galacturonic acid unit (B)

AGRICULTURE WASTE FOR SUSTAINABLE PECTIN SOURCE

Traditionally, the pectin industry has heavily relied on citrus fruits and apples as the primary sources of this valuable polysaccharide. However, the landscape of pectin sourcing is undergoing a remarkable transformation as researchers and industries alike are increasingly turning their attention to unconventional and sustainable sources. These alternative sources encompass a rich array of agro-industrial byproducts, including apple pomace (Luo et al., 2019), banana peel (Khamsucharit et al., 2018), cacao pod husk (Vriesmann et al., 2012), carrot pomace (F. Jafari et al., 2017), passion fruit peels (2014), pistachio green hull (Kazemi et al., 2019), pomegranate peel (Pereira et al., 2016), and dragon fruit (Woo et al., 2010), and various vegetables (Herrera-Rodríguez et al., 2022). This shift towards diverse sources not only offers environmental benefits by utilizing agricultural waste but also diversifies the pectin supply chain, reducing dependence on a few select fruits. Furthermore, the utility of pectin has expanded through chemical or enzymatic modifications, leading to alterations in its physicochemical characteristics. These modifications, which encompass adjustments in molecular weight, degrees of esterification (DE), and surface charges, have paved the way for enhanced functional properties and the development of novel applications (Muñoz-Almagro et al., 2017). This newfound versatility extends pectin's influence across diverse industries, ranging from food and pharmaceuticals to biotechnology, reaffirming its status as a versatile and indispensable biopolymer with a sustainable future.

EXTRACTION AND PURIFICATION OF PECTIN

Pectin, a high molecular weight polysaccharide, is a crucial structural component found in various plants, maintaining the integrity of cell structures (Flutto, 2003; Shin et al., 2021). Its diverse applications span the food industry, where it enhances viscosity in products like beverages, jams, and jellies (Chandel et al., 2022). Given the rising demand for pectin in the food, pharmaceutical, and therapeutic domains, efficient extraction processes are essential to obtain high-quality pectin.

Historically, pectin production emerged in the early 1900s when German apple juice producers began extracting pectin from dried apple pomace, a by-product of juice processing, to be used as a gelling agent (Ciriminna et al., 2015). While apple pomace and citrus peel have traditionally been the primary sources for commercial pectin production, the growing interest in utilizing by-products has led to exploration of other sources with diverse functional properties (Christiaens et al., 2015; Müller-Maatsch et al., 2016). In this day and age of sustainability, agro-industrial waste materials have also become a focus for pectin production. Pectin from various sources is extracted by a few commonly used techniques (Fig. 2). Firstly, the process involves treating alcohol insoluble material obtained from dried agro-waste with water or alcohol to remove impurities, then followed by the extraction method. After filtering the solid from the extract, the solubilized pectin should be precipitated by mixing the solubilized pectin with ethanol. The extracted pectin is then dried and further processed.

Different plant sources have been used for pectin extraction, including citrus peel (Pasandide et al., 2017), apple (Wang et al., 2014), carrot (Faeghe Jafari et al., 2017), sunflower heads (Ma et al., 2021), jackfruit peel (Xu et al., 2018), and papaya peel (Immanuel, 2018). The extraction methods vary depending on the specific plant material. Commercially, citrus peels (85%), apple pomace (14%), and sugar beet pulp (<1%) are the primary sources of pectin (Wang et al., 2018). These materials undergo thermal hydrolysis using hot mineral acids to extract pectin. However, conventional extraction processes generated acidic industrial waste and consumed significant amounts of energy. To address these concerns, innovative and non-conventional extraction techniques have been developed, aligning with the principles of "Green Chemistry." The interest in developing sustainable and efficient extraction technologies for pectin production has led to the emergence of novel techniques. These newer methods aim to achieve a cleaner process, reducing waste and environmental impact. Advances in extraction technologies have contributed to improving the overall pectin production process. Overall, the extraction of pectin from various plant-based materials involves multiple stages, including raw material pretreatment, extraction operations, and post-extraction procedures.

Conventional Extraction Methods

Conventional pectin extraction typically involves water-based methods with the addition of chemical additives. The most commonly used method is direct boiling, although it requires several hours to achieve a satisfactory pectin yield (Li et al., 2012). The extraction of pectin involves the physical-chemical hydrolysis and extraction of pectin macromolecules from plant tissues (Cui et al., 2021). Various factors, including temperature, pH, and extraction time influenced the processes. Initially, protopectin was solubilized, followed by a secondary hydrolysis reaction where the pectin was degraded. Prolonged extraction times led to reduced yield as the extractable pectin, and hydrolysis was dominant in the yield (Matsumoto et al., 2000; Turquois et al., 1999). Several parameters impacted the quantity and quality of extracted pectin, including temperature, pH, extraction duration, acid type, extraction cycle, water-to-raw material ratio, and organic solvent volumes (Chuenkaek et al., 2023; Sayed et al., 2022). Typically, the highest pectin yields were obtained using hot acid extraction methods (pH 1.5–2.5) at temperatures around 70–90 °C and acid extraction also results in pectin enriched with galacturonic acid residues (Wicker & Kim, 2016). However, this conventional acid extraction method has its drawbacks, because of leading to pectin degradation and raise concerns about hazardous chemical disposal, and driving the exploration of alternative non-conventional extraction methods. One of the commonly employed methods for extracting pectin from plant tissues is by heating the plant sample in acidified water. The addition of chelating agents, such as Ethylenediaminetetraacetic acid (EDTA) and Cyclohexanediaminetetraacetic acid (CDTA), to the extraction mixture aids in the easy release of pectin from the cell wall (Srivastava & Malviya, 2011). It is important to avoid prolonged direct heating, as it may lead to the thermal degradation of the polymer. The extraction process typically involves refluxing the plant sample in acidified water at a high temperature. The resulting hot acid extract was then filtered to remove pulp, cooled, and precipitated using double the volume of ethanol (Canteri-Schemin et al., 2005). The pectin was separated by floating it on the surface and removing it using a cheese cloth. The extracted pectin is then dried and milled to obtain powdered pectin.

Chemical extraction using acids is the most commonly employed method in pectin production and significantly impacts the content, composition, and physicochemical properties of pectin, including monomeric components such as xylose, arabinose, and galacturonic acid. Table 1 shows some examples of pectins obtained from conventional extraction methods.

Table 1. Examples of pectins obtained from conventional extraction method

Sources	Acid condition	Extraction temperature (°C)	Extraction time (min)	Yield (%)	DM (%)	References
Pomelo (*Citrus Maxima*)	Nitric acid (pH 2)	80	120	24.26 ± 0.08	59.4 ± 0.9	Methacanon et al., 2014
Ponkan (*Citrus reticulata* Blanco cv. *Ponkan*)	Nitric acid (pH 1.6)	100	100	25.8 ± 0.57	85.7	Colodel et al., 2018
Orange (*Citrus sinnensis*)	Phosphoric acid (0.6 M)	95	120	29.37 ± 4.86	83.66 ± 2.77	Tovar et al., 2019
Orange (*Citrus sinnensis*)	Hydrochloric acid (pH 1.6)	90	117	22.0 ± 0.02	Lower than 50	Tsouko et al., 2020
Orange peel waste	Citric acid (pH 2)	-	90	15.85 ± 0.77	66.79 ± 4.67	Ortiz-Sanchez et al., 2020
Pomelo (*Citrus Maxima*)	Acetic acid (pH 3)	90	90	8.2 ± 0.4	62.7 ± 2.5	Van Hung et al., 2021

Non-Conventional Extraction Methods

In recent years, there has been a growing emphasis on the development of novel and more efficient techniques for the cleaner extraction of pectin. This shift is driven by the need to address environmental concerns while simultaneously improving extraction yields. Consequently, non-conventional methods, including enzyme-assisted extraction, ultrasound-assisted extraction subcritical water extraction, and microwave-assisted extraction, have garnered significant attention. These non-conventional techniques offer numerous advantages over traditional acid-based extractions. Firstly, they contribute to a reduced environmental impact on sustainable practices within the industry. Secondly, these methods are characterized by decreased energy consumption, promoting resource efficiency. Additionally, they often result in improved extraction yields, making them economically appealing. Collectively, these non-conventional techniques signify a shift towards more sustainable and eco-friendly practices in the pectin extraction industry, aligning with the broader global trend towards greener and cleaner technologies.

Table 2. Examples of pectins obtained from enzyme-assisted extraction method

Sources	Enzyme	Extraction temperature (°C)	Extraction time (min)	Yield (%)	DM (%)	References
Lime peel	Laminex C2K (derived from *Penicillium funiculosum*) (pH 3.5)	50	240	23	82.2	Dominiak et al., 2014
Kiwifruit puree (including the skin and seed)	Celluclast	25	30	4.39	85	Yuliarti et al., 2015
Yellow passion fruit (*Passiflora edulis* f. *flavicarpa*)	Protopectinase-SE (pH 3.0)	37	45	26	67.5	Vasco-Correa & Zapata Zapata, 2017

Enzyme-Assisted Extraction: This technique employed enzyme to isolate pectin to break down the cell wall matrix and liberate pectin from plant-based materials, thereby improving the permeability of the wall and enhancing mass transfer during the extraction process. Table 2 shows some examples of pectins obtained from enzyme-assisted extraction method. The utilization of various enzymes, including cellulase, hemicellulase, xylanase, pectinase, α-amylase, β-glucanase, and endo-polygalacturonase followed specific roles in the enzymatic breakdown of cell wall components. Alternatively, crude enzymes derived from fungal cultures have been explored as viable options for this extraction process. Compared to conventional acid extraction methods, enzyme-assisted extraction generally requires lower temperatures and less solvent usage, making it more environmentally friendly. However, on this case, there are challenges associated with enzyme-assisted extraction, such as the cost of enzymes and scaling up to an industrial level. The quality and quantity of pectin obtained through enzyme-assisted extraction can vary significantly, depending on factors like enzyme composition, raw materials, and extraction conditions. Optimal enzyme selection and extraction parameter optimization are crucial to achieving desirable pectin characteristics, including the degree of esterification and methoxylation. Beyond its efficacy, enzyme-assisted extraction offers environmental advantages, requiring fewer chemical reagents and operating at lower temperatures (Marić et al., 2018). Nevertheless, it may be more expensive and time-consuming than conventional methods.

Table 3. Examples of pectins obtained from microwave-assisted extraction method

Sources	Acid condition	Power (W)	Extraction time (min)	Yield (%)	DM (%)	References
Passion fruit (*Passiflora edulis f. flavicarpa*) peel	Tartaric acid (pH 2)	628	9	18.2	50.00 ±1.48	Seixas et al., 2014
Dragon fruit (*Hylocereus* spp) peel	Nitric acid (pH 2)	450	5	6.13	57.50 ± 1.33	Tongkham et al., 2017
Orange (*Citrus reticulata*)	Sulphuric acid (pH 1.4)	422	2.8	19.24	-	Prakash Maran et al., 2013
Lemon	Hydrochloric acid	360	1	9.71	51.08 ±0.08	Karbuz & Tugrul, 2021

Microwave-Assisted Extraction: As summarized in Table 3, microwave-assisted extraction is an innovative technique that harnesses microwave radiation to heat the extraction solvent, expediting the extraction of pectin from plant materials. When subjected to microwave energy, polar molecules in both the sample matrix and solvent undergo rapid, disorganized rotation, generating heat. This localized heating facilitates the extraction of pectin by improving the mass transfer from the plant material to the solvent. Compared to conventional extraction methods, microwave heating offers several notable advantages. It significantly enhances extraction efficiency, reduces extraction time, and conserves energy. By selectively heating the solvent, microwave markedly could reduce the time required for pectin extraction and minimize the volume of solvent used, rendering it an environmentally friendly method (Koh et al., 2014). Numerous research studies delved into pectin extraction using microwave method, examining various source materials and experimenting with different extraction parameters including power, temperature, extraction time, and solid-liquid ratio. However, to exercise caution when used microwave heating was essential in prolonged extraction under harsh conditions potentially leading to the degradation of pectin chains (Spinei & Oroian). Therefore, careful control of extraction parameters is imperative to avoid excessive degradation.

Table 4. Examples of pectins obtained from ultrasound-assisted extraction method

Sources	Acid condition	Power (W)	Sonication time (min)	Yield (%)	DM (%)	References
Sour orange (*Citrus aurantium L.*) peels	Citric acid (pH 1.5)	150	10	28.07 ± 0.67	6.77 ± 0.43	Hosseini et al., 2019
Turkish grapefruit	Hydrochloric acid (pH 1.5) Temp. 70°C	450	25	17.92	75.12	Bagherian et al., 2011; Tongkham et al., 2017
Grapefruit (*Citrus paradisi Macf. cv. Changshanhuyou*) peel	Hydrochloric acid (pH 1.5) Temp. 67°C	12.56 W/cm2	28	27.34	58.78 ± 3.54	Wang et al., 2015
Lemon	Hydrochloric acid Temp. 75°C	-	45	10.11	51.13 ± 0.19	Karbuz & Tugrul, 2021
Kiwi	Hydrochloric acid Temp. 75°C	-	45	17.30	50.75 ± 0.34	Karbuz & Tugrul, 2021

Ultrasound-Assisted Extraction: Ultrasound-assisted extraction is a technique to utilize high-frequency ultrasound waves to enhance the extraction of pectin. The mechanical effects of ultrasound, including cavitation and microstreaming, are key factors in disrupting plant cell walls, thereby facilitating the release of pectin molecules. In improving extraction efficiency and reducing extracting time, when ultrasonic waves passed through a liquid medium, such as acidified water (pH 1.5), the created bubbles or cavities collapsed in a process called cavitation. This phenomenon generates high temperatures and strong stress pressures, enhancing the solubility of pectin in the solvent and increasing the rate of mass transfer. To achieve efficient cavitation, ultrasound probe systems are often used instead of conventional ultrasonic baths. Recent studies explored ultrasound application for pectin extraction from various sources (Panwar et al., 2023). Optimizing experimental conditions, such as power/temperature, extraction time, and solid-liquid ratio, crucial in ultrasound condition as listed in Table 4. Longer extraction times under more intense conditions became increase of pectin yield, but potentially reducing polymer chain length was appeared. The balancing extraction efficiency with the desired characteristics of the extracted pectin is essential. **Subcritical Water Extraction:** Subcritical water extraction, known as pressurized hot water extraction or superheated water extraction, is an innovative technique by using autoclave for efficiently extracting pectin from plant materials. This method utilizes water at temperatures above its boiling point but below its critical point, along with elevated pressures, eliminating the need for additional solvents or chemicals during the extraction process. In conventional extraction methods, organic solvents are often used at elevated temperatures and pressures to improve solvent penetration, solubility, and mass transfer while reducing solvent viscosity and surface tension. In contrast subcritical water extraction achieves similar effects by using water at temperatures above its boiling point while maintaining high pressure. The specific extraction conditions, including pressure, temperature, extraction duration, and solid-liquid ratio, significantly influence both the yield and quality of the extracted pectin, necessitating careful optimization (Yilmaz-Turan et al., 2023). In the extraction process, crude pectin was extracted typically by further processing steps for purification including filtration, centrifugation, or precipitation using solvents like methanol, ethanol, or isopropanol. Alternatively, precipitation with insoluble salts such as aluminum was utilized. The resulting precipitate was washed with alcohol to remove soluble contaminants, and the precipitated pectin and then subjected to pressing, drying, and milling to obtain powdered pectin (Table 5). In summary, subcritical water extraction offers an environmentally friendly alternative for extracting pectin from plant materials. By leveraging the unique properties of subcritical water, this technique enhances pectin extraction efficiency while reducing the reliance on additional solvents or chemicals. Combinations of different extraction methods can further optimize the extraction process, and subsequent purification steps are employed to obtain purified pectin.

Table 5. Examples of pectins obtained from subcritical water extraction

Sources	Operating conditions	Extraction temperature (°C)	Extraction time (min)	Yield (%)	DM (%)	References
Citrus peel	Autoclave, 1:30 solid-liquid ratio	120	5	21.95± 0.05	74.74 ± 1.57	Wang et al., 2014
Apple pomace		150	5	16.68± 0.20	85.99 ± 1.37	
Jackfruit peel	High-pressure reactor, 17.03 mL/g liquid-solid ratio	138	9.15	14.96	61.08±0.48	Li et al., 2019
Fresh sunflower heads	8 bar pressure, 7 mL/g liquid-solid ratio	120	20	6.57 ± 0.6	19.4 ± 0.95	Ma et al., 2020

Pressurized Carbon Dioxide and Deionized Water Method: In recent research, a novel method utilizing pressurized carbon dioxide, deionized water, and supercritical method, has emerged as an alternative approach for pectin extraction. This method capitalizes on the unique properties of pressurized carbon dioxide, which can simultaneously acidify water and generate carbonate ions (CO_3^{2-}). These carbonate ions act as natural chelating agents for calcium ions (Ca^{2+}), facilitating the extraction process. A comparative study was conducted to evaluate the efficacy of this pressurized carbon dioxide method in comparison to traditional methods involving hydrochloric acid (HCl) and a chelating agent, specifically sodium hexametaphosphate (Tsuru et al., 2021). The study revealed that while the pressurized carbon dioxide method exhibited a lower extraction efficiency for pectin from orange peel compared to the HCl and chelating agent methods, it yielded pectin with distinct advantages. Notably, pectin obtained through the pressurized carbon dioxide method exhibited higher degree of DE and a larger molecular weight when compared to the HCl method conducted under the same conditions. It is important to acknowledge that the pressurized carbon dioxide and deionized water method is still in its early stages of development. However, supercritical method holds significant promise due to its environmentally friendly nature. This method generates lower emissions in comparison to traditional extraction methods that rely on harsh chemicals. Further research and optimization of this method are imperative to fully explore its capabilities and refine the extraction process to achieve optimal pectin yield and quality.

After extraction pf pectins, the degree of esterification (DE), often refers to as the degree of methylation (DM), is a fundamental parameter for characterizing pectin and holds significant relevance in understanding its gel-forming properties. Several analytical techniques, including Fourier-transform infrared spectroscopy (FTIR), high-performance liquid chromatography (HPLC), and titrimetric, were employed to determine the DE of pectin. DE quantifies the ratio of methyl ester groups to galacturonic acid units within the pectin molecule, thereby categorizing pectin into two primary classifications based on its DE (Fig.3).

PECTIN STRUCTURES AND THEIR PROPERTIES

Pectin Chemical Structure

Pectin, a complex, negatively charged branched heteropolysaccharide, presents an intricate structural composition with up to 17 distinct monosaccharides interconnected through over 20 types of linkages (Muñoz-Almagro et al., 2021; Zhang et al., 2018). As firstly documented by Braconnot in 1825, pectin is rich in galacturonic acid (GalA) (Fig. 4), constituting up to 65% of its composition of homogalacturonan (b) (Willats et al., 2006). Pectin consists of partially methyl esterification of the carboxyl group of GalA (c) and that with dissociation of the unmethyl esterified moiety (d) are mixed in the homogalacturonan segments.

Understanding the structural intricacies and conformational flexibility of pectin chains has been a subject of extensive research, primarily employing hydrodynamic, viscometry, and modeling simulation methods to provide comprehensive insights into these molecules. The major structural regions of pectin include homogalacturonan (HG), rhamnogalacturonan I (RG-I), rhamnogalacturonan II (RG-II), xylogalacturonan (XGA) and apiogalacturonan. Here, RG-I has repeating α-1,4/1,2-bond structure of GalA and rhamnose, RG-II is a mixture part of GalA, rhamnose (e) as well as apiose, methoxylated glucuronic acid, fucose, etc. Also, the additional residues of the sugars D-xylose (f) or D-apiose (g), xylogalacturonan and apiogalacturonan moieties branch off from the D-GalA residue backbone, are also present, respectively. These are recognized as the primary structural domains, depending on the plant species of origin. Typically, HG comprises approximately 65% of pectin, RG-I constitutes 20%-35%, and RG-II makes up less than 10% (Gotoh et al., 2021). Among them, HG is the predominant polysaccharide in plant cell walls and consists of a linear structure with D-galacturonic acid units linked in an α-$(1\rightarrow 4)$ configuration. This structure plays a pivotal role in regulating plant cell wall strength and is intimately involved in fruit softening during maturation.

The biosynthesis of HG begins within plant cells, where it is initially synthesized in a highly methylesterified state within the Golgi apparatus before being secreted into the cell wall. Subsequently, enzymes from the pectin methylesterase family catalyze partial de-esterification (Wolf et al., 2009). This process involves the partial methylesterification of carboxy groups at the C-6 position of GalA residues and the partial acetylation of hydroxy groups at positions O-2 and O-3 (Willats et al., 2006).

The degree of methoxylation (DM) and degree of acetylation (DAc) express the substitution rates of methoxy and acetyl groups on the acidic form of GalA residues in pectin, which can vary significantly depending on the plant's origin and profoundly influence pectin's properties. In contrast, RG-I and RG-II, domains of pectin have branching into the pectin structure, and partial methylation and features four heteropolymer side chains. The former structure forms a ramified configuration through a repeating disaccharide unit alternating residue of GalA (galacturonic acid) and l-rhamnose (Rha). Additionally, RG-I possesses side chains attached to the C-4 position or occasionally the C-3 position of the Rha residue. The structural branching in RG-I depends on variables like the pectin source, extraction conditions, and the presence of other sugars such as xylose, fucose, and glucuronic acid, among others (Atmodjo et al., 2013). The latter structure of RG-II was characterized by approximately α-$(1\rightarrow 4)$-linked GalpA units and undergoes partial methylation and features four heteropolymer side chains (Ndeh et al., 2017; O'Neill et al., 2004). While the diversity of pectin structure results in its source and extraction conditions, the RG-II region remains relatively consistent across sources (Dranca & Oroian, 2018).

Apart from these, additionally, pectin contains functional groups beyond the carbohydrate matrix, including phenolic acids, methanol, acetic acid, and select amide groups. Methanol and acetic acid play significant roles in the methylation of galacturonic acid residues, influencing the structural characteristics of pectin. The extent of methylation (DM) serves as a valuable descriptor of pectin's structure and is distinguished by two, HM and LM, of high levels of methoxy groups having over 50% of carboxyl groups in methylated form, and lower content in low methoxy pectin, respectively. Native pectin primarily falls under the high methoxy category. In contrast, acetylation in pectin occurs sporadically in native instances, with the degree of acetylation (DA) representing the percentage of galacturonosyl residues open to acetylation per unit of monosaccharide. DA can surpass 100% and typically occurs within the branched RG regions. Different pectin sources contain varying acetylation levels; for instance, citrus and apple pectin tends to have limited acetyl groups in the HG region, unlike pectin from sugar beet and potato, which exhibit higher acetylation levels (Alpizar-Reyes et al., 2022; Ngenefeme et al., 2013). Although not naturally occurring in pectin, amide bond can be induced chemically or enzymatically to enhance functional attributes such as water solubility, gelling, and rheological properties. This transformation involves converting certain non-esterified carboxyl groups into amide groups, facilitated by diverse amino compounds (Chandel et al., 2022; Chen et al., 2020; Zheng et al., 2023).

Figure 4. Chemical structures of (a) galacturonic acid, (b) GalA-linked homogalacturonan and (c) its methylesterified state and (d) negatively charged form in homogalacturonan, (e) Rhamnose, (f) D-Xylose, and (g) D-Apiose

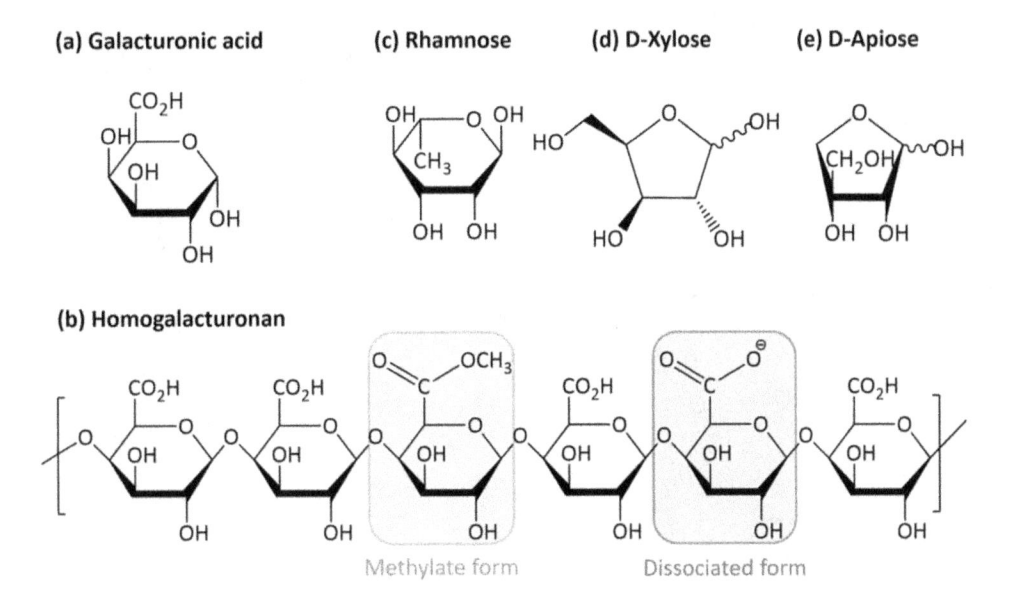

The degree of methyl esterification (DM) can generally be estimated by the titration of the COOH group in pectin, reflecting on the infrared (IR) spectrum of the sample (Chuenkaek et al., 2023). The IR spectroscopy is a fast and convenient method for the investigation of functional groups of polysaccharides. As shown in Figure 5 for Fourier Transform Infrared Spectroscopy (FT-IR), the commercial pectin from citrus fruit contains IR band; strong broad band at about 3500 cm^{-1} for O–H stretching vibration, which was mainly originated from cellulose and hemicellulose (Ma & Mu, 2016), the C–H stretching vibrations of polysaccharides at about 2950 cm^{-1}, a notable methyl-esterified C=O groups at 1740 cm^{-1}, C-H bending vibrations within carbohydrate segments around 1200-1450 cm^{-1} range and the carboxyl groups (COO-) within the galacturonic acid group of pectin at 1600 cm^{-1}. The previous research study also compares different DM% of pectins purified from *Citrus maxima* wastes (Chuenkaek et al., 2023), when purified in different acidic conditions. Changing DM, the peak intensities at 1740 cm^{-1} and 1600 cm^{-1} are varied, tending decrease the former band intensity in lower DM and increasing the OH band at 2950 cm^{-1}.

Figure 5. FTIR of commercial pectin and HM and LM pectins extracted from citrus. The pectin of HM and LM was extracted from waste peel of citrus maxima (Chuenkaek et al., 2023)

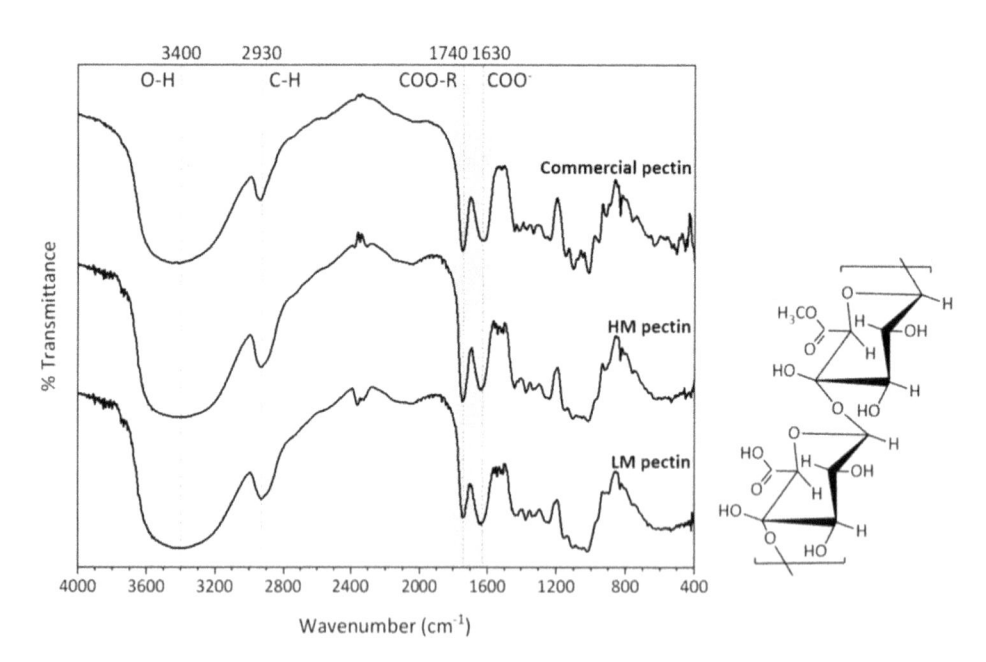

Figure 6. Structure formation mechanisms in pectin gelation. (a) hydrogen bonds between undissociated carboxyl groups, (b) hydrophobic interactions, (c) random ionic interactions (crosslinks) between dissociated carboxyl groups. ca-bridges at subsequent free dissociated carboxyl groups can form egg-box junction zones

Figure 7. Fabrication of ionic pectinate hydrogels

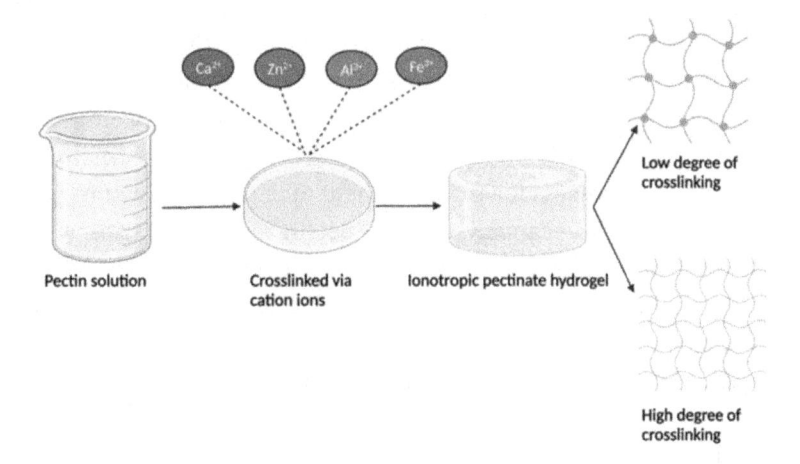

Pectin Solubility Properties

Pectin, having GalA (galacturonic acid) segments, exhibits solubility in pure water, but it can vary widely as based on several key properties like its molecular weight, and methylation degree in surrounding pH conditions.

As molecular weight of pectin increases, its solubility undergoes a notable transformation. This is because solubility depends on the degree of ionization of GalA. Higher molecular weight pectin, characterized by larger and more intricate molecular structures, exhibit a distinct propensity towards decreased solubility due to the heightened entanglement of pectin molecules. As a result, pectin molecules become less amenable to rapid dissolution in aqueous solutions. In other words, the increase in pectin concentration hinders the solubility of high molecular weight pectin, due to intricate molecular networks (Sara E. Herrera-Rodríguez et al., 2022). Mo Zhou et al. investigated the effect of Mw ($8.06 \pm 0.16 \times 10^4$ - $3.96 \pm 0.08 \times 10^5$ [g/mol]) of the extracted pectin from apple, peach, and carrot at different concentrations on their solubility. It was observed that their apparent viscosities remained relatively consistent as shear rates increased, suggesting behavior of non-pseudoplastic fluids, which behave less significantly decrease in viscosity under shear stress. Their entangled and networked molecular structures, attributed to the shear stress because of requirement of additional energy to disrupt intermolecular interactions. In other words, these substances maintain their viscosity and resist changes in flow rate due to the strength of their molecular connections. Thus, higher molecular weight of pectin presented greater solubility time (Zhou et al., 2021)

On the other hand, the solubility is highly dependent on pH. At higher pH levels, the increment of dissociation of COOH group in GalA segment exhibits negatively charged pectin (Fig. 4 (d)), particularly in lower methoxy pectin. The heightened presence of these negatively charged groups engenders electrostatic repulsion forces among neighboring pectin molecules and effectively mitigates intermolecular entanglement, thereby paving the way for enhanced solubility. In essence, the electrostatic repulsion acts as a molecular force, allowing pectin segments extended more readily throughout the solution, introducing viscus state with higher viscosity. However, particularly in both acidic and alkaline sides, undesired transformations of pectin referred to desertification and depolymerization, occurring fragmentation of pectin chains and a subsequent decrease in molecular weight (Gawkowska et al., 2018).

The degree of methylation stands out as a critical parameter in the intricate realm of pectin solubility. The high-methylation pectin containing high DM methoxy groups exhibits a unique behavior in its solubility (Gawkowska et al., 2018). Since the pectin has fewer carboxylate anions available at any given pH, reduced carboxylate anions decrease solubility and brings about a structural transformation. In contrast, low-methylation pectin possesses a higher content of carboxylate anions within their structure, promoting greater solubility. Such negatively charged pectin readily interacts with water molecules, facilitating their dissolution in aqueous solutions (Sun et al., 2023).

Temperature exerts a multifaceted influence on the solubility of pectin as mentioned before. The relationship between temperature and pectin solubility involves intricate dynamics that require careful examination (Ling et al., 2023), but temperature-induced degradation of pectin is a crucial factor, since elevated temperatures enhances the degradation of pectin structural integrity, especially at higher temperature by hydrolysis. Intriguingly, the influence of temperature on pectin solubility goes beyond mere dissolution (Ling et al., 2023).

The presence of counterions in a solution plays a pivotal role in shaping the solubility of pectin, and this influence is a subject of considerable significance. Recent studies shed light on the profound impact of counterions, particularly monovalent salts such as Na^+, on the solubility of pectin (Gawkowska et al., 2018; Gawkowska et al., 2019). The solubility of pectin was enhanced in present of monovalent salt like sodium ions (Na^+). Such ion tended to form mono-complexes with the pectin chains. The complexes served to hinder electrostatic repulsion forces between pectin molecules, effectively altering their interactions. Meanwhile, divalent salt ions (Ca^{2+}) promoted gelatinization, since divalent calcium ions bridge dissociated carboxy ions and form a gel-like structure (Zheng et al., 2020; Tran Vo et al., 2022).

Rheological Properties

Pectin finds diverse applications as a stabilizer, thickener, emulsifier, gelling agent, and glazing agent in various industries. This is because pectin can take a variety of forms, depending on conditions, from liquids to gels with both solid-liquid properties and solids. For example, when added it to a substance, pectin alters its flow characteristics, making it a vital rheology modifier, but with Ca^{2+}, gelation state is taken and then when water is evaporated, pectin takes on a solid state. Rheology delves into how substances deform and flow under different conditions of time, strain, and stress interactions. Several factors influence the rheological properties of pectin gels, including sucrose concentration, pH, temperature, and the concentration of divalent ions (Yuliarti et al., 2017). In a comprehensive viscoelastic and viscosity analyses, the behavior provides valuable insights into pectin states. As widely recognized, nearly all polymer solutions exhibit shear-thinning behavior, where apparent viscosity decreases as shear rate increases due to polymer chain disentanglement. However, the rheology of pectin solutions can be significantly influenced by co-solutes such as sugar, as well as external factors like temperature and pH. Increased pectin concentration increased solution viscosity by reducing the intermolecular distances between pectin molecules, facilitating interactions like hydrogen bonding and molecular entanglement (Guimarães et al., 2009). Several studies reported a positive correlation between pectin concentration and viscosity for intrinsic components of as galacturonic acid content, degree of esterification and molecular mass of pectin and also in external factors like pH, temperature, and ionic strength due to chemically diverse structure of pectin (Wang et al., 2016). In pectin extracted from main-harvested kiwi, enzyme-extracted pectin exhibited the lowest degree of branching compared to water-extracted pectin and acid-extracted pectin. The viscosity aligned with molecular weight and branching/linearity properties, with water-extracted pectin having the highest viscosity, suggesting that acid and enzyme treatments induced higher pectin hydrolysis. Interestingly, acid treatment showed less affecting on depolymerization, as molecular weight in cases of pectins extracted from orange and grapefruit peels using citric and sulfuric acids (Sayah et al., 2016), however these cases led higher molecular weights compared to pectin from residual peels obtained after essential oil extraction, likely due to the thermal treatment used for oil extraction.

The application of heat to pectin samples affects polymer structure and intermolecular interactions, including hydrogen bonds, electrostatic repulsion, hydrophobic association, and Van der Waals forces (Hua et al., 2015). The viscous properties were exhibited in pectin (1% w/w) extracted from kiwi fruits at different maturity by using conventional methods with and without citric acid and enzymes (Yuliarti et al., 2015). The results attributed in the higher degree of esterification of pectin for early-harvested kiwi fruits, favoring electrostatic interactions along and between pectin chains. The pectin (2% w/v) extracted from citron peel showed pseudoplastic flow behavior in the aqueous solution as shear rate increased (Pasandide et al., 2017). The viscosity of the pectin sample decreased due to chain alignment in the direction of shear flow, polymer network disentanglement, and reduced intermolecular interactions.

Fabrication of Ionotropic Pectinate Hydrogels

It was observed that the presence and concentration of divalent ions play a substantial role in influencing the rheological properties of high methoxyl pomelo pectin (HMP)in divalent ions inducing gel formation in HM pomelo pectin. Moreover, larger ions, such as $Ba^{2+} > Ca^{2+} > Mg^{2+}$, led to an increase in pectin gel modulus. Sometime, water-soluble pectin often has gelling properties. The gelation mechanism for LM pectin is often described using the "egg box" model in the presence of crosslinking agent like Ca^{2+} ion (Said et al., 2023). This model emphasizes the formation of networks through interactions between calcium ions (Ca^{2+}) and ionized carboxyl groups in the non-esterified regions of homogalacturonan (Fig. 6 (c)). Because the LM pectin contains COOH group, it readily forms cross-linkages in the presence of divalent ions. In the egg box, calcium ions fit into cavities created by the interaction between two homogalacturonan segments, resulting in the stronger gelling. In contrast, HM pectin has also calcium cross-linking for residual COOH groups, if the residues are adjacent and present in high concentration, leading to chain entanglement. At high degrees of esterification, due to non-ionic property, such as the presence of sugars and acids, play a crucial role in cross-linking of pectin segments. Co-solutes like sucrose, especially at low pH, generally promote gel formation in HM pectin and facilitate chain-to-chain interactions, reduce water activity, and protonate carboxylate residues, thereby reducing electrostatic repulsion.

Moreover, acidic conditions also encourage the formation of hydrogen bonds, as in Figure. 6 (a) and (b), between non-dissociated carboxylic groups or the COOH and alcohol groups. In the both illustrations, the stability of HM pectin gels is upheld by intermolecular hydrogen bonds and hydrophobic atmosphere created by the methyl ester type of pectin chains. Co-solutes disrupt the water molecules surrounding the methyl esters, enabling hydrophobic interactions (Kastner et al., 2012). Hydrogen bonds form between undissociated carboxyl groups or -COOH groups and adjacent pectin -OH chain, playing a significant role in forming junction zones.

In general, when the gelation occurs, pectin with higher molecular weights tends to facilitate each segment interaction. For instance, in the case of HM pectin, the cases of higher molecular weights encourage more extensive chain association. Conversely, LM pectin with higher molecular weights provides an electrostatic atmosphere. The sites have a high possibility of forming egg box structures when Ca^{2+} is present, thereby promoting the formation of interconnected network structures for gelling pectin. Consequently, pectin with higher molecular weights is likely to exhibit faster gelation, making it have a higher elastic modulus and enhanced gelation properties. On the contrary, pectin with low molecular weights, typically less than or equal to 10 kDa, the gelatinization tended to be lowered due to a diminished ability to establish these crucial interactions and structures (Cao et al., 2020).

Since the dissociation of COOH of the GalA units depends on pH, when increased from approximately pH 3.2 to around 8.0, the presence of Ca^{2+} improves by gelation. Essentially, the transition of the pectin-Ca^{2+} solution to a gel state occurred as the solution shifted towards a more alkaline pH. Conversely, at a pH of 3.2 pectin exhibited substantially lower gelation capability, due to that acidic conditions reduced dissociation of the COOH groups of the GalA segments (Tran Vo et al., 2022).

As well as pH, in another external change, temperature effects the condition of pectin. At temperatures exceeding 50 °C, abilities of water hydrogen bonding come into breaking and decreasing hydrophobic interactions surrounding the methoxy groups in pectin. But, as the system cools down, water molecules in the vicinity of the methoxy groups gather around the hydrophobic methoxy groups, enhancing aggregation of pectin segments. This shift in pectin state due to changes in temperature resulted in the development of a three-dimensional network gel structure. (Toniazzo & Fabi, 2023).

Recent research suggested that the pattern of methylation, known as the degree of homogeneity of homogalacturonan segment of pectin also influences the gelation properties of HM pectin which exhibited faster as compared with a random distributed pectin. But, these insights into the gelation behavior are still invaluable for tailoring its properties to various applications in the food and pharmaceutical industries. It possesses valuable functional properties, leading to extensive research and industrial use as thickening, gelling, and emulsifying in fields such as food, personal care, and pharmaceuticals (Freitas et al., 2021). These privileges are due to the water-soluble nature of pectin and other natural polymers like cellulose or chitin are insoluble in water. Therefore, gels solidified in a pectin solution contain large amounts of water and have low mechanical strength. In achieving high mechanical strength for diverse applications, especially in biomedical applications overcoming this obstacle was taken introducing cation ions, which act as bridges between pectin chains to form a gel network. Understanding pectin's molecular weight is crucial for assessing its gelling properties. For example, low viscosity in dragon fruit peel pectin attributed to its low molar mass (Muhammad et al., 2014) exhibited fewer junction zones, reducing cross-linking and affecting gel strength. The extraction method significantly impacts pectin's molecular weight. Depending on its degree of methoxylation (DM), HM pectin forms gels under acidic conditions (pH < 3.5) in the presence of a high concentration of sugar or a similar co-solute (Abboud et al., 2020). In LM pectin gels, with or without sugar, ionic crosslinking facilitated gelation with divalent or trivalent cations like Ca^{2+}, Zn^{2+}, Al^{3+}, and Fe^{3+}, as summarized in Figure 7.

Among more than divalent cationic ion, primarily, calcium ion is to serve as a cross-linker in the production of ionotropic LM pectin hydrogels. The gelation effected on temperature and pH tending below pH 3 to generate a reinforced gel for amidated pectin (Lootens et al., 2003). However, a weaker gel was formed at the same conditions for non-amidated pectin. Due to the amidation of carboxyl groups, the gelation was occurred as a function of pH and the amount of Ca^{2+} ions. In the pH range between pH 2 and 3.5, the greater added calcium concentration was displayed the increasingly heterogeneous gel formation (Capel et al., 2006). The important of intrinsic factors consisting of pectin degree of polymerization, degree and pattern of methoxylation was demonstrated on the properties of pectin-calcium gels (Fraeye et al., 2010). The better gel-like behavior of pectin-calcium mixture was exhibited at higher DM and lower calcium content in case of more blocked wise distribution than random distribution in GlaA segments. The excessive Ca^{2+} content partly compensated the effect of depolymerization causing the poor gel strength. Additionally, water evaporation was utilized instead of traditional mixing calcium ions (Yang et al., 2013). The results indicated that gelation of high methoxy pectin was triggered and speeded up because of non-specific binding and an electrostatic screening influence. In other words, the degree of content of the hydrophobic -OCH_3 group affected the spread of the molecular chain of pectin in aqueous solution and generated different gelling process, when cross-linked by calcium ions filling electronegative spaces in the twofold buckled ribbon structure of dissociated residues (Powell et al., 1982). The binding mechanism includes two or more chains and is cooperative, resulting in the development of junction zones rather than point-like cross-links. A minimum length of succeeding non-methyl esterified units, believed to be between 6 and 20, was required to produce a stable junction zone (Fraeye et al., 2010). Based on a molecular modeling simulation investigation, the "shifted egg-box" model for pectin-calcium gels was proposed (Cao et al., 2020). In some situations, Ca^{2+} ions engaged with single dissociated carboxyl groups, generating monocomplexes via charge reversal on a single chain and resulting in unspecific or random cross-linking between the pectin chains (Ngouémazong et al., 2012).

Contrary to Ca^{2+}, LM pectin was formed a three dimensional hydrogel network with addition of Zn^{2+}, Al^{3+}, and Fe^{3+} (Rajput et al., 2022). The dual crosslinked interpenetrating zinc-pectinate incorporated with neem gum was prepared to enhance floating properties that beneficial for stomach targeting. This pectinate hydrogel containing zinc demonstrated antibacterial properties against Staphylococcus aureus and Escherichia coli but was not conducive to facilitating cell migration. In contrast, the calcium-based hydrogel proved to be non-toxic to fibroblast cells and did not impede cell migration in any adverse way (Sarioglu et al., 2019). The low methoxy amidated pectin/sodium carboxymethylcellulose microspheres formed gels by a combination of Zn^{2+} and Al^{3+} ions used for colonic targeting of progesterone. In polyvalent cations, specifically calcium (Ca^{2+}) and aluminum (Al^{3+}), the pectin formulation was aimed for promoting gelation (El-Zahaby et al., 2014). In other studies, pectin microparticle gel was crosslinked through Zn^{2+}, Fe^{3+}, or Al^{3+} bridges. There was a significant inverse relationship observed between the Zn^{2+} content and the DM of pectin, indicating that lower DM promoted the binding of Zn^{2+} ions. Microparticle gel derived from pectin with a greater linear structure contained a higher proportion of Fe^{3+}. In comparison to the less electronegative Al^{3+} cations, both Fe^{3+} and Zn^{2+} ions possess higher electronegativity and exhibits stronger binding to pectin. Furthermore, microparticles with elevated levels of Zn^{2+}, Fe^{3+}, and Al^{3+} displayed a reduced degree of swelling when exposed to simulated digestive fluids (Günter et al., 2019). Additionally, different amount of the crosslinkers could manipulate the mechanical strength of pectin hydrogel which impacted by the degree of crosslinking. The Fe^{3+} cross-linked pectin hardness produced at a concentration of 42 mM of Fe^{3+}, was 24% greater than that of the Fe^{3+}-pectin hydrogels created with 21 mM and 84 mM of Fe^{3+}. As the concentration of the cross-linking cation increased, both the Young modulus and adhesiveness of the hydrogel also increased. Conversely, the brittleness of Fe^{3+}-pectin hydrogel decreased by 38–44% with an increase in Fe^{3+} concentration. Comparing the Ca^{2+} cross-linked hydrogel (CaP) hydrogel, at cation concentrations corresponding to 31 mM, 62 mM, and 124 mM, the Fe^{3+}-pectin hydrogel exhibited significantly higher hardness, surpassing that of the Ca^{2+}-pectin hydrogel by 4.5, 4.2, and 3.0 times, respectively. Similarly, the Young modulus of the Fe^{3+}-pectin hydrogel exceeded that of the Ca^{2+}-pectin hydrogel by 2.0, 6.4, and 4.2 times at the corresponding cation concentrations (Popov et al., 2023).

APPLICATION

In the field, the fusion of natural and synthetic polymers has opened doors to novel polymeric materials with enhanced durability and resilience (Song et al., 2018). Polymer films, in particular, play a crucial role in various applications, including the production of sponge materials, hydrogels, and drug encapsulation (Koubala et al., 2014). The increasing shift toward bio-based materials, as opposed to synthetic counterparts, is primarily driven for safety materials and then for environmental concerns and necessity for discharging non-waste treatment due to problem in the growing issue of plastic waste. This transition has been attracted in the desirable physiochemical properties and biodegradability of natural polymers (Liu et al., 2007). Among polymers, functional properties of pectin in both LM and HM, pectin can be applied for many applications in food, industrial, and biomedical sectors can be considered (Fig. 8), because pectin is known for its remarkable inertness (Kokkonen et al., 2009), biodegradability (Ren et al., 2022), and biocompatibility (Kraskouski et al., 2022), and has gained widespread utilization across diverse sectors, including textiles, the food industry as a gelling agent, pharmaceuticals, biomaterials and various other product domains (Noreen et al., 2017). Pectin finds its footing as a biomaterial in gene delivery (Katav et al., 2008), plays a role in oral drug delivery applications (Sriamornsak, 2011), contributes to food packaging as an edible coating (Norziah et al., 2000), aids biomass yield and biorefinery processes (Ishii & Matsunaga, 2001; Nakamura et al., 2002), and serves in tissue engineering as scaffolds (Ninan et al., 2013). Additionally, its applications extend to the paper and textile industries for the preparation of ultracentrifugation membranes (Swenson et al., 1953).

Pectin is widely used in the food industry as an excellent thickener agent for producing jellies and jams (Smith, 2003), a pH stabilizer in dairy products and low-calorie products (Ciriminna et al., 2022), as well as good former of edible films and coatings (Lazaridou & Biliaderis, 2020), foams (Christoph et al., 2018), and paper substitutes (ALPIZAR REYES et al., 2022). In addition with an emulsifier in pharmaceutics for the design of drugs to treat gastrointestinal disorders (Dongowski & Anger, 1996), pectin is applicable for blood cholesterol reduction (Brouns et al., 2012), and cancer treatment (Zhang et al., 2015) in medical field.

Utilization in Food Products

Pectin, as a naturally-derived polysaccharide found in the primary cell walls of edible plants, presents an appealing option for the food industry due to its renewable nature, non-toxic properties, cost-effectiveness, and biocompatibility. As a biomass compatible with both of these food constraints and significant role in the food sector, pectin has excellent natures acting as a versatile ingredient with gelling, stabilizing, and thickening capabilities. The applications in food products are diverse, ranging from traditional favorites to innovative culinary creations. One of the most iconic roles of pectin in the culinary world is to contribute it to foods of jams and jellies. In jam production, fruits undergo a carefully orchestrated cooking process to release their natural juices and pectin content. This transformation converts protopectin into more soluble forms, creating the characteristic gel-like consistency that defines jams and jellies (McNeil et al., 1980). Pectin's relevance to the food sector extends beyond traditional preserves. In the domain of sugar-free jams, it steps in as a valuable sugar substitute. For instance, low methoxy pectin showed a nature for its stability under acidic conditions (Yang et al., 2018), and was particularly well-suited for this purpose. Its application empowered the creation of delicious sugar-free jams that cater to dietary preferences while maintaining the desired texture and taste. Instant jellies, a staple in bakery production, were other examples. In HM pectin due to its remarkable thermal stability (Nesic et al.), the formulations somewhat tended to differ from those employing LM pectin, but the fundamental principle remained the same in both LM and HM, which became the key ingredient in delivering the delightful consistency of instant jellies (May, 1990).

It was noted that the influence of pectin stretches even further, embracing artificial cherries and gel puddings. In these applications, pectin extracted from fruit syrup and cold milk serves as the backbone, creating textures and mouthfeel that tantalize the palate (Albersheim et al., 1979; Norziah et al., 2000). Its role as a beverage clouding agent is evident in diabetic soft drinks, where pectin assists in achieving the desired visual opacity without compromising the beverage's quality. In the realm of dairy products, pectin plays an essential part in yogurt fruit preparations. Here, it enhances not only the softness of the fruit pieces but also contributes to the partial gel textures that are characteristic of these delightful yogurt blends (El-Shamei & El-Zoghbi, 1994; Pedrolli et al., 2009). Due to its biocompatibility, the application of pectin-based biofilms has demonstrated the potential to significantly prolong the shelf life of avocado fruits, extending it by over a month when compared to untreated fruits. This extension was achieved by reducing oxygen absorption, which in turn delays changes in texture and color (Nesic et al., 2022). The study focused on the impact of a coating of pectin and corn flour edible film, revealing the most effective in minimizing weight loss, preserving freshness, delaying respiration, and maintaining the biochemical quality of tomatoes, ultimately enhancing their shelf life (Sucheta et al., 2019).

As result, for the food industry its significance is as a versatile ingredient. From the timeless allure of jams and jellies to its innovative role in sugar-free products and bakery goods, pectin continues to shape the culinary landscape. Its ability to enhance both solid and liquid concoctions reaffirm its status as a stalwart ingredient in the world of gastronomy.

Packaging

Packaging plays a crucial role in various industries, particularly in the food sector, where it is essential for preserving food quality, safety, and extending shelf life. Petroleum-derived plastic packaging has been widely used, accounting for a significant portion of global packaging production. However, concerns have arisen regarding the environmental and health impacts of plastic polymers, which often contain various chemical additives. This has led to growing environmental pollution and waste issues.

As considered viable biomaterials for creating environmentally friendly films, pectin has feature of their film-forming, biodegradable, and non-toxic characteristics (Heba et al., 2022). In response to these challenges, pectin has emerged as a promising alternative for active food packaging due to its natural biodegradability, compatibility with living organisms, safety for consumption, and favorable chemical and physical properties. Pectin is employed as a coating material to enhance food quality and prolong the shelf life of products, making it an eco-friendly substitute for synthetic polymer-based packaging materials (Moalemiyan et al., 2012). However, pure pectin films tend to be brittle, often necessitating the use of a plasticizer to achieve the desired flexibility. Thus, the use of additives like corn flour and beetroot in commercial pectin films revealed significant improvements in tensile strength, reduced water solubility, and variations in film thickness (Rai et al., 2019). These additives enhanced the overall performance and properties of pectin-based films, making them more suitable for various food packaging applications. For instance, pectin films combined with corn flour and beetroot exhibited increased tensile strength, decreased water solubility, and improved film thickness compared to pure pectin films. Additionally, these modified films were applied as coatings for fresh produce like tomatoes (Chaturvedi et al., 2019). The combination of both components proved to be effective in reducing weight loss, delaying the respiration rate, preserving internal quality during storage, extending the shelf life of tomatoes, and minimizing shrinkage. Also, incorporating nano chitosan into pectin films was explored with varying ratios showing different effects on film properties (Ngo et al., 2020) . As blending at 50:50 ratio of nano-chitosan/pectin, for instance, increased tensile strength while simultaneously reducing water solubility, water vapor permeability, and oxygen permeability. These films even exhibited antimicrobial activity against various microorganisms, making them potentially valuable for food preservation.

In summary of packaging, pectin-based films have gained attention as eco-friendly alternatives to traditional plastic packaging due to their biodegradability and non-toxic nature. By incorporating additives and exploring novel formulations, researchers are enhancing the performance and properties of these films, making them increasingly suitable for various food packaging and preservation applications while mitigating environmental concerns associated with plastic packaging.

Utilization in Biomedical Applications

Among natural polymers, since pectin, a complex heteropolysaccharide predominantly found in its potential films for particularly notable food and pharmaceuticals (Huang et al., 2017), an illustrative example pectin films incorporated glycerol and lactic acid, conferring upon them the ability to safeguard laminated films from fungal contamination (Huang et al., 2017). Furthermore, in the development of edible pectin blend films, pectin was effectively plasticized with glycerol, exhibiting their remarkable flexibility even at room temperature (Fishman et al., 2000), expanding for biomedical applications of facilitating noteworthy application was to use for wound dressing material with healing ability (Gan & Latiff, 2011; Mishra et al., 2011). Intriguingly, such pectin films derived from citrus waste were investigated for their potential in skincare as delved into the prospect of utilizing these films for skin moisturization (Fig. 9), yielding promising results with implications for skin cell proliferation. Particularly compelling was the observation that among the tested pectin, those 70% DM% of exhibited significantly heightened levels of skin cell proliferation. A critical aspect of this study was the correlation between crystallinity and film dissolution rate: lower crystallinity was associated with quicker dissolution, while higher crystallinity resulted in extended dissolution times (Chuenkaek et al., 2023).

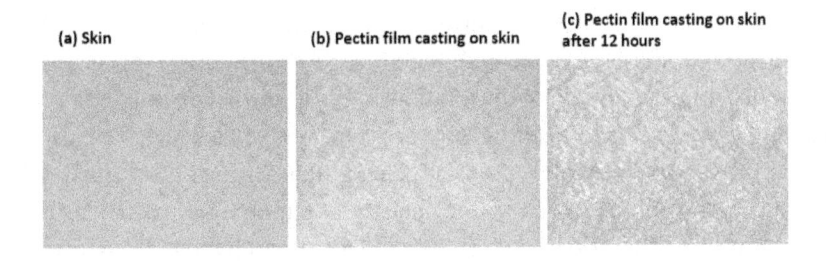

Figure 8. Applications of pectin in different area

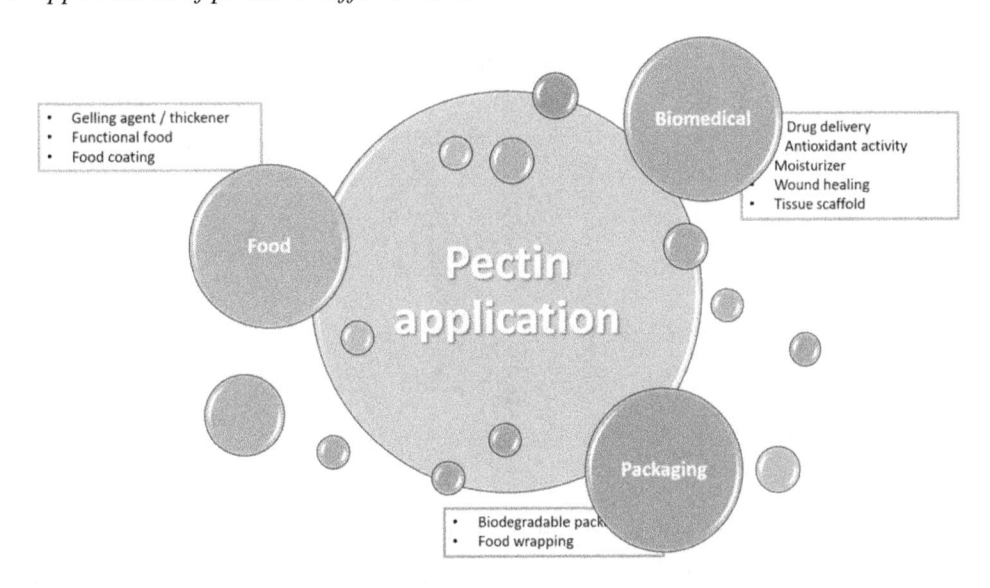

Figure 9. Pectin film for moisturization application (Reprinted from ref. (Chuenkaek et al., 2023))

Figure 10. Biomedical applications of pectin-based hydrogels

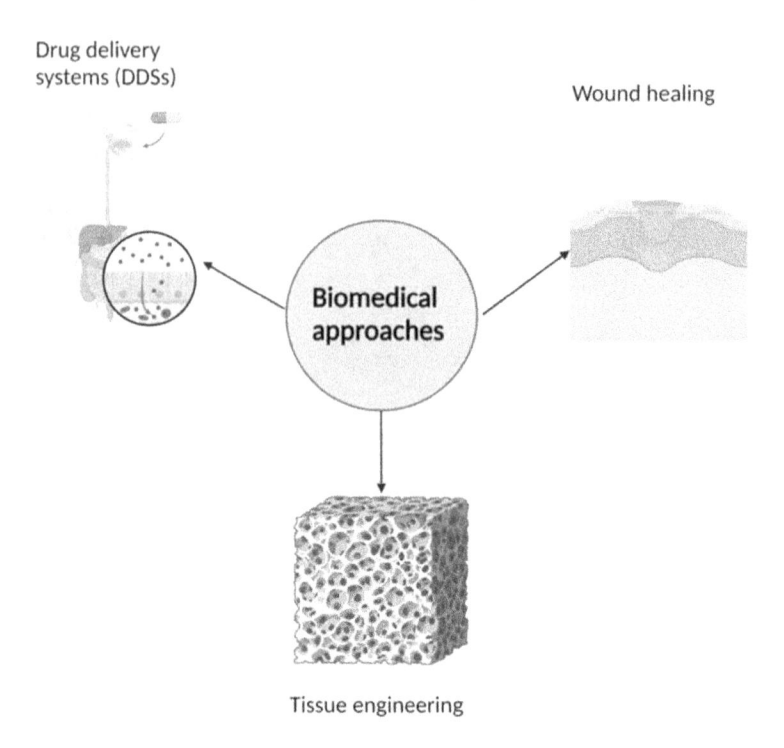

On the other hand, pectin-based hydrogels have been extensively explored for a range of potential biomedical applications, including drug delivery, wound healing, and tissue engineering, owing to their biocompatibility, biodegradability, capacity for easy gel formation, and simple manipulation, as shown in Figure 10. Recently, bioactive substances such as antibiotics, growth factors, vitamins, minerals, and more play a vital role in maintaining and enhancing health. Therefore, enhancing the effectiveness of drug delivery at the intended site while maintaining the activity of bioactive compounds, achieving the desired dosage, and minimizing potential side effects is a preferred objective. To tackle these challenges, drug delivery systems (DDS) have been created. Pectin garnered attention for its ability to form gels in acidic conditions, mucoadhesiveness, and dissolution in basic environments (Wai et al., 2010). These attributed enabled targeted and controlled drug delivery, especially in nasal and gastric environments and aided in the release of colon-related drugs and prolonged gastric drug contact (Gan & Latiff, 2011; Wai et al., 2010). The mucoadhesive properties of LM pectin was suitable for nasal drug delivery, binding to mucin via hydrogen bonding (Sriamornsak et al., 2010). The research focused on investigating hydrogels crafted from thiol-modified pectin, employing disulfide bonds for crosslinking. These hydrogels were utilized as carriers for encapsulating Acetaminophen, acting as a representative drug model. The results revealed that these manufactured hydrogels underwent swift degradation when exposed to physiological reducing agents such as L-glutathione and L-cysteine, while displaying remarkable stability in their absence (Quadrado et al., 2022). For pH-responsive hydrogel, grafting acrylamide was created onto pectin cross-linking layer with glutaraldehyde. The resulting product demonstrated pH-dependent drug release behavior, suggesting its potential for use in controlled DDS. Furthermore, the product was determined to be biocompatible, making it a promising candidate for the development of implantable medical devices (Sutar et al., 2008). Moreover, pectin recently gained recognition as a valuable substance for skin applications due to maintaining a moist wound environment, providing protection from infections, removing wound exudates, and promoting wound healing from pectin hydrogel (Nordin et al., 2022). In contrast, pectin-honey hydrogel used as a wound healing membrane compared with to liquid honey in serving a healing nature. The pectin-honey hydrogel exhibited a faster reduction in wound area as compared to the control and became effective approach to promote and expedite the wound healing process (Giusto et al., 2017). As introduced for a pectin hydrogel infused with liquid allantoin, its healing properties and hydrophilic nature are due to specific functional groups (Saucedo-Acuña et al., 2023). Allantoin is evenly distributed both on the surface and within the amorphous pectin hydrogel, which was characterized by a diverse pore distribution. This unique structure promoted effective wound drying and facilitates improved interaction between the hydrogel and the cells involved in the wound healing process.

Another use of pectin hydrogel is in bio-scaffold of tissue engineering because of its biocompatible, biodegradable, and bioactive natures. In pectin-based injectable biomaterials for bone tissue engineering, there was potential ability to mimic the natural extracellular matrix, providing an ideal environment for cells to adhere, multiply, and migrate into the matrix and facilitating the regeneration of damaged or diseased tissues effectively modified with cell adhesion protein, RGD(Arg-Gly-Asp) (Munarin et al., 2011). The pectin-based oligopeptide supported cell viability, differentiation, and the formation of mineralized extracellular matrices and improved cell adhesion and proliferation within the pectin microspheres. One of the limitations of pectin hydrogel to adapt with the requirement of tissue engineering is owing to its poor mechanical properties. For reinforcement, composite hydrogel of pectin and the poly(vinyl alcohol) was implemented and used for bone regeneration. When the scaffold had stiffness and pore size and improved proliferation and adhesion of osteoblasts, accelerated bone healing process was recognized in *in vivo* after transplantation into the femoral defect (Hu et al., 2022).

CONCLUSION

Pectin, a versatile plant-based polysaccharide, plays a pivotal role in enhancing sustainability and nutrition in plant-derived products. Recent efforts have focused on sourcing pectin sustainably from unconventional, eco-friendly sources like agro-industrial by-products and organic waste materials. From the viewpoint of circular economy, in this chapter, methods for extracting pectin from food wastes, especially fruit wastes, and their properties are presented in addition with their applications. Since pectin has been functionalized by chemical or enzyme processes that affect its physical characteristics, such as molecular weight, degree of esterification, and surface charge, the relevant topics are introduced as well as advanced techniques and applications. Emerging sources including agro-industrial sub-products and wastes, pulps, husks, hulls, peels offer promise in sustainability due to their abundance and favorable attributes. Therefore, in essence, pectin remains indispensable for sustainable food, pharmaceutical, and cosmetic applications. Future research should continue to prioritize sustainable extraction techniques from diverse raw materials, promoting sustainability throughout the supply chain. As pectin continues to be harnessed sustainably on an industrial scale, it holds the potential to drive innovation and sustainable practices for a more environmentally responsible future.

REFERENCES

Abboud, K. Y., Iacomini, M., Simas, F. F., & Cordeiro, L. M. C. (2020). High methoxyl pectin from the soluble dietary fiber of passion fruit peel forms weak gel without the requirement of sugar addition. *Carbohydrate Polymers*, 246, 116616. https://doi.org/https://doi.org/10.1016/j.carbpol.2020.116616. DOI: 10.1016/j.carbpol.2020.116616 PMID: 32747256

Albersheim, P., Brand, J., Darvill, A., Häusler, J., & McNeil, M. (1979). The structural polymers of the primary cell walls of dicots. *Fortschritte der Chemie organischer Naturstoffe / Progress in the Chemistry of Organic Natural Products. Fortschritte der Chemie organischer Naturstoffe. Progress in the Chemistry of Organic Natural Products*, 37, 191–249.

Alpizar-Reyes, E., Cortés-Camargo, S., Román-Guerrero, A., & Pérez-Alonso, C. (2022). Tamarind gum as a wall material in the microencapsulation of drugs and natural products. *In Micro-and nanoengineered gum-based biomaterials for drug delivery and biomedical applications*, 347-382.

Alpizar-Reyes, E. R. I. K., Cruz-Olivares, J. U. L. I. A. N., Cortés-Camargo, S., Rodríguez-Huezo, M. E., Macías-Mendoza, J. O., Alvarez-Ramírez, J., & Pérez-Alonso, C. E. S. A. R. (2022). Structural, physicochemical, and emulsifying properties of pectin obtained by aqueous extraction from red pitaya (Hylocereus polyrhizus) peel. Revista Mexicana de Ingeniería Química, 21(3), Alim2887-Alim2887.

Atmodjo, M. A., Hao, Z., & Mohnen, D. J. A. r. o. p. b. (2013). Evolving views of pectin biosynthesis. *64*, 747-779.

Bagherian, H., Zokaee Ashtiani, F., Fouladitajar, A., & Mohtashamy, M. (2011). Comparisons between conventional, microwave- and ultrasound-assisted methods for extraction of pectin from grapefruit. *Chemical Engineering and Processing*, 50(11), 1237–1243. https://doi.org/https://doi.org/10.1016/j.cep.2011.08.002. DOI: 10.1016/j.cep.2011.08.002

Brouns, F., Theuwissen, E., Adam, A., Bell, M., Berger, A., & Mensink, R. P. (2012). Cholesterol-lowering properties of different pectin types in mildly hyper-cholesterolemic men and women. *European Journal of Clinical Nutrition*, 66(5), 591–599.

Canteri-Schemin, M. H., Fertonani, H. C. R., Waszczynskyj, N., Wosiacki, G. J. B. a. o. b., & technology. (2005). Extraction of pectin from apple pomace. *48*, 259-266.

Cao, L., Lu, W., Mata, A., Nishinari, K., & Fang, Y. (2020). Egg-box model-based gelation of alginate and pectin: A review. *Carbohydrate Polymers*, 242, 116389. DOI: 10.1016/j.carbpol.2020.116389 PMID: 32564839

Capel, F., Nicolai, T., Durand, D., Boulenguer, P., & Langendorff, V. (2006). Calcium and acid induced gelation of (amidated) low methoxyl pectin. *Food Hydrocolloids*, 20(6), 901–907. DOI: 10.1016/j.foodhyd.2005.09.004

Chandel, V., Biswas, D., Roy, S., Vaidya, D., Verma, A., & Gupta, A. (2022). Current Advancements in Pectin: Extraction, Properties and Multifunctional Applications. *Foods, 11*(17). Chandel, V., Biswas, D., Roy, S. A.-O. X., Vaidya, D., Verma, A., & Gupta, A. Current Advancements in Pectin: Extraction, Properties and Multifunctional Applications. *Foods*, 11(17), 2683. DOI: 10.3390/foods11172683 PMID: 36076865

Chaturvedi, K., Sharma, N., & Yadav, S. K. J. I. j. o. b. m. (2019). Composite edible coatings from commercial pectin, corn flour and beetroot powder minimize post-harvest decay, reduces ripening and improves sensory liking of tomatoes. *133*, 284-293.

Chen, J., Niu, X., Dai, T., Hua, H., Feng, S., Liu, C., McClements, D. J., & Liang, R. J. F. c. (2020). Amino acid-amidated pectin: Preparation and characterization. *309*, 125768.

Christiaens, S., Uwibambe, D., Uyttebroek, M., Van Droogenbroeck, B., Van Loey, A. M., & Hendrickx, M. E. (2015). Pectin characterisation in vegetable waste streams: A starting point for waste valorisation in the food industry. *Lebensmittel-Wissenschaft + Technologie*, 61(2), 275–282. https://doi.org/https://doi.org/10.1016/j.lwt.2014.12.054. DOI: 10.1016/j.lwt.2014.12.054

Christoph, S., Hamraoui, A., Bonnin, E., Garnier, C., Coradin, T., & Fernandes, F. M. (2018). Ice-templating beet-root pectin foams: Controlling texture, mechanics and capillary properties. *Chemical Engineering Journal*, 350, 20–28. https://doi.org/https://doi.org/10.1016/j.cej.2018.05.160. DOI: 10.1016/j.cej.2018.05.160

Chuenkaek, T., Nakajima, K., & Kobayashi, T. (2023). Water-soluble pectin films prepared with extracts from Citrus maxima wastes under acidic conditions and their moisturizing characteristics. *Biomass Conversion and Biorefinery*. Advance online publication. DOI: 10.1007/s13399-023-04358-8

Ciriminna, R., Chavarría-Hernández, N., Inés Rodríguez Hernández, A., & Pagliaro, M. (2015). Pectin: A new perspective from the biorefinery standpoint. *Biofuels, Bioproducts & Biorefining*, 9(4), 368–377. https://doi.org/https://doi.org/10.1002/bbb.1551. DOI: 10.1002/bbb.1551

Ciriminna, R., Fidalgo, A., Scurria, A., Ilharco, L. M., & Pagliaro, M. (2022). Pectin: New science and forthcoming applications of the most valued hydrocolloid. *Food Hydrocolloids*, 127, 107483. https://doi.org/https://doi.org/10.1016/j.foodhyd.2022.107483. DOI: 10.1016/j.foodhyd.2022.107483

Colodel, C., Vriesmann, L. C., Teófilo, R. F., & de Oliveira Petkowicz, C. L. (2018). Extraction of pectin from ponkan (Citrus reticulata Blanco cv. Ponkan) peel: Optimization and structural characterization. *International Journal of Biological Macromolecules*, 117, 385–391. https://doi.org/https://doi.org/10.1016/j.ijbiomac.2018.05.048. DOI: 10.1016/j.ijbiomac.2018.05.048 PMID: 29753767

Crombie, H., Scott, C., & Reid, J. (2003). *Advances in Pectin and Pectinase Research* (Voragen, A. G. J., Schols, H. A., & Visser, R. G. F., Eds.). Kluwer Academic Publishers.

Cui, J., Zhao, C., Feng, L., Han, Y., Du, H., Xiao, H., & Zheng, J. (2021). Pectins from fruits: Relationships between extraction methods, structural characteristics, and functional properties. *Trends in Food Science & Technology*, 110, 39–54. DOI: 10.1016/j.tifs.2021.01.077

Dominiak, M., Søndergaard, K. M., Wichmann, J., Vidal-Melgosa, S., Willats, W. G. T., Meyer, A. S., & Mikkelsen, J. D. (2014). Application of enzymes for efficient extraction, modification, and development of functional properties of lime pectin. *Food Hydrocolloids*, 40, 273–282. https://doi.org/https://doi.org/10.1016/j.foodhyd.2014.03.009. DOI: 10.1016/j.foodhyd.2014.03.009

Dongowski, G., & Anger, H. (1996). Metabolism of pectin in the gastrointestinal tract. In *Progress in Biotechnology* (Vol. 14, pp. 659-666). DOI: 10.1016/S0921-0423(96)80300-5

Dranca, F., & Oroian, M. J. F. R. I. (2018). Extraction, purification and characterization of pectin from alternative sources with potential technological applications. *113*, 327-350.

El-Shamei, Z., & El-Zoghbi, M. J. F. N. (1994). Producing of natural clouding agents from orange and lemon peels. *38*(2), 158-166.

El-Zahaby, S. A., Kassem, A. A., & El-Kamel, A. H. (2014). Formulation and in vitro evaluation of size expanding gastro-retentive systems of levofloxacin hemihydrate. *International Journal of Pharmaceutics*, 464(1), 10–18. https://doi.org/https://doi.org/10.1016/j.ijpharm.2014.01.024. DOI: 10.1016/j.ijpharm.2014.01.024 PMID: 24472642

Fishman, M., Coffin, D., Konstance, R., & Onwulata, C. J. C. P. (2000). Extrusion of pectin/starch blends plasticized with glycerol. *41*(4), 317-325.

Flutto, L. (2003). PECTIN I Properties and Determination. In Caballero, B. (Ed.), *Encyclopedia of Food Sciences and Nutrition* (2nd ed., pp. 4440–4449). Academic Press., https://doi.org/https://doi.org/10.1016/B0-12-227055-X/00901-9 DOI: 10.1016/B0-12-227055-X/00901-9

Fraeye, I., Duvetter, T., Doungla, E., Van Loey, A., & Hendrickx, M. (2010). Fine-tuning the properties of pectin–calcium gels by control of pectin fine structure, gel composition and environmental conditions. *Trends in Food Science & Technology*, 21(5), 219–228. DOI: 10.1016/j.tifs.2010.02.001

Freitas, C. M. P., Coimbra, J. S. R., Souza, V. G. L., & Sousa, R. C. S. J. C. (2021). Structure and applications of pectin in food, biomedical, and pharmaceutical industry. *RE:view*, 11(8), 922.

Fry, S. C. (1988). *The growing plant cell wall: chemical and metabolic analysis*. Longman Group Limited.

Gamonpilas, C., Krongsin, J., Methacanon, P., & Goh, S. M. (2015). Gelation of pomelo (Citrus maxima) pectin as induced by divalent ions or acidification. *Journal of Food Engineering*, 152, 17–23. https://doi.org/https://doi.org/10.1016/j.jfoodeng.2014.11.024. DOI: 10.1016/j.jfoodeng.2014.11.024

Gan, C.-Y., & Latiff, A. A. J. C. p. (2011). Extraction of antioxidant pectic-polysaccharide from mangosteen (Garcinia mangostana) rind: Optimization using response surface methodology. *83*(2), 600-607.

Gawkowska, D., Ciesla, J., Zdunek, A., & Cybulska, J. (2019). The Effect of Concentration on the Cross-Linking and Gelling of Sodium Carbonate-Soluble Apple Pectins. *Molecules (Basel, Switzerland)*, 24(8), 1635. Advance online publication. DOI: 10.3390/molecules24081635 PMID: 31027264

Gawkowska, D., Cybulska, J., & Zdunek, A. (2018). Structure-Related Gelling of Pectins and Linking with Other Natural Compounds: A Review. *Polymers*, 10(7), 762. Advance online publication. DOI: 10.3390/polym10070762 PMID: 30960687

Giusto, G., Vercelli, C., Comino, F., Caramello, V., Tursi, M., & Gandini, M. (2017). A new, easy-to-make pectin-honey hydrogel enhances wound healing in rats. *BMC Complementary and Alternative Medicine*, 17(1), 266. DOI: 10.1186/s12906-017-1769-1 PMID: 28511700

Gotoh, S., Kitaguchi, K., & Yabe, T. (2021). Involvement of the Complex Polysaccharide Structure of Pectin in Regulation of Biological Functions. *Reviews in Agricultural Science*, 9(0), 221–232. DOI: 10.7831/ras.9.0_221

Guimarães, G., Júnior, M., & Rojas, E. (2009). Density and Kinematic Viscosity of Pectin Aqueous Solution†. *Journal of Chemical and Engineering Data -. Journal of Chemical & Engineering Data*, 0(2), 662–667. Advance online publication. DOI: 10.1021/je800305a

Günter, E. A., Popeyko, O. V., Melekhin, A. K., Belozerov, V. S., Martinson, E. A., & Litvinets, S. G. (2019). Preparation and properties of the pectic gel microparticles based on the Zn^{2+}, Fe^{3+} and Al^{3+} cross-linking cations. *International Journal of Biological Macromolecules*, 138, 629–635. https://doi.org/https://doi.org/10.1016/j.ijbiomac.2019.07.122. DOI: 10.1016/j.ijbiomac.2019.07.122 PMID: 31336115

Heba, Y., Guohua, Z., & Hassan, A. (2022). Pectin and Its Applicability in Food Packaging. In V. Işıl & U. Sinan (Eds.), *A Glance at Food Processing Applications* (pp. Ch. 5). IntechOpen. https://doi.org/ DOI: 10.5772/intechopen.101614

Herrera-Rodríguez, S. E., Pacheco, N., Ayora-Talavera, T., Pech-Cohuo, S., & Cuevas-Bernardino, J. C. (2022). Chapter 7 - Advances in the green extraction methods and pharmaceutical applications of bioactive pectins from unconventional sources: a review. In R. Atta ur (Ed.), *Studies in Natural Products Chemistry* (Vol. 73, pp. 221-264). Elsevier. https://doi.org/https://doi.org/10.1016/B978-0-323-91097-2.00015-7

Herrera-Rodríguez, S. E., Pacheco, N., Ayora-Talavera, T., Pech-Cohuo, S., & Cuevas-Bernardino, J. C. J. S. N. P. C. (2022). Advances in the green extraction methods and pharmaceutical applications of bioactive pectins from unconventional sources. *RE:view*, 73, 221–264.

Hosseini, S. S., Khodaiyan, F., Kazemi, M., & Najari, Z. (2019). Optimization and characterization of pectin extracted from sour orange peel by ultrasound assisted method. *International Journal of Biological Macromolecules*, 125, 621–629. https://doi.org/https://doi.org/10.1016/j.ijbiomac.2018.12.096. DOI: 10.1016/j.ijbiomac.2018.12.096 PMID: 30543886

Hu, Z., Cheng, J., Xu, S., Cheng, X., Zhao, J., Kenny Low, Z. W., Chee, P. L., Lu, Z., Zheng, L., & Kai, D. (2022). PVA/pectin composite hydrogels inducing osteogenesis for bone regeneration. *Materials Today. Bio*, 16, 100431. https://doi.org/https://doi.org/10.1016/j.mtbio.2022.100431. DOI: 10.1016/j.mtbio.2022.100431 PMID: 36186849

Hua, X., Wang, K., Yang, R., Kang, J., & Zhang, J. (2015). Rheological properties of natural low-methoxyl pectin extracted from sunflower head. *Food Hydrocolloids*, 44, 122–128. https://doi.org/https://doi.org/10.1016/j.foodhyd.2014.09.026. DOI: 10.1016/j.foodhyd.2014.09.026

Huang, X., Li, D., & Wang, L.-j. J. J. o. F. E. (2017). Characterization of pectin extracted from sugar beet pulp under different drying conditions. *211*, 1-6.

Immanuel, G. (2018). *EXTRACTION AND CHARACTERIZATION OF PECTIN DERIVED FROM PAPAYA (Carica papaya Linn.)*. PEEL., DOI: 10.2348/ijset07150970

Ishii, T., & Matsunaga, T. J. P. (2001). Pectic polysaccharide rhamnogalacturonan II is covalently linked to homogalacturonan. *57*(6), 969-974.

Jafari, F., Khodaiyan, F., Kiani, H., & Hosseini, S. S. (2017). Pectin from carrot pomace: Optimization of extraction and physicochemical properties. *Carbohydrate Polymers*, 157, 1315–1322. DOI: 10.1016/j.carbpol.2016.11.013 PMID: 27987838

Jafari, F., Khodaiyan, F., Kiani, H., & Hosseini, S. S. (2017). Pectin from carrot pomace: Optimization of extraction and physicochemical properties. *Carbohydrate Polymers*, 157, 1315–1322. https://doi.org/https://doi.org/10.1016/j.carbpol.2016.11.013. DOI: 10.1016/j.carbpol.2016.11.013 PMID: 27987838

Karbuz, P., & Tugrul, N. (2021). Microwave and ultrasound assisted extraction of pectin from various fruits peel. *Journal of Food Science and Technology*, 58(2), 641–650. DOI: 10.1007/s13197-020-04578-0 PMID: 33568858

Kastner, H., Einhorn-Stoll, U., & Senge, B. (2012). Structure formation in sugar containing pectin gels – Influence of Ca2+ on the gelation of low-methoxylated pectin at acidic pH. *Food Hydrocolloids*, 27(1), 42–49. https://doi.org/https://doi.org/10.1016/j.foodhyd.2011.09.001. DOI: 10.1016/j.foodhyd.2011.09.001

Katav, T., Liu, L., Traitel, T., Goldbart, R., Wolfson, M., & Kost, J. J. J. o. C. R. (2008). Modified pectin-based carrier for gene delivery: cellular barriers in gene delivery course. *130*(2), 183-191.

Kazemi, M., Khodaiyan, F., Labbafi, M., Saeid Hosseini, S., & Hojjati, M. (2019). Pistachio green hull pectin: Optimization of microwave-assisted extraction and evaluation of its physicochemical, structural and functional properties. *Food Chemistry*, 271, 663–672. DOI: 10.1016/j.foodchem.2018.07.212 PMID: 30236729

Khamsucharit, P., Laohaphatanalert, K., Gavinlertvatana, P., Sriroth, K., & Sangseethong, K. (2018). Characterization of pectin extracted from banana peels of different varieties. *Food Science and Biotechnology*, 27(3), 623–629. DOI: 10.1007/s10068-017-0302-0 PMID: 30263788

Koh, P. C., Leong, C. M., & Mohd Adzahan, N. (2014). Microwave-assisted extraction of pectin from jackfruit rinds using different power levels. *International Food Research Journal*, 21, 2091–2097.

Kokkonen, H., Niiranen, H., Schols, H. A., Morra, M., Stenbäck, F., & Tuukkanen, J. (2009). Pectin-coated titanium implants are well-tolerated in vivo. *Journal of Biomedical Materials Research. Part A*, 93(4), 1404–1409. DOI: 10.1002/jbm.a.32649 PMID: 19911385

Koubala, B. B., Christiaens, S., Kansci, G., Van Loey, A. M., & Hendrickx, M. E. J. F. r. i. (2014). Isolation and structural characterisation of papaya peel pectin. *55*, 215-221.

Kraskouski, A., Hileuskaya, K., Ladutska, A., Kabanava, V., Liubimau, A., Novik, G., Nhi, T. T. Y., & Agabekov, V. (2022). Multifunctional biocompatible films based on pectin-Ag nanocomposites and PVA: Design, characterization and antimicrobial potential. *Journal of Applied Polymer Science*, 139(42), e53023. https://doi.org/https://doi.org/10.1002/app.53023. DOI: 10.1002/app.53023

Lazaridou, A., & Biliaderis, C. G. (2020). Edible Films and Coatings with Pectin. In Kontogiorgos, V. (Ed.), *Pectin: Technological and Physiological Properties* (pp. 99–123). Springer International Publishing., DOI: 10.1007/978-3-030-53421-9_6

Li, D., Jia, X., Wei, Z., & Liu, Z.-y. (2012). Box–Behnken experimental design for investigation of microwave-assisted extracted sugar beet pulp pectin. *Carbohydrate Polymers*, 88(1), 342–346. https://doi.org/https://doi.org/10.1016/j.carbpol.2011.12.017. DOI: 10.1016/j.carbpol.2011.12.017

Li, W. J., Fan, Z. G., Wu, Y. Y., Jiang, Z. A.-O., & Shi, R. C. (2019). Eco-friendly extraction and physicochemical properties of pectin from jackfruit peel waste with subcritical water. *Journal of the Science of Food and Agriculture*, 99(12), 5283–5292. DOI: 10.1002/jsfa.9729 PMID: 30953352

Liew, S. Q., Chin, N. L., & Yusof, Y. A. (2014). Extraction and Characterization of Pectin from Passion Fruit Peels. *Agriculture and Agricultural Science Procedia*, 2, 231–236. DOI: 10.1016/j.aaspro.2014.11.033

Ling, B., Ramaswamy, H. S., Lyng, J. G., Gao, J., & Wang, S. (2023). Roles of physical fields in the extraction of pectin from plant food wastes and byproducts: A systematic review. *Food Research International*, 164, 112343. https://doi.org/https://doi.org/10.1016/j.foodres.2022.112343. DOI: 10.1016/j.foodres.2022.112343 PMID: 36737935

Liu, L., Finkenstadt, V., Liu, C. K., Jin, T., Fishman, M., & Hicks, K. J. J. o. A. P. S. (2007). Preparation of poly (lactic acid) and pectin composite films intended for applications in antimicrobial packaging. *106*(2), 801-810.

Lootens, D., Capel, F., Durand, D., Nicolai, T., Boulenguer, P., & Langendorff, V. (2003). Influence of pH, Ca concentration, temperature and amidation on the gelation of low methoxyl pectin. *Food Hydrocolloids*, 17(3), 237–244. https://doi.org/https://doi.org/10.1016/S0268-005X(02)00056-5. DOI: 10.1016/S0268-005X(02)00056-5

Luo, J., Xu, Y., & Fan, Y. (2019). Upgrading Pectin Production from Apple Pomace by Acetic Acid Extraction. *Applied Biochemistry and Biotechnology*, 187(4), 1300–1311. DOI: 10.1007/s12010-018-2893-1 PMID: 30218302

Ma, M., & Mu, T. (2016). Effects of extraction methods and particle size distribution on the structural, physicochemical, and functional properties of dietary fiber from deoiled cumin. *Food Chemistry*, 194, 237–246. https://doi.org/https://doi.org/10.1016/j.foodchem.2015.07.095. DOI: 10.1016/j.foodchem.2015.07.095 PMID: 26471550

Ma, X., Jing, J., Wang, J., Xu, J., & Hu, Z. (2020). Extraction of Low Methoxyl Pectin from Fresh Sunflower Heads by Subcritical Water Extraction. *ACS Omega*, 5(25), 15095–15104. DOI: 10.1021/acsomega.0c00928 PMID: 32637782

Ma, X., Yu, J., Jing, J., Zhao, Q., Ren, L., & Hu, Z. (2021). Optimization of sunflower head pectin extraction by ammonium oxalate and the effect of drying conditions on properties. *Scientific Reports*, 11(1), 10616. DOI: 10.1038/s41598-021-89886-x PMID: 34012041

Marić, M., Grassino, A. N., Zhu, Z., Barba, F. J., Brnčić, M., Brnčić, S. R. J. T. i. F. S., & Technology. (2018). An overview of the traditional and innovative approaches for pectin extraction from plant food wastes and by-products: Ultrasound-, microwaves-, and enzyme-assisted extraction. *76*, 28-37.

Matsumoto, T., Sugiura, Y., Kondo, A., & Fukuda, H. (2000). Efficient production of protopectinases by Bacillus subtilis using medium based on soybean flour. *Biochemical Engineering Journal*, 6(2), 81–86. https://doi.org/https://doi.org/10.1016/S1369-703X(00)00079-6. DOI: 10.1016/S1369-703X(00)00079-6 PMID: 10959081

May, C. D. J. C. p. (1990). Industrial pectins: Sources, production and applications. *12*(1), 79-99.

McNeil, M., Darvill, A. G., & Albersheim, P. J. P. p. (1980). Structure of plant cell walls: X. Rhamnogalacturonan I, a structurally complex pectic polysaccharide in the walls of suspension-cultured sycamore cells. *66*(6), 1128-1134.

Methacanon, P., Krongsin, J., & Gamonpilas, C. (2014). Pomelo (Citrus maxima) pectin: Effects of extraction parameters and its properties. *Food Hydrocolloids*, 35, 383–391. https://doi.org/https://doi.org/10.1016/j.foodhyd.2013.06.018. DOI: 10.1016/j.foodhyd.2013.06.018

Mishra, R., Majeed, A., & Banthia, A. J. I. J. o. P. T. (2011). Development and characterization of pectin/gelatin hydrogel membranes for wound dressing. *15*, 82-95.

Moalemiyan, M., Ramaswamy, H. S., & Maftoonazad, N. (2012). PECTIN-BASED EDIBLE COATING FOR SHELF-LIFE EXTENSION OF ATAULFO MANGO. *Journal of Food Process Engineering*, 35(4), 572–600. https://doi.org/https://doi.org/10.1111/j.1745-4530.2010.00609.x. DOI: 10.1111/j.1745-4530.2010.00609.x

Muhammad, K. Mohd. Zahari, N. I., Gannasin, S. P., Mohd. Adzahan, N., & Bakar, J. (2014). High methoxyl pectin from dragon fruit (Hylocereus polyrhizus) peel. *Food Hydrocolloids, 42*, 289-297. https://doi.org/https://doi.org/10.1016/j.foodhyd.2014.03.021

Müller-Maatsch, J., Bencivenni, M., Caligiani, A., Tedeschi, T., Bruggeman, G., Bosch, M., Petrusan, J., Van Droogenbroeck, B., Elst, K., & Sforza, S. (2016). Pectin content and composition from different food waste streams. *Food Chemistry*, 201, 37–45. https://doi.org/https://doi.org/10.1016/j.foodchem.2016.01.012. DOI: 10.1016/j.foodchem.2016.01.012 PMID: 26868545

Munarin, F., Guerreiro, S. G., Grellier, M. A., Tanzi, M. C., Barbosa, M. A., Petrini, P., & Granja, P. L. (2011). Pectin-Based Injectable Biomaterials for Bone Tissue Engineering. *Biomacromolecules*, 12(3), 568–577. DOI: 10.1021/bm101110x PMID: 21302960

Muñoz-Almagro, N., Montilla, A., Moreno, F. J., & Villamiel, M. J. U. S. (2017). Modification of citrus and apple pectin by power ultrasound: Effects of acid and enzymatic treatment. *38*, 807-819.

Muñoz-Almagro, N., Montilla, A., & Villamiel, M. J. F. R. I. (2021). Role of pectin in the current trends towards low-glycaemic food consumption. *140*, 109851.

Nakamura, A., Furuta, H., Maeda, H., Takao, T., Nagamatsu, Y. J. B., biotechnology,, & biochemistry. (2002). Structural studies by stepwise enzymatic degradation of the main backbone of soybean soluble polysaccharides consisting of galacturonan and rhamnogalacturonan. *66*(6), 1301-1313.

Ndeh, D., Rogowski, A., Cartmell, A., Luis, A. S., Baslé, A., Gray, J., Venditto, I., Briggs, J., Zhang, X., & Labourel, A. J. N. (2017). Complex pectin metabolism by gut bacteria reveals novel catalytic functions. *544*(7648), 65-70.

Nesic, A., Meseldzija, S., Cabrera-Barjas, G. A.-O., & Onjia, A. A.-O. (2022, January 26). Novel Biocomposite Films Based on High Methoxyl Pectin Reinforced with Zeolite Y for Food Packaging Applications. *Foods*, 11(3), 360. DOI: 10.3390/foods11030360 PMID: 35159510

Ngenefeme, F.-T. J., Eko, N. J., Mbom, Y. D., Tantoh, N. D., & Rui, K. W. (2013). *A one pot green synthesis and characterisation of iron oxide-pectin hybrid nanocomposite* (Vol. 3). Open Journal of Composite Materials.

Ngo, T. M. P., Nguyen, T. H., Dang, T. M. Q., Tran, T. X., & Rachtanapun, P. J. I. j. o. m. s. (2020). Characteristics and antimicrobial properties of active edible films based on pectin and nanochitosan. *21*(6), 2224.

Ngouémazong, D. E., Jolie, R. P., Cardinaels, R., Fraeye, I., Van Loey, A., Moldenaers, P., & Hendrickx, M. (2012). Stiffness of Ca2+-pectin gels: Combined effects of degree and pattern of methylesterification for various Ca2+ concentrations. *Carbohydrate Research*, 348, 69–76. https://doi.org/https://doi.org/10.1016/j.carres.2011.11.011. DOI: 10.1016/j.carres.2011.11.011 PMID: 22209690

Ninan, N., Muthiah, M., Park, I.-K., Elain, A., Thomas, S., & Grohens, Y. J. C. p. (2013). Pectin/carboxymethyl cellulose/microfibrillated cellulose composite scaffolds for tissue engineering. *98*(1), 877-885.

Nordin, N. N., Aziz, N. K., Naharudin, I., & Anuar, N. K. (2022). Effects of Drug-Free Pectin Hydrogel Films on Thermal Burn Wounds in Streptozotocin-Induced Diabetic Rats. *Polymers*, 14(14), 2873. DOI: 10.3390/polym14142873 PMID: 35890648

Noreen, A., Akram, J., Rasul, I., Mansha, A., Yaqoob, N., Iqbal, R., Tabasum, S., Zuber, M., & Zia, K. M. J. I. m. (2017). Pectins functionalized biomaterials; a new viable approach for biomedical applications. *RE:view*, 101, 254–272. PMID: 28300586

Norziah, M., Fang, E., & Abd Karim, A. (2000). Extraction and characterisation of pectin from pomelo fruit peels. *Gums and Stabilisers for the Food Industry*, 10, 27–36. DOI: 10.1533/9781845698355.1.27

O'Neill, M. A., Ishii, T., Albersheim, P., & Darvill, A. G. J. A. R. P. B. (2004). Rhamnogalacturonan II: structure and function of a borate cross-linked cell wall pectic polysaccharide. *55*, 109-139.

Ortiz-Sanchez, M., Solarte-Toro, J.-C., González-Aguirre, J.-A., Peltonen, K. E., Richard, P., & Cardona Alzate, C. A. (2020). Pre-feasibility analysis of the production of mucic acid from orange peel waste under the biorefinery concept. *Biochemical Engineering Journal*, 161, 107680. https://doi.org/https://doi.org/10.1016/j.bej.2020.107680. DOI: 10.1016/j.bej.2020.107680

Panwar, D., Panesar, P. S., & Chopra, H. K. (2023). Ultrasound-assisted extraction of pectin from Citrus limetta peels: Optimization, characterization, and its comparison with commercial pectin. *Food Bioscience*, 51, 102231. https://doi.org/https://doi.org/10.1016/j.fbio.2022.102231. DOI: 10.1016/j.fbio.2022.102231

Pasandide, B., Khodaiyan, F., Mousavi, Z. E., & Hosseini, S. S. (2017). Optimization of aqueous pectin extraction from Citrus medica peel. *Carbohydrate Polymers*, 178, 27–33. https://doi.org/https://doi.org/10.1016/j.carbpol.2017.08.098. DOI: 10.1016/j.carbpol.2017.08.098 PMID: 29050593

Pedrolli, D. B., Monteiro, A. C., Gomes, E., & Carmona, E. C. J. O. B. J. (2009). Pectin and pectinases: production, characterization and industrial application of microbial pectinolytic enzymes. 9-18.

Pereira, P. H., Oliveira, T. I., Rosa, M. F., Cavalcante, F. L., Moates, G. K., Wellner, N., Waldron, K. W., & Azeredo, H. M. (2016). Pectin extraction from pomegranate peels with citric acid. *International Journal of Biological Macromolecules*, 88, 373–379. DOI: 10.1016/j.ijbiomac.2016.03.074 PMID: 27044343

Popov, S., Paderin, N., Chistiakova, E., & Ptashkin, D. (2023). Serosal Adhesion Ex Vivo of Hydrogels Prepared from Apple Pectin Cross-Linked with Fe3+ Ions. *International Journal of Molecular Sciences*, 24(2), 1248. DOI: 10.3390/ijms24021248 PMID: 36674765

Powell, D. A., Morris, E. R., Gidley, M. J., & Rees, D. A. (1982). Conformations and interactions of pectins: II. Influence of residue sequence on chain association in calcium pectate gels. *Journal of Molecular Biology*, 155(4), 517–531. https://doi.org/https://doi.org/10.1016/0022-2836(82)90485-5. DOI: 10.1016/0022-2836(82)90485-5 PMID: 7086902

Prakash Maran, J., Sivakumar, V., Thirugnanasambandham, K., & Sridhar, R. (2013). Optimization of microwave assisted extraction of pectin from orange peel. *Carbohydrate Polymers*, 97(2), 703–709. https://doi.org/https://doi.org/10.1016/j.carbpol.2013.05.052. DOI: 10.1016/j.carbpol.2013.05.052 PMID: 23911504

Quadrado, R. F. N., Macagnan, K. L., Moreira, A. S., & Fajardo, A. R. (2022). Redox-responsive hydrogels of thiolated pectin as vehicles for the smart release of acetaminophen. *Reactive & Functional Polymers*, 181, 105448. https://doi.org/https://doi.org/10.1016/j.reactfunctpolym.2022.105448. DOI: 10.1016/j.reactfunctpolym.2022.105448

Rai, S. K., Chaturvedi, K., & Yadav, S. K. J. F. H. (2019). Evaluation of structural integrity and functionality of commercial pectin based edible films incorporated with corn flour, beetroot, orange peel, muesli and rice flour. *91*, 127-135.

Rajput, K., Tawade, S., Nangare, S., Shirsath, N., Bari, S., & Zawar, L. (2022). Formulation, optimization, and in-vitro-ex-vivo evaluation of dual-crosslinked zinc pectinate-neem gum-interpenetrating polymer network mediated lansoprazole loaded floating microbeads. *International Journal of Biological Macromolecules*, 222, 915–926. https://doi.org/https://doi.org/10.1016/j.ijbiomac.2022.09.216. DOI: 10.1016/j.ijbiomac.2022.09.216 PMID: 36181884

Ren, W., Qiang, T., & Chen, L. (2022). Recyclable and biodegradable pectin-based film with high mechanical strength. *Food Hydrocolloids*, 129, 107643. https://doi.org/https://doi.org/10.1016/j.foodhyd.2022.107643. DOI: 10.1016/j.foodhyd.2022.107643

Said, N. S., Olawuyi, I. F., & Lee, W. Y. (2023). Pectin Hydrogels: Gel-Forming Behaviors, Mechanisms, and Food Applications. *Gels (Basel, Switzerland)*, 9(9), 732. DOI: 10.3390/gels9090732 PMID: 37754413

Sarioglu, E., Arabacioglu Kocaaga, B., Turan, D., Batirel, S., & Guner, F. S. (2019). Theophylline-loaded pectin-based hydrogels. II. Effect of concentration of initial pectin solution, crosslinker type and cation concentration of external solution on drug release profile. *Journal of Applied Polymer Science*, 136(43), 48155. https://doi.org/https://doi.org/10.1002/app.48155. DOI: 10.1002/app.48155

Saucedo-Acuña, R. A., Meza-Valle, K. Z., Cuevas-González, J. C., Ordoñez-Casanova, E. G., Castellanos-García, M. I., Zaragoza-Contreras, E. A., & Tamayo-Pérez, G. F. (2023). Characterization and In Vivo Assay of Allantoin-Enriched Pectin Hydrogel for the Treatment of Skin Wounds. *International Journal of Molecular Sciences*, 24(8), 7377. DOI: 10.3390/ijms24087377 PMID: 37108540

Sayah, M. Y., Chabir, R., Benyahia, H., Rodi Kandri, Y., Ouazzani Chahdi, F., Touzani, H., & Errachidi, F. (2016). Yield, Esterification Degree and Molecular Weight Evaluation of Pectins Isolated from Orange and Grapefruit Peels under Different Conditions. *PLoS One*, 11(9), e0161751. DOI: 10.1371/journal.pone.0161751 PMID: 27644093

Sayed, M. A., Kumar, J., Rahman, M. R., Noor, F., & Alam, M. A. (2022). Effect of extraction parameters on the yield and quality of pectin from mango (Mangifera indica L.) peels. *Discover Food*, 2(1), 28. DOI: 10.1007/s44187-022-00029-1

Seixas, F. L., Fukuda, D. L., Turbiani, F. R. B., Garcia, P. S., Petkowicz, C. L. O., Jagadevan, S., & Gimenes, M. L. (2014). Extraction of pectin from passion fruit peel (Passiflora edulis f. flavicarpa) by microwave-induced heating. *Food Hydrocolloids*, 38, 186–192. https://doi.org/https://doi.org/10.1016/j.foodhyd.2013.12.001. DOI: 10.1016/j.foodhyd.2013.12.001

Shin, Y. A.-O., Chane, A., Jung, M., & Lee, Y. A.-O. (1712). Recent Advances in Understanding the Roles of Pectin as an Active Participant in Plant Signaling Networks. *Plants*, 10(8), 1712. Advance online publication. DOI: 10.3390/plants10081712 PMID: 34451757

Smith, D. A. (2003). JAMS AND PRESERVES | Methods of Manufacture. In Caballero, B. (Ed.), *Encyclopedia of Food Sciences and Nutrition* (2nd ed., pp. 3409–3415). Academic Press., https://doi.org/https://doi.org/10.1016/B0-12-227055-X/00660-X DOI: 10.1016/B0-12-227055-X/00660-X

Song, R., Murphy, M., Li, C., Ting, K., Soo, C., & Zheng, Z. (2018, September). Current development of biodegradable polymeric materials for biomedical applications. *Drug Design, Development and Therapy*, 12, 3117–3145. DOI: 10.2147/DDDT.S165440 PMID: 30288019

Spinei, M., & Oroian, M. (2022, July 26). Microwave-assisted extraction of pectin from grape pomace. *Scientific Reports*, 12(1), 1–17. DOI: 10.1038/s41598-022-16858-0 PMID: 35882905

Sriamornsak, P., Wattanakorn, N., & Takeuchi, H. J. C. p. (2010). Study on the mucoadhesion mechanism of pectin by atomic force microscopy and mucin-particle method. *79*(1), 54-59.

Sriamornsak, P. J. E. o. o. d. d. (2011). Application of pectin in oral drug delivery. *8*(8), 1009-1023.

Srivastava, P., & Malviya, R. (2011). Sources of pectin, extraction and its applications in pharmaceutical industry - An overview. *Indian Journal of Natural Products and Resources*, 2, 10–18.

Sucheta, C. K., Sharma, N., & Yadav, S. K. (2019). Composite edible coatings from commercial pectin, corn flour and beetroot powder minimize post-harvest decay, reduces ripening and improves sensory liking of tomatoes. *International Journal of Biological Macromolecules, 133*, 284-293. https://doi.org/https://doi.org/10.1016/j.ijbiomac.2019.04.132

Sun, R., Niu, Y., Li, M., Liu, Y., Wang, K., Gao, Z., Wang, Z., Yue, T., & Yuan, Y. (2023). Emerging trends in pectin functional processing and its fortification for synbiotics: A review. *Trends in Food Science & Technology*, 134, 80–97. https://doi.org/https://doi.org/10.1016/j.tifs.2023.03.004. DOI: 10.1016/j.tifs.2023.03.004

Sutar, P. B., Mishra, R. K., Pal, K., & Banthia, A. K. (2008). Development of pH sensitive polyacrylamide grafted pectin hydrogel for controlled drug delivery system. *Journal of Materials Science. Materials in Medicine*, 19(6), 2247–2253. DOI: 10.1007/s10856-007-3162-y PMID: 17619970

Swenson, H., Miers, J., Schultz, T., & Owens, H. J. F. T. (1953). Pectinate and pectate coatings. II. Application to nuts and fruit products. *7*(6), 232-235.

Tongkham, N., Juntasalay, B., Lasunon, P., & Sengkhamparn, N. (2017). Dragon fruit peel pectin: Microwave-assisted extraction and fuzzy assessment. *Agriculture and Natural Resources (Bangkok)*, 51(4), 262–267. https://doi.org/https://doi.org/10.1016/j.anres.2017.04.004. DOI: 10.1016/j.anres.2017.04.004

Toniazzo, T., & Fabi, J. P. (2023). Versatile Polysaccharides for Application to Semi-Solid and Fluid Foods: The Pectin Case. *Fluids (Basel, Switzerland)*, 8(9), 243. Advance online publication. DOI: 10.3390/fluids8090243

Tovar, A. K., Godínez, L. A., Espejel, F., Ramírez-Zamora, R.-M., & Robles, I. (2019). Optimization of the integral valorization process for orange peel waste using a design of experiments approach: Production of high-quality pectin and activated carbon. *Waste Management, 85*, 202-213. https://doi.org/https://doi.org/10.1016/j.wasman.2018.12.029

Tran Vo, T. M., Kobayashi, T., & Potiyaraj, P. (2022). Viscoelastic Analysis of Pectin Hydrogels Regenerated from Citrus Pomelo Waste by Gelling Effects of Calcium Ion Crosslinking at Different pHs. *Gels (Basel, Switzerland)*, 8(12), 814. Advance online publication. DOI: 10.3390/gels8120814 PMID: 36547338

Tsouko, E., Maina, S., Ladakis, D., Kookos, I. K., & Koutinas, A. (2020). Integrated biorefinery development for the extraction of value-added components and bacterial cellulose production from orange peel waste streams. *Renewable Energy*, 160, 944–954. https://doi.org/https://doi.org/10.1016/j.renene.2020.05.108. DOI: 10.1016/j.renene.2020.05.108

Tsuru, C., Umada, A., Noma, S., Demura, M., Hayashi, N. J. F., & Technology, B. (2021). Extraction of pectin from Satsuma mandarin orange peels by combining pressurized carbon dioxide and deionized water: A green chemistry method. *14*, 1341-1348.

Turquois, T., Rinaudo, M., Taravel, F. R., & Heyraud, A. (1999). Extraction of highly gelling pectic substances from sugar beet pulp and potato pulp: Influence of extrinsic parameters on their gelling properties. *Food Hydrocolloids*, 13(3), 255–262. https://doi.org/https://doi.org/10.1016/S0268-005X(99)00007-7. DOI: 10.1016/S0268-005X(99)00007-7

Van Hung, P., Anh, M. N. T., Hoa, P. N., & Phi, N. T. L. (2021). Extraction and characterization of high methoxyl pectin from Citrus maxima peels using different organic acids. *Journal of Food Measurement and Characterization*, 15(2), 1541–1546. DOI: 10.1007/s11694-020-00748-y

Vasco-Correa, J., & Zapata Zapata, A. D. (2017). Enzymatic extraction of pectin from passion fruit peel (Passiflora edulis f. flavicarpa) at laboratory and bench scale. *Lebensmittel-Wissenschaft + Technologie*, 80, 280–285. https://doi.org/https://doi.org/10.1016/j.lwt.2017.02.024. DOI: 10.1016/j.lwt.2017.02.024

Vriesmann, L. C., Teófilo, R. F., & Lúcia de Oliveira Petkowicz, C. (2012). Extraction and characterization of pectin from cacao pod husks (Theobroma cacao L.) with citric acid. *Lebensmittel-Wissenschaft + Technologie*, 49(1), 108–116. DOI: 10.1016/j.lwt.2012.04.018

Wai, W. W., Alkarkhi, A. F., Easa, A. M. J. F., & Processing, B. (2010). Effect of extraction conditions on yield and degree of esterification of durian rind pectin: An experimental design. *88*(2-3), 209-214.

Wang, M., Huang, B., Fan, C., Zhao, K., Hu, H., Xu, X., Pan, S., & Liu, F. (2016). Characterization and functional properties of mango peel pectin extracted by ultrasound assisted citric acid. *International Journal of Biological Macromolecules*, 91, 794–803. https://doi.org/https://doi.org/10.1016/j.ijbiomac.2016.06.011. DOI: 10.1016/j.ijbiomac.2016.06.011 PMID: 27283236

Wang, W., Chen, W., Zou, M., Lv, R., Wang, D., Hou, F., Feng, H., Ma, X., Zhong, J., Ding, T., Ye, X., & Liu, D. (2018). Applications of power ultrasound in oriented modification and degradation of pectin: A review. *Journal of Food Engineering*, 234, 98–107. https://doi.org/https://doi.org/10.1016/j.jfoodeng.2018.04.016. DOI: 10.1016/j.jfoodeng.2018.04.016

Wang, W., Ma, X., Xu, Y., Cao, Y., Jiang, Z., Ding, T., Ye, X., & Liu, D. (2015). Ultrasound-assisted heating extraction of pectin from grapefruit peel: Optimization and comparison with the conventional method. *Food Chemistry*, 178, 106–114. https://doi.org/https://doi.org/10.1016/j.foodchem.2015.01.080. DOI: 10.1016/j.foodchem.2015.01.080 PMID: 25704690

Wang, X., Chen, Q., & Lü, X. (2014). Pectin extracted from apple pomace and citrus peel by subcritical water. *Food Hydrocolloids*, 38, 129–137. https://doi.org/https://doi.org/10.1016/j.foodhyd.2013.12.003. DOI: 10.1016/j.foodhyd.2013.12.003

Wicker, L., & Kim, Y. (2016). Pectin and Health. In Caballero, B., Finglas, P. M., & Toldrá, F. (Eds.), *Encyclopedia of Food and Health* (pp. 289–293). Academic Press., https://doi.org/https://doi.org/10.1016/B978-0-12-384947-2.00532-8 DOI: 10.1016/B978-0-12-384947-2.00532-8

Willats, W. G., Knox, J. P., Mikkelsen, J. D. J. T. i. F. S., & Technology. (2006). Pectin: new insights into an old polymer are starting to gel. *17*(3), 97-104.

Willats, W. G., McCartney, L., Mackie, W., & Knox, J. P. (2001). Pectin: Cell biology and prospects for functional analysis. *Plant Molecular Biology*, 47(1/2), 9–27. DOI: 10.1023/A:1010662911148 PMID: 11554482

Wolf, S., Mouille, G., & Pelloux, J. (2009). Homogalacturonan Methyl-Esterification and Plant Development. *Molecular Plant*, 2(5), 851–860. https://doi.org/https://doi.org/10.1093/mp/ssp066. DOI: 10.1093/mp/ssp066 PMID: 19825662

Woo, K. K., Chong, Y. Y., Li Hiong, S. K., & Tang, P. Y. (2010). Pectin Extraction and Characterization from Red Dragon Fruit (Hylocereus polyrhizus): A Preliminary Study. *Journal of Biological Sciences (Faisalabad, Pakistan)*, 10(7), 631–636. DOI: 10.3923/jbs.2010.631.636

Xu, S.-Y., Liu, J.-P., Huang, X., Du, L.-P., Shi, F.-L., Dong, R., Huang, X.-T., Zheng, K., Liu, Y., & Cheong, K.-L. (2018). Ultrasonic-microwave assisted extraction, characterization and biological activity of pectin from jackfruit peel. *Lebensmittel-Wissenschaft + Technologie*, 90, 577–582. https://doi.org/https://doi.org/10.1016/j.lwt.2018.01.007. DOI: 10.1016/j.lwt.2018.01.007

Yang, X., Nisar, T., Liang, D., Hou, Y., Sun, L., & Guo, Y. (2018). Low methoxyl pectin gelation under alkaline conditions and its rheological properties: Using NaOH as a pH regulator. *Food Hydrocolloids*, 79, 560–571. DOI: 10.1016/j.foodhyd.2017.12.006

Yang, Y., Zhang, G., Hong, Y., Gu, Z., & Fang, F. (2013). Calcium cation triggers and accelerates the gelation of high methoxy pectin. *Food Hydrocolloids*, 32(2), 228–234. DOI: 10.1016/j.foodhyd.2013.01.003

Yilmaz-Turan, S., Gál, T., Lopez-Sanchez, P., Martinez, M. M., Menzel, C., & Vilaplana, F. (2023). Modulating temperature and pH during subcritical water extraction tunes the molecular properties of apple pomace pectin as food gels and emulsifiers. *Food Hydrocolloids*, 145, 109148. https://doi.org/https://doi.org/10.1016/j.foodhyd.2023.109148. DOI: 10.1016/j.foodhyd.2023.109148

Yuliarti, O., Hoon, A. L. S., & Chong, S. Y. (2017). Influence of pH, pectin and Ca concentration on gelation properties of low-methoxyl pectin extracted from Cyclea barbata Miers. *Food Structure*, 11, 16–23. DOI: 10.1016/j.foostr.2016.10.005

Yuliarti, O., Matia-Merino, L., Goh, K. K. T., Mawson, J., Williams, M. A. K., & Brennan, C. (2015). Characterization of gold kiwifruit pectin from fruit of different maturities and extraction methods. *Food Chemistry*, 166, 479–485. https://doi.org/https://doi.org/10.1016/j.foodchem.2014.06.055. DOI: 10.1016/j.foodchem.2014.06.055 PMID: 25053083

Zhang, H., Chen, J., Li, J., Yan, L., Li, S., Ye, X., Liu, D., Ding, T., Linhardt, R. J., & Orfila, C. J. F. H. (2018). Extraction and characterization of RG-I enriched pectic polysaccharides from mandarin citrus peel. *79*, 579-586.

Zhang, W., Xu, P., & Zhang, H. (2015). Pectin in cancer therapy: A review. *Trends in Food Science & Technology*, 44(2), 258–271. https://doi.org/https://doi.org/10.1016/j.tifs.2015.04.001. DOI: 10.1016/j.tifs.2015.04.001

Zheng, C., Zou, Y., Huang, Y., Shen, B., Fei, P., & Zhang, G. J. F. H. (2023). Biosynthesis of amidated pectins with ultra-high viscosity and low gelation restriction through ultra-low temperature enzymatic method. *134*, 108037.

Zheng, J., Chen, J., Zhang, H., Wu, D., Ye, X., Linardt, R. J., & Chen, S. (2020). Gelling mechanism of RG-I enriched citrus pectin: Role of arabinose side-chains in cation- and acid-induced gelation. *Food Hydrocolloids*, 101, 105536. https://doi.org/https://doi.org/10.1016/j.foodhyd.2019.105536. DOI: 10.1016/j.foodhyd.2019.105536

Zhou, M., Bi, J., Chen, J., Wang, R., & Richel, A. (2021). Impact of pectin characteristics on lipid digestion under simulated gastrointestinal conditions: Comparison of water-soluble pectins extracted from different sources. *Food Hydrocolloids*, 112, 106350. https://doi.org/https://doi.org/10.1016/j.foodhyd.2020.106350. DOI: 10.1016/j.foodhyd.2020.106350

Chapter 12
Biomass Hydrogel Drug and Ultrasound Delivery Therapy Technology

Harshani Iresha
https://orcid.org/0009-0006-1605-3846
University of Peradeniya, Sri Lanka

Tu Minh Tran Vo
https://orcid.org/0000-0002-6785-4188
Nagaoka University of Technology, Japan

ABSTRACT

For drug supply systems, there are various applications including the medical field and natural elements are generally polysaccharides like cellulose, chitin and others exhibit as key feature of biocompatibility, biodegradability, non-toxicity, drug-loading ability for regenerated matrix, especially for tissue regeneration and drug release. Highlighting and considering current research with biomass hydrogels of cellulose and chitin and new findings that drug delivery (DD) and drug release (DR) under external stimuli are important and becomes a symbolic area of future research. Therefore, this chapter describes present trends on biomass hydrogel drugs in the DD and DR under external smart stimulation. Among external stimuli, the control of drug release by ultrasonic external stimuli, which can be penetrated from outside the body, is mainly mentioned in this chapter because of its advanced features.

INTRODUCTION

Background of Biomass Hydrogel Composed of Cellulose and Chitins

Biomass polymer hydrogels are a modern sustainable material fabricated using naturally available polysaccharides, e.g., cellulose, hemicellulose, chitin, pectin, carrageenan, etc., that are extracted from natural sources such as sugar cane bagasse (Nakasone et al., 2016), cotton (Iresha & Kobayashi, 2021b; Song et al., 2015), agave fibers (Tovar-carrillo et al., 2013), crab shells (Iresha & Kobayashi, 2021a; Jiang & Kobayashi, 2017b), shrimp shells (Zhao et al., 2019), Sea weeds (Martín-del-Campo et al.,

DOI: 10.4018/979-8-3693-0003-9.ch012

2021), woods (Hosseinaei et al., 2012) etc. In order to utilize these biomass polymer materials, the first step in the process is the extraction process of these effective materials from biomass resources. Extraction of natural polysaccharides from available natural sources is highly systematized depending on the nature of the natural source which is another field of study for curious researchers. Here, the method of extraction of natural polymers is important for the hydrogel properties. This is because non-toxicity, biocompatibility, and material strength are important factors in the properties of these materials needed for use in the medical field, and these are highly dependent on the extraction and purification processes of the polysaccharides (Kaczmarek et al., 2020). While the properties of biomass materials depend on the polymer properties, another important factor is whether the material has characteristics similar to those of living organisms. Hydrogels, like living organisms, are materials that retain large amounts of water, and the contribution of the sugar chains that form the backbone of biomass materials, along with this water retention property, can provide non-toxic and biocompatible properties. Cellulose and chitin are materials that can impart these properties (Figure 1). However, their insolubility in solvents and difficulties in extraction, purification, and formability have hindered the development of materials with superior properties. Therefore, this is developing another wider focusing area to investigate. Interestingly, hydrogels are 3-dimensional swelling materials in their primary structure with polymer network, creating water retention condition in the inter- or intra-molecular interactions of polysaccharides (Kobayashi,2015) and also make them favorable in different fields of advanced applications including cytocompatability (Tovar-carrillo et al., 2013), biocompatibility (Nakasone et al., 2016), smart drug release (Iresha & Kobayashi, 2021b; Jiang et al., 2016). So, the excellent water adsorption capacity is the most iconic property (Pérez-Luna & González-Reynoso, 2018). Other important properties requiring hydrogel for drug releasing applications are of entrapment of drug, strength and flexibility to stimulate against external triggered force. For example, cellulose hydrogels which were fabricated from cellulose extracted from rice husks and oat husks exhibited filamentous arrangement of cellulose fibers arranged to create such appropriate network in the hydrogel (Oliveira et al., 2017). Also, chitin hydrogel with homogenous microstructure was noted with chitin hydrogels prepared super supercritical carbon dioxide (Tsioptsias & Panayiotou, 2008).

Figure 1. Chemical structure of (a) cellulose and (b) chitin

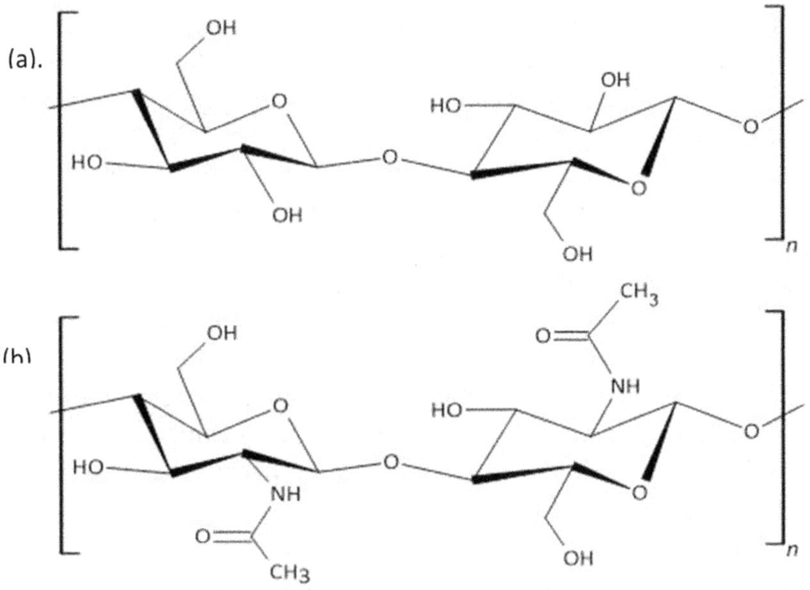

In the preparation process, chitin dissolved directly in NaOH/ Urea solution under lower temperature in which hydrogels were successfully fabricated and showed uniform porous structure (Chang et al., 2011). In this case, as following to dissolve completely in NaOH/urea aqueous solution via the freezing/thawing method to prepare transparent solutions, subsequently, hydrogels were prepared directly from the chitin solution in the NaOH/urea aqueous system, exhibiting a 293T cell viability for their excellent biocompatibility and safety. Since such natural polysaccharides bring their own characteristic chemical structures, hence unique functions are essential for their nature-assigned duties. Therefore, when consuming polysaccharides for the uses of mankind, the resulting products exhibit characteristics and qualities that align with the chemical nature of cellulose and chitin.

As seen in Figure 1 (a), cellulose structure is composed with β-(1→4) glycosidic bonds connecting D-glucose units. The presence of -OH and -O- functional groups leading to strong inter and intra-molecular hydrogen forces make cellulose fibrils rigid and crystalline (Valle et al., 2017). Thus, the original function of cellulose as a primary element which is to determine the entire cell wall structure can be strongly proposed by the rigid-rodlike crystalline cellulose fibrils. Contrary to cellulose, in its structural diference, chitin exhibits a similarity, but there is a notable difference in which the hydroxyl group on the second carbon is replaced by an acetamide group, as illustrated in Figure 1 (b). This arrangement promotes stronger hydrogen bonds than cellulose due to the existence of number of sites in chitin, i.e., -NH, -C=O, -CH_3-OH, and -OH which preferred hydrogen bonding in its networking (Kameda et al., 2005). Therefore, the exoskeletal structures of insects, including crabs and shrimps which are created by chitin as its structural element, made them stiff facilitating the original functions (Duan et al., 2018). Nonetheless, chitin's inability to dissolve in typical organic solvents is attributed to its supramolecular structure, which is characterized by numerous inter- and intramolecular hydrogen bonds. This structural feature poses challenges in processing and limits its range of applications (Zhong et al., 2020). In addition, it contains *N*-acetylglucosamine groups, with the acetyl groups partially hydrolyzed to amino groups.

Despite its polar structure, cellulose widely recognizesthat hardly soluble, but absorbs water. This characteristic is primarily attributed to the existence of an extensive network of hydrogen bonds and, as recent research has highlighted, hydrophobic interactions (Gubitosi et al., 2017). Therefore, utilizing these remarkable natural materials was a challenging task until the dissolution was achieved by using lithium chloride (LiCl) in N, N-dimethylacetamide (DMAc) as the universal solvent for the dissolution of cellulose and chitin (Kobayashi, 2015; Tovar-carrillo et al., 2013). In their film fabrication process, phase inversion of the polysaccharide solutions dissolved in LiCl/ DMAc solvent facilitates solvent exchange inducing the liquid to solid transition, making a perfectly shaped hydrogel films, which are is flexible, has excellent mechanical strength, and contains more than 1,000 times its own weight in water. Considering cellulose and chitin hydrogels, the former hydrogel is translucent compared to chitin hydrogel due to more ordered crystalline arrangement of cellulose molecules compared to chitin (Dassanayake et al., 2018). Besides, water holding capacity of cellulose and chitin also differed for almost similar cellulose and chitin loadings (Table 2) which is attributable to the different chemical nature of cellulose and chitin and alterations in the viscoelastic properties. In contrast, the latter chitin having hydroxyl group (-OH) and acetyl amide group (CH_3-O-CO-NH) caused higher water retention, and there was a report for bio- and cytocompatible properties (Kostag & El Seoud, 2021).

In the case of cellulose hydrogels, excellent cytocompatibility of agave hydrogel films (Tovar-carrillo et al., 2013), wooden pulped cellulose interpenetrated with hydroxy ethyl cellulose (Tovar-carrillo et al., 2014) and sugarcane bagasse hydrogel (Nakasone& Kobayashi, 2016) was seen by *in vitro* assays of NIH 3T3 fibroblast cells, emphasizing the prospective uses of the hydrogels as tissue engineering scaffolds due to its excellent fibroblast cell growth on the cellulose hydrogels. Furthermore, *in vivo* test using mice was conducted for cellulose hydrogels prepared from sugar cane bagasse cellulose (Nakasone et al., 2016). Their study results revealed excellent functionality of cyto and bio-compatibilities as prolonged with the mechanical strength of hydrogel over the period of 3-4 weeks duty in the living mice (Nakasone et al., 2016). Therefore, the use of biomass polymer hydrogels with such excellent skin affinity and superior biocompatibility as polymer matrices for sustained drug release represents a tremendous advantage. Accordingly, the approach of loading drugs on cellulose and chitin hydrogel matrices can be applied not only as drug delivery system but also as regenerative drugs, which is a development suitable for a sustainable society as upcycle materials recycled from food waste.

Drug Supply System With Medicines Loaded on Recycled Biomass Hydrogels

Prior of detailing about biomass derived hydrogels and its application in drug supply systems, it is worth of introducing key sectors for biomass hydrogels in biomedical therapies with number of evolving advanced technologies. The history evidences the progress of drug delivery system dating back to 1952 with the first invention of controlled drug release formulation of dextroamphetamine (Dexedrine) drug by Smith Klein Beecham (Yun et al., 2015). As introduced by Park (2014) the progression in the first generation was oral and transdermal mechanisms with basic dissolution, diffusion, osmosis and ion-exchange methods (Park, 2014). Then, in the second-generation delivery systems (1980 – 2010), smart functionalities were investigated in drug delivery as called by "smart" drug delivery system (SDDS) referring to the controlled or sustained drug administration by using smart polymers (Kabanov et al., 1992) and hydrogels (Huffman et al., 1986; Siegel et al., 1988). Such controlled and stimulated drug supply systems were achieved by the surrounding factors like pH (Dong & Hoffman, 1991; Siegel et al., 1988) and temperature (Huffman et al., 1986). Progress has been made of nanotechnology-based research commenced in the late part of this generation opening up a wide space for more research paths in the systems (Kawano et al., 2006; Ow Sullivan et al., 2003).

Briefing the history of drug supply systems, the elaboration about the technologies in the SDDS is expressed hereafter. The ultimate desires of the SDDS for a specific disease or a treatment are effective and efficient in drug administration to the targeted site with less or no drug toxicity. Thus, designing of a SDDS is a mutual combination of drug or medicine, administration route, drug carrier, potential stimulant, and drug release and delivery mechanisms (Park, 2014). Once the drug is defined for a specific therapy, by knowing physicochemical properties of the drug and the releasing kinetics of drug releasing from the matrix. Simultaneously, drug loading material behaves releasing with combining with driving mechanism by external trigger. In such SDDS, however, stimulation mechanism to drug supply depends on mode of stimulation, material, delivery route etc.

Stimulants in Smart Drug Delivery Systems

Stimulant or the trigger is the most emphasizing feature in smart drug delivery systems, when combined external stimulant with medicine loaded matrix (Iresha & Kobayashi, 2020; Nayak, 2023). The reason could be the unique stimulating mechanism exert by each stimulant in a specific SDDS which ultimately facilitate the drug administration. As shown in Figure 2, stimulants can be categorized as both exogenous and endogenous stimulants. Here, exogenous stimulant refers to the external stimuli such as temperature, electric fields, light, ultrasound, and magnetic fields, while the endogenous stimulants are the pH, enzymes, and redox changes(Mura et al., 2013). These stimulants have different trigger mechanisms for drug supply system which is thus applicable to specific treatments depending on the drug carries and targeting sites.

Figure 2. Exogenous and endogenous stimulants in triggering drug supply to human body

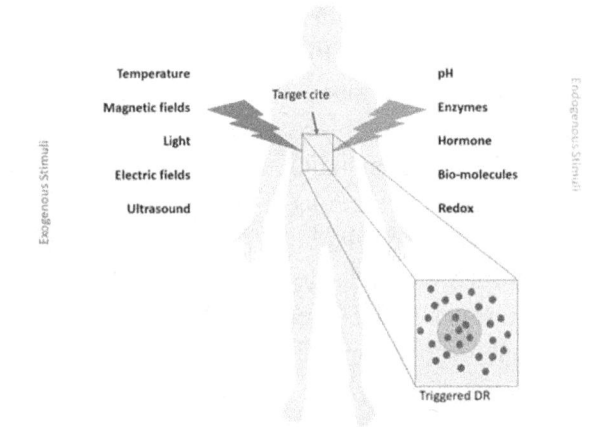

US Stimulant for SDD and SDR Applications

Among these stimuli, ultrasound (US) stimulant has tributed to its exclusive characteristics which can be coupled with green engineering applications. Compilation of positive and negative features of some existing stimulants are presented in Table 1 for readers with interest. In the medical field, US is widely established as a diagnostic tool, and it is called sonography in practical medical uses. Currently, sonography is applied for the diagnoses in vascular, abdominal, obstetrics, gynecology, neurology, cardiography, ophthalmology, etc. (Ranganayakulu, 2016). Thus, extending US for therapeutic applications is obvious, since this takes advantage of the characteristic of US to pass through the human body, a feature not found in other external stimuli.

It is known that US is pressure wave that transmitted through a tangible medium such as water by compressing and expanding the molecules (Boissenot et al., 2016; Pitt et al., 2004). The range of US frequencies is above 20 kHz. Applications of US depend on the US frequency, and such frequencies have dividing at low (20 – 200 kHz), intermediate (200 – 700 kHz), medium (0.7 – 3 MHz), and high (1 – 20 MHz) ranges (Ahmadi et al., 2012). The low-frequency US is typical for therapeutic medical applications (Ahmadi et al., 2012) as a promising stimulant for DR and DD systems because US power can be transmitted through the body organs with frequency-dependent penetrability, thereby can act on the focused biological systems. Also, US is a non-invasive stimulant to the human and animal, while it can be easily manipulated externally (Boissenot et al., 2016).

As an external trigger source, there are following characteristics of US. Under precise monitoring and controlling, the local temperature of the body for a specific time of treatment, as used as favorable and specific therapy, was elevated by the hyperthermic effects of US (Kennedy et al., 2003). For that, high intensity focused US developed mainly studies for tumor therapies. For the US drug supply system in the applications, thermo-sensitive liposomes widely utilized carriers in US-induced hyperthermic cancer treatment applications. (Lyon et al., 2021; Mylonopoulou et al., 2013; Santos et al., 2017; Tacker & Anderson, 1982). Another incident of US is collapse of cavitation bubbles, i.e., transient, or inertial cavitation, generating more microbubbles. Thus, the process continues, emitting shock waves to damage the drug carriers, enhanced drug delivery (Sirsi & Borden, 2014), and led polymer decomposition (Jiang & Kobayashi, 2017a; Kost et al., 1989). Furthermore, transient cavitation induced membrane permeabilization, thus enhancing the drug transportation (Sundaram et al., 2003). The US-induced drug permeation through the skin was also possible due to the cavitation effect at the low-frequency range (Huebsch et al., 2014; Mitragotri, 2005; Mitragotri et al., 1995; Polat et al., 2010). In addition, acoustic streaming, the circulatory motion of the liquid near to the oscillating bubbles (Nyborg, 1982), and become one of the characteristics of US inducing the drug release due to the motions of the surrounding liquid of vibrating micro-drug (Baker et al., 2001; Marmottant & Hilgenfeldt, 2003; Nyborg, 1982). Microstreaming represents a mechanically potent phenomenon of significant power. Thus, US has exerted sufficient effect on matrix permeability and stimulation cell activities (Baker et al., 2001; Rooney, 1970). Also, when the US propagates in the medium the bulk streaming was caused by US waves in the medium. Such radiation forces were capable of transporting particles near the targeted tissues (Dayton et al., 2002; Shortencarier et al., 2004). Table 1 presents in summary of these distinctive actions of US, making it more readily available than other external stimuli such as temperature, light, pH, electric fields, and magnetic fields. Therefore, in combination with the biocompatible polymer hydrogel drug, it is possible to construct an efficient drug release system that is operated with exogenous stimulants.

Table 1. Positive and negative characteristics of some common stimulants in DD and DR

Stimulant	Positive	Negative
Temperature	• Cancer cells are sensitive to hyperthermic conditions thus increased control of cell growth • Several technologies such as magnetic, US, light can be used to induce heat	• Complication in technologies • Drug carrier fabrication is difficult with complex methodologies • Risk of internal tissues or organs due to applied external heat in deep penetration • Biocompatibility of drug carriers is questionable
Magnetic fields	• Stimulated DR, magnetically guided drug targeting and contrast agent for imaging is possible with magnetic fields • Reached to clinical trials for magnetically stimulated DD and DR	• Complex equipment set for focusing, intensity controlling and deep penetration • Health risk of exposing to magnetic fields are yet unsolved • Possible toxicity by iron oxide and other chemicals used in drug carriers
Light	• Non-invasive nature • High degree of spatiotemporal precision	• Low penetrating depth due to strong scattering properties of soft tissues • Complex fabrication procedures of drug carriers • Questionable biocompatibility of drug carriers
Electric fields	• Accessibility of electric fields due to the existing iontophoresis devices in clinics • Safe levels of electric fields have been studied extensively	• Risk if damage to internal tissues by the electric sources in deep penetration • Electro-responsiveness is subjected to change due to environmental factors like presence of ionizable molecules, concentration of electrolytes.
pH	• Small changes of pH make the drug carriers responsive in DR by the changes of chemical and physical functions • Abnormal cites in body such as cancers and inflation show characteristic pH changes thus can be used as a stimulant without using outside stimulus.	• pH changes in diseased cites could be vary depend on the patient and nature of the disease • Drug carriers may react adversely by the altered compositions of the targeted due to disease
Ultrasound	• Non-invasive nature • Deep penetrability into tissues • Ultrasound is already a well-established technology in medicine thus capable of modifying for DD and DR • Several mechanisms to stimulate DD and DR by oscillatory motion, acoustic streaming, cavitation, thermal effect. • Easy to use • Outside to body manipulation ability	• Drug carriers need to be developed • Long exposure may affect for healthy tissues

(Alsawaftah et al., 2022; Boissenot et al., 2016; Golovin et al., 2021; P. Mi, 2020; Ribeiro et al., 2022)

Preparation of Biomass Hydrogels and Drug Entrapment Methods

When dealing with biomass hydrogels for drug delivery, a crucial aspect involves the fabrication of the material and the incorporation of medications. Nonetheless, drug loading remains a challenging aspect that demands careful consideration during the fabrication of drug-hydrogel systems. Most common three methods of drug encapsulation into BPHGs are shown in schematic diagrams as given in Figure 3.

In method 1, if the biomass polymer and medicine dissoluble in solvent, the drug solution is prepared in the same solvent as the natural polymer. As an illustration, in the case of cellulose and chitin, the co-solvent was DMAc/LiCl (Kobayashi, 2015), allowing the drug to attach to the hydrogel matrix via hydrogen bonds or similar linkages prior to the solidification of hydrogels, during polymer transformation. But, it ensured that DMAc is difficult in evaporation due to higher boiling temperature at 153

°C and lower vapor pressure at 0.3 kPa (20 °C). Therefore, by employing a solvent exchange process in which the solvent is eliminated through water exchange, thorough solvent removal can be achieved. In this method for wet filming, the drug also dissolves with the solvent from the solidified polymer matrix into the coagulation bath.

In the method-2, utilizing a process where the drug is adsorbed into the hydrogel after the preparation of hydrogels and subsequent immersion in a drug solution (Lei et al., 2022). In this case, when there is a robust interaction force between the drug and the gel carrier, the gel is more likely to retain the drug. Conversely, in cases where the interaction force is weaker, the drug loading tends to be less efficient. Another approach is method-3, which involves combining methods 1 and 2, showing in steps as contact of drug with the polymer prior of molding. After the gelation of drug-polymer solution, the drug encapsulation is carried out, but, limiting for special type of polymers which promotes self-gelation with time or after cooling. The most common way of drug entrapment is method-1 and then the rest of the steps depend on the gelation properties of the drug-polymer solution.

Figure 3. Common methods of drug encapsulation into biomass hydrogels

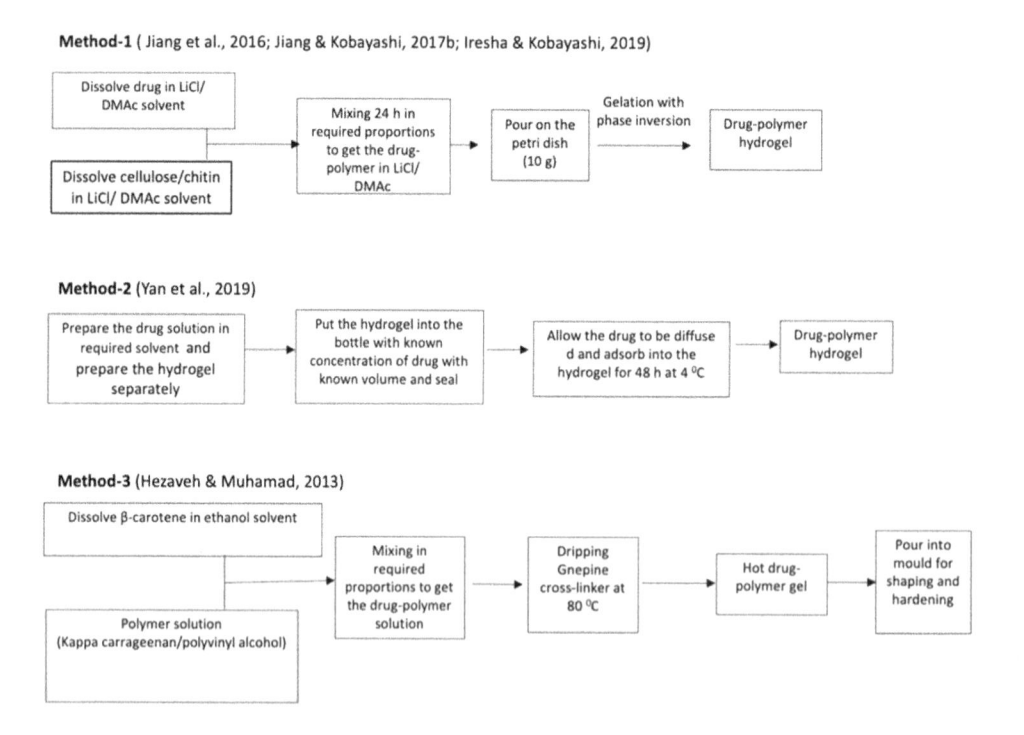

Characteristics of Cellulose and Chitin Hydrogels Loading Medicines

For emphasizing characteristic features of biomass hydrogels, combination of external triggers used in stimuli to drug release serves inevitable procedures. There are many kinds of drug carriers or matrices namely, nanoparticles (You et al., 2012), liposomes (Ranjan et al., 2012), micelles (Yu et al., 2015), hydrogels (Gutowska et al., 1992; Hezaveh & Muhamad, 2013), transdermal patches (Im et al., 2010), etc. as designed for desired applications of drug supply systems. Out of these drug carriers, hydrogels are one of the advanced materials in biomedical applications. It is possible to have characteristics of the hydrogel network at both physical and chemical levels for drug loading using interactions between the hydrogel and the drug (Li & Mooney, 2016). In particular, as seen in Figure 3 for cellulose-nicotine hydrogels (Iresha & Kobayashi, 2021b), the images of nicotine-loaded cellulose hydrogels present different yellow colors, depending on the loaded amounts of the medicine in the cellulose. Actually, the loaded amounts of nicotines were 179 ± 15, 602 ± 68, and 537 ± 68 mg/g-dry cellulose, creating with different cellulose quantities for 0.45 wt% (a), 0.9 wt% (b) and 1.8 wt% (c). There was tendency that higher cellulose concentration loaded higher nicotines in the cellulose hydrogels. As the cellulose content increases, the nicotine hydrogels became denser since the densities obtained were 1.010, 1.016, and 1.021, but, water contents % were 3075 ± 5, 1971 ± 3, 1329 ± 1 for (a), (b) and (c), respectively. The porous structure and water retention % of cellulosic hydrogels depended on the type and concentration of polysaccharides used. Secondly, to avoid adverse reactions in the body, biocompatibility is the crucial characteristic. This could be accomplished by utilizing natural polymers like pectin, chitosan, and alginate (Hezaveh & Muhamad, 2013; Huebsch et al., 2014; Jiang et al., 2016; Jiang & Kobayashi, 2017b), or by employing synthetic polymers (Brown et al., 1996; Zhang et al., 2004) that are both biodegradable and safe for the environment.

As visually confirmed by color changes in Figure 4 in the lower panel, and the UV-Vis spectra verified nicotine presence through characteristic absorption peaks. In the comparison of spectra, the yellow coloration attributed to nicotine, loading the drug in the hydrogels through hydrogen bonding and physical entrapment played in the roles. As well as nicotine hydrogel, similar color changes were noted for mimosa-loaded cellulose hydrogel (Jiang & Kobayashi, 2017a) and gallic acid-loaded chitin hydrogel (Jiang & Kobayashi, 2017b) when compared to their drug-free counterparts. For instance, the responsible wavelength for nicotine was identified as 260 nm, associated with the $\pi-\pi*$ electron transition (Clayton et al., 2013) and the corresponding peak shifted somewhat shorter wavelength side in the nicotine loaded hydrogel, meaning interacting the medicine with cellulose.

Figure 4. Over viewed pictures of nicotine-loaded cellulose hydrogels prepared in similar the medicine laoding process and their respective SEM images in freezed dryed condition (X 50 magnification) for different cellulose concentration in the phases inversion process and absorption spectra. For the hydrogels, (a) 0.45 wt%, (b). 0.9 wt% and (c). 1.8 wt% cellulose solution in DMAc/LiCl was gelled with 0.1 wt% of nicotine concentration at 25 ^0C

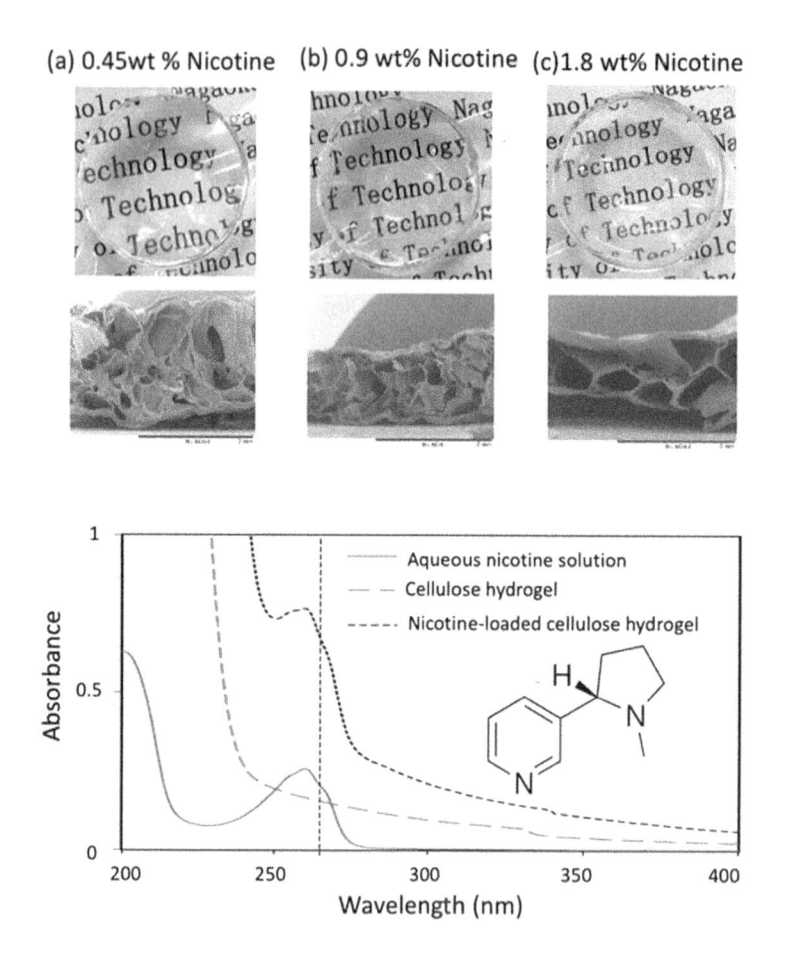

This variation in nicotine loading into the cellulose matrix was attributed to differences in hydrogel porosity, swelling index, and the hydrophilic nature of the matrix. The results show that hydrogels with lower cellulose density and higher water content tend to retain less drug. This trend suggested that drug release was also more likely in hydrogels with lower cellulose density and higher water content.

US-TRIGGERED DRUG RELEASE FROM HYDROGEL MEDICINES

Since cellulose and chitin were regenerated from food wastes like bagasse and shell, such polymer sources even in waste biomass sources have outstanding properties including biocompatibility, non-toxicity, and biodegradability. Also, US is a non-invasive green technique with premier advantages in

SDDS such as outside to body manipulation ability, deep penetrability, different mechanisms to induce drug release (thermal, oscillation, cavitation, etc.), frequency dependent wide range of drug release applications (Kobayashi, 2023)., applicability on different drug carriers, existing technology in medical field thus high capability to modify for drug release applications, and easy in use (Carovac et al., 2011; Mura et al., 2013; Pitt et al., 2004; Sirsi & Borden, 2014). In the intelligent systems for DDS, the therapeutic molecules are delivered on-demand; US power mediates drug vectors, release, and delivery drugs. In the case of microbubbles, the bubble collapse because of cavitation is the origin of drug release. Studies tabulated in *Table 3* are comprehensively discussed in this section to give reader a broad knowledge on release behaviors under US stimulations for biomass hydrogel medicines.

Table 2. Biomass hydrogel medicines used for US-triggered systems

Natural polysaccharide hydrogels	Encapsulated drug	Reference
Cellulose hydrogel	Nicotine	(Iresha & Kobayashi, 2021b)
Chitin hydrogel	Nicotine	(Iresha & Kobayashi, 2018)
Cellulose hydrogel	Nicotine	(Kobayashi et al., 2022)
Cellulose hydrogel	Mimosa	(Jiang et al., 2016)
Chitin hydrogel	Gallic acid	(Jiang & Kobayashi, 2017b)
Cellulose hydrogel		
Carrageenan hydrogel		
Agar hydrogel		
Alginate Hydrogels	Bone morphogenetic protein-2 conjugated with gold nanoparticles (BMP-2 AuNPs)	(Kearney et al., 2015)
chitosan/β-glycerophosphate hydrogel	Doxorubicin loaded into Lysolipid thermally sensitive liposomes	(López-Noriega et al., 2014)
Chitosan/ perflurohexane nanodroplets	Curcumin	(Baghbani et al., 2017)
Ca^{2+}-crosslinked alginate hydrogels	Mitoxantrone Condensed plasmid DNA (pDNA) Stromal cell-divided factor 1α (SDF-1α)	(Huebsch et al., 2014)
Cellulose hydrogel	Short-chain fatty acid	(Yan et al., 2019)

US Factors on Drug Releasing

In effect of US power and US frequency, these are two main US parameters which determine the US performance. Because the wave characteristics transmitted to the media by the US transducer, one is US frequency which is set as sound pressure frequency per second. The another one is US output power which determines the sound-wave strength. Therefore, the different waves behave differently in the transmitted medium, directly affecting. Thus, for the drug-loaded medium the circumstance is exposed to US action, resulting in enhanced drug release. Other parameters which govern the drug release are exposure time, capsule distance to transducer, capsule orientation facing the transducer, temperature of the medium, number of transducers, irradiation patterns (pulse, within time intervals or continuous). In addition to the influence of ultrasound (US), the effectiveness of drug release from the polymer matrix is also affected by the release medium, which mimics the biological environment.

For instance, these hydrogels were prepared with different cellulose loadings (0.45 wt%, 0.9 wt%, and 1.8 wt%), denoted as NC0.45, NC0.9, and NC1.8, and exposed to varying US powers (5-40 W) at a constant frequency and temperature. NC0.9 showed distinguishable deviations in the amounts of nicotine released over the range of US powers used from 5 W to 40 W, compared to the NC0.45 and NC1.8 under the same conditions. More interestingly, among the three nicotine-cellulose hydrogels, the highest nicotine release of 5 µg/ml at 60 min exposure was observed for NC0.9, at 40 W. Even though the NC1.8 contained considerably high amount of nicotine inside, the responses to different US powers were not noticeable in nicotine release, might be due to nicotines loading in tightest hydrogel matrix. Figure 5 shows time change in nicotine release for NC0.9, when exposed US in different frequency at 30 W. Relative to non-exposed case, US systems exhibited enhanced nicotine releasing. Among different US frequency, the cases of 43kHz and 23kH with lower frequency US showed higher release of nicotine. In In Fig. 5 (b) for nicotine release from NC0.9 hydrogel in physio- logical environment of PBS was compared with the release in distilled water at 37 oC. Because of in Fig. 5 (b), the nicotine release from NC0.9 hydrogel reduced in physio logical environment of PBS, as compared with the release in distilled water, due to the lower diffusion of nicotine in salt medium (Chen & Papadopoulos, 2020). However, when US at 43kHz (30W) was exposed, the releasing nicotine enhanced almost 2 times higher amounts to release nicotine from the NC0.9. Plots of US exposure time vs released mimosa concentration in the aqueous solution. The US irradiation was carried out at different frequencies of 30 W (a) and 10 W (b) for the mimosa hydrogel film, which was prepared from 1.0 wt% of cellulose solution.

Figure 5. Nicotine release behavior from NC0.9 hydrogel in (a) different frequency of US at 30W and (b) in PBS environment at 37 C. For (b), the US operated at 43 kHz frequency and 40 W output power for 60 min

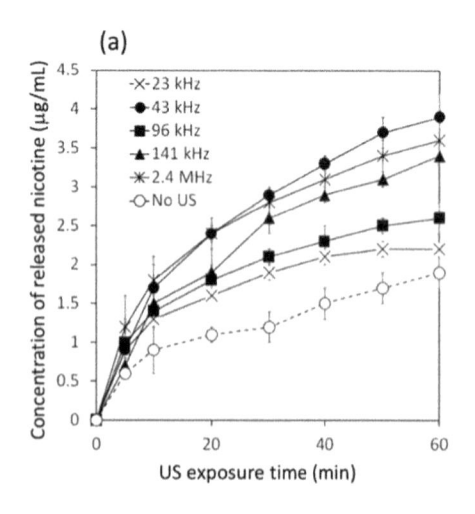

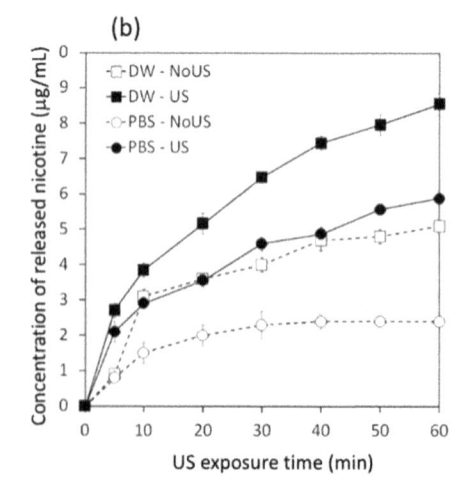

Similar releasing behavior was observed in other medicine hydrogels of cellulose- mimosa and chitin-garlic acid hydrogels. Here, it is complicated to analyze the exact fact about frequency dependency. In this situation, even though the precise cause is difficult to explain, generally, but expect for 43 kHz, at higher frequencies the nicotine release was increased. At higher frequencies the hydrogels undergo compression and rarefaction in a shorter time interval due to lower wavelength which promote enhanced drug release. Furthermore, according to Holland and Apfel, the threshold pressure for cavitation increases with the transducer frequency (Holland & Apfel, 1989). Tsutsui et al. also mentioned that at lower frequencies (20 kHz) the threshold pressure for cavitation is lower than that at higher frequencies (2 MHz) (Tsutsui et al., 2004). At higher frequencies which means at shorter transduced distance of sound cycles, it has less probability for extensive bubble growth for cavitation due to the shorter time period at rarefaction (American Institute of Ultrasound in Medicine., 2000).

In the composing of US-stimuli response on nicotine loaded cellulose hydrogel, it is important to mention the behavior of DR in an environmental condition closer to actual body conditions. There, nicotine release behaviors were studied at 37 °C while the fluid which the nicotine released during irradiation was phosphate buffer solution (PBS) as a medium analogue to body fluid. The results were summarized into Fig.5 (b). There, it is well noted that the nicotine release was suppressed when the release was done in PBS even at elevated temperature, for both US and non-US conditions. The reason could be the salt environment reduces the diffusivity of nicotine (Chen & Papadopoulos, 2020).

In such cases, the hydrogel drug must be placed near the affected area in the body and stimulated externally to stimulate drug expulsion. A case in point is for effect of US drug releasing behaviors in different hydrogel-medicine systems. Researchers have developed a novel approach using reversibly cross-linked hydrogels that can be disrupted the gel networks through US stimulation, making them easy in the delivery and the state is convenient for drug delivery to outside hydrogel (Huebsch et al., 2014). They found that cancer cells respond better to short-term via high-dose bursts of the chemotherapy drug mitoxantrone than to continuous dosages, suggesting the potential benefits of implantable devices that allow external control over dose and timing. Injectable and biocompatible alginate hydrogels demonstrated the ability to self-heal damage caused by ultrasound pulses. This enabled the on-demand release of mitoxantrone both in laboratory settings and in live subjects by US. Notably, when mitoxantrone-loaded gels were implanted near tumors and subjected to daily US pulses, they were more effective at eliminating tumor growth. In contrast, it's worth mentioning that since the release of mitoxantrone was slight when no ultrasound was applied, the combination of US had a noticeable effect.

On the other hand, the low-frequency US, below the cavitation threshold was used in the release of therapeutic molecules, as controlled medicine from a hydrogel enclosed within an acoustic horn (Gerayeli et al., 2021). This "flashing ratchet" approach released all the molecules in less than 90 seconds at low energy levels (1.5 W cm^{-2}). This method differs from traditional ultrasound-based drug delivery and can be applied to various biomedical systems involving gels, such as drug delivery and molecule separation.

For this treatment, the hydrogel needs to be a biocompatible material, and biomass hydrogel is suitable. For example, chitin was used in delivering active ingredients and developing functional foods for weight loss, lipid reduction, gastrointestinal health, and anti-aging (Lv et al., 2023). Moreover, the application of chitin-based materials was expanded into medicine. In addition, hydroxyapatite was *in situ* synthesized to obtain the chitin/hydroxyapatite composite plastics and applied for the viability of biocompatibility, hemocompatibility and *in vivo* histocompatibility of the bioplastics were evaluated (He et al., 2017). The introduction of hydroxyapatite could improve the cell adhesion, proliferation and differentiation of the osteoblast cells. Moreover, the composite exhibited good histocompatibility, hemocompatibility and in vivo biodegradability, showing potential application in the bone tissue engineering field. In contrast, the tissue biocompatibility of cellulose and its derivatives was investigated in two *in vivo* tests: absorption by living tissues and foreign body reaction (Miyamoto et al., 1989). In biocompatibility of subcutaneously implanted plant-derived cellulose biomaterials, cellulose scaffolds retained much of their original shape and did undergo a slow deformation over the 8-week length of the study, demonstrating that native cellulose scaffolds are biocompatible and exhibit promising potential as a surgical biomaterial (Modulevsky, 2016). Also, biocompatibility of cellulose hydrogel films regenerated from sugar cane bagasse waste was evaluated it's *in vivo* behavior in mice (Nakasone et al, 2016). This indicated that the implantation of the hydrogel film insignificantly affected the growth pattern of mice. In addition, there was no evidence of severe inflammation in the implantation sight, exhibiting acceptable biocompatibility and durability in living body.

US EFFECT TO VISCOELASTIC HYDROGEL MATRIX

The viscoelasticity measurements are often used to evaluate physical properties of hydrogel owing to its specific properties composing liquid-like and solid-like behavious. As shown in Figure 6 (a), a sono-rheometer incorporating an ultrasonic bath in a viscoelasticity measuring instrument has made it possible to measure the dynamic viscoelasticity of hydrogels under US irradiation (Noguchi & Kobayashi, 2020). Viscoelastic measurements reveal how materials behave under various alterations made to the sample during analysis. The amplitude sweep measurement is a type of viscoelastic analysis which test for the complex modulus change over different strain rates which the sample underwent. The sono-deviced rheometer, which enabled viscoelastic properties under ultrasound operation, was used to investigate for cellulosic hydrogels, prepared at 0.5, 1 and 2 wt% concentration in the DMAc/LiCl solution. The sono-deviced equipment could measure the effect of changes in storage modulus G' and loss modulus G" under 43 kHz ultrasound exposure, noting that the 43 kHz US significantly decreased the values of the modulus values under US condition, but, when stopped, these were immediately returned to original values. This US effect meant that the hydrogel was soften under the exposure within few seconds and then self-healing to original state in the absent of US.

Figure 6 (b) shows US cycled effect on (b) G' and G" and tanδ for composite hydrogel with cellulose and chitin when the same amount of each was added. Here, their *in-situ* viscoelastic properties were estimated under cyclic exposure of 43 kHz and 30 W US using a sono-deviced rheometer.

Figure 6. (a) Diagram of in situ sono-deviced rheometer, (b) G' and G" vs time, (c) tan δ vs time, and (d) G'/G'0 vs time behaviors of cellulose-chitin composite hydrogel with 0.2 wt% cellulose and 0.7 wt% chitin, under cyclic US on/off irradiation in 5 min intervals. US conditions were 30 W/ 43 kHz at 25°C

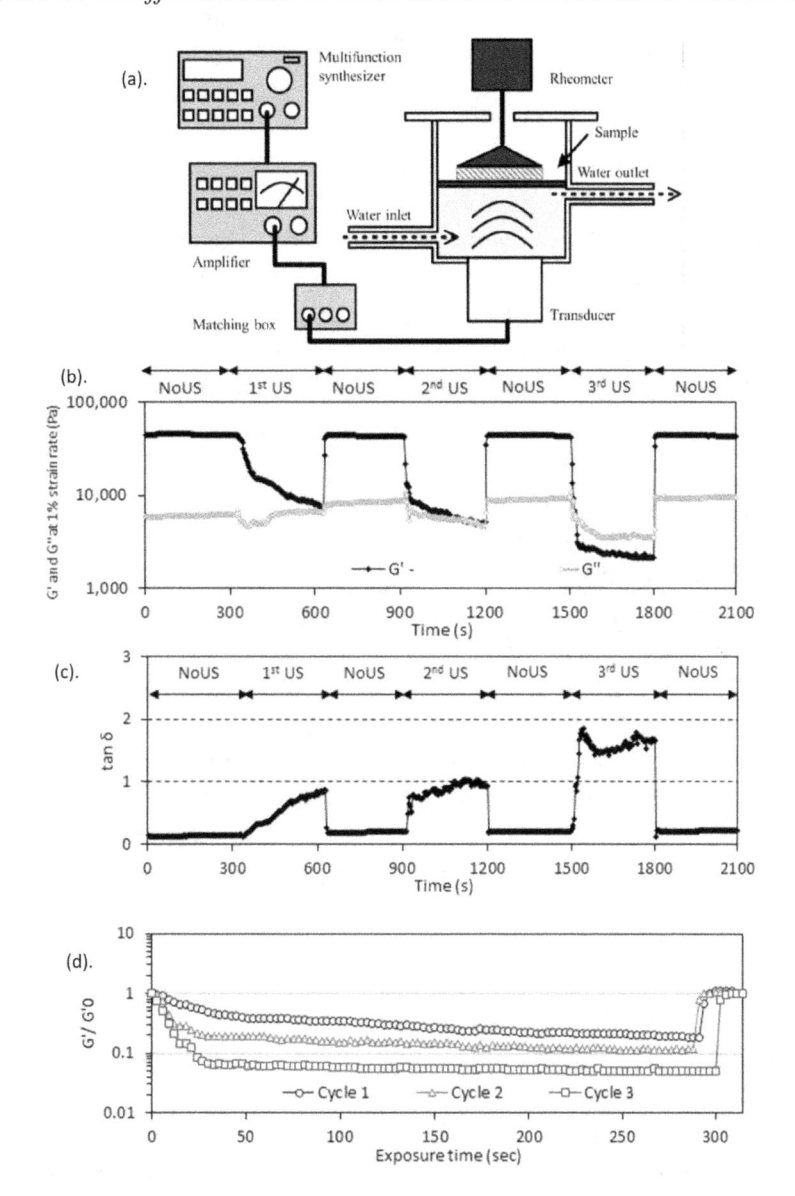

In (a), the values for G' and G" were observed during each cycle. These measurements were taken over time with a 5-minute interval, while the US was alternated on and off. Here, the US was operated at an output power of 50 W at a frequency of 43 kHz. An intriguing discovery was made the cellulose hydrogel responded rapidly to the US, resulting in a sudden drop in G' and G" values. However, after 5 minutes of irradiation, G' and G" values returned to their initial levels immediately after the US was turned off. This softening and self-healing gelation pattern, corresponding to the cycles of US activation and deactivation, was consistently observed over three cycles, indicating that US-induced hydrogen bond breakage and subsequent reformation of hydrogen bonds in the absence of US as well as the responsible behavior for cellulose hydrogels (Noguch & Kobayashi, 2020). In contrast, the cyclic US on/off operation was observed in tan δ, meaning G'/G'_0 variation, against cyclic US irradiation. The values of tan δ gradually increased, meaning that the gel state was softening toward liquid condition from solid gel state. On second cycles, the softening tendency became easily occurring within 20-25 seconds and on third cycles, immediately softened within few seconds and then the liquid state of the gel component kept during US condition for 5 min at tanδ>1. As a result, the cycled rate of softening at each cycle can be ordered as 1st US step < 2nd US step < 3rd US step, but, the self-healed gelation occurred similarly within few seconds.

Further analysis about *in situ* US irradiation was discussed for kanten agarose, κ-carrageenan, and konjac glucomannan hydrogels (Noguchi & Kobayashi, 2022). Another spotlight observations were reported in this study on how the different functional segments present in polysaccharides effect on the hydrogel responses upon US irradiation. This softening effect was not much significant for konjac glucomannan hydrogel. Further, the agarose and k-carrageenan hydrogel structures were collapsed after US irradiation under large strain rate while konjac glucomannan could withstand the US forces. The reason was explained as the differences in structure rearrangement in the hydrogel upon US irradiation for the three hydrogels explained. It is mentioned that the formation of helix aggregates upon the gelation of agarose and carrageenan hydrogels is happening due to anhydro-L-galactose and D-galactose segments present in these two polysaccharides, respectively. These helical aggregates were softened and separate into individual helical segments due to inability to tolerate the US forces by the hydrogel structures of agarose and carrageenan. Therefore, G" values were increased upon US irradiation while during the amplitude sweep test the collapse of hydrogels were seen at higher deformations for agarose and carrageenan BPHGs. For konjac hydrogels, the gelation happens due to cross-linking by deacetylation of glucomannan molecules thus the structure is comparatively stable under US exposure. Additionally, the softening effect induced by US stimulation, driven by hydrogen bond breakage, extends beyond interactions within the polymeric matrix. It is also impacted by the disturbance of hydrogen bonds between the drug and the polymer matrix, as previously detailed in the preceding section. This underscores the potential of US as a promising stimulus. Therefore, this phenomenon holds significance in the realm of BPHGs and their applications for a sustainable future.

Biomass Polymer Hydrogels in Drug Supply Systems Under Different Stimulants

Up to now, this chapter is aligned with biomass polymer hydrogels which are exactly derived by natural polysaccharides. Nevertheless, such direct natural polymers in hydrogel form are still not much common in literatures. However, there are hydrogels prepared by derivatives of such polysaccharides sometimes in combination with different substances for smart drug delivery or release under different stimuli as given in Table 3. Given the structural parallels to certain natural polymers, the adaptation of specific polymers for use in hydrogel structures, variations in drug encapsulation, and diverse responses to various stimuli in addition to their activation by US, it is essential to closely examine these hydrogels within the context of smart drug delivery and release systems.

In particular, pH and ion-sensitive DR system was developed using cationic cellulose hydrogels cross-linked by ethylenglycol diglycidylether (EGDE). Diclofenac sodium release was investigated (Rodríguez et al., 2003) from this hydrogel. Here, diclofenac sodium entrapment into the hydrogel was promoted due to ionic and hydrophobic interactions with the drug and the polymer. When the hydrogels were directly immersed in water and 0.9 wt% NaCl solution as the release mediums, the release was not permitted at the acidic conditions with pH 3. At the pH of 6, diclofenac release was noted and the release was higher in the NaCl medium than in the water medium. When the medium was a base with pH of 8, release was higher than at pH 6 both at NaCl and water mediums. When the release medium was gradually varying pH from 3 →6 →8, the drug release profiles were different compared to direct immersion into release mediums which explained earlier. From pH 3 until 6 a sustained release of drug was noted at both water and NaCl mediums and once the pH reacted to 8, the release was significant. Prominent hydrophobic bonds and ionic interactions at lower pH of 3 restrict the diclofenac release from the hydrogel. At pH 8, the drug-polymer interactions were completely vanished and higher release of drug was permitted to the medium. Moreover, the initial pH history of the hydrogel, ionic environment to release along with the release pH are essential factors to be considered in this SDR system. Here, it suggested this DR system is suitable for oral administration due to analogue in pH variation in the physiological route with the tested conditions.

Another interesting natural polymer is bacterial cellulose. Bacterial cellulose (BC) and acrylic acid (AA) crosslinked together using electron-beam radiation to make acrylic acid-grafted bacterial cellulose to form a hydrogel which is thermo- and pH- responsive for DR (Mohd Amin et al., 2012). Electron-beam radiation facilitates not only crosslinking but also sterilization and residue-free hydrogel system with excellent biocompatibility and biodegradability. Increase of electron bean dosage and AA concentration used for the BC/ AA hydrogel fabrication caused to decrease the pore size of the hydrogels. Therefore, drug entrapment, swelling behaviors and DR was higher for BC/ AA hydrogels with higher pore sizes. Bovine serum albumin (BSA) was entrapped into the hydrogel as the model drug. DR profiles were obtained for BC/ AA hydrogels fabricated with different pores sizes while the release mediums were simulated gastric fluid (SGF) and simulated intestinal fluid (SIF) in the absence of enzymes. Here, due to the high acidic environment in SGF compared to SIF, the BSA release was suspended in the SGF environment while BSA release was remarkably enhanced at the SIF medium. The maximum cumulative release of 90% was reached withing 8 h duration for the BC/ AA with largest pore size while 83% of maximum cumulative release was taken within 13 h duration for the hydrogel with smallest pore size. This drug release profile with SGF and SIF mediums suggests these BC/ AA hydrogels control the BSA release at gastric environment which is at 1.2 pH while promote release at small intestine with the pH of 6.8 for absorption in to the systemic circulation.

In another study, genipin cross-linked kappa-carrageenan/ polyvinyl alcohol hydrogel (k-C/PVA) system was proved its controlled β-carotene release with no burst release compared to its non-crosslinked k-C/PVA hydrogel. Here, genipin is a naturally derived non-toxic substance with many herbal benefits. β-carotene release was more controlled with increased inclusion of genipin into the hydrogel fabrication. However, the release behaviors were maximum when the medium pH was 7 compared to pH 1.2 and pH 12. These behaviors can be attributable to the swelling properties of hydrogels at different pH of 1.2, 7 and 12.

Hybrid hydrogel fabricated with cellulose extracted from sugarcane bagasse and green zinc oxide (ZnO) extracted from muskmelon seeds crosslinked with epichlorohydrin was studied for pH sensitive curcumin release. There, loading of curcumin drug was enhanced in the hybrid Cellulose-ZnO compared to pure cellulose hydrogel due to excellent hydrophobic interaction between ZnO and curcumin. Thus, the release of curcumin was higher in hybrid hydrogel than the pure cellulose hydrogel at each pH values tested. Furthermore, the hybrid cellulose-ZnO hydrogel shows higher curcumin release at pH 7.4 due to the curcumin stability by remarkable hydrophobic interactions between curcumin and ZnO at neutral/alkaline environment, compared to acidic pH of 1.2 and 4.0. In the pure cellulose hydrogel, even though the cellulose-curcumin complexes are possible during curcumin loading, degradation happens at neutral/alkaline medium due to less stability of curcumin complex made with only cellulose.

Lignin, due to its high molecular weight and branched structure, fabrication of hydrogels exhibits some limitations. Peroxidation reaction for alkaline lignin and hydroxymethylation reaction for organo-lignin are two chemical methods for lignin modifications which were reported by Morales et al. (2022). Their ultimate objective was to prepare lignin hydrogels and study quercetin loading and release behaviors under simulated in vitro conditions at PBS environment (Morales et al., 2022). Lignin from almond and walnuts, and commercially available lignin were used for the study. There, the modification to alkaline lignin caused to lower the swelling behaviors while the modification done to organo-lignin showed enhanced swelling. Importantly, all the samples were able to loaded with similar amount of quercetin yet the releasing profiles were different to one another. Among them hydrogel prepared with modified oregano-lignin from walnut shows the highest release at 6.5 h time. Here, modification to lignin might not significantly affected the drug entrapment however the drug interactions restrict the free diffusion of quercetin as described by the authors.

As stipulated above, there are many of such BPHGs formulated from their natural form, derivative or composites of natural polysaccharides in the form of hydrogel films, macro size hydrogels, microbeads, nanocrystals etc. In comparison, the literature-exist biomass hydrogels were showing limited studies on stimulants in smart drug delivery and release system. Mostly, the biomass hydrogel responses against pH, temperature or very recently, US. However, considering the properties which are reported suggested their applications upon many applications for range of drugs and stimulants depending on the applications.

Table 3. Summary of biomass hydrogels in smart drug delivery and release system under different stimulants except US

Biomass hydrogels	Drug	Stimulant	Reference
Cationic cellulose hydrogels	Diclofenac Sodium	pH- and ion-sensitivity	(Rodríguez et al., 2003)
Bacterial cellulose/ acrylic acid hydrogel	Bovine serum albumin	Temperature and pH	(Mohd Amin et al., 2012)
Chitin nanogels conjugated with MPA-capped-CdTe-Quantum Dots	Bovine serum albumin	pH	(Rejinold N et al., 2011)
Hydroxypropylmethyl cellulose films	Nicotine	pH and ionic strength	(Marani et al., 2015)
Kappa-carrageenan/ polyvinyl alcohol hydrogel	β-carotene	pH	(Hezaveh & Muhamad, 2013)
Bacterial Cellulose/ Acrylamide based hydrogel	Theophylline	Released to phosphate buffer (pH 7.4) at 37 °C	(Mohd Amin et al., 2012)
Cellulose nanocrystals-gelatin hydrogels	Theophylline	Released to simulated gastric fluid (HCl at pH 1.2) at 37 °C	(Ooi et al., 2016)
Carboxymethyl cellulose/ carboxymethyl β-cyclodextrin hydrogel	Tetracycline	Released to PBS buffer (pH 7.4), at 37 °C	(Jeong et al., 2018)
Carboxymethylcellulose sodium/ cellulose hydrogel	Bovine serum albumin	Released to phosphate buffer (pH 7.4) at 37 °C	(Chang et al., 2010)

continued on following page

Table 3. Continued

Biomass hydrogels	Drug	Stimulant	Reference
Chitin/PLGA blend microspheres	Bovine serum albumin	Released to PBS buffer (pH 7.4) at 37 °C	(F. L. Mi et al., 2003)
Chitosan/ alginate beads	Indomethacin	pH	(F. L. Mi et al., 2002)
Poly (ethylene glycol)/ carboxymethyl chitosan hydrogel	5-fluorouracil	pH	(El-Sherbiny & Smyth, 2010)
Carboxymethyl chitosan/ poly(N-isopropylacrylamide) semi-interpenetrating polyampholyte hydrogel	Coenzyme A	Temperature and pH	(Guo & Gao, 2007)
Thiolated hydroxypropyl cellulose nanogels		Temperature and redox	(Tan et al., 2011)
Bacterial cellulose nanofiber/ sodium alginate hybrid hydrogel	Ibuprofen	pH and electric field	(Shi et al., 2014)
Cellulose/ green Zinc oxide hybrid hydrogel crosslinked with epichlorohydrin	Curcumin	pH	(Anagha et al., 2019)
Lignin hydrogel	Quercetin	Released into phosphate buffer saline at 37 °C	(Morales et al., 2022)

Future Directions of BPHGs on DD and DR Applications

BPHGs are an emerging advanced materials which shows elevated potential in medical applications. However, this material is yet under experimental stage. There are some limitations to overcome this in order to make this remarkable material to be in the future market. One main concern is the inconsistency of yield, purity and properties of the extracted polysaccharides from naturally available sources or waste biomass sources, depending on the place of origin, season of cultivation, storage and transportation conditions, etc. These factors need to be investigated through reliable experimental studies to standardize the extracted fiber properties, purity and yield, and must be fully assessed through life cycle assessments studies.

The drug encapsulation methods are yet to be studied to ensure the constant drug amount per unit volume. Parallelly, the zero-release of drug during storage and before the stimulation must be confirmed with extended experiments. These two factors are challenging to biomass hydrogels due to the high-water content and the squeezability of biomass hydrogels upon handling, in its nature. Sterilization, drug shelf life in BPHGs, and degradation of the hydrogels are another important aspect to be studied and formalized for a standard medicinal product.

More importantly, proper drug administration route, biomass hydrogels place of body contact, method of simulation and appropriate device for an efficient stimulation are other factors to overcome the real-world applications of such a prospective material. With all these positive and challenging properties, BPHGs is emerging as a prestigious advanced material to emboss its identity in future sustainable DD and DR applications.

CONCLUSION

The chapter reveals the applicability of biomass sources on developing advanced sustainable materials for medical applications. Formulation of hydrogels from natural polysaccharides such as cellulose and chitin extracted from natural waste biomass sources like sugar cane bagasse and crab shells, etc. proven their premier properties tailor-made for medical applications especially in DR and DD. Such properties are biocompatibility, non-toxicity, biodegradability, drug loading ability, and water retention ability. As the modern level of DR and DD, 'smart' concept of drug release and delivery plays a vital role in various therapeutic applications. There, such biomass polymer hydrogels are recognized as potential sustainable capsules for smart drug delivery and release system under different stimulants including ultrasound, pH, and temperature. Among these stimulants, US-triggered effect was widely described in the chapter considering ultrasound application as a green approach. Different US output powers and frequencies make differences in drug release behaviors. Moreover, the polysaccharide content was directly affected on drug loading and the behaviors due to different morphologies formed during hydrogel fabrication. The response of the hydrogel matrix to ultrasound stimulation displayed the executive performance, as it remained intact during the operation while facilitating drug release when exposed to ultrasound. Towards the conclusion of the chapter, an examination was conducted on how cyclic ultrasound irradiation impacts the viscoelastic properties of a composite hydrogel made from cellulose and chitin. This analysis represents a specialized investigation with the potential to expand the range of applications for these hydrogels, referred to as BPHGs. Furthermore, the study recognized the potential of BPHGs as carriers for drugs, allowing for intelligent drug release and delivery under various stimulus conditions. It is emphasized that, as a promising and sustainable material for advanced medical applications, BPHGs require further research to address the limitations outlined at the end of the chapter. Without a doubt, BPHGs are expected to achieve their peak performance in the near future, given their exceptional sustainability, which sets them apart from other materials.

REFERENCES

Ahmadi, F., McLoughlin, I. V., Chauhan, S., & Ter-Haar, G. (2012). Bio-effects and safety of low-intensity, low-frequency ultrasonic exposure. *Progress in Biophysics and Molecular Biology*, 108(3), 119–138. DOI: 10.1016/j.pbiomolbio.2012.01.004 PMID: 22402278

Alsawaftah, N. M., Awad, N. S., Pitt, W. G., & Husseini, G. A. (2022). pH-Responsive Nanocarriers in Cancer Therapy. *Polymers*, 14(5), 936. DOI: 10.3390/polym14050936 PMID: 35267759

American Institute of Ultrasound in Medicine. (2000). Section 7—Discussion of the mechanical index and other exposure parameters. *Journal of Ultrasound in Medicine*, 19(2), 143–168. DOI: 10.7863/jum.2000.19.2.143 PMID: 10680619

Anagha, B., George, D., Maheswari, P. U., & Begum, K. M. M. S. (2019). Biomass Derived Antimicrobial Hybrid Cellulose Hydrogel with Green ZnO Nanoparticles for Curcumin Delivery and its Kinetic Modelling. *Journal of Polymers and the Environment*, 27(9), 2054–2067. DOI: 10.1007/s10924-019-01495-y

Baghbani, F., Chegeni, M., Moztarzadeh, F., Hadian-Ghazvini, S., & Raz, M. (2017). Novel ultrasound-responsive chitosan/perfluorohexane nanodroplets for image-guided smart delivery of an anticancer agent: Curcumin. *Materials Science and Engineering C*, 74, 186–193. DOI: 10.1016/j.msec.2016.11.107 PMID: 28254284

Baker, K. G., Robertson, V. J., & Duck, F. A. (2001). A review of therapeutic ultrasound: Biophysical effects. *Physical Therapy*, 81(7), 1351–1358. DOI: 10.1093/ptj/81.7.1351 PMID: 11444998

Boissenot, T., Bordat, A., Fattal, E., & Tsapis, N. (2016). Ultrasound-triggered drug delivery for cancer treatment using drug delivery systems: From theoretical considerations to practical applications. *Journal of Controlled Release*, 241, 144–163. DOI: 10.1016/j.jconrel.2016.09.026 PMID: 27667179

Brown, L. R., Edelman, E. R., Fischel-Ghodsian, F., & Langer, R. (1996). Characterization of Glucose-Mediated Insulin Release from Implantable Polymers. *Journal of Pharmaceutical Sciences*, 85(12), 1341–1345. DOI: 10.1021/js9600686 PMID: 8961150

Carovac, A., Smajlovic, F., & Junuzovic, D. (2011). Application of Ultrasound in Medicine. *Acta Informatica Medica*, 19(3), 168. DOI: 10.5455/aim.2011.19.168-171 PMID: 23408755

Chang, C., Chen, S., & Zhang, L. (2011). Novel hydrogels prepared via direct dissolution of chitin at low temperature: Structure and biocompatibility. *Journal of Materials Chemistry*, 21(11), 3865–3871. DOI: 10.1039/c0jm03075a

Chang, C., Duan, B., Cai, J., & Zhang, L. (2010). Superabsorbent hydrogels based on cellulose for smart swelling and controllable delivery. *European Polymer Journal*, 46(1), 92–100. DOI: 10.1016/j.eurpolymj.2009.04.033

Chen, C.-Y., & Papadopoulos, K. D. (2020). Temperature and Salting out Effects on Nicotine Dissolution Kinetics in Saline Solutions. *ACS Omega*, 5(14), 7738–7744. DOI: 10.1021/acsomega.9b02836 PMID: 32309681

Clayton, P. M., Vas, C. A., Bui, T. T. T., Drake, A. F., & McAdam, K. (2013). Spectroscopic studies on nicotine and nornicotine in the UV region. *Chirality*, 25(5), 288–293. DOI: 10.1002/chir.22141 PMID: 23494810

Dayton, P. A., Allen, J. S., & Ferrara, K. W. (2002). The magnitude of radiation force on ultrasound contrast agents. *The Journal of the Acoustical Society of America*, 112(5), 2183–2192. DOI: 10.1121/1.1509428 PMID: 12430830

de Oliveira, J. P., Bruni, G. P., Lima, K. O., El Halal, S. L. M., da Rosa, G. S., Dias, A. R. G., & Zavareze, E. da R. (2017). Cellulose fibers extracted from rice and oat husks and their application in hydrogel. *Food Chemistry*, 221, 153–160. DOI: 10.1016/j.foodchem.2016.10.048 PMID: 27979125

del Valle, L. J., Díaz, A., & Puiggal, J. (2017). Hydrogels for Biomedical Applications: Cellulose, Chitosan, and Protein/Peptide Derivatives. *Gels (Basel, Switzerland)*, 3(3), 27. DOI: 10.3390/gels3030027 PMID: 30920524

Dong, L., & Hoffman, A. S. (1991). A novel approach for preparation of pH-sensitive hydrogels for enteric drug delivery. *Journal of Controlled Release*, 15(2), 141–152. DOI: 10.1016/0168-3659(91)90072-L

Duan, Y., Freyburger, A., Kunz, W., & Zollfrank, C. (2018). Cellulose and chitin composite materials from an ionic liquid and a green co-solvent. *Carbohydrate Polymers*, 192, 159–165. DOI: 10.1016/j.carbpol.2018.03.045 PMID: 29691008

El-Sherbiny, I. M., & Smyth, H. D. C. (2010). Poly(ethylene glycol)-carboxymethyl chitosan-based pH-responsive hydrogels: Photo-induced synthesis, characterization, swelling, and in vitro evaluation as potential drug carriers. *Carbohydrate Research*, 345(14), 2004–2012. DOI: 10.1016/j.carres.2010.07.026 PMID: 20708174

Golovin, Y. I., Golovin, D. Y., Vlasova, K. Y., Veselov, M. M., Usvaliev, A. D., Kabanov, A. V., & Klyachko, N. L. (2021). Non-heating alternating magnetic field nanomechanical stimulation of biomolecule structures via magnetic nanoparticles as the basis for future low-toxic biomedical applications. In *Nanomaterials* (Vol. 11, Issue 9). MDPI. DOI: 10.3390/nano11092255

Guo, B. L., & Gao, Q. Y. (2007). Preparation and properties of a pH/temperature-responsive carboxymethyl chitosan/poly(N-isopropylacrylamide)semi-IPN hydrogel for oral delivery of drugs. *Carbohydrate Research*, 342(16), 2416–2422. DOI: 10.1016/j.carres.2007.07.007 PMID: 17669378

Gutowska, A., Bae, Y. H., Feijen, J., & Kim, S. W. (1992). Heparin release from thermosensitive hydrogels. *Journal of Controlled Release*, 22(2), 95–104. DOI: 10.1016/0168-3659(92)90194-V

He, M., Wang, X., Wang, Z., Chen, L., Lu, Y., Zhang, X., & Zhang, L. (2017). Biocompatible and biodegradable bioplastics constructed from chitin via a "green" pathway for bone repair. *ACS Sustainable Chemistry & Engineering*, 5(10), 9126–9135.

Hezaveh, H., & Muhamad, I. I. (2013). Controlled drug release via minimization of burst release in pH-response kappa-carrageenan/polyvinyl alcohol hydrogels. *Chemical Engineering Research & Design*, 91(3), 508–519. DOI: 10.1016/j.cherd.2012.08.014

Holland, C. K., & Apfel, R. E. (1989). An Improved Theory for the Prediction of Microcavitation Thresholds. *IEEE Transactions on Ultrasonics, Ferroelectrics, and Frequency Control*, 36(2), 204–208. DOI: 10.1109/58.19152 PMID: 18284969

Hosseinaei, O., Wang, S., Enayati, A. A., & Rials, T. G. (2012). Effects of hemicellulose extraction on properties of wood flour and wood–plastic composites. *Composites. Part A, Applied Science and Manufacturing*, 43(4), 686–694. DOI: 10.1016/j.compositesa.2012.01.007

Huebsch, N., Kearney, C. J., Zhao, X., Kim, J., Cezar, C. A., Suo, Z., & Mooney, D. J. (2014). Ultrasound-triggered disruption and self-healing of reversibly cross-linked hydrogels for drug delivery and enhanced chemotherapy. *Proceedings of the National Academy of Sciences of the United States of America*, 111(27), 9762–9767. DOI: 10.1073/pnas.1405469111 PMID: 24961369

Huffman, A. S., Afrassiabi, A., & Dong, L. C. (1986). Thermally reversible hydrogels: II. Delivery and selective removal of substances from aqueous solutions. *Journal of Controlled Release*, 4(3), 213–222. DOI: 10.1016/0168-3659(86)90005-2

Im, J. S., Bai, B. C., & Lee, Y. S. (2010). The effect of carbon nanotubes on drug delivery in an electro-sensitive transdermal drug delivery system. *Biomaterials*, 31(6), 1414–1419. DOI: 10.1016/j.biomaterials.2009.11.004 PMID: 19931904

Iresha, H., & Kobayashi, T. (2018). Cellulose and Chitin Hydrogels as Drug Carriers in Ultrasound-stimulated Nicotine Release System. *Proceeding of the 27th Annual Meeting of* the Japan Society of Sonochemistry, 76–77.

Iresha, H., & Kobayashi, T. (2019). Cellulose-Chitin composite hydrogel for ultrasound-triggered drug release. In M. Kubo (Ed.), *28th Annual Meeting of the Japan Society of Sonochemistry* (pp. 7–8). Japan Society of Sonochemistry.

Iresha, H., & Kobayashi, T. (2020). Smart Polysaccharide Hydrogels in Drug Delivery and Release. In Nayak, A. K., & Hasnain, M. S. (Eds.), *Advanced Biopolymeric Systems for Drug Delivery* (pp. 135–149). Springer Nature., https://doi.org/https://doi.org/10.1007/978-3-030-46923-8 DOI: 10.1007/978-3-030-46923-8_6

Iresha, H., & Kobayashi, T. (2021a). *In Situ* Viscoelasticity Behavior of Cellulose–Chitin Composite Hydrogels during Ultrasound Irradiation. *Gels (Basel, Switzerland)*, 7(3), 1–13. DOI: 10.3390/gels7030081 PMID: 34209349

Iresha, H., & Kobayashi, T. (2021b). Ultrasound-triggered nicotine release from nicotine-loaded cellulose hydrogel. *Ultrasonics Sonochemistry*, 78, 105710. DOI: 10.1016/j.ultsonch.2021.105710 PMID: 34411843

Jeong, D., Joo, S. W., Hu, Y., Shinde, V. V., Cho, E., & Jung, S. (2018). Carboxymethyl cellulose-based superabsorbent hydrogels containing carboxymehtyl β-cyclodextrin for enhanced mechanical strength and effective drug delivery. *European Polymer Journal*, 105, 17–25. DOI: 10.1016/j.eurpolymj.2018.05.023

Jiang, H., & Kobayashi, T. (2017a).. . *Ultrasound Effect on Cellulose Decomposition in Solution and Hydrogels.*, 0869(3), 45–52.

Jiang, H., & Kobayashi, T. (2017b). Ultrasound stimulated release of gallic acid from chitin hydrogel matrix. *Materials Science and Engineering C*, 75, 478–486. DOI: 10.1016/j.msec.2017.02.082 PMID: 28415488

Jiang, H., Tovar-Carrillo, K., & Kobayashi, T. (2016). Ultrasound stimulated release of mimosa medicine from cellulose hydrogel matrix. *Ultrasonics Sonochemistry*, 32, 398–406. DOI: 10.1016/j.ultsonch.2016.04.008 PMID: 27150786

Kabanov, A. V., Batrakova, E. V., Melik-Nubarov, N. S., Fedoseev, N. A., Dorodnich, T. Y., Alakhov, V. Y., Chekhonin, V. P., Nazarova, I. R., & Kabanov, V. A. (1992). A new class of drug carriers: Micelles of poly(oxyethylene)-poly(oxypropylene) block copolymers as microcontainers for drug targeting from blood in brain. *Journal of Controlled Release*, 22(2), 141–157. DOI: 10.1016/0168-3659(92)90199-2

Kameda, T., Miyazawa, M., Ono, H., & Yoshida, M. (2005). Hydrogen Bonding Structure and Stability of α-Chitin Studied by 13C Solid-State NMR. *Macromolecular Bioscience*, 5(2), 103–106. DOI: 10.1002/mabi.200400142 PMID: 15706624

Kawano, T., Yamagata, M., Takahashi, H., Niidome, Y., Yamada, S., Katayama, Y., & Niidome, T. (2006). Stabilizing of plasmid DNA in vivo by PEG-modified cationic gold nanoparticles and the gene expression assisted with electrical pulses. *Journal of Controlled Release*, 111(3), 382–389. DOI: 10.1016/j.jconrel.2005.12.022 PMID: 16487614

Kearney, C. J., Skaat, H., Kennedy, S. M., Hu, J., Darnell, M., Raimondo, T. M., & Mooney, D. J. (2015). Switchable Release of Entrapped Nanoparticles from Alginate Hydrogels. *Advanced Healthcare Materials*, 4(11), 1634–1639. DOI: 10.1002/adhm.201500254 PMID: 26044285

Kennedy, J. E., Ter Haar, G. R., & Cranston, D. (2003). High intensity focused ultrasound: Surgery of the future? *The British Journal of Radiology*, 76(909), 590–599. DOI: 10.1259/bjr/17150274 PMID: 14500272

Kobayashi, T. (2015). Fabrication of cellulose hydrogels and characterization of their biocompatible films. In F. Atta-ur-Rahman (Ed.), *Studies in Natural Products Chemistry* (Vol. 45, pp. 1–15). Elsevier B.V. https://doi.org/DOI: 10.1016/B978-0-444-63473-3.00001-0

Kobayashi, T. (2023). *Ultrasound-triggered drug delivery, 577-591. Advanced and Modern Approaches for Drug Delivery*. Academic press., DOI: 10.1016/B978-0-323-91668-4.00025-3

Kobayashi, T., Iresha, H., Noguchi, S. A., Wahab, M., & Wahab, M. A. (2022). Biomass Hydrogel Medicines for Ultrasound Drug Releasing Materials. In *Reference Module in Materials Science and Materials Engineering* (pp. 653–662). Elsevier., https://doi.org/https://doi.org/10.1016/B978-0-12-820352-1.00221-2

Kost, J., Leong, K., & Langer, R. (1989). Ultrasound-enhanced polymer degradation and release of incorporated substances. *Proceedings of the National Academy of Sciences of the United States of America*, 86(20), 7663–7666. DOI: 10.1073/pnas.86.20.7663 PMID: 2813349

Kostag, M., & El Seoud, O. A. (2021). Sustainable biomaterials based on cellulose, chitin and chitosan composites - A review. *Carbohydrate Polymer Technologies and Applications, 2*(December 2020). DOI: 10.1016/j.carpta.2021.100079

López-Noriega, A., Hastings, C. L., Ozbakir, B., O'Donnell, K. E., O'Brien, F. J., Storm, G., Hennink, W. E., Duffy, G. P., & Ruiz-Hernández, E. (2014). Hyperthermia-Induced Drug Delivery from Thermosensitive Liposomes Encapsulated in an Injectable Hydrogel for Local Chemotherapy. *Advanced Healthcare Materials*, 3(6), 854–859. DOI: 10.1002/adhm.201300649 PMID: 24436226

Lv, J., Lv, X., Ma, M., Oh, D., Jiang, Z., & Fu, X. (2023). Chitin and chitin-based biomaterials: A review of advances in processing and food applications. *Carbohydrate Polymers*, 299, 120142. DOI: 10.1016/j.carbpol.2022.120142 PMID: 36876773

Lyon, P. C., Mannaris, C., Gray, M., Carlisle, R., Gleeson, F. V., Cranston, D., Wu, F., & Coussios, C. C. (2021). Large-Volume Hyperthermia for Safe and Cost-Effective Targeted Drug Delivery Using a Clinical Ultrasound-Guided Focused Ultrasound Device. *Ultrasound in Medicine & Biology*, 47(4), 982–997. DOI: 10.1016/j.ultrasmedbio.2020.12.008 PMID: 33451816

Marani, P. L., Bloisi, G. D., & Petri, D. F. S. (2015). Hydroxypropylmethyl cellulose films crosslinked with citric acid for control release of nicotine. *Cellulose (London, England)*, 22(6), 3907–3918. DOI: 10.1007/s10570-015-0757-1

Marmottant, P., & Hilgenfeldt, S. (2003). Controlled vesicle deformation and lysis by single oscillating bubbles. *Nature*, 423(6936), 153–156. DOI: 10.1038/nature01613 PMID: 12736680

Martín-del-Campo, A., Fermín-Jiménez, J. A., Fernández-Escamilla, V. V., Escalante-García, Z. Y., Macías-Rodríguez, M. E., & Estrada-Girón, Y. (2021). Improved extraction of carrageenan from red seaweed (Chondracantus canaliculatus) using ultrasound-assisted methods and evaluation of the yield, physicochemical properties and functional groups. *Food Science and Biotechnology*, 30(7), 901–910. DOI: 10.1007/s10068-021-00935-7 PMID: 34395021

Mi, F. L., Shyu, S. S., Lin, Y. M., Wu, Y. B., Peng, C. K., & Tsai, Y. H. (2003). Chitin/PLGA blend microspheres as a biodegradable drug delivery system: A new delivery system for protein. *Biomaterials*, 24(27), 5023–5036. DOI: 10.1016/S0142-9612(03)00413-7 PMID: 14559016

Mi, F. L., Sung, H. W., & Shyu, S. S. (2002). Drug release from chitosan-alginate complex beads reinforced by a naturally occurring cross-linking agent. *Carbohydrate Polymers*, 48(1), 61–72. DOI: 10.1016/S0144-8617(01)00212-0

Mi, P. (2020). Stimuli-responsive nanocarriers for drug delivery, tumor imaging, therapy and theranostics. In *Theranostics* (Vol. 10, Issue 10, pp. 4557–4588). Ivyspring International Publisher. DOI: 10.7150/thno.38069

Mitragotri, S. (2005). Healing sound: The use of ultrasound in drug delivery and other therapeutic applications. *Nature Reviews. Drug Discovery*, 4(3), 255–260. DOI: 10.1038/nrd1662 PMID: 15738980

Mitragotri, S., Blankschtein, D., & Langer, R. (1995). Ultrasound-mediated transdermal protein delivery. *Science*, 269(5225), 850–853. DOI: 10.1126/science.7638603 PMID: 7638603

Miyamoto, T., Takahashi, S., Ito, H., Inagaki, H., & Noishiki, Y. (1989). Tissue biocompatibility of cellulose and its derivatives. *Journal of Biomedical Materials Research*, 23(1), 125–133. DOI: 10.1002/jbm.820230110 PMID: 2708402

Modulevsky, D. J., Cuerrier, C. M., & Pelling, A. E. (2016). Biocompatibility of subcutaneously implanted plant-derived cellulose biomaterials. *PLoS One*, 11(6), e0157894. Advance online publication. DOI: 10.1371/journal.pone.0157894 PMID: 27328066

Mohd Amin, M. C. I., Ahmad, N., Halib, N., & Ahmad, I. (2012). Synthesis and characterization of thermo- and pH-responsive bacterial cellulose/acrylic acid hydrogels for drug delivery. *Carbohydrate Polymers*, 88(2), 465–473. DOI: 10.1016/j.carbpol.2011.12.022

Morales, A., Labidi, J., & Gullón, P. (2022). Influence of lignin modifications on physically crosslinked lignin hydrogels for drug delivery applications. *Sustainable Materials and Technologies*, 33, e00474. Advance online publication. DOI: 10.1016/j.susmat.2022.e00474

Mukherjee, D., Wong, J., Griffin, B., Ellis, S. G., Porter, T., Sen, S., & Thomas, J. D. (2000). Ten-fold augmentation of endothelial uptake of vascular endothelial growth factor with ultrasound after systemic administration. *Journal of the American College of Cardiology*, 35(6), 1678–1686. DOI: 10.1016/S0735-1097(00)00575-1 PMID: 10807476

Mura, S., Nicolas, J., & Couvreur, P. (2013). Stimuli-responsive nanocarriers for drug delivery. *Nature Materials*, 12(11), 991–1003. DOI: 10.1038/nmat3776 PMID: 24150417

Mylonopoulou, E., Bazán-Peregrino, M., Arvanitis, C. D., & Coussios, C. C. (2013). A non-exothermic cell-embedding tissue-mimicking material for studies of ultrasound-induced hyperthermia and drug release. *International Journal of Hyperthermia*, 29(2), 133–144. DOI: 10.3109/02656736.2012.762553 PMID: 23406389

Nakasone, K., Ikematsu, S., & Kobayashi, T. (2016). Biocompatibility Evaluation of Cellulose Hydrogel Film Regenerated from Sugar Cane Bagasse Waste and Its in Vivo Behavior in Mice. *Industrial & Engineering Chemistry Research*, 55(1), 30–37. DOI: 10.1021/acs.iecr.5b03926

Nakasone, K., & Kobayashi, T. (2006). Cytocompatible cellulose hydrogels containing trace lignin. *Materials Science and Engineering C*, 64, 269–27. DOI: 10.1016/j.msec.2016.03.108 PMID: 27127053

Nayak, A. K., Hasnain, M. S., Laha, B., & Mati, S. (2023). *Advanced and Modern Approaches for Drug Delivery*. Academic Press.

Noguchi, S., & Kobayashi, T. (2022). Ultrasound Viscoelastic Properties of Biomass Polysaccharide Hydrogels as Evaluated by Rheometer Equipped with Sono-Device. *Gels (Basel, Switzerland)*, 8(3), 172. Advance online publication. DOI: 10.3390/gels8030172 PMID: 35323285

Noguchi, S., & Takaomi, K. (2020). Ultrasound response of viscoelastic changes of cellulose hydrogels triggered with Sono-deviced rheometer. *Ultrasonics Sonochemistry*, 67, 105143. DOI: 10.1016/j.ultsonch.2020.105143 PMID: 32446975

Nyborg, W. L. (1982). Ultrasonic microstreaming and related phenomena. *British Journal of Cancer*, 45(Suppl. 5), 156–160. /pmc/articles/PMC2149335/?report=abstract

Ooi, S. Y., Ahmad, I., & Amin, M. C. I. M. (2016). Cellulose nanocrystals extracted from rice husks as a reinforcing material in gelatin hydrogels for use in controlled drug delivery systems. *Industrial Crops and Products*, 93, 227–234. DOI: 10.1016/j.indcrop.2015.11.082

Ow Sullivan, M. M., Green, J. J., & Przybycien, T. M. (2003). Development of a novel gene delivery scaffold utilizing colloidal gold–polyethylenimine conjugates for DNA condensation. *Gene Therapy 2003 10:22*, *10*(22), 1882–1890. DOI: 10.1038/sj.gt.3302083

Park, K. (2014). Controlled drug delivery systems: Past forward and future back. *Journal of Controlled Release*, 190, 3–8. DOI: 10.1016/j.jconrel.2014.03.054 PMID: 24794901

Pérez-Luna, V., & González-Reynoso, O. (2018). Encapsulation of Biological Agents in Hydrogels for Therapeutic Applications. *Gels (Basel, Switzerland)*, 4(3), 61. DOI: 10.3390/gels4030061 PMID: 30674837

Pitt, W. G., Husseini, G. A., & Staples, B. J. (2004). Ultrasonic drug delivery—A general review. *Expert Opinion on Drug Delivery*, 1(1), 37–56. DOI: 10.1517/17425247.1.1.37 PMID: 16296719

Polat, B. E., Blankschtein, D., & Langer, R. (2010). Low-frequency sonophoresis: Application to the transdermal delivery of macromolecules and hydrophilic drugs. *Expert Opinion on Drug Delivery*, 7(12), 1415–1432. DOI: 10.1517/17425247.2010.538679 PMID: 21118031

Ranganayakulu, S. V. (2016). Ultrasound applications in Medical Sciences. *International Journal of Modern Trends in Engineering and Research*, 3(2), 287–293.

Ranjan, A., Jacobs, G. C., Woods, D. L., Negussie, A. H., Partanen, A., Yarmolenko, P. S., Gacchina, C. E., Sharma, K. V., Frenkel, V., Wood, B. J., & Dreher, M. R. (2012). Image-guided drug delivery with magnetic resonance guided high intensity focused ultrasound and temperature sensitive liposomes in a rabbit Vx2 tumor model. *Journal of Controlled Release*, 158(3), 487–494. DOI: 10.1016/j.jconrel.2011.12.011 PMID: 22210162

Rejinold, N. S., Chennazhi, K. P., Tamura, H., Nair, S. V., & Rangasamy, J. (2011). Multifunctional Chitin Nanogels for Simultaneous Drug Delivery, Bioimaging, and Biosensing. *ACS Applied Materials & Interfaces*, 3(9), 3654–3665. DOI: 10.1021/am200844m PMID: 21863797

Ribeiro, T. P., Moreira, J. A., Monteiro, F. J., & Laranjeira, M. S. (2022). Nanomaterials in cancer: Reviewing the combination of hyperthermia and triggered chemotherapy. *Journal of Controlled Release*, 347, 89–103. DOI: 10.1016/j.jconrel.2022.04.045 PMID: 35513211

Rodríguez, R., Alvarez-Lorenzo, C., & Concheiro, A. (2003). Cationic cellulose hydrogels: Kinetics of the cross-linking process and characterization as pH-/ion-sensitive drug delivery systems. *Journal of Controlled Release*, 86(2–3), 253–265. DOI: 10.1016/S0168-3659(02)00410-8 PMID: 12526822

Rooney, J. A. (1970). Hemolysis near an ultrasonically pulsating gas bubble. *Science*, 169(3948), 869–871. DOI: 10.1126/science.169.3948.869 PMID: 5432582

Santos, M. A., Goertz, D. E., & Hynynen, K. (2017). Focused ultrasound hyperthermia mediated drug delivery using thermosensitive liposomes and visualized with in vivo two-photon microscopy. *Theranostics*, 7(10), 2718–2731. DOI: 10.7150/thno.19662 PMID: 28819458

Shi, X., Zheng, Y., Wang, G., Lin, Q., & Fan, J. (2014). PH- and electro-response characteristics of bacterial cellulose nanofiber/sodium alginate hybrid hydrogels for dual controlled drug delivery. *RSC Advances*, 4(87), 47056–47065. DOI: 10.1039/C4RA09640A

Shortencarier, M. J., Dayton, P. A., Bloch, S. H., Schumann, P. A., Matsunaga, T. O., & Ferrara, K. W. (2004). A method for radiation-force localized drug delivery using gas-filled lipospheres. *IEEE Transactions on Ultrasonics, Ferroelectrics, and Frequency Control*, 51(7), 822–831. DOI: 10.1109/TUFFC.2004.1320741 PMID: 15301001

Siegel, R. A., Falamarzian, M., Firestone, B. A., & Moxley, B. C. (1988). pH-Controlled release from hydrophobic/polyelectrolyte copolymer hydrogels. *Journal of Controlled Release*, 8(2), 179–182. DOI: 10.1016/0168-3659(88)90044-2

Sirsi, S. R., & Borden, M. A. (2014). State-of-the-art materials for ultrasound-triggered drug delivery. *Advanced Drug Delivery Reviews*, 72, 3–14. DOI: 10.1016/j.addr.2013.12.010 PMID: 24389162

Song, B., Wu, C., & Chang, J. (2015). Ultrasound-triggered dual-drug release from poly(lactic-co-glycolic acid)/mesoporous silica nanoparticles electrospun composite fibers. *Regenerative Biomaterials*, 2(4), 229–237. DOI: 10.1093/rb/rbv019 PMID: 26816645

Sundaram, J., Mellein, B. R., & Mitragotri, S. (2003). An experimental and theoretical analysis of ultrasound-induced permeabilization of cell membranes. *Biophysical Journal*, 84(5), 3087–3101. DOI: 10.1016/S0006-3495(03)70034-4 PMID: 12719239

Tacker, J. R., & Anderson, R. U. (1982). Delivery of Antitumor Drug to Bladder Cancer by Use of Phase Transition Liposomes and Hyperthermia. *The Journal of Urology*, 127(6), 1211–1214. DOI: 10.1016/S0022-5347(17)54299-8 PMID: 7087041

Tan, J., Kang, H., Liu, R., Wang, D., Jin, X., Li, Q., & Huang, Y. (2011). Dual-stimuli sensitive nanogels fabricated by self-association of thiolated hydroxypropyl cellulose. *Polymer Chemistry*, 2(3), 672–678. DOI: 10.1039/C0PY00348D

Tovar-carrillo, K. L., Sueyoshi, S. S., Tagaya, M., & Kobayashi, T. (2013). *Fibroblast Compatibility on Scaffold Hydrogels Prepared from Agave Tequilana Weber Bagasse for Tissue Regeneration.* 52(33), 11607–11613. https://doi.org/doi.org/10.1021/ie401793w

Tovar-Carrillo, K. L., Tagaya, M., & Kobayashi, T. (2014). Biohydrogels interpenetrated with hydroxyethyl cellulose and wooden pulp for biocompatible materials. *Industrial & Engineering Chemistry Research*, 53(12), 4650–4659. DOI: 10.1021/ie403257a

Tran Vo, T. M., Kobayashi, T., & Potiyaraj, P. (2022). Viscoelastic Analysis of Pectin Hydrogels Regenerated from Citrus Pomelo Waste by Gelling Effects of Calcium Ion Crosslinking at Different pHs. *Gels (Basel, Switzerland)*, 8(12), 814. Advance online publication. DOI: 10.3390/gels8120814 PMID: 36547338

Tsioptsias, C., & Panayiotou, C. (2008). Foaming of chitin hydrogels processed by supercritical carbon dioxide. *The Journal of Supercritical Fluids*, 47(2), 302–308. DOI: 10.1016/j.supflu.2008.07.009

Tsutsui, J. M., Xie, F., & Porter, R. T. (2004). The use of microbubbles to target drug delivery. *Cardiovascular Ultrasound*, 2(1), 1–7. DOI: 10.1186/1476-7120-2-23 PMID: 15546496

You, J., Zhang, R., Xiong, C., Zhong, M., Melancon, M., Gupta, S., Nick, A. M., Sood, A. K., & Li, C. (2012). Effective Photothermal Chemotherapy Using Doxorubicin-Loaded Gold Nanospheres That Target EphB4 Receptors in Tumors. *Cancer Research*, 72(18), 4777–4786. DOI: 10.1158/0008-5472. CAN-12-1003 PMID: 22865457

Yu, P., Yu, H., Guo, C., Cui, Z., Chen, X., Yin, Q., Zhang, P., Yang, X., Cui, H., & Li, Y. (2015). Reversal of doxorubicin resistance in breast cancer by mitochondria-targeted pH-responsive micelles. *Acta Biomaterialia*, 14, 115–124. DOI: 10.1016/j.actbio.2014.12.001 PMID: 25498306

Yun, Y. H., Lee, B. K., & Park, K. (2015). Controlled Drug Delivery: Historical perspective for the next generation. *Journal of Controlled Release*, 219, 2–7. DOI: 10.1016/j.jconrel.2015.10.005 PMID: 26456749

Zhang, X. Z., Wu, D. Q., & Chu, C. C. (2004). Synthesis, characterization and controlled drug release of thermosensitive IPN-PNIPAAm hydrogels. *Biomaterials*, 25(17), 3793–3805. DOI: 10.1016/j.biomaterials.2003.10.065 PMID: 15020155

Zhao, D., Huang, W. C., Guo, N., Zhang, S., Xue, C., & Mao, X. (2019). Two-step separation of chitin from shrimp shells using citric acid and deep eutectic solvents with the assistance of microwave. *Polymers*, 11(3), 409. Advance online publication. DOI: 10.3390/polym11030409 PMID: 30960393

Chapter 13
ERS Vacuum Fermentation and Drying Bioreactor Contributing to Recycling of Organic Containing Wastes

Shinichi Shimose
JET Corporation, Japan

Tomoyuki Katayama
JET Corporation, Japan

ABSTRACT

ERS (Environmental Recycling System), a bioreactor in which microbial fermentation conducts organic waste treatment under mild-temperature heating and vacuum conditions, and in which high-speed fermentation and dehydration drying of wet wastes are carried out in a single reactor unit. The volume and weight reduction of waste containing organic materials and the conversion of the waste into compost, fertilizer, feed, livestock bedding, or biomass fuel are possible. Main features of ERS processing include that there is no emission of drain or bad odors through its processing, that the microorganisms used for fermentation in ERS are those inhabiting the surroundings of the installation site, preventing any disruption to the local microbial environment, and that there is no need to supplement ERS with microorganisms, and that unsorted general waste can be directly introduced into the ERS without sorting. This chapter introduces the current status of organic-containing waste treatment with examples of ERS implementation and also the practical applications.

INTRODUCTION

Importance and Current Status of Recycling of Organic Wastes

It is important to promote the shift to a sustainable social and economic system from the perspective of revitalizing nature and coexistence in the near future. In order to achieve this, it is necessary to change our lifestyles so that they are compatible with a recycling-oriented society (Environment Agency, 2000;

DOI: 10.4018/979-8-3693-0003-9.ch013

Fiksel, 2003; van Berkel, 2007). In addition, in order to achieve sustainable growth, it is necessary to make changes to the conventional lifestyle development through the development of innovative technologies, and to create a system in which industry and the economy are in harmony with nature. From this perspective, in addition to the sustainable use of terrestrial resources, society, the economy, and nature need to develop sustainably in symbiosis without running out of limited resources (Kobayashi and Nakajima, 2021). In other words, in order to reduce the consumption of natural resources and also to reduce the environmental burden, waste emission reduction and effective recycling technologies should be considered for realization in the future.

In the current situation, the concept of Zero-Waste Business is important first to promote the reduction of environmental burdens (Zero Waste Business Principles, 2005; Zaman, 2012; Tarannum et al., 2022), where the key concepts are "reduction of waste generation" (no waste), "cyclical use of recyclable resources" (use of waste as resources), and "proper disposal of recyclable resources" (proper disposal of waste that cannot be reused), or 3R (Figure 1a). The importance and priority of the 3Rs (reduce, reuse, recycle) is well documented and explained in the UN Environmental programme (UNEP/GC/23/INF/11, 2005; Universal Eco Pasific, 2023). In effectively operating these 3Rs and promote their cyclical use, it is essential to introduce upcycling technologies that convert waste into new products (United Nations Environment Programme, 2021). For example, the important point in 3R is to transform organic waste into a circular bioeconomy that views organic waste as a resource and contributes to more sustainable sectors. The most obvious example of this is the conversion of organic waste, such as food waste, into a resource, with the economic value of recycled materials being food, feed, fuel, and fertilizer, in that order. For example, in the conversion of food waste into fertilizer, the EU supported the training of researchers and the development of new processes for the production of biobased fertilizers in the Ferti Cycle project to contribute to a more sustainable agricultural sector without the use of natural phosphorus resources (Bayo et al., 2022). Here, collaboration with connectable sectors is essential for waste recycling and resource recovery.

As one example, in Japan, waste is classified into two categories; industrial waste and municipal waste. As a general rule, municipalities are responsible for overseeing the disposal of general waste, while businesses are responsible for the disposal of industrial waste. The municipal waste is originally generated from people's daily lives, while the industrial waste arises from business activities. Industrial wastes include cinders, sludge, waste oil, waste acids and alkalis, waste plastics, metal, glass, concrete, and ceramic scraps, debris, soot and dust, paper waste, wood waste, fiber waste, animal manure and carcasses, and animal and plant residues. Here, organic wastes in industrial wastes are wastes that are different from petroleum-based materials derived from formerly living materials (CORDIS, 2020), and examples of biodegradable and organic wastes include food wastes, food-stained paper, non-hazardous wood waste, landscape waste, and green waste such as pruning waste. Currently, food waste, biomass-derived wastes such as paper, wood, and fiber scraps, animal manure and carcasses, and animal and plant residues are often landfilled without being recycled. In Australia, it was reported that the amount of organic waste sent to landfills results in the generation of about 13 million tons of CO_2-emission (carbon dioxide equivalent), which was equivalent to about 3% of Australia's total emissions (Pickin et al., 2020). In Malaysia and Indonesia, the abandonment of empty oil palm fruit bunches (EFB) on vacant palm oil mill sites has become a common practice, and greenhouse gas (GHG) emissions due to methane generation from EFB anaerobic decomposition have become a problem (Novaviro, 2011; Fujita et al., 2019). Therefore, it is very important to know that recycling organic waste has the potential to reduce greenhouse gas emissions while conserving natural resources, making upcycling of organic waste an important approach.

According to the United State Environmental protection agency (EPA), total municipal solid waste (MSW) generation in the USA in 2018 was 292.4 million tons, of which approximately 69 million tons of MSW generated was recycled and 25 million tons was composted. Thus, almost 94 million tons of MSW were recycled and composted, corresponding to a recycling and composting rate of 32.1% (EPA, 2022). The EU generates 2.2 billion tons of waste each year, of which more than a quarter (27%) is municipal waste (European Parliament, 2023). The amount of waste and the way managed in each country vary widely among EU countries, with recycling and composting accounting for 49.6% and landfill accounting for 18%. In the future, there is a shift toward increasing recycling and reducing landfill, with the goal of a 60% recycling rate and a landfill rate of 10% or less by 2030. Among EU countries, in Germany, recycling and composting accounted for 71.1% in 2020, while landfill accounted for 9%. Meanwhile, the general waste disposal situation in Japan was 40.95 million tons, or 890 grams of waste per person, as of March 2023. Of this amount, 0.9% was disposed of in landfills, and 39.9 million tons was treated by municipalities through incineration, shredding, sorting, and other intermediate processes. The recycling rate was about 20%, but this value was significantly lower in Japan than in other EU countries, represented by Germany's 71.1%. One reason for this was the high incineration rate of 3.149 million tons, or about 79.9% (Waste Management, 2012).

In general, from the principle of manufacturing recyclables to have economic value in the 3R process, even after the 3R stage is over, energy conversion by combustion or combustion for disposal can still be performed, and these thermal treatment processes are subordinate to the economic value. Energy recovery and fertilizer conversion are important technologies for building a sustainable, resource-recycling society., however, in terms of value-added, for example, in the case of food waste, the first step is to reuse it as food, and if food use fails, then feed use comes next, and if that fails, then fuel use or fertilizer use (Figure 1). Contrary to this ideal, the actual situation in Japan is that the incineration rate is high, and efforts to improve the reuse and recycling of organic waste will be necessary.

Figure 1. 3R-cyclical use of recyclable resources and economic value of food waste

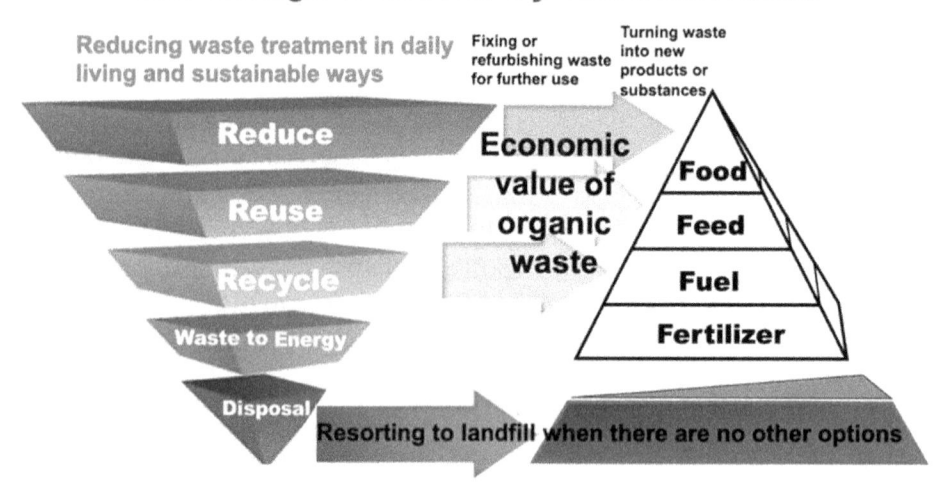

Recycling of organic wastes In order to properly treat and recycle organic waste, the following items should be properly addressed in each region where the waste is generated; the amount and characteristics of each type of waste generated, its collection and treatment, and issues related to recycling and its utilization. In other words, it is important to assume an integrated system of generation, discharge, treatment/resource conversion, and utilization, and to meet the balance of payments to form a circulatory system at an optimal scale according to the characteristics of the region and the nature of the circulating resources. By promoting the effective use of local resources, waste reduction, and activation of environmental activities, such recycling process has the effect of reducing the consumption of natural resources, thereby reducing the burden on the environment to a greater extent than before. Because of these advantages, in Japan, the Ministry of the Environment advocates the construction of a regional recycling society (MOE Japan, 2022), and in particular, soil conditioners, fertilizers, animal feeds, and heat energy applications are the main uses for organic waste in the region (European Parliament, 2023). Table 1 summarizes the main organic wastes from the related food industry, forestry, agriculture, and livestock (Professional Engineering Japan). In each of the wastes listed here, the purity of the waste is high and the feed conversion rate is high because the waste is not a mixed unit with different kinds of waste. In contrast, the high rate of combustion and landfill rates for waste from people's daily lives induces the following problems; the current situation is that the rapid increase in waste is making it difficult to incinerate waste, and it is also becoming more difficult to landfill combustible waste without incineration (Japanese recycling industry, 2014). Therefore, in the future, it is extremely important to attempt to shift from a one-way society of disposal and landfill to a recycling-oriented society of reduction and reuse.

Table 1. Major organic wastes related to agriculture in Japan, including food industry, forestry, agro-industry, and livestock production

		Type	Water content (%)	Estimated annual generation in Japan	Processing and Utilization
Food Industries		Beer dregs	80	700,000-1,050,00t	Cattle feed 95%
		Shochu dregs	94	379,712t	Marine disposal 50%, fertilizer 19%, feed 19%
		Whiskey dregs	76.4	379,712t	Feed 83.5%
		Sake dregs		160,165t	Feed 100%
		Tofu residue	80	744,600t	Feed 70%, fertilizer 16%, foodstuff 4%
	Fruit juice	Orange juice	86	22,500~153,300t	Feed 86.8%
		Canned item	84	31,900~71,600t	Feed 50.6%
		Apple juice	79.7	43,800~85,500t	Feed 64.1%
	Residue	Peach Juice		4,700~7,800t	
		Canned item		4,800t	Disposal 100%
		Coffee dregs	65	600,000t	Fuel used in factory, approx. 100,000 t/year for disposal
	Tea dregs	Green tea dregs		9,000t	
		Black tea dregs	70	14,000t	Raw material for compost
		Barley tea dregs		4,500t	
		Others		9,000t	
		Soy sauce cake	26.5	85,877t	Feed 64.1%
		Corn starch by-product	11~53	784,740t	Feed 100%
		Valeryl starch dregs	92	776,790t	Feed 70%, disposal 30%
		Corn starch residue	92	258,341t	Feed 70%, citric acid fermentation raw material 30%
Forestry products		Paper sludge	74~79	1,208,758t	Incinerated 93%
		Bark		4.6 million t (6,577,000 m3 /year) 906,458 t/year as bark compost	From sawmills burning 52%, fuel 30%, compost 7%, barn bedding 4%, 2% for ogaliiths
		Sawdust	40	2.6 million t(4,146,000m3 /year)	For livestock 62%, 13% for mushrooms 10% for ogaliiths 8% for industry
		Waste from the culture medium after mushroom cultivation	47.6 53	500,000t	

continued on following page

Table 1. Continued

Agricultural products	Beat pulp	82	1,208,758t	Feed 100%
	Bagasse	50	418,500t	Fuel used in factory 89%, compost 8%, Feed 1%
	Rice husks	11.7	2,366,408t	Burning 24.6%, barn bedding 20.9%, compost 20.1%, culvert material 12.2%, mulching 6.31%, smoked charcoal 4.61
	Rice straw	10	11,000,000t	Hoeing 61%, compost 11%, coarse feed 11%, barn bedding 6.5%, burning 4.9%, mulching 4.5%, handicraft 1.6%
	Wheat straw		728,000t	Burning 37.7%, hoeing 33.5%, compost exchange 29%
Animal husbandry	Cattle manure	80	30,369,000t	Compost 94.8%, dried 5%
	Pig manure	75	8,397,000t	Compost 50%, dried 16%
	Chicken manure	78	15,390,000t	Dried 64%, compost 12%
	Feathers		167,586t	Feed 100%
	Blood		25,228t	Feed 74.4%, the rest for disposal
	Non-edible internal organs, etc.		596,463t	Feed 97.6%

On the other hand, conversion to feed with a high resource recovery rate (Table 1), i.e., conversion of food residues to eco-feed, is not only an effective use of food resources, but also an important initiative to improve feed self-sufficiency, etc. in the country. In this initiative, the form of eco-feed is powdered or flaked, which requires a process to dry and grind food residues with high moisture contents.

Composting of organic wastes. As mentioned above, the European and U.S. responses have equated recycling and composting, and have responded with a policy of increasing this ratio while reducing landfills (EPA, 2022; European Parliament, 2023). Although composting is known to be slow because it uses microbial metabolism, it is possible to convert organic waste into high-value-added products through methane conversion (EESI, 2017 a; Dhimana et al., 2018), carbonization (Le et al 2022; Ischia and Fiori, 2021), and conversion of resources into organic acids (Thygesen et al., 2021; Dharmalingam et al., 2022). In the composting of organic waste, it is necessary to ensure aerobic conditions and to create conditions for active microorganisms in order for composting to proceed smoothly (Chavan et al., 2022; JEMAI, 2014). However, due to the moisture content varying with the type of waste, livestock manure with a high moisture content generally becomes anaerobic, because air cannot penetrate to the interior if these left untreated. Therefore, composting requires a moisture regulation of about 55-77%. In general, composting of organic waste has a standard process configuration of pretreatment, followed by primary fermentation and secondary fermentation due to composting process can be used for biological decomposition. Here, pretreatment includes crushers, vibrating screens, wind sorting machines, etc. After that, the first fermentation takes about one month, and the second fermentation is maturation, which spends about one to two months. In addition, the fermentation process

Figure 2. Reclamation of organic waste for regional development through sustainable resource utilization

Formation of a Recycling-Oriented Society

Collecting organic wastes
1) Household food waste, industrial food waste including food processing residues
2) Sewage sludge, sewage sludge, septic tank sludge
3) Livestock manure including excrement of cows, pigs and chickens
4) Prunings

↓

Conversion of waste into valuable resources

↓

Renewable energy (steam/electricity/water)

↓

Produce and give back to sustainable community

requires a turnover operation of the waste. On the other hand, many organic wastes have a highwater content; for example, animal manure has a water content of about 80%, and sewage sludge has a water content of about 80-97%. Therefore, improper disposal and management of organic wastes can cause bad odors, which may flow into rivers and groundwater, resulting in water pollution. Such waste also produces methane gas, a greenhouse gas, when it is landfilled, and if it is incinerated, the high water content requires extra auxiliary fuel, which inhibits energy recovery.

Conversion of organic waste into energy. It has been reported by that organic wastes are potential high-impact resources for the domestic production of biofuels, bioproduct precursors, heat, and electricity (Sridhar et al., 2021; Ochieng et al., 2022; Bioenergy Technologies Office, 2023). The waste-to-energy nexus based on the 5R principle (Reduce, Reuse, Recycle, Recovery, and Restore) (Sharma et al., 2020) is for energy production from the agricultural residues. In general, general waste, including organic waste, undergoes thermal conversion, composting by microbial conversion, and landfill treatment (Varma 2014). For example, waste plastic could be converted into fuel through pyrolysis (Jha and Kannan, 2021) and catalytic action (Sivagami et al., 2022). Landfilling is also converting organic components into biogas fuel through decomposition, and composting is also being considered for conversion into fuel through biogas generation (Jaber and Noguchi, 2007; EESI, 2017b; Wang et al., 2023). In terms of thermal conversion, energy is generally converted to steam vapor through incineration (Kumar and Samadder, 2017; IGES, 2020; Vatavuk, 2000) and used with turbine generator (FEMP, 2016) and fuel cell (Dessie and Tadesse, 2022) for bioelectricity.

The use of organic waste as a resource, if not adapted to the 3Rs, it ultimately involves the combustion of hydrocarbons in organic materials or microbial decomposition with energy release, which eventually is converted to finally CO_2. Therefore, there are various resource conversion methods or technologies depending on the form in which the carbon in the waste is converted. In all of these technologies, most organic wastes have a high moisture content, as shown in Table 1, which is problematic because of the high energy consumption for the initial incineration for drying. In general, when food wastes are reused as food products, it is necessary to dry wastes with high moisture content for value-added food commercialization of food wastes. In this case, the product can be dried by natural drying, draft drying, freeze drying, or sporadic drying. In terms of equipment prices, natural drying is less expensive, while freeze drying and spray drying require expensive equipment. Then, in the pulverization process, it is necessary to powderize by impact, airflow, ball milling, or other equipment. Thus, even when simply converting organic waste with high water content into dry powder, diversification of equipment is inevitable. Therefore, there are problems of breakdowns and higher costs, and there is a demand for technology that enables resource processing in a single pot.

As shown in Table 1, the wastes represent in the food, forestry, agriculture, and livestock sectors: 1) commercial food waste, including household food waste, agricultural waste, and food processing residues; 2) human waste, sewage sludge, and septic tank sludge; 3) livestock manure, including cattle, pig, and chicken excreta; and 4) pruning branches. In any cases, for the reuse of these materials, it is necessary to take the following steps:

Waste → Dried powder → Powder function evaluation → Product utilizing an organic waste by recycling plant.

However, equipment that includes a drying process for organic waste containing water is required, and for efficiency, plants are typically accompanied by crushing, heating, and draining equipment. An anaerobic fermentation layer and aerobic fermentation are essential for methane conversion (Western Australia, 2022; EPA,2023) and composting (Faniyi and Oyatokun, 2021), respectively. In the case of methane conversion, a typical plant configuration consists of following processes;

1 feedstock supply process → 2 crushing process → 3 storage process → 4 fermentation process → 6 dehydration process.

These recycling plants are systems that combine multiple unit operation processes of crushing, fermentation, and dehydration, and the composting of organic waste is a future challenge because it takes a long time to reach a product and involves labor-intensive operations, including turnaround operations of organic waste.

Organic Waste Treatment by Vacuum Fermentation Bioreactor (ERS)

Vacuum fermentation bioreactor (ERS). The vacuum fermentation bioreactor ERS is an organic waste treatment system that can perform fermentation, agitation, and dehydration in a single reactor. This bioreactor has the advantage that the fermentation process is carried out under vacuum conditions, while the dehydration process can be carried out in a single pot without transferring the contents. In principle, environmental microorganisms that are ubiquitous in soil can be cultured and fixed in the reactor, and organic waste can be treated through microbial decomposition while dehydration is performed. Generally, an ERS system consists of a main bioreactor chamber, an associated hopper, and a deodorization cooling system (Fig. 3), and can treat waste by its fermentation, stirring, heating, and dehydration process in a single unit under vacuum. Environmental microorganisms (FERMBP-7504, FERMBP-7505, and FERMBP-7506, deposited at the National Institute of Advanced Industrial Science and Technology (NITE)) are planted in the reactor for fermentation and the fermentation reaction on organic waste can be carried out at 60-70°C in a heated state with stirring, utilizing heat from steam of boilers. After the fermentation process, the organic waste with high water content can be further powdered in the dehydration process by vacuum without transferring the fermented material from the reactor. Therefore, ERS has the following features which are not found in other treatment plants;

1) Organic turns into a valuable resource such as compost within 1 day.
2) The reactor unit is simple with few components and easy to maintain.
3) No complicated pre-processing (moisture adjustment, etc.) and post-processing (ripening, etc.) are required because of the dehydration function.
4) No drain is generated (no need for equipping drainage facility).
5) Space saving for storage.

This is an advantage for daily waste disposal, allowing for a significant reduction in waste volume (saving on transportation costs) and easier handling of dried waste. In addition, the fermented and dried waste can be used as various resources such as compost and fuel.

Figure 3. Diagram of the ERS bioreactor composed of reactor chamber with processes of fermentation, dehydration, heating and stirring, hopper for feeding waste, and deodorization cooling system

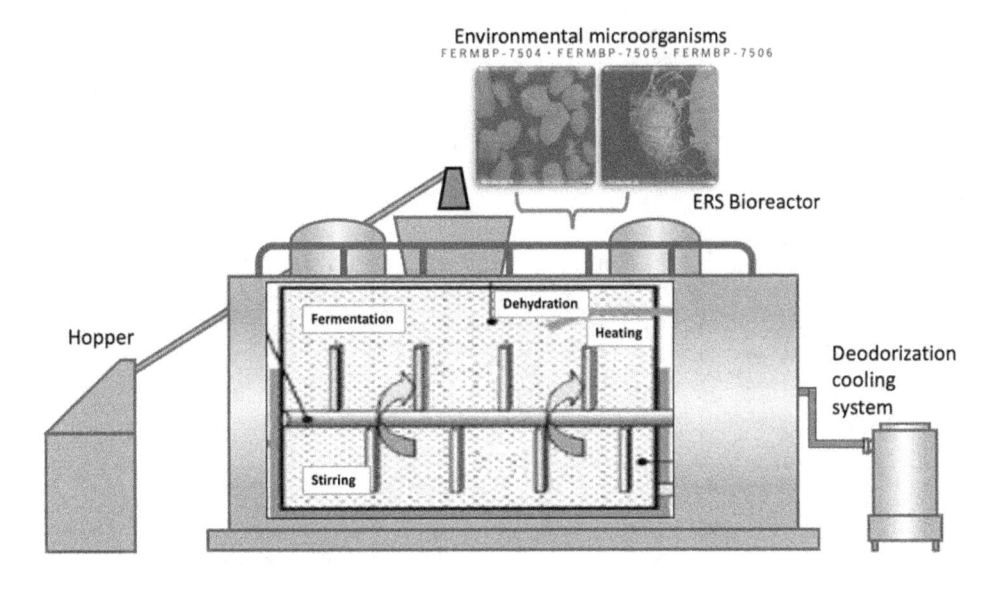

There are various examples of organic waste treatment by ERS (Table 2). In these actual cases, the processed materials include livestock manure, biomass such as palm residues, and food waste such as waste milk, shochu dregs, tofu residue, and tea and coffee dregs, as well as industrial and general wastes. The deliverables generated through ERS processing can be reused mainly in agriculture and livestock farming as fertilizers, livestock feed and bedding materials, while some could be used as bio-fuel.

Table 2. Examples of ERS adapted implementation

Treated material	Installation Location	Processing quantity (tons/day)	Introduction date	Use of deliverables	Reference
Beef cattle manure	Tokoro-gun, Hokkaido	75 tons/day	Mar 2022	Barn bedding	Subsidies utilized: Livestock Production and Soil Composting Production and Distribution System Support Project
Animal manure and feed scraps	Nishimuro-gun, Wakayama	1.4 tons/day	Mar 2023	Fertilizer	
Beef cattle manure	Omura City, Nagasaki Prefecture	25 tons/day	Mar 2022	Barn bedding	Special measures project for livestock and dairy farming profit enhancement (facility improvement project)
Pig manure and wash water	Miyakonojo City, Miyazaki Prefecture	2 tons/day	Apr 2021	Barn bedding and compost	
Beef cattle manure	Kawakami-gun, Hokkaido	28 tons/day	Sep 2020	Barn bedding	
Dairy cattle manure, parlor water, waste milk	Fujinomiya City, Shizuoka	25 tons/day	Apr 2020	Fuel	Special measures project for livestock and dairy farming profit enhancement and improvement
Dairy cattle manure, pig manure, chicken manure, shochu dregs and volcanic ash	Kagoshima Prefecture	10 tons/day	1997	Feed and barn bedding	
Wild game (Deer, boar, etc.)	Kihoku Town, Ehime Prefecture	2 tons/day	Mar 2022	Fertilizer and feed	
Wild game (Deer, boar, etc.)	Mimasaka City, Okayama Prefecture	2 tons/day	Sep 2019	Fertilizer	
Barley tea dregs/coffee dregs	Gunma Prefecture	1.8 tons/day	May 2020	Feed, fertilizer and fuel	
Tofu including packed tofu	Gunma Prefecture	25 tons/day	Mar 2019	Feed	
Industrial waste (food residues)	Yuzawa City, Akita Prefecture	2 tons/day	Dec 2016	Feed and fertilizer	
Palm residue (POME and EFB)	Penang, Malaysia	2 tons/day	Dec 2018	Fuel	
Digested liquid after methane fermentation	Fujinomiya City, Shizuoka Prefecture	15 tons/day	Apr 2018	Fertilizer (special fertilizer certification acquired)	
Industrial and general waste	Chikusei City, Ibaraki	15 tons/day	Apr 2018	Fertilizer	
Municipal and landfill waste	Bali, Indonesia	2 tons/day	Dec 2017	Fuel and fertilizer	

"Regional Recycling and Symbiosis Zone" project for volume reduction and fuel conversion of general waste in Minami-Izu Town. As an attempt to reduce the volume of general waste that is difficult to dispose of, and to convert it into fuel in a one-pot process, there is an example of a demonstration experiment in Minami-Izu Town (Shizuoka, Japan) using ERS to ferment and dry general waste and convert it into fuel being used at a municipal incineration facility in 2023.

In the cooperation with the Minami-Izu Town's local government for the treatment of general waste in the town, the ERS shown in Figure 3 has been installed in the incineration facility (15 t/day) at the Minami-Izu Town Cleaning Center to demonstrate the system. In addition to the ERS treatment for the co-combustion experiment with the general waste-derived ERS' deliverables and general waste, only disposable diapers were extracted from the general waste, and the ERS treatment for the co-combustion experiment with the diaper-derived ERS treated products and general waste was conducted. In the co-combustion experiment, the raw materials were converted to fuel by ERS processing and burned at the incinerator in the cleaning center.

ERS has the advantage of being able to perform the crushing, storage, fermentation, and dehydration processes in a single equipment, in one pot, after feeding the raw materials, and because the vacuum fermentation bioreactor functions to ferment and dry while dehydrating, less energy is required to reduce the moisture content than for a dehydration reaction to simply burn. The ERS system of Minami-Izu Town with such performance consists of an ERS bioreactor, raw material hopper, heat exchanger, and deodorization cooling system. The illustration in Figure 4 shows the configuration of the ERS system installed at the Minami-Izu Town Cleaning Center as part of this regional recycling symbiosis project. By processing general waste collected from the community with the ERS system at the processing site, not only the volume of general waste can be reduced, but also the cost and labor of transportation can be reduced. In addition, the water content of waste fermented under vacuum conditions using ERS is much lower than that of untreated general waste. Therefore, damage to the incinerator is expected to be reduced compared to untreated waste by ERS. It is also expected to be possible to store the ERS deliverables without generating odors and to reduce its volume, thus making it possible to secure storage space and adjust the amount of waste incinerated on a daily basis.

The final goal is to use the ERS' deliverables as fuel to be supplied to the incinerator, The main general wastes are food waste, disposal diapers, plastics, pruned branches, etc., all of which are household general wastes with high water content, and increases the use of fossil fuels for incineration. Therefore, the volume reduction and dehydration using ERS and the use of ERS' deliverables as fuel can reduce the amount of the fossil fuel required for initial combustion in the incinerator. In addition, the incinerator's furnace life can be extended through stable combustion by using ERS' deliverables as fuel and the general waste generated can be used as a refuse-derived fuel instead of incinerated waste.

Figure 4. Treatment flow of general waste by the ERS system in the regional recycling and symbiosis zone project in Minami-Izu Town

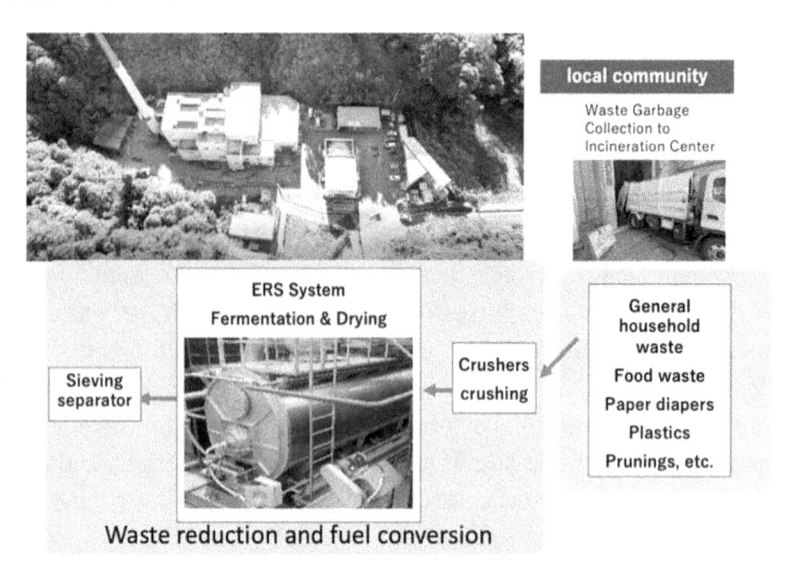

The ERS system installed here consists of a material input hopper - ERS Bioreactor - heat exchanger – deodorization cooling system (Figure 5). The actual experimental flow is shown below.

Collected material is brought in → the material is fed into the hopper → fermentation and drying by ERS → discharge of deliverables → storage of the deliverables → combustion by the municipal incinerator

The material input to the ERS was mixed waste containing organic and plastic wastes. After feeding the material to the ERS, the fermentation temperature was set at 70°C and the fermentation processing time was changed in the experiment, after which the material was dried for 1 hour and flake-like organic matter containing plastics were obtained. The moisture content of the ERS' deliverables was 11.0, 13.6%, and 5.6% at 0.5, 1, and 2 hours of fermentation time, respectively, compared to 82% moisture content in the unprocessed material which is a general waste before being fed to the hopper. Since the deliverables were dry enough, they could be stored until being burned at the incinerator. The calorific values for each fermentation processing time were determined. The heating value including the latent heat of condensation obtained when the product steam in the combustion gas condenses is referred to as the higher heating value, while the heating value remaining as steam and not including the latent heat of condensation is referred to as the lower heating value. Table 3 summarizes these heating values. The unprocessed material had higher and lower heating values of 870 kcal/kg and <50 kcal/kg, respectively, while ERS' deliverables had higher heating values of 3400-3600 kcal/kg and 2900-3200 kcal/kg, respectively, for the deliverables.

Table 3. Calorific value of ERS treated materials

	Unprocessed material	ERS's deliverables		
Fermentation time		0.5 hour	1 hour	2 hours
Water content (%)	82.8	11	13.6	5.6
Higher heating value (kcal/kg)	870	3650	3420	3530
Lower calorific value (kcal/kg)	<50	3240	2990	3190

Figure 5. Configuration of ERS for general waste combustion treatment in Minami-Izu Town and general waste treatment flow

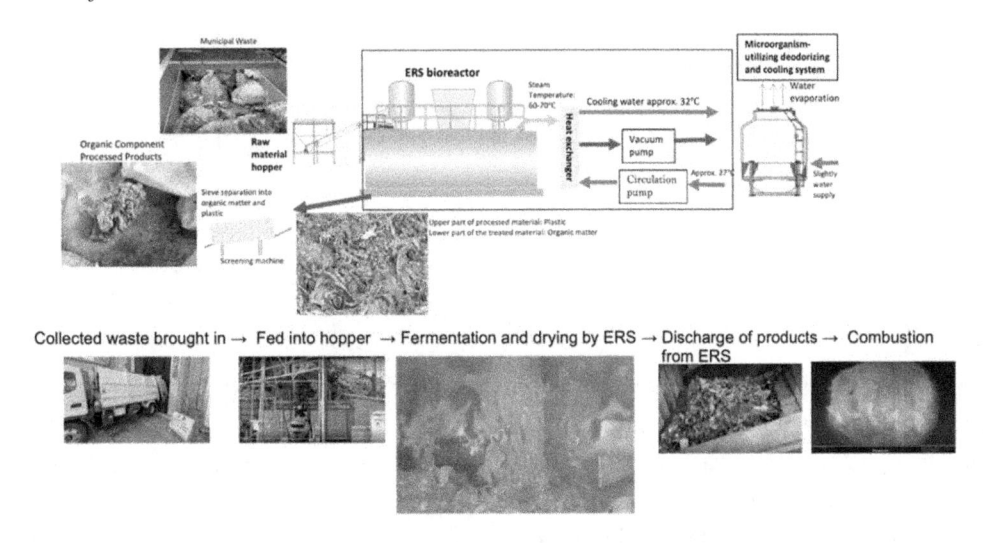

The results of thermogravimetry (TG) and differential thermal analysis (DTA) of these products are shown in Figure 6. According to the DTA results, the weight loss around 100°C is the endothermic peak due to water evaporation, resulting in a caloric loss of 158 kcal/kg. The weight loss at around 100°C is the endothermic peak due to water evaporation, resulting in an endothermic loss of 158 kcal/kg. The exothermic peak at around 300°C is followed by an exothermic peak at around 49.4 kcal/kg, and at 400-500°C, an exothermic value of 688 kcal/kg is observed. On the other hand, in the ERS product, the weight loss around 100°C was almost unobservable, and the weight loss mainly around 300°C was 50-60%, with a calorific value of 287-475 kcal/kg. The weight loss around 460-470°C was also observed, with a calorific value of 287-475 kcal/kg. A sharp exothermic peak was observed at 460-470°C, and about 20% of the weight loss was observed during this combustion, resulting in 239 kcal/kg at 0.5 hours of fermentation treatment and 382 kcal/kg at 1 and 2 hours.

As a result, the heat required for evaporation of water at around 100°C is about 158 kcal/kg for the untreated sample, which causes a loss of about 20% of the total calorific value of about 737 kcal/kg. On the other hand, for the 0.5-hour ERS treated sample, the total calorific value is 668 cal/g, which is about 7% of the heat loss of 51.3 kcal/kg of heat absorption for dewatering. In this comparison, the ERS sample was found to be more efficient in the case of fuel conversion of general waste.

Figure 6. Thermogravimetric (TG) and differential thermal analysis (DTA) profiles from thermogravimetric differential thermal analyzer

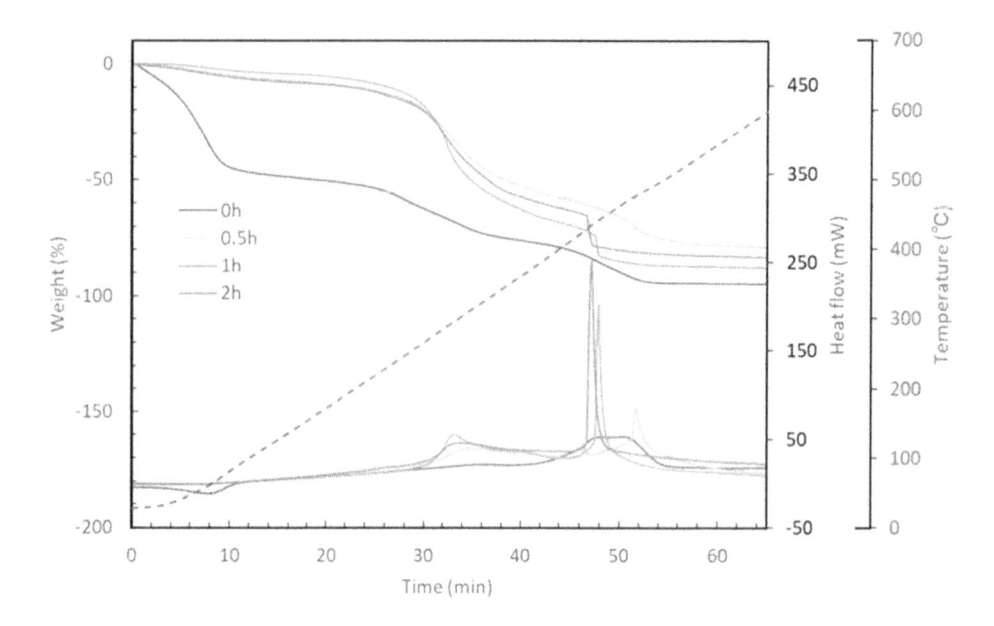

Organic Waste Treatment by ERS. As shown in the ERS treatment examples in Table 2, one of the features of ERS with vacuum drying capability is its ability to dry and convert liquid wastes into solid resources. The moisture in the waste can be evaporated into the atmosphere after being deodorized by the deodorization cooling system, so there is no need to install drainage facilities (Figure 3). For example, in the case of Fujinomiya City in Table 2, liquid manure from dairy cattle is processed by ERS used solid fuel in combination with a biomass boiler. The manure-derived ERS' deliverables are also used as solid compost and livestock bedding. Furthermore, using it as livestock bedding in the barn has the advantage that no bad odor emission in the barn, and there is no need to replace the bedding or even to add sawdust for the bedding. Also, the ERS system can complete the fermentation and drying of the input material in one day without releasing odors (Figure 7), whereas the typical slurry storage and fermentation system produces odors and requires several months to complete fermentation and drying.

Figure 7. Image of livestock manure recycling by ERS system

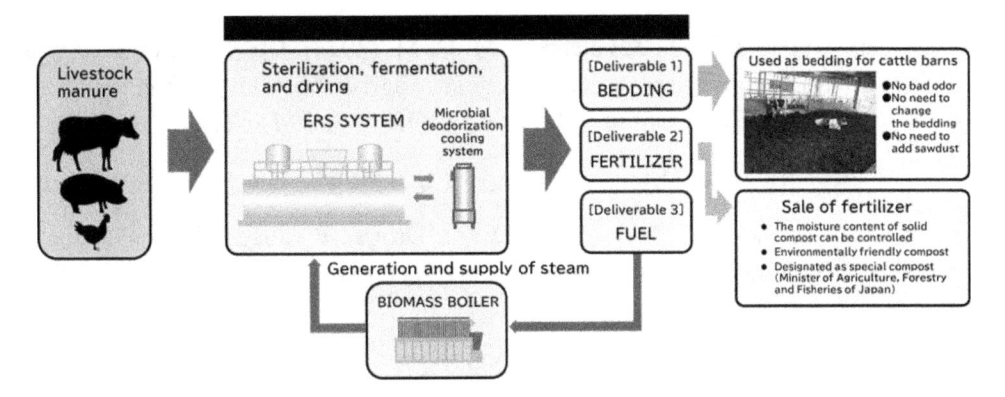

On the other hand, in the case of beef cattle in Saroma Town, Hokkaido, as shown in Table 2, when the increasing of breeding volumes was planned, there were concerns that due to high breeding volumes, the large amount of cow manure would keep being generated and waste treatment would not catch up the volume of such manure, and that due to neglected cow manure, groundwater contamination would occur.

At the farm, about 75 tons of beef cow manure is treated by the ERS on the daily basis, and the ERS' deliverables is used as fuel for biomass boilers, resulting in the reduction of fossil fuel consumption by more than 1/3 compared to heavy oil or gas boilers.

CONCLUSION

In this chapter, we have outlined the treatment of waste and especially the importance of the 3Rs in building a recycling-oriented society. When applied the ERS system for treating organic wastes having higher water contents, the ERS system can sterilize, ferment, and dry, and solidify the wastes within one day without generating odors and drain in the entire treatment process, and can convert the wastes into valuable resources such as fuel, compost, fertilizer, livestock bedding, and feed. Through this process, the volume of waste can be reduced and the treated material can be stored for a long time. Therefore, it contributes to systematic waste disposal and is a powerful method of waste validation. Thus, there are high expectations for the ERS system for reducing environmental impact through waste recycling and energy recovery with society's growing awareness of the need for recovery.

Author Note

Portions of these findings for data and analysis were obtained at the facility where the ERSs were installed. Experiments at a general waste combustion facility were conducted with assistance from the town of Minami Izu (Japan) and the support of Nagaoka University of Technology in Regional Circulation and Symbiosis Project of JET. We have no conflicts of interest to disclose.

Correspondence concerning this article should be addressed to the Public Relations, JET Corporation, NOSAI Bldg. 4F, 19 Ichibancho, Chiyoda, Tokyo 102-0082, Japan.

Email: info@jet-e.jp

REFERENCES

Bayo, J. F., Achmon, Y., Gioia, F. D., & Guerrero, M. D. M. (2022). Upcycling organic waste for the sustainable management of soilborne pests and pathogens in agri-food systems. *Frontiers in Sustainable Food Systems*, 6, 1012789. Advance online publication. DOI: 10.3389/fsufs.2022.1012789

Bioenergy Technologies Office. (2023): https://www.energy.gov/eere/bioenergy/2023-project-peer-review

Chavan, S., Yadav, B., Atmakuri, A., Wongc, J. W. C., & Drogui, P. (2022). Bioconversion of organic wastes into value-added products: A review. *Bioresource Technology*, 344, 12639828. DOI: 10.1016/j.biortech.2021.126398 PMID: 34822979

CORDIS. (2020): https://cordis.europa.eu/project/id/860127

Dessie, Y., & Tadesse, S. (2022). Advancements in Bioelectricity Generation Through Nanomaterial-Modified Anode Electrodes in Microbial Fuel Cells. *Frontiers in Nanotechnology*, 4, 876014. Advance online publication. DOI: 10.3389/fnano.2022.876014

Dharmalingam, B., Tantayotai, P., Panakkal, E. J., Cheenkachorn, K., Kirdponpattara, S., Gundupalli, M. P., Y Cheng, U. S., Sriariyanun, M. (2022). Organic Acid Pretreatments and Optimization Techniques for Mixed Vegetable Waste *Biomass Conversion into Biofuel Production*: https://doi.org/DOI: 10.1007/s12155-022-10517-y

Dhimana, S. S., Shresthad, N., Davida, A., Basotra, N., Johnson, G. R., Chadha, B. S., Gadhamshetty, V., & Sani, R. K. (2018). Producing methane, methanol and electricity from organic waste of fermentation reaction using novel microbes. *Bioresource Technology*, 258, 270–278. DOI: 10.1016/j.biortech.2018.02.128 PMID: 29544100

EESI. (2017a); tpps://www.eesi.org/papers/view/fact-sheet-biogasconverting-waste-to-energy

EESI. (2017b): https://www.eesi.org/papers/view/fact-sheet-biogasconverting-waste-to-energy

EPA. (2022); https://www.epa.gov/facts-and-figures-about-materials-waste-and-recycling/national-overview-facts-and-figures-materials

EPA. (2023); https://www.epa.gov/anaerobic-digestion/basic-information-about-anaerobic-digestion-ad

European Parliament. (2023); https://www.europarl.europa.eu/news/en/headlines/society/20180328STO00751/waste-management-in-the-eu-infographic-with-facts-and-figures

Faniyi, T. O., & Oyatokun, O. S. (2021). *Fermentation in the Perspective of Agriculture, Fermentation - Processes, Benefits and Risks* (Laranjo, M., Ed.)., DOI: 10.5772/intechopen.97608

FEMP. (2016); https://www.wbdg.org/resources/biomass-electricity-generation

Fujita, H., Nakano, K., Chaisomphob, T., & Hambali, E. (2019). Greenhouse Gas Emission of Electricity Generation and Field Abandonment Scenarios Using Empty Fruit Bunches at a Palm Oil Mill, Surat Thani Province, Thailand. *Journal of the Japan Institute of Energy*, 98(5), 119–123. DOI: 10.3775/jie.98.119

IGES 2(020); https://www.iges.or.jp/en/publication_documents/pub/policysubmission/en/10877/WtEI_guideline_web_200615.pdf

Ischia, G., & Fiori, L. (2021). Hydrothermal Carbonization of Organic Waste and Biomass: A Review on Process, Reactor, and Plant Modeling. *Waste and Biomass Valorization*, 12(6), 2797–2824. DOI: 10.1007/s12649-020-01255-3

Jaber, N., & Noguchi, N. (2007). Literature Review of Recent Research on Biogas and its Usage in Diesel Engines. *Nogyo Kikai Gakkaishi*, 69(1), 89–98.

Japan, M. O. E. (2022): https://www.env.go.jp/en/press/press_00629.html

Japanese recycling industry (2014): https://www.env.go.jp/content/900453392.pdf

JEMAI. (2014); https://www.cjc.or.jp/cjc/plant05.html

Jha, K. K., & Kannan, T. T. M. (2021). Recycling of plastic waste into fuel by pyrolysis - a review. *Materials Today: Proceedings*, 37(2), 3718–3720.

Kobayashi, T., & Nakajima, L. (2021). Sustainable development goals for advanced materials provided by industrial wastes and biomass sources. *Current Opinion in Green and Sustainable Chemistry*, 28, 100439. DOI: 10.1016/j.cogsc.2020.100439

Kumar, A., & Samadder, S. R. (2017). A review on technological options of waste to energy for effective management of municipal solid waste. *Waste Management (New York, N.Y.)*, 69, 407–422. DOI: 10.1016/j.wasman.2017.08.046 PMID: 28886975

Le, H. S., Chen, W. H., Ahmed, S. F., Said, Z., Rafa, N., Le, A. T., Ibham Veza, Ü. A., Nguyen, X. P., Duong, X. Q., Huang, Z., & Hoan, A. T. (2022). Hydrothermal carbonization of food waste as sustainable energy conversion path. *Bioresource Technology*, 363, 127958. DOI: 10.1016/j.biortech.2022.127958 PMID: 36113822

Novaviro. (2011). Novaviro Technology Sdn. Bhd. [Internet]. Specifications of Novaviro-KS anaerobic digester technology and scope of KS anaerobic digester system. Available from: http://novaviro.com.my/ [accessed 10 December 2011].

Ochieng, R., Gebremedhin, A., & Sarker, S. (2022). Integration of Waste to Bioenergy Conversion Systems: A Critical Review. *Energies*, 15(7), 2697. DOI: 10.3390/en15072697

Pickin, J., Wardle, C., O'Farrell, K., Nyunt, P., & Donovan, S. (2020). National Waste Report, *Blue Environment Pty Ltd.,* www.blueenvironment.com.au

Professional Engineering Japan. https://www.engineer.or.jp/c_cmt/cpd/topics/008/attached/attach_8458_1.pdf

Sharma, S., Basu, S., Shetti, N. P., Kamali, M., Walvekar, P., & Aminabhavi, T. M. (2020). Waste-to-energy nexus: A sustainable development. *Environmental Pollution*, 267, 115501. DOI: 10.1016/j.envpol.2020.115501 PMID: 32892013

Sivagami, K., Kumar, K. V., Tamizhdurai, P., Govindarajan, D., Kumarc, M., & Nambi, I. (2022). Conversion of plastic waste into fuel oil using zeolite catalysts in a bench-scale pyrolysis reactor. *RSC Advances*, 12(13), 7612–7620. DOI: 10.1039/D1RA08673A PMID: 35424760

Sridhar, A., Kapoor, A., Senthil Kumar, P., Ponnuchamy, M., Balasubramanian, S., & Prabhakar, S. (2021). Conversion of food waste to energy: A focus on sustainability and life cycle assessment. *Fuel*, 302, 11069. DOI: 10.1016/j.fuel.2021.121069

Tarannum, N., Kumar, N., & Pooja, K. (2022). Zero Waste as an Approach to Develop a Clean and Sustainable Society, *Handbook of Smart Materials, Technologies, and Devices Applications of Industry 4.0*, 381-423. https://link.springer.com/referencework/10.1007/978-3-030-84205-5

Thygesen, A., Tsapekos, P., Alvarado-Morales, M., Angelidaki, I. (2021). Valorization of municipal organic waste into purified lactic acid, Bioresource Technology, 342, 125933. UNEP/GC/23/INF/11. (2004). Reduce, reuse and recycle concept (the "3Rs") and life-cycle economy, Governing Council of the United Nations Environment Programme.

United Nations Environment Programm. (2021):https://wedocs.unep.org/handle/20.500.11822/40346

Universal Eco Pasific. (2023): https://www.universaleco.id/blog/detail/prinsip-3r-reduce-reuse-dan-recycle/156

Varma, V. S., & Kalamdhad, A. S. (2014). Bio-conversion of organic waste using composting technologies- A review. *International Journal of Environmental Technology and Management*, 17(6), 483. Advance online publication. DOI: 10.1504/IJETM.2014.066518

Vatavuk, W. M. (2000). Chapter 2 Incinerators, EPA/452/B-02-001; https://www.epa.gov/sites/default/files/2020-07/documents/cs3-2ch2.pdf

Wang, Q., Xia, C., Alagumalai, K., Le, T. T. N., Yuan, Y., Khademi, T., Berkani, M., & Lu, H. (2023). Biogas generation from biomass as a cleaner alternative towards a circular. *Fuel*, 333(2), 126456. DOI: 10.1016/j.fuel.2022.126456

Waste Management. (2012): https://www.env.go.jp/content/900453393.pdf

Western Australia. (2022); https://www.agric.wa.gov.au/climate-change/composting-avoid-methane-production----western-australia

Zaman, A.U. (2012). Developing a Social Business Model for Zero Waste Management Systems: A Case Study Analysis, Journal of Environmental Protection, 3(11)(2012), Article ID:24726,12 pages DOI:DOI: 10.4236/jep.2012.311163

Zero Waste Business Principles. (2005): https://zwia.org/zero-waste-business-principles/

Chapter 14
Pollutant Remediation Using Inorganic Polymer-Based Fibrous Composite Adsorbents

Anh Phuong Le Thi
Nagaoka University of Technology, Japan

Ngan Phan Thi Thu
Nagaoka University of Technology, Japan

Takaomi Kobayashi
Nagaoka University of Technology, Japan

ABSTRACT

In the prospective of building sustainable development society, environmental remediation technology is important for reconstruction of polluted nature. This chapter highlights new remediation technology of fibrous adsorbents consist of inorganic-polymer composites having several merits on combining both properties of polymer and inorganic components. These materials possess a fibrous structural arrangement, providing them with a substantial surface area that facilitates efficient pollutant capture. Their characters address several topics of a wide range of environmental issues for removal of heavy metals, organic compounds, and gases, as customized to specific pollutant removal requirements. Becasue environmental concerns are increasingly issued, it is clear for decontamination process that fibrous adsorbents have great potential in reducing the negative effects of pollutants on both the environment and human health.

INTRODUCTION

Global Pollution and Decontamination Processes

While human prosperity, global pollution is serious, and one has the problem of pollutant decontamination, since toxic pollutants discharged into global environmental systems has become an increasingly worrisome global risk. A variety of toxic substances including H_2S, CO_x, nitrogen compounds, sulfur-containing compounds, and metal compounds is annually discharged into both water and air environment (Kampa & Castanas, 2008; Pöschl & Shiraiwa, 2015; J. Tang et al., 2017; Z. Zhang et al., 2018). These toxic emissions primarily originate from human activities, such as the inadvertent release of toxic gases and wastewater from industrial processes, posing a significant threat to the environment, ecology, and human health (Chou & Harper, n.d.; Stejskal, 2022; Ukaogo et al., 2020). As a result, the effective removal of toxic pollutants from the environment has garnered significant attention and become a prominent topic of discussion (Wen et al., 2019). On the other hand, utilizing these technologies, mankind has made efforts to remove pollution sources from nature. Pollutants and contaminants pose a significant threat to our environment and human health due to their ability to easily spread and traverse between the various components of the Earth's ecosystem, including the atmosphere, pedosphere (soil), and hydrosphere (water) (Figure 1). Consequently, if we do not decontaminate a contaminated site, the source of contamination has the ability to spread to other locations. This is because the source of contamination has the ability to move within the environment, rather than remaining at the contaminated site, due to external natural forces. The external action by rain, wind, water flow, and soil spreading has such an effect that it can spread to the entire earth.

Figure 1. Overview of the harmfulness of environmental pollutions on ecosystems

The successful elimination of these toxic substances is paramount for safeguarding the environment and the lives of those on site who are exposed to such dangerous compounds. To attenuate and overcome this contamination problem, various technologies are employed for environmental conservation (Sidhikku Kandath Valappil et al., 2021) in terms of chemical precipitation (Oncel et al., 2013), membrane filtration (Hube et al., 2020), ion exchange (Swanckaert et al., 2022) and adsorption flotation (Qasem et al., 2021), which involves adsorption processes for pollutants. In actual decontamination sites, however, many of these treatment methods encounter challenges such as high costs, limited effectiveness, and the generation of secondary pollutants. Among these technologies adsorbents have recognized comparative in potential effect on generating less sludge, introducing alternatively without special equipment (Zito

& Shipley, 2015), offering advantages such as cost-effectiveness, minimal production of secondary by-products and then equipping an easily manageable separation process, corresponding to environmental counter-plan (Chen et al., 2012).

Of them, the shape and properties of the adsorbent are particularly important and also affect the effectiveness of the decontamination process in addition with offering the advantages of easy handling and low cost, for example, in chelating polymers (Tyagi & Jacob, 2020), activates carbon (Dlamini et al., 2020), biosorbents (Bhatnagar et al., 2010), metal oxides (Yang, 2021), zeolites (Shah et al., 2011), etc.

Table 1. Inorganic minerals used for adsorbents of some pollutants

Inorganic minerals	Adsorbate	Maximum adsorption capacity q_m (mg/g)	References
Mordenite X- or Y- type	Benzen Toluene	904.8 995.11	(K.-J. Kim & Ahn, 2012)
Natural zeolite	Cd	197.5	(M. Wang et al., 2018)
Chabazite	Pb	6.6	(Yuna, 2016)
	Zn	5.5	
	Cu	5.1	
	Cr	3.6	
Zeolite Na-X	Toluene Xylene	383.67	(Bandura et al., 2016)
Clinoptilolite	Methylene blue	25.52	(S. Wang & Zhu, 2006)
	Cd^{2+}	4.48	(Sprynskyy et al., 2006)
	Pb^{2+}	26.9	(Llanes-Monter et al., 2007)
	Cu^{2+}	26.24	(Sprynskyy et al., 2006)
Halloysite	Benzene	62.7	(Deng et al., 2020)
	Pb^{2+}	14.9	(Nguyen et al., 2019)
Bentonite	O-xylene	110.6	(Zaitan et al., 2008)
Kaolinitic clays	CO C_2H_2 SO_2	1.95 6.11 34.7	(Volzone & Ortiga, 2011)
$\gamma\text{-}Fe_2O_3$	As(V)	90	(Hei et al., 2014)
Fe-Co-MOF-74	As(III) As(V)	266 292	(Sun et al., 2019)
$(Fe_3O_4\text{-}ED)/MIL\text{-}101(Fe)$	Cd(II)	155	(Babazadeh et al., 2015)
ZIF-8	Cr(VI)	0.15	(Niknam Shahrak et al., 2017)
$NH_2\text{-}Zr\text{-}MOFs$	Pb(II)	166.74	(K. Wang et al., 2017)
ZIF-90-SH	Hg(II)	22.4	(Bhattacharjee et al., 2015)

Mineral Adsorbents

In the context of environmental applications, certain waste materials, such as fly ash and sludge, have been identified as valuable resources. They can serve as effective adsorbents for various pollutants and can also be used as precursors for the synthesis of aluminosilicate source materials like various types of zeolites (Bandura et al., 2016; Hosseini Asl et al., 2019). Table 1 lists some inorganic minerals with properties that can adsorb contaminants. Among them, zeolite is known as a highly selective mineral. Basically, if we enable pollutant elimination from dirty water medium, such adsorbent materials first come into contact with the contaminant and then hold onto the material. The adsorption process for capturing substances relies on the adsorption capacity and selectivity of adsorbent (Guo et al., 2015; Lee et al., 2021; Lee et al., 2017; Wong et al., 2018). Here, obeying Langmuir adsorption model, maximum adsorption capacities of q_m means that adsorbate behaves at isothermal conditions to finally saturate the adsorption sites of adsorbent. This model is expressed as the following equation: $q_e = \frac{K_L q_m C_e}{1 + K_L C_e}$, where q_e denotes equilibrium adsorption amount [mg/L] ate equilibrium concentration C_e of substance, K_L is the Langmuir constant [L/mg], and q_m is maximum adsorption capacity [mg/g-adsorbent].

Accordingly, these materials enable specific toxic compounds to access the surface with some driving forces either through van der Waals forces (physical adsorption) (Reimerink et al., 1999), electro-statical force (Ma et al., 2021) or chemical bonding between the material sites and pollutant (Saleh et al., 2022). Therefore, selection and designing the ability of adsorption to specific pollutant is necessary, even if inorganic mineral adsorbents are already known to adsorb contaminants.

On the other hand, it is worth noting that such adsorbents are initially found in a powdery form, are favored for their strong adsorption abilities, primarily due to their large surface area and numerous adsorption sites (Dlamini et al., 2020). For example, zeolite NaY derived from rice husk and commercial zeolite NaY were modified with hexadecyltrimethylammonium (HDTMA) in quantities equivalent to 50%, 100%, and 200% of their external cation exchange capacity and then tested to assess their capacity for removing Cr(VI) and As(V) from aqueous solutions (Yusof & Malek, 2009). Removal of organic pollutants such as phenol (Shah et al., 2011; Wang et al., 2011), textile dye (Ozdemir et al., 2009) by surfactant-modified zeolite gained notable success in recent investigations. Through these examples for pollutant adsorbents, it seems to be an important factor of practically used absorbent having several shapes of adsorbents. Typically, powdery adsorbents are favored due to their high adsorption performance, primarily because they offer a large surface area with numerous adsorption sites (Hu & Srinivasan, 2001). Within the category of powdery adsorbents, nanoparticles, which are defined as having at least one dimension within the range of 1-100 nm, exhibit even higher surface areas and superior adsorption performance compared to bulk particles of the same materials (Zito & Shipley, 2015). For example, Tang et al. reported using nanoparticles for water treatment, specifically synthesizing ultrafine iron oxide (α-Fe_2O_3) nanoparticles with a size of approximately 5 nm (W. Tang et al., 2011). These nanoparticles showed notable maximum adsorption capacities for arsenite and arsenate, measuring at 96.2 and 45.88 mg/g, respectively. In contrast, Fe_2O_3@MIL-101(Cr) achieved notable maximum adsorption capacities of q_m=121 mg/g for As(III) and 80 mg/g for As(V) (Folens et al., 2016). Also, zirconium-based metal-organic composite was used for the removal of protein-bound uremic toxin from human serum albumin (Kato et al., 2019).

There are many different types of mineral-derived adsorbents, and it is possible to select a material that is compatible with the pollutant, but many of these are often in the form of powders, which presents several problems when used in the actual decontamination process. For example, powders are cohesive as particularly agglomerated in water, cannot efficiently utilize all retained adsorbent sites, causing a loss of efficiency. In the other hands, in large-scale continuous adsorption treatment, wastewater is typically directed through adsorbents packed within a filtration tank, often in the form of a column. The configuration of the packed adsorbent significantly influences the efficiency and reliability of the treatment process. While powdery adsorbents exhibit excellent adsorption capabilities, it is meaningless if adsorbents cannot sustainably exert their action in the field.

To solve this problem, there have been already ways to composite a polymer with good moldability and a powdered adsorbent. That is, polymers composited with inorganic minerals can maintain their adsorption properties and introduce polymer properties such as formability into the adsorbents. As known, polymers are organic substances with remarkable attributes, including impressive mechanical strength, exceptional flexibility, chemical stability, and a generous surface area. These qualities have made polymers valuable for hosting a wide array of both organic and inorganic materials. Notably, polymer composites gained substantial attention, particularly in the realms of water treatment and air pollution control (Berber, 2020).

One notable advantage of polymer composites is their ability to precisely adjust their adsorption properties through techniques like blending, cross-linking, and surface modification. This category of composite materials offers several benefits, including ease of preparation and application, strong chemical stability under harsh operation (Hsissou et al., 2021). Nano iron oxide/Pd/Cu composites have proven highly effective in removing nitrate from contaminated underground water, operating under pseudo-first-order kinetics (Jieun Lee et al., 2016; Liu et al., 2014). Another noteworthy approach involving nano composites is the magnetization of carbon nanotubes with iron, optimizing adsorption behavior for the removal of nitrate and chlorinated organic contaminants from water (Azari et al., 2014).

Fibrous Adsorbents

Recently, there has been a lot of research on fiber adsorbents, the name of which utilizes electrospinning to give adsorbents a larger surface area. Elkady et al. conducted a synthesis of a copolymer consisting of styrene and acrylonitrile using a solution polymerization process. Subsequently, this copolymer was transformed into nanofibers employing an electrospinning technique. To enhance its dye absorption capabilities, the nanofiber surface was chemically modified by the addition of carboxylic acid groups. These structural modifications significantly improved the adsorption capacity for the basic violet dye, reaching an impressive $q_m = 67.1$ mg/g in less than 30 minutes following the surface modification of the nanofiber (Elkady et al., 2016). Furthermore, using metal–organic framework-based electrospun fibers, zeolitic imidazolate framework/polyacrylonitrile composite fibers was characterized by interconnected mesopores, resulted in rapid adsorption kinetics, substantial capturing capacity, and excellent cycling stability in the removal of tetracycline (R. Zhao et al., 2021). This exceptional performance was primarily attributed to the incorporation of imidazolate zeolite and the creation of mesopores within the composite fibers. Both the pseudo-second-order kinetic model and the Langmuir isotherm model confirmed maximum adsorption capacity of 885.2 mg/g. Notably, recyclability tests indicated that the removal efficiency remained at 97% even after 10 cycles (Awe et al., 2017; R. Zhao et al., 2021).

On the other hand, the following examples of fibrous adsorbents other than electrospinning were found. For example, Young and colleagues embarked on the synthesis of chelating fibers to enhance the adsorption capacity for heavy metals, with their findings revealing that these chelating fibers exhibited superior contact efficiencies with water in comparison to traditional resin beads (Ko et al., 2011). Furthermore, Fe_2O_3-Al_2O_3 nanocomposite characterized its fine fibrous structure designed for efficient removal of heavy metals (Mahapatra et al., 2013). The primary advantage of these fibrous adsorbents lies in their exceptional removal efficiency due to their fine fiber filaments. Furthermore, research by Rabiul Awual explored zirconium-loaded bifunctional fibers containing both phosphonate and sulfonate as a highly selective ligand exchange adsorbent for the trace phosphate removal from water (Awual, 2019). In the past years, amidoxime containing polymeric fibers were shown to be the most efficient adsorbent for uranium extraction in aqueous solution because of their favorable adsorption capacity and certain selectivity with competing ion including vanadium copper, iron, etc (Kuo et al., 2015; Ladshaw et al., 2017; Ling et al., 2017; Neti et al., 2017; A. Zhang et al., 2003). The amidoxime group was usually prepared by treating the nitrile group with hydroxylamine hydrochloride, therefore, polyacrylonitrile fibers are the most suitable candidate to modified with amidoxime groups for uranium adsorption.

In other example, polypropylene fibers are another type of universal fibrous polymer adsorbent for uranium. To achieve amidoximated polypropylene fiber, GMA is first grafted onto polypropylene fibers via a radiation polymerization route (H. Zhang et al., 2018). Then an aminoxylation process is realized by the reaction with dicyandiamide followed with hydroxylamine hydrochloride. The adsorption of uranium equilibrium is achieved within 5 h at pH 8, presenting $q_m = 112$ mg/g. In addition, amine and phosphonic groups, there are some other functional groups modified polymeric fibers such as carboxylic fiber, chelating fiber containing amine and hydroxyl groups etc. (He et al., 2018; Lei et al., 2012; Wu et al., 2017). Recently, polyacrylonitorile (PAN) fibers modified with lysine (PAN-Lys) were also prepared for uranium adsorption (W. Li et al., 2017). It was well known that the lysine is a typical amino acid with a carboxylic amine and two amine groups, which can easily interact with uranyl ions. The adsorption capacity reached as high as 405 mg/g and the kinetic process are well fitted with pserdo-second order model.

In the case of ε-polycaprolactone (PCL) and beta-cyclodextrin-based polymer (PCD) composite fiber materials, they demonstrated remarkable selective adsorption capabilities for methylene blue and 4-aminoazobenzene solutions. The inclusion of a PCD component in the composite fibers not only improved the mechanical strength of the membranes but also altered the adsorption behavior due to the cavity molecular structure enabled by host-guest interactions. The dye removal reached $q_m = 24.1$ mg/g for 4-aminnoazobenzene (R. Guo et al., 2019). Additionally, modified polyurethane and biomass-based Juncus effusus (JE) fibers exhibited impressive adsorption capacities for three different anionic dyes, surpassing many existing the modified adsorbents. These fibers conformed to the Langmuir and pseudo-second-order models for equilibrium and kinetics studies. Furthermore, a vertically oriented JE polymer composite fiber was fabricated for continuous dye filtration, demonstrating high flow flux and rapid filtration performance without the need for added pressure (Xia et al., 2020).

Inorganic Composited Polymeric Adsorbents

Adsorbents in the form of membranes rather than fiber shapes have also been studied extensively for inorganic-polymer composites. Polyacrylonitrile-alumina hollow fiber membranes demonstrated their capacity for adsorptive nitrate ion removal with achieving q_m = 15 mg/g. The results indicated that the hydrophilicity, mechanical strength, and zeta potential of the membrane increased with alumina concentration. Among the isotherm models, the Langmuir isotherm exhibited the best fit for nitrate adsorption. It was worth noting that the presence of co-ions like chloride, sulfate, and carbonate has a detrimental effect on nitrate removal (Mukherjee & De, 2014). For example, composite membranes comprising nylon 6 and mordenite zeolite with different loading amounts were effectively employed for selectively isolating heavy metal ions-specifically, Pb^{2+}, Cu^{2+}, and Cd^{2+} from aqueous solutions (Ton Nu & Kobayashi, 2020a), especially supporting the preferential adsorption to Pb^{2+} over Cd^{2+} and Cu^{2+}. Particularly noteworthy was the separation factor achieved by the composite membrane containing 30% mordenite loading, denoted as NZC 30, which reached an impressive value of 9.2 for Pb^{2+} relative to Cu^{2+}. In addition to these, this material was useful for adsorbing hormonal gases, emitted from fruits such as bananas, and was useful as a long-term preservation material without hastening the ripening process (Ton Nu & Kobayashi, 2020b). In this case, the mordenite powder agglomerated with moisture during long-term storage and showed a decrease in adsorption function, but the nylon-mordenite composite membrane retained its air content during the 10-day banana retention period and retained its freshness at room temperature.

For heavy metal retention, halloysite nano clay was mixed with chitin hydrogel films and adsorption experiments of the ternary ion of Pb^{2+}, Cu^{2+}, and Cd^{2+} were taken (Nguyen et al., 2019). The removal capacity of Pb(II) was highly retained in Pb^{2+} relative to other ions.

Figure 2. Picture of zeolite-PES composite fiber(a) and SEM images of cross section of the fibrous adsorbents (b)-(d)

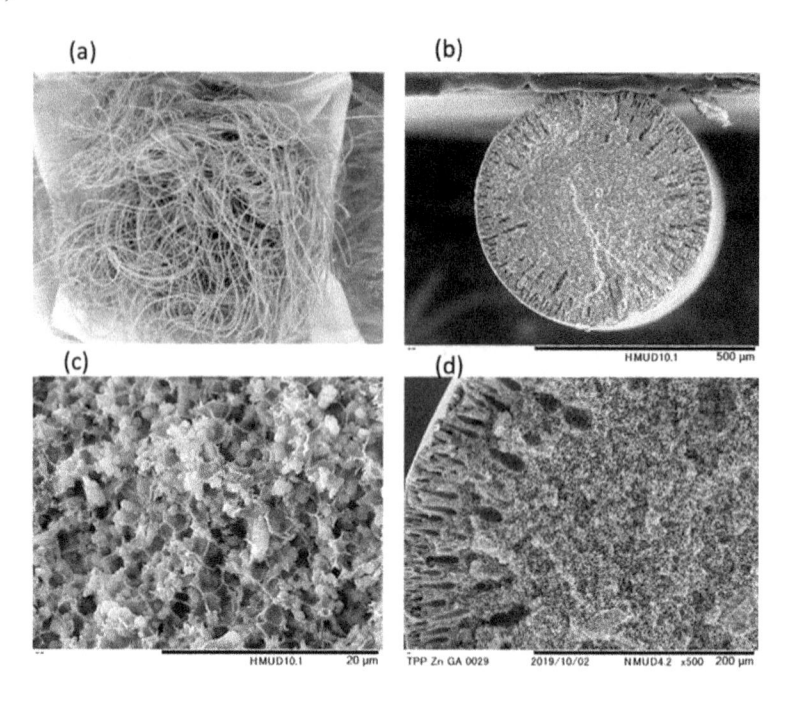

However, natural water sources, where pollutants are present, are often very complex environments and do not provide a pure, clean environment and contain with certainty suspended solids components and organic plant algae, which tend to cover the adsorbent sites. Thus, in the cases of gel films and membranes that perform filtration, deposits on the such flat surface inevitably tend to degrade its performance. Especially in the case of filtration, although selective adsorption can be achieved, the degradation of filtration performance due to fouling is still a problem. Over time, the entrapment of these obstructive substances leads to increased pressure loss in the operation and reduced removal efficiency.

Consequently, there is a pressing need for fibrous adsorbents capable of sustaining efficient performance in challenging environmental conditions of this nature. To address this challenge, as an example of a pioneering material, zeolite-polymer composite fiber utilizing a phase inversion technique with polyether sulfone (PES) is presented in detail below.

As shown in Figure 2, the resultant fibers had approximately 300 μm diameter. Their length was adjustable in about 50-100 cm. The scanning electron microscope (SEM) images (Figure 2b) of the cross sections for the zeolite composite fibers presented a porous structure having mese pores. In actual outdoor decontamination (Kobayashi et al., 2016), the fibers are simply placed in a nylon net and left in the decontamination source (Figure 3), allowing effective outdoor decontamination of radioactive cesium for over a year. This persistence was due to the fact that the entanglement of the fibers trapped external soil and organic plant components at the surface entrapped, allowing adsorption without disturbing the adsorption sites inside the fiber adsorbent. In the passive decontamination to prevent radioactive Cs penetration to downstream residential areas, the innovation was aimed at facilitating the decontamination of radioactive substances with low maintenance, presenting a potential solution to the issue of maintaining optimal adsorption performance even in the presence of suspended solids in the water.

Figure 3. Illustration of extension of radioactive cesium flowing downstream in natural field and passive type adsorbent for the decontamination. Photo shows decontamination of a stream in Date City, Fukushima Prefecture, using zeolite polymer fiber and illustration of changes in absorbed radioactive cesium over a 26-day period

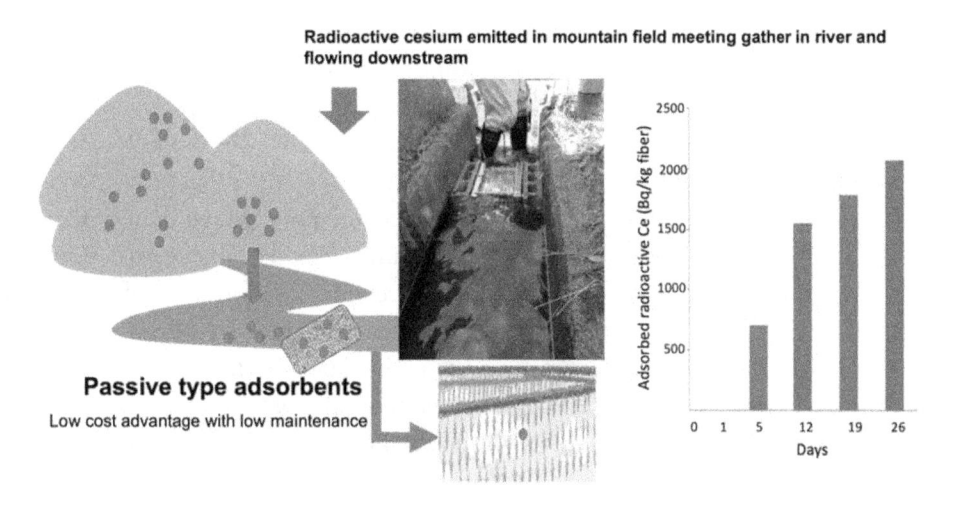

Since then, this type of fibrous adsorbent has been used for a variety of applications. These fibrous materials hold great potential in the field of adsorption composite due to fibrous shape, adjustable porosity, especially the fibrous shape could solve powder disadvantage, making them potential substitutes adsorbents (Nakamoto, 2018). In many researches the fibrous adsorbents were prepared by combining polymer and integrating inorganic compounds like mordenite, limonite (Nakamoto et al., 2017; Ngan at al., 2023; Phuong et al., 2023).

INORGANIC POLYMER BASED FIBROUS ADSORBENTS

As shown in Figure 1, this interconnectedness of the earth's environment presents a complex challenge, as efforts to mitigate contamination should be tailored to specific problems in each site. To effectively reduce the spread of contaminants, it is essential to employ strategies that facilitate decontamination in each of these distinct locations. Adsorbents can be used to adsorb contaminants in the air, soil, and water and purify each environment, halting the spread of contaminants at their respective locations to reduce the impact on people. For example, this approach was particularly valuable in addressing environmental disasters, such as the Fukushima nuclear incident in Japan on March 11, 2011. During the Fukushima disaster, the collapse of the nuclear power plant led to the release of radioactive cesium into the surrounding environment, including water sources. Since water moves freely, radioactive cesium spreads with it. In addition, in adsorptive soils, there are localized areas of high concentrations, called hot spots. To combat this crisis, innovative solutions were required, and one such solution involved the use of polymer composite fibers containing zeolite. These fibers effectively served as a decontamination method, specifically aimed at reducing the dispersion of radioactive cesium in water.

Inorganic polymer composites based on fibrous materials offer a versatile and efficient platform for various adsorption mechanisms, since these composites combine advantages of both embedded inorganic powders with specific adsorptive ability and high surface area and polymer matrix having fibrous shape, chemical stability with the structural integrity and flexibility especially in PES for stability against chemicals and thermal changes. As illustrated in Figure 4, for example, zeolite-PES composite fibers have several functionalities in the fibers and are suitable for environmental pollutant remediation, since the composite fibers combine distinctive adsorption performance including ion exchange, chemisorption, and van der Waals forces (hydrogen bonding), electrostatic attraction, making them effective for selective capture of molecules based on size, charge, and chemical affinity. The effectiveness of this approach can be attributed to the unique structure of the composite material within the polymer matrix, where zeolite powders were dispersed and embedded in spongy PES in a manner that prevented its dissipation (Figure 2c). This structural arrangement formed with zeolite and polymer, as seen in the SEM images indicated enabled efficient adsorption of radioactive cesium. This composite structure significantly enhanced the outdoor adsorption capacity for radioactive cesium, leading to a substantial increase in overall decontamination efforts. Zeolites are particularly well-suited for adsorbing radioactive cesium due to their three-dimensional framework structure, which includes interconnected channels and large extra-framework sites capable of hosting exchangeable cations and water molecules (Bish, 2013). This inherent property of zeolites makes them a valuable tool in the ongoing battle to contain and reduce the impact of environmental contaminants.

Figure 4. The proposed mechanism for inorganic-polymer based composite fibrous materials

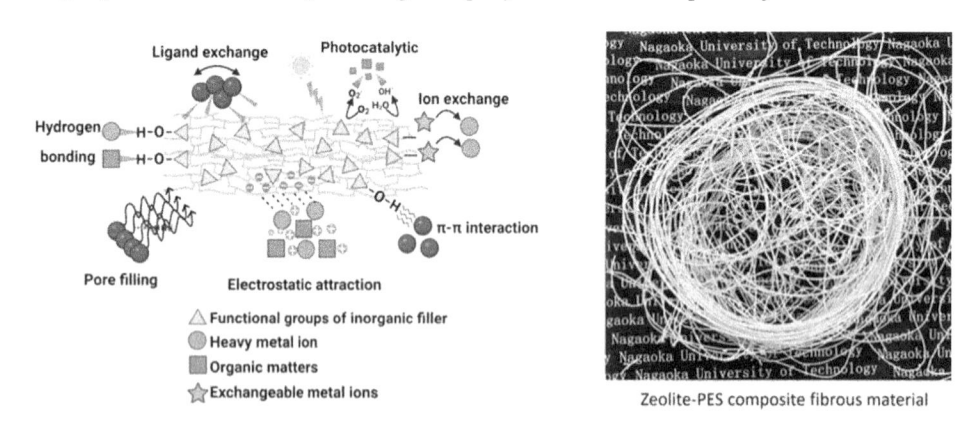

Fabrication of Inorganic-Polymer Composite Adsorbent Fibers

Inorganic-polymer composite materials have been synthesized through various methods, including the sol-gel process (Pandey & Mishra, 2011), *in-situ* interface polymerization (Chazot et al., 2020), deposition of inorganic materials onto polymer surfaces (M. Zhang et al., 2023), melt-mixing (Wijesinghe et al., 2020), and phase inversion (Duraikkannu et al., 2021). The phase inversion technique, often referred to as the wet coagulation method for creating asymmetric polymeric membranes (Chai et al., 1998; Kobayashi et al., 2002), involves casting a polymer solution onto a substrate, such as glass plates, followed by immersion in a coagulation bath containing water. During this coagulation process, there is an exchange of solvent and non-solvent at the interfaces. Excess non-solvent, like water, can solvate the solvent within the polymer, leading to the coagulation of the polymer dissolved in the solution. This phase-inversed polymer from liquid to solid state exhibits a range of configurations, boasting high porosity as a key advantage, resulting in a high surface area that enhances permeability and flux properties for water treatment membranes. However, in order to provide further functionality, it was required to use materials that imparted functionality to polymers (Kobayashi et al., 1998; H. Y. Wang et al., 1997), and a simpler method was thought to be to composite functional powders with polymers that have membrane refining functionality for environmental hormones (Takeda & Kobayashi, 2006). Subsequently, composite adsorbents of the hollow-fiber type appeared, taking advantage of the moldability of polymers and introducing porous sponge structure via phase inversion process (Son et al., 2011).

In contrast, although there are not many examples of inorganic-polymer composite fibers, polypropylene- with montmorillonite was fabricated by melt processing for investigating its adsorption capacity for Pb^{2+}(Motsa et al., 2011) . However, the adsorption abilities of composite were limited, with Pb^{2+} adsorption capacities of only q_m = 1.59 mg/g, primarily due to that the melting process only provided composite-montmorillonite embedded in the non-porous polymer fiber, which could not accommodate montmorillonite adsorption sites. Namely, the polymer phase is homogeneous, i.e., the montmorillonite embedded in the polymer phase cannot participate in adsorption, and adsorption is attributed only to the exposed sites on the fiber surface. Therefore, the phase inversion method providing porous sponge structure of the polymer layer stands out as a favorable approach to inorganic-polymer composite fibers, particularly for water treatment adsorbents. Apart from the melting method, the membrane creation technique was applied to the wet spinning method to produce polymer fibers with a porous structure. As seen in Figure 5, polymer solution dispersed with zeolite powders was applied by phase inversion method and wet spinning process is convenient for easily preparation of fibrous inorganic-polymer composite fiber for applicable to water treatment adsorbents (Kobayashi et al., 2016). However, there is a notable scarcity of research focused on the utilization of the phase wet spinning inversion method for the production of fibrous inorganic-polymer composites. The wet spinning procedure involves several sequential stages, encompassing spinning, coagulation, and immersion processes. In this method, a solution containing a mixture of inorganic powder and polymer is extruded into a coagulation bath in a fibrous form, facilitated by the application of air pressure. Then, within the coagulation and immersion processes, coagulation of the polymer occurs through a solvent exchange mechanism between the initial solvent and a non-solvent. This intricate process ultimately yields a composite material wherein inorganic particles are seamlessly integrated into a porous polymer matrix.

Figure 5. Fabrication of inorganic polymer based fibrous absorbent by wet spinning process

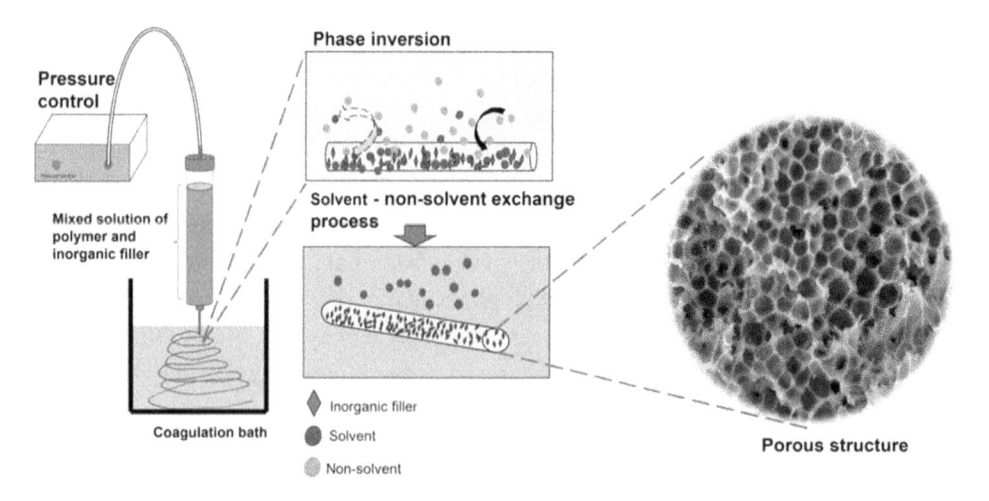

As mentioned before, the zeolite-polymer composite fibers were fabricated by modified phase inversion of the PES-NMP solution to solid PES in water (Kobayashi et al., 2016; Nakamoto & Kobayashi, 2017). In this procedural sequence, polyethersulfone (PES) was dissolved in N-Methyl-2-pyrrolidone (NMP) for preparation of PES-NMP solution. Then, predetermined quantities of zeolite powders were mixed and dispersed in the PES-NMP solution at various addition amounts. Following this, the mixed solution was subjected to agitation for a designated period. After the preparation, the zeolite-polymer composite solution was extruded through a cylindrical needle using air pressure and immersed into coagulation bath. The extruded mixture isolation was subsequently solidified in form of fiber shape in water. Since the PES and zeolite are insoluble, when NMP dissolves in water, the fibrous zeolite-PES fibers solidify in the coagulation bath. In this method, the structure of the polymer depends on the degree of water dissolution in the solvent phase containing the polymer (Kobayashi et al., 1998). As a result, the polymer phase precipitates and solidifies in the water in a reticulate, porous structure with embedding zeolite powders.

WASTEWATER TREATMENT APPLICATIONS

Wastewater, characterized by the presence of heavy metals, organic compounds, and inorganic substances, constitutes a critical category of pollutants that pose substantial threats to ecological equilibrium (Briffa et al., 2020; Murugesan et al., 2006; Toda et al., 2014; Q. Wang et al., 2022). Consequently, the removal of these pollutants from aqueous pollutant represents a urgent global environmental concern. In addressing this concern of decontamination, which is part of "Remediation and Cleanup", the adsorption process has emerged as a prominent method due to its simplicity and cost-effectiveness. In addition to environmental purification, the cleaned water can be reused in industry, etc., contributing to the construction of a recycling-oriented society (EPA, 2023). Therefore, extensive efforts have been devoted to the development of highly effective and selective adsorbents endowed with exceptional properties (C. Wang et al., 2019). For exploration of inorganic polymer-based adsorbents materials and assesses the adsorption efficacy of various developed composite materials, designing and implementation for the removal of pollutant from aqueous solutions mainly carried out in both batch-water and continuous-water systems.

Inorganic-Polymer Composite Fiber Adsorbent Tailor-Made for Pollutant Capture

One remarkable property is that the introduced zeolite in the composite has the property of efficiently retaining heavy metals. Thus, zeolite-polymer composite fibers also have the ability to efficiently capture heavy metals. This is largely dependent on the properties of the mainly inorganic components. Therefore, the inorganic powder can be changed to match the desired pollutant. This approach has the advantage of allowing tailor-made inorganic-polymer composite fiber adsorbents, i.e., inorganic materials selected for heavy metal capture. In the combined mordenite zeolite with polymer fibers through a wet spinning process, employing varying loadings ranging from 27.0% to 69.0% by weight, resulting in the formation of solid composite fibers. These composites exhibited an increase in surface area, reaching a maximum of 145 m^2/g with a zeolite loading of 69.0% by weight. Additionally, they demonstrated robust thermal stability and highly efficient heavy metal adsorption capabilities. Notably, the fibers containing 58.8% and 69.0% zeolite content displayed a remarkable maximum adsorption capacity of 0.67 138.7 mg/g, fitting well with the Langmuir isotherm model. In batch solutions featuring single or mixed heavy metal ions of Pb^{2+}, Cd^{2+}, Cu^{2+}, and Ni^{2+} at a pH 5, Pb^{2+} exhibited the highest adsorption selectivity in the order of $Pb^{2+} > Cd^{2+} > Cu^{2+} > Ni^{2+}$ (Nakamoto et al., 2017).

Furthermore, the research delved into continuous flow column adsorption of mordenite composite fibrous materials for with specifically focusing on Pb^{2+} adsorption (Nakamoto et al., 2018). Figure 6 shows breakthrough curves of Pb^{2+} at different Volumatic flow rates. The solid line for Adams-Bohart model and dash line for Wolborska model. Here, 0.2M Pb^{2+} solution was flowed at 5, 10 and 15 mL/min. When decreased flow rate, the value of C_t/C_0 meaning ratio of Pb^{2+} concentration at initial and the observed time t became increased slowly to spend longer operation time, obeying Adams-Bohart mode. For the breakpoint defined as $C_t/C_0 = 1$ in the curves, the breaking time for 5, 10 and 15 ml/min were 43, 10 and 4 hours, respectively. Also. in quaternary mixed solutions comprising Pb^{2+}, Cd^{2+}, Cu^{2+}, and Ni^{2+}, the adsorbent exhibited a selective removal of Pb^{2+}, leaving Cu^{2+} and Ni^{2+} desorbed during column operation.

Figure 6. Breakthrough curves of Pb2+ at different Volumatic flow rates

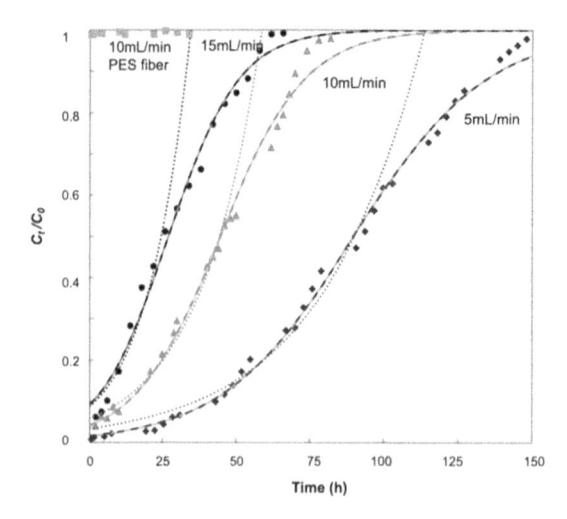

As well as similar approach of the zeolite- PES composite fibers, alumina (Al_2O_3) powder was composited with PES by wet spinning method. The Al_2O_3 PES fibrous adsorbents were then treated by depositing CeO_2 on the surface by oxidation of $CeCl_3$ in an H_2O_2 solution. The modified adsorbent exhibited superior performance in removing antimony (Sb), with adsorption behavior following the Langmuir isotherm and maximum capacity q_m = 7.7 mg/g (Thu et al., 2023). This capacity was approximately twice as high as that of the non-deposited Al_2O_3-polymer fiber and the composite fibers also packed in column for continuous flow system (Thu et al., 2023). In another research of Al_2O_3-PES composite fibers were developed to prevent silica contamination in water. Various amounts of Al_2O_3 loaded (10%, 20%, and 30%) were incorporated into the fiber using a wet-spinning process. Higher Al_2O_3 content led to improved silica adsorption, obeying Langmuir isotherm model with a maximum capacity of 69.9 mg/g at pH 6. Impressively, even after undergoing five regeneration cycles with 75% removal, the 30% Al_2O_3 loaded fiber maintained its high silica adsorption performance, demonstrating for industrial silica-related challenges (Ngan et al., 2023).

As well as heavy metal pollution, arsenic contamination has attracted public concern due to the potential restriction of soil for agricultural purposes (Tóth et al., 2016). When used the cerium composited activated alumina in the fibrous shapes, the adsorption capacities increased with high loading amounts of CeO_2 - Al_2O_3 with 65 wt% in the fiber and showed maximum adsorption capacities of 10.4 mg/g for arsenate and 7.4 mg/g for arsenite. However, powdered CeO_2 - Al_2O_3 still exhibited superior adsorption abilities compared to composite fibers. As then CeO_2 was deposited on Al_2O_3-polymer composite fiber in Ce^{3+} solutions, the adsorption capacities for both arsenate and arsenite were significantly enhanced with q_m = 13.32 and 10.36 mg/g, respectively (Nakamoto, 2018).

Other work in As remediation processing, the column systems with limonite composite fibers containing 60 wt% limonite powders and varying PES concentrations, effective adsorption of As(III) and As(V) at pH 6.3 occurred, following Langmuir-type adsorption. The maximum adsorption capacities were 1.5 mg/g for As(III) and 3.2 mg/g for As(V). Varying column conditions (length, loading ratio, flow rate) showed that slowing flow velocity and increasing fiber height improved arsenic removal, consistent with the Adams-Bohart model (Phuong at al. 2023).

Numerous composite materials have been synthesized and proposed for the purpose of removing dyes from wastewater. For instance, the mordenite zeolite-polymer composite fibers also acted their performance to retain organic dyes. With a loading of 58.8% mordenite zeolite at varying fiber diameters, the maximum adsorption capacity was found to be 18.16 mg/g methylene blue and, obeying Langmuir adsorption isotherm (Nakamoto & Kobayashi, 2017). This trend shows that zeolites exhibit good adsorption properties for positively charged ions and dyes. Therefore, ammonia was applied as pollutant for the elimimination. The equilibrium isotherm (a) indicated Langmuir typed adsorption having $q_m =$ 9.1 mg/g and 16.8 mg/g, respectively for composite fiber and mordenaite. Reduction of ammonia concentration is important in fish rearing, and the results of adapting the fibrous adsorbent to rainbow trout rearing are shown in (b). In this rearing facility, rainbow trout were reared for about six months without water exchange, maintaining a water volume of 200 L with only water addition. (c) is a photograph of the actual farm.

Figure 7. Ammonia isotherms of mordenite zeolite powder and mordenite -PES fibers (a) and picture and illustration and picture of aquaculture system in Nagaoka University of Technology for Rainbow Trout breeding (b) and breeding facility in Niigata (c)

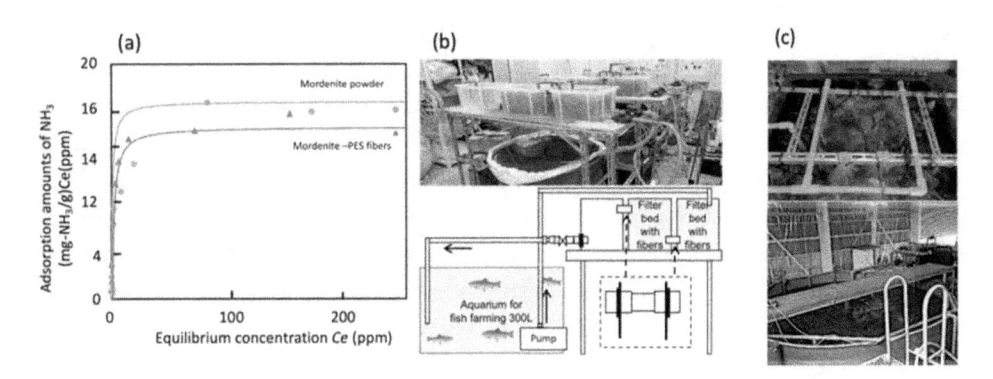

On the other hand, zeolite-polymer composite fiber adsorbent was transformed in pellets which were reported in uses for storage of radioactive Cs after decontamination processes (Ooshiro et al., 2017). The immobilization matrix was prepared by using compressed heat treatment at various temperatures (100–800 °C) for the zeolite polymer composite fibers. This process reduced the matrix volume to one-sixth of its original size. When tested with non-radioactive and radioactive Cs, minimal Cs leakage (0.05%) occurred with heat treatment at 300 °C. Below 300 °C, no pollutants were released. However, at 500 °C, the matrix released radioactive Cs due to the decomposition of the plastic component. Comparatively, this method proved effective for compactly immobilizing radioactive Cs when compared to natural zeolite.

Air Pollution Control Application of Inorganic-Polymer Based Adsorbents

With the release of various gaseous pollutants from industrial and anthropogenic activities, there is a growing focus on removing these pollutants from the environment to safeguard living organisms. In recent times, adsorption has emerged as a promising technology for remediation, offering distinct advantages such as low cost, low energy consumption, and operational simplicity. Fibrous adsorbent has the advantages over membrane adsorbent in that it has a lower pressure drop than membrane adsorbent when air containing contaminants is blown through the adsorbent and, like zeolite-PES composite fiber, the polymer portion is a porous fiber that can easily contact the inorganic portion of the adsorption site.

In particular, inorganic-polymer-based composites have garnered significant attention due to their superior adsorption capacity compared to alternative materials such as porous silica, zeolite, porous alumina, and organometallic compounds for CO_2 removal. Researchers are actively synthesizing and studying these materials to determine their potential for efficiently capturing and reducing CO_2 emissions. However, the majority of these efforts have focused on developing laboratory-scale films of TR membranes, with only a few studies reporting the fabrication of hollow fiber TR membranes (S. Kim et al., 2012; Sunghoon Lee et al., 2017; Woo et al., 2015). In a separate study, Li et al. synthesized cobalt tetraphenyl porphyrin (CoTPP) and prepared a PDMS coating solution containing CoTPP. This coating solution was applied to PES hollow fibers to create TFC-HFMs. The resulting membranes exhibited an O_2/N_2 selectivity of 2.8, an O_2/CO_2 selectivity of 1.5, and an oxygen permeance of 53 GPU at 0.05 bar. The inclusion of CoTPP facilitated oxygen transport through the membranes, resulting in higher O_2/N_2 and O_2/CO_2 selectivity compared to pristine membranes (H. Li et al., 2016).

On the other hand, hazardous gases such as SO_x, NO_x, CO, NH_3, H_2S and halogens, mostly released by the burning of fossil fuels, are big threats to environmental and public safe (Hiraide et al., 2020; Woellner et al., 2018). The mitigation of ambient gases, owing to their potential to induce deleterious effects on respiratory health through toxic gas exposure, constitutes a paramount facet of pollution control. For the gas elimination, a porous inorganic composite material was used in practical applications, where the removal is mainly governed by the physical sorption on the adequate-sized pores (An & Rosi, 2010; Marchesini et al., 2017; Yu et al., 2017; J. Zhao et al., 2016).

In contrast, limonite, which is general term for either or both aggregates of needle iron ore (goethite, α-FeOOH) or lepidocrocite (lepidocrocite, γ-FeOOH), is known to adsorb H_2S (Zhou et al., 2016). Thus, the powders were composited with PES by wet spinning method which used for hydrogen sulfide suppression in the anaerobic sludge environment et al., 2023). Figure 8 depicts decontamination of H_2S emitted from microbe soil sourced from the Ariake Sea, located in Kumamoto Prefecture, Japan. In this region, issues such as red tides, hypoxic waters, and declining fishery yields have been evident since 1981, and H_2S poisoning escalated significantly in recent years. The depletion of oxygen in the seawater also has led to a series of environmental problems, including a reduction in fishery production (Jia et al., 2018). In this study, limonite composite fibers provided to diminish the H_2S gas emanating from the sludge. The resultant limonite-PES composite fibers were efficient in reduction of H_2S generation from anaerobic microbial sludge (Phuong et al., 2023). They exhibited strong H_2S adsorption, following a Langmuir-type isotherm with a 3.7-4.4 g/g maximum capacity, indicating mesopore adsorption. In an *in vitro* H_2S fermentation environment with anaerobic microbial sludge, these fibers suppressed H_2S production and reduced CO generation, impacting microorganism metabolism. During anaerobic digestion, particularly between 672-840 hours (28-35 days), the mesopores of fibers transformed into macropore adsorption sites. Limonite reacted with H_2S, forming pyrite (FeS_2) and iron sulfate ($Fe_2(SO_4)_3$) products.

Figure 8. The coexistence system of limonite composite fibers and anaerobic sludge (taken from the Ariake Sea, Kumamoto Prefecture, Japan) for hydrogen sulfide suppression

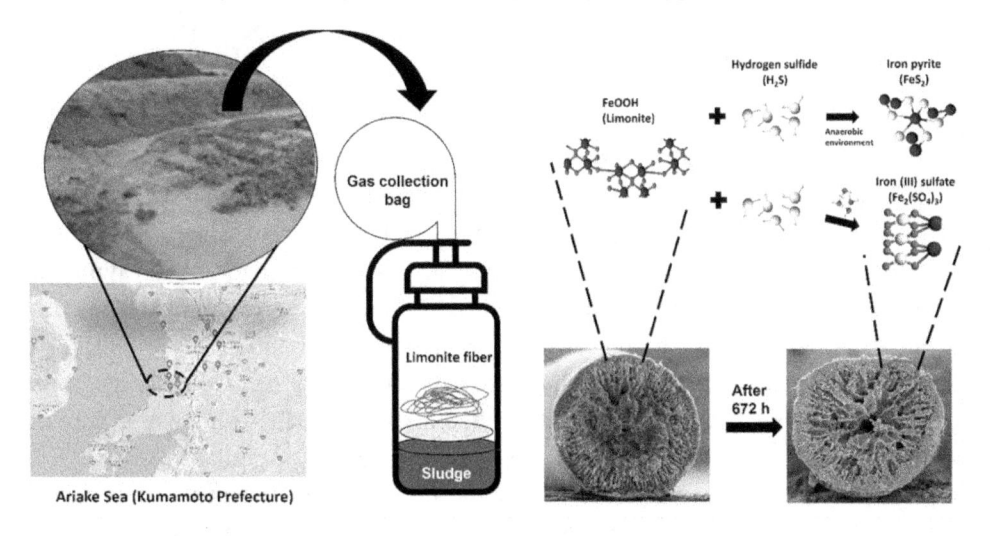

In the study conducted by Jinwook, the report delves into the incorporation of zeolite imidazole framework-8 (ZIF-8) into polypropylene nonwoven fibers for the purpose of SO_2 adsorption (Jinwook Lee et al., 2021). The research discerned that the sol-gel impregnation method proved to be the most efficient for loading ZIF-8 onto the fibers. In case of incorporated nafion/TiO_2 nanoparticles into a Polyimide-polyether (PEBAX) solution, which was subsequently coated on the inner surface of a polyethersulfone support hollow fiber to remove SO_2 gas (Park et al., 2019). This addition of nafion/TiO_2 nanoparticles was found to enhance membrane performance by introducing sulfonate and hydroxyl functional groups to the membrane. Consequently, this led to increased SO_2 permeation and selectivity. The achieved ideal selectivities for SO_2/N_2 and SO_2/CO_2 were 2928 and 72, respectively.

Volatile organic compounds (VOCs) pose significant threats to both the environment and human health. Over the past three decades, adsorption technology has been employed for VOC abatement and has consistently proven its effectiveness (X. Li et al., 2020; Roberts et al., 2020). Into recent research advancements in one such VOC adsorption material—the inorganic-based fibrous adsorbent. The mechanisms underlying VOCs adsorption within the adsorbent primarily involve electrostatic attraction between polar VOCs and hydrophilic sites, interactions of nonpolar VOCs and hydrophobic sites, partitioning in non-carbonized regions. As specific surface area, pore volume, and surface chemical functional groups increased, while pore size tended to decrease, the adsorption capacity enhanced, meaning that the volume of narrow micropores played a pivotal role in controlling VOC adsorption. For example, active carbon fiber was typically synthesized through a process involving spinning, carbonization, and activation, utilizing precursors such as viscose, phenolic polymer, and polyacrylonitrile (Baur et al., 2015). Tsai conducted a study on the adsorption characteristics of chloroform, acetone, and acetonitrile on commercial activated carbons. The diffusion coefficient of VOCs in the microporous structure (1.72 nm) was determined from the adsorption kinetic curve (Jiun-horng et al., 2008). In another study, the porosity distribution and adsorption capacity of various activated carbons were investigated (Lillo-Ródenas et al., 2007). These carbons were prepared through heat treatments in a nitrogen atmosphere using 16 different carbonaceous precursors (including coals, lignocellulosic materials, and carbon fiber) and two different alkaline hydroxides (KOH and NaOH) for activation.

CONCLUSION

In summary, the utilization of inorganic-polymer-based adsorbents stands as a prominent and evolving trend in the realm of pollutant remediation. These innovative materials have demonstrated their efficacy in addressing a diverse array of environmental challenges, providing a range of advantages, including enhanced adsorption efficiency, versatility, and facile modification. The symbiosis between inorganic constituents, such as zeolites, metal-organic frameworks, and nanoparticles, and polymer matrices has engendered the development of highly efficient adsorbents. These materials exhibit the capacity to target a wide spectrum of pollutants, encompassing heavy metals, organic compounds, and gaseous contaminants. Their adjustable attributes, encompassing surface chemistry, pore dimensions, and mechanical robustness, offer tailored solutions to meet specific pollutant removal requirements. Furthermore, the fibrous and composite nature of these adsorbents augments their practical utility. Their fibrous morphology affords an expanded surface area and improved accessibility to pollutants, while their compatibility with large-scale processes renders them suitable for real-world applications in pollutant remediation. As the global community grapples with pressing environmental challenges, the emergence of inorganic-polymer-based adsorbents holds promise for ameliorating the repercussions of pollutants on ecosystems and human health. Ongoing research and development endeavors in this field harbor the potential to further refine and optimize these materials, establishing them as indispensable tools in the quest for a cleaner and more sustainable environment.

REFERENCES

An, J., & Rosi, N. L. (2010). Tuning MOF CO2 Adsorption Properties via Cation Exchange. *Journal of the American Chemical Society*, 132(16), 5578–5579. DOI: 10.1021/ja1012992 PMID: 20373762

Awe, O. W., Minh, D. P., Lyczko, N., Nzihou, A., & Zhao, Y. (2017). Laboratory-scale investigation of the removal of hydrogen sulfide from biogas and air using industrial waste-based sorbents. *Journal of Environmental Chemical Engineering*, 5(2), 1809–1820. DOI: 10.1016/j.jece.2017.03.023

Awual, M. R. (2019). Efficient phosphate removal from water for controlling eutrophication using novel composite adsorbent. *Journal of Cleaner Production*, 228, 1311–1319. https://doi.org/https://doi.org/10.1016/j.jclepro.2019.04.325. DOI: 10.1016/j.jclepro.2019.04.325

Azari, A., Babaei, A.-A., Rezaei-Kalantary, R., Esrafili, A., Moazzen, M., & Kakavandi, B. (2014). Nitrate removal from aqueous solution using carbon nanotubes magnetized by nano zero-valent iron. *Journal of Mazandaran University of Medical Sciences*, 23(2), 14–27. https://www.scopus.com/inward/record.uri?eid=2-s2.0-84907289460&partnerID=40&md5=70ac412be5fe3e608f606ef278b68fd5

Babazadeh, M., Hosseinzadeh-Khanmiri, R., Abolhasani, J., Ghorbani-Kalhor, E., & Hassanpour, A. (2015). Solid phase extraction of heavy metal ions from agricultural samples with the aid of a novel functionalized magnetic metal–organic framework. *RSC Advances*, 5(26), 19884–19892. DOI: 10.1039/C4RA15532G

Bandura, L., Panek, R., Rotko, M., & Franus, W. (2016). Synthetic zeolites from fly ash for an effective trapping of BTX in gas stream. *Microporous and Mesoporous Materials*, 223, 1–9. https://doi.org/https://doi.org/10.1016/j.micromeso.2015.10.032. DOI: 10.1016/j.micromeso.2015.10.032

Baur, G. B., Beswick, O., Spring, J., Yuranov, I., & Kiwi-Minsker, L. (2015). Activated carbon fibers for efficient VOC removal from diluted streams: The role of surface functionalities. *Adsorption*, 21(4), 255–264. DOI: 10.1007/s10450-015-9667-7

Berber, M. R. (2020). Current Advances of Polymer Composites for Water Treatment and Desalination. *Journal of Chemistry*, 2020, 1–19. Advance online publication. DOI: 10.1155/2020/7608423

Bhatnagar, A., Vilar, V. J. P., Botelho, C. M. S., & Boaventura, R. A. R. (2010). Coconut-based biosorbents for water treatment — A review of the recent literature. *Advances in Colloid and Interface Science*, 160(1), 1–15. https://doi.org/https://doi.org/10.1016/j.cis.2010.06.011. DOI: 10.1016/j.cis.2010.06.011 PMID: 20656282

Bhattacharjee, S., Lee, Y.-R., & Ahn, W.-S. (2015). Post-synthesis functionalization of a zeolitic imidazolate structure ZIF-90: A study on removal of Hg(ii) from water and epoxidation of alkenes. *CrystEngComm*, 17(12), 2575–2582. DOI: 10.1039/C4CE02555E

Bish, D. L. (2013). Chapter 14.2 - Parallels and Distinctions Between Clay Minerals and Zeolites. In F. Bergaya & G. B. T.-D. in C. S. Lagaly (Eds.), *Handbook of Clay Science* (Vol. 5, pp. 783–800). Elsevier. https://doi.org/https://doi.org/10.1016/B978-0-08-098258-8.00026-2

Briffa, J., Sinagra, E., & Blundell, R. (2020). Heavy metal pollution in the environment and their toxicological effects on humans. *Heliyon*, 6(9), e04691. DOI: 10.1016/j.heliyon.2020.e04691 PMID: 32964150

Chai, X., Kobayashi, T., & Fujii, N. (1998).. . *Quaternized Ammonium Groups for Ultrafiltration.*, (January), 1821–1828.

Chazot, C. A. C., Jons, C. K., & Hart, A. J. (2020). In Situ Interfacial Polymerization: A Technique for Rapid Formation of Highly Loaded Carbon Nanotube-Polymer Composites. *Advanced Functional Materials*, 30(52), 2005499. https://doi.org/https://doi.org/10.1002/adfm.202005499. DOI: 10.1002/adfm.202005499

Chen, H., Nanayakkara, C. E., & Grassian, V. H. (2012). Titanium Dioxide Photocatalysis in Atmospheric Chemistry. *Chemical Reviews*, 112(11), 5919–5948. DOI: 10.1021/cr3002092 PMID: 23088691

Chou, C.-H. S. J. (Chi-H. S. J. & Harper, C. (n.d.). *Toxicological profile for arsenic* (A. for T. S. and D. R. United States & S. R. Corporation. (eds.)). . https://stacks.cdc.gov/view/cdc/11481DOI: 10.15620/cdc:11481

Deng, H., Pan, T., Zhang, Y., Wang, L., Wu, Q., Ma, J., Shan, W., & He, H. (2020). Adsorptive removal of toluene and dichloromethane from humid exhaust on MFI, BEA and FAU zeolites: An experimental and theoretical study. *Chemical Engineering Journal*, 394, 124986. https://doi.org/https://doi.org/10.1016/j.cej.2020.124986. DOI: 10.1016/j.cej.2020.124986

Dlamini, D. S., Tesha, J. M., Vilakati, G. D., Mamba, B. B., Mishra, A. K., Thwala, J. M., & Li, J. (2020). A critical review of selected membrane- and powder-based adsorbents for water treatment: Sustainability and effectiveness. *Journal of Cleaner Production*, 277, 123497. https://doi.org/https://doi.org/10.1016/j.jclepro.2020.123497. DOI: 10.1016/j.jclepro.2020.123497

Duraikkannu, S. L., Castro-Muñoz, R., & Figoli, A. (2021). A review on phase-inversion technique-based polymer microsphere fabrication. *Colloid and Interface Science Communications*, 40, 100329. https://doi.org/https://doi.org/10.1016/j.colcom.2020.100329. DOI: 10.1016/j.colcom.2020.100329

Elkady, M. F., El-Aassar, M. R., & Hassan, H. S. (2016). Adsorption Profile of Basic Dye onto Novel Fabricated Carboxylated Functionalized Co-Polymer Nanofibers. In *Polymers* (Vol. 8, Issue 5). DOI: 10.3390/polym8050177

EPA. (2021). *Publications on Remediation Technologies for Cleaning Up Contaminated Sites*. United States Environmental Protection Agency. https://www.epa.gov/remedytech/publications-remediation-technologies-cleaning-contaminated-sites

Folens, K., Leus, K., Nicomel, N. R., Meledina, M., Turner, S., Van Tendeloo, G., Du Laing, G., & Van Der Voort, P. (2016). Fe3O4@MIL-101 – A Selective and Regenerable Adsorbent for the Removal of As Species from Water. *European Journal of Inorganic Chemistry*, 2016(27), 4395–4401. https://doi.org/https://doi.org/10.1002/ejic.201600160. DOI: 10.1002/ejic.201600160

Guo, L., Ye, P., Wang, J., Fu, F., & Wu, Z. (2015). Three-dimensional Fe3O4-graphene macroscopic composites for arsenic and arsenate removal. *Journal of Hazardous Materials*, 298, 28–35. DOI: 10.1016/j.jhazmat.2015.05.011 PMID: 26001621

Guo, R., Wang, R., Yin, J., Jiao, T., Huang, H., Zhao, X., Zhang, L., Li, Q., Zhou, J., & Peng, Q. (2019). Fabrication and Highly Efficient Dye Removal Characterization of Beta-Cyclodextrin-Based Composite Polymer Fibers by Electrospinning. *Nanomaterials (Basel, Switzerland)*, 9(1), 127. Advance online publication. DOI: 10.3390/nano9010127 PMID: 30669533

Hang, C., Li, Q., Gao, S., & Shang, J. K. (2012). As(III) and As(V) adsorption by hydrous zirconium oxide nanoparticles synthesized by a hydrothermal process followed with heat treatment. *Industrial & Engineering Chemistry Research*, 51(1), 353–361. DOI: 10.1021/ie202260g

He, Y., Wang, X., Nie, X., Zou, H., & Pan, N. (2018). Uranium(VI) removal from aqueous solutions by a chelating fiber. *Journal of Radioanalytical and Nuclear Chemistry*, 317(2), 1005–1012. DOI: 10.1007/s10967-018-5956-4

Hei, S., Jin, Y., & Zhang, F. (2014). Fabrication of γ-Fe_2O_3 Nanoparticles by Solid-State Thermolysis of a Metal-Organic Framework, MIL-100(Fe), for Heavy Metal Ions Removal. *Journal of Chemistry*, 546956, 1–6. Advance online publication. DOI: 10.1155/2014/546956

Hiraide, S., Sakanaka, Y., Kajiro, H., Kawaguchi, S., Miyahara, M. T., & Tanaka, H. (2020). High-throughput gas separation by flexible metal–organic frameworks with fast gating and thermal management capabilities. *Nature Communications*, 11(1), 3867. DOI: 10.1038/s41467-020-17625-3 PMID: 32747638

Hosseini Asl, S. M., Javadian, H., Khavarpour, M., Belviso, C., Taghavi, M., & Maghsudi, M. (2019). Porous adsorbents derived from coal fly ash as cost-effective and environmentally-friendly sources of aluminosilicate for sequestration of aqueous and gaseous pollutants: A review. *Journal of Cleaner Production*, 208, 1131–1147. https://doi.org/https://doi.org/10.1016/j.jclepro.2018.10.186. DOI: 10.1016/j.jclepro.2018.10.186

Hsissou, R., Seghiri, R., Benzekri, Z., Hilali, M., Rafik, M., & Elharfi, A. (2021). Polymer composite materials: A comprehensive review. *Composite Structures*, 262, 113640. https://doi.org/https://doi.org/10.1016/j.compstruct.2021.113640. DOI: 10.1016/j.compstruct.2021.113640

Hu, Z., & Srinivasan, M. P. (2001). Mesoporous high-surface-area activated carbon. *Microporous and Mesoporous Materials*, 43(3), 267–275. https://doi.org/https://doi.org/10.1016/S1387-1811(00)00355-3. DOI: 10.1016/S1387-1811(00)00355-3

Hube, S., Eskafi, M., Hrafnkelsdóttir, K. F., Bjarnadóttir, B., Bjarnadóttir, M. Á., Axelsdóttir, S., & Wu, B. (2020). Direct membrane filtration for wastewater treatment and resource recovery: A review. *The Science of the Total Environment*, 710, 136375. https://doi.org/https://doi.org/10.1016/j.scitotenv.2019.136375. DOI: 10.1016/j.scitotenv.2019.136375 PMID: 31923693

Jia, R., Lei, H., Hino, T., & Arulrajah, A. (2018). Environmental changes in Ariake Sea of Japan and their relationships with Isahaya Bay reclamation. *Marine Pollution Bulletin*, 135, 832–844. https://doi.org/https://doi.org/10.1016/j.marpolbul.2018.08.008. DOI: 10.1016/j.marpolbul.2018.08.008 PMID: 30301105

Jiun-horng, T., Hsiu-mei, C., Guan-yinag, H. & Hung-lung, C. (2008). *Adsorption characteristics of acetone, chloroform and acetonitrile on sludge-derived adsorbent, commercial granular activated carbon and activated carbon fibers. 154*, 1183–1191. DOI: 10.1016/j.jhazmat.2007.11.065

Kampa, M., & Castanas, E. (2008). Human health effects of air pollution. *Environmental Pollution*, 151(2), 362–367. https://doi.org/https://doi.org/10.1016/j.envpol.2007.06.012. DOI: 10.1016/j.envpol.2007.06.012 PMID: 17646040

Kato, S., Otake, K. I., Chen, H., Akpinar, I., Buru, C. T., Islamoglu, T., Snurr, R. Q., & Farha, O. K. (2019). Zirconium-Based Metal-Organic Frameworks for the Removal of Protein-Bound Uremic Toxin from Human Serum Albumin. *Journal of the American Chemical Society*, 141(6), 2568–2576. DOI: 10.1021/jacs.8b12525 PMID: 30707010

Kim, K.-J., & Ahn, H.-G. (2012). The effect of pore structure of zeolite on the adsorption of VOCs and their desorption properties by microwave heating. *Microporous and Mesoporous Materials*, 152, 78–83. https://doi.org/https://doi.org/10.1016/j.micromeso.2011.11.051. DOI: 10.1016/j.micromeso.2011.11.051

Kim, S., Han, S. H., & Lee, Y. M. (2012). Thermally rearranged (TR) polybenzoxazole hollow fiber membranes for CO2 capture. *Journal of Membrane Science, 403–404*, 169–178. https://doi.org/https://doi.org/10.1016/j.memsci.2012.02.041

Ko, Y. G., Chun, Y. J., Kim, C. H., & Choi, U. S. (2011). Removal of Cu(II) and Cr(VI) ions from aqueous solution using chelating fiber packed column: Equilibrium and kinetic studies. *Journal of Hazardous Materials*, 194, 92–99. https://doi.org/https://doi.org/10.1016/j.jhazmat.2011.07.088. DOI: 10.1016/j.jhazmat.2011.07.088 PMID: 21871730

Kobayashi, T., Fukaya, T., Abe, M., & Fujii, N. (2002). Phase Inversion Molecular Imprinting by Using Template Copolymers for High Substrate Recognition. *Langmuir*, 18(7), 2866–2872. DOI: 10.1021/la0106586

Kobayashi, T., Ohshiro, M., Nakamoto, K., & Uchida, S. (2016). Decontamination of Extra-Diluted Radioactive Cesium in Fukushima Water Using Zeolite-Polymer Composite Fibers. *Industrial & Engineering Chemistry Research*, 55(25), 6996–7002. DOI: 10.1021/acs.iecr.6b00903

Kobayashi, T., Wang, H. Y., Fukaya, T., & Fujii, N. (1998). Molecular Imprinted Membranes Prepared by Phase Inversion of Polyacrylonitrile Copolymers Containing Carboxylic Acid Groups. In *Molecular and Ionic Recognition with Imprinted Polymers* (Vol. 703, pp. 13–188). American Chemical Society., https://doi.org/doi:10.1021/bk-1998-0703.ch013 DOI: 10.1021/bk-1998-0703.ch013

Kuo, L. J., Janke, C. J., Wood, J. R., Strivens, J. E., Das, S., Oyola, Y., Mayes, R. T., & Gill, G. A. (2015). Characterization and testing of amidoxime-based adsorbent materials to extract uranium from natural seawater. *Industrial & Engineering Chemistry Research*, 55(15), 4285–4293. DOI: 10.1021/acs.iecr.5b03267

Ladshaw, A. P., Wiechert, A. I., Das, S., Yiacoumi, S., & Tsouris, C. (2017). Amidoxime Polymers for Uranium Adsorption: Influence of Comonomers and Temperature. In *Materials* (Vol. 10, Issue 11). DOI: 10.3390/ma10111268

Lee, J., Lee, K., & Kim, J. (2021). Fiber-Based Gas Filter Assembled via in Situ Synthesis of ZIF-8 Metal Organic Frameworks for an Optimal Adsorption of SO2: Experimental and Theoretical Approaches. *ACS Applied Materials & Interfaces*, 13(1), 1620–1631. DOI: 10.1021/acsami.0c19957 PMID: 33395254

Lee, J., Ye, Y., Ward, A. J., Zhou, C., Chen, V., Minett, A. I., Lee, S., Liu, Z., Chae, S.-R., & Shi, J. (2016). High flux and high selectivity carbon nanotube composite membranes for natural organic matter removal. *Separation and Purification Technology*, 163, 109–119. DOI: 10.1016/j.seppur.2016.02.032

Lee, S., Binns, M., Lee, J. H., Moon, J. H., Yeo, J. G., Yeo, Y. K., & Kim, J. K. (2017). Membrane separation process for CO2 capture from mixed gases using TR and XTR hollow fiber membranes: Process modeling and experiments. *Journal of Membrane Science*, 541, 224–234.

Lee, S., Lee, T., & Kim, D. (2017). Adsorption of Hydrogen Sulfide from Gas Streams Using the Amorphous Composite of α-FeOOH and Activated Carbon Powder. *Industrial & Engineering Chemistry Research*, 56(11), 3116–3122. DOI: 10.1021/acs.iecr.6b04747

Lei, Y., Cao, W., Lin, K., & Jin, Y. (2012). Adsorption kinetics of uranium(VI) on carboxylic acid fiber. *Advanced Materials Research*, 550–553, 2625–2628. . DOI: 10.4028/www.scientific.net/AMR.550-553.2625

Li, H., Choi, W., Ingole, P. G., Lee, H. K., & Baek, I. H. (2016). Oxygen separation membrane based on facilitated transport using cobalt tetraphenylporphyrin-coated hollow fiber composites. *Fuel*, 185, 133–141. https://doi.org/https://doi.org/10.1016/j.fuel.2016.07.097. DOI: 10.1016/j.fuel.2016.07.097

Li, W., Liu, Q., Liu, J., Zhang, H., Li, R., Li, Z., Jing, X., & Wang, J. (2017). Applied Surface Science Removal U (VI) from artificial seawater using facilely and covalently grafted polyacrylonitrile fibers with lysine. *Applied Surface Science*, 403, 378–388. DOI: 10.1016/j.apsusc.2017.01.104

Li, X., Zhang, L., Yang, Z., Wang, P., Yan, Y. & Ran, J. (2020). Adsorption materials for volatile organic compounds (VOCs) and the key factors for VOCs adsorption process: A review. *Separation and Purification Technology, 235*(October 2019), 116213. DOI: 10.1016/j.seppur.2019.116213

Lillo-Ródenas, M. A., Marco-Lozar, J. P., Cazorla-Amorós, D., & Linares-Solano, A. (2007). Activated carbons prepared by pyrolysis of mixtures of carbon precursor/alkaline hydroxide. *Journal of Analytical and Applied Pyrolysis*, 80(1), 166–174. https://doi.org/https://doi.org/10.1016/j.jaap.2007.01.014. DOI: 10.1016/j.jaap.2007.01.014

Ling, C., Liu, X., Yang, X., Hu, J., Li, R., Pang, L., Ma, H., Li, J., Wu, G., Lu, S., & Wang, D. (2017). Uranium adsorption tests of amidoxime-based ultrahigh molecular weight polyethylene fibers in simulated seawater and natural coastal marine seawater from different locations. *Industrial & Engineering Chemistry Research*, 56(4), 1103–1111. DOI: 10.1021/acs.iecr.6b04181

Liu, H., Guo, M., & Zhang, Y. (2014). Nitrate removal by Fe0/Pd/Cu nano-composite in groundwater. *Environmental Technology*, 35(7), 917–924. DOI: 10.1080/09593330.2013.856926 PMID: 24645474

Llanes-Monter, M. M., Olguín, M. T., & Solache-Ríos, M. J. (2007). Lead sorption by a Mexican, clinoptilolite-rich tuff. *Environmental Science and Pollution Research International*, 14(6), 397–403. DOI: 10.1065/espr2006.10.357 PMID: 17993223

Ma, Y., Kuhn, A. N., Gao, W., Al-Zoubi, T., Du, H., Pan, X., & Yang, H. (2021). Strong electrostatic adsorption approach to the synthesis of sub-three nanometer intermetallic platinum–cobalt oxygen reduction catalysts. *Nano Energy*, 79, 105465. https://doi.org/https://doi.org/10.1016/j.nanoen.2020.105465. DOI: 10.1016/j.nanoen.2020.105465

Mahapatra, A., Mishra, B. G., & Hota, G. (2013). Electrospun Fe2O3–Al2O3 nanocomposite fibers as efficient adsorbent for removal of heavy metal ions from aqueous solution. *Journal of Hazardous Materials, 258–259*, 116–123. https://doi.org/https://doi.org/10.1016/j.jhazmat.2013.04.045

Marchesini, S., McGilvery, C. M., Bailey, J., & Petit, C. (2017). Template-Free Synthesis of Highly Porous Boron Nitride: Insights into Pore Network Design and Impact on Gas Sorption. *ACS Nano*, 11(10), 10003–10011. DOI: 10.1021/acsnano.7b04219 PMID: 28892607

Motsa, M. M., Thwala, J. M., Msagati, T. A. M., & Mamba, B. B. (2011). The potential of melt-mixed polypropylene–zeolite blends in the removal of heavy metals from aqueous media. *Physics and Chemistry of the Earth Parts A/B/C*, 36(14), 1178–1188. https://doi.org/https://doi.org/10.1016/j.pce.2011.07.072. DOI: 10.1016/j.pce.2011.07.072

Mukherjee, R., & De, S. (2014). Adsorptive removal of nitrate from aqueous solution by polyacrylonitrile-alumina nanoparticle mixed matrix hollow-fiber membrane. *Journal of Membrane Science*, 466, 281–292. DOI: 10.1016/j.memsci.2014.05.004

Murugesan, G. S., Sathishkumar, M., & Swaminathan, K. (2006). Arsenic removal from groundwater by pretreated waste tea fungal biomass. *Bioresource Technology*, 97(3), 483–487. https://doi.org/https://doi.org/10.1016/j.biortech.2005.03.008. DOI: 10.1016/j.biortech.2005.03.008 PMID: 16216732

Nakamoto, K. (2018). *Inorganic - Polymer Composite Fibers Developed for Decontamination of Contaminated Water. March.*

Nakamoto, K. & Kobayashi, T. (2017). *Fibrous mordenite zeolite - polymer composite adsorbents to methylene blue dye. 0869*(12), 131–136.

Nakamoto, K., Ohshiro, M., & Kobayashi, T. (2017). Mordenite zeolite - Polyethersulfone composite fibers developed for decontamination of heavy metal ions. *Journal of Environmental Chemical Engineering*, 5(1), 513–525. DOI: 10.1016/j.jece.2016.12.031

Nakamoto, K., Ohshiro, M., & Kobayashi, T. (2018). Continuous flow column adsorption of mordenite zeolite–polymer compositfibers for lead removal. *Desalination and Water Treatment*, 109, 297–306. DOI: 10.5004/dwt.2018.22115

Neti, V. S., Das, S., Brown, S., Janke, C. J., Kuo, L. J., Gill, G. A., Dai, S., & Mayes, R. T. (2017). Efficient Functionalization of Polyethylene Fibers for the Uranium Extraction from Seawater through Atom Transfer Radical Polymerization. *Industrial & Engineering Chemistry Research*, 56(38), 10826–10832. DOI: 10.1021/acs.iecr.7b00482

Nguyen, K. D., Trang, T. T. C., & Kobayashi, T. (2019). Chitin-halloysite nanoclay hydrogel composite adsorbent to aqueous heavy metal ions. *Journal of Applied Polymer Science*, 136(11), 47207. https://doi.org/https://doi.org/10.1002/app.47207. DOI: 10.1002/app.47207

Niknam Shahrak, M., Ghahramaninezhad, M., & Eydifarash, M. (2017). Zeolitic imidazolate framework-8 for efficient adsorption and removal of Cr(VI) ions from aqueous solution. *Environmental Science and Pollution Research International*, 24(10), 9624–9634. DOI: 10.1007/s11356-017-8577-5 PMID: 28247275

Oncel, M. S., Muhcu, A., Demirbas, E., & Kobya, M. (2013). A comparative study of chemical precipitation and electrocoagulation for treatment of coal acid drainage wastewater. *Journal of Environmental Chemical Engineering*, 1(4), 989–995. https://doi.org/https://doi.org/10.1016/j.jece.2013.08.008. DOI: 10.1016/j.jece.2013.08.008

Ooshiro, M., Kobayashi, T., & Uchida, S. (2017). Fibrous zeolite-polymer composites for decontamination of radioactive waste water extracted from radio-Cs fly ash. *International Journal of Engineering and Technical Research*, 7(4).

Ozdemir, O., Turan, M., Turan, A. Z., Faki, A., & Engin, A. B. (2009). Feasibility analysis of color removal from textile dyeing wastewater in a fixed-bed column system by surfactant-modified zeolite (SMZ). *Journal of Hazardous Materials*, 166(2), 647–654. https://doi.org/https://doi.org/10.1016/j.jhazmat.2008.11.123. DOI: 10.1016/j.jhazmat.2008.11.123 PMID: 19136207

Pandey, S., & Mishra, S. B. (2011). Sol–gel derived organic–inorganic hybrid materials: Synthesis, characterizations and applications. *Journal of Sol-Gel Science and Technology*, 59(1), 73–94. DOI: 10.1007/s10971-011-2465-0

Park, H. J., Bhatti, U. H., Nam, S. C., Park, S. Y., Lee, K. B., & Baek, I. H. (2019). Nafion/TiO2 nanoparticle decorated thin film composite hollow fiber membrane for efficient removal of SO2 gas. *Separation and Purification Technology*, 211, 377–390. https://doi.org/https://doi.org/10.1016/j.seppur.2018.10.010. DOI: 10.1016/j.seppur.2018.10.010

Phuong, A., Thi, L., Wakasugi, R. & Kobayashi, T. (2023). *Suppression of Hydrogen Sulfide Generation via the Coexistence of Anaerobic Sludge and Goethite-Rich Limonite / Polyethersulfone Composite Fibers*. DOI: 10.1021/acsomega.3c04540

Pöschl, U., & Shiraiwa, M. (2015). Multiphase Chemistry at the Atmosphere–Biosphere Interface Influencing Climate and Public Health in the Anthropocene. *Chemical Reviews*, 115(10), 4440–4475. DOI: 10.1021/cr500487s PMID: 25856774

Qasem, N. A. A., Mohammed, R. H., & Lawal, D. U. (2021). Removal of heavy metal ions from wastewater: A comprehensive and critical review. *NPJ Clean Water*, 4(1), 36. DOI: 10.1038/s41545-021-00127-0

Reimerink, W. M. T. M., & Kleut, D. v. (1999). Air pollution control by adsorption. In A. B. T.-S. in S. S. and C. Dąbrowski (Ed.), *Adsorption and its Applications in Industry and Environmental Protection* (Vol. 120, pp. 807–819). Elsevier. https://doi.org/https://doi.org/10.1016/S0167-2991(99)80380-2

Roberts, A. D., Lee, J. S. M., Magaz, A., Smith, M. W., Dennis, M., Scrutton, N. S., & Blaker, J. J. (2020). Hierarchically porous silk/activated-carbon composite fibres for adsorption and repellence of volatile organic compounds. *Molecules (Basel, Switzerland)*, 25(5), 5–7. DOI: 10.3390/molecules25051207 PMID: 32156015

Saleh, T. A. (2022). Chapter 2 - Adsorption technology and surface science. In T. A. B. T.-I. S. and T. Saleh (Ed.), *Surface Science of Adsorbents and Nanoadsorbents* (Vol. 34, pp. 39–64). Elsevier. https://doi.org/https://doi.org/10.1016/B978-0-12-849876-7.00006-3

Shah, B., Tailor, R., & Shah, A. (2011). Sorptive sequestration of 2-chlorophenol by zeolitic materials derived from bagasse fly ash. *Journal of Chemical Technology and Biotechnology*, 86(10), 1265–1275. https://doi.org/https://doi.org/10.1002/jctb.2646. DOI: 10.1002/jctb.2646

Sidhikku Kandath Valappil, R., Ghasem, N., & Al-Marzouqi, M. (2021). Current and future trends in polymer membrane-based gas separation technology: A comprehensive review. *Journal of Industrial and Engineering Chemistry*, 98, 103–129. https://doi.org/https://doi.org/10.1016/j.jiec.2021.03.030. DOI: 10.1016/j.jiec.2021.03.030

Son, L. T., Katagawa, K., & Kobayashi, T. (2011). Using molecularly imprinted polymeric spheres for hybrid membranes with selective adsorption of bisphenol A derivatives. *Journal of Membrane Science*, 375(1), 295–303. https://doi.org/https://doi.org/10.1016/j.memsci.2011.03.054. DOI: 10.1016/j.memsci.2011.03.054

Son, L. T., & Takaomi, K. (2011). Hollow-fiber membrane absorbents embedded molecularly imprinted polymeric spheres for bisphenol A target. *Journal of Membrane Science*, 384(1–2), 117–125. DOI: 10.1016/j.memsci.2011.09.013

Sprynskyy, M., Buszewski, B., Terzyk, A. P., & Namieśnik, J. (2006). Study of the selection mechanism of heavy metal (Pb2+, Cu2+, Ni2+, and Cd2+) adsorption on clinoptilolite. *Journal of Colloid and Interface Science*, 304(1), 21–28. https://doi.org/https://doi.org/10.1016/j.jcis.2006.07.068. DOI: 10.1016/j.jcis.2006.07.068 PMID: 16989853

Stejskal, J. (2022). Recent Advances in the Removal of Organic Dyes from Aqueous Media with Conducting Polymers, Polyaniline and Polypyrrole, and Their Composites. *Polymers*, 14(19), 4243. Advance online publication. DOI: 10.3390/polym14194243 PMID: 36236189

Sun, J., Zhang, X., Zhang, A., & Liao, C. (2019). Preparation of Fe–Co based MOF-74 and its effective adsorption of arsenic from aqueous solution. *Journal of Environmental Sciences (China)*, 80, 197–207. https://doi.org/https://doi.org/10.1016/j.jes.2018.12.013. DOI: 10.1016/j.jes.2018.12.013 PMID: 30952337

Swanckaert, B., Geltmeyer, J., Rabaey, K., De Buysser, K., Bonin, L., & De Clerck, K. (2022). A review on ion-exchange nanofiber membranes: Properties, structure and application in electrochemical (waste)water treatment. *Separation and Purification Technology*, 287, 120529. https://doi.org/https://doi.org/10.1016/j.seppur.2022.120529. DOI: 10.1016/j.seppur.2022.120529

Takeda, K., & Kobayashi, T. (2006). Hybrid molecularly imprinted membranes for targeted bisphenol derivatives. *Journal of Membrane Science*, 275(1), 61–69. https://doi.org/https://doi.org/10.1016/j.memsci.2005.09.004. DOI: 10.1016/j.memsci.2005.09.004

Tang, J., An, T., Xiong, J., & Li, G. (2017). The evolution of pollution profile and health risk assessment for three groups SVOCs pollutants along with Beijiang River, China. *Environmental Geochemistry and Health*, 39(6), 1487–1499. DOI: 10.1007/s10653-017-9936-3 PMID: 28315117

Tang, W., Li, Q., Gao, S., & Shang, J. K. (2011). Arsenic (III,V) removal from aqueous solution by ultrafine α-Fe2O3 nanoparticles synthesized from solvent thermal method. *Journal of Hazardous Materials*, 192(1), 131–138. DOI: 10.1016/j.jhazmat.2011.04.111 PMID: 21684075

Toda, K., Tanaka, T., Tsuda, Y., Ban, M., Koveke, E. P., Koinuma, M., & Ohira, S.-I. (2014). Sulfurized limonite as material for fast decomposition of organic compounds by heterogeneous Fenton reaction. *Journal of Hazardous Materials*, 278, 426–432. https://doi.org/https://doi.org/10.1016/j.jhazmat.2014.06.033. DOI: 10.1016/j.jhazmat.2014.06.033 PMID: 24997258

Ton Nu, P. T., & Kobayashi, T. (2020a). Methanol/CaCl2 Wet Phase Inversion for Selective Separation of Heavy Metal Ions by Nylon 6-Mordenite Zeolite Composite Membranes. *International Journal of Environmental Research*, 14(6), 667–683. DOI: 10.1007/s41742-020-00291-0

Ton Nu, P. T., & Kobayashi, T. (2020b). Nylon-6–Mordenite Composite Membranes for Adsorption of Ethylene Gas Released from Chiquita Bananas. *Industrial & Engineering Chemistry Research*, 59(17), 8212–8222. DOI: 10.1021/acs.iecr.9b06149

Tyagi, R., & Jacob, J. (2020). Design and synthesis of water-soluble chelating polymeric materials for heavy metal ion sequestration from aqueous waste. *Reactive & Functional Polymers*, 154, 104687. https://doi.org/https://doi.org/10.1016/j.reactfunctpolym.2020.104687. DOI: 10.1016/j.reactfunctpolym.2020.104687

Ukaogo, P. O., Ewuzie, U., & Onwuka, C. V. (2020). *21 - Environmental pollution: causes, effects, and the remedies* (P. Chowdhary, A. Raj, D. Verma, & Y. B. T.-M. for S. E. and H. Akhter (eds.); pp. 419–429). Elsevier. https://doi.org/https://doi.org/10.1016/B978-0-12-819001-2.00021-8

Volzone, C., & Ortiga, J. (2011). SO2 gas adsorption by modified kaolin clays: Influence of previous heating and time acid treatments. *Journal of Environmental Management*, 92(10), 2590–2595. https://doi.org/https://doi.org/10.1016/j.jenvman.2011.05.031. DOI: 10.1016/j.jenvman.2011.05.031 PMID: 21696883

Wang, C., Luan, J., & Wu, C. (2019). Metal-organic frameworks for aquatic arsenic removal. *Water Research*, 158, 370–382. DOI: 10.1016/j.watres.2019.04.043 PMID: 31055017

Wang, H.-Y., Huang, H.-F., & Jiang, J.-Q. (2011). The effect of metal cations on phenol adsorption by hexadecyl-trimethyl-ammonium bromide (hdtma) modified clinoptilolite (Ct.). *Separation and Purification Technology*, 80(3), 658–662. https://doi.org/https://doi.org/10.1016/j.seppur.2011.06.030. DOI: 10.1016/j.seppur.2011.06.030

Wang, H. Y., Kobayashi, T., Fukaya, T., & Fujii, N. (1997). Molecular Imprint Membranes Prepared by the Phase Inversion Precipitation Technique. 2. Influence of Coagulation Temperature in the Phase Inversion Process on the Encoding in Polymeric Membranes. *Langmuir*, 13(20), 5396–5400. DOI: 10.1021/la970114x

Wang, K., Gu, J., & Yin, N. (2017). Efficient Removal of Pb(II) and Cd(II) Using NH2-Functionalized Zr-MOFs via Rapid Microwave-Promoted Synthesis. *Industrial & Engineering Chemistry Research*, 56(7), 1880–1887. DOI: 10.1021/acs.iecr.6b04997

Wang, M., Cai, H., & Zhang, J. (2018). Application research on the adsorption of cadmium ion in wastewater by zeolite molecular sieve. *Chemical Engineering Transactions*, 71, 403–408. DOI: 10.3303/CET1871068

Wang, Q., Zhu, S., Xi, C., & Zhang, F. (2022). A Review: Adsorption and Removal of Heavy Metals Based on Polyamide-amines Composites. *Frontiers in Chemistry*, 10(March), 1–15. DOI: 10.3389/fchem.2022.814643 PMID: 35308790

Wang, S., & Zhu, Z. H. (2006). Characterisation and environmental application of an Australian natural zeolite for basic dye removal from aqueous solution. *Journal of Hazardous Materials*, 136(3), 946–952. https://doi.org/https://doi.org/10.1016/j.jhazmat.2006.01.038. DOI: 10.1016/j.jhazmat.2006.01.038 PMID: 16504394

Wen, M., Li, G., Liu, H., Chen, J., An, T., & Yamashita, H. (2019). Metal-organic framework-based nanomaterials for adsorption and photocatalytic degradation of gaseous pollutants: Recent progress and challenges. *Environmental Science. Nano*, 6(4), 1006–1025. DOI: 10.1039/C8EN01167B

Wijesinghe, W. P. S. L., Mantilaka, M. M. M. G. P. G., Ruparathna, K. A. A., Rajapakshe, R. B. S. D., Sameera, S. A. L., & Thilakarathna, M. G. G. S. N. (2020). 4 - Filler matrix interfaces of inorganic/biopolymer composites and their applications. In K. L. Goh, A. M.K., R. T. De Silva & S. B. T.-I. in P. and F. R. C. Thomas (Eds.), *Woodhead Publishing Series in Composites Science and Engineering* (pp. 95–112). Woodhead Publishing. https://doi.org/https://doi.org/10.1016/B978-0-08-102665-6.00004-2

Woellner, M., Hausdorf, S., Klein, N., Mueller, P., Smith, M. W., & Kaskel, S. (2018). Adsorption and Detection of Hazardous Trace Gases by Metal–Organic Frameworks. *Advanced Materials*, 30(37), 1704679. https://doi.org/https://doi.org/10.1002/adma.201704679. DOI: 10.1002/adma.201704679 PMID: 29921016

Wong, S., Lim, Y., Ngadi, N., Mat, R., Hassan, O., Inuwa, I. M., Mohamed, N. B., & Low, J. H. (2018). Removal of acetaminophen by activated carbon synthesized from spent tea leaves: Equilibrium, kinetics and thermodynamics studies. *Powder Technology*, 338, 878–886. https://doi.org/https://doi.org/10.1016/j.powtec.2018.07.075. DOI: 10.1016/j.powtec.2018.07.075

Woo, K. T., Lee, J., Dong, G., Kim, J. S., Do, Y. S., Hung, W.-S., Lee, K.-R., Barbieri, G., Drioli, E., & Lee, Y. M. (2015). Fabrication of thermally rearranged (TR) polybenzoxazole hollow fiber membranes with superior CO_2/N_2 separation performance. *Journal of Membrane Science*, 490, 129–138. https://doi.org/https://doi.org/10.1016/j.memsci.2015.04.059. DOI: 10.1016/j.memsci.2015.04.059

Wu, F., Pu, N., Ye, G., Sun, T., Wang, Z., Song, Y., Wang, W., Huo, X., Lu, Y., & Chen, J. (2017). Performance and Mechanism of Uranium Adsorption from Seawater to Poly(dopamine)-Inspired Sorbents. *Environmental Science & Technology*, 51(8), 4606–4614. DOI: 10.1021/acs.est.7b00470 PMID: 28332830

Xia, L., Zhou, S., Zhang, C., Fu, Z., Wang, A., Zhang, Q., Wang, Y., Liu, X., Wang, X., & Xu, W. (2020). Environment-friendly Juncus effusus-based adsorbent with a three-dimensional network structure for highly efficient removal of dyes from wastewater. *Journal of Cleaner Production*, 259, 120812. https://doi.org/https://doi.org/10.1016/j.jclepro.2020.120812. DOI: 10.1016/j.jclepro.2020.120812

Yang, J. H. (2021). Hydrogen sulfide removal technology: A focused review on adsorption and catalytic oxidation. *Korean Journal of Chemical Engineering*, 38(4), 674–691. DOI: 10.1007/s11814-021-0755-y

Yu, J., Mu, C., Yan, B., Qin, X., Shen, C., Xue, H., & Pang, H. (2017). Nanoparticle/MOF composites: Preparations and applications. *Materials Horizons*, 4(4), 557–569. DOI: 10.1039/C6MH00586A

Yuna, Z. (2016). Review of the Natural, Modified, and Synthetic Zeolites for Heavy Metals Removal from Wastewater. *Environmental Engineering Science*, 33(7), 443–454. DOI: 10.1089/ees.2015.0166

Yusof, A. M., & Malek, N. A. N. N. (2009). Removal of Cr(VI) and As(V) from aqueous solutions by HDTMA-modified zeolite Y. *Journal of Hazardous Materials*, 162(2), 1019–1024. https://doi.org/ https://doi.org/10.1016/j.jhazmat.2008.05.134. DOI: 10.1016/j.jhazmat.2008.05.134 PMID: 18632204

Zaitan, H., Bianchi, D., Achak, O., & Chafik, T. (2008). A comparative study of the adsorption and desorption of o-xylene onto bentonite clay and alumina. *Journal of Hazardous Materials*, 153(1), 852–859. https://doi.org/https://doi.org/10.1016/j.jhazmat.2007.09.070. DOI: 10.1016/j.jhazmat.2007.09.070 PMID: 17977653

Zhang, A., Uchiyama, G., & Asakura, T. (2003). Dynamic-State Adsorption and Elution Behaviour of Uranium(VI) Ions from Seawater by a Fibrous and Porous Adsorbent Containing Amidoxime Chelating Functional Groups. *Adsorption Science and Technology*, 21(8), 761–773. DOI: 10.1260/026361703773581812

Zhang, H., Zhang, L., Han, X., Kuang, L., & Hua, D. (2018). Guanidine and Amidoxime Cofunctionalized Polypropylene Nonwoven Fabric for Potential Uranium Seawater Extraction with Antifouling Property. *Industrial & Engineering Chemistry Research*, 57(5), 1662–1670. DOI: 10.1021/acs.iecr.7b04687

Zhang, M., Wang, X., Xue, Y., Li, J., Wang, J., Fang, C., & Zhu, L. (2023). Robust and Scalable In Vitro Surface Mineralization of Inert Polymers with a Rationally Designed Molecular Bridge. *ACS Applied Materials & Interfaces*, 15(6), 8730–8741. DOI: 10.1021/acsami.2c21286 PMID: 36735823

Zhang, Z., Chen, J., Gao, Y., Ao, Z., Li, G., An, T., Hu, Y., & Li, Y. (2018). A coupled technique to eliminate overall nonpolar and polar volatile organic compounds from paint production industry. *Journal of Cleaner Production*, 185, 266–274. https://doi.org/https://doi.org/10.1016/j.jclepro.2018.03.037. DOI: 10.1016/j.jclepro.2018.03.037

Zhao, J., Lee, D. T., Yaga, R. W., Hall, M. G., Barton, H. F., Woodward, I. R., Oldham, C. J., Walls, H. J., Peterson, G. W., & Parsons, G. N. (2016). Ultra-Fast Degradation of Chemical Warfare Agents Using MOF–Nanofiber Kebabs. *Angewandte Chemie International Edition*, 55(42), 13224–13228. https://doi.org/https://doi.org/10.1002/anie.201606656. DOI: 10.1002/anie.201606656 PMID: 27653957

Zhao, R., Shi, X., Ma, T., Rong, H., Wang, Z., Cui, F., Zhu, G., & Wang, C. (2021). Constructing Mesoporous Adsorption Channels and MOF-Polymer Interfaces in Electrospun Composite Fibers for Effective Removal of Emerging Organic Contaminants. *ACS Applied Materials & Interfaces*, 13(1), 755–764. DOI: 10.1021/acsami.0c20404 PMID: 33373204

Zhou, Q., Jiang, X., Li, X., & Jiang, W. (2016). The control of H2S in biogas using iron ores as in situ desulfurizers during anaerobic digestion process. *Applied Microbiology and Biotechnology*, 100(18), 8179–8189. DOI: 10.1007/s00253-016-7612-7 PMID: 27209038

Zito, P., & Shipley, H. J. (2015). Inorganic nano-adsorbents for the removal of heavy metals and arsenic: A review. *RSC Advances*, 5(38), 29885–29907. DOI: 10.1039/C5RA02714D

Chapter 15
Removal of Metal Ions With Biomasses and Bioremediation

Minoru Satoh

National Institute of Technology, Ibaraki College, Japan

ABSTRACT

Eco-friendly technologies using biomasses and bioremediation are presented, since remediation technologies for serious heavy metal pollution are necessary to protect the global environmental. In the perspective of the 3Rs (Recycle, Reuse, Reduce), one of such approaching is used with biomasses, plants and microbe resource derived from living organisms. The removal of metal ions via biomasses and bioremediation, especially with waste biomasses materials, becomes environmentally and friendly technologies in addition with another merit of waste reduction and effective utilization of unused biomass. Thus, bioremediation is gaining attention as a sustainable and effective method of cleaning up environmental pollutants using certain plants and animals. In particular, the method using plants as called phytoremediation, plant roots absorb water and nutrients from contaminated soil, sediment, surface water, and groundwater, removing metals and other toxic chemicals.

INTRODUCTION

The term "heavy metals" generally refers to metals and metal compounds with a density of 5 g/cm^3 or greater, including arsenic (As), cadmium (Cd), chromium (Cr), copper (Cu), iron (Fe), lead (Pb), mercury (Hg), silver (Ag), and zinc (Zn) (Duruibe et al., 2007). These substances are widely involved in human activities such as fossil fuel combustion, mining, and electroplating and released into the environment in large amounts daily through wastewater and other pathways (Zakhama et al. 2011), but, also, bioaccumulation is causing serious environmental and health problems (Yang et al., 2015). Since heavy metal and smelting metals, use is inevitable in industrial development through mining, it is also a frequent source of pollution in our lives, and soil, water, and food have become increasingly contaminated. Under these circumstances, certain plants are adapted to high concentrations of metals and can also accumulate toxic metals in their bodies to very high levels, but other plant species cannot survive

DOI: 10.4018/979-8-3693-0003-9.ch015

in such environment. Therefore, phytoremediation using such plants is now gaining attention as a low maintenance decontamination method (Peterson, 1983).

On the other hand, heavy metal contamination of wastewater is a problem of widespread concern. Because effective removal of heavy metals from water and wastewater is of great ecological importance, many techniques have been reported and established to recover metals from wastewater by using chemical precipitation, flotation, biosorption, electrolytic recovery and membrane separation. However, negative decontamination operations do not generate profit, so low cost and low maintenance are desired, but the more advanced technology to remove them efficiently tends to be higher cost and much maintenance required. Thus, the use of low-cost like biomass and bioremediation adsorption of heavy metals appears to be an appropriate choice for wastewater treatment.

In the criteria for heavy metal contamination status, according to U.S. Environmental Protection Agency regulations, the maximum contaminant limits (MCLs) for As, Cd, Cr, Hg, and Pb are 0.01, 0.005, 0.1, 0.002, and 0.015 mg/L, respectively (USEPA, 2016). These concentration levels of heavy metals and their compounds are regulated due to highly toxic, carcinogenic, mutagenic, and teratogenic even at very low concentrations. In Japan, until the Small Home Appliance Recycling Law enforced in 2013, and the recycling of small electronic devices, etc., that are no longer in use are to be handled in accordance with the Law, since the metals such as gold and copper, except for some metals such as iron and aluminum were disposed of as landfill waste without being recycled. In response to the requirement to ensure proper disposal of waste and effective use of resources, about 102,000 tons of used small electronic devices, etc. were disposed of under the Small Home Appliance Recycling Law in 2020, and the weight of metals recycled became about 52,000 tons. By type of metal recycled, about 45,000 tons of iron, 4,000 tons of aluminum, 3,000 tons of copper, 340 kg of gold, and 3,700 kg of silver were recycled. In order to further promote the use of useful metals contained in used products and contribute to securing resources and reducing the consumption of natural resources, it is necessary to continue to ensure the volume of recovery and improve the efficiency of recycling for all major products, including rare metals.

The realization of a low-carbon society is a result of reducing the concentration and emission of greenhouse gases in the atmosphere. In addition to reducing energy consumption, the REMEDIATION process must be reviewed and led in the direction of low energy consumption. To achieve this, it is necessary to build a recycling-oriented society in which waste emissions are reduced as much as possible, waste is reused as much as possible, and recycled with as little energy consumption as possible. In this regard, the developments of metal ion removal technologies by using 1) biomasses and 2) bioremediation are important in their harvestable efforts (Figure 1).

Figure 1. Metal ion removal technologies by using biomasses and bioremediation

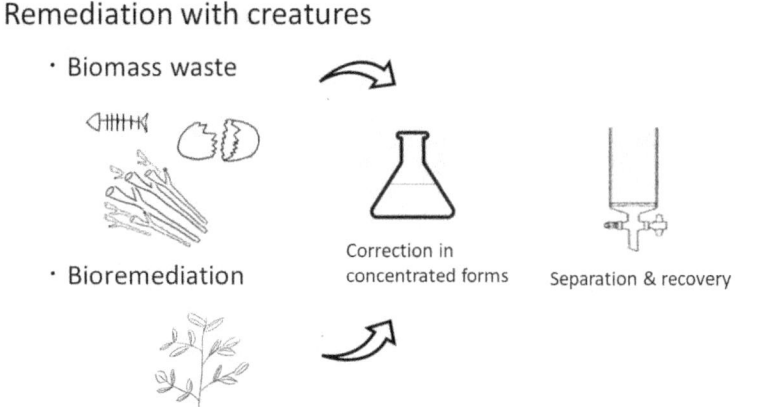

Here, remediation takes biomass, which is organic matter from living organisms and is produced by the growth of living organisms. Furthermore, bioremediation utilizes the natural purification effects of living organisms. Both are technologies that effectively utilize the power of nature.

The world's accelerated industrial revolution and the use of natural resources in metal mining and industry have had a major impact on the environment through heavy metal pollution. One of the most devastating impacts the world faces today is pollution by heavy metals reaching the air, soil, and water (Asha & Sandeep, 2013; Raghunandan et al. 2014, 2018). To concentrate contaminants from soil and water, remediation materials are adapted to the contaminated area and the concentrates are collected. Then the recovered contaminants can then be separated and recovered as resources.

REMOVAL OF METAL IONS BY USING BIOMASSES

As included for biomass materials to which remediation technologies can be adapted, there are various types of biomasses, which are classified into waste biomass, unutilized biomass, and resource crops (plants grown for energy or product manufacturing purposes), as followed in sources and uses;

(1) waste biomass includes waste paper, livestock manure, food waste, construction wood, lumber residues, black liquor (pulp mill effluent), sewage sludge, and human waste sludge,
(2) unused biomass includes rice straw, wheat straw, rice husks, and forest residues (thinned wood, damaged wood, etc.)
(3) resource crops including starch and sugar crops such as sugarcane and corn, and oilseed crops such as rapeseed.

Figure 2. Three types of biomasses

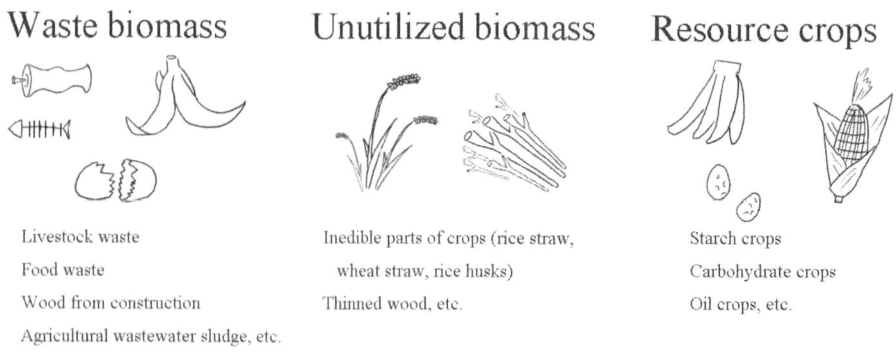

Waste biomass
Livestock waste
Food waste
Wood from construction
Agricultural wastewater sludge, etc.

Unutilized biomass
Inedible parts of crops (rice straw, wheat straw, rice husks)
Thinned wood, etc.

Resource crops
Starch crops
Carbohydrate crops
Oil crops, etc.

As illustrated in Figure 2, these of waste biomass, utilized biomass and resource crops are plant sources grown primarily for the purpose of being used as energy sources or product materials. In particular, waste biomass and unused biomass are important in the removal of metal ions using biomass waste in a recycling-based society.

In general technologies like chemical precipitation, flotation, electrolytic recovery, membrane separation, adsorption process by adsorbents, their methods of separation of metal ions and the recovery have disadvantages such as incomplete removal of metal ions, generation of large amounts of harmful sludge, and high costs involving. In contract, agricultural biomasses derived from peat moss, coffee beans, sawdust, and crab shell etc., (Table 1), have recently been attracting attention in treatment and purification of wastewater, since such wastes are its abundance and sustainability.

Table 1. Removal of metal ions by using biomasses

Materals	Metals	Maximum Adsorption Capacity /mgg^{-1}	References
Wheat shell	Cu	8.34	Basci et al., 2004
Wheat straw	Cd Cu	4.20 4.16	Dang et al., 2009
Rice husk	Pb Zn	0.62 19.61	Elham et al., 2010
Rice straw	Cd	13.9	Yang et al., 2012
Peat moss	Co	34.2 (treated with NaOH) 29.75 (un-treated) 20.85 (treated with HNO$_3$)	Caramalău, et al., 2009
Peat moss	Pb Hg	117.58 81.97	Bulgariu et al., 2008
Peat moss	Cu	16.1	Gardea-Torresdey et al., 1995

continued on following page

Table 1. Continued

Materals	Metals	Maximum Adsorption Capacity /mgg⁻¹	References
Biochar of peat moss	Pb Cd Cu	81.3 39.8 18.2	Lee et al., 2015
Coffee beans	Cd Cu Zn Pb Fe	6.72 - - - -	Kaikake et al., 2007
Cicer arientinum (chickpea var. black gram)	Pb Cd Zn Cu Ni	49.97 39.99 33.81 25.73 19.56	Saeed et al., 2005
Polyporus tenuiculus	Cu Pb Cd	14.7 92 11.4	Grassi, et al., 2011
Tea husk	Cu Pb	48 65	Amarasinghe et al., 2007
Eggshell	Cr	21-160	Chojnacka, 2005
Coconut tree sawdust (CTS)	Cu Pb Zn	3.89 25.00 23.81	Putra et al., 2014
Eggshell (ES)	Cu Pb Zn	34.48 90.90 35.71	
Sugarcane bagasse (SB)	Cu Pb Zn	3.65 21.28. 40.00	
Papaya (Carica papaya)	Cu Cd Zn	19.88 17.22 13.45	Saeed et al., 2005
Treated sawdust	Cr Pb Hg Cu	111.61 52.38 20.62 5.64	Meena et al., 2008
Crab cell	Cu Co	243.9 322.6	Vijayaraghavan et al., 2006
Salmon milt	Nd La, Ce, Pr, Pm, Sm, Eu, Gd, Tb, Dy, Y, Ho, Er, Tm, Yb, Lu	50.1 - -	Takahashi et al., 2014
DNA	Nd La, Ce, Pr, Sm, Eu, Gd, Tb, Dy, Ho, Er, Tm, Yb, Lu	0.182 - -	Takahashi et al., 2012

continued on following page

Table 1. Continued

Materials	Metals	Maximum Adsorption Capacity /mgg⁻¹	References
Pomegranate peel	Cu Pb	1.31 13.87	El-Ashtoukhy et al., 2008
Chlorella vulgaris	Cu Cd Ni Zn	25 45 21 31	Fraile et al., 2005
Coir fiber	Cu Pb Ni Fe	9.43 29.41 8.84 11.11	Pushkar et. al., 2013
Dye-loaded peanut shell	Cu Ni Zn	7.60 7.49 9.57	Shukla & Pai, Separation and Purification Technology, 2005
Dye-loaded sawdust	Cu Ni Zn	8.07 9.87 17.09	
Dye-loaded jute	Cu Ni Zn	8.4 5.26 5.95	Shukla & Pai, Bioresource Technology, 2005
Natural cotton fiber modified with citric acid (NCFCA)	Cu Zn Cd Pb	6.12 4.53 8.22 21.62	Paulino et al., 2013
Succinylated Okra	Cu Zn Cd Pb	72.72 57.11 121.51 273.97	Singha & Guleria, 2015

Waste biomass.

On those biomasses, waste sorbents are used as the main approach to decontaminate metal ions. Among waste biomass, wheat shell was adsorbed well Cu(II) (Basci et al., 2004) at various ranges of pH 2–7. On this case, pH played an important role in the efficiency and at pH 5 maximum efficiency of 99% was observed, when used for the wheat hull of 12 g/L. Onto wheat straw, *Triticum aestivum*, in aqueous systems (Dang et al., 2009), the biosorption equilibria were observed for both Cd(II) and Cu(II) and best fitted by the Langmuir model. The capacity of Cd(II) was about 27% and higher than that of Cu(II). At an initial concentration of 50 mg/L, approximately 80% of their metal ions were removed after 2.5 hours and 87% was achieved at equilibrium after approximately 3.5 hours. After adsorption, the metal-containing wheat straw was desorbed and reused for adsorption. In the effect of pH, increasing the pH from 4.0 to 7.0 increased the retention capacity at the equilibrium adsorption by about 130% for Cd(II) and 60% for Cu(II). However, when the temperature was increased from 25 to 30°C, the capacity became to be relatively small. As well as wheat hulls, rice husks produced as agricultural waste in large quantities removed Zn(II) and Pb(II) ions from dairy wastewater (Elham et al., 2010). The main parameters affecting the adsorption on these ions for rice husk bio-adsorbent were the amount of adsorbent, contact time, and the pH value of the wastewater; the optimal pH was in the pH range from 7.0 to 9.0, and the maximum adsorption capacities of Zn(II) and Pb(II) ions were 19.6 mg/g and 0.62 mg/g, respectively, with removal rates of 70% and 96.8%. Against other heavy metals, cadmium which is the most common toxic metal, rice straw was used to investigate the potential to remove Cd (Yang et al., 2012). For large amounts of wastewater contaminated with Cadmium, the biosorption process occurred within 5 min of equilibrium time with high biosorption capacity of 13.9 mg/g in the pH range of 2.0-6.0. The main Cd biosorption mechanism that Cd(II) was ion exchange of K(I), Na(I), Mg(II), and Ca(II) in rice straw with functional groups such as C=C, C-O, O-H, and carboxylic acids for chelation with Cd(II). When 0.5% (w/v) rice straw was shaken in a 50 mg/mL $CdSO_4$ solution with shaking at 150 r/min for 3 h, about 80% of the aquatic Cd was absorbed with the Cd amount of 8-10 mg/g. The metal-enriched rice straw contained industrial mining grade metals and was easily recovered for metal recovery, resulting in high quality with bio-ecoprocess.

In order to adsorb metal ions in aqueous solutions, various polar functional groups (-COOH, -OH, -NH_2, etc.) becomes important in the framework included, for example, in case of peat moss (Caramalău et al., 2009). The adsorption behavior of Co (II) on peat moss collected from Poiana Stampei, Romania, was influenced by initial pH. When the peat moss dosed in aqueous initial Co (II) concentration of 240 mg Co(II)/L with contact time in batch experiments, showing optimal conditions at pH=6.0 (acetate buffer), the adsorbent amount was 5 g/L (Caramalău et al., 2009). Such bio-sorbent peat moss also behaved sorption of Pb(II) and Hg(II) ions onto peat moss (Bulgariu et al., 2008), obeying Langmuir model with the maximum uptake capacities of Pb(II) and Hg(II) for 117.58 mg/g and 81.97 mg/g, respectively. The Cu(II) binding properties with humic acid and humin were in Canadian Sphagnum peat moss (Gardea-Torresdey et al. 1995). The optimum adsorption of Cu(II) depended on pH dependent and appeared at pH 4.0 by very rapid binding with binding capacities, when used with 16.1 mg/g-peat moss, of 28.2 mg /g-humic acid, and 17.9 mg/g-humin. But, in the case of esterified carboxyl groups of the biomass, less adsorption was observed, meaning that the carboxyl groups of the peat moss played a role in Cu(II) adsorption. In de-adsorption process on the biomasses, more than 90% Cu(II) adsorbed was recovered by treatment with 0.1 M HCl. Then, the used peat moss was converted to biochar and taken to use for heavy metal adsorption, again. The heavy metal capacity of biochar could be controlled by its carbonization (Lee et al., 2015) under pyrolysis of the peat moss at 400-1000°C for 30-90 min. The adsorbed heavy metals Pb, Cu, and Cd were under different carbonization conditions. Biochar produced at 800°C for 90 min was the most effective in removing Pb, Cd and Cu with maximum adsorption of 81.3 mg/g, 39.8 mg/g and 18.2 mg/g, respectively.

Furthermore, coffee beans have a possibility for bio-adsorbents. Approximately 8 million tons of coffee are produced annually worldwide, and most of the coffee beans become residual waste after being dripped and defatted (Potts et al., 2014). The residue, coffee beans after being dripped and degreased, were used as an adsorbent for base metals such as Cu(II), Zn(II), Pb(II), Fe(III) and Cd(II) (Kaikake et al., 2007). Analysis using fluorescent X-ray revealed that the residual beans contained sulfur and calcium and the micro-morphology had porous structure in the plant walls and specific surface area was found to be 1.2 m^2/g. In Gu(II), Zn(II), Pb(II), Fe(III), and Cd(II) the adsorption was dependent on pH. Even though no adsorption occurred at pH 2, about 100% was adsorbed at above pH 4. Of particular interest was the adsorption characteristics of Cd(II) on the residual coffee beans. The adsorption isotherm of Cd(II) at pH 8 obeyed Langmuir equation, yielding an adsorption equilibrium constant of 55.2 mmol/L and an adsorption capacity of 5.98×10^{-2} mmol/g. Also the desorption of cadmium(II) retained was easily achieved above 90% by batch treatment in the presence of HCl or HNO_3 solutions at more than 0.01 M concentration, indicating that cation exchange reaction between metal ions and the residual beans.

This material has a possibility of removing harmful metal ions from drinking water, groundwater, and wastewater.

In the cases of agro-waste, *cicer arientinum* (chickpea var. black gram) was useful as a bio-sorbent naturally occurring in certain biomass (Saeed et al., 2005). The bio-sorbent efficiently removed heavy metal ions from aqueous solution in the order of Pb>Cd>Zn>Cu>Ni. When the equilibriums were established within 30 minutes, following Langmuir adsorption isotherms, and the maximum amounts (q_m) were 49.97, 39.99, 33.81, 25.73, and 19.56 mg/g for Pb, Cd, Zn, Cu, and Ni, respectively. The Pb removal efficiency of the bio-sorbent from binary and ternary solutions containing Cd, Cu, Ni, and Zn was comparable to that of single each ion. In the desorption of Pb and other metals in single and multi-metallic solutions 0.1 M HCl was used for both experiments of batch and column releasing.

Papaya trees (Carica papaya) grown in tropical and subtropical countries become waste when they bear fruit. It is a very inexpensive, readily available, and abundant source of biomaterials. Papaya wood obtained from the felled trunks of adult trees has been reported to be a bio-sorbent for heavy metals (Saeed et al., 2005). In the treatment procedure, the bark of the trunk was stripped, cut into small pieces, and soaked in boiling water for 30 minutes. It was then thoroughly washed with tap water and soaked in distilled water for 2 to 3 hours. The washed wood pieces were crushed and oven-dried at 80°C. Biosorption was performed in 100 ml batch medium with contacting 10 mg/L of Cu(II), Cd(II), and Zn(II), as the 5 g/L papaya wood was added and the contact time was 60 min. The resulting metal removal rates were 97.8%, 94.9%, and 66.8% for Cu(II), Cd(II), and Zn(II), respectively, at pH 5. The advantages of this bio-sorbent are its rapid and high adsorption/desorption properties and its reusability in repeated cycles.

Polyporus tenuiculus (naturally occurring giant mushroom) with large fruiting bodies from Central and South America grows easily on straw or sawdust composing of lignocellulose and used for removal of Cu, Pb, and Cd from aqueous solutions (Grassi, et al., 2011). Obeying Langmuir model introduced the values of qe (metal mg/g-biomass) of 14.7 ± 0.3 mg/g for Cu, 11.4 ± 0.4 mg/g for Cd, and 92 ± 13 mg/g for Pb, respectively, in the order of Pb(II) > Cu(II) > Cd(II). Pb was removed less efficiently, and Cd was removed more efficiently than Pb. In all cases, equilibrium was reached in about 30 minutes and the pattern was highly dependent on pH. Also tea leaves were tested in aqueous copper or lead solution (Amarasinghe et al., 2007) and the husks adsorbed both metal ions significantly in the pH range of 5-6. The highest metal adsorption was 48 mg/g for Cu and 65 mg/g for Pb. Under all experimental conditions, Pb showed higher affinity and adsorption kinetics than Cu. Kinetic studies revealed that Pb and Cu adsorption was fast, with more than 90% of adsorption occurring within the first 15-20 min of contact time.

In addition to food plant waste, there is also sawdust as industrial plant waste. The effectiveness of coconut tree sawdust (CTS), and sugarcane bagasse (SB) as low-cost alternative bio-sorbents for removing Cu(II), Pb(II), and Zn(II) ions from aqueous solutions was reported (Putra et al., 2014). The optimal biosorption condition was pH 6.0, when used as biomass addition of 0.1 g, at equilibrium time of 90 minutes and the maximum adsorption capacities (q_m) for Cu(II), Pb(II) and Zn(II) of 3.89, 25.00 and 23.81 mg/g for CTS, and 3.65, 21.28 and 40.00 mg/g for SB, respectively.

The sawdust was chemically treated to remove lignin and the treated sawdust was used to remove heavy metals from wastewater and aqueous solutions (Meena et al., 2008) for Cr(VI), Pb(II), Hg(II) and Cu(II). The chemically treated sawdust found maximum removal rates of 99.4, 92.2, 94.6 and 99.3% for Hg(II), Cu(II), Pb(II) and Cr(VI) at pH 6, respectively. Especially the adsorption capacity was high in Cr(VI) (111.61 mg/g), and observed in Pb(II) (52.38 mg/g), Hg(II) (20.62 mg/g) and Cu(II) (5.64 mg/g). Also, the adsorption rate increased with pH, reaching a maximum at pH 6, and then decreased as pH increased further. This meant that the possible adsorption sites for complexation with metal ions on sawdust might be -COOH, -C_6H_5-OH, and amide groups in the treated sawdust.

Animal waste biomass.

There are several resources that are consumed in large quantities in animal food wastes as opposed to unutilized plant waste. Among them the world's egg supply was 92.66 million tons (Food and Agricultural Organization of the United Nations, 2020) and most of the eggshells were waste. But, it was reported that the crushed eggshells presented relatively high sorption capacity to Cr(III) ions (Chojnacka, 2005). The dried eggshells consist of approximately 85-95% calcium carbonate, 1.4% magnesium carbonate, 0.76% phosphate, 4% organic matter, and trace amounts of sodium, potassium, zinc, manganese, iron, and copper (Schaafsma et al, 2000; Daengprok et al., 2002). The efficient sorbent experimentally determined for maximum sorption capacities range from 21 to 160 mg/g, depended on pH and temperature. Experimentally determined maximum sorption capacities were obtained at 20 oC, pH 5.

For another animal waste, salmon milt is interesting as an environmentally friendly biomaterial as well as a recycled waste product from the fishing industry. The use of milt is a cost-effective way to recover rare earths and has great potential for the extraction of rare earths (REEs) from liquid ore waste (Takahashi et al, 2014). Adsorption of rare earths on milt was effective at an adsorption capacity of 1.04 mEq/g, with phosphate in milt showing sufficiently high affinity. In addition, the milt powder was used in batch-type REE recovery and separation experiments for lanthanides, excepting for both Y and Pm (total 15 elements mixed at 0.10 mg/L each REE; milt: 2.0 g/L) at pH 4 and 25°C for 1 hour. Batch experiments showed that more than 80% of all rare earth elements were adsorbed on the milt at pH 4, especially for Tm, Yb, and Lu. When the pH was adjusted to values between 2 and 4, less than 20% of the adsorbed rare earth elements remained in the milt solution, except for Yb and Lu.

DNA is a very interesting material extracted from marine wastes in marine products, since the extract is a relatively low-cost and environmentally friendly biomaterial. Adhered cellulose-DNA-filter hybrids were successfully used for separation and recovery of rare earth elements by a column method (Takahashi et al., 2012). The binding site of the rare earth elements was due to the presence of phosphoric acid group in the DNA. The reacting amines of DNA with *N,N'*-disuccinimidyl groups, bound in the filter paper as a cross-linking reagent. When DNA was fixed on the fibrous cellulose filter, it was able to sufficiently collect rare earths on the modified filter. The adsorption capacity of 0.182 mg/g of Nd was observed, meaning that the affinity of rare earths for DNA was stronger for the rare earths with larger atomic numbers. From the viewpoint of practical application, it was useful to be able to separate Nd and Fe(III) from a synthetic solution of Nd magnet waste using a column packed with DNA-filter hybrids.

Food wastes for bio-sorbents.

Pomegranate fruit is widely consumed fresh or as processed products such as juice, jam, and wine and pomegranate peel are by-products of the juice industry but, remain unused. There were studies using the peel for removal of lead (II) and copper (II) in aqueous solutions (El-Ashtoukhy et al., 2008). As obtained after firing peels, the activated carbon prepared from pomegranate peel and activated carbon was also tested. There were four types of materials induced the peel for raw peel, anaerobic firing, chemically treated peel and acidic treated one. The raw part was used as bio-sorption of heavy metals and in firing part, then, for the ample of anaerobic firing, the raw prepared was calcined in stainless-steel tube without air and placed in the center of a muffle furnace heated at 500°C for 1 hour. Furthermore, the activated carbon produced was cooled in a desiccator. The chemical treatment was carried out by immersing the peel in a solution of phosphoric acid and zinc chloride in a 1:1 ratio for 24 hours. After that, similar process under the anaerobic firing was taken place for drying and carbonization. In the case of chemically acidic treaded peel, immersing in nitric acid (10% wt) was implemented for 24 hours, then the sample was dried and carbonized in the same manner. In these samples, chemically treated peel was effective in the removal of Pb and Cu ions from aqueous solutions at the optimum pH around 5.6 and 5.8, respectively.

Chlorella (*Chlorella vulgaris*) is a unicellular green alga that is primarily marine but can also grow in freshwater. After the algae were grown, the cells were collected from the cultured cells by centrifugation. The recovered biomass pellets were washed with distilled water and centrifuged again. They were then dried at 50-60°C for 24 hours. The sorption capacity of the recovered biomass pellets was examined using different metals (Cu, Zn, Cd, and Ni) in both monometallic and binary metal solutions (Fraile et al., 2005). In the case of copper, pretreatment at acidic pH 3 was necessary for the alga sample. This was due to avoid subsequent Cu colloid precipitation in the removal test. In the pH effects the basic pH showed greater in the metal uptake. The maximum amounts of heavy metals adsorbed at equilibrium were 0.40, 0.40, 0.36, and 0.48 mmol/g for Cu, Cd, Ni and Zn, respectively, and exhibited very similar value for the four metals. The controlling pH also was significant and high recovery yields were obtained at the lowest pH of the eluent solution.

Coir, a fruit fiber obtained from the tropical coconut palm (Cocos nucifera), is inexpensive and abundantly available. The main components of coir are cellulose (36-43%), lignin (41-45%), and hemicellulose (18-20%), with pectin (2-3%) as a trace component. Batch adsorption of Cu(II), Pb(II), Ni(II), and Fe(II) ions on coconut coir was studied (Pushkar et. al, 2013). Alkali treatment (18% (w/v) NaOH) of coir fiber increased the metal adsorption capacity and adsorption rate. The maximum metal ion adsorption capacities of Cu(II), Pb(II), Ni(II), and Fe(II) increased about 3-fold to 9.43, 29.41, 8.84, and 11.11 mg/g, respectively, after alkaline treatment of coir fiber. Alkali treatment removed hemicellulose and others from the fibers, resulting in porous and increasing higher surface area of the coir fibers. After the treatments, the produced functional groups were available for metal ion adsorption. The pH of the solution played an important role in metal adsorption, and the optimum pH for maximum adsorption was different for each metal ion. Repeated desorption cycles showed that the alkali-treated coir fiber could be reused up to three times, with negligible loss of adsorption capacity.

Figure 3. Chemical modification of the materials with a specific dye, C.I. Reactive Orange 13

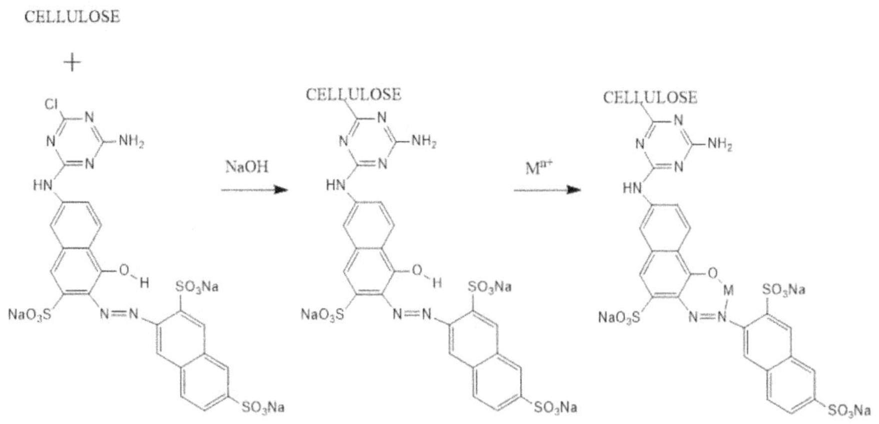

In the case of peanut shells, Cu(II), Ni(II), and Zn(II) adsorptions were examined (Shukla & Pai, 2005), showing good adsorption capacity and depending on the combination of adsorbent material and metal ions. Chemical modification of the materials with a specific dye, C.I. Reactive Orange 13 (Figure 3), furthermore enhanced the adsorption capacity due to the chelating effect of the dye. The maximum metal ion adsorption of Cu(II), Ni(II), and Zn(II) for the dye-loaded peanut shells were 7.60 mg/g, 7.49 mg/g, and 9.57 mg/g, respectively, and for the dye-unloaded peanut shells were 4.46 mg/g, 3.83 mg/g, and 7.62 mg /g, respectively. Similarly, the dye-loaded sawdust adsorbed 8.07 mg/g of Cu(II), 9.87 mg/g of Ni(II), and 17.09 mg/g of Zn(II) and the adsorption amount was higher than the values of 4.94 mg/g, 8.05 mg/g, and 10.96 mg/g, respectively, for sawdust adsorbent without the dye. A decrease in adsorption capacity was observed for all adsorbent materials when the pH of the metal cation aqueous solution was lowered. When the pH of the aqueous metal cation solutions was lowered to be pH 1.5, the adsorption capacities reached very low values in all cases. Even at low pH, the adsorption of metal ions on the dye-loaded adsorbent was relatively high. This was used in the desorption studies. Regeneration and reusability of the adsorbents were evaluated after three consecutive adsorption-desorption cycles, and it was found that the adsorption capacity was retained.

As well as peanut shells, Jute, whose main production areas are India and Bangladesh, is composed of wood fibers, has a reticular structure, and is traditionally used for packaging and inexpensive decorative purposes. For Cu(II), Ni(II) and Zn(II), jute lignocellulose fiber was evaluated (Shukla & Pai, Bioresource Technology, 2005) by dying with two different chemically modified fibers for loading of C.I. reactive orange 13, a dye with a specific structure, and oxidation with hydrogen peroxide (Figure 3). In the dying treatment process, hydrogen peroxide oxidizes the hydroxyl groups of the cellulose of the jute to carboxyl groups, making it a weak cationic ion exchanger. Thus, both modified jute fibers showed high metal ion adsorption with 8.4, 5.26, and 5.95 mg/g of Cu(II), Ni(II), and Zn(II), respectively, while the oxidized jute fiber adsorbed 7.73, 5.57, and 8.02 mg/g, as against 4.23, 3.37 and 3.55 mg/g for unmodified jute fibers. The adsorption amounts decreased with decreasing pH. The desorption efficiency, regenerative capacity, and recycling capacity of these adsorbents were evaluated in three consecutive adsorption-desorption cycles.

Figure 4. Natural cotton fiber modified with citric acid

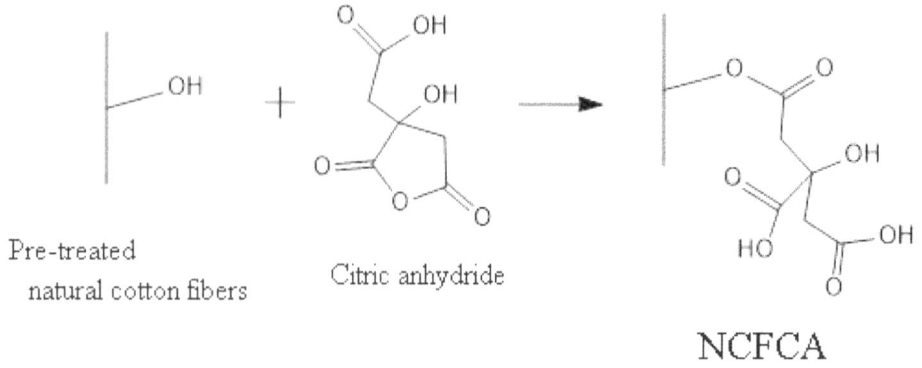

The adsorption of Cu(II), Zn(II), Cd(II), and Pb(II) ions by natural cotton fiber modified with citric acid was reported by using both batch and continuous systems (Paulino et al., 2013). The modified un-hydrated cottonwas obtained as shown in Figure 4, the adsorbent adsorbed metal ions within short time and maximum adsorption capacities were 6.12, 4.53, 8.22, and 21.62 mg/g for Cu(II), Zn(II), Cd(II), and Pb(II) ions at 25°C and pH 5, respectively. In a fixed-bed column packed with the modified cotton was able to efficiently remove heavy metals under flowing system for their metal remediation.

Chemical modified of okra biomass exhibited to remove Cu(II), Zn(II), Cd(II), and Pb(II) from aqueous solutions (Singha, & Guleria, 2015), when the biomass was treated by succinylation. The adsorption behavior followed Langmuir isotherm model and resultant q_m values for Cu(II), Zn(II), Cd(II) and Pb(II) metal ions were 72.72, 57.11, 121.51 and 273.97 mg/g, respectively.

Table 2. Removal of metal ions by using chitin and chitosan

Materials	Metals	Maximum Adsorption Capacity /mg g^{-1}	References
R_1= -H, R_2= -H Chitosan	Cd	5.93	Jha et al., 1988
	Cu	4.7	Ngah & Isa, 1998
R_1= -H, R_2= (2-pyridyl-methyl) N-(2-pyridyl-methyl)chitosan	Cu	104	Rodrigues et al., 1998.
R_1= -H, R_2= (4-pyridyl-methyl) N-(4-pyridyl-methyl)chitosan		54	
R_1= -H, R_2= -CH$_2$-(pyridyl) N-[3-(methylthio)propyl]chitosan	Hg	420	Baba et al., 1998
R_1= -H, R_2= GLA R_1= -CH$_2$-CH(OH)-CH$_2$-, R_2= -H ECH R_1= -H, R_2= -CH$_2$-CH$_2$-S-CH$_3$ EGDE	Cu	60 62 46	Ngah et al., 2002.
R_1= -CH$_2$-CH(OH)-(aminophenyl), R_2= -(CH$_2$CH$_2$CH$_3$)$_m$ APTMC	Pd Pt	304 569	Yamashita et al., 2015

Note. GLA: glutaraldehyde chitosan derivative, ECH: epichlorohydrine chitosan derivative, EGDE: ethylene glycol diglycidyl ether chitosan derivative, APTMC: aminophenyl thiomethyl chitosan

Relative to lingo cellulose waste, chitin $(C_8H_{13}O_5N)_n$ occurs in nature as a major component of the exoskeletons of arthropods such as crabs, lobsters, shrimp, and other crustaceans, insects, and mollusk caecilians and is biomass polymer with *N*-acetylglucosamine. A partial derivative of glucose group is obtained from chitin by treating the shells of crustaceans such as shrimp with the alkali sodium hydroxide (No et al., 1995). The polymer is characterized as being depletion-free, biodegradable, safe, and bio-compatible. In animal wastes, chitin and chitosan composed of crab cells were reported as bio-sorbents for elimination of metal ions. Table 2 lists examples of bio-sorbents of chitin and chitosan. Chitosan adsorbed Cd in batch experiments (Jha et al., 1988) and the adsorption capacity was 5.93 mg/g. The presence of ethylenediaminetetraacetic acid (EDTA) reduced the Cd adsorption. Crab cells adsorbed Cu(II) and Co(II) (Vijayaraghavan et al., 2006) for the acidic treated sample pretreated by washing with 0.1 M HCl for 4 h to remove $CaCO_3$. Then, the treated shells were washed with water and dried naturally. The biosorption capacity of crab shells for copper and cobalt depended on the particle size (0.456-1.117 mm), bio-sorbent dosage (1-10 g/L), initial metal concentration (500-2000 mg/L) and solution pH value (3.5-6). In batch mode, 243.9 mg/g and 322.6 mg/g were adsorbed for Cu(II) and Co(II), respectively, when the optimum particle size was 0.767 mm at pH 6. The presence of heavy metal cations of Ni(II), Cd(II), and Zn(II) caused desorption of Cu(II) and Co(II) from the modified crab shells. The presence of EDTA (pH 3.5, in HCl) worked well for the eluting agents of Cu(II) and Co(II) from crab shells, caused little bio-sorbent damage.

Because chitosan is easy to make many derivatives due to the presence of its amino group, many chitosan derivatives for recovering metal ions have been reported (Table 2). In an interesting application to bio-sorbents, the amino phenyl thiomethyl chitosan derivative was used to selectively separate precious metals such as Au(III), Pd(II), and Pt(IV) from base metal solutions (Yamashita et al., 2015). The modified chitosan adsorbed Au(III), Pd(II), and Pt(IV) depending on the HCl concentration, especially with increasing HCl. It is important to notice that unmodified chitosan itself is a unique cationic polysaccharide with very good complexing properties for a number of metal ions, but its sorption capacity and selectivity can be sufficiently increased by chemical modification and altered to meet specific application requirements. The review by Pestov et al. presents the results of studies on the binding of metal ions to chitosan and a very large number of chitosan derivatives (Pestov & Bratskaya, 2016).

REMOVAL OF METAL IONS BY BIOREMEDIATION

As mentioned in the previous section, bio-sorbents could efficiently remove heavy metals and could be adapted to many pollutants. These are mostly contaminated effluents. In contrast, physicochemical methods such as solidification/insolubilization, chemical oxidation, incineration, and washing are mainly used as methods for remediation of contaminated soil. While some metals at trace concentrations have important effects on biological health, high levels of heavy metals also represent toxicity (Ahemad, 2019; Ahuti, 2015). Furthermore, unlike organic contaminants, heavy metals are rarely degraded in soil. In this case, since purification costs are high for these (Atlas 2005), bioremediation technology, which is relatively inexpensive compared to conventional technology and is a detoxification treatment technology, is attracting attention (Table 3).

Table 3. Cost of soil remediation

Purification technology	Cost dollar/m^3
Thermal decomposition	178 - 715
Excavation and removal	53 - 134
Soil cleaning	26 - 71
Soil covering	26 - 62

continued on following page

Table 3. Continued

Purification technology	Cost dollar/m³
Containment	71 - 178
Chemical oxidation	71 - 152
Bioremediation	2 - 268

Bioremediation is an environmental restoration technology that utilizes living organisms such as microorganisms, plants, and animals. *In situ* bioremediation is a passive technology, generally requiring little equipment installation, except in some cases where elaborate irrigation systems are required, and implementation costs are generally low compared to other more aggressive technologies.

The following are the advantages of Bioremediation.

(1) Low cost.
(2) Possible to clean up contamination over a wide area.
(3) In-situ and in-operation remediation is possible.
(4) Low energy requirements due to normal temperature and pressure.
(5) Low environmental impact from remediation

Bioremediation of heavy metals using various microorganisms such as bacteria, microalgae, yeast and fungi was widely applied as an alternative to traditional methods due to its environmental friendliness and cost-effectiveness at low heavy metal concentrations.

In particular, phytoremediation using plants is one of the most promising soil remediation technologies because it is low-cost, has almost no environmental impact such as CO_2 emissions. In addition, such plant inevitably uses solar energy, and is effective in preventing the spread of contaminated soil and in afforestation (Anderson, 1993; Cunningham & Berti, 1993; Fiorenza et al., 2000; Li & Zang, 2003). The use of microalgae for phytoremediation of heavy metals is gaining momentum due to their abundant availability, low cost, excellent metal removal efficiency, and environmental friendliness.

Phytoremediation processes.

More than 500 plant species, or about 0.2% of all angiosperms, can accumulate heavy metals (Sarma, H., 2011). Hyperaccumulators, in particular, are very important in the phytoremediation of heavy metals. Hyperaccumulators are generally defined as plants that accumulate more than 1000 mg/kg of Cu, Co, Cr, Ni, and Pb and 1000 mg/kg of Fe, Mn, and Zn in shoot dry matter (Baker & Brooks, 1989: Market, 2003). The phytoremediation processes utilize plants to remove or detoxify contaminants and metals from soil and water and essentially the plant utilization decontaminates or stabilizes contaminated areas near the plants. In many studies on hyperaccumulators, about 75% of them was based on Ni accumulation, and other metals such as Zn, Cd, Cu, Pb, Co, Tl, Mn, Al, As, and Se were also hyperaccumulated (Brooks, 1998; Baker et al., 2000; Jansen et al., 2003). In New Zealand, treatment with ammonium thiocyanate of gold from ores with high gold yields ranged from 9.27 to 19.34 μg/g (Anderson et al., 1998). The cultivation of Brassicaceae plants, specifically Alyssum, allows these metals to be recovered from the soil from soils containing nickel/cobalt (Chaney et al., 1998). Accumulation of nickel in aboveground tissues was achieved on the order of 2.5% or higher, and by harvesting, drying, and subsequently burning the aboveground plants. Through these processes, the metal was recovered in concentrations 10 to 20 times higher than in the soil.

Copper roofs of Buddhist temples in Asian countries reported corrosion product by forming copper oxide, which inhibits most of the moss growth as the roof due to exposing to washed rain (Nomura et al., 2015; Shaw, 1994). Scopelophila cataractae (Mitt.) Broth is a copper moss, often found in copper-enriched environments, but unable to reproduce under normal conditions in nature. The moss Scopelophila cataractae (Mitt.) Broth is a typical hyperaccumulator and is suitable for phytoremediation of heavy metal-contaminated soils, which usually found copper-rich. Such polluted environments accumulate high levels of copper in the moss tissues. Quantitatively on concentrations of copper and 15 other elements (Na, Mg, Al, P, K, Ca, Ti, V, Cr, Mn, Fe, Co, Ni, Zn, and Pb) in copper-rich samples, copper concentrations in rainwater were detected in the order of ppm, while copper concentration in shoots was as high as 1-3% dry weight (Satake et al., 1988). The cell wall was particularly important in accumulation site for Cu. Thus, this moss species could be recognized as a hyperaccumulator (Rascio & Navari-Izzo, 2011).

Athyrium yokoscense is a member of the fern family Athyaceae, which is known as Hebino-negoza in Japan. It is metal tolerant and grows even in soil contaminated with high concentrations of heavy metals such as zinc, cadmium, lead, and copper (Morishita & Boratynski, 1992) and Athyrium yokoscense showed the highest cadmium concentrations in leaf fragments in the range of 451-996 mg/kg. Arabis geminifera had the second highest Cd accumulation, with a concentration of 94.8 mg/kg in leaves. Also, Arabis geminifera had the second highest Cd accumulation, with a leaf concentration of 94.8 mg/kg and *Miscanthus ssp.* accumulated in leaves 1.75 mg/kg with lower than others.

Among plants sampled from 17 locations in their native Japan provenance, *Chengiopanax sciadophylloides* Franch. et Sav. had a maximum foliar Mn concentration of 23,000 mg/kg dry weight (Mizunoa et al., 2013). Mn concentrations in soil were low to normal and a strong positive correlation was observed between foliar Mn concentrations and calcium (Ca) concentrations. Half of the foliar Mn was readily extractable with water, and almost all Mn was extractable with HCl. Furthermore, as recovery, it was known that high purity Mn compounds were precipitated from aqueous leaf ash solutions when pH was adjusted to 8-10.

Three wild aquatic macrophytes of the genera Ipomea, Eclipta, and Marsilea accumulated several heavy metal ions such as Fe, Cr, Mn, Zn, Cu, and Cd in their roots and shoots (Gupta et al., 2008).

The pattern of metal accumulation in all three species was in the order Fe > Cr > Mn > Zn > Cu > Cd. The metal accumulation capacity of the genus Marsilea was inferior to that of the genera Eclipta and Ipomea. The heavy metal concentrations in roots, stems, leaves, and fruits of these three species indicated that the soil-to-plant part concentration ratio (TFS) is important for selecting plant species for phytoremediation of soil contaminated with high levels of heavy metals. Thus, values of TFS> 1 mean higher metal accumulation in the plant body than in the soil (Barman et al., 2000). Ipomea sp. showed TFS$\geq$1 for Cd, Cu, Mn, and Zn, while Eclipta sp. and Marsilea sp. did TFS$\geq$1 for Fe, Cu, and Cd.

Water spinach (Ipomoea aquatica) and okra (Abelmoschus esculentus) are tropical vegetables found in equatorial regions. The metal tolerance and metal ion accumulation characteristics of water spinach and okra were investigated under different contaminated soils: 50 mg Pb/kg soil, 50 mg Zn/kg soil, and 50 mg Cu/kg soil (Ng et al., 2016). The concentration and accumulation of Pb, Zn, and Cu metal ions from soil to roots and shoots (edible parts) showed the highest accumulation of Pb in the roots of okra (80.2 mg/kg), followed by Zn in the roots (35.7 mg/kg) and shoots (34.8 mg/kg) of water spinach mustard. Different accumulation orders were observed for okra and water spinach: Pb > Zn > Cu in okra and Zn

> Pb > Cu in water spinach. Both water spinach and okra have excellent potential as phytoremediators of Pb and Zn.

Biosorption, which is adsorption process of using biological functions, has adsorption ability of metals using biological functions concerning with actions of various types of proteins and polysaccharides which exist in the cell wall and cell membrane. Here, the membrane separates the inside and outside of cells, and many functional groups such as phosphate, sulfonate, and carboxyl groups exist on them. Since these functional groups have negative charges, they can electrostatically adsorb positively charged metal ions (Tsezos, 2013) in Figure 5. When used microorganisms as adsorbents, the cells can naturally multiply and replicate adsorbents for metal ions simply by providing nutrients, so biosorption is expected to be a low-cost and environmentally friendly metal recovery system.

Figure 5. Biosorption of metal ions on cell surface

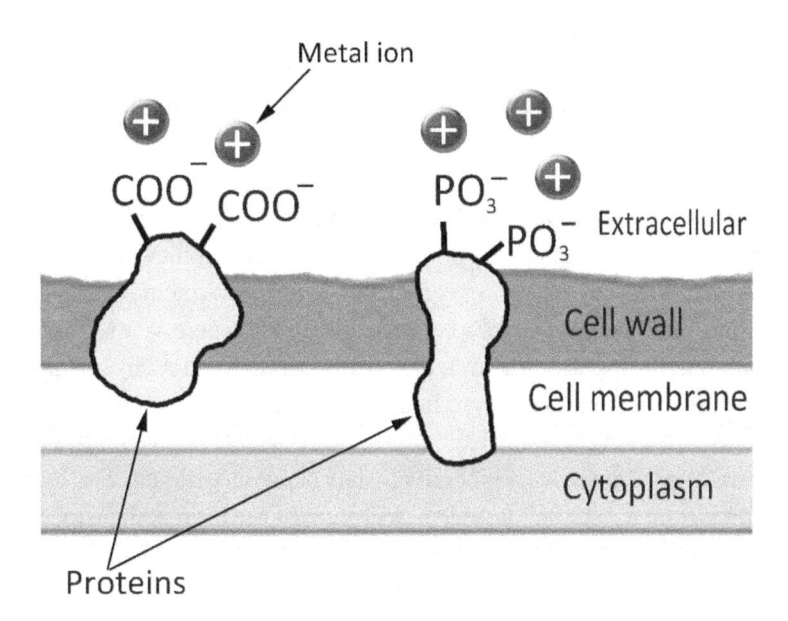

The biosorption behavior of Cr, Cd, Co, Fe, and As by Chlorella colonialis algae was examined (Jaafari & Yaghmaeian, 2019) and the heavy metals Cr, Cd, Co, Fe, and As by Chlorella colonialis algae increased as reaction time increased from 30 to 100 hours and then decreased steadily. The biosorption capacity of the Chlorella colonials increased when the initial Cd concentration was increased from 5 mg/L to 12 mg/L. At low concentrations of heavy metals, Chlorella colonialis was effective in bioaccumulating Cr, Co, Fe, and As, but at higher concentrations, the bioaccumulation efficiency decreased; the maximum removal efficiencies for Cr, Cd, Co, Fe, and As were 97.8, 97.05, 95.15, 98.6, and 96.5%, respectively, suggesting good alternative to current expensive methods of removing heavy metals from aqueous solutions.

Chromium is widely used in chromium plating because of its superior luster, hardness, and corrosion resistance. Alloys containing more than 10.5% iron, nickel, and chromium (ferrochrome) are called stainless steels. Since chromium forms a passive film, it is virtually rust-free and has a wide range of applications from heavy industrial products such as automobiles and machinery to kitchen utensils such as sinks and knives. However, chromium metal and Cr(III) are weakly toxic, but Cr(VI) are highly toxic. In the past, Cr(VI) was often used for plating, but this was often considered as problem, causing soil contamination. Although the use of chromium has decreased, plating solutions using Cr(VI) are still the mainstream for chrome plating at this time. The main mechanisms of the remediation include bio-adsorption, bioaccumulation, complexation, electrostatic attraction, reduction of Cr(VI) to Cr(III), and ion exchange, which reduce Cr(VI) concentrations and convert Cr(VI) to Cr(III), making them less toxicity (Shaopan, et al.,2019). Microbial reduction of Cr(VI) to Cr(III) was first reported in the late 1970s (Romanenko & Koren'kov, 1977), when Pseudomonas dechromaticans, isolated from sewage sludge, presented to be capable of reducing Cr(VI) under anaerobic conditions.

Several bacterial strains, Arthrobacter sp. (Rosaleset, al.,2012), Intrasporangium sp. (Liu et al., 2012), Microbacterium sp. (Soni et al., 2014), Bacillus TE9 and TE5 (Thilakarathne & Rathnayake, 2016), Pseudomonas GT2, GT3, and GT7 (Zhang et al., 2016), Pseudomonas putida (Kamran et al., 2017), Sporosarcina saromensis (Zhao et al., 2016), Bacillus sp. MNU16 (Upadhyay et al., 2017), Bacillus sp. FY1 (Xiao et al., 2017), Cellulosimicrobium sp. (Bharagava & Mishra, 2018), and Serratia sp. S2 (He et al., 2018), can be used to purify soil and wastewater contaminated with Cr(VI) by reducing Cr(VI) to Cr(III).

A method for stable purification of mine wastewater containing heavy metals using the activity of sulfate-reducing bacteria fed by rice bran as a nutrient source was reported (Sato et. al., 2020). Acidic mine wastewater containing heavy metals may be generated at former mine sites. At such sites, wastewater treatment continues in order to prevent adverse effects on the environment. Generally, mine wastewater is neutralized using specialized equipment and chemicals, but in recent years, attention has focused on low-cost, low environmental impact treatment technologies that utilize microbial activity. The development of equipment that precipitates and removes heavy metals through the action of *Deltaproteobacteria*, sulfate-reducing bacteria, utilizing rice husks and rice bran, which are agricultural wastes, as carriers and nutrient sources for microorganisms, respectively, has been undertaken. The coexistence of sulfate-reducing bacteria is important for stable long-term operation of the bioreactor. Since this technology can purify wastewater containing heavy metals at low cost and with low environmental impact, it is expected to be applied not only to mine wastewater but also to industrial wastewater.

Figure 6. EPR analysis using (a) existence process of Cu(II) ions incorporated within the microorganisms for EPR analysis and (b) EPR spectrum of Rhodotorula rubra

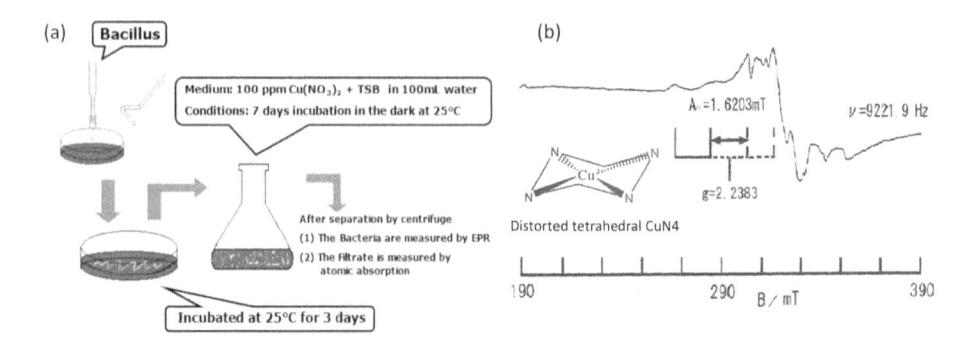

Performing bioremediation with using microorganisms from aqueous solutions containing Cu(II) ions, the mode of existence of Cu(II) ions incorporated within the microorganisms was investigated using EPR analysis (Ahiko & Satoh, our unpublished data). Each yeast was inoculated with 100 ppm Cu and Tryptic Soy Broth (TSB) as a medium and incubated at 25°C in the dark for 7 days (Figure 6 (a)). After centrifugation, the bacteria were measured for EPR in a frozen solution state at 77 K, and the filtrate was measured for atomic absorption spectrophotometry. The EPR spectrum of *Rhodotorula rubra* after culture shows in Figure 6 (b) having typical EPR pattern for Cu(II) ions. The EPR parameters ($g_{//}$ and $A_{//}$) are summarized in Table 4 along with data for other bacteria. These parameters revealed the coordination environment of the Cu(II) ion, which was found to be a distorted tetrahedral CuN4 (Yokoi, 1974; Yokoi & Addison, 1977). The EPR spectra of only the medium containing Cu(II) have similar EPR parameters, indicating that Cu(II) is present in the bacteria while maintaining the same state as in the medium. As shown in Table 4, some bacteria are not tolerant to Cu(II) ions, some are tolerant but do not absorb much Cu(II) ions, and some absorb Cu(II) ions. The Cu(II) ion exists in the bacteria that absorb Cu(II) ions is a distorted tetrahedral CuN4 having chemical structure in Figure 6 (b).

Table 4. EPR parameters of yeast fungus

Bacteria	$g_{//}$	$A_{//} / 10^{-2} cm^{-1}$	Existence form of Cu(II)	Cu(II) absorption rate / %
Candida utilis	2.2426	1.5299	distorted tetrahedral CuN4	4.8
Rhodotorula rubra	2.2383	1.6203	distorted tetrahedral CuN4	17.8
Cryptococcus albidus	——	——	EPR not observed	0

As described above, the detailed mechanism of bioremediation will be elucidated in the process of heavy metal adsorption in the future. Yoong et al. have compiled a review article that aims to provide useful insights for the future development of efficient and commercially viable technologies for microalgae-based heavy metal bioremediation (Yoong, 2020).

CONCLUSION

Mineral resources, each with its own unique characteristics, are indispensable materials for industry. For example, in automobiles, many mineral resources are used in every part of the car, from the body to the motor and battery inside. Today, industrial products would not be possible without mineral resources. From the perspective of building a recycling-oriented society, it is becoming increasingly important to (1) save resources and develop alternative materials and (2) recycle mineral resources. Technologies for recovering mineral resources from discarded products and for restoring the pollution that occurs in the refining process of mineral resources are issues that scientists and engineers must address. For example, the Ashio Copper Mine poisoning incident in Japan was an environmental disaster caused by the Ashio Mine operations in the late 19th and early 20th centuries. In emerging countries, such problems are still being reported in developing countries.

We must take these problems into account when developing technologies. The removal of metals using environmentally friendly biomass and the remediation of contaminated soil and water through bioremediation are sustainable and will become increasingly important in the future.

REFERENCES

Ahemad, M. (2019). Remediation of metalliferous soils through the heavy metal resistant plant growth promoting bacteria: Paradigms and prospects. *Arabian Journal of Chemistry*, 12(7), 1365–1377. DOI: 10.1016/j.arabjc.2014.11.020

Ahuti, S. (2015). Industrial growth and environmental degradation. *International Education and Research Journal*, 1(5), 5–7.

Amarasinghe, B. M. W. P. K., & Williams, R. A. (2007). Tea Waste as a Low Cost Adsorbent for the Removal of Cu and Pb from Wastewater. *Chemical Engineering Journal*, 132(1-3), 299–309. DOI: 10.1016/j.cej.2007.01.016

Anderson, C. W. N., Brooks, R. R., Stewart, R. B., & Simcock, R. (1998). Harvesting a crop of gold in plants. *Nature*, 395(6702), 553–554. DOI: 10.1038/26875

Anderson, T. A., Guthrie, E. A., & Walton, B. T. (1993). Bioremediation in the rhizosphere. *Environmental Science & Technology*, 27(13), 2630–2635. DOI: 10.1021/es00049a001

Asha, L. P., & Sandeep, R. S. (2013). Review on bioremediation-potential tool for removing environmental pollution. *Int J Basic Appl Chem Sci*, 3(3), 21–33.

Atlas, R. M., & Philp, J. (2005). *Bioremediation*. ASM Press., DOI: 10.1128/9781555817596

Baba, Y., Kubota, F., Kamiya, N., & Goto, M. (2011). Selective recovery of dysprosium and neodymium ions by a supported liquid membrane based on ionic liquids. *Solvent Extraction Research and Development, Japan*, 18(0), 193–198. DOI: 10.15261/serdj.18.193

Baker, A. J. M., & Brooks, R. R. (1989). Terrestrial higher plants which hyperaccumulate metallic elements—A review of their distribution, ecology and phytochemistry. *Biorecovery*, 1, 81–126.

Baker, A. J. M., McGrath, S. P., Reeves, R. D., & Smith, J. A. C. (2000). Metal hyperaccumulator plants: A review of the ecology and physiology of a biological resource for phytoremediation of metal-polluted soils. In Terry, N., & Bañuelos, G. (Eds.), *Phytoremediation of Contaminated Soil and Water* (pp. 85–108). CRC Press.

Barman, S. C., Sahu, R. K., Bhargava, S. K., & Chatterjee, C. (2000). Distribution of heavy metals in wheat, mustard, and weed grown in fields irrigated with industrial effluents. *Bulletin of Environmental Contamination and Toxicology*, 64(4), 489–496. DOI: 10.1007/s001280000030 PMID: 10754044

Basci, N., Kocadagistan, E., & Kocadagistan, B. (2004). Biosorption of copper (II) from aqueous solutions by wheat shell. *Desalination*, 164(2), 1, 135–1400. DOI: 10.1016/S0011-9164(04)00172-9

Bharagava, R. N., & Mishra, S. (2018). Hexavalent chromium reduction potential of Cellulosimicrobium sp. isolated from common effluent treatment plant of tannery industries. *Ecotoxicology and Environmental Safety*, 147, 102–109. DOI: 10.1016/j.ecoenv.2017.08.040 PMID: 28841524

Chaney, R. L., Angle, R. S., Baker, A. J. M., & Li, Y. M. (1998). Method for phytomining of nickel cobalt and other metals from soil. *U.S. Patent No. 5711784*.

Chojnacka, K. (2005). Biosorption of Cr(III) ions by eggshells. *Journal of Hazardous Materials*, 121(1-3), 167–173. DOI: 10.1016/j.jhazmat.2005.02.004 PMID: 15885418

Cunningham, S. D., & Berti, W. R. (1993). Phytoremediation of contaminated soils: Progress and Promise. 205th National Meeting, *American Chemical Society*, 265-268.

Daengprok, W., Garnjanagoonchorn, W., & Mine, Y. (2002). Fermented pork sausage fortified with commercial or hen eggshell calcium lactate. *Meat Science*, 62(2), 199–204. DOI: 10.1016/S0309-1740(01)00247-9 PMID: 22061412

Dang, V. B. H., Doan, H. D., Dang-Vu, T., & Lohi, A. (2009). Equilibrium and kinetics of biosorption of cadmium(II) and copper(II) ions by wheat straw. *Bioresource Technology*, 100(1), 211–219. DOI: 10.1016/j.biortech.2008.05.031 PMID: 18599289

Duruibe, J. O., Ogwuegbu, M. O. C., & Egwuruhwu, J. N. (2007). Heavy metal pollution and human biotoxic effect. *International Journal of Physical Sciences*, 2(5), 112–118.

El-Ashtoukhy, E. S., Amin, N. K., & Abdelwahab, O. (2008). Removal of Lead(II) and Copper(II) from Aqueous Solution Using Pomegranate Peel as a New Adsorbent. *Desalination*, 223(1-3), 162–173. DOI: 10.1016/j.desal.2007.01.206

Elham, A., Hossein, T., & Mahnoosh, H. (2010). Removal of Zn(II) and Pb(II) Ions Using Rice Husk in Food Industrial Wastewater. *Journal of Applied Science & Environmental Management*, 14(4), 159–162. DOI: 10.4314/jasem.v14i4.63306

Fiorenza, S., Oubre, C. L., & Ward, C. H. (2000). *Phytoremediation of Hydrocarbon-contaminated Soil*. CRC Press LLC.

Food and Agricultural Organization of the United Nations, *Food Balances*, (2020).

Fraile, A., Penche, S., Gonzalez, F., Blazquez, M. L., Munoz, J. A., & Ballester, A. (2005). Biosorption of Copper, Zinc, Cadmium and Nickel by Chlorella vulgaris. *Chemistry and Ecology*, 21(1), 61–75. DOI: 10.1080/02757540512331334933

Gardea-Torresdey, J. L., Tang, L., & Salvador, J. M. (1995). Copper Adosoption by Sphagnum Peat Moss and Its Different Humic Fractions. *Proceedings of the 10th Annual Conference on Hazardous Waste Research*, 249-260.

Grassi, D. A., Galicio, M., & Fernandez Cirelli, A. (2011). A Homogeneous and Low-Cost Biosorbent for Cd, Pb and Cu Removal from Aqueous Effluents. *Chemistry and Ecology*, 27(4), 297–309. DOI: 10.1080/02757540.2011.565750

Gupta, S., Nayek, S., Saha, R. N., & Satpati, S. (2008). Assessment of heavy metal accumulation in macrophyte, agricultural soil, and crop plants adjacent to discharge zone of sponge iron factory. *Environmental Geology (Berlin)*, 55(4), 731–739. DOI: 10.1007/s00254-007-1025-y

He, Y., Dong, L. L., Zhou, S. M., Jia, Y., Gu, R. J., Bai, Q. H., Gao, J., Li, J., & Xiao, H. (2018). Chromium resistance characteristics of Cr(VI) resistance genes ChrA and ChrB in Serratia sp. S2. *Ecotoxicology and Environmental Safety*, 157, 417–423. DOI: 10.1016/j.ecoenv.2018.03.079 PMID: 29655157

Jaafari, J., & Yaghmaeian, K. (2019). Optimization of heavy metal biosorption onto fresh water algae (Chlorella coloniales) using response surface methodology (RSM). *Chemosphere*, 217, 447–455. DOI: 10.1016/j.chemosphere.2018.10.205 PMID: 30439657

Jansen, S., Broadley, M. R., Robbrecht, E., & Smets, E. (2003). Aluminium hyperaccumulation in angiosperms: A review of its phylogenetic significance. *Botanical Review*, 68(2), 235–269. DOI: 10.1663/0006-8101(2002)068[0235:AHIAAR]2.0.CO;2

Jha, I. N., Iyengar, L., & Rao, Λ. V. S. P. (1988). Removal of Cadmium Using Chitosan. *Journal of Environmental Engineering*, 114(4), 962–974. DOI: 10.1061/(ASCE)0733-9372(1988)114:4(962)

Kaikake, K., Hoaki, K., Sunada, H., Dhakal, R. P., & Baba, Y. (2007). Removal characteristics of metal ions using degreased coffee beans: Adsorption equilibrium of cadmium(II). *Bioresource Technology*, 98(15), 2787–2791. DOI: 10.1016/j.biortech.2006.02.040 PMID: 17400448

Kamran, M. A., Bibi, S., Xu, R. K., Hussain, S., Mehmood, K., & Chaudhary, H. J. (2017). Phyto-extraction of chromium and influence of plant growth promoting bacteria to enhance plant growth. *Journal of Geochemical Exploration*, 182, 269–274. DOI: 10.1016/j.gexplo.2016.09.005

Lee, S.-J., Park, J. H., Ahn, Y.-T., & Chung, J. W. (2015, February). Comparison of Heavy Metal Adsorption by Peat Moss and Peat Moss-Derived Biochar Produced Under Different Carbonization Conditions. *Water, Air, and Soil Pollution*, 226(2), 9. DOI: 10.1007/s11270-014-2275-4

Leong, Y. K., & Chang, J.-S. (2020). Bioremediation of heavy metals using microalgae: Recent advances and Mechanisms. *Bioresource Technology*, 303, 122886. DOI: 10.1016/j.biortech.2020.122886 PMID: 32046940

Li, F., Zang, S., & Yi, L. (2003). Bioremediation of contaminated soil. *Shengtaixue Zazhi*, (1), 35–39. PMID: 12916207

Liu, H., Wang, H., & Wang, G. (2012). Intrasporangium chromatireducens sp. nov., a chromate-reducing actinobacterium isolated from manganese mining soil, and emended description of the genus Intrasporangium. *Journal of Hazardous Materials*, 62(2), 403–408. DOI: 10.1099/ijs.0.030528-0 PMID: 21441371

Meena, A. K., Kadirvelu, K., Mishra, G. K., Rajagopal, C., & Nagar, P. N. (2008). Adsorptive Removal of Heavy Metals from Aqueous Solution by Treated Sawdust (Acacia arabica). *Journal of Hazardous Materials*, 150(3), 604–611. DOI: 10.1016/j.jhazmat.2007.05.030 PMID: 17600619

Mizunoa, T., Emorib, K., & Itoa, S. (2013). Manganese hyperaccumulation from non-contaminated soil in Chengiopanax sciadophylloides Franch. et Sav. and its correlation with calcium accumulation. *Soil Science and Plant Nutrition*, 59(4), 591–602. DOI: 10.1080/00380768.2013.807213

Morishita, T., & Boratynski, J. K. (1992). Accumulation of cadmium and other metals in organs of plants growing around metal smelters in Japan. *Soil Science and Plant Nutrition*, 38(4), 781–785. DOI: 10.1080/00380768.1992.10416712

Ng, C. C., Rahman, M. M., Boyce, A. N., & Abas, M. R. (2016). Heavy metals phyto-assessment in commonly grown vegetables: Water spinach (I. aquatica) and okra (A. esculentus). *SpringerPlus*, 5(1), 469. DOI: 10.1186/s40064-016-2125-5 PMID: 27119073

Ngah, W. S. W., & Isa, I. M. (1998). Comparison Study of Copper Ion Adsorption on Chitosan, Dowex A-1 and Zerolit 225. *Journal of Applied Polymer Science*, 67(6), 1067–1070. DOI: 10.1002/(SICI)1097-4628(19980207)67:6<1067::AID-APP14>3.0.CO;2-Y

Ngah, W. W. S., Endud, C. S., & Mayanar, R. (2002). Removal of copper(II) ions from aqueous solution onto chitosan and cross-linked chitosan beads. *Reactive & Functional Polymers*, 50(2), 181–190. DOI: 10.1016/S1381-5148(01)00113-4

No, H. K., & Meyers, S. P. (1995). Preparation and Characterization of Chitin and Chitosan – A review. *Journal of Aquatic Food Product Technology*, 4(2), 27–52. DOI: 10.1300/J030v04n02_03

Nomura, T., Itouga, M., Kojima, M., Kato, Y., Sakakibara, H., & Hasezawa, S. (2015). Copper mediates auxin signalling to control cell differentiation in the copper moss Scopelophila cataractae. *Journal of Experimental Botany*, 66(5), 1205–1213. DOI: 10.1093/jxb/eru470 PMID: 25428998

Paulino, A. L. G., Cunha, A. J. D., Alfaya, R. V. D. S., & Alfaya, A. A. D. S. (2013). Chemically Modified Natural Cotton Fiber: A Low-Cost Biosorbent for the Removal of the Cu(II), Zn(II), Cd(II), and Pb(II) from Natural Water. *Desalination and Water Treatment*, •••, 1–11. DOI: 10.1080/19443994.2013.804451

Pestov, A., & Bratskaya, S. (2016). Chitosan and Its Derivatives as Highly Efficient Polymer Ligands. *Molecules (Basel, Switzerland)*, 21(3), 330. DOI: 10.3390/molecules21030330 PMID: 26978343

Peterson, P. J. (1983). Adaptation to toxic metals. In Robb, D. A., & Pierpoint, W. S. (Eds.), *Metals and Micronutrients* (pp. 51–69). Academic Press. DOI: 10.1016/B978-0-12-589580-4.50010-9

Pushkar, M., Shukla, P. M., & Shukla, S. R. (2013). Biosorption of Cu(II), Pb(II), Ni(II), and Fe(II) on Alkali Treated Coir Fibers. *Separation Science and Technology*, 48(3), 421–428. DOI: 10.1080/01496395.2012.691933

Putra, W. P., Kamari, A., Yusoff, S. N. M., Ishak, C. F., Mohamed, A., Hashim, N., & Isa, I. M. (2014). Biosorption of Cu(II), Pb(II) and Zn(II) Ions from Aqueous Solutions Using Selected Waste Materials: Adsorption and Characterisation Studies. *Journal of Encapsulation and Adsorption Sciences*, 4(1), 25–35. DOI: 10.4236/jeas.2014.41004

Raghunandan K, Kumar A, Kumar S, Permaul K, Singh S (2018). Production of gellan gum, an exopolysaccharide, from biodiesel-derived waste glycerol by Sphingomonas spp. *3 Biotech* 8(1):71 https://doi.org/DOI: 10.1007/s13205-018-1096-3

Raghunandan, K., Mchunu, S., Kumar, A., Kumar, K. S., Govender, A., Permaul, K., & Singh, S. (2014). Biodegradation of glycerol using bacterial isolates from soil. *Journal of Environmental Science and Health. Part A, Toxic/Hazardous Substances & Environmental Engineering*, 49(1), 85–92. DOI: 10.1080/10934529.2013.824733 PMID: 24117087

Rascio, N., & Navari-Izzo, F. (2011). Heavy metal hyperaccumulating plants: How and why do they do it? And what makes them so interesting? *Plant Science*, 180(2), 169–181. DOI: 10.1016/j.plantsci.2010.08.016 PMID: 21421358

Rodrigues, C. A., Laranjeira, M. C. M., de Fávere, V. T., & Stadler, E. (1998). Interaction of Cu(II) on N-(2-pyridylmethyl) and N-(4-pyridylmethyl) chitosan. *Polymer*, 39(21), 5121–5126. DOI: 10.1016/S0032-3861(97)10190-2

Romanenko, V. I., & Koren'kov, V. N. (1977). Pure culture of bacteria using chromates and bichromates as hydrogen acceptors during development under anaerobic conditions. *Mikrobiologiia*, 46, 414–417. PMID: 895551

Rosales, E., Pazos, M., Sanroman, M. A., & Tavares, T. (2012). Application of zeolite Arthrobacter viscosus system for the removal of heavy metal and dye: Chromium and Azure B. *Desalination*, 284, 150–156. DOI: 10.1016/j.desal.2011.08.049

Saeed, A., Akhter, M. W., & Iqbal, M. (2005). Removal and recovery of heavy metals from aqueous solution using papaya wood as a new biosorbent. *Separation and Purification Technology*, 45(1), 25–31. DOI: 10.1016/j.seppur.2005.02.004

Saeed, A., Iqbal, M., & Akhtar, M. W. (2005). Removal and recovery of lead (II) from single and multimetal (Cd, Cu, Ni, Zn) solutions by crop milling waste (black gram husk). *Journal of Hazardous Materials*, 117(1), 65–73. DOI: 10.1016/j.jhazmat.2004.09.008 PMID: 15621354

Sarma, H. (2011). Metal hyperaccumuulation in plants: A Review focusing on phytoremediation technology. *Journal of Environmental Science and Technology*, 4(2), 118–138. DOI: 10.3923/jest.2011.118.138

Satake, K., Shibata, K., Nishikawa, M., & Fuwa, K. (1988). Copper accumulation and location in the moss Scopelophila cataractae. *Journal of Bryology*, 15(2), 353–376. DOI: 10.1179/jbr.1988.15.2.353

Sato, Y., Hamai, T., Hori, T., Aoyagi, T., Inaba, T., Hayashi, K., Kobayashi, M., Sakata, T., Habe, H., (2022). Optimal start-up conditions for the efficient treatment of acid mine drainage using sulfate-reducing bioreactors based on physicochemical and microbiome analyses. *Journal of hazardous materials* 423(Pt B) 127089-127089. https://doi.org/DOI: 10.1016/j.jhazmat.2021.127089

Schaafsma, A., Pakan, I., Hofstede, G. J. H., Muskiet, F. A. J., Van Der Veer, E., & De Vries, P. J. F. (2000). Mineral, amino acid, and hormonal composition of chicken eggshell powder and the evaluation of its use in human nutrition. *Poultry Science*, 79(12), 1833–1838. DOI: 10.1093/ps/79.12.1833 PMID: 11194049

Shaopan, X., Zhaoliang, S., Paramsothy, J., Sabry, M. S., Jörg, R., Yong, S. O., Nanthi, B., & Hailong, W. (2019). A critical review on bioremediation technologies for Cr(VI)-contaminated soils and wastewater. *Critical Reviews in Environmental Science and Technology*, 49(12), 1027–1078. DOI: 10.1080/10643389.2018.1564526

Shaw, A. J. (1994). Adaptation to metals in widespread and endemic plants. *Environmental Health Perspectives*, 102(suppl 12), 105–108. DOI: 10.1289/ehp.94102s12105 PMID: 7713025

Shukla, S. R., & Pai, R. S. (2005). Adsorption of Cu(II), Ni(II) and Zn(II) on Dye Loaded Groundnut Shells and Sawdust. *Separation and Purification Technology*, 43(1), 1–8. DOI: 10.1016/j.seppur.2004.09.003

Shukla, S. R., & Pai, R. S. (2005). Adsorption of Cu(II), Ni(II) and Zn(II) on Modified Jute Fibres. *Bioresource Technology*, 96(13), 1430–1438. DOI: 10.1016/j.biortech.2004.12.010 PMID: 15939269

Singha, A. S., & Guleria, A. (2015). Utility of chemically modified agricultural waste okra biomass for removal of toxic heavy metal ions from aqueous solution. *Engineering in Agriculture, Environment and Food*, 8(1), 52–60. DOI: 10.1016/j.eaef.2014.08.001

Soni, S. K., Singh, R., Awasthi, A., & Kalra, A. (2014). A Cr (VI)-reducing Microbacterium sp. strain SUCR140 enhances growth and yield of Zea mays in Cr (VI) amended soil through reduced chromium toxicity and improves colonization of arbuscular mycorrhizal fungi. *Environmental Science and Pollution Research International*, 21(3), 1971–1979. DOI: 10.1007/s11356-013-2098-7 PMID: 24014225

Takahashi, Y., Kondo, K., Miyaji, A., Umeo, M., Honma, T., & Asaoka, S. (2012). Recovery and separation of rare earth elements using columns loaded with DNA-filter hybrid. *Analytical Sciences*, 28(10), 985–992. DOI: 10.2116/analsci.28.985 PMID: 23059995

Takahashi, Y., Kondo, K., Miyaji, A., Watanabe, Y., Fan, Q., Honma, T., & Tanaka, K. (2014). Recovery and Separation of Rare Earth Elements Using Salmon Milt. *PLoS One*, 9(12), e114858. Advance online publication. DOI: 10.1371/journal.pone.0114858 PMID: 25490035

Thilakarathne, P. R., & Rathnayake, I. V. N. (2016). Comparison of hexavelant chromium tolerance and removal capacity of two Bacillus species isolated from tannery effluent. *International Symposium on Water Quality and Human Health: Challenges Ahead*.

Tsezos, M. (2013). Biosorption: A Mechanistic Approach. In Schippers, A., Glombitza, F., Sand, W. (eds) *Geobiotechnology I. Advances in Biochemical Engineering/Biotechnology*, vol 141, 173–209, Springer. https://doi.org/DOI: 10.1007/10_2013_250

Upadhyay, N., Vishwakarma, K., Singh, J., Mishra, M., Kumar, V., Rani, R., Mishra, R. K., Chauhan, D. K., Tripathi, D. K., & Sharma, S. (2017). Tolerance and reduction of chromium(VI) by Bacillus sp. MNU16 isolated from contaminated coal mining soil. *Frontiers in Plant Science*, 8, 778. DOI: 10.3389/fpls.2017.00778 PMID: 28588589

USEPA. (2016). National Primary Drinking Water Regulations. in: EPA 816-F-09-004, (Ed.) U.S.E.P. Agency, United State Environmental Protection Agency. United State.

Vijayaraghavan, K., Palanivelu, K., & Velan, M. (2006). Biosorption of copper(II) and cobalt(II) from aqueous solutions by crab shell particles. *Bioresource Technology*, 97(12), 1411–1419. DOI: 10.1016/j.biortech.2005.07.001 PMID: 16112568

Xiao, W. D., Ye, X. Z., Yang, X. E., Zhu, Z. Q., Sun, C. X., Zhang, Q., & Xu, P. (2017). Isolation and characterization of chromium (VI)-reducing Bacillus sp. FY1 and Arthrobacter sp. WZ2 and their bioremediation potential. *Bioremediation Journal*, 21(2), 100–108. DOI: 10.1080/10889868.2017.1282939

Yamashita, A., Yoshida, K., Oshima, T., & Baba, Y. (2015). Synthesis of an Aminophenylthiomethyl Chitosan Derivative and Its Adsorption of Precious Metals. *Journal of Chemical Engineering of Japan*, 48(11), 897–902. DOI: 10.1252/jcej.15we025

Yang, D., Jing, D., Gong, H., Zhou, L., Yang, X. (2012). Biosorption of aquatic cadmium(II) by unmodified rice straw. *Bioresour Technol*, 114(none), 20–25. https://doi.org/DOI: 10.1016/j.biortech.2012.01.110

Yang, J., Cao, J., Xing, G., & Yuan, H. (2015). Lipid production combined with biosorption and bioaccumulation of cadmium, copper, manganese and zinc by oleaginous microalgae Chlorella minutissima UTEX2341. *Bioresource Technology*, 175, 537–544. DOI: 10.1016/j.biortech.2014.10.124 PMID: 25459865

Yokoi, H. (1974). ESR and Optical Absorption Studies of Various Bis(N-salicylidenealkylaminato) copper(II) Complexes with Tetrahedrallydistorted Coordination Geometry. *Bulletin of the Chemical Society of Japan*, 47(12), 3037–3040. DOI: 10.1246/bcsj.47.3037

Yokoi, H., & Addison, A. W. (1977). Spectroscopic and redox properties of pseudotetrahedral copper(II) complexes. Their relation to copper proteins. *Inorganic Chemistry*, 16(6), 1341–1349. DOI: 10.1021/ic50172a018

Zakhama, S., Dhaouadi, H., & M'Henni, F. (2011). Nonlinear modelisation of heavy metal removal from aqueous solution using Ulva lactuca algae. *Bioresource Technology*, 102(2), 786–796. DOI: 10.1016/j.biortech.2010.08.107 PMID: 20855200

Zhang, J. K., Wang, Z. H., & Ye, Y. (2016). Heavy metal resistances and chromium removal of a novel Cr(VI)-reducing Pseudomonad strain isolated from circulating cooling water of iron and steel plant. *Applied Biochemistry and Biotechnology*, 180(7), 1328–1344. DOI: 10.1007/s12010-016-2170-0 PMID: 27350052

Zhao, R., Wang, B., Cai, Q. T., Li, X. X., Liu, M., Hu, D., & Fan, C. (2016). Bioremediation of hexavalent chromium pollution by Sporosarcina saromensis M52 isolated from offshore sediments in Xiamen, China. *Biomedical and Environmental Sciences*, 29, 127–136. DOI: 10.3967/bes2016.014 PMID: 27003170

Chapter 16
Extracting Technology Upcycling Toward Useful Metallic Materials From Mineral Wastes and Pollutant Soil by Ultrasound Washing

Tri Phuoc Phan
JET Corporation, Japan

ABSTRACT

The increasing of industrial waste and soil contamination has raised so many concerns for the environment and the sustainable development of society. Recently, ultrasound (US) technology has emerged as the new tool for enhancing the recovery process of mineral waste and also for the soil washing process, making US highly efficient cleaning soil or extracting pollutants from waste soils. The unique and highly efficient US action is attributed to acoustic cavitation phenomena in ultrasound circumstance, cleaning and extraction can be more efficient than usual one. This chapter reviews the application of high-power US for eliminating pollutants from contaminant soil and US-extracting valuable metals from mineral wastes. Since US washing technology enhances the recovery process from unusual wastes, this also describes an upcycle technology that uses US cleaning to convert them into valuable resources.

INTRODUCTION

After several decades of rapid industrial development with population growth, The consumption of natural resources for the development of human society shows no sign of abating. The underutilization of limited natural wealth and the unabated expansion of industry are both at odds with each other, and the difficulty of balancing economic growth while coexisting with nature is a major challenge for human civilization. Against this background, the importance of building a recycling-oriented society has been increasingly recognized in recent years, and the recycling of waste, efficient purification of contaminated materials, and effective utilization of their residues should be promoted in the future. Thus, trends in

DOI: 10.4018/979-8-3693-0003-9.ch016

the last decade have seen a growing trend to reduce waste and pollution and to recycle used resources to extend their life in the economic cycle, especially metallic elements to help reduce the use of natural resources (Phiri et al., 2021; Tayebi-Khorami et al., 2019). Annually, industrial activities of countries around the world have produced millions of tons of liquid and solid waste, which contain many valuable metal elements that could be recovered inside the economic systems (Abdel-Shafy & Mansou, 2018; Krishnan et al, 2021; Phiri et al., 2021). Generally, the waste usually ends up in a landfill or incinerator, which not only causes the waste of land resources but also poses an environmental threat to the nearby environment due to the leaking of contaminants to the outside (Ettler et al., 2023; Nyirenda & Mwansa, 2022). In addition to ferrous metals, other precious metals such as copper (Cu), zinc (Zn), manganese (Mn), lead (Pb), titanium (Ti), chromium (Cr), aluminum (Al), silver (Ag), gold (Au) and platinum (Pt) are included in scrap (Cotty et al., 2023; Gorai et al., 2003a; Sun et al., 2017). Many disasters have been documented beginning with the Industrial Revolution and continuing after the Industrial Revolution due to lack of proper management and treatment. For example, the coal waste spill in England (Lacey, 1972), the fly ash slurry spill in the US (Korol et al., 2004), the Gold King mine wastewater spill (Clausen et al., 2023), the Minamata disaster in Japan (Murata & Sakamoto, 2013), the contaminating arsenic disaster in Bangladesh and India (Gupta & Chatterjee, 2017; Shrivastava et al., 2015). The areas affected by the disaster remained polluted for long periods, depending on the type of waste and the effort of the government to solve the problem. Those disasters caused thousands of deaths and affected millions of people's health.

On the other hand, there is a similar pattern of environmental disasters from over-accumulating waste generated during the mining or refining process without proper treatment. With the right treatment, these wastes could potentially become secondary sources of metal or urban mine, if people can successfully separate valuable metal from each source. In industrial waste from mining and refining industries, like tailing, low-grade ore, slag, anode slime, furthermore, the increase of the annual dispose of electronic waste also raising the same concern for recycling treatment to recover the precious metal in those waste for the future demand. In some cases, wastes generated during production are left without effective utilization methods and without being recovered. Industrial activities that produce wastes cause soil contamination in the surrounding areas, so these must also be cleaned up and the utilization of the residues must be considered. (Hu et al., 2018; V. Kumar et al., 2019; Tóth et al., 2016). However, in many cases, the final solution is landfill or other disposal.

Industrial Mining Wastes

The potential industrial wastes for recovery process are summarized in Figure 1. In the present landscape of industrial mining, encompassing approximately 3500 active mining sites, the mining sector processes an estimated 100 billion tons of solid waste and sludge annually. This waste includes tailing residues from ore processing, slag resulting from smelting byproducts, and other materials discarded during extraction and refining processes (Rampacek, 1982; Rankin, 2015; Tayebi-khorami et al., 2019). A notable incident occurred in 2019 when the tailing dam of the Brumadinho iron mine in Brazil suffered a catastrophic failure, leading to the loss of more than 250 lives (Silva Rotta et al., 2020), drawing global attention to the peril of poorly managed mining waste. Copper, recognized as one of the most versatile metals globally, has been mined and refined since ancient times. With a copper slag production rate of 2.2 per ton of refined copper smelting (Gorai et al., 2003a), the improper management and treatment of copper slag could pose substantial environmental risks. Although recycling technologies emerged to mitigate these concerns, some waste materials have been repurposed for secondary applications, such as in construction materials like sand replacements for cement, granules, and pavement (Gorai et al., 2003a; Soni et al., 2022; Yaswanth et al., 2022). Moreover, copper slag also contains trace amounts of valuable metals like Zn, Cu, and Pb that could be targeted through chemical leaching processes (Khalid et al., 2019). Recently, despite the great comfort and convenience brought to our daily life, the waste coming from electrical and electronic equipment (EEE) is raising a lot of concern internationally. There are e-waste sources coming from daily life equipment such as smartphones, washing machines, television, personal computer, refrigerators, and medical devices. Depending on the recycling employed for e-waste, a potentially hazardous material could be introduced into the environment. These materials comprise heavy metals and persistent organic pollutants, encompassing persistent organic compounds (like polycyclic aromatic hydrocarbons (PAHs), polychlorinated biphenyls (PCBs), brominated flame retardants (BFRs), perfluoroalkyl and polyfluoroalkyl substances (PFASs), along with polychlorinated dibenzo-p-dioxins (PCDDs) and polychlorinated dibenzofurans (PCDFs) (Kumar et al., 2019; Orlins & Guan, 2016; Quan et al., 2015; Zheng et al., 2015). Newly published global e-waste statistics revealed that in 2019, approximately 53.6 million tons (Mt) of e-waste was generated worldwide (Ahirwar & Tripathi, 2021). This marked an escalation from 41.8 Mt in 2014 and 44.7Mt in 2016. Forecasts indicate that, with the roughly 2 Mt annual growth rate, the amount of e-waste predicts to reach 74.7Mt by the year 2030 (Ahirwar & Tripathi, 2021).

Figure 1. Summarizing of the potential industrial waste for recovery process

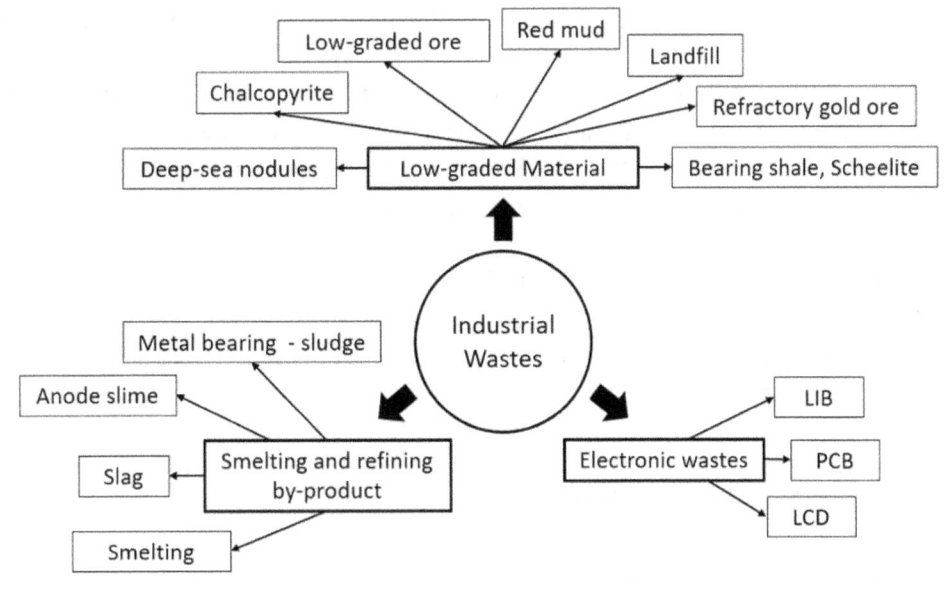

Mine Contamination in Soils

Besides the solid wastes generated by industrial activities, industrial and infrastructure development activities have emerged as significant contributors to soil contamination, posing serious environmental and human health concerns. From 2005 to 2013, the Ministry of Land and Resources (MLR) and the Ministry of Environmental Protection (MEP) conducted the soil analyzing for 13 inorganic contaminants (As, vanadium (V), Zn, mercury (Hg), Co, Cr, Cu, fluoride (F), Cd, Mn, Ni, Pb, selenium (Se),) and 3 types of organic (hexachlorocyclohexane, dichlorodiphenyltrichloroethane, and polyaromatic hydrocarbons). According to the report of MEP, 16.1% of the sample exceed the environmental regulation and also in the case of agricultural soils, exceeding 19.4% (account for 26 million ha presenting in the sample). In these cases, 82.4% belonged to the metal and metalloids contamination and inorganic account for the rest (The Ministry of Environmental Protection; The Ministry of Land and Resources Report, 2014). Furthermore, most of the contaminated areas are the surroundings of the mining and industrial activities.

In Japan, the infrastructure endeavored from the early 2000s onward, necessitated the extensive excavation of soil and rock. The Japan Ministry of Land, Infrastructure, Transport and Tourism (MLIT) reported that the nation amassed around 140 million cubic meters of such soil in 2012 (Ministry of Land, Infrasture, Transport and Tourism, 2012). If left unaddressed, these naturally contaminated rocks can pose considerable environmental risks for adjacent areas and the individuals residing in proximity to construction and waste disposal zones. Li *et al.* documented an arsenic leaching test on four soil samples excavated from diverse construction sites in Tokyo, revealing arsenic concentrations of 9.22, 79.4, 6.75, and 11.7 mg.kg^{-1}(Li et al., 2017). Despite the relatively modest levels of arsenic in these samples, the substantial yearly volume of construction soil raises noteworthy apprehensions. In addition to industrial pollution, certain regions, like aquifers in India and Bangladesh, face potential contamination concerns. These areas exhibited groundwater arsenic levels that ranged from 50 to 3200 µg/l, significantly surpassing standards set by entities like the Indian national limit (50 µg/l) and the WHO guideline (10 µg/l) (Bhattacharya et al., 1997; Bhattacharyya et al., 2003; WHO & World Health Organization, 2019). The presence of high arsenic concentrations in soil represents a looming contamination issue for the region, jeopardizing the well-being of local inhabitants. The escalating soil arsenic contamination is raising multiple alarms, and it appears that this contamination may be influenced by tainted water sources. Numerous studies highlight the potential for crops, particularly rice, grown in arsenic-contaminated soil to accumulate the toxic element, posing a threat to consumers in countries with substantial rice consumption rates such as China, Japan, Vietnam, and India (Chen et al., 2022; Meharg & Rahman, 2003; Nguyen et al., 2021).

ULTRASOUND WASHING TECHNOLOGY

Ultrasound Wave and its Action

Among the many technologies, ultrasound processing technology can provide a distinctive method and has emerged as a promising and innovative solution for addressing the challenges of contaminated soil treatment and recovery of valuable elements from industrial waste. The physical and chemical effects generated from the intense acoustic waves can create such unique processes such as particles dispersion, contamination breakdown, and modification of soil properties. This has the potential to significantly enhance the efficiency of soil remediation efforts, mitigating the adverse effects of pollutants on ecosystems and human health. Moreover, ultrasound technology holds immense promise in the realm of industrial waste management, offering an efficient recovery of valuable elements from various waste streams. By utilizing ultrasound-assisted techniques in the leaching process of metals species into water medium the remaining precious elements in the industrial waste like slag, tailing, or electronic waste can be extracted and reclaimed.

Figure 2. Illustration of ultrasound at different frequencies and cavitation behaving with compressing or depressing sound pressures in aqueous medium when ultrasound transduces

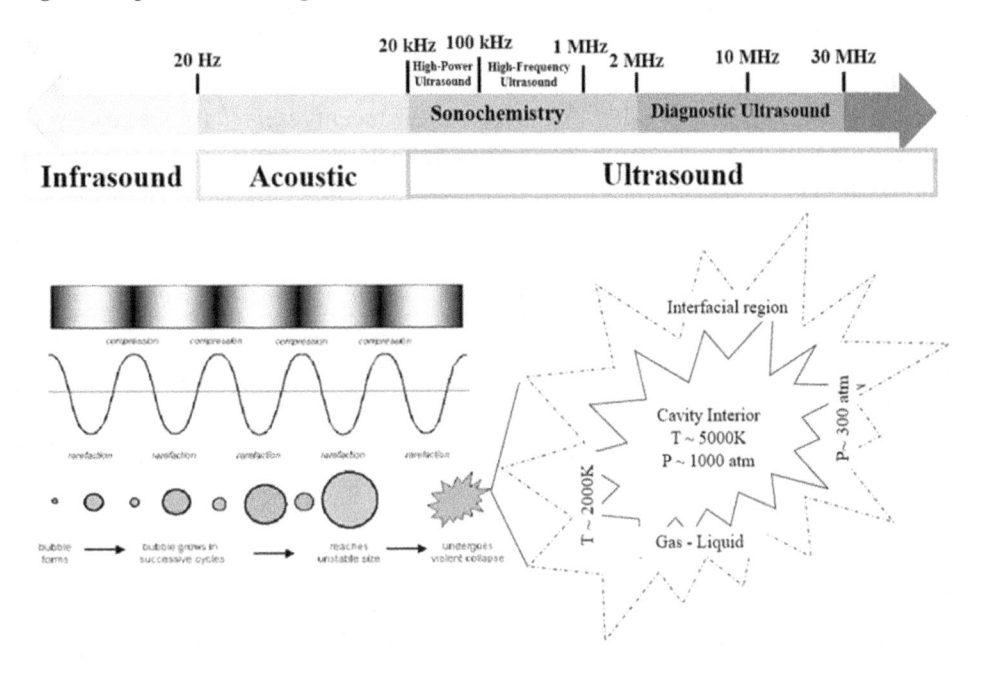

As seen in Figure 2, ultrasonic waves, often referring to US waves, encompass sound waves that possess frequencies exceeding the upper threshold of human hearing, typically surpassing 20 kHz. Lazzaro documented the discovery of ultrasonic waves during his investigation into the navigational abilities of bats (Kane et al., 2004). Based on the frequency, it's widely accepted that ultrasound can be classified into two type: (1) low frequency ultrasound (ranging between 20 – 2 MHz) (Ojha et al., 2017), where acoustic cavitation is generated and applied to induce and enhance chemical reaction and industrial process, (2) high frequency ultrasound (ranging between 2 – 10 MHz), which are principally adopted in the field of diagnostic and medical filed (Verweij et al., 2014). Like other sound waves, ultrasonic waves travel through a medium by inducing particles to oscillate back and forth along the wave's propagation, creating cycles of compression and rarefaction. In the compression phase, particles within the medium are compressed, leading to heightened pressure. Conversely, during rarefaction, particles are distanced, resulting in reduced pressure. When the pressure fluctuations stemming from ultrasonic waves are sufficiently potent to overcome the cohesive forces binding liquid molecules, tiny bubbles form and subsequently collapse- a phenomenon termed "acoustic cavitation." Within the bubble's implosion, both pressure and temperature undergo rapid elevation (reaching over 5000 K and exceeding 1000 atm) over a brief span (Suslick & Price, 1999), generating chemical species.

ULTRASOUND-ASSISTED EXTRACTION OF VALUABLE MINES

Ultrasound is often used as an assistive technology for cleaning and extraction. Such ultrasound-assisted extraction is based on the principle of acoustic cavitation due to the propagation of mechanical waves formed by a set of cycles of high and low pressures called compressions and rarefactions (Rayess et al., 2022; Linares & Rojas, 2022). In the US-assisted leaching system to value-added substances, generally, the distinct ultrasonic cavitation generates specific outcomes, including thermal, mechanical, and sonochemical effects. Mechanical impacts induce the formation of microcracks or the removal of the passive layer on particle surfaces, thereby enhancing lixiviant mass transfer. Thermal effects raise temperatures, leading to accelerated leaching reaction rates or extents. For example, ultrasound-assisted extraction was applied for rare-earth elements extracted from carbonatite rocks (Diehl et al., 2018) and also heavy metals (Bao et al., 2023). Table 1 summarizes the recent research for the recovery of valuable metal from various types of wastes, from the low-grade material, smelting and refining by-products, and electronic wastes.

Table 1. Summary of ultrasound assisted recovery from various of industrial wastes

Raw material (Targeted metal)	Ultrasound Condition	Additive and condition	Efficiency	Reference
Low-graded material				
Chalcopyrite (Cu)	24 kHz – 400W	$Fe_2(SO_4)_3$, 300 min, 80°C	57.5%	(Wang et al., 2020)
Low-grade Zinc (Zn)	20 kHz – 600W	NH_3-$(NH_4)_2SO_4$, 60 min, 30°C	83.33%	(S. Li et al., 2015)
Deep-sea nodules (Ni, Co)	28 kHz – 100W	$(NH_4)_2S_2O_3$, 90 min, 85°C	71% (Ni) 32% (Co)	(Knaislová et al., 2018)
Red mud (Ti)	20 kHz – 400W	H_2SO_4, 5 hours, 90°C	20% increase	(Şayan & Bayramo lu, 2004)
Aluminum dross (Al)	40 kHz – 100 W	NaOH, 4 hours, 50°C	60%	(Nguyen & Lee, 2019)
Landfill (Pb)	40 kHz – 20 W	H_2SO_4/NaCl, 30 min, 25°C	26%	(John et al., 2020)
Refractory gold ores (Au)	1500W	NaOH/Na_2S, 2 hours, 55°C	68.9%	(Guo et al., 2019)
Silica sand (Fe)	20 kHz – 300 W	$C_2H_2O_4$, 30 min, 95°C	75.4%	(Du et al., 2011)
Vanadium-bearing shale (V)	20 kHz – 1200 W	H_2SO_4, CaF_2, 30 min, 95°C	92.93%	(B. Chen et al., 2020)
Scheelite (W)	23, 39-42 kHz – 300 W	NaOH, 6 hours, 80°C	71.5%	(Johansson et al., 2021)
Smelting and refining by-product				
Copper slag (Cu, Zn, Co, Fe)	20 kHz – 400W	H_2SO_4-$Fe_2(SO_4)_3$, 180 min, 65°C	Cu (89.28%) Zn (51.32%) Co (69.81%) Fe (13.73%)	(Beşe, 2007)
Copper-cadmium slag (Zn, Cd, Mg)	20 kHz – 300W	H_2SO_4, 120 min, 40°C	Over 97%	(H. Liu et al., 2022)
Copper slag (Si)	26 kHz – 600W	HCl, 120 min, 60°C	90%	(Phuoc Tri et al., 2023)
Copper-bearing tailing (Cu)	40 kHz – 200W	$NH_4(OH)$, 6 days, 25°C	89.5%	(Zhang et al., 2008)

continued on following page

Table 1. Continued

Raw material (Targeted metal)	Ultrasound Condition	Additive and condition	Efficiency	Reference
Mine tailing (Zn, Al, Mn, Cu, Fe)	37/80 kHz – 130/100 W	pH =2, 30 min, 30°C	Zn (+150%) Al (+95%) Mn (+48%) Cu (+38%) Fe (+28%)	(Hernández et al., 2023)
Lead-Zinc tailing (Ag)	20 kHz – 200 W	$Na_2S_2O_3$, 5 min, 30°C	71.5%	(H. Li et al., 2020)
Zinc smelting (Ge)	20.21 kHz – 700 W	$HCl/CaCl_2$, 40 min, 80°C	92.7%	(J. P. Wang & Erdenebold, 2020)
Zinc slag (In)	20.21 kHz – 700 W	$HCl-CaCl_2/Ca(ClO)_2$, 10 min, 70°C	93%	(Zou et al., 2020)
Germanium containing slag (Zn,Ge)	19.5 kHz – 300 W	$H_2SO_4 – H_2O_2$, 60 min, 90°C	Zn (96.66%) Ge (92.96%)	(Xin et al., 2022)
Lead-rich and Antimony rich slag (Sb, Pb)	20-25 kHz – 400 W	HCl-NaCl, 6 hours, 95°C	Sb (85.56%) Pb (75.37%)	(R. L. Zhang et al., 2015)
EAF dust (Zn)	20 kHz – 60 W	H_2SO_4, 30 min, 80°C	55%	(Brunelli & Dabalà, 2015)
Corundum flue dust (Ga, Zn)	20 kHz – 900 W	H_2SO_4, 50 min, 90°C	Ga (82,56%) Zn (99.57%)	(Ding et al., 2022)
Sintering Dust (Ag)	20 kHz – 400 W	CH_4N_2S, 90 min, 50°C	95%	(Chang et al., 2017)
Copper anode slime (Cu)	19 kHz – 800 W	$H_2SO_4/Na_2S_2O_8$, 50 min, 50°C	98.4%	(J. Liu et al., 2022)
Lab-synthesized metal bearing sludge (Cu, Cr)	20 kHz – 48W	H_2SO_4, 60 min, 25°C	Cu (97.47%) Cr (97.33%)	(P. Zhang et al., 2013)
Electronic wastes				
PCBs (Cu)	40 kHz – 600 W	HCL, 180 min, 60°C	99%	(Thi Hong Hoa et al., 2023)
PCBs (Au, Ag, Pd, Pt)	20 kHz – 700 W	Water, 30 min, 25°C	Over 90%	(Jadhao et al., 2020)
PCBs sludge (Cu)	20 kHz – 160 W	H_2SO_4, 60 min, 25°C	90%	(Xie et al., 2009)
PCB sludge + etching solution (Cu)	20 kHz – 300 W	Lime, 1 hour, 25°C	93.76%	(Huang et al., 2011)
LIB (Co, Li)	20 kHz – 360 W	H_2SO_4, 30 min, 30°C	Co (94.36%) Li (98.62%)	(Jiang et al., 2018)
LIB black mass (Li)	40 kHz	Water, 3h, 30°C	92.6%	(Makuza et al., 2021)
LIB cathode material (Ni, Li, Co, Mn)	90 W	DL-malic acid – H_2O_2, 30 min, 80°C	Ni (97.8%) Co (97.6%) Mn (97.3%) Li (98%)	(Ning et al., 2020)
LIB cathode (Co)	20 kHz – 800 W	H_2SO_4, 4 hours, 65°C	98%	(W. Y. Wang et al., 2020)
LIB (Li, Mn, Co, Ni)	40 kHz – 110 W	Citric acid, 24 hours, 50°C	Li, Mn (96%) Co (99%) Ni (99%)	(Xiao et al., 2021)
LCD (In)	Microwave-ultrasound 650 W	HCl, 142s, 45°C	100%	(W. Zhang et al., 2020)
LCD (In)	20 kHz – 300 W	H_2SO_4, 60 min, 25°C	96.8%	(K. Zhang et al., 2017)

continued on following page

Table 1. Continued

Raw material (Targeted metal)	Ultrasound Condition	Additive and condition	Efficiency	Reference
LCD (In)	20 kHz – 360 W	H_2SO_4, 20 min, 60°C	92%	(Souada et al., 2018)
LCD (Gd,Pr)	90 kHz – 120 W	PPi, 60 min, 25°C	Gd (87%) Pr (85%)	(Toache-Pérez et al., 2020)
LCD (Er, In, Sn, Gd)	90 kHz – 120 W	$Na_4P_2P_7 - H_2O_2$, 2 hours, 25°C	Er (93%) In (97%) Sn (72%) Gd (99%)	(Toache-Pérez et al., 2022)

Ultrasound Assisted Ore Recovery

Recovery of raw material resources from low-grade ores is possible with conventional technologies, but has been neglected because of high capital investment and low economics (Pradhan et al., 2006). However, in recent years, with the rise of symbiosis with nature and resource recovery, a trend has emerged to be considered with the introduction of technology equipped with efficiency. As summarized in Table 1, low-grade ores are effectively efficient in the special environment of applying ultrasound techniques due to enhancement of the beneficial recovery aspects. For copper (Cu) leaching from Chalcopyrite in acidic ferric sulfate media only increase the efficiency from 50.4% to 57.5% (Wang et al., 2020), and for low-grade zinc (Zn) ore using NH_3-$(NH_4)_2SO_4$ as the additive (S. Li et al., 2015). In the latter case, ultrasound demonstrated profound leaching efficiency at low NH_3 concentration. When increased the power from 200W to 600W the final leaching rate was increased to 83.33% within 60 minutes. Nickel (Ni) and Cobalt (Co) were targeted for the recovery from deep-sea nodules by comparing microwave and ultrasound systems with the presence of ammonium thiosulfate (($NH_4)_2S_2O_3$) as reducing agent (Knaislová et al., 2018). Using 28kHz ultrasound at 100 W leached Ni and Co in efficiency of 71% and 32% for 90 minutes but, microwave heating made in comparison had an extraction efficiency of 48% and 8% at 60 W for 210 minutes, respectively. Red mud was utilized at 20 kHz – 400 W of ultrasound powers to recover titanium (Ti) in the presence of sulfuric acid (H_2SO_4) (Şayan & Bayramo lu, 2004), resulting at 20% more of Ti in the recovery. Also the presence of 40 kHz ultrasound at 100W enhanced the leaching % of aluminum (Al) from aluminum dross and increased from 35 to 60%, when added NaOH in 4 hours (Nguyen & Lee, 2019). As optimized for lead (Pb) recovery from landfill, sonicated system was used with H_2SO_4 and sodium chloride (NaCl) under 40kHz ultrasound operation at 20 W (John et al., 2020), improving from 19% to 26% efficiency. ultrasound at 15000W for the antimony (Sb) removal of refractory for gold ores (Guo et al., 2019) reached 94.5% leaching after 1 hours operation in the presence of sodium hydroxide (NaOH) and sodium sulfide (Na_2S). In this case, if no ultrasonic treatment was performed, the efficiency was reached to 58.7% at the same condition. The recovery yield of gold (Au) increased from 13.35% to 68.9% with the ultrasound assist extraction process. Iron (Fe) was leached process from silica sand when oxalic acid ($C_2H_2O_4$) was present with the 20kHz ultrasound assist at 300 W by 75.4% yield within 30 minutes (Du et al., 2011) . Also, the enhancing ability of 20 kHz ultrasound assisted at 1200 W leaching Vanadium (V) for vanadium-bearing shale, when used in H_2SO_4 and calcium fluoride (CaF_2) (B. Chen et al., 2020). The ultrasound operation showed V leaching efficiency from 87.86% to 92.93%, within an 8-fold shorten time for 30 min compared to 240 min. In recovery of tungsten (W) from scheelite (Johansson et al., 2021; Yang et al., 2018), the presence of ultrasound boosted the effectiveness of leaching process from 36.7% to 71.5% .

Ultrasound Assisted Treatment of Smelting and Refining By-Products

Refining technology can enable the recovery of high-value-added metals from low-grade ores that have not previously been subject to smelting and refining. Therefore, smelting and refining technologies are contributing to the construction of a sustainable, recycling-oriented economy. On the other hand, a large amount of unused slag is generated, and its disposal is also a problem. Slag is a residual product resulting from extractive pyrometallurgical operations, possesses distinctive chemical and physical attributes influenced by a range of factor (Gao et al., 2020; Gorai et al., 2003b; Piatak et al., 2015). Generally, the chemical composition of slag, especially in instances of legacy slag was categorized broadly into ferrous and nonferrous types in steelmaking (Gorai et al., 2003b; Piatak et al., 2015; Shen & Forssberg, 2003). In this case, apart from slag, additional by-products generated during both ferrous and nonferrous ore processing, encompassing waste rock, tailings, sludges, slimes, and dust (Liu et al., 2021b; P. Zhang et al., 2013; Y. Zhang et al., 2019). Moreover, slag can be generated not only from refining, but also from recycling activities such as lead scrap and alkaline battery recycling, and from the vitrification of municipal and nuclear waste. It is essential to realize the usefulness of mining by-products that are constantly generated in the course of industrial industries, such as slag, and to pursue them as a sustainable material use through technology. One of the most researched approaches to recycling this type of industrial waste is to use it as a secondary resource in hydrometallurgical processes.

Copper Slag Treated With US

As of the most available slag in the world, copper slag has been widely researched for the recovery of previous metal remaining inside. Metallurgical solid waste residue is produced after the extraction of valuable metals (Zue et al., 2021). Copper slag is a type of nonferrous metal smelting slag produced in pyrometallurgy and is a type of secondary resource with a large quantity and high quality. The typical composition of copper slag is Cu 0.42–4.6%; Fe 29–40%; SiO_2 30–40%; CaO $\leq$ 11%; Al_2O_3 $\leq$ 10% (Piatak et al., 2015). Thus, in European Union, Cu slag is freely traded between nations without legally sanction for the metal leaching potential (Duester et al., 2016). The growth rate in recent years would predict that the output of copper slag will exceed 24 million tons by 2020 (Fuentes et al., 2020).

Therefore, researches have also been conducted using ultrasonic extraction techniques in order to leach out valuable mineral species. For instance, Beşe et al. reported about ultrasound-assisted leaching system for extracting Cu, Zn, Co and Fe from copper slag using H_2SO_4 and ferric sulphate ($Fe_2(SO_4)_3$) (Beşe, 2007). The presence of 20 kH ultrasound at 400 W output power increased the leaching efficiency into 89,28%, 51.32%, 69.87% and 13.73% for Cu, Zn, Co and Fe, respectively. Recently, the improvement of the leaching process of copper-cadmium slag was investigated with 20 kHz ultrasound operated at 300 W in H_2SO_4 medium (H. Liu et al., 2022), demonstrating that, over 97% of Zn, cadmium (Cd), and magnesium (Mg) was leached after 2 hours treatment. As utilized with 26 kHz ultrasound at 600 W for the silica extraction (Phuoc Tri et al., 2023), ultrasound process showed high purity silica deposition in acidic media like HCl, H_2SO_4 and HNO_3. Depending on their concentration, the purification degree was varied especially in 6M concentration of HCl, reaching over 90% purity of silica deposited without other element contaminations like iron oxide. Figure 3 shows pictures of copper slag and treated ones in 6M HCl without and with ultrasound exposure. It was discovered in the research that ultrasound had the ability to silica gelation purified in the presence of acidic environment, leading to the separation between the silica gel and the residue copper slag. In these cases, 90% of silica was recovered from copper slag with only 0.04% of Fe contamination. In the original slag, the elemental analysis was indicated as CuO 0.98%; Fe_2O_3 53.5%; SiO_2 26.5%%; CaO 3.79%; Al_2O_3 5.64%, resulting in high silica content secondly only to the iron component. These results show that silica in waste slag can easily contribute to industry as a resource, since silica is well known as a necessary material for daily life and for use in diverse industries, such as semiconductors, glass, composites, cosmetics, ceramics, and concrete (Zhuang & Chen, 2019). In these cases, sand is a very common silica source, although its overexploitation has caused environmental impacts [28] and crises worldwide, especially in developing countries (Devi, et al., 2021; Duker et al., 2020; Bisht,2022).

Figure 3. Silica gel recovered from copper slag with and without the 26 kHz – 600 W ultrasound irradiation using HCl 6M

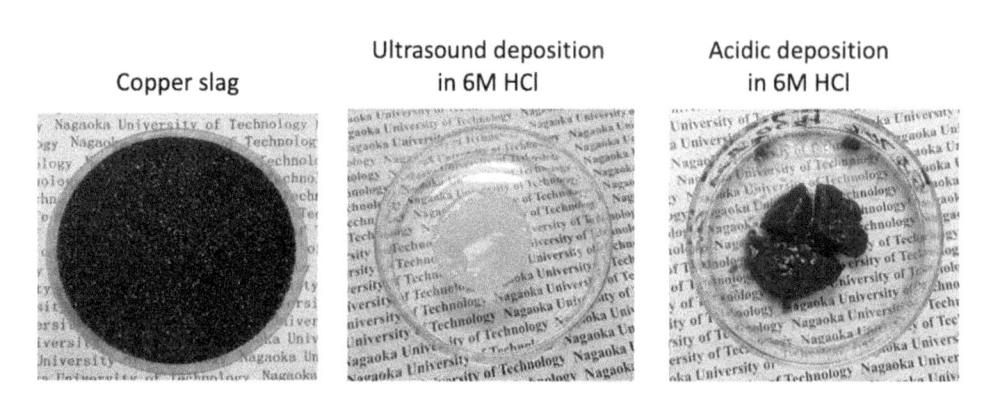

Furthermore, a mixture of crushed rock and processing liquid originating from mills, washeries, or concentrated ores could remain after the extraction of economic metals, minerals fuels or coal from mine resource (Kossoff et al., 2014), showing that, ultrasound leaching of Cu from copper-bearing tailings with 40 kHz at 200 W) with ammonia (NH_4OH) increased the final leaching rate up to 13.5% (Zhang et al., 2008). Not only for the direct treatment, the ultrasound technique (37/80 kHz-130/100 W) was implemented for the pre-treatment of the bioleaching of metal from mine tailing (Hernández et al., 2023). Under 80 kHz-100 W ultrasound, the tailing mine particles was observed to be severely fragmented, leaching to the increase of mass transfer rates, resulting in increased in the metal leaching rate: Zn (150%), Al (95%), Mn (48%), Cu (38%), Fe (28%). Li et al. studied the enhancing effects of ultrasound (20 kHz -200 W) to the recovery of Ag from the lead-zinc tailings using $Na_2S_2O_3$ (H. Li et al., 2020). These enhancements of ultrasound were explained due to the fact that ultrasonic irradiation destroyed the surface areas of the tailing, accelerating the mass transfer and diffusion rate.

Other Industrial Wastes With Ultrasound Leaching Processes

On the other hand, the remaining precious metal in by-product of Zn smelting and refining process was appealing to the recovery as a secondary source. When used HCl-CaCl$_2$ mixed solution to extract germanium (Ge) from a by-product of lead and zinc metallurgical process with the assist of 20 kHz ultrasound at 700 W (L. Zhang et al., 2016). The ultrasound shortened the time by 60% (from 100 min to 40 min) and yielded a 92.7% efficiency. With the same ultrasound system, indium (In) was extracted with using HCl-CaCl$_2$ and Ca(ClO)$_2$ with a 700 W ultrasound resulting at nearly 93% after 10 min operation(Zou et al., 2020). They indicated that the efficiency was due to both mechanical and oxidation induced by ultrasound generated bubbles. Also in 19.5 kHz ultrasound at 300 W, the recovery rate of Zn and Ge was investigated for germanium-containing slag dust using H$_2$SO$_4$ (Xin et al., 2022) and the lead-rich and antimony-rich oxidizing slag was recovered for Sb and Pb in HCl-NaCl medium as ultrasound was operated at 20-25 kHz – 400 W (Zhang et al., 2015).

An Electric Arc Furnace (EAF) for steel-making furnace generates around 1.6 billion tons of steel every year, which account for 40% of process steel from recycled waste-feed (Antuñano et al., 2019). In the by-product of the EAF - sintering dust and corundum flue dust, valuable metal like Fe, Zn, Pb, gallium (Ga) was recovered by 20 kHz ultrasound at 60 W (Brunelli & Dabalà, 2015). At low concentration of 0.5 M of H$_2$SO$_4$, ultrasound demonstrated an efficacious enhance in the leaching of Zn, resulting in a 55% improvement compared to the original leaching due to the dissolution of franklinite. Similar ultrasound effects were investigated for the recovery of Ga and Zn from corundum flue dust (Ding et al., 2022), Ag recovery from sintering dust in the presence of thiourea acid (CH$_4$N$_2$S) (Chang et al., 2017). The ultrasound leaching of Cu from anode slime also recorded positive result when combining with H$_2$SO$_4$ or sodium persulfate (Na$_2$S$_2$O$_8$), with the improvement from 66.64% to 98,11% (P. Zhang et al., 2013), or reduce the content of copper in the waste from 17.8% to 2.64% (J. Liu et al., 2022).

Ultrasound Assists in Recovery Valuable From E-Waste

Advancements in the economy and rapid technological growth have influenced how people live and various industries. In recent years, the electronics manufacturing sector has seen remarkable progress, and continuously improving electronic devices has become essential to stay competitive and meet the demands of consumers (Kaliyavaradhan et al., 2022; Rene et al., 2021). Electronic waste contains valuable materials, particularly precious and scarce metals (Mir & Dhawan, 2022). Proper recycling of e-waste can generate economic advantages and simultaneously decrease the release of hazardous substances resulting from improper or inadequate disposal techniques (Frazzoli et al., 2022; Zhu et al., 2022). The reason for this is that printed circuit boards (PCBs), which are used in most of today's electronic devices, are composed of a variety of materials, including non-metallic polymeric substrates (resins), metal foils, laminates, and electronic components, and it is extremely important to view these as resources. (Mir & Dhawan, 2022). Typically, 30-35% of PCBs weight is consisted of metals, 35-42% of refractories and the rest 24-30% is resins (A. Kumar et al., 2018). In term of metallic components, the metal component of PCB varied depend on the type of device, encompassing 8-38% Fe, 10-27% Cu, 2-19% Al, 1-3% Pb, 0.3-2% Ni, as well as traces of Ag ranging from 200 to 3000 ppm, Au from 20 to 1200 ppm, and Pd from 10 to 300 ppm (Arya & Kumar, 2020). When producing the same quality of gold, silver, copper, aluminum, and steel, the original mining cost is approximately seven times higher than the cost of recycling PCBs, which made it become a compelling secondary metal source (Zhou et al., 2021). As the demand for portable devices increases exponentially, the large amount of annual disposal of batteries has become a problem due to the end-of-life phase of the product. Lithium-ion batteries (LIBs) are currently the most common chosen for daily appliances because of its reliability, efficiency in energy storage due to high energy density, long cycle life, low self-discharge, etc. (C. Li et al., 2023). Furthermore, as the shift of vehicle from fossil fuel combustion engines to electric vehicles (EVs), the requirement for LIBs to EVs and hybrid electric vehicles (HEVs) are predicted to raise every single year (Gao et al., 2018; Maisel et al., 2023). Beside the battery, liquid crystal display LCD, is one of the most common parts of today's electronic devices (for instance PC monitors, laptops, tablets, mobile phones, televisions). This kind of waste contain a variety of substances, including highly toxic elements like mercury (Hg), Cr, Cd, and Pb (Savvilotidou et al., 2014) and organic component like cellulose triacetate, polyvinyl alcohol, triphenyl phosphate (R. Wang & Xu, 2016). Besides that, there amount of precious metal inside waste LCDs (like In and Sn) are appealing for urban mining (Nakashima & Kumahara, 2002).

Consisting of many different materials like ceramics, polymer, and variety of metals (Cu, Al, Fe, Pb, Ni, Ag, Au, palladium (Pd), platinum (Pt)) (Awasthi et al., 2016; H. Wang et al., 2017), the high amount of annual disposal of PCBs makes the recycling process to retrieve the precious metals inside become compelling in both economic and environmental side (Pariatamby & Victor, 2013). 2.5 fold of Cu recovery efficiency from mobile phone PBCs was recorded with the presence of 40 kHz – 600 W ultrasound with HCl (Hoa et al., 2023). Jadhao et al. employed the 20 kHz ultrasound at 700 W to separate Au, Ag, Pd, and Pt from the pyrolyzed PCBs only using water (Jadhao et al., 2020), result at 1.6g out of 1.7g of metal fraction was recovered after 30 min exposure. Beside the PCBs, the sludge generated from the rinsewater of PCBs also contained valuable metals like Cu, thus ultrasound was implemented at 20 kHz – 160 W to recover Cu from PCBs waste sludge using H_2SO_4 (Xie et al., 2009). Results showed that, at only 25°C, the ultrasound delivery the result at nearly 98%, which is much higher than even 80°C (78%) and 25°C (56%) at conventional method. Beside PCB waste sludge, the making of PCBs also require acid etching process, which will create a waste that contain significant amounts of copper due to direct contact between copper foil from PCBs with acid (Yu et al., 2016). As creative recovery effort, combining by leaching PCB waste sludge using PCBs etching solution was treated with 20 kHz ultrasound at 300W (Huang et al., 2011).

Ultrasound Leaching to Lithium-Ion Batteries

For the continued development of LIBs, the raw material requirement for future generations much be secured. One of the best ways to limit the use of raw material is extracting the precious metal in the end-of-life phase LIBs battery (Baum et al., 2022). For 30 minutes leaching, the used LIBs was treated with 20 kHz ultrasound at 360 W in aqueous $H_2SO_4 – H_2O_2$. This reached to Co and Li leaching with 91.65% and 92.70% recovery rate, respectively (Jiang et al., 2018). Water leaching of "black mass" – a mixed-roasted of cathode and anode material, with 40 kHz ultrasound ended up with 92.6% of Li recovery (Makuza et al., 2021). More than 97% of Ni, Li, Co, and Mn was leached out from waste cathode material of spent LIB with the leaching system using DL-malic acid and 90W ultrasound pool after 30 minutes (Ning et al., 2020). Ultrasound not only reported for the enhancing of dissolution of the desired metal, but also enhancing the precipitation of precious metal for easier to recover. When applying with 20kHz ultrasound at 800 W, the recovery of Co from cathode material was investigated for waste LIBs (W. Y. Wang et al., 2020). Under combination of ultrasound and H_2SO_4, while other metal like Li, Mn, and Ni was leaching at more than 80%, almost no Co was detected in the leachate, which is very useful for the recovery of Co in the next step, resulting in 98% of Co recovery after the process. Also, the 40 kHz ultrasound at 110 W was used for the recovery of Li, Mn, Co, and Ni from discarded LIBs with citric acid ($C_2H_4O_2$) (Xiao et al., 2021). After 4 hours leaching with ultrasound resulting at more than 99% recovery rate of Ni and Co and 94 – 96% of Li and Mn, while the recovery rate of conventional leaching rate is around 89%. To extract Li, Co, and Ni from spent LIBs by using fresh lemon juice and H_2O_2 (Esmaeili et al., 2020), the ultrasonic irradiation enhanced the leaching rate of Li, Co, and Ni from 75%, 58,2%, and 74,2% to nearly 100%, respectively. The presence of all three components: lemon juice, H_2O_2 and ultrasonic wave made significant contribution to the efficiency of the process.

Ultrasound Applied for Liquid Crystal Display

As the demand for LCDs will keep rising in the future, recycling the precious metal inside the waste, or spent LCDs for the secondary sources should be considered for the sustainable future. Zhang et al. proposed an innovative system of microwave and ultrasound leaching for recovery of In from LCDs panels with various kind of acid (HCl, H_3PO_4, and $HClO_4$) (W. Zhang et al., 2020). At the optimal condition, 100% of In can be leached from the crushing panel using HCl only after 142s. Based on the structure of the indium tin oxide glass (ITO), Zheng et al. developed a not crushing-panel method for the recovery of In from LCD waste with the assistance of the 20 kHz – 300 W ultrasound system and low concentration HCl (K. Zhang et al., 2017). The research team stated that, at 300W, 96.8% of In was leaching out of a 4 x 4 cm of glass panel from the with 0.8M of HCl without any heating requirement. Souada et al. implemented the ultrasound technique (20 kHz – 320 W) for extracting In from ITO from LCD waste screen by using H_2SO_4 (Souada et al., 2018). The research team stated that, leaching with ultrasound at 4 min resulted at 82% In removing from the panel while the result from conventional method only reach 13%, demonstrated greatly enhance in both leaching effiency and shorten the time. Finally, 92% of it was recovered after 20 min leaching with ultrasound and H_2SO_4 at 60°C. Beside the In, LCD also contained a lot of other precious metals. Toache-Pérez et al. applied the utlrasound technology for the extraction of rare earth elements (REE) from spent LCD using pyrophosphate anion ($P_2O_7^{4-}$, PPi) with 90 kHz – 120 W ultrasound aid (Toache-Pérez et al., 2020). The data showed that, after 60 min leaching with ultrasound at room temperature, 87% of gadolinium (Gd) and 85% of praseodymium (Pr) was extracted from the screen waste. In 2022, the same research team continued their ultrasound-assist recovery research for the recovery of REE from LCD screen waste with more element was targeted. With the same utrasonic system, using sodium pyrophosphate ($Na_4P_2O_7$) – H_2O_2 to leach out Er, In, Sn and Gd from LCD screens (Toache-Pérez et al., 2022). As the result, 2h leaching of screen waste at room temperature with the assist of ultrasound result as 93, 97, 72, and 99% of Er, In Sn and Gd recovery, respectively.

ULTRASOUND ASSISTS IN CONTAMINATED SOIL REMEDIATION

Soil, an irreplaceable natural asset, stands as a fundamental constituent of terrestrial ecosystems. Regrettably, soil serves as a prominent reservoir for pollutants of diverse types, notably heavy metals and organic contaminant, which persist as contaminants due to their innate resistance to degradation or elimination. These elements exist naturally in the Earth's crust, manifesting in different forms within the solid phase of soils and sediments, as well as dissolved in water. Some heavy metals, such as As, Fe, Cu, Co, Zn, Ni, Mn, selen (Se), and molybdenum (Mo), are required in low concentrations for metabolic processes. However, elevated concentrations of these elements pose toxicity risks to humans, plants, and microorganisms.

Conventional Method for Contaminated-Soil Treatment

In this context, various methods of decontamination have been proposed and implemented. The primary goals of soil remediation technologies are to remove and stabilize contaminants, protect against the spread of contaminants, not harm other organisms, and alter ecosystems. These methods can be categorized into two main approaches: in-situ and off-site. In-situ means at the original site, while off-site means outside the original site. In other words, in terms of soil treatment, the in-situ method means that the contaminated soil is treated directly at the site of contamination, while the off-site method means that the contaminated soil is removed from its original location and transported to a specialized facility for treatment. For the on-site treatment, there are three main techniques that can be used to treat the contaminant from the soil: bioremediation – improving the soil quality by extracting or stabilizing the contaminants through harnessing the metabolic microorganism or animal (Vishwakarma et al., 2020), phytoremediation – using certain type of plant to absorb, accumulate, degrade, or render the contaminant on the polluting site (McCutcheon & Jørgensen, 2008), and electrokinetic extraction – an electrical field is created in a soil matrix by applying a low-voltage direct current (DC) which will mobilize or concentrated the contaminant at the electrodes (Cameselle & Gouveia, 2018).

Beside extraction, in-situ technique also includes the immobilization/solidification of the contaminant to prevent the spread out of the unwanted substance (L. Wang et al., 2019). In terms of ex-situ soil technique, after excavation, there is also 2 types of approaches extraction or immobilization. While the solidification approach is quite the same as the in-situ technique, the biggest different lying in the extraction process. The ex-situ extraction technique usually call as "soil washing ", which is removal of the contaminant by using one or several treatment step involve with one or many different technique. Depend on the type of contaminant, the treatment process can only use physical process depend on the difference in physical properties of the soil like size, density, and magnetic (Anca et al., 2011). However, the combination between both physical and chemical seem to be more versatile options due to much more treatment option and combination for different type of contamination (Dermont, Bergeron, & Mercier, 2008; Trellu et al., 2016).

ULTRASOUND-ASSIST SOIL WASHING

Ultrasound-Assist Soil Washing on Metal-Contaminated Soil

Recently, the application of ultrasound technique to enhance the efficiency of the soil washing technique have gathered a lot attention in the soil treatment community due to high increase in efficiency, shorten the time and also report to have high removing rate even at disadvantages condition (Cho et al., 2020; Choi et al., 2021a; Kobayashi & Phuoc Tri, 2022; Son et al., 2019). Not only showing effect on single treatment, ultrasound could also play an enhancing step for the multi-stage treatment of the soil (Fraiese et al., 2020; Kwedi-Nsah et al., 2019). Furthermore, ultrasound also demonstrate great enhance in treating two of the most common type of soil contamination which is heavy-metal contamination and organic-substances contamination.

As the most powerful tool in the *ex-situ* soil remediation technology, soil washing techniques are recognized as the most highly effective ex-situ remediation methods, primarily relying on the release of contaminants from excavated soil at the specialized facilities (Dermont et al., 2008). This achieved through vigorous agitation and scrubbing of soil particles with a washing additives in a slurry phase. Recently, ultrasound technology was applied as the main washing mechanism for removing contaminant from soil, especially heavy metal-contaminated soil. There is a group of dedicated researcher from Koreas to focus on developing this specific soil washing technique. As utilized as large lab-scaled 28 kHz ultrasound, pairing with a 200 rpm mechanical stirring could treat contaminated-soil containing heavy metal of Cu, Pb, and Zn in acidic HCl media (Park & Son, 2017). As result, the presence of the ultrasound not only greatly improved the washing efficiency, but also reduced the required concentration of the washing additive to reached the local regulation. The removal rate of conventional washing of Cu, Pb and Zn with 0.5M are 25.1%, 24% and 29.3%, while ultrasound washing results in 74.5%, 76%, and 71,1%, respectively. When the 28 kHz ultrasound at 170 W was combined with mechanical agitation to eliminate contaminated metals from the soil with mildly HCl acid (Son et al., 2019), ultrasound-assisted washing demonstrated better result even with lower liquid to solid ratio, with less concentration of HCl. Furthermore, the sonochemist-luminescence, testing the formation of cavitation bubble, indicated that in the presence of smaller particles having about 75 µm-2mm sizes, active cavitation activity occurred throughout most of the vessel area due to nucleation of suspended particles. As using soil-washing process applied to remove Cu, Pb, and Zn from contaminated soil, the aid of 28 kHz ultrasound with mechanic stirring effected in the presence of HCl/EDTA (Choi et al., 2021b). Not only ultrasound aide improved the average leaching rate of the optimal condition from 44.8% to 57.9%, it also demonstrated the working ability in less favorite. At 5:1 liquid to solid condition, ultrasound irradiation enhanced the removal efficiency from 17.1% to 50.3%, which was quite comparable to the optimal one. Furthermore, the team conducted a experiment for better understanding of ultrasound washing mechanism, by ultrasonic washing different size painted beads at 4 mm, 2 mm and 1 mm. It was reported that, visually, the smaller the beads, the more violent it vibrated, resulting as more paint was removed from the smaller beads compared with the bigger one. In the paint removal effect of ultrasound, cavitation bubble effected to create the microjet and shockwave, when applying two different type of ultrasound, namely 20 kHz – 50 W probe type ultrasound and 24 kHz – 200 W cup-horn sonoreactor, for the remedy of many different heavy metals element (Cr, Ni, Cu, Zn, and Pb) using EDTA and acetic acid (De La Calle et al., 2013). A drastically decrease in time, from 1 hour to 2 min (ultrasonic probe) or 10 min (cup-horn sonoreactor) in the EDTA system, resulting at 83-118% recovery rate in all five metal. The time required for the acetic acid system also reduce from 16 hours to 6 min (ultrasonic probe) or to 30 min (cup-horn sonoreactor), reaching 83-122% for all the metals.

Figure 4. SEM images of As contaminanted soil at 5,000 magnification before and after ultrasound treatment with distilled water and aqueous NaOH (1M and 2M), HCl (1 M and 2M) and the particle size distribution respectively using US 1,200 W at 22°C for 60min treatment

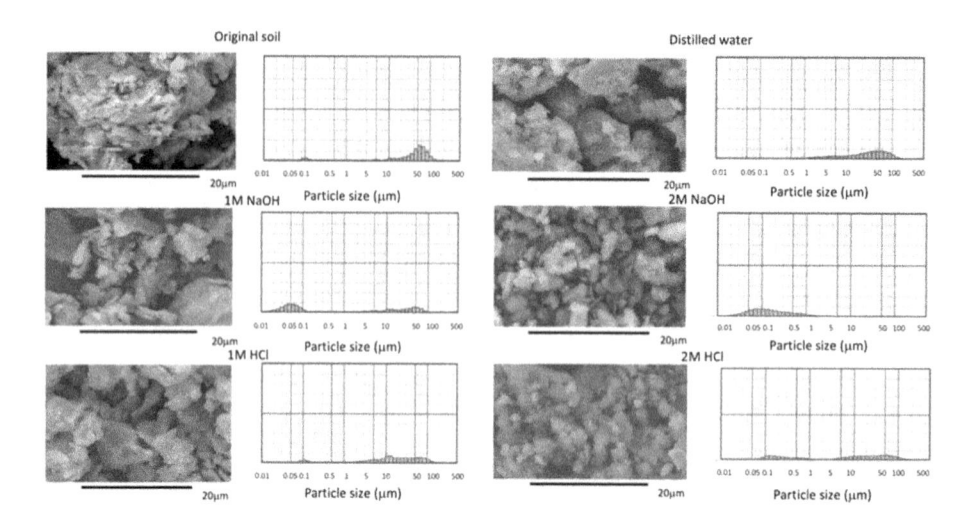

In order to purify arsenic (As) contaminated soil by usng 28 kHz ultrasound at 1200 W, effect of HCl and NaOH was investigated(Kobayashi & Phuoc Tri, 2022). Around 96 to 99% of As remained in the soil was eliminated by 1 hour washing with just ultrasound technique only. From SEM images of each soil situations (Figure 4), the contaminated soils became smaller particle sizes when ultrasound was applied with acid an alkali as indicated by size distribution data. the soil particles was heavily pulverized under high-power ultrasound, leading to mechanically crush the soil particles and resulting as enhancing in the As washing process. Particularly with NaOH, the particle size distribution was reduced from almost 50-100 μm size to 0.05 μm size at 1 M concentration. This size reduction effect facilitates contact between the contained arsenic and the alkaline water, thus increasing clean efficiency. As a result, arsenic containing 9 ppm could be cleaned to 0.45 ppm.

Not only for the elimination of heavy metals, but ultrasound also demonstrated the ability to eliminate other organic contaminants from the soil. When 35/130 kHz ultrasound system was deployed to combine with electrokinetic soil for the extraction of benzo pyrene, benzo anthracene, Cd, Pb, and Zn with Na_2SO_4 and nitric acid as washing additive (Fraiese et al., 2020), the ultrasound enebled for 10 min treatment of the electrokinetic waste to remove almost all the organic contaminants from the soil. In terms of metal contaminants, nearly 100% of Cd and Zn were removed about the 2-step treatment, while the ultrasonic pre-treatment improved the elimination rate of Pb from 29.44% to 64.64%.

The contamination of the environment with petroleum poses a significant and widespread risk due to the toxic nature of petroleum hydrocarbons, which is harmful to all life forms. This environmental contamination is a common occurrence due to the extensive use of petroleum and its associated disposal methods, as well as accidental spills (Zahed et al., 2010). Annually, there are around 1,300,000 tons of petroleum hydrocarbons introduced into the oceans from various sources. These leaked oils are intricate blends primarily composed of aliphatic and aromatic hydrocarbons, along with relatively small amounts of asphaltenes, resins, and heavy metals. Since numerous components found in petroleum are recognized as potential toxicity, mutagenicity, or carcinogenic properties, many research have demonstrated (Abdel-Shafy & Mansour, 2016). Among them, ultrasound process showed tremendous effect in eliminating various of contaminants from soil, including petroleum related substances with a 20 kHz – 100 W ultrasound system (Kim & Wang, 2003). The presence of ultrasound increased the oil removal efficiency from around 44% to 61% at the optimal condition. Son et al. proposed using a 35 kHz – 100 W ultrasound system pairing with mechanical stirring to remove waste diesel from the contaminated sand (Son et al., 2011). With only 1 min washing with ultrasound and mechanism stirring, around 74.7% of the waste diesel was removed from the sand. While 1 min of conventional washing required sodium dodecyl sulfate to reach 31% of removal rate, ultrasound-assisted process did not need any additional chemical to reach 74.7% at the same condition, proving that ultrasound process can be considered as green process in this scenario. In ultrasound system with 40kHz at 200W, using additive $Na_2S_2O_8$, heat, and ultrasound efficiently degraded petroleum hydrocarbons from contaminated soil (Y.-T. Li et al., 2022). The combination of those three factors played an important role in the elimination of petroleum hydrocarbons. The assist of ultrasound improved the efficiency from 72% to 78%, while only 2 factors like ultrasound and $Na_2S_2O_8$, only reached 56.4%. The effectiveness of ultrasound come from the ability to degrade persulfate to sulfate for the degradation of petroleum hydrocarbons, and to deliver the washing additive to the macropores and micropores to hard-to-reach locations of the contaminated soil. In contrast, the ultrasound system to enhance the degradation of petroleum hydrocarbons was implemented with pairing Fe and $Na_2S_2O_8$ (Y. T. Li et al., 2020) for removal efficiency of $Na_2S_2O_8$. The presence of ultrasound nearly triple the efficiency at 50%, and the combination of all three factors resulted at 70% at the same condition. The result indicated that the effect of ultrasound came from many factors like free-radical generated from the cavitation bubble, the ability to enhance the solid to liquid transfer and also the ensure the continuous release of ferrous iron to activate $Na_2S_2O_8$. By using the same mechanism for the degradation of petroleum hydrocarbons, two different ultrasound systems of 43 kHz – 220W and 20 kHz – 550 W were used to treat contaminated soil sample (Lei et al., 2020), showing that, at single factor like only $Na_2S_2O_8$, high-frequency ultrasound, low-frequency ultrasound or raising temperature alone results in less than 10% of petroleum hydrocarbons degradation. But, combining two or three more factors together exhibited much greater result. For example, the dual ultrasound yielded a 38.7% degradation rate compared to 5.2% of single ultrasound frequency, and when coupling dual frequencies ultrasound system with $Na_2S_2O_8$ and evading the temperature, 88.9% of contaminant was degraded after 180 min process. It was noted from the research team that, both $SO_4^{\bullet-}$ and $\bullet OH$ contributed to the degradation, $SO_4^{\bullet-}$ took the leading position.

Beside total petroleum hydrocarbons (TPHs), other organic pollutants found in soil encompass pesticides, phenols, cyanides, microplastic, halogenated hydrocarbons (HOPs), and polycyclic aromatic hydrocarbons (PAHs). Among these, TPHs, HOPs, and PAHs are frequently detected in urban contaminated-soil (Trellu et al., 2016). These substances were characterized by their persistence, non-degradabilitym hydrophobic nature, and tendency to accumulate over time, as emphasized by Zheng at el. (Zheng et al., 2022). Their presence poses a risk to all living organisms because of the associated toxic, mutagenic, teratogenic, and carcinogenic hazards. The direct use of such contaminated soil for construction without remediation poses a serious threat to human health and urban development (Han et al., 2016; Shi et al., 2020). Traditional physical remediation technique such as excavation, landfill disposal, and soil washing offer high effectiveness but come with high-price and slow-operation pace (Cheng et al., 2021; Santos et al., 2018). Furthermore, the use of chemical in the washing of those organic contamination poses a threat of secondary pollutant from the process, and also required high amount of chemical for efficiency washing process. By generating strong fluid-solid shear forces, facilitating the mechanical detachment and elimination of contaminations, the soil washing process with the assist of ultrasound showed improvement in efficiency, shorten the washing time and also required additives (Mason et al., 2004; Newman et al., 1997). An innovative approach to eliminate dioxin-contaminated soil like poly-chlroinated dibenzo-p-dioxins and dibenzo-furans PCDFs was proposed by using the anaerobic compost tea from 1-year compost of seafood waste, combined wit 40 kHz ultrasound system and mechanical stirring (Hung et al., 2017). Efficiencies exceeding 85% for moderately contaminated soils and over 95% for highly contaminated soils were attained following 6 and 10 washing cycles, respectively, at ambient temperatures conditions with the assistance of ultrasound and mechanical stirring. While the mechanical stirring ensured the macro-scale mixing, ultrasound cavitation enabled micro-scale mixing which improved the washing efficiency of the process. The team observed and recorded that different alcoholic and acidic compounds from the tea is not only suitable for eliminating the dioxin from the soil, it also enriches the bioavailability of and bioactivity of the soil. In the experiment to treat the dioxin-contaminated soil, the same ultrasonic system was used with another solvent with fish oil (Vu et al., 2017). The moderately contaminated soil required 5 cycles (60 min/ultrasonic washing) for 94.12% removal rate, and 10 cycles for highly contaminated-soil. Fish oil in this case proved to be effective in removing dioxin from long time contaminated-soil. Furthermore, the environmentally friendly solvent also provided the bio-attenuation for the soil, which is a huge advantage compared with other synthesis surfactants. As applied the 28 kHz ultrasound system with mechanical stirring, eliminating polychlorinated biphenyls (PCB) from fresh soil (soil A) and weathered soil (soil B) was investigated in n-hexane and methanol (Lee & Son, 2021). The presence of ultrasound enhanced the outcome of the process from 13% to 20% at the same conditions. At the optimal condition, 90% of PCB from soil A was eliminated after washing with methanol and ultrasound, while the result for soil B was 70%. At 100 W ultrasound with 30kHz frequency, eliminating hexachlorobenzene (HCBs) and phenanthrene (PHE) was applied for different type of soils, namely synthetic clay, natural farm clay, and kaolin (Shrestha et al., 2009). Ultrasound washing worked best on normal soil with the removal rate of HCBs and PHEs around 57% and 71.% respectively. One point to be noted from this research is that, the increase of power will result in increase of efficiency to a certain extent, after that point the result will stable or slightly decrease, and higher liquid/solid also usually yield higher result but the ratio should be considered for the optimal condition.

Figure 5. Illustration of effect of ultrasound on soil washing

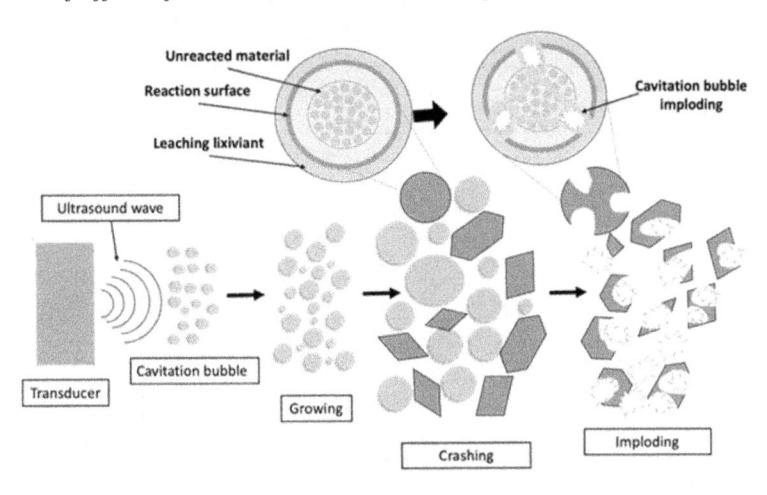

In conclusion for soil washing with ultrasound, as shown in Figure 5, mechanical effect of ultrasound, which is highly attributed with action of cavitation bubbles, behaves to pulverize soil particles. The effect of ultrasonic waves is not possible with ordinary cleaning or chemical cleaning, and a major advantage of ultrasonic cleaning is that it allows for higher efficiency in a shorter time. Finally, Table 2 summarizes the list of ultrasound-related research papers on soil cleaning. These examples use low-frequency band ultrasound in the frequency range of 20 kHz to 40 kHz. In particular, many studies have used acidic water and chelating agents in coexistence for the cleaning of heavy metals and toxic arsenic. On the other hand, in the decomposition of hydrocarbon-based organic substances, the direction is toward decontamination by decomposition with the mechanical pulverizing power of low-frequency ultrasound.

Table 2. Summarize of ultrasound-assisted soil washing

Contamination	Ultrasound Condition	Additive and condition	Efficiency	Reference
Cu, Pb, Zn	28 kHz – 250 W	HCl, 30 min, 25°C	Cu (74.5%) Pb (76%) Zn (71.1%)	(Park & Son, 2017)
Cu, Pb, Zn	28 kHz – 170 W	HCl/ EDTA, 20 min, 25°C	54.7%	(Choi et al., 2021b)
Cu, Pb, Zn	20 kHz – 170 W	HCl, 20 min, 25°C	62%	(Son et al., 2019)
As	28 kHz – 1200 W	HCl/NaOH, 1 hour, 25°C	96-99%	(Kobayashi & Phuoc Tri, 2022)
PAHs/ Cd, Pb, Zn	35/130 kHz – 110 W	Na_2SO_4/Nitric Acid, 10 min, 30°C	B(a)P (97%) B(a)A (91%) Cd (100%) Pb (63%) Zn (100%)	(Fraiese et al., 2020)
Cr, Ni, Cu, Zn, Pb	20 kHz – 50 W (Probe type) 24 kHz – 200 W (Cup-horn style)	EDTA. 2 – 10 min Acetic Acid, 6 – 30 min	EDTA (81-118%) Acetic acid (83-122%)	(De La Calle et al., 2013)

continued on following page

Table 2. Continued

Contamination	Ultrasound Condition	Additive and condition	Efficiency	Reference
Petroleum hydrocarbons	40 kHz – 200 W	$N_2S_2O_8$/Fe, 20°C, 3 days	70%	(Y. T. Li et al., 2020)
Petroleum hydrocarbons	40 kHz – 200 W	$N_2S_2O_8$, 50°C, 12 min	78.2%	(Y.-T. Li et al., 2022)
Petroleum hydrocarbons	43 kHz – 220 W 20 kHz – 550 W	$Na_2S_2O_8$, 180 min	88.9%	(Lei et al., 2020)
Oil-contaminated	20 kHz – 100 W	Water, 22°C	61%	(Kim & Wang, 2003)
PCBs	28 kHz – 480 W	Methanol, n-hexane, 20 – 25 min	84.5%	(F. Chen et al., 2016)
Dioxin (PCDDs, PCDFs)	40 kHz – 200 W)	Anaerobic compost tea, 25°C, 60 min/cycle	Moderated contaminated (6 cycles 85%) Highly contaminated (10 cycles 90%)	(Hung et al., 2017)
Dioxin (PCDDs, PCDFs)	40 kHz – 200 W	Fish oil, 25°C, 60 min/cycle	Moderated contaminated (5 cycles 94.12%) Highly contaminated (10 cycles 94.51%)	(Vu et al., 2017)
HCBs PHE	30 kHz – 100 W	Water, 60 min, 60°C	HCBs (57%) PHE (71%)	(Shrestha et al., 2009)

Author Note

Portions of these findings for data and analysis were obtained at Nagaoka university of Technology. Experiment samples of pollutant soil and slug wastes were provided with Fukuda construction and Onahama smelting and refining Co., LTD (Japan). We have no conflicts of interest to disclose.

Correspondence concerning this article should be addressed to the Nagaoka University of Technology, Japan. Email: p-tri@jet-e.jp

REFERENCES

Abdel-Shafy, H.-I., & Mansou, M. S. M. (2018). Solid waste issue: Sources, composition, disposal, recycling, and valorizatio. *Egyptian Journal of Petroleum*, 27(4), 1275–1290. DOI: 10.1016/j.ejpe.2018.07.003

Abdel-Shafy, H. I., & Mansour, M. S. M. (2016). A review on polycyclic aromatic hydrocarbons: Source, environmental impact, effect on human health and remediation. *Egyptian Journal of Petroleum*, 25(1), 107–123. DOI: 10.1016/j.ejpe.2015.03.011

Anca, F., Brindusa, C., Robu, M., & Gavrilescu, M. (2011). *Soil and groundwater cleanup : benefits and limits of emerging technologies*. 241–268. DOI: 10.1007/s10098-010-0319-z

Antuñano, N., Cambra, J. F., & Arias, P. L. (2019). Hydrometallurgical processes for Waelz oxide valorisation – An overview. *Process Safety and Environmental Protection*, 129, 308–320. DOI: 10.1016/j.psep.2019.06.028

Arya, S., & Kumar, S. (2020). E-waste in India at a glance: Current trends, regulations, challenges and management strategies. *Journal of Cleaner Production*, 271, 122707. DOI: 10.1016/j.jclepro.2020.122707

Awasthi, A. K., Zeng, X., & Li, J. (2016). Integrated bioleaching of copper metal from waste printed circuit board—A comprehensive review of approaches and challenges. *Environmental Science and Pollution Research International*, 23(21), 21141–21156. DOI: 10.1007/s11356-016-7529-9 PMID: 27678000

Bao, S., Chen, B., Zhang, Y., Ren, L., Xin, C., Ding, W., Yang, S., & Zhang, W. (2023). A comprehensive review on the ultrasound-enhanced leaching recovery of valuable metals: Applications, mechanisms and prospects. *Ultrasonics Sonochemistry*, 98, 106525. DOI: 10.1016/j.ultsonch.2023.106525 PMID: 37453257

Baum, Z. J., Bird, R. E., Yu, X., & Ma, J. (2022). Lithium-Ion Battery Recycling—Overview of Techniques and Trends. *ACS Energy Letters*, 7(2), 712–719. DOI: 10.1021/acsenergylett.1c02602

Beşe, A. V. (2007). Effect of ultrasound on the dissolution of copper from copper converter slag by acid leaching. *Ultrasonics Sonochemistry*, 14(6), 790–796. DOI: 10.1016/j.ultsonch.2007.01.007 PMID: 17383213

Bisht, A. (2022). Sand futures: Post-growth alternatives for mineral aggregate consumption and distribution in the global south. *Ecological Economics*, 191, 107233. DOI: 10.1016/j.ecolecon.2021.107233

Brunelli, K., & Dabalà, M. (2015). Ultrasound effects on zinc recovery from EAF dust by sulfuric acid leaching. *International Journal of Minerals Metallurgy and Materials*, 22(4), 353–362. DOI: 10.1007/s12613-015-1080-4

Cameselle, C., & Gouveia, S. (2018). Electrokinetic remediation for the removal of organic contaminants in soils. *Current Opinion in Electrochemistry*, 11, 41–47. DOI: 10.1016/j.coelec.2018.07.005

Chang, J., Zhang, E. D., Zhang, L. B., Peng, J. H., Zhou, J. W., Srinivasakannan, C., & Yang, C. J. (2017). A comparison of ultrasound-augmented and conventional leaching of silver from sintering dust using acidic thiourea. *Ultrasonics Sonochemistry*, 34, 222–231. DOI: 10.1016/j.ultsonch.2016.05.038 PMID: 27773239

Chen, B., Bao, S., Zhang, Y., & Li, S. (2020). A high-efficiency and sustainable leaching process of vanadium from shale in sulfuric acid systems enhanced by ultrasound. *Separation and Purification Technology, 240*(December 2019), 116624. DOI: 10.1016/j.seppur.2020.116624

Chen, F., Tan, M., Ma, J., Li, G., & Qu, J. (2016). Restoration of manufactured gas plant site soil through combined ultrasound-assisted soil washing and bioaugmentation. *Chemosphere*, 146, 289–299. DOI: 10.1016/j.chemosphere.2015.12.050 PMID: 26735729

Cheng, Y., Sun, H., Yang, E., Lv, J., Wen, B., Sun, F., Luo, L., & Liu, Z. (2021). Distribution and bioaccessibility of polycyclic aromatic hydrocarbons in industrially contaminated site soils as affected by thermal treatment. *Journal of Hazardous Materials*, 411(January), 125129. DOI: 10.1016/j.jhazmat.2021.125129 PMID: 33486229

Cho, K., Myung, E., Kim, H., Park, C., Choi, N., & Park, C. (2020). *Effect of Soil Washing Solutions on Simultaneous Removal of Heavy Metals and Arsenic from Contaminated Soil.*

Choi, J., Lee, D., & Son, Y. (2021a). Ultrasound-assisted soil washing processes for the remediation of heavy metals contaminated soils: {The} mechanism of the ultrasonic desorption. *Ultrasonics Sonochemistry*, 74, 105574. DOI: 10.1016/j.ultsonch.2021.105574 PMID: 33975185

Choi, J., Lee, D., & Son, Y. (2021b). Ultrasound-assisted soil washing processes for the remediation of heavy metals contaminated soils: {The} mechanism of the ultrasonic desorption. *Ultrasonics Sonochemistry*, 74, 105574. DOI: 10.1016/j.ultsonch.2021.105574 PMID: 33975185

De La Calle, I., Cabaleiro, N., Lavilla, I., & Bendicho, C. (2013). Ultrasound-assisted single extraction tests for rapid assessment of metal extractability from soils by total reflection X-ray fluorescence. *Journal of Hazardous Materials*, 260, 202–209. DOI: 10.1016/j.jhazmat.2013.05.021 PMID: 23770487

Dermont, G., Bergeron, M., & Mercier, G. (2008). *Soil washing for metal removal : A review of physical / chemical technologies and field applications. 152*, 1–31. DOI: 10.1016/j.jhazmat.2007.10.043

Dermont, G., Bergeron, M., Mercier, G., & Richer-Laflèche, M. (2008). Soil washing for metal removal: A review of physical/chemical technologies and field applications. *Journal of Hazardous Materials*, 152(1), 1–31. DOI: 10.1016/j.jhazmat.2007.10.043 PMID: 18036735

Devi, R. L., Sangeetha, K., & Shamini, S. S. (2021). Protection of sand silicamaterial from unauthorized intruders in semiconductor industries. *Materials Today: Proceedings*, 46, 3828–3831. DOI: 10.1016/j.matpr.2021.02.048

Diehl, O., Gatiboni, T., L., Mello,P., L., Muller, E., I., Duarte, F., A., Flores, E., M., M. (2018). *Ultrasonics Sonochemistry*, 40 (B), 24-29. DOI: 10.1016/j.ultsonch.2017.04.012

Ding, W., Bao, S., Zhang, Y., & Xiao, J. (2022). Mechanism and kinetics study on ultrasound assisted leaching of gallium and zinc from corundum flue dust. *Minerals Engineering*, 183(April), 107624. DOI: 10.1016/j.mineng.2022.107624

Du, F., Li, J., Li, X., & Zhang, Z. (2011). Improvement of iron removal from silica sand using ultrasound-assisted oxalic acid. *Ultrasonics Sonochemistry*, 18(1), 389–393. DOI: 10.1016/j.ultsonch.2010.07.006 PMID: 20674453

Duester, L., Brinkmann, C., Ternes, T. A., & Heininger, P. (2016). Commentary. *Critical Reviews in Environmental Science and Technology*, 46(4), 434–437. DOI: 10.1080/10590501.2015.1131559

Duker, A. E. C., Mawoyo, T. A., Bolding, A., de Fraiture, C., & van der Zaag, P. (2020). Shifting or drifting? the crisis-driven advancement and failure of private smallholderirrigation from sand river aquifers in southern arid Zimbabwe. *Agricultural Water Management*, 241, 106342. DOI: 10.1016/j.agwat.2020.106342

Esmaeili, M., Rastegar, S. O., Beigzadeh, R., & Gu, T. (2020). Ultrasound-assisted leaching of spent lithium ion batteries by natural organic acids and H2O2. *Chemosphere*, 254, 126670. DOI: 10.1016/j.chemosphere.2020.126670 PMID: 32325352

Fraiese, A., Cesaro, A., Belgiorno, V., Sanromán, M. A., Pazos, M., & Naddeo, V. (2020). Ultrasonic processes for the advanced remediation of contaminated sediments. *Ultrasonics Sonochemistry*, 67(March), 105171. DOI: 10.1016/j.ultsonch.2020.105171 PMID: 32446202

Frazzoli, C., Ruggieri, F., Battistini, B., Orisakwe, O. E., Igbo, J. K., & Bocca, B. (2022). E-WASTE threatens health: The scientific solution adopts the one health strategy. *Environmental Research, 212*(PA), 113227. DOI: 10.1016/j.envres.2022.113227

Fuentes, I., Ulloa, C., Jimenez, R., & Garcia, X. (2020). The Reduction of Fe-Bearing Copper Slag for its Use as a Catalyst in Carbon Oxide Hydrogenation to Methane. A Contribution to Sustainable Catalysis. *Journal of Hazardous Materials*, 387, 121693. DOI: 10.1016/j.jhazmat.2019.121693 PMID: 31787399

Gao, W., Liu, C., Cao, H., Zheng, X., Lin, X., Wang, H., Zhang, Y., & Sun, Z. (2018). Comprehensive evaluation on effective leaching of critical metals from spent lithium-ion batteries. *Waste Management (New York, N.Y.)*, 75, 477–485. DOI: 10.1016/j.wasman.2018.02.023 PMID: 29459203

Guo, P., Wang, S., & Zhang, L. (2019). Selective removal of antimony from refractory gold ores by ultrasound. *Hydrometallurgy*, 190(October), 105161. DOI: 10.1016/j.hydromet.2019.105161

Han, L., Qian, L., Yan, J., Liu, R., Du, Y., & Chen, M. (2016). A comparison of risk modeling tools and a case study for human health risk assessment of volatile organic compounds in contaminated groundwater. *Environmental Science and Pollution Research International*, 23(2), 1234–1245. DOI: 10.1007/s11356-015-5335-4 PMID: 26354114

Hernández, I. A., Díaz, H. L. M., Morales, F. J. F., Romero, L. R., & Camacho, J. V. (2023). Bioleaching of metal polluted mine tailings aided by ultrasound irradiation pretreatment. *Environmental Technology & Innovation*, 31, 103192. DOI: 10.1016/j.eti.2023.103192

Hu, W., Wang, H., Dong, L., Huang, B., Borggaard, O. K., Bruun Hansen, H. C., He, Y., & Holm, P. E. (2018). Source identification of heavy metals in peri-urban agricultural soils of southeast {China}: {An} integrated approach. *Environmental Pollution*, 237, 650–661. DOI: 10.1016/j.envpol.2018.02.070 PMID: 29529426

Huang, Z., Xie, F., & Ma, Y. (2011). Ultrasonic recovery of copper and iron through the simultaneous utilization of Printed Circuit Boards (PCB) spent acid etching solution and PCB waste sludge. *Journal of Hazardous Materials*, 185(1), 155–161. DOI: 10.1016/j.jhazmat.2010.09.010 PMID: 20932641

Hung, W., Huang, W. Y., Lin, C., Vu, C. T., Yotapukdee, S., Kaewlaoyoong, A., Chen, J. R., & Shen, Y. H. (2017). The use of ultrasound-assisted anaerobic compost tea washing to remove poly-chlorinated dibenzo-p-dioxins (PCDDs), dibenzo-furans (PCDFs) from highly contaminated field soils. *Environmental Science and Pollution Research International*, 24(23), 18936–18945. DOI: 10.1007/s11356-017-9517-0 PMID: 28656572

Jadhao, P. R., Ahmad, E., Pant, K. K., & Nigam, K. D. P. (2020). Environmentally friendly approach for the recovery of metallic fraction from waste printed circuit boards using pyrolysis and ultrasonication. *Waste Management (New York, N.Y.)*, 118, 150–160. DOI: 10.1016/j.wasman.2020.08.028 PMID: 32892092

Jiang, F., Chen, Y., Ju, S., Zhu, Q., Zhang, L., Peng, J., Wang, X., & Miller, J. D. (2018). Ultrasound-assisted leaching of cobalt and lithium from spent lithium-ion batteries. *Ultrasonics Sonochemistry*, 48(May), 88–95. DOI: 10.1016/j.ultsonch.2018.05.019 PMID: 30080590

Johansson, Ö., Pamidi, T., & Shankar, V. (2021). Extraction of tungsten from scheelite using hydrodynamic and acoustic cavitation. *Ultrasonics Sonochemistry, 71*(November 2020). DOI: 10.1016/j.ultsonch.2020.105408

John, J. J., De Houwer, V., Van Mechelen, D., & Van Gerven, T. (2020). Effect of ultrasound on leaching of lead from landfilled metallurgical residues. *Ultrasonics Sonochemistry*, 69(July), 2–10. DOI: 10.1016/j.ultsonch.2020.105239 PMID: 32645663

Joo, S. H., Kang, J., Woong, K., & Shin, S. M. (2013). Production of chemical manganese dioxide from lithium ion battery ternary cathodic material by selective oxidative precipitation of manganese. *Materials Transactions*, 54(5), 844–849. DOI: 10.2320/matertrans.M2012379

Kaliyavaradhan, S. K., Prem, P. R., Ambily, P. S., & Mo, K. H. (2022). Effective utilization of e-waste plastics and glasses in construction products - a review and future research directions. *Resources, Conservation and Recycling, 176*(September 2021), 105936. DOI: 10.1016/j.resconrec.2021.105936

Kim, Y. U., & Wang, M. C. (2003). Effect of ultrasound on oil removal from soils. *Ultrasonics*, 41(7), 539–542. DOI: 10.1016/S0041-624X(03)00168-9 PMID: 12919689

Knaislová, A., Vu, H. N., & Dvořák, P. (2018). Microwave and ultrasound effect on ammoniacal leaching of deep-sea nodules. *Minerals (Basel)*, 8(8), 1–16. DOI: 10.3390/min8080351

Kobayashi, T., & Phuoc Tri, P. (2022). Effect of High-Power Ultrasound Washing on Arsenic-Polluted Soil. *JOURNAL OF CHEMICAL ENGINEERING OF JAPAN, 55*(9), 22we027. DOI: 10.1252/jcej.22we027

Kossoff, D., Dubbin, W. E., Alfredsson, M., Edwards, S. J., Macklin, M. G., & Hudson-Edwards, K. A. (2014). Mine tailings dams: Characteristics, failure, environmental impacts, and remediation. *Applied Geochemistry*, 51, 229–245. DOI: 10.1016/j.apgeochem.2014.09.010

Krishnan, S., Zulkapli, N. S., Kamyab, H., Taib, S. M., Din, M. F. B., Majid, Z. A., Chaiprapat, S., Kenzo, I., Ichikawa, Y., Nasrullah, M., Chelliapan, S., & Othman, S. (2021). Current technologies for recovery of metals from industrial wastes: An overview. *Environmental Technology & Innovation*, 2, 101525. DOI: 10.1016/j.eti.2021.101525

Kumar, A., Holuszko, M. E., & Janke, T. (2018). Characterization of the non-metal fraction of the processed waste printed circuit boards. *Waste Management (New York, N.Y.)*, 75, 94–102. DOI: 10.1016/j.wasman.2018.02.010 PMID: 29449113

Kumar, V., Sharma, A., Kaur, P., Singh Sidhu, G. P., Bali, A. S., Bhardwaj, R., Thukral, A. K., & Cerda, A. (2019). Pollution assessment of heavy metals in soils of India and ecological risk assessment: A state-of-the-art. In *Chemosphere* (Vol. 216, pp. 449–462). Elsevier Ltd., DOI: 10.1016/j.chemosphere.2018.10.066

Kwedi-Nsah, L.-M., Watanabe, Y., & Kobayashi, T. (2019). *Soil Recycling Geopolymers Fabricated from High Power Ultrasound Treated Soil Slurry in the Presence of Ammonia.*

Lee, D., & Son, Y. (2021). Ultrasound-assisted soil washing processes using organic solvents for the remediation of {PCBs}-contaminated soils. *Ultrasonics Sonochemistry*, 80, 105825. DOI: 10.1016/j.ultsonch.2021.105825 PMID: 34775161

Lei, Y. J., Zhang, J., Tian, Y., Yao, J., Duan, Q. S., & Zuo, W. (2020). Enhanced degradation of total petroleum hydrocarbons in real soil by dual-frequency ultrasound-activated persulfate. *The Science of the Total Environment*, 748, 141414. DOI: 10.1016/j.scitotenv.2020.141414 PMID: 32814296

Li, C., Zhang, H., Ding, P., Yang, S., & Bai, Y. (2023). Deep feature extraction in lifetime prognostics of lithium-ion batteries: Advances, challenges and perspectives. *Renewable & Sustainable Energy Reviews*, 184(April), 113576. DOI: 10.1016/j.rser.2023.113576

Li, H., Zhang, L., Xie, H., Yin, S., Peng, J., Li, S., Yang, K., & Zhu, F. (2020). Ultrasound-Assisted Silver Leaching Process for Cleaner Production. *Journal of the Minerals Metals & Materials Society*, 72(2), 766–773. DOI: 10.1007/s11837-019-03843-8

Li, S., Chen, W., Yin, S., Ma, A., Yang, K., Xie, F., Zhang, L., & Peng, J. (2015). Impacts of ultrasound on leaching recovery of zinc from low grade zinc oxide ore. *Green Processing and Synthesis*, 4(4), 323–328. DOI: 10.1515/gps-2015-0036

Li, Y. T., Li, D., Lai, L. J., & Li, Y. H. (2020). Remediation of petroleum hydrocarbon contaminated soil by using activated persulfate with ultrasound and ultrasound/Fe. *Chemosphere*, 238, 124657. DOI: 10.1016/j.chemosphere.2019.124657 PMID: 31473526

Li, Y.-T., Zhang, J.-J., Li, Y.-H., Chen, J.-L., & Du, W.-Y. (2022). Treatment of soil contaminated with petroleum hydrocarbons using activated persulfate oxidation, ultrasound, and heat: {A} kinetic and thermodynamic study. *Chemical Engineering Journal*, 428, 131336. DOI: 10.1016/j.cej.2021.131336

Linares, G., & Rojas, L. (2022). Ultrasound-Assisted Extraction of Natural Pigments From Food Processing By-Products: A Review. *Frontiers in Nutrition*, 9, 891462. Advance online publication. DOI: 10.3389/fnut.2022.891462 PMID: 35685880

Liu, H., Wang, S., Fu, L., Zhang, G., Zuo, Y., & Zhang, L. (2022). Mechanism and kinetics analysis of valuable metals leaching from copper-cadmium slag assisted by ultrasound cavitation. *Journal of Cleaner Production*, 379(P2), 134775. DOI: 10.1016/j.jclepro.2022.134775

Liu, J., Wang, S., Zhang, Y., Zhang, L., & Kong, D. (2022). Synergistic mechanism and decopperization kinetics for copper anode slime via an integrated ultrasound-sodium persulfate process. *Applied Surface Science*, 589(February), 153032. DOI: 10.1016/j.apsusc.2022.153032

Maisel, F., Neef, C., Marscheider-Weidemann, F., & Nissen, N. F. (2023). A forecast on future raw material demand and recycling potential of lithium-ion batteries in electric vehicles. *Resources, Conservation and Recycling*, 192(February), 106920. DOI: 10.1016/j.resconrec.2023.106920

Makuza, B., Yu, D., Huang, Z., Tian, Q., & Guo, X. (2021). Dry Grinding - Carbonated Ultrasound-Assisted Water Leaching of Carbothermally Reduced Lithium-Ion Battery Black Mass Towards Enhanced Selective Extraction of Lithium and Recovery of High-Value Metals. *Resources, Conservation and Recycling*, 174(June), 105784. DOI: 10.1016/j.resconrec.2021.105784

Mason, T. J., Collings, A., & Sumel, A. (2004). Sonic and ultrasonic removal of chemical contaminants from soil in the laboratory and on a large scale. *Ultrasonics Sonochemistry*, 11(3–4), 205–210. DOI: 10.1016/j.ultsonch.2004.01.025 PMID: 15081982

McCutcheon, S. C., & Jørgensen, S. E. (2008). Phytoremediation. *Encyclopedia of Ecology*, 568–582. DOI: 10.1016/B978-0-444-63768-0.00069-X

Ministry of Land, Infrasture, T. and T. (2012). *Investigation results of construction by-product in 2012*.

Mir, S., & Dhawan, N. (2022). A comprehensive review on the recycling of discarded printed circuit boards for resource recovery. *Resources, Conservation and Recycling, 178*(November 2021), 106027. DOI: 10.1016/j.resconrec.2021.106027

Nakashima, K., & Kumahara, Y. (2002). Effect of tin oxide dispersion on nodule formation in ITO sputtering. *Vacuum*, 66(3–4), 221–226. DOI: 10.1016/S0042-207X(02)00145-8

Newman, A. P., Lorimer, J. P., Mason, T. J., & Hutt, K. R. (1997). An investigation into the ultrasonic treatment of polluted solids. *Ultrasonics Sonochemistry*, 4(2), 153–156. DOI: 10.1016/S1350-4177(97)00020-5 PMID: 11237034

Nguyen, T. T. N., & Lee, M. S. (2019). Improvement of alumina dissolution from the mechanically activated dross using ultrasound-assisted leaching. *Journal of Korean Institute of Metals and Materials*, 57(3), 154–161. DOI: 10.3365/KJMM.2019.57.3.154

Ning, P., Meng, Q., Dong, P., Duan, J., Xu, M., Lin, Y., & Zhang, Y. (2020). Recycling of cathode material from spent lithium ion batteries using an ultrasound-assisted DL-malic acid leaching system. *Waste Management (New York, N.Y.)*, 103, 52–60. DOI: 10.1016/j.wasman.2019.12.002 PMID: 31865035

Pariatamby, A., & Victor, D. (2013). Policy trends of e-waste management in Asia. *Journal of Material Cycles and Waste Management*, 15(4), 411–419. DOI: 10.1007/s10163-013-0136-7

Park, B., & Son, Y. (2017). Ultrasonic and mechanical soil washing processes for the removal of heavy metals from soils. *Ultrasonics Sonochemistry*, 35, 640–645. DOI: 10.1016/j.ultsonch.2016.02.002 PMID: 26867953

Phuoc Tri, P., Takaomi, K., & Syuji, U. (2023). Ultrasound effects on restricted silica gelation during silica extraction from Pyro-Metallurgical copper slag under acidifying conditions. *Ultrasonics Sonochemistry*, 97(April), 106447. DOI: 10.1016/j.ultsonch.2023.106447 PMID: 37245264

Piatak, N. M., Parsons, M. B., & Seal, R. R.II. (2015). Characteristics and Environmental Aspects of Slag: A Review. *Applied Geochemistry*, 57, 236–266. DOI: 10.1016/j.apgeochem.2014.04.009

Pradhan, N., Das, B., Gahan, C. S., Kar, R. N., & Sukla, L. B. (2006). Beneficiation of iron ore slime using Aspergillus niger and Bacillus circulans. *Bioresource Technology*, 97(15), 1876–1879. DOI: 10.1016/j.biortech.2005.08.010 PMID: 16531043

Rayess, Y. E., Dawra, M., & Beyrouthy, M. E. (2022). Chapter 17 - Modern extraction techniques for herbal bioactives, *Herbal Bioactive-Based Drug Delivery Systems Challenges and Opportunities*, 437-455. DOI: 10.1016/B978-0-12-824385-5.00002-9

Rene, E. R., Sethurajan, M., Kumar Ponnusamy, V., Kumar, G., Bao Dung, T. N., Brindhadevi, K., & Pugazhendhi, A. (2021). Electronic waste generation, recycling and resource recovery: Technological perspectives and trends. *Journal of Hazardous Materials,* 416(May 2020), 125664. DOI: 10.1016/j.jhazmat.2021.125664

Santos, A., Fernandez, J., Rodriguez, S., Dominguez, C. M., Lominchar, M. A., Lorenzo, D., & Romero, A. (2018). Abatement of chlorinated compounds in groundwater contaminated by HCH wastes using ISCO with alkali activated persulfate. *The Science of the Total Environment*, 615, 1070–1077. DOI: 10.1016/j.scitotenv.2017.09.224 PMID: 29751410

Savvilotidou, V., Hahladakis, J. N., & Gidarakos, E. (2014). Determination of toxic metals in discarded Liquid Crystal Displays (LCDs). *Resources, Conservation and Recycling*, 92, 108–115. DOI: 10.1016/j.resconrec.2014.09.002

Şayan, E., & Bayramo lu, M. (2004). Statistical modeling and optimization of ultrasound-assisted sulfuric acid leaching of TiO2 from red mud. *Hydrometallurgy*, 71(3–4), 397–401. DOI: 10.1016/S0304-386X(03)00113-0

Shi, Z., Liu, J., Tang, Z., Zhao, Y., & Wang, C. (2020). Vermiremediation of organically contaminated soils: Concepts, current status, and future perspectives. *Applied Soil Ecology,* 147(February 2019), 103377. DOI: 10.1016/j.apsoil.2019.103377

Shrestha, R. A., Pham, T. D., & Sillanpää, M. (2009). *Effect of ultrasound on removal of persistent organic pollutants (POPs) from different types of soils.* 170, 871–875. DOI: 10.1016/j.jhazmat.2009.05.048

Son, Y., Cha, J., Lim, M., Ashokkumar, M., & Khim, J. (2011). Comparison of {Ultrasonic} and {Conventional} {Mechanical} {Soil}-{Washing} {Processes} for {Diesel}-{Contaminated} {Sand}. *Industrial \& Engineering Chemistry Research, 50*(4), 2400–2407. DOI: 10.1021/ie1016688

Son, Y., Lee, D., Lee, W., Park, J., Hyoung Lee, W., & Ashokkumar, M. (2019). Cavitational activity in heterogeneous systems containing fine particles. *Ultrasonics Sonochemistry*, 58, 104599. DOI: 10.1016/j.ultsonch.2019.05.016 PMID: 31450376

Souada, M., Louage, C., Doisy, J. Y., Meunier, L., Benderrag, A., Ouddane, B., Bellayer, S., Nuns, N., Traisnel, M., & Maschke, U. (2018). Extraction of indium-tin oxide from end-of-life LCD panels using ultrasound assisted acid leaching. *Ultrasonics Sonochemistry*, 40, 929–936. DOI: 10.1016/j.ultsonch.2017.08.043 PMID: 28946505

Thi Hong Hoa, N., Hoan, N. T., Nguyen Trong, N., Linh, N. T. N., Quy, B. M., Pham, T. T. H., Nguyen, V. Q., Nguyen Van, P., & Nguyen, V. D. (2023). Enhancement of Leaching Copper from Printed Circuit Boards of Discarded Mobile Phones Using Ultrasound–Ozone Integrated Approach. *Metals*, 13(6), 1145. DOI: 10.3390/met13061145

Toache-Pérez, A. D., Bolarín-Miró, A. M., Sánchez-De Jesús, F., & Lapidus, G. T. (2020). Facile method for the selective recovery of Gd and Pr from LCD screen wastes using ultrasound-assisted leaching. *Sustainable Environment Research*, 30(1), 1–14. DOI: 10.1186/s42834-020-00060-w

Toache-Pérez, A. D., Lapidus, G. T., Bolarín-Miró, A. M., & De Jesús, F. S. (2022). Selective Leaching and Recovery of Er, Gd, Sn, and in from Liquid Crystal Display Screen Waste by Sono-Leaching Assisted by Magnetic Separation. *ACS Omega*, 7(36), 31897–31904. DOI: 10.1021/acsomega.2c02729 PMID: 36119989

Tóth, G., Hermann, T., Da Silva, M. R., & Montanarella, L. (2016). Heavy metals in agricultural soils of the {European} {Union} with implications for food safety. *Environment International*, 88, 299–309. DOI: 10.1016/j.envint.2015.12.017 PMID: 26851498

Trellu, C., Mousset, E., Pechaud, Y., Huguenot, D., van Hullebusch, E. D., Esposito, G., & Oturan, M. A. (2016). Removal of hydrophobic organic pollutants from soil washing/flushing solutions: A critical review. *Journal of Hazardous Materials*, 306, 149–174. DOI: 10.1016/j.jhazmat.2015.12.008 PMID: 26707974

Vishwakarma, G. S., Bhattacharjee, G., Gohil, N., & Singh, V. (2020). Current status, challenges and future of bioremediation. *Bioremediation of Pollutants: From Genetic Engineering to Genome Engineering*, 403–415. DOI: 10.1016/B978-0-12-819025-8.00020-X

Vu, C. T., Lin, C., Hung, W., Huang, W. Y., Kaewlaoyoong, A., Yotapukdee, S., Chen, J. R., & Shen, Y. H. (2017). Ultrasonic Soil Washing with Fish Oil Extract to Remove Polychlorinated Dibenzo-p-dioxins (PCDDs), Dibenzofurans (PCDFs) from Highly Contaminated Field Soils. *Water, Air, and Soil Pollution*, 228(9), 343. Advance online publication. DOI: 10.1007/s11270-017-3534-y

Wang, H., Zhang, S., Li, B., Pan, D., Wu, Y., & Zuo, T. (2017). Recovery of waste printed circuit boards through pyrometallurgical processing: A review. *Resources, Conservation and Recycling*, 126(April), 209–218. DOI: 10.1016/j.resconrec.2017.08.001

Wang, J., Faraji, F., & Ghahreman, A. (n.d.). *minerals Effect of Ultrasound on the Oxidative Copper Leaching from Chalcopyrite in Acidic Ferric Sulfate Media*. DOI: 10.3390/min10070633

Wang, J. P., & Erdenebold, U. (2020). A study on reduction of copper smelting slag by carbon for recycling into metal values and cement raw material. *Sustainability (Basel)*, 12(4), 1421. Advance online publication. DOI: 10.3390/su12041421

Wang, L., Cho, D., Tsang, D. C. W., Cao, X., Hou, D., Shen, Z., Alessi, D. S., Sik, Y., & Sun, C. (2019). Green remediation of As and Pb contaminated soil using cement-free clay- based stabilization / solidi fi cation. *Environment International, 126*(November 2018), 336–345. DOI: 10.1016/j.envint.2019.02.057

Wang, R., & Xu, Z. (2016). Pyrolysis characteristics and pyrolysis products separation for recycling organic materials from waste liquid crystal display panels. *Journal of Hazardous Materials*, 302, 45–56. DOI: 10.1016/j.jhazmat.2015.09.038 PMID: 26444486

Wang, S., Cui, W., Zhang, G., Zhang, L., & Peng, J. (2017). Ultra fast ultrasound-assisted decopperization from copper anode slime. *Ultrasonics Sonochemistry*, 36, 20–26. DOI: 10.1016/j.ultsonch.2016.11.013 PMID: 28069202

Wang, W. Y., Yen, C. H., & Hsu, J. K. (2020). Selective recovery of cobalt from the cathode materials of NMC type Li-ion battery by ultrasound-assisted acid leaching and microemulsion extraction. *Separation Science and Technology*, 55(16), 3028–3035. DOI: 10.1080/01496395.2019.1665071

Xiao, X., Hoogendoorn, B. W., Ma, Y., Ashoka Sahadevan, S., Gardner, J. M., Forsberg, K., & Olsson, R. T. (2021). Ultrasound-assisted extraction of metals from Lithium-ion batteries using natural organic acids. *Green Chemistry*, 23(21), 8519–8532. DOI: 10.1039/D1GC02693C

Xie, F., Li, H., Ma, Y., Li, C., Cai, T., Huang, Z., & Yuan, G. (2009). The ultrasonically assisted metals recovery treatment of printed circuit board waste sludge by leaching separation. *Journal of Hazardous Materials*, 170(1), 430–435. DOI: 10.1016/j.jhazmat.2009.04.077 PMID: 19457612

Xin, C., Xia, H., Jiang, G., Zhang, Q., Zhang, L., Xu, Y., & Cai, W. (2022). Mechanism and kinetics study on ultrasonic combined with oxygen enhanced leaching of zinc and germanium from germanium-containing slag dust. *Separation and Purification Technology*, 302(July), 122167. DOI: 10.1016/j.seppur.2022.122167

Yang, J., He, L., Liu, X., Ding, W., Song, Y., & Zhao, Z.YANG. (2018). Comparative kinetic analysis of conventional and ultrasound-assisted leaching of scheelite by sodium carbonate. [English Edition]. *Transactions of Nonferrous Metals Society of China*, 28(4), 775–782. DOI: 10.1016/S1003-6326(18)64710-7

Yu, M., Zeng, X., Song, Q., Liu, L., & Li, J. (2016). Examining regeneration technologies for etching solutions: A critical analysis of the characteristics and potentials. *Journal of Cleaner Production*, 113, 973–980. DOI: 10.1016/j.jclepro.2015.10.131

Zahed, M. A., Aziz, H. A., Isa, M. H., & Mohajeri, L. (2010). Effect of initial oil concentration and dispersant on crude oil biodegradation in contaminated seawater. *Bulletin of Environmental Contamination and Toxicology*, 84(4), 438–442. DOI: 10.1007/s00128-010-9954-7 PMID: 20224975

Zhang, J., Wu, A., Wang, Y., & Chen, X. (2008). Experimental research in leaching of copper-bearing tailings enhanced by ultrasonic treatment. *Journal of China University of Mining and Technology*, 18(1), 98–102. DOI: 10.1016/S1006-1266(08)60021-8

Zhang, K., Li, B., Wu, Y., Wang, W., Li, R., Zhang, Y. N., & Zuo, T. (2017). Recycling of indium from waste LCD: A promising non-crushing leaching with the aid of ultrasonic wave. *Waste Management (New York, N.Y.)*, 64, 236–243. DOI: 10.1016/j.wasman.2017.03.031 PMID: 28347586

Zhang, L., Guo, W., Peng, J., Li, J., Lin, G., & Yu, X. (2016). Comparison of ultrasonic-assisted and regular leaching of germanium from by-product of zinc metallurgy. *Ultrasonics Sonochemistry*, 31, 143–149. DOI: 10.1016/j.ultsonch.2015.12.006 PMID: 26964934

Zhang, P., Ma, Y., & Xie, F. (2013). Impacts of ultrasound on selective leaching recovery of heavy metals from metal-containing waste sludge. *Journal of Material Cycles and Waste Management*, 15(4), 530–538. DOI: 10.1007/s10163-013-0131-z

Zhang, R. L., Zhang, X. F., Tang, S. Z., & Huang, A. D. (2015). Ultrasound-assisted HCl-NaCl leaching of lead-rich and antimony-rich oxidizing slag. *Ultrasonics Sonochemistry*, 27, 187–191. DOI: 10.1016/j.ultsonch.2015.05.020 PMID: 26186836

Zhang, W., Lin, Y. C., Chien, S. K., Wu, T. Y., Chen, S. C., Cheng, P. C., & Lai, C. N. (2020). Efficient indium leaching and recovery from waste liquid crystal displays panels using microwave and ultrasound-assisted heating system. *Separation and Purification Technology*, 250(May), 117154. DOI: 10.1016/j.seppur.2020.117154

Zheng, W., Cui, T., & Li, H. (2022). Combined technologies for the remediation of soils contaminated by organic pollutants. A review. *Environmental Chemistry Letters*, 20(3), 2043–2062. DOI: 10.1007/s10311-022-01407-y

Zhou, W., Liang, H., & Xu, H. (2021). Recovery of gold from waste mobile phone circuit boards and synthesis of nanomaterials using emulsion liquid membrane. *Journal of Hazardous Materials, 411*(December 2020). DOI: 10.1016/j.jhazmat.2020.125011

Zhu, M., Yuan, Y., Yin, H., Guo, Z., Wei, X., Qi, X., Liu, H., & Dang, Z. (2022). Environmental contamination and human exposure of polychlorinated biphenyls (PCBs) in China: A review. *The Science of the Total Environment*, 805, 150270. DOI: 10.1016/j.scitotenv.2021.150270 PMID: 34536863

Zhuang, C., & Chen, Y. (2019). The effect of nano-SiO2 on concrete properties: A review. *Nanotechnology Reviews*, 8(1), 562–572. DOI: 10.1515/ntrev-2019-0050

Zou, J., Luo, Y., Yu, X., Li, J., Xi, Y., Zhang, L., Guo, W., & Lin, G. (2020). Extraction of Indium from By-products of Zinc Metallurgy by Ultrasonic Waves. *Arabian Journal for Science and Engineering*, 45(9), 7321–7328. DOI: 10.1007/s13369-020-04471-0

Zuo, Z., Feng, Y., Luo, S., Dong, X., Li, X., Ren, D., Yu, Q., & Guo, J. (2021). Element Distribution and Migration Behavior in the Copper Slag Reduction and Separation Process. *Frontiers in Energy Research*, 9, 760312. Advance online publication. DOI: 10.3389/fenrg.2021.760312

Chapter 17
Assessing the Photocatalytic Performance of Carbon-Based Semiconductors in the Degradation of Pharmaceutical Wastes

Nursarah Sofea Ismail
Universiti Sains Malaysia, Malaysia

Ahmad Fadhil Bin Rithwan
Universiti Sains Malaysia, Malaysia

Sirikanjana Thongmee
Kasetsart University, Thailand

Siti Fairus Mohd Yusoff
Universiti Kebangsaan Malaysia, Malaysia

Rohana Adnan
Universiti Sains Malaysia, Malaysia

Noor Haida Mohd Kaus
https://orcid.org/0000-0002-6257-7302
Universiti Sains Malaysia, Malaysia

ABSTRACT

This review highlights recent research endeavors in the field, emphasizing the critical role of carbon-based photocatalysts for their superior performance under visible light irradiation and their capacity to efficiently photodegrade pharmaceutical contaminants. The integration of carbon materials with semiconductors has demonstrated remarkable synergistic effects, leading to enhanced photocatalytic efficiency. The exploration of carbon-based semiconductor materials represents a promising avenue for addressing the persistent issue of pharmaceutical waste in the environment, offering a greener and more efficient solution for water treatment.

INTRODUCTION

One of the emerging pollutants that piqued the researcher's attention was pharmaceutical compounds. Pharmaceutical compounds have garnered significant attention as emerging pollutants due to their indispensable roles in disease treatment and prevention in both human and animal health. The widespread

DOI: 10.4018/979-8-3693-0003-9.ch017

use of pharmaceuticals has led to their increasing presence in the environment. These pharmaceutical residues originate from various sources, including inadequate wastewater treatment systems, hospitals, pharmaceutical industries, and even from the excretion of unmetabolized drugs in urine (Gupta et al., 2017; Hapeshi et al., 2010). Concentrations of pharmaceuticals ranging from micrograms per liter (µg/L) to milligrams per liter (mg/L) have been detected in both groundwater and surface water (Cabeza et al., 2012; Roberts et al., 2016; Wang and Wang, 2016).

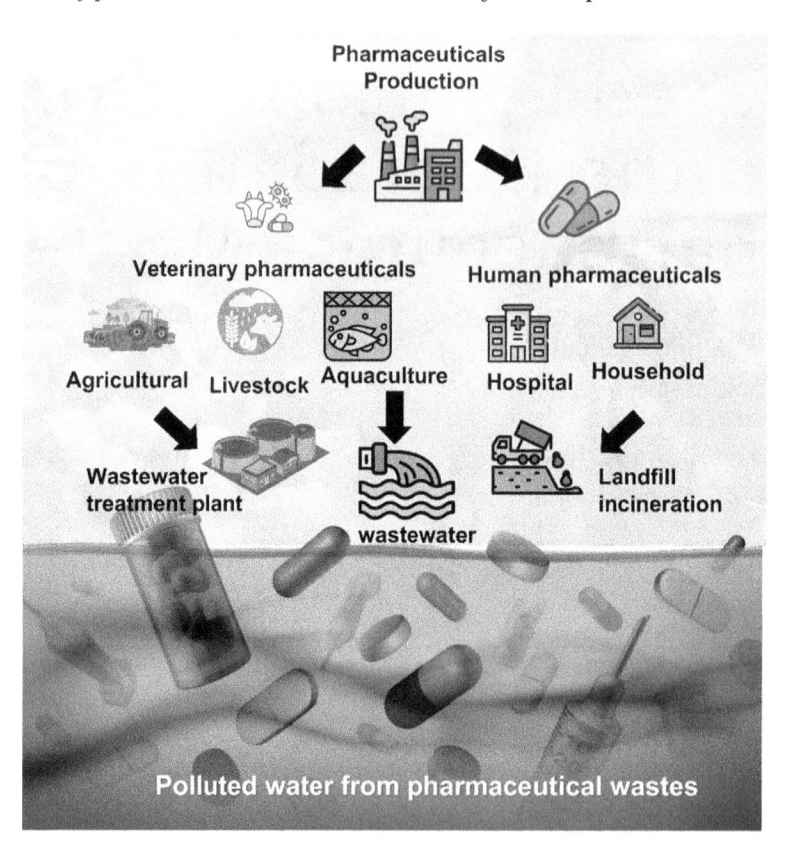

Figure 1. The sources of polluted water to the environment from the pharmaceutical wastes

Despite the multiple routes through which pharmaceuticals enter the water environment, pharmaceutical industry effluents remain a major point source of pollution. Improper disposal of unused or expired drugs in households and hospitals, which often find their way into sewage systems, contributes further to the presence of pharmaceuticals in aquatic environments. Additionally, runoff from agricultural activities, containing herbicides, pesticides, and veterinary medicines, can frequently contaminate water bodies (Fick et al., 2009; Larsson et al., 2007 and Mompelat et al., 2009). Figure 1 summarizes the sources of pharmaceutical waste that become hazardous substances in water due to inappropriate disposal and treatment practices. As global pollution levels rise, so does the demand for pharmaceutical products, resulting in an increase in pharmaceutical-contaminated water and, consequently, water stress. Unfortunately, the presence of pharmaceuticals in aquatic environments has been linked to various negative biological and economic impacts. Prolonged exposure to pharmaceutically contaminated water can lead to the development of drug-resistant bacterial strains in both humans and animals (Danner et al., 2019). Recent research indicates a substantial increase in antibiotic consumption, with a projected 200% increase by 2030 (Polianciuc et al., 2020). According to Ambrosetti et al. (2015), antibiotics can persist in the environment as parent molecules, breakdown products, and partially metabolized forms, further exacerbated by their propensity to bioaccumulate and transform within solid and aqueous phases (Ambrosetti et al., 2015; Manzetti & Ghisi, 2014). This persistence raises concerns about antimicrobial resistance in the environment, necessitating the development of new drugs to combat this issue. To mitigate the adverse consequences of pharmaceuticals in aquatic environments and enhance water quality for human consumption, there is a growing need for more effective techniques to remove these pollutants (Ferri et al., 2017).

Regardless the frequent detection of pharmaceuticals in the environment at extremely low concentrations (ng/L - µg/L), their non-biodegradable and pseudo-persistent nature renders conventional wastewater treatment approaches, both chemical and biological, largely ineffective (Karpi´nska and Kotowska, 2019; Xiang et al., 2018). Conventional wastewater treatment plants (WWTPs), which rely on traditional procedures, lack the capability to efficiently remove pharmaceuticals from contaminated water. Consequently, influents and effluents from WWTPs commonly retain residual pharmaceuticals. In response, researchers have conducted numerous studies over the years to explore various methods and materials for pharmaceutical removal from wastewater (Deng et al., 2016). Conventional techniques, such as coagulation, sedimentation, and adsorption, have proven inefficient in degrading antibiotics. Adsorption and membrane technologies transport pollutants from water to the pore structure of adsorbent materials or the reject stream of membrane filtration, leaving behind residues that necessitate further treatment or disposal, often in landfills. Photocatalysis, a subset of advanced oxidation processes (AOPs), offers a promising solution for breaking down recalcitrant contaminants, both organic and inorganic, in wastewater treatment. Photocatalysis generates oxidizing agents like hydroxyl radicals, superoxide radicals, and sulfate radicals, which play a crucial role in the degradation process (Schneider et al., 2020).

In contemporary practice, the integration of carbonaceous materials such as carbon dots, fullerene, biochar, activated carbon, and graphene into photocatalytic systems can enhance their efficiency, particularly in the adsorption of pollutants before photocatalytic treatment. Carbon materials are employed to augment the performance of semiconductor photocatalysts due to their ability to significantly accelerate electron transfer rates while reducing the rate of recombination of photogenerated electron-hole pairs (Orimolade et al., 2019). Furthermore, carbon materials possess exceptional mechanical properties, high thermal conductivity, remarkable charge carrier mobility, and substantial specific surface area (Allen et al., 2010). The incorporation of carbon-based materials into multifunctional systems has opened doors to future applications in areas such as batteries, supercapacitor electrodes, and catalysts. This concise review provides insight into the current status and recent developments concerning the incorporation of carbonaceous metal-based catalysts for the photocatalytic degradation of pharmaceutical wastes. It also delves into the mechanisms underlying the role of carbon materials in enhancing photocatalytic efficiency.

Emerging Pollutant

The global development of infrastructure and technology has greatly facilitated human life, addressing various challenges. However, these advancements have also brought about adverse consequences for both the environment and human health, including issues such as climate change, air pollution, noise pollution, and water contamination. Anthropogenic activities associated with these concerns encompass practices such as open burning, deforestation, urban construction, untreated industrial and hospital wastewater discharge, and illegal waste disposal. In contemporary times, a multitude of contaminants, including dyes (Tkaczyk et al., 2020), heavy metals (Mokarram et al., 2020), microplastics (Viaroli et al., 2022), insecticides, herbicides (El-Nahhal & El-Nahhal, 2021), oil spillage (Singh et al., 2020), and pharmaceuticals (Khan et al., 2020), have been detected in aquatic environments. Emerging pollutants refer to non-regulated trace substances that have recently entered or been identified in the environment (Khan et al., 2020). Examples of emerging contaminants include personal care products, hormones, endocrine-disrupting chemicals, and pharmaceuticals commonly used in daily life, as well as various industrial additives (Stefanakis & Becker, 2019).

Numerous studies have investigated the presence of emerging pollutants. For instance, Chaudary and Walker (2019) identified contamination in the Ganga River, attributing it to compounds, both organic and inorganic, originating from agricultural waste, sewage, and religious practices. In Malaysia, Praveena et al. (2019) detected trace amounts of pharmaceutical residues in drinking water sourced from the Semenyih River. Despite the relatively low concentrations of these pollutants in water bodies, ranging from µg/L to ng/L, their impact on human health and ecosystems is significant. As reviewed by Patel et al. (2020) and Gonsioroski et al. (2020), the adverse effects associated with emerging pollutants include low sperm count, birth defects, breast cancer, and delayed brain development. These pollutants exhibit persistence in the environment due to their adsorption onto solid surfaces like soil, manure, and sludge, enabling their transmission to other locations (Manzetti & Ghisi, 2014; Zhi et al., 2019). Consequently, the elimination or minimization of pharmaceutical waste in water bodies is imperative.

Pharmaceutical wastes.

Pharmaceutical industries play a crucial role in addressing diseases affecting both humans and animals, especially in modern times. For instance, the emergence of methicillin-resistant Staphylococcus aureus (MRSA) necessitated the global adoption of vancomycin as a treatment option. However, alarming reports indicate the emergence of vancomycin-resistant S. aureus (VRSA) strains in the United States and Indonesia (Hamzah et al., 2019). Additionally,, as documented between 2011 and 2016, Salmonella Typhimurium, responsible for salmonellosis, has developed resistance to multiple drugs, including ampicillin, streptomycin, sulfonamides, tetracycline, and chloramphenicol (Wang et al., 2019). This underscores the rapid development of antimicrobial resistance in pathogens against existing drugs. Given the swift resistance development in Gram-positive and Gram-negative bacteria as well as various pathogens, the development of novel drugs is imperative (Ezelarab et al., 2018).

Fluoroquinolones, including drugs like ciprofloxacin, marbofloxacin, orbifloxacin, norfloxacin, and ofloxacin are widely used for treating bacterial infections affecting the skin, bone, soft tissue, respiratory tract, and urinary tract (Madaan et al., 2015). However, the presence of pharmaceutical waste in surface water has become a concern due to direct discharges from factories (Scott et al., 2018), landfills (Lu et al., 2016), farms (Jaffrézic et al., 2017), and inefficient removal in wastewater treatment plants (Yang et al., 2017). The chemical structures of these prominent fluoroquinolones are illustrated in Figure 2.

Figure 2. The chemical structure of prominent fluoroquinolones pharmaceutical wastes

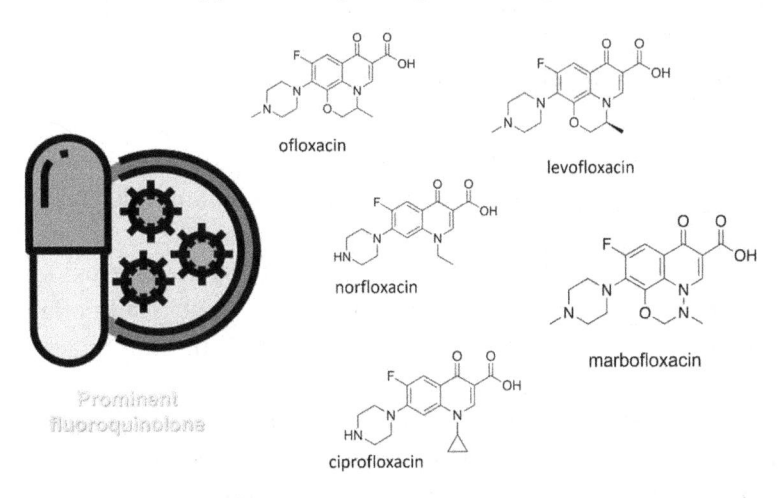

Ofloxacin and norfloxacin have been frequently detected in water bodies worldwide. Ofloxacin, effective against a broad spectrum of gram-negative and gram-positive bacteria, is commonly employed for treating respiratory, urinary, and gastrointestinal infections in humans and animals (Jaswal et al., 2021; Singh et al., 2020). Meanwhile, norfloxacin, which is another fluoroquinolone, is used against bacterial infections like Bacillus subtilis, Escherichia Coli, and Staphylococcus aureus (Suaifan et al., 2022). Moreover, Deng et al. (2016) reported 0.7 ng/L of ofloxacin in a river in Hong Kong, with a detection rate of 69.6%. He et al. (2015) recorded ofloxacin concentrations of 0.29-0.57 µg/L in raw water, 0.10-0.15 µg/L downstream, and 0.009-0.04 µg/L in effluent from municipal wastewater treatment plants in Maryland, USA. In India, Gothwal and Shashidhar (2017) detected elevated concentrations of ofloxacin, ranging from 1.55 to 542.45 µg/L, in the Musi River, suggesting potential ecological risks. Moreover, Teglia et al. (2019) identified ofloxacin contamination in animal wastewater and rivers in Argentina, with concentrations ranging from 0.71 to 1.78 µg/L.

Norfloxacin was detected in the urine of preschool children in Hong Kong at a concentration of 0.037 ng/mL, (Li et al., 2017). Adachi et al. (2013) revealed the presence of norfloxacin, ranging from undetected levels to 33 ng/L, in surface water in Japan. Furthermore, norfloxacin was found in sewage treatment plant effluents in Canada and China, with concentrations of 50 and 96 ng/L, respectively (Miao et al., 2004; Tong et al., 2011). Wagil et al. (2014) reported the presence of norfloxacin that range from undetected to 442.8 ng/L in surface water in Poland. Ilie et al. (2020) reported the detection of paracetamol (PAR) and oxytetracycline (OXY) in surface waters in Romania. However, the concentrations detected were extremely low and did not pose an ecological risk to the aquatic ecosystem. Nonetheless, it is essential to address the removal of OXY and PAR to prevent potential consequences, as the use of these medications is expected to increase in the future. Conversely, the misuse of drugs can lead to increased bacterial resistance. The World Health Organization (WHO) has identified antimicrobial resistance (AMR) as a significant risk to both humans and animals (Neelam & Megha, 2019). The presence of pharmaceutical waste in water bodies can contribute to AMR, as microbes exposed to antibiotics can become more resistant, rendering these medications less effective in treating infections in both humans and animals (Ferri et al., 2017). These findings underscore the prevalence of pharmaceutical waste in water bodies worldwide, even at low concentrations, emphasizing the need for its elimination to mitigate potential risks to human health and ecosystems.

Advanced Oxidation processes (AOPs).

Advanced Oxidation Processes (AOPs) represent a viable approach for remediating pollutants and can be categorized into several processes, including photolysis (Abramović et al., 2021), sonochemical treatment (Arvaniti et al., 2020), photo-Fenton reaction (Chen et al., 2022), and photocatalysis (Huang et al., 2021). These methods leverage reactive oxygen species (ROS) like superoxide anion radicals (O_2^-) and hydroxyl radicals (•OH) to effectively degrade both organic and inorganic pollutants. The efficiency of pollutant remediation relies on factors such as the specific AOP utilized, ambient conditions, and the physicochemical properties of the pollutants (Deng & Zhao, 2015).

AOPs offer several advantages, including high and non-selective degradation capabilities (Hayati et al., 2018), suitability for small-scale applications (Ponraj et al., 2017), and rapid reaction rates (Lafi et al., 2009). These processes contribute to the mineralization of contaminants by forming intermediates, ultimately resulting in the production of less harmful compounds. In a recent study by Zambrano et al. (2022), photolysis was employed to achieve substantial degradation of sulfamethoxazole, sulfadiazine, ciprofloxacin, and tetracycline, with degradation percentages ranging between 98 to 99% after 240 h for sulfamethoxazole, 212 h for sulfadiazine, 122 h for ciprofloxacin, and 100 h for tetracycline. Choi et al. (2020) investigated the sonolysis of Eosin B, achieving an 88% degradation efficiency with 50% ultrasound intensity during a 60-minute experiment. Conversely, in a study by Rahmani et al. (2014), sonolysis alone exhibited a relatively low removal efficiency, 8.5%, for tinidazole. On the other hand, González-Rodríguez et al. (2021) applied the photo-Fenton process using Fe_3O_4/SBA15 nanocomposites and achieved 100% and 90% degradation of sulfamethoxazole and orange II, respectively, within 120 minutes. These findings underscore the effectiveness of AOPs in wastewater treatment, enabling the breakdown and mineralization of a broad spectrum of contaminants within a short timeframe. Catalytic materials play a crucial role in the effective removal of pharmaceutical pollutants from wastewater. These materials leverage various mechanisms to enhance the degradation and removal efficiency of complex organic compounds.

Photocatalysis

Photocatalysis, a well-established method for treating wastewater, effectively removes various organic and inorganic contaminants. It boasts several advantages over alternative water remediation approaches, including cost-effectiveness, complete mineralization, environmental friendliness, reusability, and the absence of secondary waste. There were several roles of catalytic materials in removing pharmaceutical pollutants (Table 1), as such the generation of reactive species. Photocatalysts such as TiO_2, ZnO, and g-C_3N_4 generate reactive oxygen species (ROS) like hydroxyl radicals (•OH) and superoxide radicals (•O_2^-) under light irradiation. These ROS are highly reactive and can degrade pharmaceutical compounds into less harmful or benign substances (Wang et al., 2021; Jiang et al., 2022). Another role is for extending light absorption by doping and composite formation with materials like graphene oxide, carbon nanotubes, or metal ions extend the light absorption range of photocatalysts into the visible spectrum. This allows for more efficient use of natural sunlight or artificial visible light sources (Zhang et al., 2022; Wei et al., 2023). Furthermore, the enhancement of charge separation where catalytic materials often suffer from rapid recombination of photogenerated electron-hole pairs, reducing efficiency. Composites like g-C_3N_4/graphene oxide improve charge separation, prolonging the lifetime of reactive species and enhancing degradation efficiency (Liu et al., 2021; Chen et al., 2021).

It's also exhibited the behaviour of improved adsorption with synergistic effect. Materials like activated carbon have high surface areas and porosity, providing ample adsorption sites for pharmaceutical molecules. When combined with photocatalytic materials, they facilitate close contact between pollutants and reactive species, enhancing degradation rates (Sun et al., 2022; Ma et al., 2021). Synergistic effects can be observed by combining different catalytic materials, where the properties of each component complement and enhance overall photocatalytic performance. For example, the combination of g-C_3N_4 with carbon nanotubes enhances both light absorption and charge mobility (Wei et al., 2023; Li et al., 2022).

Catalytic degradation can convert pharmaceutical pollutants into less toxic or non-toxic compounds. This is particularly important for persistent pollutants like antibiotics, which can contribute to the development of antibiotic resistance (Jiang et al., 2022; Qiu et al., 2023) and these materials are designed to be stable under operational conditions and can be reused multiple times without significant loss of activity. This is crucial for practical applications in wastewater treatment (Madima et al., 2022; Sekar et al., 2022).

Table 1. Role of several catalysts for removing pharmaceutical pollutants

Catalyst	Roles and Mechanisms	References
TiO_2	Generates ROS under UV light, high stability, prone to recombination	Madima et al., 2022; Kang et al., 2021
ZnO	Similar to TiO_2, but with better charge separation, prone to photocorrosion	Singh et al., 2023; Sekar et al., 2022
g-C_3N_4	Visible light absorption, generates ROS, often combined with other materials for synergy	Jiang et al., 2022; Wang et al., 2021
Graphene Oxide	Enhances charge separation and light absorption when used in composites	Zhang et al., 2022; Liu et al., 2021
Activated Carbon	High surface area for adsorption, enhances contact between pollutants and photocatalysts	Sun et al., 2022; Ma et al., 2021
Carbon Nanotubes	Improved electron transfer, structural stability, enhances light absorption in composites	Wei et al., 2023; Chen et al., 2021
Fullerene	Enhances light absorption and charge separation, works well in g-C_3N_4 composites	Qiu et al., 2023; Li et al., 2022

Photocatalysis utilizes photo-functional materials as catalysts and light energy to generate radicals. This process triggers the formation of electron-hole pairs when the energy of light surpasses the semiconductor's bandgap energy, resulting in the generation of radicals such as hydroxyl and superoxide (Mestre and Calvalho, 2019; Ma et al., 2021; Liu et al., 2020; Yang & Wang, 2018). Equations 1-5 describe the principal reactions that occur during photocatalysis:

$$\text{Photocatalysis} + h\nu = e^- C_B + h+ V_B \tag{1}$$

$$O_2 + e^- C_B =$$

$$O_2^{\bullet -} \tag{2}$$

$$H_2O + h+ V_B = \bullet OH + H^+ \quad (3)$$

$$O_2^{\bullet-} + \text{Pollutants} = \text{degradation products} \quad (4)$$

$$\bullet OH + \text{Pollutants} = \text{degradation products} \quad (5)$$

In the realm of photocatalysis, a semiconductor material is exposed to light possessing an excitation energy surpassing its band gap energy. This light exposure induces electronic transitions from the valence band (V_B) to the conduction band (C_B) of the semiconductor, giving rise to the generation of electrons (e^-) and positive holes (h^+). These charge carriers can follow two paths: they may recombine or migrate to the surface of the catalyst, impacting the efficiency of photocatalysis. When charge carriers migrate to the photoactive material's surface and interact with adsorbed species, they initiate redox reactions. To enhance the photo-oxidation of target pollutants, it is imperative to prevent the recombination of e^-/h^+ pairs.

The photo-generated hole functions as a potent oxidizing agent, directly reacting with the target pollutant or water to produce hydroxyl radicals ($\bullet OH$), depending on the redox potential of the pollutant. Conversely, electrons act as strong reducing agents, engaging with electron acceptor species like dissolved oxygen to generate superoxide radicals ($O_2^{\bullet-}$) or species with a higher redox potential than the photocatalyst's conduction band. These radicals serve as potent oxidizing agents used for the mineralization of pollutants in water, ultimately yielding less harmful by-products such as carbon dioxide and water. However, the lifespan of these generated radicals is limited, necessitating their continuous regeneration to sustain the degradation process. Some research endeavours concentrate on extending radical lifespans through the modification of the photocatalyst, including the integration of additional substances into the system.

Heterogeneous semiconductor as photocatalyst.

In the realm of photocatalysis, various semiconductors such as metal oxides, metal ferrites, metal sulfides, and metal titanates serve as photocatalysts. Each of these photocatalysts possesses distinctive features that play a pivotal role in determining the suitability of photocatalysis. Key factors include the bandgap energy, surface area, morphology, and particle size of the photocatalysts. Notable semiconductors utilized in recent photocatalysis studies include $BiVO_4$ (Raja et al., 2020), CuO (Arunadevi et al., 2018), SnO_2 (Zinatloo-Ajabshir et al., 2019), ZnO (Mohamed, 2018), and TiO_2 (Ruidíaz-Martínez et al., 2020). TiO_2, the first semiconductor employed in photocatalysis, remains a prominent choice with ongoing modifications to enhance its performance. It stands out as a non-toxic photocatalyst known for its high photo-efficiency and rapid reactivity, even in low-light conditions. However, its utilization is limited to UV light irradiation due to its wide bandgap energy (Lee & Park, 2013). Efforts to optimize photocatalysts have primarily centered on improving light absorption efficiency, particularly by searching for visible-light-responsive photocatalysts and mitigating electron-hole recombination (Tong et al., 2012; Pan et al., 2010).

In the case of nanoscale semiconductors, the particles have demonstrated superior performance which are attributed to their vast surface area, excellent dispersion capabilities, and quantum size effects (Tong et al., 2012; Bell, 2003). However, the widespread application of semiconductor nanoparticles faces significant challenges in suspension because aggregation leads to a rapid loss of active sites, resulting in reduced photocatalytic efficiency and giving rise to the challenges to separate, recover, and reuse the nanoparticles (Tong et al., 2012). Owing to these characteristics, the efficiency of pollutant removal varies among different types of semiconductors, contingent upon numerous factors, especially the properties of the photocatalysts and the nature of the pollutants. Consequently, ongoing research in photocatalysis aims to identify versatile and high-performance photocatalysts for addressing waterborne pollutants. For instance, Zhang et al. (2018) reported promising results in the photocatalytic degradation of phenol, methyl orange, and rhodamine B using TiO_2 with degradation efficiencies of 54%, 80.7%, and 84.2% under a UV lamp for phenol, methyl orange, and rhodamine B, respectively was reported (Zhang et al., 2018).

In a separate study, ZnO was employed for the degradation of flumequine under a UV lamp although the performance was relatively low, with only 11% of flumequine being degraded within 240 minutes (Vaizogullar et al., 2017). Tammina et al. investigated the use of SnO_2 as a photocatalyst for the complete degradation of methylene blue under a UV lamp, achieving a 100 percent removal efficiency in just 30 minutes (Tammina et al., 2018). Meanwhile, Arshad et al. utilized CuO for the photodegradation of methylene blue under solar light conditions and achieved a 75% removal efficiency within 80 minutes (Arshad et al., 2017).

On the other hand, heterogeneous photocatalysts offer several advantages for use in photocatalysis, including cost-effectiveness, non-selective catalytic activity, non-toxicity, ease of separation, and chemical stability in liquid media (Tahir et al., 2020). These catalysts can be categorized as mixed metal oxides when they contain two or more types of metal cations in the oxide structure, such as binary or ternary metal oxides (Gawande et al., 2012). Mixed metal oxides are appealing to researchers due to their active acidic or basic sites and redox properties, and they can have crystal structures like perovskites, scheelites, and spinels (Gawande et al., 2012). In photocatalytic applications, the B sites of the photocatalyst play a crucial role in the adsorption process as active sites for pollutant molecules' interaction with the photocatalyst's surface, while the A site stabilizes the structure and enhances its activity (Singh et al., 2019; Hu et al., 2015; Li et al., 2019).

Perovskite oxides containing bismuth, known as Bi-contained perovskite oxides, are of particular interest for their excellent optical properties. Among them, bismuth ferrite with perovskite structure and an unusual compound of bismuth, iron, and oxygen (BFO) has gained attention as an alternative to traditional semiconductor photocatalysts like TiO_2 and ZnO, especially in wastewater treatment. Bismuth ferrite has three phases: mullite ($Bi_2Fe_4O_9$), sillenite ($Bi_{25}FeO_{40}$), and perovskite ($BiFeO_3$) with the perovskite phase having a chemical formula of ABO_3 (Hu et al., 2015). BFO nanocatalysts are particularly appealing due to their narrow bandgap of 2.4 eV (Li et al., 2019), allowing the photocatalyst to absorb a broader spectrum of UV and visible light regions, compared to photocatalysts with wider bandgaps that only respond to UV light.

For instance, Dumitru et al., utilized BFO as a photocatalyst for the degradation of doxorubicin, a pharmaceutical waste in wastewater (Dumitru et al.,2019). They achieved a 79% photodegradation efficiency under UV lamp irradiation for 180 minutes. Similarly, Basith et al. (2018) successfully degraded rhodamine B using the sillenite phase of BFO, achieving a photocatalytic efficiency of 62% under high-power mercury-xenon lamp irradiation for 240 minutes (Basith et al., 2018). However, some phases of BFO, such as mullite, exhibited lower photocatalytic efficiency for degradation of methyl violet under visible light and the obtained a removal efficiency was only 10.6% (Liu et al., 2017). The limited performance of some semiconductors may be attributed to the rapid recombination rate of electron-hole pairs

(Liu et al., 2017). To overcome these limitations, modifications like doping, heterojunction formation, and the incorporation of other materials as support have been explored to enhance the photocatalytic degradation ability.

One promising approach to improve photocatalysis involves creating heterojunction catalysts composed of nanocomposites of two distinct materials, which can enhance photocatalytic photodegradation while minimizing recombination rates (Raza et al., 2018). Additionally, carbon-based materials have emerged as intriguing photocatalysts for breaking down organic pollutants, showing improved photocatalytic activity compared to their pure counterparts.

CARBON-BASED MATERIALS

Figure 3. The remarkable properties of carbonaceous materials with various types of carbon

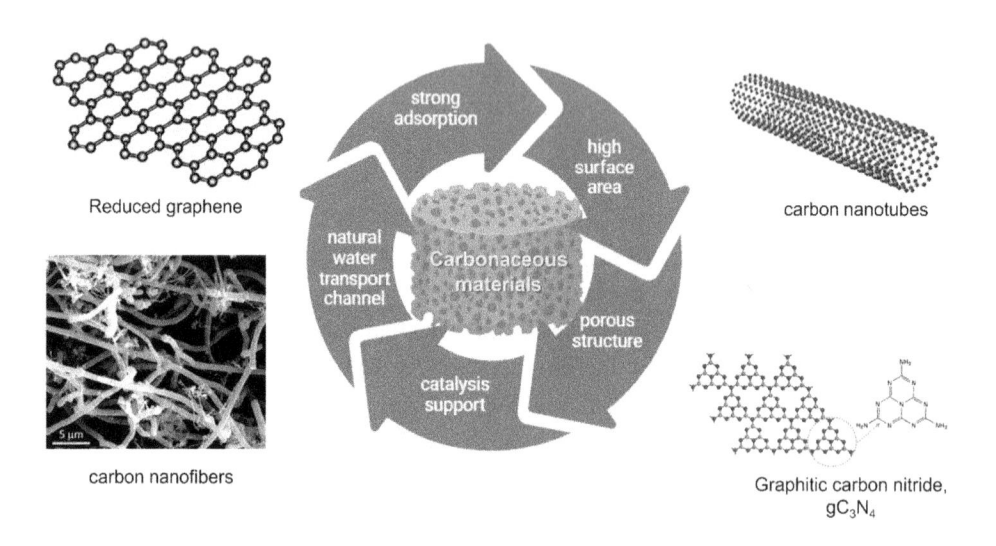

Carbon-based semiconductor materials hold great promise as prospective photocatalysts, offering a viable alternative to address the challenges and limitations faced in this field (Minella et al., 2017; Shan et al., 2010). The extensive surface area, black body properties, and high charge carrier mobility have led to extensive exploration of carbon materials as dopants, hybridizing materials, or supports for semiconductors across various applications, including electrical devices (Wang et al., 2021), solar cells (Fagiolari & Bella, 2019), batteries (Yuan et al., 2020), and photocatalysis (Liu et al., 2018). Carbon-based materials (CBMs), including activated carbon (AC), diamond, fullerene (C60), carbon nanotubes (CNTs), carbon quantum dots, and graphene, are anticipated to play a significant role in semiconductor systems, particularly in wastewater treatment (Mohapatra et al., 2023). CBMs offer several advantages (Figure 3), of their stability under chemical and thermal conditions, making them practical for various reactions. The carbonaceous matrix possesses several advantages as follows: a high specific surface area, large porosity, catalyst loading ability, water transport channel and high adsorption capacity for water contaminants. These materials exhibit resistance to corrosive environments, including extremes of pH, and are readily available with excellent mechanical strength, making them easily obtainable and suitable as supporting materials (Mohammad et al., 2021; Mohapatra et al., 2023). Furthermore, the integration of carbon materials with semiconductors caused substantial synergistic effects and enhanced photocatalytic performance (Minella et al., 2017; Mohapatra et al., 2023).

Figure 4. Simplified photocatalysis schemes using semiconductor and carbon-based materials.

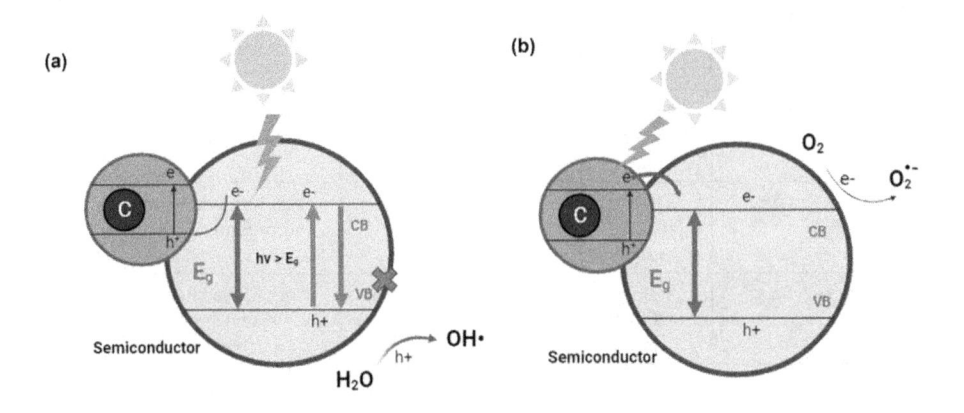

(a) charge separation in the presence of carbon, and (b) carbon acting as photosensitizer

Notably, carbon materials can serve as efficient and cost-effective photocatalysts when their carbon skeletons were carefully tailored (Rhimi et al., 2020). Two strategies have been employed to enhance photocatalytic performance when incorporating carbon into a semiconductor. Figure 4 shows illustration of photocatalysis schemes using semiconductor and carbon-based materials for (a) charge separation in the presence of carbon, and (b) carbon acting as photosensitizer. In the former case (a), this improved charge separation, where carbon helps trap generated electrons, preventing recombination. The latter case (b) of carbon acting photosensitizer can transfer electrons to the semiconductor's conduction band. Woan et al. proposed a charge separation mechanism for carbon-semiconductor composites and suggested that carbon materials with metallic conductivity can trap electrons generated in the semiconductor, thus preventing recombination (Woan et al., 2009). In contrast, Wang et al., proposed a mechanism in which carbon materials act as photosensitizers by transferring electrons to the semiconductor's conduction band (Wang et al.,2005). Numerous studies have investigated the incorporation of carbon-based materials with semiconductors for the removal of pharmaceutical wastes.

Table 2 summarizing the application conditions and efficiencies of various carbon-based photocatalysts used in general and pharmaceutical (including antibiotic) in wastewater treatment. The table includes information on the type of photocatalyst, the optimal pH range, light source, and reported efficiencies, along with recent studies available.

Table 2. Efficiencies of carbon based photocatalyst with different pH and light sources

Photocatalyst	Optimal pH Range	Light Source	Efficiency (%Degradation)	Observations	References
g-C_3N_4 (Graphitic Carbon Nitride)	5-8	Visible (420-450 nm)	50-80% for antibiotics and hormones	Visible light active, lower efficiency than TiO_2 under UV	Jiang et al., 2022; Wang et al., 2021
g-C_3N_4/Graphene Oxide	5-8	Visible (420-450 nm)	60-90% for various pharmaceuticals	Enhanced charge separation and visible light absorption	Zhang et al., 2022; Liu et al., 2021
g-C_3N_4/Carbon Nanotubes	5-9	Visible (400-450 nm)	65-85% for antibiotics and NSAIDs	Improved electron transfer properties	Wei et al., 2023; Chen et al., 2021
g-C_3N_4/Activated Carbon	4-8	Visible (400-500 nm)	70-90% for various pharmaceuticals	High surface area and adsorption capacity	Sun et al., 2022; Ma et al., 2021
g-C_3N_4/Fullerene	4-8	Visible (420-450 nm)	60-85% for antibiotics and hormones	Enhanced light absorption and charge separation	Qiu et al., 2023; Li et al., 2022

Reduced graphene oxide.

Reduced graphene oxide (rGO) is a relatively recent carbon compound gaining significant attention for its integration with semiconductors in photocatalysis applications. rGO is derived from the reduction of graphene oxide through processes like thermal treatment, photoreduction, or the use of chemical reducing agents such as sodium borohydride, ascorbic acid, and hydrazine hydrate (Mohd Kaus et al., 2021; Yuan et al., 2020). Its outstanding properties, including a large specific surface area, excellent adsorption capacity, and high conductivity, make it an attractive choice as a supporting material for semiconductors in wastewater treatment (Johra & Jung, 2015).

The large specific surface area enhanced the adsorption of pollutants onto the photocatalyst before mineralization (Zheng et al., 2020). Furthermore, rGO serves as an electron reservoir due to its high conductivity, reducing the recombination rate of electron-hole pairs by transferring excited electrons to the rGO surface, thereby promoting the production of radicals (ref). Oxygen functional groups on rGO's surface provide anchor points for metal ions and serve as nucleation sites for photocatalyst formation. This feature enables control over the morphology and size of the photocatalyst, ultimately improving photoactivity performance (Hu et al., 2013; Raja et al., 2020).

For instance, Mangalam et al. employed a silver/reduced graphene oxide (Ag/rGO) nanocomposite for methyl orange degradation, achieving a remarkable 91% degradation within 45 minutes compared to 62% with pristine GO (Mangalam et al., 2019). In another study, Naraginti et al. utilized a Zr/Ag-TiO_2@rGO photocatalyst for p-bromophenol photodegradation and achieved 95% removal in 180 minutes. Reactive oxygen species (ROS) such as hydroxyl radicals (•OH), superoxide anion radicals ($O_2^{•-}$), and holes (h^+) were identified as contributing to the photodegradation process (Naraginti & Yong, 2019). Aleksandrzak et al. also explored the impact of rGO incorporation into a zinc oxide (ZnO) photocatalyst. ZnO/rGO efficiently photodegraded 94% of methylene blue (MB) within 45 minutes of UV irradiation, outperforming ZnO, which required 105 minutes to achieve the same degradation percentage (Aleksandrzak et al., 2015). Hydroxyl radicals were identified as the primary ROS in this photocatalytic activity (Labhane et al., 2016).

A novel hydrothermal photocatalyst involving reduced graphene oxide with nitrogen doping and $BiVO_4$ was developed for the degradation of metronidazole (MTZ) and chloramphenicol (CAP) under visible light exposure (Sun et al., 2022). This hybrid photocatalyst achieved impressive degradation efficiencies, degrading 95% of MTZ in 4 h and 93% of CAP in 5 h. The presence of nitrogen-doped rGO led to the formation of •OH and $O_2^{•-}$ radicals before recombination, resulting in superior catalytic activity compared to the bare counterparts. Kadi and colleagues constructed mesoporous nanocomposites of $BiFeO_3$/graphene with varying $BiFeO_3$ contents and assessed their photocatalytic capabilities for ciprofloxacin degradation (Kadi et al., 2020). The 20% $BiFeO_3$/graphene nanocomposites exhibited an impressive photocatalytic efficiency of 98.8% within 120 minutes, which is four times greater than the 50% efficiency achieved with pure mesoporous $BiFeO_3$ material (Kadi et al., 2020). Feng et al. employed a straightforward one-pot mild hydrothermal synthesis method to fabricate boron-doped reduced graphene oxide (B-rGO) supported by bismuth ferrite ($BiFeO_3$) nanocomposites 8 Feng et al., 2020). This group reported the degradation efficiency of tetracycline (TC) reached 86.7%, with excellent photostability observed for B-rGO/$BiFeO_3$ after four consecutive reaction cycles.

Meanwhile, Peng et al. used ZnO/rGO in the photocatalytic degradation of methylene blue (MB) and acid red 249 (AR249) (Peng et al., 2015). The optimized ZnO/rGO achieved impressive degradation rates of 92.9% for MB and 94.8% for AR249, surpassing the performance of ZnO alone, which achieved 87.4% and 78.6% degradation, respectively, under UV light for 120 minutes. The primary reactive oxygen species (ROS) involved in this photocatalytic degradation were holes (h^+) (Peng et al., 2015). Soltani et al. (2016a) investigated $BiFeO_3$/rGO for the photocatalytic degradation of Bisphenol A. The photocatalytic degradation efficiency increased from 78% for pure BFO to 98% for BFO/rGO nanocomposites after 70 minutes of irradiation with a 55 W fluorescent lamp.

Ju et al. demonstrated that the BiOBr/Ag_3PO_4@rGO composite efficiently degraded TC, achieving an elimination rate of 98.5% within 24 minutes (Ju et al., 2020). The synergistic effects of rGO in combination with BiOBr/Ag_3PO_4 enhanced both the photostability and recyclability of TC decomposition (Ju et al., 2020). Ma et al. (yr) conducted the photocatalytic degradation of norfloxacin using hydrothermally prepared Bi_2WO_6/TiO_2/rGO nanocomposites under visible light. They achieved the removal of over 87% of the pharmaceutical waste. The presence of heterojunctions and reduced graphene oxide (rGO) improved the visible light absorption and promoted efficient charge separation compared to the Bi_2WO_6/TiO_2 photocatalyst. The degradation mechanism involved photogenerated holes and hydroxyl radicals as the major oxidants responsible for norfloxacin breakdown. A similar trend in the photocatalytic removal of norfloxacin was observed using Bi_2WO_6/$Bi_{25}FeO_{40}$/rGO composite (Zhang et al., 2020). Although only 78% removal of norfloxacin was achieved, it still outperformed the unmodified Bi_2WO_6/$Bi_{25}FeO_{40}$ photocatalyst.

Figure 5. Photocatalytic degradation of emerging pollutants by using semiconductor with reduced graphene oxide (rGO) under light irradiation

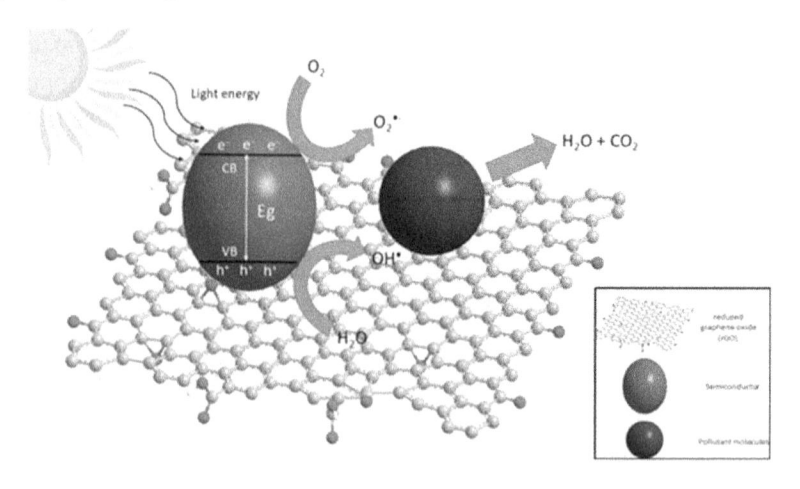

The proposed model for the photocatalytic mechanism involving semiconductors integrated with reduced graphene oxide (rGO) for the degradation of pharmaceutical pollutants is illustrated in Figure 5. The incorporation of a heterostructure semiconductor with rGO facilitates the photodegradation of pollutant molecules under visible and/or solar light irradiation. This suggests that rGO enhances the absorption of visible light and mitigates the recombination of electron-hole pairs during photocatalysis. Furthermore, the observed enhancement in the photodegradation of the pollutant is attributed to the formation of a large interfacial contact between semiconductor and the rGO heterostructures, leading to superior interfacial charge carrier dynamics. This model elucidates that the main active species for the efficient pollutant degradation are the superoxide radical $\left(O_2^{\bullet-}\right)$ and hydroxyl radicals ($\bullet$OH). These radicals play crucial roles in degrading the pollutants and several intermediates were formed through the various potential reaction pathways. In this review, rGO extended the visible light responsiveness, increased the surface area of the photocatalysts, and reduced the rapid recombination of photogenerated electron-hole pairs. Collectively, these studies underscore the advantages of incorporating rGO as a support material for semiconductors to enhance photocatalytic degradation performance.

Other carbon-based photocatalytic materials

According to Khalid and colleagues, the incorporation of carbon-based materials into photocatalytic systems, including activated carbons (AC), fullerenes, carbon nanotubes (CNTs), or graphene, can yield several advantages (Khalid et al., 2007). This includes enhanced photocatalytic activity, increased adsorption capacity, improved electron scavenging ability, sensitization properties, extended visible light absorption range, and simplified separation processes. It's important to consider the synergistic effects resulting from improved adsorption when evaluating the enhanced photocatalytic performance.

When TiO_2/C70 was employed for the treatment of sulfathiazole under a 300 W Xe lamp for 180 minutes, an impressive elimination efficiency (80%) was achieved and surpassed the 10% performance of pure TiO_2 (Wang et al., 2015). Additionally, Gar Alalm et al. found that TiO_2/AC outperformed pure TiO_2 in the degradation of amoxicillin, diclofenac, paracetamol, and ampicillin (Gar Alalm et al., 2016).

Under solar light exposure for 180 minutes, TiO_2/AC achieved complete degradation of amoxicillin and ampicillin (89%), while pure TiO_2 achieved only 83% degradation. For diclofenac and paracetamol decomposition, TiO_2/AC achieved 85% and 70% removal, respectively, while pure TiO_2 achieved 75% and 63% removal (Gar Alalm et al., 2016).

Furthermore, Xu et al. reported that $CNT/LaVO_4$ outperformed pristine $LaVO_4$ in the degradation of tetracycline under visible light irradiation for 180 minutes, with $CNT/LaVO_4$ degrading 81% of the pollutant, whereas $LaVO_4$ achieved only 40% removal (Xu et al., 2019). Moreover, Pankaj et al. investigated a ternary nanocomposite consisting of g-carbon nitride (GCN) and Bi_2O_3/Fe_3O_4 for the removal of oxytetracycline (OTC) (Pankaj et al., 2019). Under solar light irradiation, the $GCN/Bi_2O_3/Fe_3O_4$ photocatalyst efficiently degraded and mineralized the antibiotic OTC, outperforming GCN and Bi_2O_3 alone. Within 8 h of visible light exposure, OTC was completely mineralized (Pankaj et al., 2019). These studies highlight the advantages of incorporating carbon-based materials into semiconductor systems for the photocatalytic degradation of waterborne pollutants.

Hong et al. demonstrated the enhanced photocatalytic activity of Bi_2MoO_6 photocatalysts modified with carbon quantum dots (CQDs) for the degradation of ciprofloxacin and tetracycline hydrochloride (Hong et al.,2022). The incorporation of 2 wt percent CQDs content into Bi_2MoO_6 resulted in a five-fold improvement in photocatalytic performance compared to bare Bi_2MoO_6 after 120 minutes of light exposure (Hong et al., 2022). Moreover, Gao et al. developed a $C/Fe_3O_4/Bi_2O_3$ composite with doubly conductive photocatalytic properties for the degradation of tetracycline. The authors reported the composite achieved a photocatalytic decomposition of 91% in 90 minutes, surpassing the performance of pure Bi_2O_3 (Gao et al., 2016).

Furthermore, Li et al. developed a novel C_3N_4/carbon nanotube/$Bi_{25}FeO_{40}$ (CN/CNT/BFO) nanocomposite with a Z-scheme heterostructure (Li et al., 2022). This composite demonstrated superior photocatalytic degradation of tetracycline hydrochloride (92.2%) compared to individual components. The presence of carbon nanotubes facilitated electron transfer between BFO and CN, enhancing photocatalytic performance (Li et al., 2022). Lastly, Wu and colleagues reported the degradation of ciprofloxacin using a $Bi_2MoO_6/C_3N_4/Ti_3C_2$ composite photocatalyst (Wu et al., 2020b). The composite exhibited a significantly faster degradation rate of approximately five times that of Bi_2WO_6 without. (Wu et al., 2020b). This shows the positive impact of incorporating carbon materials with various nanocrystal formations, high electrical conductivity, inherent hydrophobicity, and customizable surface properties into photocatalytic systems for the degradation of pharmaceutical waste and other waterborne pollutants.

CONCLUSION

The rapid expansion within the realm of carbon-based nanomaterials over recent decades has undeniably unveiled novel and intriguing properties, marked notably by significantly enhanced catalytic activities. These materials manifest exceptional thermal conductivity, mechanical strength, and optical characteristics. Carbon-based nanomaterials (CBMs) find extensive applications as supporting substrates in catalysis. This review delves into diverse innovative carbon-based catalysts, presenting unique attributes and advantages. The evolution of distinct CBMs is governed by three pivotal overarching factors: the augmentation of surface area, mitigation of photogenerated electron-hole pair recombination, and augmentation of active sites on the supported material's surface. Guided by these criteria, various dimensions of CBMs emerge as a promising class of materials for modifying hybrid structures, distinguished by their cost-effectiveness, superior electrical conductivity, and appealing optical properties with immense potential in water treatment and purification, particularly in the removal of pharmaceutical and industrial pollutants. In summary, CBMs photocatalysts represent an unexplored frontier, and the pursuit of high activity, stability, and efficiency stands as the ultimate objective in the realm and future research endeavors in the domain of wastewater treatment. This advancement holds promise for identifying prospective applications in the next generation of photodegradation technology.

Author Note

This research was sponsored by Fundamental Research Grant Scheme (FRGS) [FRGS/1/2019/STG07/USM/02/7/] from Ministry of Higher Education Malaysia and Universiti Sains Malaysia. We have no conflicts of interest to disclose.

Correspondence concerning this article should be addressed to Noor Haida Mohd Kaus, School of Chemical Sciences, Universiti Sains Malaysia, 11800, Penang, Malaysia. Email: noorhaida@usm.my

REFERENCES

Abramović, B. F., Uzelac, M. M., Armaković, S. J., Gašić, U., Četojević-Simin, D. D., & Armaković, S. (2021). Experimental and computational study of hydrolysis and photolysis of antibiotic ceftriaxone: Degradation kinetics, pathways, and toxicity. *The Science of the Total Environment*, 768, 144991. Advance online publication. DOI: 10.1016/j.scitotenv.2021.144991 PMID: 33736306

Adachi, F., Yamamoto, A., Takakura, K. I., & Kawahara, R. (2013). Occurrence of fluoroquinolones and fluoroquinolone-resistance genes in the aquatic environment. *The Science of the Total Environment*, 444, 508–514. DOI: 10.1016/j.scitotenv.2012.11.077 PMID: 23291652

Ahmed, S., Khan, F. S. A., Mubarak, N. M., Khalid, M., Tan, Y. H., Mazari, S. A., Karri, R. R., & Abdullah, E. C. (2021). Emerging pollutants and their removal using visible-light responsive photocatalysis – A comprehensive review. *Journal of Environmental Chemical Engineering*, 9(6), 106643. DOI: 10.1016/j.jece.2021.106643

Aleksandrzak, M., Adamski, P., Kukułka, W., Zielinska, B., & Mijowska, E. (2015). Effect of graphene thickness on photocatalytic activity of TiO2 -graphene nanocomposites. *Applied Surface Science*, 331, 193–199. DOI: 10.1016/j.apsusc.2015.01.070

Allen, M. J., Tung, V. C., & Kaner, R. B. (2010). Honeycomb carbon: A review of graphene. *Chemical Reviews*, 110(1), 132–145. DOI: 10.1021/cr900070d PMID: 19610631

Arshad, A., Iqbal, J., Siddiq, M., Ali, M. U., Ali, A., Shabbir, H., Nazeer, U., & Saleem, M. S. (2017). Solar light triggered catalytic performance of graphene- CuO nanocomposite for waste water treatment. *Ceramics International*, 43(14), 10654–10660. DOI: 10.1016/j.ceramint.2017.03.165

Arunadevi, R., Kavitha, B., Rajarajan, M., Suganthi, A., & Jeyamurugan, A. (2018). Investigation of the drastic improvement of photocatalytic degradation of Congo red by monoclinic Cd, Ba-CuO nanoparticles and its antimicrobial activities. *Surfaces and Interfaces*, 10, 32–44. DOI: 10.1016/j.surfin.2017.11.004

Arvaniti, O. S., Frontistis, Z., Nika, M. C., Aalizadeh, R., Thomaidis, N. S., & Mantzavinos, D. (2020). Sonochemical degradation of trimethoprim in water matrices: Effect of operating conditions, identification of transformation products and toxicity assessment. *Ultrasonics Sonochemistry*, 67(April), 105139. DOI: 10.1016/j.ultsonch.2020.105139 PMID: 32348950

Atiqah, H., Azmy, M., Razuki, N. A., & Aziz, A. W. (2017). Visible Light Photocatalytic Activity of BiFeO 3 Nanoparticles for Degradation of Methylene Blue. *The Journal of Physiological Sciences; JPS*, 28(2), 85–103. DOI: 10.21315/jps2017.28.2.6

Basith, M. A., Ahsan, R., Zarin, I., & Jalil, M. A. (2018). Enhanced photocatalytic dye degradation and hydrogen production ability of Bi25FeO40-rGO nanocomposite and mechanism insight. *Scientific Reports*, 8(1), 33–35. DOI: 10.1038/s41598-018-29402-w PMID: 29311608

Bell, A. T. (2003). The impact of nanoscience on heterogeneous catalysis. *Science*, 299(5613), 1688–1691. DOI: 10.1126/science.1083671 PMID: 12637733

Berkani, M., Smaali, A., Kadmi, Y., Almomani, F., Vasseghian, Y., Lakhdari, N., & Alyane, M. (2022). Photocatalytic degradation of Penicillin G in aqueous solutions: Kinetic, degradation pathway, and microbioassays assessment. Journal of Hazardous Materials, 421(July 2021), 126719. https://doi.org/ DOI: 10.1016/j.jhazmat.2021.126719

Bychko, I., Abakumov, A., Didenko, O., Chen, M., Tang, J., & Strizhak, P. (2022). Differences in the structure and functionalities of graphene oxide and reduced graphene oxide obtained from graphite with various degrees of graphitization. *Journal of Physics and Chemistry of Solids*, 164(January), 110614. DOI: 10.1016/j.jpcs.2022.110614

Cabeza, Y., Candela, L., Ronen, D., & Teijon, G. (2012). Monitoring the occurrence of emerging contaminants in treated wastewater and groundwater between 2008 and 2010. The Baix Llobregat (Barcelona, Spain). *Journal of Hazardous Materials*, 239–240, 32–39. DOI: 10.1016/j.jhazmat.2012.07.032 PMID: 22877748

Causanilles, A., Ruepert, C., Ib'a˜nez, M., Emke, E., Hern'andez, F., & de Voogt, P. (2017). Occurrence and fate of illicit drugs and pharmaceuticals in wastewater from two wastewater treatment plants in Costa Rica. *The Science of the Total Environment*, 599–600, 98–107. Advance online publication. DOI: 10.1016/j.scitotenv.2017.04.202 PMID: 28472697

Chaudhary, M., & Walker, T. R. (2019). River Ganga pollution: Causes and failed management plans (correspondence on Dwivedi et al. 2018. Ganga water pollution: A potential health threat to inhabitants of Ganga basin. Environment International 117, 327–338). In Environment International (Vol. 126, Issue February, pp. 202–206). https://doi.org/DOI: 10.1016/j.envint.2019.02.033

Chen, J., Wang, X., & Liu, H. (2021). Enhanced photocatalytic performance of g-C_3N_4/carbon nanotubes for pharmaceutical degradation under visible light. *Applied Surface Science*, 567, 150847.

Chen, Y., Li, Y., Luo, N., Shang, W., Shi, S., Li, H., Liang, Y., & Zhou, A. (2022). Kinetic comparison of photocatalysis with H2O2-free photo-Fenton process on BiVO4 and the effective antibiotic degradation. Chemical Engineering Journal, 429(September 2021), 132577. https://doi.org/DOI: 10.1016/j.cej.2021.132577

Choi, Y., Lee, D., Hong, S., Khan, S., Darya, B., Lee, J. Y., Chung, J., & Cho, S. H. (2020). Investigation of the synergistic effect of sonolysis and photocatalysis of titanium dioxide for organic dye degradation. *Catalysts*, 10(5), 500. Advance online publication. DOI: 10.3390/catal10050500

Cycoń, M., Mrozik, A., & Piotrowska-Seget, Z. (2019). Antibiotics in the soil environment—Degradation and their impact on microbial activity and diversity. *Frontiers in Microbiology*, 10(Mar), 338. Advance online publication. DOI: 10.3389/fmicb.2019.00338 PMID: 30906284

Dai, J. F., Xian, T., Di, L. J., & Yang, H. (2013). Preparation of BiFeO3 -graphene nanocomposites and their enhanced photocatalytic activities. *Journal of Nanomaterials*, 2013(1), 642897. Advance online publication. DOI: 10.1155/2013/642897

Danner, M. C., Robertson, A., Behrends, V., & Reiss, J. (2019). Antibiotic pollution in surface fresh waters: Occurrence and effects. *The Science of the Total Environment*, 664, 793–804. DOI: 10.1016/j.scitotenv.2019.01.406 PMID: 30763859

Deng, W., Li, N., Zheng, H., & Lin, H. (2016). Occurrence and risk assessment of antibiotics in river water in Hong Kong. *Ecotoxicology and Environmental Safety*, 125, 121–127. DOI: 10.1016/j.ecoenv.2015.12.002 PMID: 26685784

Di, L. J., Yang, H., Hu, G., Xian, T., Ma, J. Y., Jiang, J. L., Li, R. S., & Wei, Z. Q. (2014). Enhanced photocatalytic activity of BiFeO3 particles by surface decoration with Ag nanoparticles. *Journal of Materials Science Materials in Electronics*, 25(6), 2463–2469. DOI: 10.1007/s10854-014-1896-0

Dumitru, R., Ianculescu, A., Păcurariu, C., Lupa, L., Pop, A., Vasile, B., Surdu, A., & Manea, F. (2019). BiFeO3-synthesis, characterization and its photocatalytic activity towards doxorubicin degradation from water. *Ceramics International*, 45(2), 2789–2802. DOI: 10.1016/j.ceramint.2018.07.298

El-Nahhal, I., & El-Nahhal, Y. (2021). Pesticide residues in drinking water, their potential risk to human health and removal options. *Journal of Environmental Management*, 299(August), 113611. DOI: 10.1016/j.jenvman.2021.113611 PMID: 34526283

Ezelarab, H. A. A., Abbas, S. H., Hassan, H. A., & Abuo-Rahma, G. E. D. A. (2018). Recent updates of fluoroquinolones as antibacterial agents. *Archiv der Pharmazie*, 351(9), 1–13. DOI: 10.1002/ardp.201800141 PMID: 30048015

Feng, Z., Zeng, L., Zhang, Q., Ge, S., Zhao, X., Lin, H., & He, Y. (2020). In situ preparation of g-C3N4/Bi4O5I2 complex and its elevated photoactivity in methyl orange degradation under visible light. *Journal of Environmental Sciences (China)*, 87, 149–162. DOI: 10.1016/j.jes.2019.05.032 PMID: 31791488

Ferri, M., Ranucci, E., Romagnoli, P., & Giaccone, V. (2017). Antimicrobial resistance: A global emerging threat to public health systems. *Critical Reviews in Food Science and Nutrition*, 57(13), 2857–2876. DOI: 10.1080/10408398.2015.1077192 PMID: 26464037

Fick, J., Söderström, H., Lindberg, R. H., Phan, C., Tysklind, M., & Larsson, D. G. J. (2009). Contamination of surface, ground, and drinking water from pharmaceutical production. *Environmental Toxicology and Chemistry*, 28(12), 2522–2527. DOI: 10.1897/09-073.1 PMID: 19449981

Gao, N., Lu, Z., Zhao, X., Zhu, Z., Wang, Y., Wang, D., Hua, Z., Li, C., Huo, P., & Song, M. (2016). Enhanced photocatalytic activity of a double conductive C/Fe3O4/Bi2O3 composite photocatalyst based on biomass. *Chemical Engineering Journal*, 304, 351–361. DOI: 10.1016/j.cej.2016.06.063

Gao, T., Chen, Z., Zhu, Y., Niu, F., Huang, Q., Qin, L., Sun, X., & Huang, Y. (2014). Synthesis of BiFeo3 nanoparticles for the visible-light induced photocatalytic property. *Materials Research Bulletin*, 59, 6–12. DOI: 10.1016/j.materresbull.2014.06.022

Gar Alalm, M., Tawfik, A., & Ookawara, S. (2016). Enhancement of photocatalytic activity of TiO2 by immobilization on activated carbon for degradation of pharmaceuticals. *Journal of Environmental Chemical Engineering*, 4(2), 1929–1937. DOI: 10.1016/j.jece.2016.03.023

Gautam, S., Agrawal, H., Thakur, M., Akbari, A., Sharda, H., Kaur, R., & Amini, M. (2020). Metal oxides and metal organic frameworks for the photocatalytic degradation: A review. *Journal of Environmental Chemical Engineering*, 8(3), 103726. DOI: 10.1016/j.jece.2020.103726

Gawande, M. B., Pandey, R. K., & Jayaram, R. V. (2012). Role of mixed metal oxides in catalysis science - Versatile applications in organic synthesis. *Catalysis Science & Technology*, 2(6), 1113–1125. DOI: 10.1039/c2cy00490a

Gonsioroski, A., Mourikes, V. E., & Flaws, J. A. (2020). Endocrine disruptors in water and their effects on the reproductive system. *International Journal of Molecular Sciences*, 21(6), 1929. Advance online publication. DOI: 10.3390/ijms21061929 PMID: 32178293

González-Rodríguez, J., Fernández, L., Vargas-Osorio, Z., Vázquez-Vázquez, C., Piñeiro, Y., Rivas, J., Feijoo, G., & Moreira, M. T. (2021). Reusable Fe_3O_4/SBA15 nanocomposite as an efficient photo-fenton catalyst for the removal of sulfamethoxazole and orange II. *Nanomaterials (Basel, Switzerland)*, 11(2), 1–19. DOI: 10.3390/nano11020533 PMID: 33669767

Gothwal, R., & Shashidhar, . (2017). Occurrence of high levels of fluoroquinolones in aquatic environment due to effluent discharges from bulk drug manufacturers. *Journal of Hazardous, Toxic and Radioactive Waste*, 21(3), 05016003. DOI: 10.1061/(ASCE)HZ.2153-5515.0000346

Hamzah, A. M. C., Yeo, C. C., Puah, S. M., Chua, K. H., & Chew, C. H. (2019). Staphylococcus aureus infections in Malaysia: A review of antimicrobial resistance and characteristics of the clinical isolates, 1990–2017. *Antibiotics (Basel, Switzerland)*, 8(3), 1990–2017. DOI: 10.3390/antibiotics8030128 PMID: 31454985

Haque, M. M., & Muneer, M. (2007). Photodegradation of norfloxacin in aqueous suspensions of titanium dioxide. *Journal of Hazardous Materials*, 145(1–2), 51–57. DOI: 10.1016/j.jhazmat.2006.10.086 PMID: 17223263

Hayati, F., Isari, A. A., Fattahi, M., Anvaripour, B., & Jorfi, S. (2018). Photocatalytic decontamination of phenol and petrochemical wastewater through ZnO/TiO2 decorated on reduced graphene oxide nanocomposite: Influential operating factors, mechanism, and electrical energy consumption. *RSC Advances*, 8(70), 40035–40053. DOI: 10.1039/C8RA07936F PMID: 35558237

Hong, X., Li, Y., Wang, X., Long, J., & Liang, B. (2022). Carbon nanosheet/MnO2/BiOCl ternary composite for degradation of organic pollutants. *Journal of Alloys and Compounds*, 891, 162090. DOI: 10.1016/j.jallcom.2021.162090

Hu, C., Lu, T., Chen, F., & Zhang, R. (2013). A brief review of graphene–metal oxide composites synthesis and applications in photocatalysis. *Journal of the Chinese Advanced Materials Society*, 1(1), 21–39. DOI: 10.1080/22243682.2013.771917

Hu, Z. T., Liu, J., Yan, X., Da Oh, W., & Lim, T. T. (2015). Low-temperature synthesis of graphene/Bi2Fe4O9 composite for synergistic adsorption-photocatalytic degradation of hydrophobic pollutant under solar irradiation. *Chemical Engineering Journal*, 262, 1022–1032. DOI: 10.1016/j.cej.2014.10.037

Huang, S., Zhang, J., Qin, Y., Song, F., Du, C., & Su, Y. (2021). Direct Z-scheme SnO2/Bi2Sn2O7 photocatalyst for antibiotics removal: Insight on the enhanced photocatalytic performance and promoted charge separation mechanism. Journal of Photochemistry and Photobiology A: Chemistry, 404(July 2020), 112947. https://doi.org/DOI: 10.1016/j.jphotochem.2020.112947

Ilie, M., Deák, G., Marinescu, F., Ghita, G., Tociu, C., Matei, M., Covaliu, C. I., Raischi, M., & Yusof, S. Y. (2020). Detection of emerging pollutants oxytetracycline and paracetamol and the potential aquatic ecological risk associated with their presence in surface waters of the Arges-Vedea, Buzau- Ialomita, Dobrogea-Litoral River Basins in Romania. *IOP Conference Series. Earth and Environmental Science*, 616(1), 012016. Advance online publication. DOI: 10.1088/1755-1315/616/1/012016

Irfan, S., Zhuanghao, Z., Li, F., Chen, Y. X., Liang, G. X., Luo, J. T., & Ping, F. (2019). Critical review: Bismuth ferrite as an emerging visible light active nanostructured photocatalyst. *Journal of Materials Research and Technology*, 8(6), 6375–6389. DOI: 10.1016/j.jmrt.2019.10.004

Jaffari, Z. H., Lam, S. M., Sin, J. C., & Zeng, H. (2019). Boosting visible light photocatalytic and antibacterial performance by decoration of silver on magnetic spindle-like bismuth ferrite. *Materials Science in Semiconductor Processing*, 101(April), 103–115. DOI: 10.1016/j.mssp.2019.05.036

Jaffrézic, A., Jardé, E., Soulier, A., Carrera, L., Marengue, E., Cailleau, A., & Le Bot, B. (2017). Veterinary pharmaceutical contamination in mixed land use watersheds: From agricultural headwater to water monitoring watershed. *The Science of the Total Environment*, 609, 992–1000. DOI: 10.1016/j.scitotenv.2017.07.206 PMID: 28783916

Jaswal, A., Kaur, M., Singh, S., Kansal, S. K., Umar, A., Garoufalis, C. S., & Baskoutas, S. (2021). Adsorptive removal of antibiotic ofloxacin in aqueous phase using rGO-MoS2 heterostructure. Journal of Hazardous Materials, 417(November 2020), 125982. https://doi.org/DOI: 10.1016/j.jhazmat.2021.125982

Jeyasubramanian, K., Muthuselvi, M., Hikku, G. S., & Muthusankar, E. (2019). Improving electrochemical performance of reduced graphene oxide by counteracting its aggregation through intercalation of nanoparticles. *Journal of Colloid and Interface Science*, 549, 22–32. DOI: 10.1016/j.jcis.2019.04.046 PMID: 31015053

Ji, B., Zhao, W., Duan, J., Fu, L., & Yang, Z. (2020). Antibiotic under visible light. Journal name? 4427–4435. https://doi.org/DOI: 10.1039/C9RA08678A

Jiang, R., Lu, G., Yan, Z., Wu, D., Liu, J., & Zhang, X. (2019). Enhanced photocatalytic activity of a hydrogen bond-assisted 2D/2D Z-scheme SnNb2O6/Bi2WO6 system: Highly efficient separation of photoinduced carriers. *Journal of Colloid and Interface Science*, 552, 678–688. DOI: 10.1016/j.jcis.2019.05.104 PMID: 31176051

Jiang, W., Chen, X., Zhang, P., & Li, Y. (2022). Enhanced photocatalytic degradation of antibiotics by graphitic carbon nitride-based composites under visible light: A review. *Chemical Engineering Journal*, 429, 132262.

Jin, X., Zhou, X., Sun, P., Lin, S., Cao, W., Li, Z., & Liu, W. (2019). Photocatalytic degradation of norfloxacin using N-doped TiO2: Optimization, mechanism, identification of intermediates and toxicity evaluation. *Chemosphere*, 237, 124433. Advance online publication. DOI: 10.1016/j.chemosphere.2019.124433 PMID: 31352100

Johra, F. T., & Jung, W. (2015). RGO-TiO2-ZnO composites : Synthesis, characterization, and application to photocatalysis. Journal name. 491, 52–57.

Ju, B., Yang, F., Huang, K., & Wang, Y. (2020). Fabrication, Characterization and photocatalytic mechanism of a novel Z-scheme BiOBr/Ag3PO4@rGO composite for enhanced visible light photocatalytic degradation. *Journal of Alloys and Compounds*, 815, 151886. DOI: 10.1016/j.jallcom.2019.151886

Kadi, M. W., Mohamed, R. M., & Ismail, A. A. (2020). Facile synthesis of mesoporous BiFeO3/graphene nanocomposites as highly photoactive under visible light. *Optical Materials*, 104, 109842. DOI: 10.1016/j.optmat.2020.109842

Kanakaraju, D., Glass, B. D., & Oelgemöller, M. (2018). Advanced oxidation processmediated removal of pharmaceuticals from water: A review. *Journal of Environmental Management*, 219, 189–207. DOI: 10.1016/j.jenvman.2018.04.103 PMID: 29747102

Kang, S., Choi, W., & Lee, J. (2021). Photocatalytic degradation of pharmaceuticals and personal care products using TiO_2-based materials: A review. *Journal of Hazardous Materials*, 402, 123890.

Karpi'nska, J., & Kotowska, U. (2019). Removal of organic pollution in the water environment. *Water (Basel)*, 11(10), 2017. Advance online publication. DOI: 10.3390/w11102017

Khalid, N. R., Majid, A., Tahir, M. B., Niaz, N. A., & Khalid, S. (2017). Carbonaceous-TiO2 nanomaterials for photocatalytic degradation of pollutants: A review. *Ceramics International*, 43(17), 14552–14571. DOI: 10.1016/j.ceramint.2017.08.143

Khan, H. K., Rehman, M. Y. A., & Malik, R. N. (2020). Fate and toxicity of pharmaceuticals in water environment: An insight on their occurrence in South Asia. *Journal of Environmental Management*, 271(July), 111030. DOI: 10.1016/j.jenvman.2020.111030 PMID: 32778310

Khan, N. A., Khan, S. U., Ahmed, S., Farooqi, I. H., Yousefi, M., Mohammadi, A. A., & Changani, F. (2020). Recent trends in disposal and treatment technologies of emerging-pollutants- A critical review. *Trends in Analytical Chemistry*, 122, 115744. Advance online publication. DOI: 10.1016/j.trac.2019.115744

Krishnan, S., Rawindran, H., Sinnathambi, C. M., & Lim, J. W. (2017). Comparison of various advanced oxidation processes used in remediation of industrial wastewater laden with recalcitrant pollutants. *IOP Conference Series. Materials Science and Engineering*, 206(1), 012089. Advance online publication. DOI: 10.1088/1757-899X/206/1/012089

Kumar, A., Sharma, G., Naushad, M., Ahamad, T., Veses, R. C., & Stadler, F. J. (2019). Highly visible active Ag2CrO4/Ag/BiFeO3@RGO nano-junction for photoreduction of CO2 and photocatalytic removal of ciprofloxacin and bromate ions: The triggering effect of Ag and RGO. *Chemical Engineering Journal*, 370(March), 148–165. DOI: 10.1016/j.cej.2019.03.196

Labhane, P. K., Patle, L. B., Huse, V. R., Sonawane, G. H., & Sonawane, S. H. (2016). Synthesis of reduced graphene oxide sheets decorated by zinc oxide nanoparticles: Crystallographic, optical, morphological and photocatalytic study. *Chemical Physics Letters*, 661, 13–19. DOI: 10.1016/j.cplett.2016.08.041

Lafi, W. K., Shannak, B., Al-Shannag, M., Al-Anber, Z., & Al-Hasan, M. (2009). Treatment of olive mill wastewater by combined advanced oxidation and biodegradation. *Separation and Purification Technology*, 70(2), 141–146. DOI: 10.1016/j.seppur.2009.09.008

Lapworth, D. J., Baran, N., Stuart, M. E., & Ward, R. S. (2012). Emerging organic contaminants in groundwater: A review of sources, fate and occurrence. *Environmental Pollution*, 163, 287–303. DOI: 10.1016/j.envpol.2011.12.034 PMID: 22306910

Larsson, D. G. J., De Pedro, C., & Paxeus, N. (2007). Effluent from Drug Manufactures Contains Extremely High Levels of Pharmaceuticals. *Journal of Hazardous Materials*, 148(3), 751–755. DOI: 10.1016/j.jhazmat.2007.07.008 PMID: 17706342

Lee, S. Y., & Park, S. J. (2013). TiO2 photocatalyst for water treatment applications. *Journal of Industrial and Engineering Chemistry*, 19(6), 1761–1769. DOI: 10.1016/j.jiec.2013.07.012

Li, D., & Shi, W. (2016). Recent developments in visible-light photocatalytic degradation of antibiotics. Cuihua Xuebao. *Chinese Journal of Catalysis*, 37(6), 792–799. DOI: 10.1016/S1872-2067(15)61054-3

Li, J., Chen, G., & Zhang, R. (2022). Efficient photocatalytic degradation of pharmaceuticals using g-C_3N_4/fullerene nanostructures. *Environmental Research*, 204, 112050.

Li, J., Wang, Y., Ling, H., Qiu, Y., Lou, J., Hou, X., Bag, S. P., Wang, J., Wu, H., & Chai, G. (2019). Significant enhancement of the visible light photocatalytic properties in 3d BiFeO3/graphene composites. *Nanomaterials (Basel, Switzerland)*, 9(1), 65. Advance online publication. DOI: 10.3390/nano9010065 PMID: 30621245

Li, K., Ji, M., Chen, R., Jiang, Q., Xia, J., & Li, H. (2020). Construction of nitrogen and phosphorus co-doped graphene quantum dots/Bi5O7I composites for accelerated charge separation and enhanced photocatalytic degradation performance. *Chinese Journal of Catalysis*, 41(8), 1230–1239. DOI: 10.1016/S1872-2067(20)63531-8

Li, N., Ho, K. W. K., Ying, G. G., & Deng, W. J. (2017). Veterinary antibiotics in food, drinking water, and the urine of preschool children in Hong Kong. *Environment International*, 108(August), 246–252. DOI: 10.1016/j.envint.2017.08.014 PMID: 28889029

Li, X., Qiu, Y., Zhu, Z., Zhang, H., & Yin, D. (2022). Novel recyclable Z-scheme g-C3N4/carbon nanotubes/Bi25FeO40 heterostructure with enhanced visiblelight photocatalytic performance towards tetracycline degradation. *Chemical Engineering Journal*, 429, 132130. DOI: 10.1016/j.cej.2021.132130

Liu, B., Wu, H., & Parkin, I. P. (2020). New Insights into the Fundamental Principle of Semiconductor Photocatalysis. *ACS Omega*, 5(24), 14847–14856. DOI: 10.1021/acsomega.0c02145 PMID: 32596623

Liu, W., He, T., Wang, Y., Ning, G., Xu, Z., Chen, X., Hu, X., Wu, Y., & Zhao, Y. (2020). Synergistic adsorption-photocatalytic degradation effect and norfloxacin mechanism of ZnO/ZnS@BC under UV-light irradiation. *Scientific Reports*, 10(1), 1–12. DOI: 10.1038/s41598-020-68517-x PMID: 32681000

Liu, Y., Zhang, R., & Zhang, M. (2021). Visible-light-driven g-C_3N_4/graphene oxide for photocatalytic degradation of pharmaceuticals: A review. *Catalysis Today*, 371, 32–47.

Lu, H., Du, Z., Wang, J., & Liu, Y. (2015). Enhanced photocatalytic performance of Ag-decorated Bi-FeO3 in visible light region. *Journal of Sol-Gel Science and Technology*, 76(1), 50–57. DOI: 10.1007/s10971-015-3749-6

Lu, M. C., Chen, Y. Y., Chiou, M. R., Chen, M. Y., & Fan, H. J. (2016). Occurrence and treatment efficiency of pharmaceuticals in landfill leachates. *Waste Management (New York, N.Y.)*, 55, 257–264. DOI: 10.1016/j.wasman.2016.03.029 PMID: 27026494

Ma, J., Miao, T. J., & Tang, J. (2022). Charge carrier dynamics and reaction intermediates in heterogeneous photocatalysis by time-resolved spectroscopies. *Chemical Society Reviews*, 51(14), 5777–5794. DOI: 10.1039/D1CS01164B PMID: 35770623

Ma, L., Duan, J., Ji, B., Liu, Y., Li, C., Li, C., Zhao, W., & Yang, Z. (2021). Ligand-metal charge transfer mechanism enhances TiO2/Bi2WO6/rGO nanomaterials photocatalytic efficient degradation of norfloxacin under visible light. *Journal of Alloys and Compounds*, 869, 158679. DOI: 10.1016/j.jallcom.2021.158679

Ma, Y., Li, Z., & Zhang, Z. (2021). g-C_3N_4/activated carbon for photocatalytic degradation of pharmaceuticals: Recent progress and future directions. *Journal of Environmental Management*, 287, 112324.

Ma, Z. P., Zhang, L., Ma, X., Zhang, Y.-H., & Shi, F.-N. (2022). Design of Zscheme g-C3N4/BC/Bi-25FeO40 photocatalyst with unique electron transfer channels for efficient degradation of tetracycline hydrochloride waste. *Chemosphere*, 289, 133262. DOI: 10.1016/j.chemosphere.2021.133262 PMID: 34906528

Madaan, A., Verma, R., Kumar, V., Singh, A. T., Jain, S. K., & Jaggi, M. (2015). 1,8- Naphthyridine Derivatives: A Review of Multiple Biological Activities. *Archiv der Pharmazie*, 348(12), 837–860. DOI: 10.1002/ardp.201500237 PMID: 26548568

Madima, N., Mishra, A. K., & Arotiba, O. A. (2022). TiO_2-based photocatalysts for water purification: A review. *Journal of Environmental Chemical Engineering*, 10(2), 107143.

Mangalam, J., Kumar, M., Sharma, M., & Joshi, M. (2019). High adsorptivity and visible light assisted photocatalytic activity of silver/reduced graphene oxide (Ag/rGO) nanocomposite for wastewater treatment. Nano-Structures and Nano- Objects, 17, 58–66. https://doi.org/DOI: 10.1016/j.nanoso.2018.11.003

Manzetti, S., & Ghisi, R. (2014). The environmental release and fate of antibiotics. *Marine Pollution Bulletin*, 79(1–2), 7–15. DOI: 10.1016/j.marpolbul.2014.01.005 PMID: 24456854

Mboya, E. A., Sanga, L. A., & Ngocho, J. S. (2018). Irrational use of antibiotics in the moshi municipality Northern Tanzania: A cross sectional study. *The Pan African Medical Journal*, 31, 1–10. DOI: 10.11604/pamj.2018.31.165.15991 PMID: 31086618

Mestre, A. S., & Carvalho, A. P. (2019). Photocatalytic degradation of pharmaceuticals carbamazepine, diclofenac, and sulfamethoxazole by semiconductor and carbon materials: A review. *Molecules (Basel, Switzerland)*, 24(20), 3702. DOI: 10.3390/molecules24203702 PMID: 31618947

Miao, X. S., Bishay, F., Chen, M., & Metcalfe, C. D. (2004). Occurrence of antimicrobials in the final effluents of wastewater treatment plants in Canada. *Environmental Science & Technology*, 38(13), 3533–3541. DOI: 10.1021/es030653q PMID: 15296302

Minella, M., Fabbri, D., Calza, P., & Minero, C. (2017). Selected hybrid photocatalytic materials for the removal of drugs from water. *Current Opinion in Green and Sustainable Chemistry*, 6, 11–17. DOI: 10.1016/j.cogsc.2017.05.002

Mohamed, H. H. (2018). Sonochemical synthesis of ZnO hollow microstructure/reduced graphene oxide for enhanced sunlight photocatalytic degradation of organic pollutants. *Journal of Photochemistry and Photobiology A Chemistry*, 353, 401–408. DOI: 10.1016/j.jphotochem.2017.11.052

Mohammad, A., Khan, M. E., Cho, M. H., & Yoon, T. (2021). Graphitic-carbon nitride based mixed-phase bismuth nanostructures: Tuned optical and structural properties with boosted photocatalytic performance for wastewater decontamination under visible-light irradiation. *NanoImpact*, 23, 100345. DOI: 10.1016/j.impact.2021.100345 PMID: 35559846

Mohd Kaus, N. H., Rithwan, A. F., Adnan, R., Ibrahim, M. L., Thongmee, S., & Mohd Yusoff, S. F. (2021). Effective strategies, mechanisms, and photocatalytic efficiency of semiconductor nanomaterials incorporating rGO for Environmental contaminant degradation. *Catalysts*, 11(3), 302. DOI: 10.3390/catal11030302

Mokarram, M., Saber, A., & Sheykhi, V. (2020). Effects of heavy metal contamination on river water quality due to release of industrial effluents. *Journal of Cleaner Production*, 277, 123380. DOI: 10.1016/j.jclepro.2020.123380

Mompelat, S., Le Bot, B., & Thomas, O. (2009). Occurrence and fate of pharmaceutical products and by-products, from resource to drinking water. *Environment International*, 35(5), 803–814. DOI: 10.1016/j.envint.2008.10.008 PMID: 19101037

Mukherjee, A., Chakrabarty, S., Kumari, N., Su, W. N., & Basu, S. (2018). Visible-Light-mediated electrocatalytic activity in reduced graphene oxide-supported bismuth ferrite. *ACS Omega*, 3(6), 5946–5957. DOI: 10.1021/acsomega.8b00708 PMID: 30023934

Naraginti, S., & Yong, Y. C. (2019). Enhanced detoxification of p-bromophenol by novel Zr/Ag-TiO2@rGO ternary composite: Degradation kinetics and phytotoxicity evolution studies. Ecotoxicology and Environmental Safety, 170(December 2018), 355–362. https://doi.org/DOI: 10.1016/j.ecoenv.2018.12.001

Neelam, T., & Megha, S. (2019). Antimicrobial resistance in the environment: The Indian scenario. *The Indian Journal of Medical Research*, 149(2), 119–128. DOI: 10.4103/ijmr.IJMR_331_18 PMID: 31219076

Niu, F., Chen, D., Qin, L., Gao, T., Zhang, N., Wang, S., Chen, Z., Wang, J., Sun, X., & Huang, Y. (2015). Synthesis of Pt/BiFeO3 heterostructured photocatalysts for highly efficient visible-light photocatalytic performances. *Solar Energy Materials and Solar Cells*, 143, 386–396. DOI: 10.1016/j.solmat.2015.07.008

Orimolade, B. O., Koiki, B. A., Zwane, B. N., Peleyeju, G. M., Mabuba, N., & Arotiba, O. A. (2019). Interrogating solar photoelectrocatalysis on an exfoliated graphite–BiVO4/ZnO composite electrode towards water treatment. *RSC Advances*, 9(29), 16586–16595. DOI: 10.1039/C9RA02366F PMID: 35516409

Pan, J. H., Dou, H., Xiong, Z., Xu, C., Ma, J., & Zhao, X. S. (2010). Porous photocatalysts for advanced water purifications. *Journal of Materials Chemistry*, 20(22), 4512–4528. DOI: 10.1039/b925523k

Pankaj, R., Anita, S., Singh, V. P., Gupta, V. K., Hosseini-Bandegharaei, A., Rajesh, K., & Pardeep, S. (2019). Solar light assisted degradation of oxytetracycline from water using Bi2O3/Fe3O4 supported graphitic carbon nitride photocatalyst. *Desalination and Water Treatment*, 148, 338–350. DOI: 10.5004/dwt.2019.23831

Patel, N., Khan, Z. A., Shahane, S., Rai, D., Chauhan, D., Kant, C., & Chaudhary, V. K. (2020). Emerging pollutants in aquatic environment: Source, effect, and challenges in biomonitoring and bioremediation-A review. *Pollution*, 6(1), 99–113. DOI: 10.22059/POLL.2019.285116.646

Peng, Y., Ji, J., & Chen, D. (2015). Ultrasound assisted synthesis of ZnO/reduced graphene oxide composites with enhanced photocatalytic activity and anti- photocorrosion. *Applied Surface Science*, 356, 762–768. DOI: 10.1016/j.apsusc.2015.08.070

Ponraj, C., Vinitha, G., & Daniel, J. (2017). A review on the visible light active BiFeO3 nanostructures as suitable photocatalyst in the degradation of different textile dyes. In *Environmental Nanotechnology* (Vol. 7, pp. 110–120). Monitoring and Management., DOI: 10.1016/j.enmm.2017.02.001

Praveena, S. M., Mohd Rashid, M. Z., Mohd Nasir, F. A., Sze Yee, W., & Aris, A. Z. (2019). Occurrence and potential human health risk of pharmaceutical residues in drinking water from Putrajaya (Malaysia). *Ecotoxicology and Environmental Safety*, 180(February), 549–556. DOI: 10.1016/j.ecoenv.2019.05.051 PMID: 31128553

Preetha, S., Pillai, R., Ramamoorthy, S., Mayeen, A., Archana, K. M., Kalarikkal, N., Narasimhamurthy, B., & Lekshmi, I. C. (2022). TiO2–rGO nanocomposites with high rGO content and luminescence quenching through green redox synthesis. Surfaces and Interfaces, 30(June 2021). https://doi.org/DOI: 10.1016/j.surfin.2022.101812

Qiu, B., Luo, W., & Wang, Y. (2023). Fullerene-based photocatalysts for the removal of pharmaceutical contaminants: A review. *Journal of Environmental Chemical Engineering*, 11(1), 108566.

Rahmani, H., Gholami, M., Mahvi, A. H., Alimohammadi, M., Azarian, G., Esrafili, A., Rahmani, K., & Farzadkia, M. (2014). Tinidazole removal from aqueous solution by sonolysis in the presence of hydrogen peroxide. *Bulletin of Environmental Contamination and Toxicology*, 92(3), 341–346. DOI: 10.1007/s00128-013-1193-2 PMID: 24420343

Raja, A., Rajasekaran, P., Selvakumar, K., Arunpandian, M., Kaviyarasu, K., Asath Bahadur, S., & Swaminathan, M. (2020). Visible active reduced graphene oxide- BiVO4-ZnO ternary photocatalyst for efficient removal of ciprofloxacin. *Separation and Purification Technology*, 233, 115996. DOI: 10.1016/j.seppur.2019.115996

Raza, W., Bahnemann, D., & Muneer, M. (2018). A green approach for degradation of organic pollutants using rare earth metal doped bismuth oxide. *Catalysis Today*, 300, 89–98. DOI: 10.1016/j.cattod.2017.07.029

Rhimi, B., Wang, C., & Bahnemann, D. W. (2020). Latest progress in g-C3N4 based heterojunctions for hydrogen production via photocatalytic water splitting: A mini review. *Journal of Physics: Energy*, 2(4), 042003. DOI: 10.1088/2515-7655/abb782

Roberts, J., Kumar, A., Du, J., Hepplewhite, C., Ellis, D. J., Christy, A. G., & Beavis, S. G. (2016). Pharmaceuticals and personal care products (PPCPs) in Australia's largest inland sewage treatment plant, and its contribution to a major Australian river during high and low flow. *The Science of the Total Environment*, 541, 1625–1637. DOI: 10.1016/j.scitotenv.2015.03.145 PMID: 26456435

Ruidíaz-Martínez, M., Álvarez, M. A., López-Ramón, M. V., Cruz-Quesada, G., Rivera-Utrilla, J., & Sánchez-Polo, M. (2020). Hydrothermal synthesis of RGO- TiO2 composites as high-performance UV photocatalysts for ethylparaben degradation. *Catalysts*, 10(5), 1–25. DOI: 10.3390/catal10050520

Sárközy, G. (2001). Quinolones: A class of antimicrobial agents. *Veterinární Medicína*, 46(9–10), 257–274. DOI: 10.17221/7883-VETMED

Scott, T. M., Phillips, P. J., Kolpin, D. W., Colella, K. M., Furlong, E. T., Foreman, W. T., & Gray, J. L. (2018). Pharmaceutical manufacturing facility discharges can substantially increase the pharmaceutical load to U.S. wastewaters. *The Science of the Total Environment*, 636, 69–79. DOI: 10.1016/j.scitotenv.2018.04.160 PMID: 29704718

Sekar, K., Anandan, S., & Sivasankar, T. (2022). ZnO nanomaterials for water treatment applications: A review. *Environmental Nanotechnology, Monitoring & Management*, 17, 100616.

Shahid, M. K., Kashif, A., Fuwad, A., & Choi, Y. (2021). Current advances in treatment technologies for removal of emerging contaminants from water – A critical review. *Coordination Chemistry Reviews*, 442, 213993. DOI: 10.1016/j.ccr.2021.213993

Shan, A. Y., Ghazi, T. I. M., & Rashid, S. A. (2010). Immobilisation of titanium dioxide onto supporting materials in heterogeneous photocatalysis: A review. *Applied Catalysis A, General*, 389(1-2), 1–8. DOI: 10.1016/j.apcata.2010.08.053

Shang, E., Li, Y., Niu, J., Li, S., Zhang, G., & Wang, X. (2018). Photocatalytic degradation of perfluorooctanoic acid over Pb-BiFeO3/rGO catalyst: Kinetics and mechanism. *Chemosphere*, 211, 34–43. DOI: 10.1016/j.chemosphere.2018.07.130 PMID: 30071434

Singh, D., Tabari, T., Ebadi, M., Trochowski, M., Baris Yagci, M., & Macyk, W. (2019). Efficient synthesis of BiFeO3 by the microwave-assisted sol-gel method: "A" site influence on the photoelectrochemical activity of perovskites. Applied Surface Science, 471(December2018), 1017–1027. https://doi.org/DOI: 10.1016/j.apsusc.2018.12.082

Singh, H., Bhardwaj, N., Arya, S. K., & Khatri, M. (2020). Environmental impacts of oil spills and their remediation by magnetic nanomaterials. Environmental Nanotechnology, Monitoring and Management, 14(September 2019), 100305. https://doi.org/DOI: 10.1016/j.enmm.2020.100305

Singh, J., Kumar, A., & Bhati, M. S. (2023). ZnO nanostructures for environmental remediation: A review on photocatalytic degradation of organic pollutants. *Materials Science in Semiconductor Processing*, 151, 106996.

Singh, K., Mohan, M., & Nautiyal, S. (2020). Comparing cefixime, cefpodoxime and ofloxacin as antimicrobial agents and their effects on gut microbiota. *International Journal of Scientific Research in Science and Technology*, (November), 131–152. DOI: 10.32628/IJSRST207619

Soltani, T., & Lee, B.-K. (2016a). Sono-synthesis of nanocrystallized BiFeO3/reduced graphene oxide composites for visible photocatalytic degradation improvement of bisphenol A. *Chemical Engineering Journal*, 306, 204–213. DOI: 10.1016/j.cej.2016.07.051

Stefanakis, A. I., & Becker, J. A. (2019). A review of emerging contaminants in water: Classification, sources, and potential risks. *Waste Management: Concepts, Methodologies, Tools, and Applications*, (January), 177–202. DOI: 10.4018/978-1-7998-1210-4.ch008

Strankowski, M., Włodarczyk, D., Piszczyk, Ł., & Strankowska, J. (2016). Polyurethane nanocomposites containing reduced graphene oxide, FTIR, Raman, and XRD Studies. *Journal of Spectroscopy*, 2016, 1–6. Advance online publication. DOI: 10.1155/2016/7520741

Suaifan, G. A. R. Y., Mohammed, A. A. M., & Alkhawaja, B. A. (2022). Fluoroquinolones' Biological Activities against Laboratory Microbes and cancer cell lines. *Molecules (Basel, Switzerland)*, 27(5), 1658. Advance online publication. DOI: 10.3390/molecules27051658 PMID: 35268759

Sun, C., Liu, H., & Zhang, Y. (2022). Photocatalytic degradation of pharmaceutical contaminants using g-C_3N_4/activated carbon: A review. *Chemosphere*, 292, 133414.

Sun, W., Hu, Q., Wu, T., Wang, Z., Yi, J., & Yin, S. (2022). Construction of 0D/3D carbon quantum dots modified PbBiO2Cl microspheres with accelerated charge carriers for promoted visible-light-driven degradation of organic contaminants, Colloids Surface andPhysicochemical Engineering. *Colloids and Surfaces. A, Physicochemical and Engineering Aspects*, 642, 128591. DOI: 10.1016/j.colsurfa.2022.128591

Tahir, M. B., Iqbal, T., Rafique, M., Rafique, M. S., Nawaz, T., & Sagir, M. (2020). Nanomaterials for photocatalysis. In *Nanotechnology and Photocatalysis for Environmental Applications*. Elsevier Inc., DOI: 10.1016/B978-0-12-821192-2.00005-X

Tammina, S. K., Mandal, B. K., & Kadiyala, N. K. (2018). Photocatalytic degradation of methylene blue dye by nonconventional synthesized SnO2 nanoparticles. *Environmental Nanotechnology, Monitoring & Management*, 10, 339–350. DOI: 10.1016/j.enmm.2018.07.006

Teglia, C. M., Perez, F. A., Michlig, N., Repetti, M. R., Goicoechea, H. C., & Culzoni, M. J. (2019). Occurrence, distribution, and ecological risk of fluoroquinolones in rivers and wastewaters. *Environmental Toxicology and Chemistry*, 38(10), 2305–2313. DOI: 10.1002/etc.4532 PMID: 31291022

Tkaczyk, A., Mitrowska, K., & Posyniak, A. (2020). Synthetic organic dyes as contaminants of the aquatic environment and their implications for ecosystems: A review. *The Science of the Total Environment*, 717, 137222. DOI: 10.1016/j.scitotenv.2020.137222 PMID: 32084689

Tong, C., Zhuo, X., & Guo, Y. (2011). Occurrence and risk assessment of four typical fluoroquinolone antibiotics in raw and treated sewage and in receiving waters in Hangzhou, China. *Journal of Agricultural and Food Chemistry*, 59(13), 7303–7309. DOI: 10.1021/jf2013937 PMID: 21630710

Tong, H., Ouyang, S. X., Bi, Y. P., Umezawa, N., Oshikiri, M., & Ye, J. H. (2012). Nano-photocatalytic materials: Possibilities and challenges. *Advanced Materials*, 24(2), 229–251. DOI: 10.1002/adma.201102752 PMID: 21972044

Vaizoğullar, A. I. (2017). TiO2/ZnO supported on sepiolite: Preparation, structural characterization, and photocatalytic degradation of flumequine antibiotic in aqueous solution. *Chemical Engineering Communications*, 204(6), 689–697. DOI: 10.1080/00986445.2017.1306518

Velempini, T., Prabakaran, E., & Pillay, K. (2021). Recent developments in the use of metal oxides for photocatalytic degradation of pharmaceutical pollutants in water—A review. *Materials Today. Chemistry*, 19, 100380. DOI: 10.1016/j.mtchem.2020.100380

Viaroli, S., Lancia, M., & Re, V. (2022). Microplastics contamination of groundwater: Current evidence and future perspectives. A review. *The Science of the Total Environment*, 824, 153851. DOI: 10.1016/j.scitotenv.2022.153851 PMID: 35176372

Wagil, M., Kumirska, J., Stolte, S., Puckowski, A., Maszkowska, J., Stepnowski, P., & Białk-Bielińska, A. (2014). Development of sensitive and reliable LC-MS/MS methods for the determination of three fluoroquinolones in water and fish tissue samples and preliminary environmental risk assessment of their presence in two rivers in northern Poland. *The Science of the Total Environment*, 493, 1006–1013. DOI: 10.1016/j.scitotenv.2014.06.082 PMID: 25016107

Wang, J., & Wang, S. (2016). Removal of pharmaceuticals and personal care products (PPCPs) from wastewater: A review. *Journal of Environmental Management*, 182, 620–640. DOI: 10.1016/j.jenvman.2016.07.049 PMID: 27552641

Wang, W., Serp, P., Kalck, P., & Faria, J. L. (2005). Visible light photodegradation of phenol on MWNT-TiO2 composite catalysts prepared by a modified sol–gel method. *Journal of Molecular Catalysis A Chemical*, 235(1-2), 194–199. DOI: 10.1016/j.molcata.2005.02.027

Wang, X., Biswas, S., Paudyal, N., Pan, H., Li, X., Fang, W., & Yue, M. (2019). Antibiotic resistance in salmonella typhimurium isolates recovered from the food chain through national antimicrobial resistance monitoring system between 1996 and 2016. *Frontiers in Microbiology*, 10(MAY), 1–12. DOI: 10.3389/fmicb.2019.00985 PMID: 31134024

Wang, Y., He, D., & Sun, H. (2021). Graphitic carbon nitride (g-C_3N_4)-based materials for photocatalytic degradation of pharmaceuticals: Recent progress and future challenges. *Applied Catalysis B: Environmental*, 297, 120463.

Wei, X., Huang, L., & Wang, L. (2023). N-doped g-C_3N_4/carbon nanotubes: A review on their applications in environmental purification. *Journal of Cleaner Production*, 398, 136308.

Woan, K., Pyrgiotakis, G., & Sigmund, W. (2009). Photocatalytic carbon-nanotube–TiO2 composites. *Advanced Materials*, 21(21), 2233–2239. DOI: 10.1002/adma.200802738

Wu, K., Song, S., Wu, H., Guo, J., Zhang, L., 2020b. Facile synthesis of Bi2WO6/C3N4/ Ti3C2 composite as Z-scheme photocatalyst for efficient ciprofloxacin degradation and H2 production. Applied. Catalysis A-. General. 608, 117869. https://doi.org/. apcata.2020.117869.DOI: 10.1016/j

Xiang, J., Wu, M., Lei, J., Fu, C., Gu, J., & Xu, G. (2018). The fate and risk assessment of psychiatric pharmaceuticals from psychiatric hospital effluent. *Ecotoxicology and Environmental Safety*, 150, 289–296. DOI: 10.1016/j.ecoenv.2017.12.049 PMID: 29289864

Xu, Y., Liu, J., Xie, M., Jing, L., Xu, H., She, X., Li, H., & Xie, J. (2019). Construction of novel CNT/LaVO4 nanostructures for efficient antibiotic photodegradation. Chemical Engineering Journal, 357(July 2018), 487–497. https://doi.org/DOI: 10.1016/j.cej.2018.09.098

Yang, X., & Wang, D. (2018). Photocatalysis: From fundamental principles to materials and applications. *ACS Applied Energy Materials*, 1(12), 6657–6693. DOI: 10.1021/acsaem.8b01345

Yang, Y., Ok, Y. S., Kim, K. H., Kwon, E. E., & Tsang, Y. F. (2017). Occurrences and removal of pharmaceuticals and personal care products (PPCPs) in drinking water and water/sewage treatment plants: A review. *The Science of the Total Environment*, 596–597, 303–320. DOI: 10.1016/j.scitotenv.2017.04.102 PMID: 28437649

Yi, F., Ma, J., Lin, C., Wang, L., Zhang, H., Qian, Y., & Zhang, K. (2020). Insights into the enhanced adsorption/photocatalysis mechanism of a Bi4O5Br2/g-C3N4 nanosheet. *Journal of Alloys and Compounds*, 821, 153557. DOI: 10.1016/j.jallcom.2019.153557

Zambrano, J., García-Encina, P. A., Jiménez, J. J., López-Serna, R., & Irusta-Mata, R. (2022). Photolytic and photocatalytic removal of a mixture of four veterinary antibiotics. *Journal of Water Process Engineering*, 48(April), 102841. Advance online publication. DOI: 10.1016/j.jwpe.2022.102841

Zhang, G., Song, A., Duan, Y., & Zheng, S. (2018). Enhanced photocatalytic activity of TiO2/zeolite composite for abatement of pollutants. *Microporous and Mesoporous Materials*, 255, 61–68. DOI: 10.1016/j.micromeso.2017.07.028

Zhang, J., Fu, D., & Wu, J. (2012). Photodegradation of Norfloxacin in aqueous solution containing algae. *Journal of Environmental Sciences (China)*, 24(4), 743–749. DOI: 10.1016/S1001-0742(11)60814-0 PMID: 22894111

Zhang, N., & Chu, D. (2020). Fabricating erythrocyte-like BiOI/Bi2WO6 heterostructures with enhancing the visible-light photocatalytic performance. *Journal of Nanoparticle Research*, 22(7), 206. Advance online publication. DOI: 10.1007/s11051-020-04938-z

Zhang, X., Li, P., & Wang, Z. (2022). Enhanced photocatalytic degradation of pharmaceuticals using g-C_3N_4/graphene oxide composites. *Journal of Environmental Chemical Engineering*, 8(4), 104200.

Zhao, Y., Liang, X., Hu, X., Fan, J., 2021. rGO/Bi2WO6 composite as a highly efficient and stable visible-light photocatalyst for norfloxacin degradation in aqueous environment. Journal of Colloid Interface Sci. 589, 336–346. https://doi.org/. jcis.2021.01.016.DOI: 10.1016/j

Zheng, X., Yuan, J., Shen, J., Liang, J., Che, J., Tang, B., He, G., & Chen, H. (2019). A carnation-like rGO/Bi2O2CO3 /BiOCl composite: Efficient photocatalyst for the degradation of ciprofloxacin. *Journal of Materials Science Materials in Electronics*, 30(6), 5986–5994. DOI: 10.1007/s10854-019-00898-w

Zheng, Y., Liu, Y., Guo, X., Chen, Z., Zhang, W., Wang, Y., Tang, X., Zhang, Y., & Zhao, Y. (2020). Sulfur-doped g-C3N4/rGO porous nanosheets for highly efficient photocatalytic degradation of refractory contaminants. *Journal of Materials Science and Technology*, 41, 117–126. DOI: 10.1016/j.jmst.2019.09.018

Zhi, D., Yang, D., Zheng, Y., Yang, Y., He, Y., Luo, L., & Zhou, Y. (2019). Current progress in the adsorption, transport and biodegradation of antibiotics in soil. *Journal of Environmental Management*, 251(September), 109598. DOI: 10.1016/j.jenvman.2019.109598 PMID: 31563054

Zinatloo-Ajabshir, S., Morassaei, M. S., & Salavati-Niasari, M. (2019). Facile synthesis of Nd2Sn2O7-SnO2 nanostructures by novel and environment-friendly approach for the photodegradation and removal of organic pollutants in water. Journal of Environmental Management, 233(December 2018), 107–119. DOI: 10.1016/j.jenvman.2018.12.011

Chapter 18
Geopolymers Prepared From Unused Resources and Their Applications

Yuta Watanabe Nikaido
Tama Chemicals Co. Ltd., Japan

Sujitra Onutai
https://orcid.org/0000-0003-1334-5369
Japan Atomic Energy Agency, Japan

Sirithan Jiemsirilers
Chulalongkorn University, Thailand

Takaomo Kobayashi
Nagaoka University of Technology, Japan

ABSTRACT

As an alumino-silicate material, geopolymers present several of beneficial properties, including high strength, low permeability, high acid resistance, the capacity to immobilize poisonous compounds, and the ability to resist hazardous contaminants. In this chapter, geopolymer materials using raw materials derived from unutilized waste are discussed, especially the case of using metakaolin-based and fly ash-based raw materials, and the preparation and properties of geopolymers and their applications are explained. The main point is that geopolymers are materials that can contribute to the construction of a sustainable society as a technology that can create high value-added products through the effective use of unused waste materials.

INTRODUCTION

In building a recycling-oriented society, the use of waste and unused products is accelerating. But, waste and unused products are being generated at an increasingly rapid pace, and these are left untreated, with harmful effects on the ecosystem in some respects. For instance, disposal of wastes that are

DOI: 10.4018/979-8-3693-0003-9.ch018

Copyright ©2025, IGI Global. Copying or distributing in print or electronic forms without written permission of IGI Global is prohibited.

not properly treated can contaminate the ground. In particular, wastes containing heavy metals have a negative impact on the ecosystem. Therefore, it is important to recover waste from industry and use it to solve the world's environmental problems. In general, the principle is that the 3R process (reduce, reuse, recycle) produces recyclables that have economic value, but even after the 3R stage is completed, unused waste is eventually disposed of in landfills. For example, kaolin processing industry generates large amounts of waste around the world (Menezes et al., 2008; Menezes et al., 2009) and generates large amounts of waste materials, so, kaolin mining present significant impact toward the surrounding ecology. During the clay production process, miners use powerful water jets to wash, and this powerful watering process produces a slurry consisting of a mixture of kaolin clay and other components, most of which are unutilized products. On the other hand, municipal solid waste includes food scraps, paper, plastic, wood, and textiles, and its volume has been increasing with rapid urbanization and improvement of people's living standards. The final product is finally incinerated and the residue remains as fly ash. The final use of the remaining fly ash is to mix it with building materials such as concrete, but most of it is treated as waste. The highest percentage of incineration was used to dispose of municipal solid waste in Japan (78.5%), followed by the UK (38.4%) and the EU (27.4%) (Y. Zhang et al., 2021). Therefore, effective utilization of fly ash residues is an important issue. One solution to these problems is upcycling toward geopolymers, also known as inorganic composites like ceramics, since kaolin slurry residue and fly ash components contain silica and alumina, and thus, alkali-catalyzed inorganic condensation products can be gained for such waste sources. Figure 1 shows pictures of metakaolin clay and fly ash and Table shows components of oxidative elements for both waste minerals. Included table has elemental analysis data by XRF for metakaolin (Prasanphan et al., 2019) and fly ash (Onutai et al., 2016).

Figure 1. Pictures of metakaolin (a and b) and fly ash (c and d) and their elemental analysis data

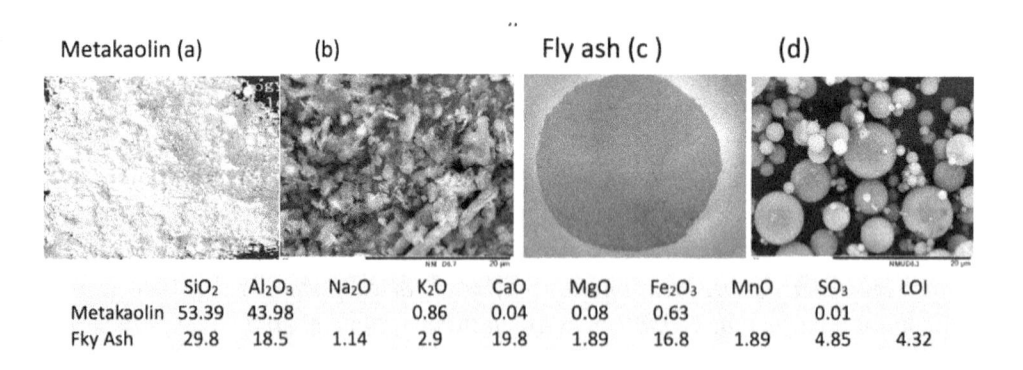

	SiO_2	Al_2O_3	Na_2O	K_2O	CaO	MgO	Fe_2O_3	MnO	SO_3	LOI
Metakaolin	53.39	43.98		0.86	0.04	0.08	0.63		0.01	
Fky Ash	29.8	18.5	1.14	2.9	19.8	1.89	16.8	1.89	4.85	4.32

Geopolymer materials are synthesized by aluminosilicate composites and alkali hydroxide and alkali silicate. Generally, alumino-silicate materials such as slag [12], metakaolin [13], and fly ash [14] enable to convert to geopolymers, since such row materials require containing silica and alumina and an alkaline solution. Then, the alkali paste slurry can form and transform quickly into a rigid geopolymer under mild condition without sintering process. Then, sufficient strength of ceramic-like geopolymers can be gained. As examples for geopolymer sources (Figure 1), here is fly ash from Mae Moh coal fired power plant, Lampang, Thailand. Ten 300 MW units compose the electricity produced by the lignite coal-fired Mae Moh coal power station, which has a daily capacity of more than 2500 MW. The process generates fly ash and bottom ash as by-products. Six of the ten units—the fly ashes—are given to businesses that manufacture cement. Fly ash from four leftover units is thrown in an abandoned mine. Table 1 displays a summary of output and consumption for coal power plants (Brigden et al., 2002).

According to the American Society for Testing Materials (ASTMC618) (American society for testing and Materials, 2002), class C fly ash is characterized as coal fly ash and raw or calcined natural pozzolan (Ahmaruzzaman, 2010), both of which contain a total mass of SiO_2, Al_2O_3, and Fe_2O_3 indicating that the fly ash of Mae Moh Lignite is class C type. The maximum free calcium level is 2.0%, and the maximum sulfur content (SO_4) is 4.0%, but the natural origin of the coal used in power plants renders the chemical composition of fly ash unstable as shown in Table 1 in the periods of 1985 to 2001. The ratios of SiO_2 and Al_2O_3 vary from year to year, but these components are present in high proportions in fly ash each year.

Table 1. Chemical composition of Mae Moh lignite fly ash from 1985-2001

Year	Composition %								
	SiO_2	Al_2O_3	Fe_2O_3	CaO	MgO	Na_2O	K_2O	SO_3	LOI
1985	12.0	5.9	17.3	39.5	4.6	2.0	0.8	11.5	6.3
1990	37.8	20.5	14.2	17.4	3.3	0.9	2.1	3.9	0.8
1992	40.3	24.0	15.0	11.2	2.8	1.0	2.6	3.1	0.5
1997	41.5	28.1	12.3	10.0	1.2	0.6	3.3	2.0	0.8
2001	39.9	18.2	13.6	17.2	2.4	1.3	2.7	1.5	0.1

Geopolymers in its Background

Victor Glukhovsky and others described soil silicate in the 1950s for alkaline alumino-silicate cementitious systems which included calcium silicate hydrated and alumino-silicate phases and were used to construct a tall building (Komnitsas & Zaharaki, 2007). Joseph Davidovits coined the term "Geopolymer" in 1972 to describe a new material with cementitious properties made of silicon and aluminum activated with a high alkali solution and having a three-dimensional structure made of alumino-silicates. Figure 1 shows illustration scheme of geopolymerization. The structure of geopolymers ranges from amorphous to semi-crystalline. Additionally, the three-dimensional Si-O-Al polymeric structure of geopolymer links together SiO_4 and AlO_4 tetrahedra by sharing all of their oxygen atoms, with Al in four-fold coordination. According to Si/Al ratios, geopolymers can be found in three main forms: poly(sialate), poly(sialate-siloxo), and poly(sialate-disiloxo) (Liew et al., 2016).

Figure 2. Geoplymerization processes including dissolution, oligomerization and polymerization for SiO$_4$ and AlO$_4$ units. M+ refers to alkali cations (usually Na+ and/or K+). SiO44- refers to dissolved silicate

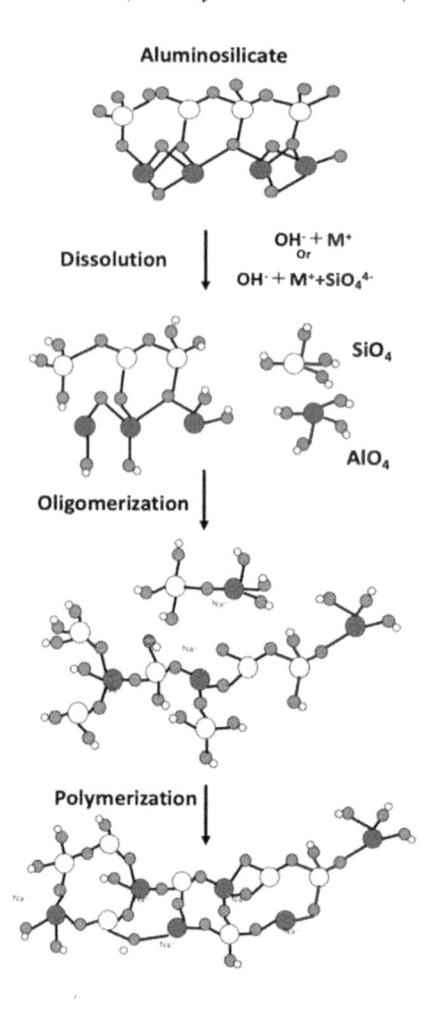

Alkali activators and alumino-silicate polymerize to form a geopolymer. In addition, using alumino-silicate sources produced from natural minerals and industrial wastes such as fly ash, metakaolin, and blast furnace slag with a high alkaline concentration solution, geopolymer can be synthesized at low pressure and ambient temperature (Abdulkareem et al., 2014; Rovnaník, 2010). The experimental reports of geopolymer that have been reported as summarized in Table 2.

Table 2. Main classes of currently developed geopolymers

Class of geopolymer materials	References
- Metakaolin MK-750-based geopolymer, poly(sialate-siloxo) Si:Al=2:1	(Cui et al., 2010; Medri et al., 2010)
- Waterglass-based geopolymer, poly(siloxonate), soluble silicate, Si:Al=1:0	(Barbosa, 2000)
- Rock-based geopolymer, poly(sialate-multisiloxo) 1< Si:Al<5	(Li et al., 2014; Tchadjié et al., 2016)

continued on following page

Table 2. Continued

Class of geopolymer materials	References
- Calcium-based geopolymer, (Ca, K, Na)-sialate, Si:Al=1, 2, 3	(Wang, 2000)
- Fly ash-based geopolymer	(Soutsos et al., 2016; Zhuang et al., 2016)
- Silica-based geopolymer, sialate link and siloxo link in poly(siloxonate) Si:Al>5	(Deb et al., 2015; Rattanasak et al., 2010)
- Phosphate-based geopolymer, AlPO$_4$-based geopolymer	(Douiri et al., 2014; Le-ping et al., 2010)

As mentioned before, the sialate (silicon-oxo-aluminate) network is a geopolymer structure composed of tetrahedral SiO$_4$ and AlO$_4$ linked alternately by sharing all oxygen atoms. Al3+'s negative charge is balanced by positive ions or alkali cations such as Ca^{++}, Na^{++}, and K^{++} in four-fold coordination (Davidovits, 1994). The main chain of the geopolymeric structure is generated from a polymeric structure of (poly)sialate, and the formula for polysialates is as follows (Hanjitsuwan et al., 2014):

Mn [(-Si-O$_2$)z-Al-O$_2$-]n, wH$_2$O

where M is monovalent cation such as sodium or potassium, n is degree of polycondensation, z is amount of Si-O$_2$ (1, 2, 3 or higher), w is amount of water. A small unit of a three-dimensional macromolecular structure known as an oligomer, which interacts at ambient temperature or at curing temperature, is the basis for geopolymerization. Three stages of the geopolymerization process can be identified: deconstruction, polymerization, and stabilization (Blackstock et al., 2012; Muñiz-Villarreal et al., 2011; Yao et al., 2009). In geopolymer, the polymerized links are bonded together in a sialate (Si-O-Al) and siloxo (Si-O-Si) type of structure. The term Si/Al atomic ratio divides the classification of polysialate in the chemical structure of geopolymer into four units. (Davidovits, 1982;Davidovits,1984; Davidovits, 1994). The first unit is sialate and poly(sialate) at Si: Al = 1. For second units, the ratio of Si: Al at 2 is sialate- siloxo and poly(sialate- siloxo). Third, at Si: Al at 3 is sialate link and poly(sialate- multisiloxo) unit. The last elementary unit is sialate link and poly(sialate- multisiloxo) unit at Si: Al > 3 which are chains from connected Si-O-Al between two poly(siloxonate), poly(silanol) or poly(sialate). Finally, in geopolymer framework, the elementary units (poly(sialate) chain and ring) are coordinated together with an oxygen atom to build a 3D network on the Si/Al ratio as

poly(sialate) (S-type, **-Si-O-Al-O-**)

poly(sialate-siloxo) (PSS-type, **-Si-O-Al-O-Si-O-**)

and poly(sialate-disiloxo) (PSDS-type, **-Si-O-Al-O-Si-O-Si-O-**)

As shown in Fig. 2, the geopolymerization procedure consists of three phases. In step one of the deconstruction process (I), alkali hydrolysis is used to leach the solid alumino-silicate components onto the surface, producing aluminate and silicate species (Comrie & Kriven, 2006). The Al-O-Al, Si-O-Si, and Si-O-Al of alumino-silicates source are dissolved in strong alkali solution (Somna et al., 2011). The metal cations from alkali solution could balance the charge of aluminium in form of four-fold co-ordination with oxygen (Lecomte et al., 2006). Following that, the solution undergoes a process of (II) polymerization. As Si-O-Na and O-H-Al and silicate molecules from alkali silicate are reacted together and then an additional polycondensation reaction occurs to obtain a 3D framework of geopolymer, the nearby reactive groups are generated. The next phase, called (III) Stabilization, involves forming preliminary units by a polycondensation process that can produce rigid chains or networks of oxygen bound to aluminate and silicate tetrahedra between room temperature and 90 °C. The stronger geopolymers formed with higher temperature curing make macromolecules' hydroxyl groups more capable of condensing with nearby molecules.

Metakaolin-Based and Fly Ash-Based Geopolymers Made From Unused Pozzolanic Waste

In the production of geopolymers, the raw materials containing silicate and alumina units are important and the sources are composed of raw material fillers and alkali activator in water medium. These raw materials solidify in the form of an aqueous solution-based paste like a geopolymer liquid by the time geopolymer is made. One important property of the suitable raw material and inactive filler are pozzolanic property (Seco et al., 2012). Raw materials used for synthesis geopolymer could be aluminosilicate natural minerals or industrial wastes for example slag (Karakoç et al., 2014), rice husk ash (He et al., 2013), waste glass (Tchakouté et al., 2016) and fly ash (Ryu et al., 2013). As shown in Fig. 1, furthermore, kaolinite or metakaolinite (Alshaaer, 2013) is an inactive filler which is used for supply Al^{3+} ions. The main composition of the pozzolanic materials are SiO_2, Al_2O_3, and CaO as inorganic fillers. Each raw material has an original oxide composition, which affects the reaction's potential. In addition, the geopolymerization reaction is impacted by the particle size of the raw materials or the active filler (Kumar & Kumar, 2011).

Water Based Alkali Activator

Alkali hydroxide and/or silicate solutions are two examples of chemical activators used in geopolymer liquor. An alkali hydroxide solution is required to dissolve raw materials and sodium or potassium silicate is used as a binder, dispersion, or activator for alkalis (Abdullah et al., 2012; Nazari et al., 2011). Sodium hydroxide solution and potassium hydroxide solution with high concentration was used as an activator for the geopolymer synthesis (Fernández-Jiménez et al., 1999; Palomo et al., 1999; Swanepoel & Strydom, 2002; Xu & Van Deventer, 2002). NaOH solution can leach Si and Al ions from raw materials and has a higher reaction rate than KOH solution, it is the most commonly used hydroxide activator (van Jaarsveld & van Deventer, 1999). When the raw material is mixed with alkali solutions, the geopolymer paste transforms into a solid with increased strength. Alkali concentration and curing temperature are only a few of the many variables that affect the strength of geopolymer samples. A higher concentration of NaOH increased the strength of the geopolymer sample because it could accelerate the rate at which the Si and Al ions would dissolve from the raw material (Chindaprasirt et al., 2009). When a high concentration of NaOH was used, a dense matrix in the microstructure increased the compressive strength of the geopolymer (Görhan & Kürklü, 2014; Hanjitsuwan et al., 2014). The highest compressive geopolymer was produced using 10 M NaOH as an alkali activator (Rattanasak & Chindaprasirt, 2009). When using NaOH concentrations greater than 14 M, geopolymer strength was reduced. In excess of the hydroxide ion concentration, the polycondensation reacted rapidly (Rattanasak & Chindaprasirt, 2009; Zuhua et al., 2009). This result demonstrated that a high alkali concentration enhanced the dissolution of Si and Al on the surface of raw material particles and increased the strength of the geopolymer. However, an excess of hydroxide ions caused rapid precipitation of alumino-silicate species at an early stage, which decreased the strength of the geopolymer sample (Lecomte et al., 2006; Lee & van Deventer, 2002).

Figure 3 shows the time variation of the shear viscosity of the metakaolin-based geopolymer paste slurry when NaOH and NH_3 activator were used. The temperature of the rheometer stage shown in the figure was fixed at 80°C, and the viscosity change of the slurry paste was measured at concentrations from 0.5 to 10 M alkaline activator. Here, the increase in viscosity of the slurry paste is seen as the initial process of forming and solidifying three-dimensional Si-O-Si or Si-O-Al networks over time through dehydration-condensation reactions (Davidovits, 2008; Rifaai et al., 2019). In both cases, an increase in viscosity was observed with time; in the case of NH_3, the degree of viscosity increase was greater at concentrations of 5M and 10M, and the viscosity value decreased rapidly after about 300 seconds. This phenomenon indicates that the T-type rotor of the viscometer slips due to solidification of the paste, making measurement impossible. In the case of 2M and 0.5M, which are low alkali concentrations, the time required for the viscometer to become unable to measure due to slippage increased with decreasing concentration, reaching approximately 400 seconds and 430 seconds, respectively. In contrast, in the case of NaOH, slip occurred at concentrations of 10M and 5M in about 30 and 60 seconds, respectively, making the measurement impossible; at 2M, slip occurred in as little as 300 seconds, and at 0.5M, almost no rapid increase in viscosity occurred within this measurement, even at 80°C. It is important to notice the difference between the large and small viscosity values at the time the viscosity value decreased due to rotor slippage. In the case of ammonia, this occurs at about 40-60 Pa s, while in the case of NaOH it occurs at 10 Pa s. This suggests that in the case of a high concentration of NaOH activator, it rapidly solidifies near the rotor, making it impossible to measure. In the case of ammonia, however, the condensation occurs gradually, indicating that it does not solidify rapidly and render the measurement impossible.

Figure 3. Viscosity changes in metakaolin based aqueous slurry pastes for NaOH or ammonia activators and illustration of the measurement using viscometer. For the time dependent measurements, share rate was conducted at 100 s^{-1} and time change of the viscosity were measured for each sample at 80 °C

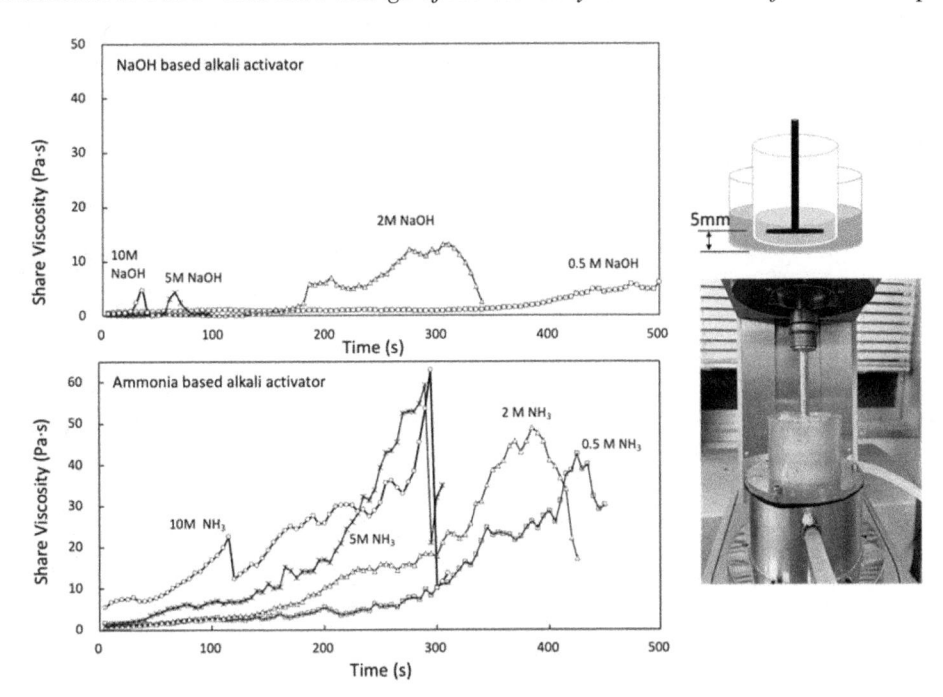

The above results were obtained at 80°C and changes could be observed, but at 25°C, viscosity changes almost never occur within this measurement time. In other words, the temperature change is one of the factors affecting the process of geopolymerization like a curing temperature. The setting and hardening times for a geopolymer sample are influenced by the curing temperature, and the strength increases as the curing time (Djwantoro Hardjito & Rangan) and temperature increase (Chindaprasirt et al., 2007). However, geopolymer sample strength reduced at temperatures over 80 °C due to over-hydration, and over-shrinking occurred in geopolymer sample curing at temperatures above 100 °C (Abdulkareem et al., 2014; Bakharev, 2005; Görhan & Kürklü, 2014).

The results of infrared spectral analysis of metakaolin-based geopolymers obtained when heated at 80°C and NaOH are shown in Fig. 4. When comparing NaOH concentrations of 10 M and 1 M. In both cases, characteristic O-H and Si-O-Si group stretching vibrations around 3400 cm^{-1} and 1000 cm^{-1} peaks were strongly observed in the vicinity of 3400 cm^{-1} and 1000 cm^{-1}. At 1M concentration, the peak intensities of O-H and Si-O-Si groups were observed around 3330 cm^{-1} and 1080 cm^{-1}, and at 10M, these peaks were 3340 cm^{-1} and 990 cm^{-1}. Peak separation of these IR absorption peaks showed that in the case of O-H stretching vibrations, the ratio of 3322cm^{-1} and 2977cm^{-1} was higher at 1M concentration, and 3443cm^{-1} was higher at 10M. This suggests that at 10M, unreacted NaOH was incorporated into the geopolymer through hydrogen bonds, whereas at 1M, the peak intensity above 3400 cm^{-1} was low, indicating that the OH groups contributed to the condensation reaction, were incorporated into the geopolymer via intermolecular hydrogen bonds. In the 1000 cm^{-1} region, the degree of -Si-O-Si- bonding was dominated by condensed components in the 1100-1080 cm^{-1} region at 1M, while the peak intensity at 1079 cm^{-1} was higher at 10M, indicating that there are many components at 960-996 cm^{-1} with low -Si-O-Si- condensation.

Figure 4. FT-IR spectra of metakaolin based geopolymer prepared with 10M or 1M NaOH activator and peak deconvolution of 2500cm^{-1}–4000 cm^{-1} and 800 cm^{-1}–1400 cm^{-1} regions for -O-H group and -Si-O-Si- group of their stretching vibration modes. The solid lines represent the original peak while dotted lines represent the fitted curves

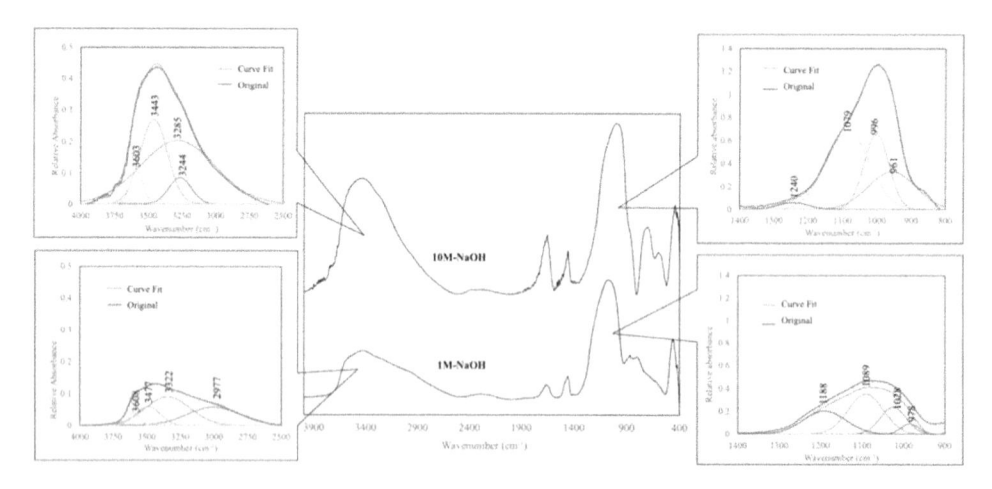

Importance of Geopolymers Made From Unused Resources as Inorganic Raw Materials

The utilization of unused resources is very important for contributing to the construction of a sustainable society. In this sense, the utilization of soil, which is an infinite mineral, in geopolymers is important in terms of mineral resource conservation and utilization. However, soils contain high levels of organic matter, which tends to inhibit geopolymer formation, making their cleaning important. Soil slurry was recycled to prepare a geopolymer after treatment with high-power ultrasound (US) in the presence of NH_3, HCl, and NaOH. Under 28 kHz US, 0.1 M NH_3 additives effectively decarbonized the slurry, eliminating 72.2% of the carbon content from the original soil. The US-treated soils were used as raw materials for the geopolymer (Kwedi-Nsah et al., 2019). This study stated that the ultrasonic cleaning with ammonia could lower carbon content and increased the strength of soil based geopolymer. On the other hand, recently, we prepared a geopolymer from white sandy soil, a volcanic soil in Kagoshima Prefecture (Japan) (Figure 5). This was confirmed by XRF elemental analysis of the white sandy soil material, so called as Shirasu Daichi in Japanese, which showed that the analyzed values (%) were 17.5% for Al_2O_3, 65.1% for SiO_2, 6.84% for CO_2, 2.23% for Na_2O, 0.314% for MgO, 3.28% for K_2O, 1.49% for CaO and 2.64% for Fe_2O_3, possessing low in carbon and iron contents, but high in silica and aluminum moisture in the soil.

Figure 5. Pictures of white sandy soil and XRD and its geopolymers for SEM when changed in solid/liquid ratio and NaOH concentrations. Solid/liquid ratio=weight of Fly ash or metakaolin/weights of sodium silicate and NaOH solution, when S /L=1.25

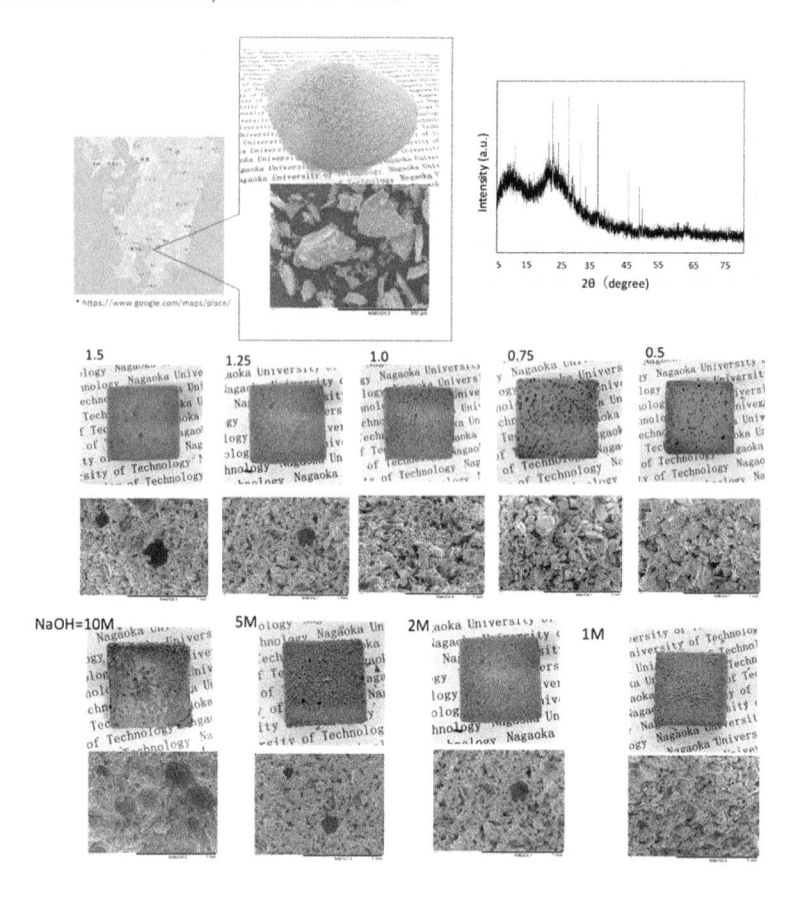

Figure 5 also shows photographs and SEM results of the geopolymer samples obtained by changing the solid/liquid ratio from 0.5 to 1.5 and the NaOH concentration from 1 to 10 M. The solid/liquid ratio of the geopolymer samples was 0.5-1.5, and the NaOH concentration was 1-10 M, respectively.

Characteristics of Geopolymers

Geopolymer has many attractive properties such as high temperature resistance, high chemical resistance, long durability, low shrinkage and fast solidification with high-strength etc. As demonstrated in Table 3, the geopolymer materials have a wide range of applications (Davidovits, 2016). In the geopolymers for concrete made from fly ash, there were several publications that studied and synthesized geopolymer for a particular construction use (Pavithra et al., 2016), geopolymer concrete created with crumb rubber and fly ash (Park et al., 2016). Geopolymers can also be used as an adsorbent to remove heavy metals or dyes from water or wastewater. Recently, a fly ash-based geopolymer was used to remove heavy metal ions from aqueous solutions (Al-Zboon et al., 2011; Wang et al., 2007; Li et al., 2006). Additionally, the geopolymers enabled to immobilize radioactive and heavy metal waste in the matrix with the 3D framework of geopolymer (Van Jaarsveld et al., 1997). Moreover, the production of hybrid composite materials composed of geopolymer and epoxy melamine resin demonstrated that these materials were high-temperature resistant and could be used for flooring, coatings, and kitchenware (Roviello et al., 2015). The ferronickel (FeNi) slag was mixed with pure alumina to create the potassium-based geopolymers (K-geopolymer), which are good fire protection materials. (Sakkas et al., 2014).

Table 3. Summary of geopolymer applications

Geopolymer applications	Sources	References
Concrete and cement	Fly ash	Pavithra et al., 2016
	Fly ash	Singh et al., 2020
	Review	Parathi et al., 2021
Adsorbents	Fly ash	Li et al., 2006;
	Rice Husk/metakaolin	Francisco et al., 2014
	Zeolite	Udaibah, 2020
	Review	Liang et al., 2021
Immobilisation of toxic pollutants	Fly ash	Watanabe et al., 2023
	Metakaolin	Kim et al., 2021
	Fly ash	Zhang et al., 2008
	Review	Vu & Gowripalan, 2018
Composites	Fly ash/polyethersulfone	Onutai et al., 2019
	Cementitious fly ash	Ohno & Li, 2023
	Epoxy melamine/metakaolin	Roviello et al., 2015
	Polypropylene fibers/ground granulated blast furnace slag	Farhan et al., 2023
Coating	Fly ash	Hamidi et al., 2022
	Metakaolin/Aluminium alloy	Novotný et al., 2022
	Metakaolin/Moroccan clays and sands	Khoms et al., 2021
	Review	Jiang, 2020
Fire resistant and protection matrix	Fly-Ash	Razak et al., 2022
	Ferronickel (FeNi) slag	Sakkas et al., 2014
	Review	Lahoti et al., 2019

APPLICATIONS OF GEOPOLYMERS

Since Davidowitz presented geopolymers in 1976 (IUPAC, 1976), they have already been applied in all industrial fields as a new generation of materials. Due to the problem of greenhouse gases emitted from the production of Portland cement, One reason for the development is that geopolymer has attracted attention as a pozzolanic raw material to replace all or part of the by-product, so geopolymer has been focused on as used to reduce or eliminate greenhouse gases. In addition to contributing to energy savings and environmental preservation in the production of raw materials, the fact that geopolymer raw materials are waste and unused inorganic materials also contributes to building a sustainable society. To date, as new materials, geopolymers have applications in refractory and heat-resistant coatings and adhesives, pharmaceutical applications, high-temperature ceramics, new binders for refractory fiber composites, encapsulation of toxic and radioactive waste, and new cements for concrete (D'Angelo, 2015; Davidovits, 2016; Raja et al., 2023). As seen Figure 6, the illustration explains several applications of geopolymers with Bulk boday, foam body, coating and composites. In recent years, not only geopolymer materials alone are being considered, but they are also increasingly being developed as composite materials with other materials.

Figure 6. Illustration explains several applications of geopolymers with Bulk boday, foam body, coating and composites

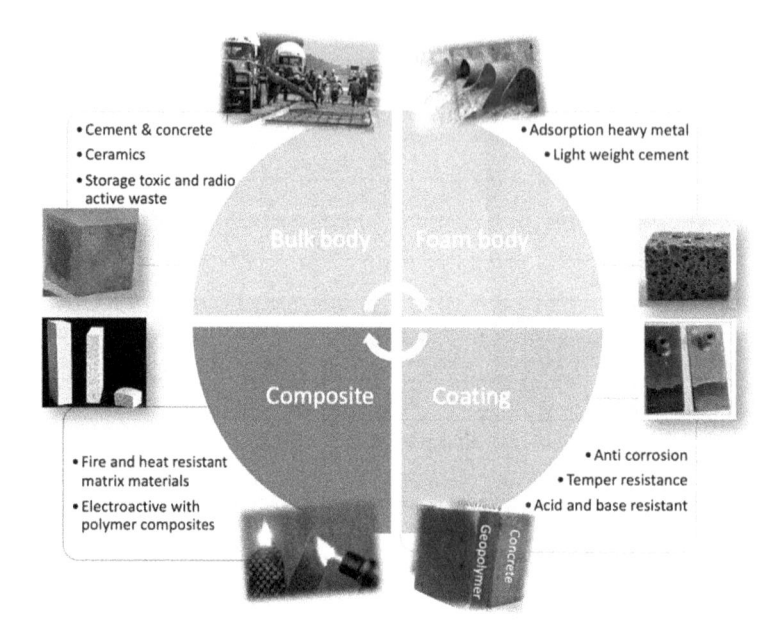

In addition to this, various, not only rely on thermal condensation reactions of alkaline activators, but also consider accelerating and controlling condensation reactions by external stimuli.

Enhanced and Controlled Geopolymerization With External Forces

In general, the geopolymer can be prepared with paste like the sources of metakaolin or fly ash dissolved in highly alkaline materials via hydrolysis in thermal process. Such thermal curing has been applied to improve the reaction and strength of geopolymer materials, and recently innovative heating tools have been reported that adapt microwave heating to geopolymerization and control the properties of the resulting geopolymer (Chindaprasirt et al., 2013; Watanabe&Kobayashi,2023). Microwave curing of fly ash based geopolymer was effective to prepare porous geopolymers with higher NaOH concentrations (Onutai et al., 2016). At high NaOH concentrations, microwave energy was efficiently absorbed and instantaneous heating was possible. At 850 W, the geopolymer slurry cured within 130°C and 1 min, resulting in porous fly ash, which was effective in forming aluminosilicate bonds in the porous geopolymer foams (Figure 7). In contrast, before microwave cure, when pre-cured at 30°C for 2 hours, then the microwave was exposed 1 min, providing dense structured fly ash geopolymers (Watanabe et al., 2023).

Figure 7. Pictures of fly ash geopolymers in comparison with conventional oven and household microwave oven

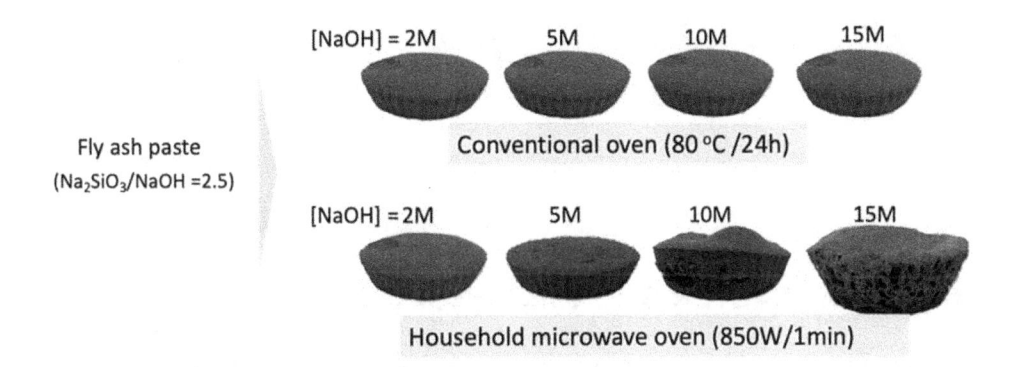

On the other hand, in geopolymerization, condensation reactions occur under high pressure to obtain densified geopolymers. Forming the geopolymer by the pressing method led to the particles being close together in a compact matrix (Prasanphan et al., 2019). Subsequently, their surfaces were dissolved with NaOH solution, inducing the geopolymerization reaction. This reaction and continuous reaction resulted in the compressive strength and microstructure of the metakaolin geopolymer. Porous properties were observed from the shape of nitrogen isotherms according to the International Union of Pure and Applied Chemistry (IUPAC) (IUPAC, 1994; Leofanti, 1998) According to the IUPAC classification, the isotherms of all geopolymer samples in this study corresponded to type IV isotherms of the hysteresis loop.

Compared to porous geopolymers, dense geopolymers are obtained when pressure is applied during conditioning (Prasanphan et al., 2019). Figure 8 shows Illustration of the effect of pressure on geopolymerization; the application of a hydraulic pressure of 20 MPa removed internal bubbles, and metakaolin particles were also present in the geopolymer in a crushed form on a flat surface (SEM photograph). Especially, flattened and molten-looking metakaolin particles were tended to observe at higher alkali concentrations, when the alkaline activator concentration was changed from 4M-10M. the compressive strength of geopolymers formed with different NaOH concentrations at curing time of 7 days. The compressive strength of the geopolymer increased with an increase in NaOH concentration. The compressive strengths of the geopolymer prepared from NaOH concentrations of 4, 6, 8, and 10 M were 22.01, 22.92, 25.70, and 26.98 MPa, respectively. The highest compressive strength was obtained from a high NaOH concentration of 10 M, which was approximately 22.58% higher compared with that obtained at a NaOH concentration of 4 M. this was due to higher dissolution abilities of Si^{4+} and Al^{3+} ions from aluminosilicate particles of the calcined waste under higher pressure.

Figure 8. Pictures of metakaolin geopolymer block and SEM images of geopolymers having different NaOH concentrations and schematic illustration for the pressing metakaolin particles

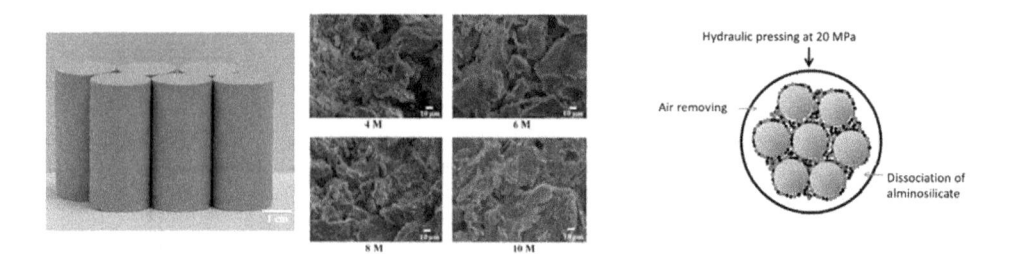

APPLICATIONS OF WASTER SOURCED GEOPOLYMERS

Utilization of Waste Rice Husk Ash for Sustainable Geopolymer

Global development and population growth tend to raise demand for construction infrastructure and buildings. In general, this construction sector accelerates energy consumption, consumes large amounts of natural resources, and generates large amounts of waste through the synthesis and construction of materials. Therefore, it is necessary to develop technologies that do not have adverse ecological effects and that do not result in a shortage of natural resources or fuels, or emissions of greenhouse gases. In agricultural countries, the rice milling process generates a large amount of by-product, namely rice husks (RH), which are usually used in boilers as fuel to produce solid waste, but there are attempts to utilize them in geopolymers (Hossain et al., 2021). This is because that rice husk contains a high quantity (85–90 wt%) of amorphous silica. In addition, with its pozzolanic characteristics (Das et al., 2020; Das et al., 2022), the utilization of rice husk ash in fly ash was applied for blast furnace slag based geopolymer concrete. In rice husk ash based geopolymer mortar was incorporated with sewage sludge ash to reduce or completely eliminate the use of cement in concrete (Zaidahtulakmal et al., 2019) and the use of rice husk ash enhanced fly ash-based self-compacted geopolymer concrete (sari et al., 2021). Due to the negative environmental impacts of cement production, there are sustainable options that can reduce the carbon footprint of the cement industry as followed (Ogwang et al., 2021); 1) improving energy efficiency, 2) waste heat recovery, 3) carbon capture and storage, 4) replacing fossil fuels with renewable energy, and 5) producing low-carbon cement to replace ordinary Portland cement. Therefore, the need for low-carbon cement is calling for geopolymers such as rice husk, which is made from unused waste.

Immobilization of Hazardous Waste

As hazardous waste or ions can be immobilized in a 3D framework of geopolymer. The fly ash-based geopolymer had a significant potential for immobilizing Pb ions (Lee et al., 2016). Geopolymer is suitable for composite materials as well as for the adsorption or immobilization of ions or hazardous waste. At this time, heavy metal solid wastes have been mainly produced by metal processes in industries such as battery manufacturing, electroplating, dye and pigment production, and mining. After generation of such waste, they are usually stabilized and solidified with Portland cement before disposal. This is because that Portland cement has as an ability immobilization of such heavy metal waste (Portland Cement Association 1991; Vu & Gowripalan, 2018). However, heavy metal immobilization using Portland cement has several problems due to those heavy metals immobilized in Portland cement leached easily under acidic conditions (Bonen & Sarkar 1995; Fan et al. 2023). Alternatively, to Portland cement which practically used other materials to propose waste fly ash geopolymers for the heavy metal immobilization. Thus, effective utilization of fly ash waste for the amount and treatment of toxic metal like Pb become important for symbiosis environment economy and society. This three-dimensional network structure becomes an immobilizing matrix for toxic heavy metals can physically encapsulate solid heavy metal contaminants (Deja 2002; Zhang et al., 2008). In addition, there were reports for highly efficient adsorption of lead in fly ash based geopolymers (Onutai et al. 2019), indicating strong immobilization on Pb to FA geopolymers. Currently, microwave cure was used for the quick curing process of fly ash (FA) geopolymer fabrication for heavy metal immobilization purposes (Watanabe et al., 2023). As applied for heavy metal immobilization of Pb ion in the matrix, fly ash and alkaline activator of sodium silicate and sodium hydroxide (NaOH) solution were fixed at mass ratio of 2.5 for geopolymerization. The geopolymer pastes precured at 30°C for 2h and then was treated under microwave irradiation at 300 and 700 W for 1 min exposure. The geopolymer without precure became porous, but the precure geopolymer densified and efficiently fixed lead after 1 minute of microwave irradiation. It was apparent that the immobilization efficiency was high in tested samples. According to them, especially, the geopolymers with more than 99% immobilization efficiency were 300 W microwave irradiated samples, as prepared with 6 M and 10 M alkaline activator. However, the values were in the range of 97–98% for the geopolymers prepared by 700 W microwave irradiation.

Adsorbents

Adsorption is one of the most efficient methods for removing heavy metal ions from wastewater, and it is also one of the least expensive(Papageorgiou et al., 2009). Recently, natural mineral or industrial waste such as mud, silica, ash, farm waste, and solid waste have been employed as sorbent materials (Al-Harahsheh et al., 2015). The concept of using waste for waste removal has become increasingly popular in the geopolymers. The sample was crushed into smaller than 200 µm particles by then grinding and sieving and then removal of Cu^{2+} ions was investigated on the effects of pH, contact time, temperature, and initial Cu^{2+} concentration. When pH increased from 1 to 6, the geopolymer absorbed more Cu^{2+}, from 5.6% to 88.21% and the removal efficiency increased from 45% to 88.2% when the dosage increased from 0.03 g to 0.15 g. The result demonstrated that as the initial concentration of Cu^{2+} increased, the adsorption efficiency decreased. Additionally, the pseudo-second-order was successfully suited to the kinetic data, and Langmuir isotherm model was shown to be more feasible than the Freundlich isotherm model.

Fly ash in Mae Moh power plant in Lampang, Thailand, provided the fly ash source for geopolymer for heavy metal removals (Onutai et al., 2018). The effectiveness of fly ash based geopolymer used as an adsorbent for heavy metals had specific surface area of 85.01 m^2/g and about ten times higher than 0.83 m^2/g for the original fly ash. The highest heavy metal removal capacity was obtained at pH 5, showing in the order of Pb^{2+} > Cu^{2+} > Cd^{2+} > Ni^{2+}. For the study a interaction between adsorbate molecules and active sites on the adsorbent surface, the adsorption isotherm was calculated. The equilibrium data for the adsorption of Pb^{2+}, Cu^{2+}, Ni^{2+}, and Cd^{2+} was fitted by using Langmuir and Freundlich equations in Eqs. (1) and (2), respectively.

Langmuir model:

$$q_e = \frac{K_L q_m C_e}{1 + K_L C_e} \tag{1}$$

Freundlich model:

$$q_e = K_F C_e^{1/n} \tag{2}$$

where q_e is the amount of heavy metal ions adsorbed per unit mass at equilibrium (mg/g), C_e is the equilibrium concentration of heavy metal ions (mg/L), K_L is the Langmuir adsorption constant (L/mg), q_m is the monolayer adsorption capacity (mg/g), while K_F is the Freundlich adsorption constant [(mg/g)(mg/L) -n], and n is adsorption intensity.

By affecting the contact time from 5 to 180 minutes, the influence of contact time on the solution of heavy metal ions (Pb^{2+}, Cu^{2+}, Cd^{2+}, and Ni^{2+}) was investigated (Al-Zboon et al., 2011). Accordingly heavy metal ions adsorb more strongly to geopolymer powder at higher temperatures to the fly ash geopolymers. The improvement of the micro-cavities in the geopolymer powder was found to cause the vaporization of water, increasing the adsorption capacity at high temperatures (Andini et al., 2008). As compared with results of Langmuir and Freundlich analyses (Table 4), the results well fitted in Langmuir model. It was found that the fly ash geopolymer had extremely higher q_m value of 434.78 mg/g for Pb^{2+}.

Table 4. Constants of Langmuir and Freundlich equations at different mono-cation system (test condition: 0.1 g geopolymer in 40 mL at pH 5, temperature at 25 °C, contact time 120 min)

Metal ion type	Langmuir constants			Freundlich constants		
	K_L (L/mg)	q_m (mg/g)	R^2	K_F [(mg/g)(mg/L)-n]	N	R^2
Pb^{2+}	0.021	434.78	0.928	15.169	4.386	0.677
Cu^{2+}	0.330	9.017	0.970	4.662	6.230	0.916
Cd^{2+}	0.066	12.41	0.994	3.350	3.205	0.561
Ni^{2+}	0.013	18.02	0.990	1.808	1.669	0.978

In contrast, coal fly ash geopolymers used removal of some organic dyes like methylene blue and crystal violet, from aqueous solution (Li et al., 2006). The adsorption capacity of the coal fly ash adsorbents depended on the preparation conditions such as NaOH:fly-ash ratio and fusion temperature and the optimal conditions was at 1.2:1 weight ratio of Na:fly-ash at 250-350 °C. Noteworthy, the synthesized geopolymer exhibited much higher adsorption capacity than fly ash itself and natural zeolite. Other porous and reusable geopolymer adsorbents for dye wastewater treatment using fly ash and silica fume effectively removed methylene blue dye from wastewater (Li et al, 2022).

On the other hand, in synthesis of zeolite based geopolymers from rice husk ash, kaolin and NaOH activators, ground water contaminated with Ca^{2+} and Mg^{2+} ions, as commonly known as hard water, was aimed to treat using the potential of zeolite-based geopolymer adsorbents for desalination (Udaibah, 2020).

Also, metakaolin based geopolymer as an adsorbent for removal Zn^{2+} and Ni^{2+} from aqueous solution was studied (Kara et al., 2017), producing with compositions of $SiO_2/Al_2O_3 = 3.2$; $Na_2O/SiO_2 = 0.2$; $Na_2O/Al_2O_3 = 0.7$; and $H_2O/Na_2O = 13.8$ and then crushing and sieving to produce particles of about 150 μm in size. The BET specific surface area of the metakaolin geopolymer was 39.24 m²/g. In the adsorption test, pH, geopolymer content, contact time, temperature, and initial concentration were examined as parameters, and the metakaolin geopolymer content affected the retention of attached Zn^{2+} and Ni^{2+} ions. Also, metakaolin-fume silica geopolymer adsorbents were prepared from silica and metakaolin in different Al and Si components and applied for removal of metal ions, Cs^+ and Pb^{2+}, from other heavy metal ions mixture (Lopez et al, 2014b). The geopolymer was optimized at Si/Al = 2, targeting to Cs^+ and Pb^{2+} separation. The binding behavior was well fitted to Langmuir model, which proving that the metakaolin-based geopolymer had multi-binding to adsorb ions. As be noted, the effective adsorption was also observed independent of NaCl concentration for the Cs^+ and Pb^{2+}, meaning that the non-ion exchanged adsorption occurred. In addition, metakaolin based-rice husk silica was used to be porous foam geopolymers (Lopez et al., 2014 a), which were applied for Cs^+ ion. The experimental results of foamed geopolymers fitted well with the Langmuir equation with correlation coefficients of 0.99, showing the maximum amount of cesium ion adsorption, q_m of 11.6 mg/g, 41.3 mg/g and 50.8 mg/g for Si/Al ratios of 2.5, 5 and 10, respectively (Lopez et al., 2014 c). When the Si/Al was 2.5, the Langmuir isotherms expressed saturation isotherm for each metal ions, but, the geopolymer showed highly adsorbed behavior to cesium, lead and copper (Figure 9).

Figure 9. Langmuir isotherms of rice husk silica-metakaolin geopolymer for the adsorption of a) individual metal ions solution and b) mixture of multicomponent of metal ions

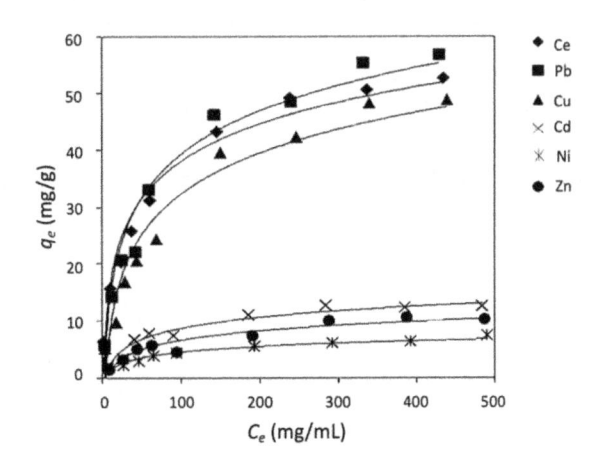

The use of geopolymers for the adsorption of CO_2 is a recent emergent technology aimed at the removal of CO_2 from gaseous streams (Minelli et al., 2016; Pei et al., 2019; Papa et al., 2021). Application of geopolymers produced with fly ash and rice husk ash was performed as an adsorbent material for the capture of CO_2 (Freire et al., 2020). The calcined rice husk ash activated by NaOH was found to be the most suitable precursor material to produce a geopolymer and CO_2 was effectively adsorbed with a capacity of 24.4% higher than the best geopolymer adsorbent reported in the literature to date.

Geopolymer Composites and Coating

Geopolymer has been developed in many different kinds of forms and shapes for the adsorption of heavy metals, including powder, spheres (Tang et al., 2015), beads (Peretz et al., 2015), and sponge (Bai et al., 2017; Kovářík et al., 2017), which requires a complex procedure and resistance to use them. Geopolymer composites have recently become a promising ecological alternative to conventional cementitious materials and becomes cost-effective, environmentally friendly, and require relatively little energy to manufacture., since such composites also have excellent compressive strength, durability, and heat resistance, and are resistant to flame and heat. The mechanical properties of the geopolymer were Reinforced as based on fly ash with short natural fibers such as cotton, sisal, raffia and coconut (Korniejenko et al., 2016) and also wo types of fibers of namely steel and polyvinyl alcohol (PVA) ones (Shaikh, 2013) for the class F fly ash as used instead of portland cement. In reinforcement of concrete, tensile properties of a novel fiber reinforced geopolymer composite enhanced strain hardening characteristics (Al-Majidi, 2017) and reviewed for engineered geopolymer composite with a low-carbon ultra-high-performance concrete (Elmesalami & Celik, 2022).

Mordenite zeolite-polyethersulfone composite fiber was developed (Nakamoto et al., 2017). N-Methyl-2-pyrrolidone (NMP) was used as a solvent to dissolve the polyethersulfone (PES) in order to produce the solvent that contained 30% PES. Then wet spinning was applied for preparation of fibrous zeolite-PES composites and the products were used as heavy metal elimination adsorbents, effectively adsorbing Pb^{2+}, Cd^{2+}, Cu^{2+}, and Ni^{2+} ions in the following order: $Pb^{2+}>Cu^{2+}>Cd^{2+}>Ni^{2+}$. For the preparation process of phase inversion technique for synthesizing fiber (Kobayashi et al., 2016; Nakamoto et al., 2017), fly ash geopolymer powders were composite to be fibrous shape and used for heavy metal adsorbent (Onutai et al., 2019). The phase inversion approach was used to create the geopolymer composite fiber and porous fly ash-based geopolymer composite fiber effected as an adsorbent for removal of heavy metal ions from wastewater. The fly ash taken from the Mae Moh power plant was used as fly ash-based geopolymer. Then PES was dissolved in NMP solvent and the resultant geopolymer powders were mixed. As shown in Figure 10, the wet spinning process enabled fiber composite formation in water medium. The composite fiber had BET surface area of $168.3 m^2/g$, showing adsorbed heavy metal cations in order of $Pb^{2+} > Cu^{2+} > Cd^{2+} > Ni^{2+}$ and also obeying Langmuior model on the isotherm (Table 5).

Figure 10. Illustration flow to fabrication of fly ash based geopolymer composite fiber via wet spinning process

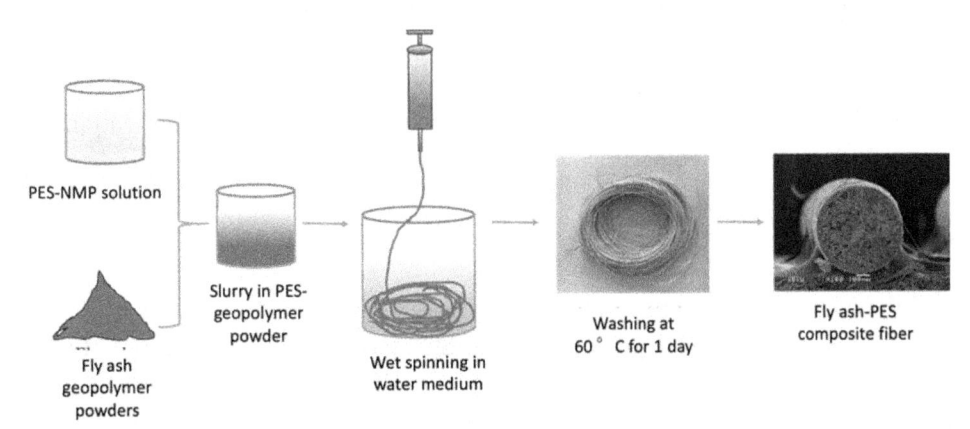

Table 5. Parameters on Langmuir and Freundlich isotherm of fly ash based geopolymer composite fiber for several heavy metal ions

Isotherm model	Parameters	Metal ions			
		Pb^{2+}	Cd^{2+}	Cu^{2+}	Ni^{2+}
Langmuir	q_m	0.74	0.60	0.96	0.71
	K_L	10.57	2.87	1.86	2.95
	R^2	0.960	0.985	0.982	0.927
Freudlich	K_F	1.71	2.16	3.59	3.99
	$1/n$	3.13	14.47	6.22	9.17
	R^2	0.762	0.311	0.858	0.348

On the other hand, geopolymer coatings protected concrete from chemical attack and corrosion, when the use of geopolymer mortar made of local waste materials and sodium silicate as a chemo-resistive coating was applied for concrete (Shiroka et al., 2018). When investigated the resistance of concrete samples coated with geopolymer to 10% inorganic/organic acids and saturated solution of sodium chloride, the coating composite of concrete sample showed great resistance to organic and inorganic acids. Also, geopolymer coating effected on mild steel (Zainal et al., 2018), piezo-resistivity of PEDOT: PSS conductive polymer layer by metakaolin geopolymer (Chen et al., 2019), and also sub-ambient daytime radiative cooling by coating with metakaolin geopolymer (Yang et al, 2023).

CONCLUSION

In this chapter, geopolymer materials using raw materials derived from unutilized waste, especially metakaolin-based and fly ash-based raw materials, are discussed. In particular, the use of inorganic fillers containing silica and alumina components as waste materials is discussed as an upcycling technology that can create high value-added products from unutilized resources, and geopolymerization is a technique that should be paid attention in the future.

Author Note

One of results were obtained on the 100th Anniversary Chulalongkorn University Fund sponsored data collection and preliminary analysis for Doctoral Scholarship, Graduate School of Chulalongkorn University and the Cultivation of Global Engineering by Pacific Rim Green Innovation project of Nagaoka University of Technology. We have no conflicts of interest to disclose.

Correspondence concerning this article should be addressed to Sujitra Onutai, Nuclear Fuel Cycle Engineering Laboratories, Japan Atomic Energy Agency, 4-33 Muramatsu, Tokai-mura, Naka-gun, Ibaraki 319-1194, Japan Email: onutai.sujitra@jaea.go.jp, sujitra.onutai@gmail.com

REFERENCES

Abdulkareem, O. A., Mustafa Al Bakri, A. M., Kamarudin, H., Khairul Nizar, I., & Saif, A. A. (2014). Effects of elevated temperatures on the thermal behavior and mechanical performance of fly ash geopolymer paste, mortar and lightweight concrete. *Construction & Building Materials*, 50, 377–387. https://doi.org/http://dx.doi.org/10.1016/j.conbuildmat.2013.09.047. DOI: 10.1016/j.conbuildmat.2013.09.047

Abdullah, M. M. A. B., Jamaludin, L., Hussin, K., Bnhussain, M., Ghazali, C. M. R., & Ahmad, M. I. (2012). Fly Ash Porous Material using Geopolymerization Process for High Temperature Exposure. *International Journal of Molecular Sciences*, 13(4), 4388–4395. https://www.mdpi.com/1422-0067/13/4/4388. DOI: 10.3390/ijms13044388 PMID: 22605984

Ahmaruzzaman, M. (2010). A review on the utilization of fly ash. *Progress in Energy and Combustion Science*, 36(3), 327–363. https://doi.org/http://doi.org/10.1016/j.pecs.2009.11.003. DOI: 10.1016/j.pecs.2009.11.003

Al-Harahsheh, M. S., Al Zboon, K., Al-Makhadmeh, L., Hararah, M., & Mahasneh, M. (2015). Fly ash based geopolymer for heavy metal removal: A case study on copper removal. *Journal of Environmental Chemical Engineering*, 3(3), 1669–1677. https://doi.org/http://doi.org/10.1016/j.jece.2015.06.005. DOI: 10.1016/j.jece.2015.06.005

Al-Majidi, M. H., Lampropoulos, A., & Cundy, A. B. (2017). Tensile properties of a novel fibre reinforced geopolymer composite with enhanced strain hardening characteristics. *Composite Structures*, 168, 402–427. DOI: 10.1016/j.compstruct.2017.01.085

Al-Zboon, K., Al-Harahsheh, M. S., & Hani, F. B. (2011). Fly ash-based geopolymer for Pb removal from aqueous solution. *Journal of Hazardous Materials*, 188(1–3), 414–421. https://doi.org/http://dx.doi.org/10.1016/j.jhazmat.2011.01.133. DOI: 10.1016/j.jhazmat.2011.01.133 PMID: 21349635

Alshaaer, M. (2013). 12//). Two-phase geopolymerization of kaolinite-based geopolymers. *Applied Clay Science*, 86, 162–168. https://doi.org/http://dx.doi.org/10.1016/j.clay.2013.10.004. DOI: 10.1016/j.clay.2013.10.004

Alvarez-Ayuso, E., Querol, X., Plana, F., Alastuey, A., Moreno, N., Izquierdo, M., Font, O., Moreno, T., Diez, S., Vazquez, E., & Barra, M. (2008). Environmental, physical and structural characterisation of geopolymer matrixes synthesised from coal (co-)combustion fly ashes. *Journal of Hazardous Materials*, 154(1-3), 175–183. DOI: 10.1016/j.jhazmat.2007.10.008 PMID: 18006153

Andini, S., Cioffi, R., Colangelo, F., Grieco, T., Montagnaro, F., & Santoro, L. (2008). Coal fly ash as raw material for the manufacture of geopolymer-based products. *Waste Management*, 28(2), 416-423. https://doi.org/http://doi.org/10.1016/j.wasman.2007.02.001

Bai, C., Conte, A., & Colombo, P. (2017). Open-cell phosphate-based geopolymer foams by frothing. *Materials Letters*, 188, 379–382. https://doi.org/https://doi.org/10.1016/j.matlet.2016.11.103. DOI: 10.1016/j.matlet.2016.11.103

Bakharev, T. (2005). Geopolymeric materials prepared using Class F fly ash and elevated temperature curing. *Cement and Concrete Research*, 35(6), 1224–1232. https://doi.org/http://dx.doi.org/10.1016/j.cemconres.2004.06.031. DOI: 10.1016/j.cemconres.2004.06.031

Barbosa, V., MacKenzie, K. J. D., & Thaumaturgo, C. (2000). Synthesis and characterisation of materials based on inorganic polymers of alumina and silica: Sodium polysialate polymers. *International Journal of Inorganic Materials*, 2(4), 309–317. DOI: 10.1016/S1466-6049(00)00041-6

Bonen, D., & Sarkar, S. L. (1995). The effects of simulated environmental attack on immobilization of heavy metals doped in cement-based materials. *Journal of Hazardous Materials*, 40(3), 321–335. DOI: 10.1016/0304-3894(94)00091-T

Brigden, K., Santillo, D., & Stringer, R. (2002). Hazardous emissions from Thai coal-fired power plants. *Toxic and potentially toxic elements in fly ashes collected from the Mae Moh and Thai Petrochemical industry coal-fired power plants in Thailand.*

Chen, S., Li, Y., Yan, D., Wu, C., & Leventis, N. (2019). Piezoresistive geopolymer enabled by crack-surface coating. *Materials Letters*, 255, 126582. DOI: 10.1016/j.matlet.2019.126582

Chindaprasirt, P., Chareerat, T., & Sirivivatnanon, V. (2007). Workability and strength of coarse high calcium fly ash geopolymer. *Cement and Concrete Composites*, 29(3), 224–229. https://doi.org/http://dx.doi.org/10.1016/j.cemconcomp.2006.11.002. DOI: 10.1016/j.cemconcomp.2006.11.002

Chindaprasirt, P., Jaturapitakkul, C., Chalee, W., & Rattanasak, U. (2009). Comparative study on the characteristics of fly ash and bottom ash geopolymers. *Waste Management,* 29(2), 539-543. https://doi.org/http://dx.doi.org/10.1016/j.wasman.2008.06.023

Chindaprasirt, P., Rattanasak, U., & Taebuanhuad, S. (2013). Resistance to acid and sulfate solutions of microwave-assisted high calcium fly ash geopolymer. *Materials and Structures*, 46(3), 75–381. DOI: 10.1617/s11527-012-9907-1

Comrie, D. C., & Kriven, W. M. (2006). Composite Cold Ceramic Geopolymer in a Refractory Application, 211-225. In *Advances in Ceramic Matrix Composites.* John Wiley & Sons, Inc., DOI: 10.1002/9781118406892.ch14

Cui, X., Liu, L., Zheng, G., Wang, R., & Lu, J. (2010). Characterization of chemosynthetic Al2O3–2SiO2 geopolymers. *Journal of Non-Crystalline Solids*, 356(2), 72–76. https://doi.org/http://dx.doi.org/10.1016/j.jnoncrysol.2009.10.008. DOI: 10.1016/j.jnoncrysol.2009.10.008

D'Angelo, J. (2015). Elsevier-geopolymer institute research focus on geopolymer science, Materials today. https://www.materialstoday.com/polymers-soft-materials/comment/elsevier-geopolymer-science/

Das, S. K., Adediran, A., Kaze, C. R., Mustakim, S. M., & Leklou, N. (2022). Production, characteristics, and utilization of rice husk ash in alkali activated materials: An overview of fresh and hardened state properties. *Construction & Building Materials*, 345, 128341.

Das, S. K., Mishra, J., Singh, S. K., Mustakim, S., Patel, A., Das, S. K., & Behera, U. (2020). Characterization and utilization of rice husk ash (RHA) in fly ash – Blast furnace slag based geopolymer concrete for sustainable future. *Materials Today: Proceedings*, 33, 33. DOI: 10.1016/j.matpr.2020.02.870

Davidovits, J. (1994). Properties of geopolymer cements. *First international conference on alkaline cements and concretes, 1*, 131-149.

Davidovits, J. (2008). *GEOPOLYMER Chemistry&Applications*. Institut Géopolymère 16 rue Galilée F-02100 Saint-Quentin France.

Davidovits, J. (2016). Geopolymers based on natural and synthetic metakaolin - A critical review, *Materialstodays*. https://www.materialstoday.com/polymers-soft-materials/features/geopolymers-natural-and-synthetic-metakaolin/

Deb, P. S., Sarker, P. K., & Barbhuiya, S. (2015). Effects of nano-silica on the strength development of geopolymer cured at room temperature. *Construction & Building Materials*, 101(Part 1), 675–683. https://doi.org/http://dx.doi.org/10.1016/j.conbuildmat.2015.10.044. DOI: 10.1016/j.conbuildmat.2015.10.044

Deja, J. (2002). Immobilization of Cr6+, Cd2+, Zn2+ and Pb2+ in alkaliactivated slag binders. *Cement and Concrete Research*, 32(12), 1971–1979. DOI: 10.1016/S0008-8846(02)00904-3

Djwantoro Hardjito, S. E. W. D. M. J. S., & Rangan, B. V. (2004). On the Development of Fly Ash-Based Geopolymer Concrete. *ACI Materials Journal*, 101(6). Advance online publication. DOI: 10.14359/13485

Douiri, H., Louati, S., Baklouti, S., Arous, M., & Fakhfakh, Z. (2014). Structural, thermal and dielectric properties of phosphoric acid-based geopolymers with different amounts of H3PO4. *Materials Letters*, 116, 9–12. https://doi.org/http://dx.doi.org/10.1016/j.matlet.2013.10.075. DOI: 10.1016/j.matlet.2013.10.075

Elmesalami, N., & Celik, K. (2022). A critical review of engineered geopolymer composite : A low-carbon ultra-high-performance concrete. *Construction & Building Materials*, 346, 128491. DOI: 10.1016/j.conbuildmat.2022.128491

Fan, J., Yan, J., Zhou, M., Xu, Y., Lu, Y., Duan, P., Zhu, Y., Zhang, Z., Li, W., Wang, A., & Sun, D. (2023). Heavy metals immobilization of ternary geopolymer based on nickel slag, lithium slag and metakaolin. *Journal of Hazardous Materials*, 453, 131380. DOI: 10.1016/j.jhazmat.2023.131380 PMID: 37043859

Fernández-Jiménez, A., Palomo, J. G., & Puertas, F. (1999). Alkali-activated slag mortars: Mechanical strength behaviour. *Cement and Concrete Research*, 29(8), 1313–1321. https://doi.org/http://dx.doi.org/10.1016/S0008-8846(99)00154-4. DOI: 10.1016/S0008-8846(99)00154-4

Freire, A., Moura-Nickel, C. D., Scaratti, G., De Rossi, A., Araújo, M. H., De Noni Júnior, A., Rodrigues, A. E., Castellón, E. R., & de Fátima Peralta Muniz Moreira, R. (2020). Geopolymers produced with fly ash and rice husk ash applied to CO_2 capture. *Journal of Cleaner Production*, 273, 122917. DOI: 10.1016/j.jclepro.2020.122917

Görhan, G., & Kürklü, G. (2014). The influence of the NaOH solution on the properties of the fly ash-based geopolymer mortar cured at different temperatures. *Composites. Part B, Engineering*, 58, 371–377. https://doi.org/http://dx.doi.org/10.1016/j.compositesb.2013.10.082. DOI: 10.1016/j.compositesb.2013.10.082

Hanjitsuwan, S., Hunpratub, S., Thongbai, P., Maensiri, S., Sata, V., & Chindaprasirt, P. (2014). Effects of NaOH concentrations on physical and electrical properties of high calcium fly ash geopolymer paste. *Cement and Concrete Composites*, 45, 9–14. https://doi.org/http://dx.doi.org/10.1016/j.cemconcomp.2013.09.012. DOI: 10.1016/j.cemconcomp.2013.09.012

He, J., Jie, Y., Zhang, J., Yu, Y., & Zhang, G. (2013). Synthesis and characterization of red mud and rice husk ash-based geopolymer composites. *Cement and Concrete Composites*, 37, 108–118. https://doi.org/http://dx.doi.org/10.1016/j.cemconcomp.2012.11.010. DOI: 10.1016/j.cemconcomp.2012.11.010

Hossain, S. S., Roy, P. K., & Bae, C. J. (2021). Utilization of waste rice husk ash for sustainable geopolymer: A review. *Construction & Building Materials*, 310, 125218.

IUPAC. (1976). https://www.geopolymer.org/library/technical-papers/20-milestone-paper-iupac-76/

Kara, İ., Yilmazer, D., & Akar, S. T. (2017). 2017/04/01/). Metakaolin based geopolymer as an effective adsorbent for adsorption of zinc(II) and nickel(II) ions from aqueous solutions. *Applied Clay Science*, 139, 54–63. https://doi.org/https://doi.org/10.1016/j.clay.2017.01.008. DOI: 10.1016/j.clay.2017.01.008

Karakoç, M. B., Türkmen, İ., Maraş, M. M., Kantarci, F., Demirboğa, R., & Uğur Toprak, M. (2014). Mechanical properties and setting time of ferrochrome slag based geopolymer paste and mortar. *Construction & Building Materials*, 72, 283–292. https://doi.org/http://dx.doi.org/10.1016/j.conbuildmat.2014.09.021. DOI: 10.1016/j.conbuildmat.2014.09.021

Kobayashi, T., Ohshiro, M., Nakamoto, K., & Uchida, S. (2016). *Decontamination of Extra-Diluted Radioactive Cesium in Fukushima Water Using Zeolite-Polymer Composite Fibers* (Vol. 55). https://doi.org/DOI: 10.1021/acs.iecr.6b00903

Komnitsas, K., & Zaharaki, D. (2007). Geopolymerisation: A review and prospects for the minerals industry. *Minerals Engineering*, 20(14), 1261–1277. https://doi.org/http://dx.doi.org/10.1016/j.mineng.2007.07.011. DOI: 10.1016/j.mineng.2007.07.011

Korniejenko, K., Frączek, E., Pytlak, E., & Adamski, M. (2016). Mechanical Properties of Geopolymer Composites Reinforced with Natural Fibers. *Procedia Engineering*, 151, 388–393. DOI: 10.1016/j.proeng.2016.07.395

Kovářík, T., Křenek, T., Rieger, D., Pola, M., Říha, J., Svoboda, M., Beneš, J., Šutta, P., Bělský, P., & Kadlec, J. (2017). Synthesis of open-cell ceramic foam derived from geopolymer precursor via replica technique. *Materials Letters*, 209, 497–500. https://doi.org/https://doi.org/10.1016/j.matlet.2017.08.081. DOI: 10.1016/j.matlet.2017.08.081

Kumar, S., & Kumar, R. (2011). Mechanical activation of fly ash: Effect on reaction, structure and properties of resulting geopolymer. *Ceramics International*, 37(2), 533–541. https://doi.org/http://dx.doi.org/10.1016/j.ceramint.2010.09.038. DOI: 10.1016/j.ceramint.2010.09.038

Ladewig, B., & Al-Shaeli, M. N. Z. (2017). Fundamentals of Membrane Processes. In Ladewig, B., & Al-Shaeli, M. N. Z. (Eds.), *Fundamentals of Membrane Bioreactors: Materials, Systems and Membrane Fouling* (pp. 13–37). Springer Singapore., DOI: 10.1007/978-981-10-2014-8_2

Le-ping, L., Xue-min, C., Shu-heng, Q., Jun-li, Y., & Lin, Z. (2010). Preparation of phosphoric acid-based porous geopolymers. *Applied Clay Science*, 50(4), 600–603. https://doi.org/http://dx.doi.org/10.1016/j.clay.2010.10.004. DOI: 10.1016/j.clay.2010.10.004

Lecomte, I., Henrist, C., Liégeois, M., Maseri, F., Rulmont, A., & Cloots, R. (2006). (Micro)-structural comparison between geopolymers, alkali-activated slag cement and Portland cement. *Journal of the European Ceramic Society*, 26(16), 3789–3797. https://doi.org/http://dx.doi.org/10.1016/j.jeurceramsoc.2005.12.021. DOI: 10.1016/j.jeurceramsoc.2005.12.021

Lee, S., van Riessen, A., Chon, C.-M., Kang, N.-H., Jou, H.-T., & Kim, Y.-J. (2016). Impact of activator type on the immobilisation of lead in fly ash-based geopolymer. *Journal of Hazardous Materials*, 305, 59–66. https://doi.org/http://dx.doi.org/10.1016/j.jhazmat.2015.11.023. DOI: 10.1016/j.jhazmat.2015.11.023 PMID: 26642447

Lee, W. K. W., & van Deventer, J. S. J. (2002). The effects of inorganic salt contamination on the strength and durability of geopolymers. *Colloids and Surfaces. A, Physicochemical and Engineering Aspects*, 211(2–3), 115–126. https://doi.org/http://dx.doi.org/10.1016/S0927-7757(02)00239-X. DOI: 10.1016/S0927-7757(02)00239-X

Li, C., Zhang, T., & Wang, L. (2014). Mechanical properties and microstructure of alkali activated Pisha sandstone geopolymer composites. *Construction & Building Materials*, 68, 233–239. https://doi.org/http://dx.doi.org/10.1016/j.conbuildmat.2014.06.051. DOI: 10.1016/j.conbuildmat.2014.06.051

Li, C., J., Zhang, Y., J., Chen, H., He, P. Y., Meng, Q. (2022). Development of porous and reusable geopolymer adsorbents for dye wastewater treatment, *Journal of Cleaner Production*, 348, 10, 131278

Li, L., Wang, S., & Zhu, Z. (2006). Geopolymeric adsorbents from fly ash for dye removal from aqueous solution. *Journal of Colloid and Interface Science*, 300(1), 52–59. DOI: 10.1016/j.jcis.2006.03.062 PMID: 16626729

Liew, Y.-M., Heah, C.-Y., Mohd Mustafa, A. B., & Kamarudin, H. (2016). Structure and properties of clay-based geopolymer cements: A review. *Progress in Materials Science*, 83, 595–629. https://doi.org/https://doi.org/10.1016/j.pmatsci.2016.08.002. DOI: 10.1016/j.pmatsci.2016.08.002

Medri, V., Fabbri, S., Dedecek, J., Sobalik, Z., Tvaruzkova, Z., & Vaccari, A. (2010). Role of the morphology and the dehydroxylation of metakaolins on geopolymerization. *Applied Clay Science*, 50(4), 538–545. https://doi.org/http://dx.doi.org/10.1016/j.clay.2010.10.010. DOI: 10.1016/j.clay.2010.10.010

Minelli, M., Medri, V., Papa, E., Miccio, F., Landi, E., & Doghieri, F. (2016). Geopolymers as solid adsorbent for CO2 capture. *Chemical Engineering Science*, 148, 267–274. Advance online publication. DOI: 10.1016/j.ces.2016.04.013

Muñiz-Villarreal, M. S., Manzano-Ramírez, A., Sampieri-Bulbarela, S., Gasca-Tirado, J. R., Reyes-Araiza, J. L., Rubio-Ávalos, J. C., Pérez-Bueno, J. J., Apatiga, L. M., Zaldivar-Cadena, A., & Amigó-Borrás, V. (2011). The effect of temperature on the geopolymerization process of a metakaolin-based geopolymer. *Materials Letters*, 65(6), 995–998. https://doi.org/http://dx.doi.org/10.1016/j.matlet.2010.12.049. DOI: 10.1016/j.matlet.2010.12.049

Nakamoto, K., Ohshiro, M., & Kobayashi, T. (2017). Mordenite zeolite—Polyethersulfone composite fibers developed for decontamination of heavy metal ions. *Journal of Environmental Chemical Engineering*, 5(1), 513–525. https://doi.org/https://doi.org/10.1016/j.jece.2016.12.031. DOI: 10.1016/j.jece.2016.12.031

Nazari, A., Bagheri, A., & Riahi, S. (2011). Properties of geopolymer with seeded fly ash and rice husk bark ash. *Materials Science and Engineering A*, 528(24), 7395–7401. https://doi.org/http://dx.doi.org/10.1016/j.msea.2011.06.027. DOI: 10.1016/j.msea.2011.06.027

Ogwanga, G., Olupota, P. W., Kaseddea, H., Menya, H., Storzc, H., & Kiro, Y. (2021). Experimental evaluation of rice husk ash for applications in geopolymer mortars. *Journal of Bioresources and Bioproducts*, 6(2), 160–167. DOI: 10.1016/j.jobab.2021.02.008

Onutai, S., Jiemsirilers, S., Thavorniti, P., & Kobayashi, T. (2016). Fast microwave syntheses of fly ash based porous geopolymers in the presence of high alkali concentration. *Ceramics International*, 42(8), 866–9874. DOI: 10.1016/j.ceramint.2016.03.086

Onutai, S., Jiemsirilers, S., Thavorniti, P., & Kobayashi, T. (2018). Removal of Pb^{2+}, Cu^{2+}, Ni^{2+}, Cd^{2+} from Wastewater using Fly Ash Based Geopolymer as an Adsorbent. *Key Engineering Materials*, 773, 373–378. DOI: 10.4028/www.scientific.net/KEM.773.373

Onutai, S., Kobayashi, T., Thavorniti, P., & Jiemsirilers, S. (2019). Porous fly ash-based geopolymer composite fiber as an adsorbent for removal of heavy metal ions from wastewater. *Materials Letters*, 236, 30–33. DOI: 10.1016/j.matlet.2018.10.035

Palomo, A., Grutzeck, M. W., & Blanco, M. T. (1999). Alkali-activated fly ashes: A cement for the future. *Cement and Concrete Research*, 29(8), 1323–1329. DOI: 10.1016/S0008-8846(98)00243-9

Papa, E., Landi, E., Murri, A. M., Miccio, F., Vaccari, A., & Medri, V. (2021). CO_2 adsorption at intermediate and low temperature by geopolymer-hydrotalcite composites. *Open Ceramics*, 5, 00048. DOI: 10.1016/j.oceram.2020.100048

Papageorgiou, S. K., Katsaros, F. K., Kouvelos, E. P., & Kanellopoulos, N. K. (2009). Prediction of binary adsorption isotherms of Cu(2+), Cd(2+) and Pb(2+) on calcium alginate beads from single adsorption data. *Journal of Hazardous Materials*, 162(2-3), 1347–1354. DOI: 10.1016/j.jhazmat.2008.06.022 PMID: 18653278

Park, Y., Abolmaali, A., Kim, Y. H., & Ghahremannejad, M. (2016). Compressive strength of fly ash-based geopolymer concrete with crumb rubber partially replacing sand. *Construction & Building Materials*, 118, 43–51. https://doi.org/http://dx.doi.org/10.1016/j.conbuildmat.2016.05.001. DOI: 10.1016/j.conbuildmat.2016.05.001

Pavithra, P., Srinivasula Reddy, M., Dinakar, P., Hanumantha Rao, B., Satpathy, B. K., & Mohanty, A. N. (2016). A mix design procedure for geopolymer concrete with fly ash. *Journal of Cleaner Production*, 133, 117–125. https://doi.org/http://dx.doi.org/10.1016/j.jclepro.2016.05.041. DOI: 10.1016/j.jclepro.2016.05.041

Pei, Y., Choi, G., Asahina, S., Yang, J.-H., Vinu, A., & Choy, J.-H. (2019). A novel geopolymer route to porous carbon: High CO2 adsorption capacity. *Chemical Communications*, 55(22), 3266–3269. DOI: 10.1039/C9CC00232D PMID: 30810144

Peretz, S., Anghel, D. F., Vasilescu, E., Florea-Spiroiu, M., Stoian, C., & Zgherea, G. (2015). Synthesis, characterization and adsorption properties of alginate porous beads. *Polymer Bulletin*, 72(12), 3169–3182. DOI: 10.1007/s00289-015-1459-4

Portland Cement Association. (1991). Solidification and stabilization of wastes using Portland cement. *The United States of America*: Report 7355.

Prasanphan, S., Wannagon, A., Kobayashi, T., & Jiemsirilers, S. (2019). Reaction mechanisms of calcined kaolin processing waste-based geopolymers in the presence of low alkali activator solution. *Construction & Building Materials*, 221, 409–420. DOI: 10.1016/j.conbuildmat.2019.06.116

Raja, V. K. B., Selvarani, V., Kannan, S., Sujan, S., Sahas, S., Padmapriya, R., Kumar, V. G., & Baalamurugan, J. (2023). Geopolymer based foams from steel slag – A green technology. *Materials Today: Proceedings*. Advance online publication. DOI: 10.1016/j.matpr.2023.07.156

Rattanasak, U., & Chindaprasirt, P. (2009). Influence of NaOH solution on the synthesis of fly ash geopolymer. *Minerals Engineering*, 22(12), 1073–1078. https://doi.org/http://dx.doi.org/10.1016/j.mineng.2009.03.022. DOI: 10.1016/j.mineng.2009.03.022

Rattanasak, U., Chindaprasirt, P., & Suwanvitaya, P. (2010). Development of high volume rice husk ash alumino silicate composites [journal article]. *International Journal of Minerals Metallurgy and Materials*, 17(5), 654–659. DOI: 10.1007/s12613-010-0370-0

Rifaai, Y., Yahia, A., Mostafa, A., Aggoun, S., & Kadri, E. H. (2019). Rheology of fly ash-based geopolymer: Effect of NaOH concentration. *Construction & Building Materials*, 223(30), 583–594. DOI: 10.1016/j.conbuildmat.2019.07.028

Roviello, G., Ricciotti, L., Ferone, C., Colangelo, F., & Tarallo, O. (2015). Fire resistant melamine based organic-geopolymer hybrid composites. *Cement and Concrete Composites*, 59, 89–99. https://doi.org/http://dx.doi.org/10.1016/j.cemconcomp.2015.03.007. DOI: 10.1016/j.cemconcomp.2015.03.007

Rovnaník, P. (2010). Effect of curing temperature on the development of hard structure of metakaolin-based geopolymer. *Construction & Building Materials*, 24(7), 1176–1183. DOI: 10.1016/j.conbuildmat.2009.12.023

Ryu, G. S., Lee, Y. B., Koh, K. T., & Chung, Y. S. (2013). The mechanical properties of fly ash-based geopolymer concrete with alkaline activators. *Construction & Building Materials*, 47, 409–418. https://doi.org/http://dx.doi.org/10.1016/j.conbuildmat.2013.05.069. DOI: 10.1016/j.conbuildmat.2013.05.069

Sakkas, K., Panias, D., Nomikos, P. P., & Sofianos, A. I. (2014). Potassium based geopolymer for passive fire protection of concrete tunnels linings. *Tunnelling and Underground Space Technology*, 43, 148–156. https://ui.adsabs.harvard.edu/abs/2014TUSTI.43.148S/abstract. DOI: 10.1016/j.tust.2014.05.003

Sari, A. N., Srisunarsih, E., & Sucipto, T. L. A. (2021). The Use of Rice Husk Ash in Enhancing the Material Properties of Fly Ash-Based Self Compacted Geopolymer Concrete. *Journal of Physics: Conference Series*, 1808(1), 012011. DOI: 10.1088/1742-6596/1808/1/012011

Seco, A., García, B., Prieto, E., Ramirez, F., Miqueleiz, L., Urmeneta, P., & Oroz, V. (2012). *Types of Waste for the Production of Pozzolanic Materials-A Review*. INTECH Open Access Publisher. DOI: 10.5772/36285

Shaikh, F. U. A. (2013). Deflection hardening behaviour of short fibre reinforced fly ash based geopolymer composites. *Materials & Design*, 50, 674–682. DOI: 10.1016/j.matdes.2013.03.063

Sikora, S., Gapys, E., Michalowski, B., Horbanowicz, T., Hynowski, M. (2018). Geopolymer coating as a protection of concrete against chemical attack and corrosion, *E3S Web of Conferences*, 49, 00101. DOI: 10.1051/e3sconf/20184900101

Somna, K., Jaturapitakkul, C., Kajitvichyanukul, P., & Chindaprasirt, P. (2011). NaOH-activated ground fly ash geopolymer cured at ambient temperature. *Fuel*, 90(6), 2118–2124. https://doi.org/http://dx.doi.org/10.1016/j.fuel.2011.01.018. DOI: 10.1016/j.fuel.2011.01.018

Soutsos, M., Boyle, A. P., Vinai, R., Hadjierakleous, A., & Barnett, S. J. (2016). Factors influencing the compressive strength of fly ash based geopolymers. *Construction & Building Materials*, 110, 355–368. https://doi.org/http://dx.doi.org/10.1016/j.conbuildmat.2015.11.045. DOI: 10.1016/j.conbuildmat.2015.11.045

Swanepoel, J. C., & Strydom, C. A. (2002). Utilisation of fly ash in a geopolymeric material. *Applied Geochemistry*, 17(8), 1143–1148. https://doi.org/http://dx.doi.org/10.1016/S0883-2927(02)00005-7. DOI: 10.1016/S0883-2927(02)00005-7

Tang, Q., Ge, Y.-y., Wang, K., He, Y., & Cui, X. (2015). Preparation of porous P-type zeolite spheres with suspension solidification method. *Materials Letters*, 161, 558–560. https://doi.org/https://doi.org/10.1016/j.matlet.2015.09.062. DOI: 10.1016/j.matlet.2015.09.062

Tchadjié, L. N., Djobo, J. N. Y., Ranjbar, N., Tchakouté, H. K., Kenne, B. B. D., Elimbi, A., & Njopwouo, D. (2016). Potential of using granite waste as raw material for geopolymer synthesis. *Ceramics International*, 42(2, 2, Part B), 3046–3055. https://doi.org/http://dx.doi.org/10.1016/j.ceramint.2015.10.091. DOI: 10.1016/j.ceramint.2015.10.091

Tchakouté, H. K., Rüscher, C. H., Kong, S., Kamseu, E., & Leonelli, C. (2016). Geopolymer binders from metakaolin using sodium waterglass from waste glass and rice husk ash as alternative activators: A comparative study. *Construction & Building Materials*, 114, 276–289. https://doi.org/http://dx.doi.org/10.1016/j.conbuildmat.2016.03.184. DOI: 10.1016/j.conbuildmat.2016.03.184

Udaibah, W. (2020). Zeolite based Geopolymer from Biomass: A Sustainable Adsorbent for Water Softener. *Journal of Physics: Conference Series*, 1539(1), 012004. DOI: 10.1088/1742-6596/1539/1/012004

van Jaarsveld, J. G. S., & van Deventer, J. S. J. (1999). Effect of the Alkali Metal Activator on the Properties of Fly Ash-Based Geopolymers. *Industrial & Engineering Chemistry Research*, 38(10), 3932–3941. DOI: 10.1021/ie980804b

Van Jaarsveld, J. G. S., Van Deventer, J. S. J., & Lorenzen, L. (1997). The potential use of geopolymeric materials to immobilise toxic metals: Part I. Theory and applications. *Minerals Engineering*, 10(7), 659–669. https://doi.org/http://dx.doi.org/10.1016/S0892-6875(97)00046-0. DOI: 10.1016/S0892-6875(97)00046-0

Vu, T. H., & Gowripalan, N. (2018). Mechanisms of heavy metal immobilisa- tion using geopolymerisation techniques – a review. *Journal of Advanced Concrete Technology*, 16(3), 124–135. DOI: 10.3151/jact.16.124

Wang, S., Li, L., & Zhu, Z. H. (2007). Solid-state conversion of fly ash to effective adsorbents for Cu removal from wastewater. *Journal of Hazardous Materials*, 139(2), 254–259. https://doi.org/http://doi.org/10.1016/j.jhazmat.2006.06.018. DOI: 10.1016/j.jhazmat.2006.06.018 PMID: 16839666

Wang, S.-D. (2000). The role of sodium during the hydration of alkali-activated slag. *Advances in Cement Research*, 12(2), 65–69. https://doi.org/doi:10.1680/adcr.2000.12.2.65. DOI: 10.1680/adcr.2000.12.2.65

Watanabe, Y., Jiemsirilers, S., & Kobayashi, T. (2023). Lead Immobilized Fly Ash-Based Geopolymer Ceramics Fabricated by Microwave Quick Cure. *Journal of Chemical Engineering of Japan*, 56(1), 2222780. DOI: 10.1080/00219592.2023.2222780

Watanabe, Y., & Kobayashi, T. (2023). *Geopolymers Prepared by Microwave Treatments, Advanced ceramics*. Springer.

Yang, N., Fu, Y., Xue, X., Lei, J. D., & Dai, J.-G. (2023). Geopolymer-based sub-ambient daytime radiative cooling coating. *EcoMat*, 12284. Advance online publication. DOI: 10.1002/eom2.122

Yao, X., Zhang, Z., Zhu, H., & Chen, Y. (2009). Geopolymerization process of alkali–metakaolinite characterized by isothermal calorimetry. *Thermochimica Acta*, 493(1–2), 49–54. DOI: 10.1016/j.tca.2009.04.002

Zaidahtulakmal, M. Z., Kartini, K., & Hamidah, M. S. (2019). Rice husk ash (RHA) based geopolymer mortar incorporating sewage sludge ash (SSA). *Journal of Physics: Conference Series*, 1349(1), 012022. DOI: 10.1088/1742-6596/1349/1/012022

Zainal, F. F., Fazill, M. F., Kamarudin, H., Rahmat, A., Bakri Abdullah, M., M., A., Wazien, W. (2018). Effect of Geopolymer Coating on Mild Steel, Solid State Phenomena (Volume 273), 175-180. DOI: 10.4028/www.scientific.net/SSP.273.175

Zhang, J., Provis, J. L., Feng, D., & van Deventer, J. S. J. (2008). Geopolymers for immobilization of Cr6+, Cd2+, and Pb2+. *Journal of Hazardous Materials*, 157(2-3), 587–598. DOI: 10.1016/j.jhazmat.2008.01.053 PMID: 18313213

Zhuang, X. Y., Chen, L., Komarneni, S., Zhou, C. H., Tong, D. S., Yang, H. M., Yu, W. H., & Wang, H. (2016). Fly ash-based geopolymer: Clean production, properties and applications. *Journal of Cleaner Production*, 125, 253–267. https://doi.org/http://dx.doi.org/10.1016/j.jclepro.2016.03.019. DOI: 10.1016/j.jclepro.2016.03.019

Zuhua, Z., Xiao, Y., Huajun, Z., & Yue, C. (2009). Role of water in the synthesis of calcined kaolin-based geopolymer. *Applied Clay Science*, 43(2), 218–223. https://doi.org/http://dx.doi.org/10.1016/j.clay.2008.09.003. DOI: 10.1016/j.clay.2008.09.003

Chapter 19
Rapid and Easy Colorimetric Detection for Specific Heavy Metal Ions Contaminated in Environmental Soil

Reiko Wakasugi
National Institute of Technology, Kumamoto College, Japan

Ryo Shoji
https://orcid.org/0000-0003-4451-7872
National Institute of Technology, Tokyo College, Japan

Hitomi Fukaura
Limited Company Sakamoto Lime Industry, Japan

Yasunori Takaki
Limited Company Sakamoto Lime Industry, Japan

Hiroyuki Kono
https://orcid.org/0000-0001-5675-3157
National Institute of Technology, Tomakomai College, Japan

ABSTRACT

Among the environmental pollution, soil is making it not only easy for analysis of contamination to remain, but also difficult to clean up. It is therefore necessary to develop a tool that can be used to easily and quickly determine the necessity of soil contamination at the on sites. In this chapter, a method developed to quickly detect soil pollutant species due to four specific types of heavy metals as hexavalent chromium, fluorine, boron, and lead. The detection kit has a sensitivity achieved to the elution standard value stipulated by Japan's Soil Contamination Countermeasures Law. The detection agent consists of a composite of water-absorbing polymer and coloring reagent that specifically reacts with such pollutant. This kit has already been sold under the trade name OCTES® (OCTES, 2014) and can now be used at soil contaminated sites without requiring any particularly difficult operations or electric power sources.

DOI: 10.4018/979-8-3693-0003-9.ch019

This OCTES is expected to be used as a screening method to easily determine the need for detailed soil contamination assessment and purification.

INTRODUCTION

Soil becomes fertile through the decomposition of organic matter by resident microorganisms. Such nature's cleansing action has been responsible for the preservation of the global environment from time immemorial to the present, but human prosperity is now releasing pollutants and going against the natural cycle. Therefore, as the construction of a sustainable society is strongly called, the management of pollutants has become extremely important. However, in today's industrialized world, deforestation, the increasing coverage of land with concrete and cement, and the disposal of waste containing pollutants through landfilling have weakened the regenerative capacity of soil, leading to a serious issue of soil contamination. Efforts to remediate contaminated soil and restore it are essential.

However, the wide range of pollutants present in contaminated soil makes it challenging, and in many cases, even the involved parties struggle to fully comprehend the extent of the contamination. Therefore, the emergence of a method that allows rapid analysis at the detection site is eagerly awaited. This chapter presents the recent progress of the soil environment and efforts to investigate and understand heavy metal contaminants in soil as part of efforts to sustainably rehabilitate soils.

Current State of Soil Environment.

The soil environment plays an important role in the circulation of materials and the maintenance of ecosystems and performs functions such as water purification and the production of food and wood. Land use varies, including industrial, agricultural, residential, and forest areas; Unfortunately, soil contamination has also been reported in Japan, in recent years, so-called "urban-type" (non-agricultural) soil pollution and this has increased rapidly. Currently, the main sources of soil contamination are the chemical and electroplating industries, with the main pollutants being lead, hexavalent chromium, and trichloroethylene (Conservation of Soil Environment, 2004). The Environmental Quality Standards (EQSs) for soil pollution were issued in August 1991. At present, the EQSs regulates 25 substances, including heavy metals like cadmium, lead, chromium (VI), copper and arsenic, volatile organic compounds, and pesticides. Even though eighty percent of cases exceeding the regulatory concentration, limits for soil contamination are attributed to heavy metals and other substances (Ministry of the Environment, Government of Japan [MOE], 2021).

Furthermore, complex contamination involving all nine heavy metals is rare, and in most cases, contamination occurs due to one to three types of heavy metals associated with factories handling specific hazardous substances. For example, specific harmful substances regulated for cadmium, copper and arsenic have target levels of soil quality examined through leaching test and content test for 0.01 mg/l in sample solution and less than 0.4 mg/kg in rice for agricultural land, less than 125 mg/kg in soil for agricultural land and 0.01 mg/l or less in sample solution, and less than 15 mg/kg in soil, respectively.

When soil becomes contaminated, rainwater can infiltrate from the soil, leading to groundwater contamination (Xu et al., 2024). As a result, if the contaminant mixes with drinking water, it can pose significant health risks. Japan has also experienced significant environmental pollution incidents in the past (Aoshima, 2012).

In the fiscal year 2021, the most frequently reported specific hazardous substances after the dismantling of facilities with their usage were "fluorine and its compounds," followed by "boron and its compounds," and "hexavalent chromium compounds." "Hexavalent chromium compounds" were commonly used not only in cleaning facilities of workplaces engaged in research, testing, inspection, or specialized education but also in facilities for surface treatment using acids or alkalis, electroplating facilities, and cleaning facilities in hospitals (Chen et al., 2024). Furthermore, "fluorine and its compounds" were commonly used not only in the aforementioned facilities but also in exhaust gas cleaning ones of the metal product manufacturing industry or machinery and equipment manufacturing industry, including the weapons manufacturing industry (MOE, 2021).

Furthermore, in many countries abroad, there are no standards for soil and water pollution, and contaminated areas exceeding Japan's leachate standards by more than ten times are widely distributed. However, there is a shortage of analytical facilities, and there are limited personnel with specialized knowledge. Furthermore, extensive financial resources are required for analysis and investigations. Consequently, there is a situation where surveys of contamination areas are not being conducted. In many cases, wastewater and sludge containing heavy metals are being discharged into rivers and fields, resulting in serious and irreversible damage to human bodies, nerves, organs, and so on in the river basins. These contaminants are soluble and migrated to the deep in the soil (Hou et al., 2022).

In urban areas, many former heavy industrial zones have seen factories relocate to rural areas or overseas, leaving behind prime real estate locations with significant value. However, there are often instances of heavy metal contamination on these former factory sites, necessitating costly investigations and remediation efforts. In order to facilitate the regeneration of land and its subsequent use, it is essential to conduct rapid analysis and apply prompt and appropriate treatments to contaminated soil. Minimizing the associated costs is desirable in this process.

Soil Analysis.

Soil analysis is necessary for various purposes, including emergency accident investigations, identifying the extent of contaminated water leakage, and soil surveys during civil engineering projects. It involves analyzing whether the concentration of heavy metals in the soil exceeds the national standards set by the government. Since soil is a solid and non-flowable medium, contamination is typically localized to specific areas. Extensive soil sample analysis is necessary to fully determine the extent of soil contamination. However, it is not uncommon for the area requiring remediation to be limited to a relatively small area. In Japan, soil contamination surveys are mandatory for real estate transactions with a land area of 3,000 m^2 or more. The analysis involves measuring the concentration distribution of one to three specific heavy metals at numerous locations to identify the extent of contamination. Efficiently defining the sampling scope for analysis can help reduce the cost associated with the analysis. Sampling can be made more efficient by dividing the survey area into a grid evaluating representative points, and conducting further detailed assessments for areas where contamination is observed, as well as their surroundings, significantly reducing the overall workload. In Japan, the Ministry of the Environment has established official methods under Environmental Agency notifications for analyzing soil, particularly targeting heavy metals and similar substances.

In terms of the situation of soil pollution in China, according to nationwide survey of mainland China conducted in 2014, approximately 16% of China's land area is said to have soil contaminated with fertilizers, pesticides, heavy metals, plastics, and other chemicals. In August 2018, Soil Contamination

Countermeasures Law has established the "Soil Environment Mass Construction Land Soil Contamination Risk Control Standard (Trial)", which includes 85 soil contamination standards. In addition, in China, not only industrial land and construction land such as housing, but also all mines, agricultural land, and construction land are regulated under the Soil Contamination Law.

In the United States, the first soil pollution law, the Comprehensive Environmental Measures, Compensation, and Liability Act (commonly known as the Superfund Act), (the United States Environmental Protection Agency [EPA], 2023) was enacted in 1980 and has achieved great results in combating soil pollution problems. The US Superfund Act specifies approximately 800 types of substances as hazardous substances. The law's main purpose is to prevent the spread of environmental pollution. Also, this law requires current and former landowners to clean up their soil at all costs.

Development of reaction reagents and water-absorbing resins.

Currently, in soil analysis, test solutions are prepared using the methods specified according to these official methods, and the concentration of heavy metal ions in the test solution is measured using analytical equipment for assessment. However, there is still insufficient preparation for *in situ* analysis to rapidly measure these elements. Therefore, it is necessary to wait for the results of the analysis over time to determine if the soil contains the designated heavy metals. Rapid *in-situ* analytical methods still require urgent development for the 10 designated elements, which are designated as specific hazardous substances under the Soil Contamination Countermeasures Act (MOE, 2017) for cadmium, hexavalent chromium, cyanide, total mercury, alkyl mercury, selenium, lead, arsenic, fluorine, and boron.

Since these are characteristic of leaching out as ions and accumulating in the soil and tissues of animals and plants, immediate countermeasures through rapid analysis and diagnosis are essential. Among these, hexavalent chromium is an extremely dangerous substance that has been listed by the International Agency for Research on Cancer (IARC) and the EPA as one of the two major carcinogens, along with asbestos, which has become a serious social problem. The effects on the human body include chronic disorders such as allergic dermatitis and lung cancer, and sudden death disorders such as skin burns, necrosis, diarrhea, and vomiting. Also, lead is a typical toxic substance that accumulates in bones. If body ingests, there are concerns about anemia and effects on the central nervous system. Based on the results of animal experiments related to lead, the International Agency for Research on Cancer (IARC) has determined that it is "possibly carcinogenic to humans" (rank 2B). In addition, fluorine and boron were classified as substances that pose a risk to human health when soil environmental standards were established in 2001, but the International Agency for Research on Cancer (IARC) has not evaluated the carcinogenicity of fluorine. However, because fluorine sometime exists in high concentrations in nature, there are many cases in which soil environmental standard values are exceeded. For these pollutants, with the aim of developing a detection agent that can quickly and easily detect soil contaminants such as heavy metals at low cost, there is a simple detection agent that combines a contaminant detection reagent and a polymer resin. Existing technology only allows coloration, but this colorant is composed of an optimized water-absorbing resin, which not only produces color, but also has properties that are not affected by the coloring reagent or soil pH. Compared to conventional easy analysis method, the methodology has the advantage of omitting the complicated work of preparing test solutions.

Development of reaction reagents

Colorants that specifically adsorb and react with certain ions have long been studied. In contrast, it is necessary to consider detection agents that not only change color but are not affected by soil pH or coexisting substances such as metal cations or oxioanions. Therefore, detection agent that specifically adsorbs and reacts with the target substance is developed. Table 1 summarizes the target substances for Cr (VI), F, B and Pb, coloring principles, and the targeted detection limits. The Japanese soil elution standard values are also shown in Table 1.

Here is an example of fluorine coloration. Alizarin is often used for fluorine coloration. In La-ALC solution, $(La-ALC)_4$ complex is formed (Figure 1). This structure is a cyclic structure with consecutive (La-ALC) units. This corresponds to a structure in which lanthanum is combined with two fluoride ions and one proton removed $(La-ALC)_4$. Even in the absence of fluoride ions, protons are eliminated under basic condition, and the color changes to blue-purple (Figure 2). In this study, we prepared a complex compound of alizarin complexone and lanthanum (III) and a buffering agent supported on a water-absorbing polymer (described later) as a coloring reagent. In terms of the principle of the fluorine detection using the lanthanum alizarin complex, a complex of lanthanum (III) and Alizarin Complexone is added, and this complex reacts with fluoride ions, resulting in a blue color.

For hexavalent chromium, boron, and lead as well, it would be informative to refer to specific reagents listed in Table 1.

Table 1. Target substances, coloring principles and detection limits

Heavy metals	Colorization principle	Coloring agent	Required time for the detection (min)	Color Change	Target detection limit (mg/L)	Environemental Standard, Japan (mg/L)
Cr (VI)	Diphenylcarbazide	N',2-Diphenylhydrazine-1-carbohydrazide	30	Colorless to purple	0.05	0.05
F*	Lanthanum alizarin complexone	3-[N,N-Bis(carboxymethyl)aminomethyl]-1,2-dihydroxyanthraquinone and Lanthanum	40	Purple to blue-purple	0.8	0.8
B*	Azomethine H	8-Hydroxy-1-(salicylideneamino)naphthalene-3,6-disulfonic Acid Monosodium Salt	90	Colorless to yellow	1.0	1.0
Pb	Pyridylazoresorcin (PAR)	Pyridylazoresorcin	60	Yellow to red	0.01	0.01

*In Japanese environmental standard for soil pollution, F and B are classified as heavy metals.

Figure 1. Structure of alizarin complexon (ALC)

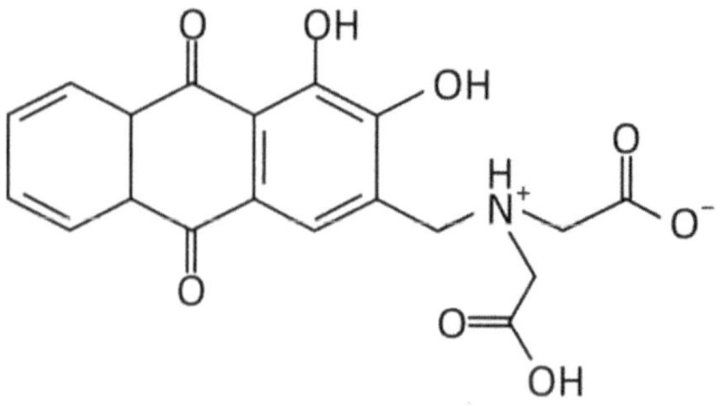

Figure 2. Alizarin blue

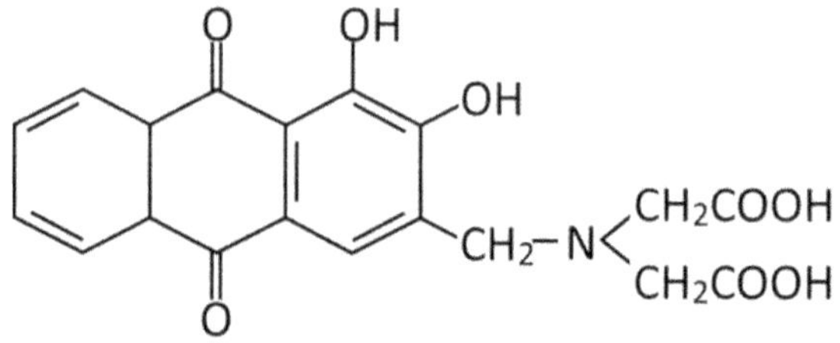

Figure 3. Structure of lanthanide-alizarin complexon

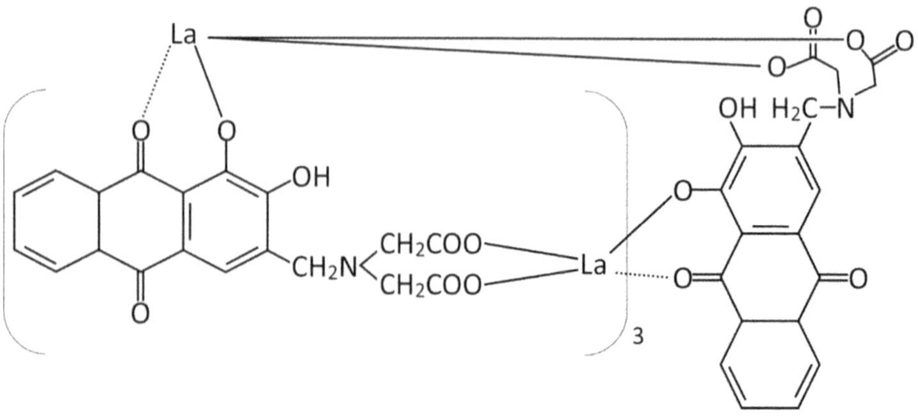

The speciation of fluorine compounds in the environment is complex; in addition to ions, they form fluoro-complexes with metal elements such as iron and aluminum. They exist in the form of fluoride precipitates with alkaline earth metals. Therefore, pretreatment such as distillation is normally required to quantify fluorine compounds. In this study, only fluoride ions that are easily eluted into water are targeted. Since fluoride ions react with many cations, there are many elements that interfere with this detection.

For the determination of Cr(VI), diphenylcarbazide spectrophotometry, atomic absorption spectrometry, inductively coupled plasma (ICP) emission, and ICP mass spectrometry are available. Among these methods, diphenylcarbazide spectrophotometry is suitable for the purpose of this study because it uses an in-situ color reaction, and even if Cr(III) coexists, Cr(VI) is not separated. It is excellent in that it can be quantified.

In terms of the principle of the hexavalent chromium detection, the complex formed by the reaction of diphenylcarbazide and hexavalent chromium exhibits a reddish-purple color, but, it does not react with trivalent chromium, so if redox substances coexist, the detection of hexavalent chromium becomes ambiguous.

On the other hand, apart from color detection reagents, when adsorbent to adsorb pollutant is present near the chromophore, support that these coloring reagents can achieve a coloring reaction. With a material that has such an integrated adsorption and coloration site, testing can be performed on site and without weighing, without the use of coloring reagents or pH adjusters, which are generally crystalline powders or solutions.

In terms of boron detection, a method using Azomethine H has already been proposed as a chelation of boron in aquatic solution (Oxspring et al., 1995). The OCTES we propose applies the same principle to soil eluates. In fact, a promising technique has been examined (Boonkanon et al., 2020). By using starch thin film, their colorimetric sensor can detect 0.052 mg/L of boron in aquatic solution. However, compared to the environmental standard for boron (1.0 mg/L in Japan), this method is too sensitive, and concerns about its stability still remain.

For example, focusing on the fact that heavy metals exist as cations and some oxonium ions as anions in soil aqueous solutions, silica gel and alumina can be used as adsorbents with high affinity for ions and as supporting materials for coloring substances. When a detection material made by impregnating silica gel or alumina with a solution of a coloring reagent and drying it at 40°C or lower is brought into contact with the test solution, the coloring reagent gradually dissolves and the color on the surface of the substrate becomes lighter over time. This combination successfully increased the reactivity of the heavy metal ions dissolved in the aqueous soil solution with the coloring substance, allowing the solution to be colored in several tens of minutes without forced agitation, becoming possible to complete the color reaction until it reached an equilibrium state.

Additionally, the color changes can be analyzed by using some image analysis technique. RGB (Red-Green-Blue) analysis was performed highlighting the changes before and after detection. Figure 4 shows the relationship between fluoride ion concentration and changes in R values before and after exposed to fluoride solution by using OCTES for F.

Figure 4. Relationship between fluoride ion concentration and changes in R values before and after exposed to fluoride solution

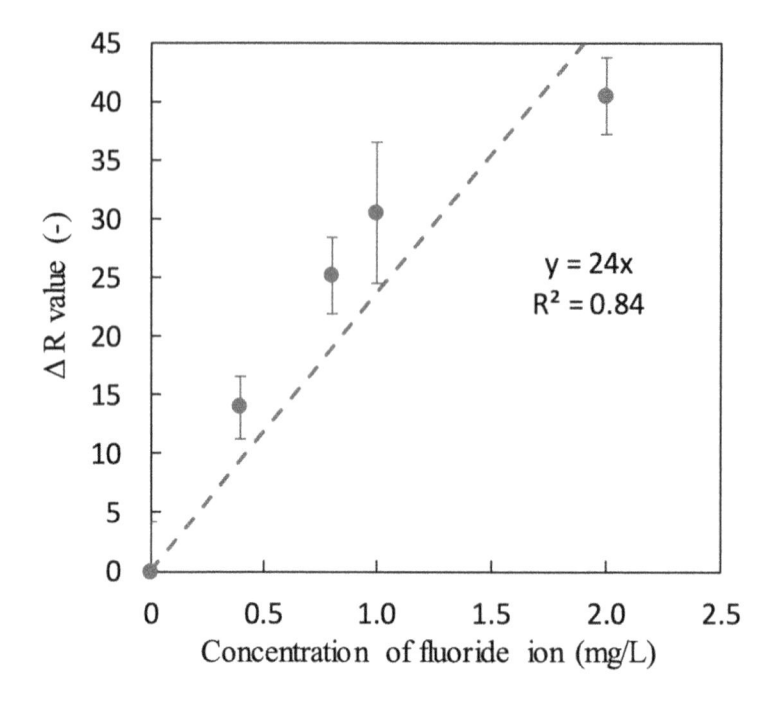

$y = 24x$
$R^2 = 0.84$

As can be seen in Figure 4, there is a significant positive proportional relationship between coloring and concentration. So, the OCTES can be used as the quantitative analysis as well.

Water-absorbing resins for heavy metals in soil

Hydrogels are generally cross-linked hydrophilic polymers that can absorb large amounts of water, as much as 100-500 times their own weight. Among the many commercially available hydrogels, cross-linked sodium polyacrylate, synthesized by copolymerization of acrylic acid with various monomers, is the most widely used superabsorbent polymer (Hua & Qian, 2001; Liu & Guo 2001). However, since crosslinked sodium polyacrylate has anionic functional site of -COO⁻ group in its structure, its water absorption performance strongly depends on the pH and ionic species of the aqueous solution in which it absorbs water (Kono & Fujita, 2012). In particular, the negative charges of heavy metals, calcium ions, and other multivalent ions and cross-linked polyacrylic acid are ionically complexed and then electrostatically neutralized, resulting in less water absorption ability at all (Kono & Zakimi, 2013). Therefore, nonionic water absorbent resins are desirable for absorbing leachate from soil and aqueous solutions containing relatively high concentrations of ionic species.

As predicted for heavy metal ion detection reagent, when it is immobilized on the surface of an absorbent resin, the coloring heavy metals in soil leachate achieves in the absorbent resin. The difference in coloration could be used to develop a device for easy visual determination of heavy metal concentrations. However, most detection reagents are based on chemical reactions, and the presence of ionic functional groups in the water-absorbent resin itself, which serves as the immobilization carrier, could affect the coloration. The policy of doing non-ionic absorbents that do not affect the coloration of heavy metal detection reagents is reasonable. Nonionic water-absorbent resins were prepared from hydroxylethyl cellulose (HEC), a typical nonionic cellulose derivative, by applying ethylene glycol diglycidiyl ether (EGDE) as crosslinking agent according to method previously reported (Kono, 2014; Kono et al., 2014). Epoxides are highly reactive and form ether bonds with hydroxyl groups in the presence of alkaline catalysis. Therefore, bifunctional crosslinking agents with epoxy groups at both ends, such as EGDE, can be used to form intermolecular cross-links between HEC chains (Figure 5). The HEC molecules crosslinked by EGDE prepared in this method us a polar molecule with a number of hydroxyl groups and thus has an extremely high affinity for water. In addition, the intermolecular crosslinking between HEC molecules forms a three-dimensional network structure, which permit to retain water molecules in the structure. Therefore, the HEC molecules crosslinked by EGDE can be used as superabsorbent resin in the same way as the crosslinked sodium polyacrylate resin, even though it is nonionic.

Figure 5. Scheme for nonionic water-absorbent resin synthesis via crosslinking of hydroxyethyl cellulose (HEC) with ethylene glycol diglycidiyl ether (EGDE). EGDE can react with the hydroxyl groups of HEC to form ether cross-linking between HEC molecules

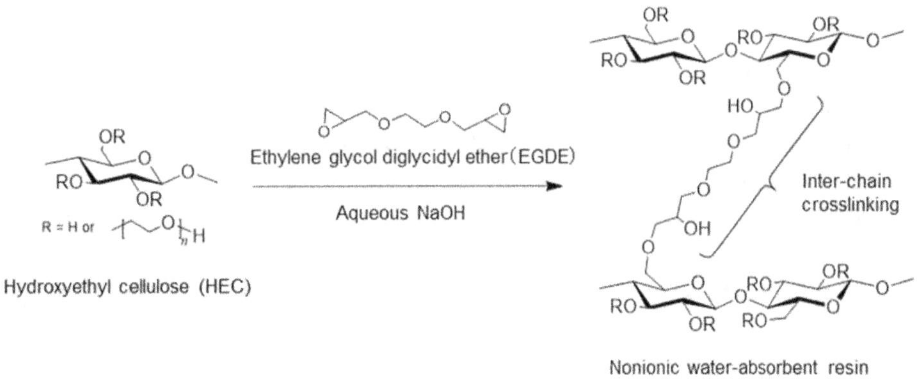

The water absorption property of the HEC-based hydrogel was compared with that of the commercially available polyacrylic acid water-absorbent resin (SA60N type II, Sumitomo Seika Co., Ltd., Japan) as a reference sample (Figure 6). The polyacrylic acid water-absorbent resin showed a rapid decrease in water absorbency with increasing NaCl concentration. The polyacrylic acid water-absorbent resin did not absorb any water when the water absorption performance was evaluated by use of acid aqueous solutions. In addition, the amount of water absorbed to polyacrylic acid water-absorbed resin dropped rapidly even in the alkaline range. On the other hand, the nonionic HEC-based water-absorbent resin maintained stable water absorption without being affected by pH as well as salt concentration.

Figure 6. Effect of salt concentration (a) and pH (b) on the relative water-absorbency of the HEC-based nonionic resins and commercially available polyacrylic acid-based resin

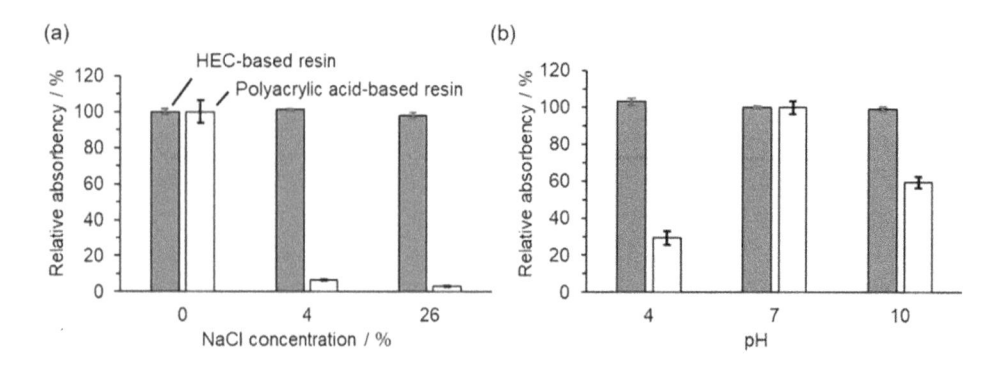

Nonionic HEC-based hydrogels have no ionic functional groups in their structure, and the hydrophilicity of the hydroxyl groups is responsible for water absorption. Because it is nonionic, it is not affected by foreign substances, and once retained, water molecules can be stably retained inside the structure due to the well-developed crosslinked structure. Therefore, it is expected to show stable water absorption without fluctuation even in soil leachate containing various substances, and to function effectively as a support agent. As a result, increasing the reactivity of heavy metal ions was achieved in dissolved aqueous soil medium with the coloring substance and this completed the coloring reaction to equilibrium within several tens of minutes, even without forcing the solution to be stirred. Therefore, the nonionic HEC-based hydrogels were applied as supporting agents for the heavy metal detection agent in soil.

Principle of detection of heavy metals in soil using water-absorbent polymer

Resins for detecting heavy metals in soil can be obtained by loading a heavy metal detection reagent onto an absorbent resin. Specifically, heavy metal detection resins are obtained by swelling the water-absorbent resin with an aqueous solution containing a completely dissolved soil heavy metal detection reagent and then lyophilizing the swollen resin (Figure 7a). The heavy metal detection resin swells again when it absorbs water from the soil elution solution containing heavy metals, and the detection reagent reacts with the heavy metals to produce a color (Fig. 6b). In other words, the presence or absence of coloration can be used to identify the presence of heavy metals in the soil. Furthermore, since the degree of coloration depends on the concentration of the heavy metal, it is possible to estimate the concentration of the heavy metal from the shade of coloration.

Figure 7. Preparation of heavy metal detection agents in soils using water absorbent-resins and colorants (a) and principles of heavy metal detection using the agents (b).

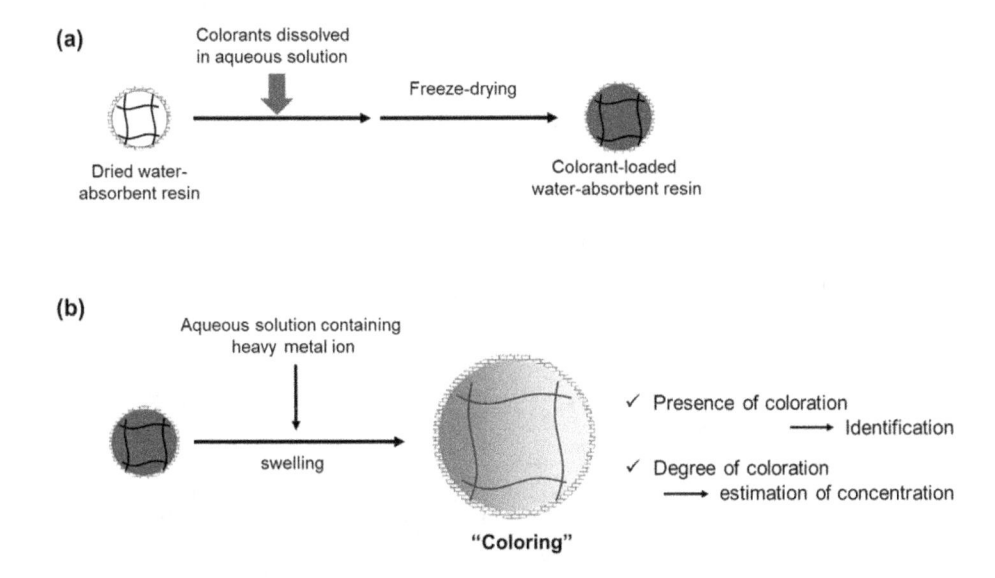

Advantages of nonionic water-absorbent resin in detection of pollutants in soil

When using a colorant immobilized on an absorbent resin to estimate the concentration of a target contaminant, the interaction between the resin and the target contaminant strongly affects the accuracy of detection (Situ et al., 2023). Figure 8(a) shows the lead ion (Pb^{2+}) standard solutions at the concentration from 0 to 5 ppm adsorbed on a detection agent in which Pb^{2+}-detection reagents (Table 1) are immobilized on an anionic polyacrylic acid resin and a nonionic HEC resin. When the anionic polyacrylic resin was used as the immobilization carrier, all chromophores turned orange regardless of Pb^{2+} concentration, making it difficult to estimate the concentration from the difference in coloration. On the other hand, when the nonionic HEC-based resin was used as the carrier, the chromophore responded in a stepwise manner from yellow to orange with increasing Pb^{2+} concentration, allowing to the concentration to be determined in the concentration range (0–5 ppm).

Figure 8(b) shows the coloring state of a fluorine ion detection reagent (Table 1) immobilized on the anionic and nonionic water-absorbent resins that were subjected to fluorine standard solutions in the concentration of 0–5 ppm. In the case of the fluorine ion detection, the detection reagent changed its chromophore in response to fluorine ion concentration, regardless of the water absorbent resins used for the carrier. Therefore, the fluorine ion concentration can be estimated by the difference in the color change of these detection materials. However, it can be confirmed in all concentration ranges that the chromogenicity when the anionic water-absorbent resin is used as the carrier is much lighter than that when nonionic water-absorbent resin is used as a carrier. Therefore, visual concentration estimation was more sensitive when nonionic water-absorbent resin was used as a carrier for fluorine-detection reagent than anionic polyacryrate water-absorbent resin.

Figure 8. The state of coloration by the lead cation (Pb^{2+}, a) and fluorine ion (F^-, b) detection reagents (Table 1) immobilized on the water absorbent resin, respectively. Anionic polyacrylic resin and nonionic HEC-based water-absorbent resin (Figure 5) were used as water absorbent resins

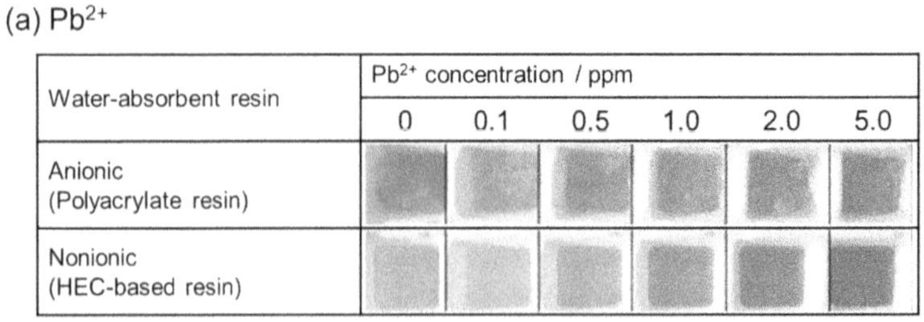

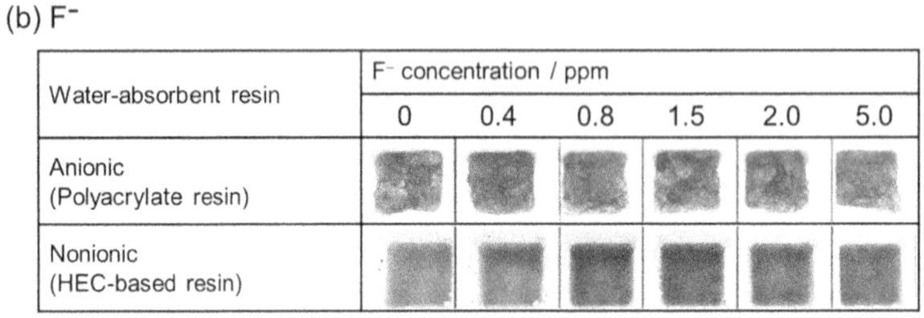

(a) Pb^{2+}

Water-absorbent resin	Pb^{2+} concentration / ppm					
	0	0.1	0.5	1.0	2.0	5.0
Anionic (Polyacrylate resin)						
Nonionic (HEC-based resin)						

(b) F^-

Water-absorbent resin	F^- concentration / ppm					
	0	0.4	0.8	1.5	2.0	5.0
Anionic (Polyacrylate resin)						
Nonionic (HEC-based resin)						

The dependence of the coloring state of lead and fluorine ions on the absorbent resin indicates that there is an interaction between the absorbent resin and the ions (Zhao et al., 2019). In the case of polyacrylate resin, the sodium carboxylate salt groups of the resin dissociate upon contact with water, changing to carboxylate anions. As a result, polyacrylate resin exhibit anionic property. When water contains metal cations such as Pb^{2+}, electrostatic attraction occurs between the carboxylate anions of the resin and the metal cations (Figure 9, top). The electrostatic attraction prompts preferential incorporation of the cations into the resin (Mingnon et al., 2019). The metal cations that penetrate into the resin not only react with the coloring reagent, but also electrostatically neutralize the carboxylate anions of the resin. As a result, the metal cation concentration inside the resin differs from the actual cation concentration of test solution, making accurate estimation of the metal ion concentration difficult. On the other hand, in the case of nonionic HEC-based resin, the ionic nature of the resin does not change upon water absorption and remains neutral because of on ionic functional groups (Figure 9, bottom). The concentration of metal ions inside of resin is almost identical to that of the test solution because there is no interaction between the resin and metal ion. As a result, the detection reagent inside the resin develops a color corresponding to the cation concentration of test solutions, allowing accurate estimation of metal ion concentration.

Figure 9. Effect of ionic properties of water-absorbent resins on the chromogenic detection of metal cations. Anionic polyacrylate resin(top) and nonionic HEC-based resin (bottom)

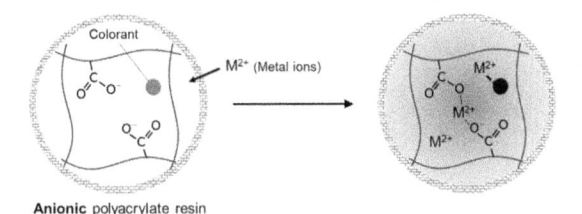

Anionic polyacrylate resin

✓ Electrostatic attraction facilitates the penetration of metal ions into the resin.
✓ In addition to reacting with the chromophore, metal ions undergo electrostatic neutralization with the resin.
✓ Accurate estimation of metal ion concentration is difficult.

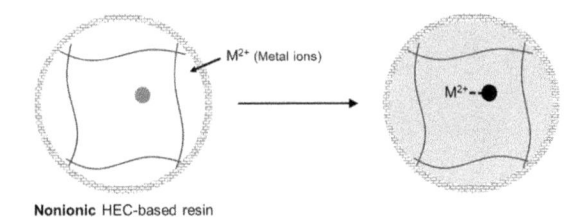

Nonionic HEC-based resin

✓ No electrostatic interaction between metal cations and resin.
✓ Metal ions react with the chromophore
✓ **Concentration of metal ions can be accurately estimated.**

In the detection of anion species such as fluorine with the detection agent loaded on the anionic polyacrylate resin, electrostatic repulsion between the carboxylate groups of resin and the target anion is considered, contrary to metal cation detection (Figure 9, top). The electrostatic repulsion between the resin and the anion species prevents the anions from penetrating into the resin (Mingnon et al., 2019). As a result, the target anion concentration inside of the resin is lower than the actual concentration of the test solution. In the detection of fluorine ions by a chromophore immobilized on a water-absorbing resin, the degree of coloration with an anionic resin was much lower than with a nonionic resin. This indicates that the penetration of fluorine ions into the resin was prevented by electrostatic repulsion with the polyacrylate resin. When nonionic resin is used as a carrier, there is no electrostatic interaction with the target anion species, allowing accurate anion concentration estimation (Figure 10 bottom). Therefore, in order to estimate ion concentrations visually using chromophores immobilized on water-absorbent resins, it is necessary to use nonionic water-absorbent resins to suppress the interaction between the resin and the ion species to be measured.

Figure 10. Effect of ionic properties of water-absorbent resins on the chromogenic detection of fluorine ions. Anionic polyacrylate resin(top) and nonionic HEC-based resin (bottom)

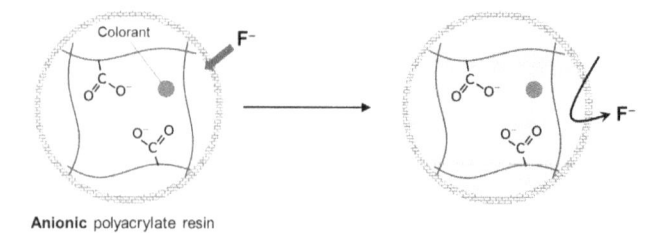

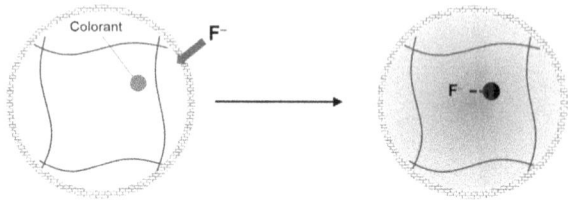

ADDRESSING SOIL CONTAMINATION

Examples of Simple Detection by OCTES®

In introducing one of the simple detection materials, OCTES®, this is composed of simple detection material units having reactive chromophore with the target heavy metals to exhibit color change and adsorbent parts. The product of OCTES® (Figure 11) is packaged in a transparent film and a filtration filter, integrating with a filtration filter. So, this kit can be directly placed on the suspended leachate, when detected in *in situ* situation. The detection material changes color depending on the concentration of the target heavy metals. Compare with the color swatch prepared in advance, the concentration of the pollutant can be determined from the change in color of OCTES®. The development of a detection agent that enables anyone to easily determine the presence or absence of heavy metals in a short time, without the use of analytical equipment on site, has made it possible to respond quickly to the cleanup of contaminated soil. However, this simplified detection method cannot accurately quantify the concentration of heavy metals in soil. Therefore, it cannot be used for final determination until it is superseded by an official method, but it has the advantage of quickly confirming measurement results as a monitoring tool for target groups where relatively high concentrations are of concern. It also has the advantage of allowing efficient on-the-spot evaluation of sampling at survey sites; no expensive analytical equipment, no specialized knowledge required and quick and easy measurement for everyone.

Figure 11. Picture of OCTES® kit for in situ detection for pollutants. The inset shows an example of fluorine detection, where the color tone changes from reddish purple to bluish purple due to fluorine

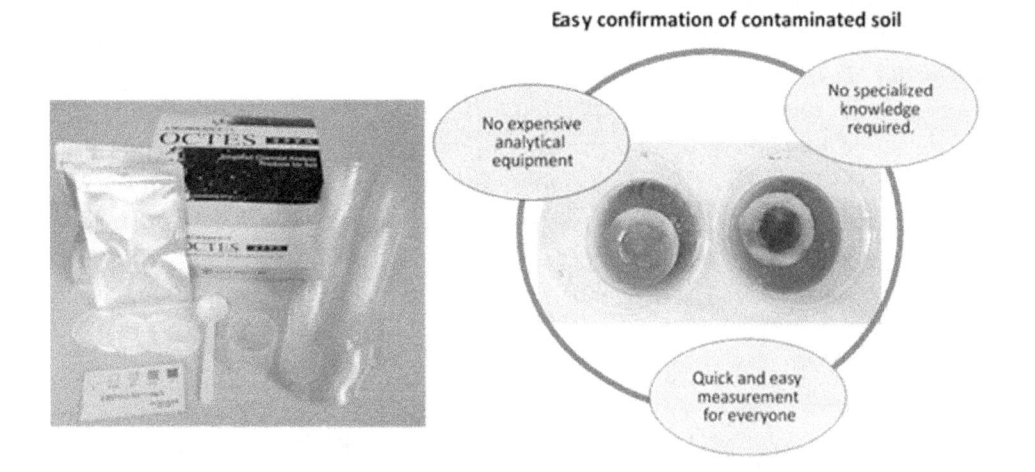

Influence of Coexisting Ions

Soil pH and moisture levels in soil can vary depending on the characteristics of the land and climatic conditions, impacting the leaching state of substances within the soil. Detection materials typically react only to specific substances that are present and allow for qualitative and quantitative analysis. However, when substances coexist with ions in the soil environment, such obstructions interact with each other, interfering with detection. Soil is composed of two main components: inorganic and organic components. The components that primarily influence color are ionized inorganic substances.

Minimizing interference from coexisting ions is desirable; however, a significant challenge in simple methods that rely on color change for determination is the presence of many substances that can ionize and leach out due to variations in pH. In the case of the simplified detection material OCTES® discussed in the previous section, let's consider the example of fluoride and hexavalent chromium. This illustrates the impact of coexisting ions on the detection materials for fluoride and hexavalent chromium. In the fluoride detection material, when fluoride is detected, as explained earlier, it changes color to blue. However, when iron ions (Fe^{3+}) coexist, if the concentration of iron ions exceeds 0.8 ppm, the color of fluoride detection changes from the blue indicating fluoride to red (as shown in the diagram). Furthermore, similar to iron, the coexistence of manganese ions (Mn^{2+}) with a high oxidation state also significantly influences the detection, as shown in the diagram. The influence of manganese ions (Mn^{2+}) results in a reddish tint at around 10 mg/L, causing a bluish-purple color, and the typical blue color indicative of fluoride is not displayed. This can impact the determination of whether fluoride is present or not. Subsequently, when the concentration of manganese ions exceeds 20 mg/L, the detection material undergoes a discoloration, transitioning from the bluish-purple color to a yellowish tint. Figure 12 shows influence of (a)Fe^{3+} (b) Mg^{2+} and (c)Zn^{2+} on OCTES® induced coloration at 2 mg/L fluoride concentration. Even in hexavalent chromium detection as shown in Figure 13, the color changes can be influenced by the presence of lead (Pb^{2+}), tin (Sn^{2+}), and other substances. In soil, ionization of components occurs due to the influence of pH, leading to ionization tendencies among coexisting components. In measurements, it is desirable to obtain stable results even in complex environmental conditions. However, there are still many challenges in simple detection methods, and improvements are needed.

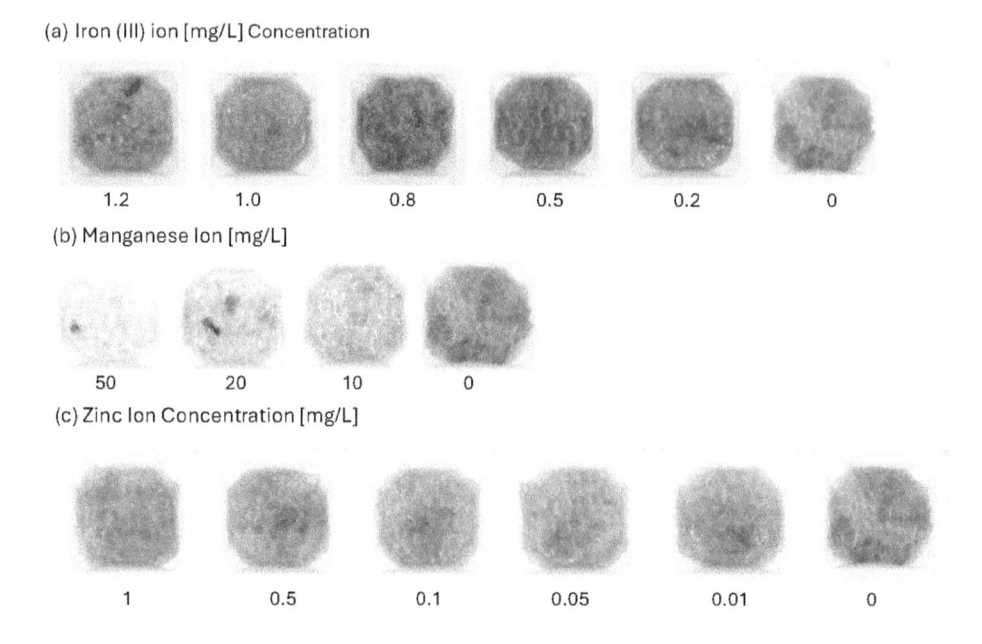

Figure 12. Influence of (a)Fe^{3+}, (b)Mg2+ and (c)Zn2+ on OCTES®-induced coloration at 2 mg/l fluoride concentration (a) In the presence of iron ions, a reddish tint starts to appear at around 0.8 ppm of iron ions, shifting the color from blue. (b) In the presence of manganese ions, the bluish tint of the detection material fades at around 10 ppm of manganese ions. Subsequently, as the concentration of manganese ions increases, the color changes from transparent to a yellowish tint, with a significant impact from coexisting ions. (c) The influence of zinc ions is significant, and the detection material turns from blue to red with the mere coexistence of 0.01 ppm of zinc ions

Figure 13. Influence of (a)Pb^{2+} and (b)Sn^{2+} on OCTES®-Induced coloration at 0.15 mg/l hexavalent chromium concentration (a) When lead ions coexist at around 500 ppm, the detection material takes on a pinkish tint from a flesh color.(b) When Tin Ion coexists at around 10 ppm, the detection material takes on a pinkish tint. As the concentration of Tin Ion increases to 500 ppm or higher, the material exhibits a pink color with a reddish tint, and the impact becomes significant

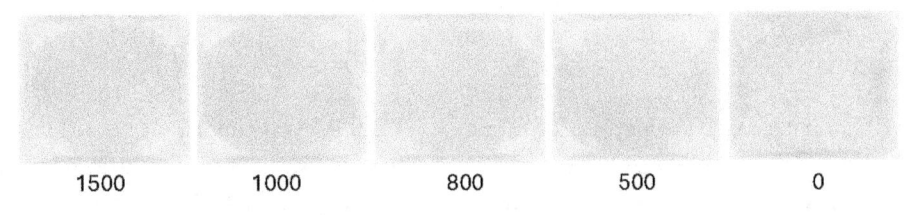

(a) Lead Ion Concentration [mg/L] Concentration

| 1500 | 1000 | 800 | 500 | 0 |

(b) Tin Ion Concentration [mg/L] [mg/L]

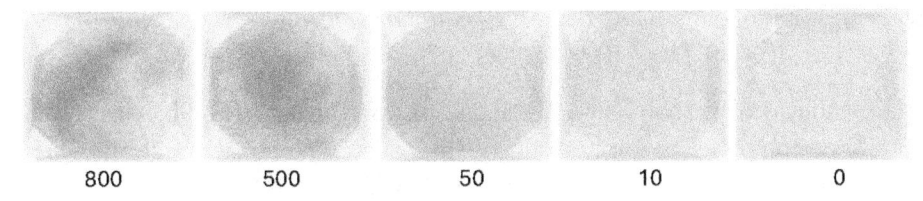

| 800 | 500 | 50 | 10 | 0 |

CONCLUSION

Soil is a natural resource, and humans utilize it for their livelihoods, industrial activities, and development. Unfortunately, soil pollution, primarily caused by heavy metals and other substances, occurs to some extent and is contributing to environmental degradation. Fortunately, soil contains microorganisms and has its own self-purification processes under the atmosphere. However, to protect sustainable natural resources, it is essential to employ the right knowledge and technology for regeneration and treatment. In developing a simple and quick method for detecting heavy metals I contaminant soil, important part is a detection agent in which a highly selective coloring reagent. To enable this, the chromophore part coexistences with nonionic water-absorbing polymer. In the composited medium, the adsorption and filtration are at the same time, occurred in the vicinity of chromophore in the OCTES®, providing easy and quick determination regardless of the presence or absence of serious soil contamination. This system improves the detection accuracy and sensitivity of the detection method.

REFERENCES

Aoshima, K. (2012). Itai-itai disease: Cadmium-induced renal tubular osteomalacia–Current situations and future perspectives-. *Nihon Eiseigaku Zasshi. Japanese Journal of Hygiene*, 67(4), 455–463. DOI: 10.1265/jjh.67.455 PMID: 23095355

Boonkanon, C., Phatthanawiwat, K., Wongniramaikul, W., & Choodum, A. (2020). Curcumin nanoparticle doped starch thin film as a green colorimetric sensor for detection of boron. *Spectrochimica Acta. Part A: Molecular and Biomolecular Spectroscopy*, 224, 117351. DOI: 10.1016/j.saa.2019.117351 PMID: 31336322

Chebotarev, A., Demchuk, A., Bevziuk, K., & Snigur, D. (2020). Mixed ligand complex of lanthanum(III) and alizarine-complexone with fluoride in micellar medium for spectrophotometric determination of total fluorine. *Chemistry and Chemical Technology*, 14(1), 1–6. DOI: 10.23939/chcht14.01.001

Chen, L., Tan, K., Wang, X., & Chen, Y. (2024). A rapid soil Chromium pollution detection method based on hyperspectral remote sensing data. *International Journal of Applied Earth Observation and Geoinformation*, 128, 103759. DOI: 10.1016/j.jag.2024.103759

Chikuma, M., Okabayashi, Y., Nakagawa, T., Inoue, A., & Tanaka, H. (1987). Separation and determination of fluoride Ion by using ion exchange resin loaded with alizarin fluorine blue. *Chemical & Pharmaceutical Bulletin*, 35(9), 3734–3739. DOI: 10.1248/cpb.35.3734

Hou, G., Zheng, J., Cui, X., He, F., Zhang, Y., Wang, Y., Li, X., Fan, C., & Tan, B. (2022). Suitable coverage and slope guided by soil and water conservation can prevent non-point source pollution diffusion: A case study of grassland. *Ecotoxicology and Environmental Safety*, 241, 113804. DOI: 10.1016/j.ecoenv.2022.113804 PMID: 35753270

Hua, F., & Qian, M. (2001). Synthesis of self-crosslinking sodium polyacrylate hydrogel and water-absorbing mechanism. *Journal of Materials Science*, 36(3), 731–738. DOI: 10.1023/A:1004849210718

Kono, H. (2014). Characterization and properties of carboxymethyl cellulose hydrogels crosslinked by polyethylene glycol. *Carbohydrate Polymers*, 106, 84–93. DOI: 10.1016/j.carbpol.2014.02.020 PMID: 24721054

Kono, H., & Fujita, S. (2012). Biodegradable superabsorbent hydrogels derived from cellulose by esterification crosslinking with 1, 2, 3, 4-butanetetracarboxylic dianhydride. *Carbohydrate Polymers*, 87(4), 2582–2588. DOI: 10.1016/j.carbpol.2011.11.045

Kono, H., Hara, H., Hashimoto, H., & Shimizu, Y. (2014). Nonionic gelation agents prepared from hydroxypropyl guar gum. *Carbohydrate Polymers*, 117, 636–643. DOI: 10.1016/j.carbpol.2014.09.085 PMID: 25498682

Kono, H., & Zakimi, M. (2013). Preparation, water absorbency, and enzyme degradability of novel chitin-and cellulose/chitin-based superabsorbent hydrogels. *Journal of Applied Polymer Science*, 128(1), 572–581. DOI: 10.1002/app.38217

Liu, M., & Guo, T. (2001). Preparation and swelling properties of crosslinked sodium polyacrylate. *Journal of Applied Polymer Science*, 82(6), 1515–1520. DOI: 10.1002/app.1990

Martínez-Vargas, S., Gómez-Tagle, P., & Yatsimirsky, A. K. (2011). Alizarin complexone-lanthanide(III)-fluoride system: Revised speciation and the origin of the analytical signal. *Inorganica Chimica Acta*, 373(1), 226–232. DOI: 10.1016/j.ica.2011.04.024

Mingnon, A., De Belie, N., Dubruel, P., & Van Vlierberghe, S. (2019). Superabsorbent polymers: A review on the characteristics and applications of synthetic, polysaccharide-based, semi-synthetic and 'smart' derivatives. *European Polymer Journal*, 117, 165–178. DOI: 10.1016/j.eurpolymj.2019.04.054

Ministry of the Environment, Government of Japan. (2017). Soil contamination countermeasures act, Amendment of act No. 45 of June 2, 2017. https://www.env.go.jp/content/900452902.pdf

Ministry of the Environment, Government of Japan. (2021). Survey results on the implementation status of the soil contamination countermeasures act and case studies of soil contamination surveys and remediation. Report, 40–44. https://www.env.go.jp/water/report/r3-01/index_r4.html

Ministry of the Environment, Government of Japan. (2021). Survey results on the implementation status of the soil contamination countermeasures act and case studies of soil contamination surveys and remediation. Report, 33–36. https://www.env.go.jp/water/report/r3-01/index_r4.html

Ministry of the Environment, Government of Japan. (2023). Areas requiring measures under the soil contamination countermeasures act and areas requiring notification of changes in characteristics under the act. Report, 1–140. https://www.env.go.jp/content/000162541.pdf

Oxspring, D. A., McClean, S., O'Kane, E., & Smyth, W. F. (1995). Study of the chelation of boron with Azomethine H by differential pulse polarography, liquid chromatography and capillary electrophoresis and its analytical applications. *Analytica Chimica Acta*, 317(1-3), 295–301. DOI: 10.1016/0003-2670(95)00394-0

Situ, Y., Yang, Y., Huang, C., Liang, S., Mao, X., & Chen, X. (2023). Effects of several superabsorbent polymers on soil exchangeable cations and crop growth. *Environmental Technology & Innovation*, 30, 103126. DOI: 10.1016/j.eti.2023.103126

Xu, D., Wang, Z., Tan, X., Xu, H., Zhu, D., Shen, R., Ding, K., Li, H., Xiang, L., & Yang, Z. (2024). Integrated assessment of the pollution and risk of heavy metals in soils near chemical industry parks along the middle Yangtze River. *The Science of the Total Environment*, 917, 170431. DOI: 10.1016/j.scitotenv.2024.170431 PMID: 38301773

Yamashita, K., Nakada, T., Kawasaki, H., & Arakawa, R. (2016). Characterization of chemical species in lanthanide(III)-alizarin complexone (ALC)-fluoride solution using ESI-MS. [Lexical characteristics of Japanese language]. *Bunseki Kagaku*, 65(1), 39–43. DOI: 10.2116/bunsekikagaku.65.39

Zhao, C., Zhang, M., Liu, Z., Guo, Y., & Zhang, Q. (2019). Salt-tolerant superabsorbent polymer with high capacity of water-nutrient retention derived from sulfamic acid-modified starch. *ACS Omega*, 4(3), 5923–5930. DOI: 10.1021/acsomega.9b00486 PMID: 31459741

About the Contributors

Rohana Adnan graduated from Clarkson University, New York in 1994. She then pursued her Doctoral study in University of Southampton, UK. She is currently a Professor in Physical Chemistry at the School Chemical Sciences, Universiti Malaysia (USM). Her research interest is on wastewater and wastewater treatment focusing mainly on adsorption, Fenton oxidation process, photocatalysis especially using visible light active photocatalysts, and biodegradation. Dyes, antibiotics, heavy metals and phenolic compounds are among the pollutants that she has worked on. Currently she and her research team is working on different methods and materials for the treatment of palm oil mill effluent.

Tolga Aytug is a senior research staff member of the Chemical Sciences Division and distinguished UT-Battelle inventor at ORNL. He has more than 25 years of experience in processing of advanced materials using both physical and chemical vapor deposition approaches as well as advanced materials characterization and property correlations for process optimization. His primary research interests are fabrication of thin film heterostructures using both physical vapor and chemical deposition approaches, thermodynamic and kinetic effects on phase nucleation and structure formation as well as development of unique nanostructured material systems and composites for various energy technologies, and novel approaches for the development of transparent and non-transparent nanostructured superhydrophobic coatings. He has authored and co-authored over 100 publications and has written two book chapters. He has 18 issued and 25 more pending US patents. He has commercialized and licensed his technologies to 6 companies. He has received numerous awards of excellence including 2007, 2010, 2012, 2014, 2015, and 2021 R&D100 Awards; UT-Battelle Inventor of the Year Award (2019); 2008, 2010, 2017 National Federal Laboratory Consortia Excellence in Technology Transfer Awards; 2007 Southeast Federal Laboratory Consortia Award; 2014 ORNL Research Accomplishment Award; and 2006 and 2017 ORNL Excellence in Technology Transfer Awards.

Siti Baidurah received her PhD degree in bioscience and biotechnology in 2015 from Chubu University in Japan. Upon graduation, she has been working in Universiti Sains Malaysia at the Bioprocess Technology Division, School of Industrial Technology. Her research focuses on the production and characterization of biodegradable polymers such as polyhydroxyalkanoates and polylactic acids. Recent research is focusing on the bio-valorization of wastes such as palm oil wastes, molasses, and other agro-industrial residue into value-added products such as biodegradable plastics and solid biomass fuel via fermentation technology. In addition to research activities, she has peer-reviewed papers over 50 manuscripts related to the fields of biopolymer and fermentation technology including biomass waste valorization, which are aligned with SDGs. Two monographs are published related to palm oil waste treatment in Malaysia. She participated in local and international conferences as an invited, keynote, and plenary speaker at 11 events from 2012-2023. One completed MoU with Nagaoka University of

Technology and Takasago Thermal Engineering which focused on treatment of palm oil wastes. Research grants obtained are as follows: 1 ongoing grant, 5 completed grant. Graduated 16 students and current postgraduate supervision is 4 students. Honors and Awards are followed: 1) Presenter Award, International Conference on Polymer Analysis and Characterization 2010, 2) Research Encouragement Award 2011, Research Conference of Materials Life Society, Japan, 3) Yokoyama International Scholarship 2013, Japan, 4)Certificate of Commendation 2011 and 2012, Chubu University, Japan.

Ahmad Fadhil Bin Rithwan is currently working as MyStep in the Domestic, Asset Management and Maintenance at Institut Latihan Kehakiman Dan Perundangan (ILKAP), Bandar Baru Bangi. He completed his BSc. (Hons.) with a major in Chemistry and minor in Food Technology from Universiti Sains Malaysia, Malaysia. His final year project is on "The Photocatalytic Degradation of Antibiotic Using AgI/ZnO". Besides, he is waiting for his Master's degree Viva Voce in Universiti Sains Malaysia with research work entitled "Kinetics and Mechanism of Photocatalytic Degradation of Emerging Pollutants Using Silver Metal /Reduced Graphene Oxide/Bismuth Ferrite Nanocomposite. He received the Best Research Presentation Award in 5th STI-Gigaku 2020 conference organized by Nagaoka University of Technology, Japan. He has been focusing on these past few years were photocatalytic degradation, antibiotic pollutants, and nanomaterials.

Karla Lizette Tovar Carrillo received her Bachelor Degree in Chemistry in Universidad Autónoma de Cd. Juárez (1998-2003) and then Master Degree in Materials Science (2005-2008). She held her Doctor Degree in Engineering at Nagaoka University of Technology (2010-2014) as title of "Study on Cellulose hydrogel films regenerated from natural plant bagasses and evaluation of their cyto-compatible properties for tissue engineering". She started her research in Universidad Autónoma de Cd. Juárez from 2014 to present in the fields of medical materials concerning with cellulose hydrogels.

Tapanee Chuenkaek received her Bachelor's Degree from the Department of Science with a major in Materials Science, focusing on Polymer and Textiles, at Chulalongkorn University, Thailand, in 2018. Following her undergraduate success, she belongs currently a Ph.D. course for the 5-year integrated program in the Department of Science of Technology Innovation at Nagaoka University of Technology, Japan. She is a member of the Environment-friendly Material Function and Engineering Laboratory. Her research interests lie in the area of biomaterials, especially in pectin functional materials.

Shuvodeep De is Postdoctoral Research Associate of the Multiscale Materials Group in the Computational Sciences and Engineering Division in Oak Ridge National Laboratory. Shuvo received his doctorate degree from Virginia Tech in Engineering Science and Mechanics with a focus on Finite Element Modeling and Multidisciplinary Design Optimization. Subsequently, he worked as a Postdoctoral Fellow in the Department of Chemical and Biological Engineering at The University of Alabama (Tuscaloosa). At Alabama his research focus was Multiphysics Simulation, Electrodeposition and Superconducting materials. He is currently focusing on Additive Manufacturing, Hybrid Manufacturing, Nanocomposite Materials, Physics-informed Machine Learning and Optimization. Apart from academics, Shuvo has interest in music, arts, photography and traveling.

Hitomi Fukaura received her BS from National Institute of Technology, Ariake College. She researched the recovery and reuse of phosphoric acid in wastewater. She was involved in quality control at a paint company, and now she is engaged in quality management and product development at her current company.

Maki Horikawa achieved her Bachelor of Engineering degree and master's degree in Department of Applied Chemistry & Biochemistry from Kumamoto University, Japan, in 2003 and 2005 respectively. She received her Ph. D degree from Kumamoto University, Japan, in 2016. She has been working at Kumamoto Industrial Research Institute, Japan, since 2010. Her research interests include the synthesis of functional polymers for industrial applications such as conductive polymers and thermoelectric conversion polymers.

Ayano Ibaraki received her bachelor's degree in Materials Science and Bioengineering from the Faculty of Engineering at Nagaoka University of Technology in 2018. After completing her undergraduate studies, she pursued a 5-year integrated doctoral degree program in the Department of Science of Technology Innovation at Nagaoka University of Technology. Her current research focuses on the utilization of natural polymers extracted from waste materials for the development of bioplastics and hydrogel materials.

Harshani Iresha is a young academic in Sri Lanka's number one university, the University of Peradeniya. She works attached to the Department of Chemical and Process Engineering at the faculty of Engineering. Previously, she was a Lecturer (on contract) in the Department of Chemical and Process Engineering, University of Moratuwa, Katubedda, Sri Lanka, from September 2014 to August 2016. Iresha graduated with a Doctor of Engineering, in March 2022 from Nagaoka University of Technology, Nagaoka, Japan. Her study was based on 'ultrasound-stimulated behaviors on biomass hydrogels'. There, she mainly focused on drug release and viscoelasticity behavior of biomass hydrogels upon ultrasound trigger. Her Master of Engineering degree was obtained from the same university in August 2018. During her studies in Nagaoka, Japan, her studies were published as reputed journal articles, a few book chapters, and many conference proceedings nationally and internationally which led to conference awards. Iresha's basic degree was obtained from the Department of Chemical and Process Engineering, University of Moratuwa, Sri Lanka, in March 2014. With her curiosity about how products come to our hands and how the amazing material 'polymer' behaves in a diverse range of applications, she selected chemical and process engineering for her undergraduate studies. Her undergraduate research was basically on biomass filler for natural rubber. As a young academic, her research focus is laying towards the development of sustainable materials for advanced engineering applications such as ultrasound drug release, the natural rubber industry, ayurvedic medicine in Sri Lanka, and many other applications in Chemical Engineering.

Nursarah Sofea Binti Ismail currently holds the position of Protégé in the Research and Development division at the Sime Darby Plantation Technology Centre. She attained a BSc. (Hons) major in Chemistry and a minor in Management, from Universiti Sains Malaysia (USM). Her final year project focused on the topic of "Visible Light-Induced Photocatalytic Degradation of Norfloxacin Antibiotic using Silver/Bismuth Ferrite/Reduced Graphene Oxide (Ag/BFO/rGO) Nanocomposites." In her current role, Nursarah is actively engaged in using a variety of equipment, including homogenizers, high-pressure homogenizers (HPH), spray dryers and particle size analyzer (PSA), for the encapsulation of active ingredients. Nursarah also has exposure to basic chromatographic systems and sample preparation techniques for the determination of small molecules, particularly amino acids and is highly proactive in participating in the preparation of samples for metabolomics. Nursarah expresses a keen interest in environmental, safety, and health (ESH) tasks that related to safety data sheet (SDS) documentation and updating Hazard Identification, Risk Assessment, and Control measures (HIRAC).

Huixin Jiang is a Postdoctoral Research Associate in the Chemical Science Division at Oak Ridge National Laboratory (ORNL). She received a Ph.D. in Energy and Environment Science from Nagaoka University of Technology, Japan in 2017. Her research interests are material development and characterization, fabrication of ultraconductive film composites, fabrication of biocidal filters to improve indoor air quality, biomass conversion, and CO_2 capture. She has received numerous awards of excellence including the Honors Scholarship of Japan Student Services Organization (JASSO), First place in Donghua University Comprehensive Scholarship, First place in Donghua University Scholarship, Excellent student, Academic Excellence Award.

Tomoyuki Katayama, born in Japan in 1959, majored in machine engineering and received the B.E. degree from Kyoto Institute of Technology, Kyoto, Japan in 1982. He started his career in 1982 as a mechanical engineer for ShinMaywa Industries, Ltd., Kanagawa, Japan. He was mainly engaged in design of waste treatment device during his 35-year tenure in the company. After leaving ShinMaywa, he started working with Mr. Shinichi Shimose, the implementor of ERS (Environmental Recycling System) manufactured by JET Corporation, Tokyo, Japan, and has been engaging in design of ERS as the Chief Operating Officer of JET. He is the Professional Engineer in Environmental Engineering certified by the Institution of Professional Engineers, Japan.

Noor Haida Mohd Kaus received her Ph.D. degree in Chemistry from University of Bristol, United Kingdom, in 2012 and currently an Associate Professor in the School of Chemical Sciences, Universiti Sains Malaysia (USM). She has been cross appointed as an Associate Professor at Nagaoka University of Technology (NUT), Japan with substantial involvement in teaching and research with the Department of Science of Technology Innovation. Her current research focuses mainly on the development of eco-friendly nanomaterials for environmental protection and energy generation, securing over RM3.0 million in international and national research grants. She and her team has published numerous works in reputable journals, obtained copyrights, and delivered keynotes and invited speeches at conferences worldwide.

Hiroyuki Kono received his BD and MD from Hokkaido University in 1993 and 1995, respectively, joined the Hokkaido Research Organization, Industrial Research Institute, and then moved to Bruker BioSpin Co. Ltd. During that time, he received his PhD from Hokkaido University in 2003. He was appointed Associate Professor at the National Institute of Technology in 2007 and was promoted to Professor in 2019. He engages in functionalization of structural polysaccharides (mainly cellulose, chitin, and chitosan) and elucidation of their structure-function relationship. Awards: Hayashi Jisuke Award from the Cellulose Society of Japan (2017), Technology Development Award from the Japanese Society of Applied Glycoscience (2020).

Anh PhuongLe Thi received her bachelor's degree in Environmental Technology, Ho Chi Minh University of Science in 2018. After graduation, she continued with her 5th years Doctoral Degree program in Science of Technology Innovation, Nagaoka University of Technology. Currently, she has worked environmental remediation technology in fibrous composites consisting of inorganic mineral powder and polymer and implemented research works on water treatment and elimination technology of toxic gaseous chemicals.

Kai Li is a R&D associate staff in Multifunctional Equipment Integration Group in Buildings and Transportation Science Division. He received his M.S. degree (2013) in Polymer Chemistry and Physics from Zhengzhou University, and a Ph.D. degree (2016) in Energy and Environment Science from Nagaoka University of Technology. His research focuses material and process development on the carbon capture, moisture management, heat exchangers, biocomposites, and advanced conductor. He has more than 10 years of experience in material and process development for various applications to reduce carbon footprint and provide energy-saving solutions. Dr. Li has co-authored more than 50 peer reviewed journal articles. Li is also the recipient of two R&D 100 awards (2020 and 2021) and one UT-Battelle awards.

Shoji Nagaoka achieved his Bachelor of Engineering degree and master's degree in Department of Applied Chemistry & Biochemistry from Kumamoto University, Japan, in 1985 and 1988 respectively. He received his Ph. D degree from Kumamoto University, Japan, in 1994. He has been working at Kumamoto Industrial Research Institute, Japan, since 1995. He is engaged in the development of carbon-neutral materials through various collaborative researchwith industry, aca demia and government. He is particularly focused on cellulose and its related materials, the main element of plants, and is actively conducting research on the development of attractive materials that contribute to the SDGs. He is currently a senior researcher at the Kumamoto Industrial Research Institute, a visiting professor in Innovative Development Organization of Kumamoto University, and a part-time lecturer at Nagaoka University of Technology.

Keita Nakajima received his Bachelor's degree in Materials Science from Nagaoka University of Technology in 2021. After graduation, he continued with his Ph.D. degree in Science of Technology Innovation at Nagaoka University of Technology. He is studying about biomaterial engineering. His research interests are biomass materials, cosmetics, ultrasound, and biomaterials. The major application of his study is medical industry.

Kashif Nawaz (PI) is the section head of Building Technologies Research and has 15 years of experience in fundamental and applied energy conservation science and technology. He is widely recognized for his work in building heating, cooling, and dehumidification systems, including novel heat exchangers and enhanced phase change processes. Nawaz has authored or coauthored more than 170 peer-reviewed publications including journal articles, conference papers and reports. Nawaz is the recipient of the ASHRAE Distinguished Service Award (2018) and Exceptional Service Award (2022), as well as two R&D 100 awards (2021 and 2022). He has authored over 150 publications on topics related to advanced heat pumps, low–global warming potential (GWP) refrigerants, novel heat exchangers, moisture management, and direct air capture.

Dang Khoa Nguyen received BSc. Degree in Environmnetal Technology, Ho Chi Minh University of Science in 2014 and earned PhD degree in Science of Technology Innovation, Nagaoka University of Technology in 2020 under the supervision of Prof. Dr. Takaomi Kobayashi. Currently, he is Head of Department of Master degree program in Environment at Faculty of Environment, Van Lang University and his research interests is focus on industial and by-product or agro-waste regeneration to functional materials for environmental treatments. Till now, he has done 6 projects, published 15 papers, and reviewed for different prestigious journals indexed in ISI/ Scopus.

Yuta (Watanabe) Nikaido currently has worked in Tama Chemical Co. Ltd, since 2023 and been researching applications of inorganic materials like colloidal silica for semiconductor fields. He received his doctoral degree of Engineer at the Department of Science and Technology Innovation, Nagaoka University of Technology, on 2023 on the basis of "Effect of Alkali Activated Agents on Geopolymers and Their Properties and Application for Lead Immobilization."

Sujitra Onutai received her Bachelor's in Materials Science from Chiangmai University in 2009. After graduation, she continued with her Master's Degree in Materials Science at the Faculty of Science, Chulalongkorn University. In 2013, she was a doctoral student in the Double Doctoral Degree Program in collaboration with Chulalongkorn University and Nagaoka University of Technology. She graduated doctoral degree in 2016 and 2018. She studied geopolymer materials and holds Master's and Doctoral degrees. She is researching the structural transition of alkali-activated materials at Japan Atomic Energy Agency, Japan.

Soydan Ozcan is Distinguished Research Scientist and the Group Leader for the Sustainable Manufacturing Technologies Group at Oak Ridge National Laboratory (ORNL). His research aims to addresses the broad and vital issue of identifying novel, high-value biomaterials from renewable sources and viable processes for preparing composite and additive manufacturing applications for zero waste, and develop manufacturing techniques and exploring new materials to improve energy efficiency during composite manufacturing, decrease material waste, and improve material cost and performance. Ozcan is also a prolific researcher who has published over 100 papers, holds 24 issued and pending patents, has published 11 book chapters, and has been an active speaker with over a hundred presentations and short courses on sustainable manufacturing materials-related topics and research. Ozcan is also the recipient of 2023 TAPPI Nano Technical Award, two R&D 100 awards (2020 and 2021), 2021 Battelle Memorial Institute Distinguished Inventor.

Saurabh Prakash Pethe is a PhD candidate at The University of Tennessee Oak Ridge Innovation Institute's (UT-ORII's) Bredesen center for Interdisciplinary Research and Graduate Education. He is working under the guidance of Dr. Parans Paranthaman at the Chemical Sciences Division at Oak Ridge National Laboratory (ORNL). Before joining the Bredesen center he earned his master's degree in mechanical engineering at the University of Tennessee under the guidance of Dr. Uday Vaidya. His research during the masters was focused on thermoplastic impregnation of textile-grade carbon fiber. Prior to that he worked in the composite pressure vessel and hazardous area electrical equipment industries for 4 years. He obtained his B.S from Pune University in mechanical engineering. His research focus is on recycling of lithium-ion batteries, lithium extraction, high voltage cathodes and sodium-ion batteries.

Phuoc Tri Phan received his bachelor's degree from the Environment Faculty of the University of Science, Vietnam National University Ho Chi Minh City, Vietnam in 2017. Since April 2018, he continued to pursue his graduated study in the 5-year Integrated Doctoral Program from the department of Science of Technology Innovation from the Nagaoka University of Technology, Japan. He finished his doctoral course from the Nagaoka University of Technology on 2023. During his doctoral studies, his main research was focused on utilizing low-frequency, high-power ultrasound for solid waste treatment and valuable elemental recovery from industrial waste. On 2023, he has started worked in JET Corporation and currently, his work is focused on applying high-speed composting technology for various types of waste treatment.

NganPhan Thi Thu completed her Bachelor of Science degree in Environmental Technology from Ho Chi Minh University of Science, Viet Nam. Following her undergraduate studies, she pursued her Ph.D. as a candidate in the 5-year integrated program at the Department of Science and Technology Innovation at Nagaoka University of Technology in Japan. She is an active member of the Environment-friendly Material Function and Engineering Laboratory in Nagaoka University of technology. Her research primarily focuses on fabricating inorganic-polymer composite fibers and exploring their potential applications as environmentally friendly adsorbents for remediation projects.

Minoru Satoh received his Doctor of Science from Tohoku University in 1991. He worked at Yamagata Research Institute of Technology, as a Researcher, and Ibaraki National College of Technology, Department of Industrial Chemistry, as an assistant professor, an associate professor, and then a professor. He became a professor in the Department of Materials Science and Technology, Nagaoka University of Technology in 2014. Since 2015 he has been working as a professor in the Division of Department of Industrial Engineering, National Institute of Technology (KOSEN), Ibaraki College. He has been studying the removal of heavy metal ions by using some ligands.

Shinichi Shimose received the B.E. degree from Kitasato University, Kanagawa, Japan in 1978 with his academic research themes "Removal of lipid from high-catch fish, TLC (Thin Layer Chromatography) separation identification test on fish oil, and quantitative test using a spectrometer on fish oil." He started his career in 1978 at Shimose Foods, Co., Ltd., a fish processing company, Yamaguchi, Japan. During his tenure at Shimose Foods, he collaborated with Kyushu University on joint research related to the recovery of proteins and oils from fish broth using UF (Ultra Filtration) and RO (Reverse Osmosis) membranes, and the concentration of fish extract using RO membranes. He was engaged in efforts to produce edible products from fish processing residues. In 1984, he worked with several food processing companies as a technical advisor and engaged in the engineering work for establishing an environmentally-friendly food processing plant. Using all the knowledge and experiences accumulated through his academic and business career, he started environmental recycling business, developing ERS (Environmental Recycling System), a high-speed bioreactor using vacuum drying technology. Since 2001 to 2006, he also worked as an environmental engineering expert of Asian Productivity Organization to develop the treatment and application method of wastes in Singapore and Fiji. In Singapore, a joint research project with Nanyang Technological University was conducted for the efficient and environmentally friendly processing of unsorted urban waste containing plastics. Also, he provided a lecture on the mechanism and case studies on ERS at a workshop in Indonesia. He founded JET Corporation in 2013 and Shimose Microbes Laboratory Corporation in 2016, and has been working as the Chief Technical Officer for the development of environmentally-friendly recycling technologies and apparatus.

Tomohiro Shirosaki achieved his Bachelor of Engineering degree and master's degree in Department of Applied Chemistry & Biochemistry from Kumamoto University, Japan, in 2000 and 2002 respectively. He received his Ph. D degree from Kumamoto University, Japan, in 2005. He has been working at

Kumamoto Industrial Research Institute, Japan, since 2009. His research interests include the preparation of functional organic materials for industrial applications such as drug delivery system, cosmetics and organic solar cells.

Ryo Shoji received his PhD from University of Tokyo in 2000. His doctoral thesis was entitled as "Development of rapid and easy detection system for cytotoxicity of environmental water". He joined the Institute of Industrial Science, University of Tokyo as a postdoctoral researcher. He then was appointed Associate Professor at the National Institute of Technology, Tokyo College in 2000 and was promoted to Professor in 2021. During that time, he spent one year as an overseas researcher at the Department of Environmental and Urban Engineering at the University of Delaware, USA. Now, he engages in bioassay for evaluating the quality of environment (mainly soil and water) and remediation using eco-friendly materials. Awards: Excellent paper award from the Society of Bioengineering Japan (2003), etc.

Hannah Snider is a chemical engineering and chemistry undergraduate student who is anticipating her graduation in May of 2024. Throughout the summer of 2023, Snider had the opportunity to conduct a 10-week summer internship with Oak Ridge National Laboratory, where she learned more about cellulose nanofibers and biocomposites. Her past research experiences have focused on green chemistry and materials science.

Yasunori Takaki was researching material development for motors at university. After graduating, he joined an automotive-related company and was involved in launching and improving product production lines. At his current company, he is involved in developing new products and building production lines.

Sirikanjana Thongmee now is an Assistant professor of Physics, She was in Physics Department Faculty of Science, Kasetsart University. She got her B.Sc in Physics at Prince of Songkla University, M. Sc. in Chemical Physics at Mahidol University and Ph.D. (Materials Science) at National University of Singapore. Dr. Sirikanjana Thongmee got the Thesis Presentation Award, Mahidol University, Thailand, 1999 and Outstanding Research of the Year 2nd Class Award, Office of the National Research Council of Thailand, Thailand, 2003. Currently Dr. Sirikanjana Thongmee's researches focus on the metal doped ZnO for spintronics and gas sensors applications, magnetic nanomaterials, graphene oxide for different applications and activated carbon from agricultural wasted.

Tu Tran Vo has expertise in hydrogel materials using biomass materials, especially her doctoral degree from Chulalongkorn University in COLLAGEN Nanogel-CHITOSAN Hydrogel Composites and a degree in radiation-enhanced green hydrogels.

Truong Thi Cam Trang was a lecturer in University of Economic and Technology Industry (2002-2011) and then received PhD degree in Energy and Environment Science, Nagaoka University of Technology in 2011. Then she started her education and research study in Vietnam National University-Ho Chi Minh (VNUHCM) - University of Science as lecture and now she is senior lecturer in department of Environment. She has been teaching in VNUHCM - Ho Chi Minh University of Science for 20 years at Faculty of Environment. She is working on researches related to biomass functional materials of environmental water treatments, especially in chitin and chitosan composites for remediation technologies to environmental pollution. Till now, she has done 6 projects, published 25 in international and domestic journals.

Reiko Wakasugi received her D.Eng. from Kumamoto University in 2004. Her doctoral thesis was entitled as "Recovery of Volatile Organic Compounds by Dual Reflux Pressure Swing Adsorption". In 2006, she was employed as an assistant at Yatsushiro National College of Technology (later reorganized as Kumamoto National College of Technology in 2009), and she was promoted to associate professor in 2021. She specializes in the field of chemical engineering, particularly in the area of adsorption. Her research focuses on adsorbents and adsorption separation processes, especially the Pressure Swing Adsorption process. In recent years, she has been conducting research and development on the application of natural resources to environmental materials. Additionally, she is working on enhancing the added value of local specialties and resources in the region. She has received the Poster Award from the Japan Adsorption Society in 2003.

Siti Fairus Mohd Yusoff, Associate Professor Dr., serves as a Senior Lecturer in the Department of Chemical Sciences, Faculty of Science and Technology at Universiti Kebangsaan Malaysia (UKM). She attained a Bachelor's Degree in Chemistry with First Class Honors from Universiti Teknologi Malaysia (UTM) and subsequently earned a Doctor of Philosophy degree from the University of Bristol, United Kingdom. With expertise in Inorganic Polymer Chemistry, she has been a faculty member at UKM since 2007. Dr. Siti Fairus is actively involved in rubber research, having secured research grants exceeding RM2 million. Additionally, she frequently contributes to scientific articles, research books, general books, and Science Magazines.

Guillermo Zarzosa received his Bachelor's Degree in Medicine from the Autonomous University of Juarez City (UACJ) in 2020. During his bachelor's degree, he was awarded with the national prize in Mental Health hosted by FCCyTAC in 2017. In 2020, he was accepted into the 5 year integrate doctoral program of Science of Technology Innovation, currently in its 4th year. During this time he participated and was awarded in different innovation contests around the world USA/Mexico (TrepCamp), Canada (Social Business Creation Competition), and Japan (Business Model Competition). Currently, his objective is to develop bioproducts extracted from insects to improve society's well-being. His current research field is insect chitin, having published a paper on Zophobas Morio's chitin properties during its life cycle.

Xianhui Zhao is an R&D Associate Staff member of Oak Ridge National Laboratory, USA. He received a PhD in Agricultural, Biosystems and Mechanical Engineering from South Dakota State University in 2015. His research interests and expertise fall in the general areas of material development and characterization, biomass conversion to polymer composite and biofuel, waste recycling and upcycling, circular economy, CO2 utilization, 3D printing, and applied catalysis. He holds numerous prestigious awards including National Aspiration Scholarship, Outstanding Service Award, and Excellent Graduate Leader.

Index

A

adsorbent 19, 24, 30, 179, 180, 191, 192, 193, 195, 197, 220, 231, 371, 372, 373, 374, 376, 377, 378, 381, 382, 383, 384, 385, 387, 388, 389, 390, 392, 396, 397, 404, 405, 409, 410, 418, 419, 457, 496, 501, 503, 504, 510, 511, 512, 514, 523, 530

adsorption 19, 20, 22, 23, 28, 47, 48, 53, 106, 148, 170, 172, 178, 179, 180, 181, 182, 183, 184, 186, 188, 189, 190, 191, 192, 193, 194, 195, 196, 198, 200, 204, 207, 219, 220, 228, 233, 239, 240, 263, 322, 370, 371, 372, 373, 374, 375, 376, 377, 378, 379, 380, 381, 382, 383, 384, 385, 386, 387, 388, 389, 390, 391, 392, 393, 394, 395, 396, 397, 400, 402, 404, 405, 406, 407, 408, 409, 410, 412, 415, 416, 417, 420, 421, 422, 423, 457, 458, 461, 463, 464, 466, 468, 474, 477, 484, 500, 501, 502, 503, 504, 510, 512, 523, 533

Agriculture waste 285, 288

Alumino-silicate 487, 488, 489, 490, 491, 492

B

Bacterial Cellulose 12, 59, 93, 96, 109, 115, 119, 120, 122, 124, 132, 133, 139, 140, 141, 142, 143, 144, 145, 146, 147, 148, 149, 150, 151, 152, 153, 154, 179, 192, 318, 337, 338, 339, 346, 347

Biocompatibility 14, 27, 39, 45, 47, 48, 49, 50, 51, 57, 106, 136, 170, 171, 172, 174, 204, 209, 210, 211, 212, 213, 222, 227, 230, 238, 270, 273, 274, 302, 303, 306, 321, 322, 323, 324, 327, 329, 330, 333, 334, 337, 340, 341, 345, 346

bioconversion 167, 261, 269, 277, 366

biodegradable 7, 25, 34, 35, 39, 41, 49, 57, 58, 62, 63, 101, 105, 109, 138, 143, 147, 149, 156, 157, 158, 160, 162, 164, 165, 166, 167, 168, 176, 177, 184, 188, 190, 193, 194, 195, 199, 204, 207, 208, 219, 233, 235, 269, 270, 274, 278, 281, 282, 283, 284, 304, 307, 316, 317, 329, 342, 345, 352, 411, 457, 534

Biological Process 120, 249, 271

Biomass fuel 168, 283, 351

Biomass Hydrogels 203, 321, 324, 327, 328, 338, 339

Biomass Materials 11, 12, 21, 322, 401

Biomass Waste 1, 155, 156, 157, 164, 272, 402

Biomedical Applications 57, 100, 169, 171, 199, 209, 210, 217, 231, 233, 236, 238, 301, 304, 306, 308, 315, 317, 329, 342

Biopolymer 55, 119, 120, 139, 144, 162, 163, 166, 167, 170, 195, 219, 242, 255, 257, 272, 288, 396

Biorefineries 12

Bioremediation 20, 22, 23, 399, 400, 401, 412, 413, 417, 418, 420, 422, 423, 424, 438, 452, 480

Biosorption 181, 189, 400, 404, 405, 406, 412, 415, 418, 419, 420, 421, 423, 424

Biosynthesis 31, 55, 96, 120, 122, 123, 124, 139, 142, 150, 152, 161, 168, 246, 283, 294, 308, 320

biovalorization 158, 162

C

carbon-based materials 174, 457, 464, 465, 468, 469

carbon neutral 11, 62, 88, 271

Cellulose fibers 13, 14, 32, 33, 35, 37, 38, 39, 40, 42, 43, 50, 58, 59, 60, 61, 79, 138, 272, 322, 342

Cellulose Hydrogels 25, 33, 46, 47, 49, 50, 55, 56, 57, 226, 263, 322, 324, 329, 332, 336, 338, 344, 346, 347, 534

Chemical Composition 52, 433, 489

Chemical Treatment 37, 39, 40, 96, 101, 103, 120, 249, 408

Chitin Biomass 241, 245

Chitosan 12, 26, 32, 45, 46, 52, 53, 56, 57, 64, 88, 93, 97, 98, 112, 159, 160, 166, 168, 169, 170, 171, 172, 173, 174, 175, 176, 177, 178, 179, 180, 181, 182, 183, 184, 185, 186, 187, 188, 189, 190, 191, 192, 193, 194, 195, 196, 197, 198, 199, 200, 201, 203, 204, 205, 206, 207, 208, 209, 210, 211, 212, 213, 214, 216, 217, 218, 219, 220, 221, 222, 223, 224, 226, 227, 228, 229, 230, 231, 232, 233, 234, 235, 236, 237, 238, 239, 249, 253, 254, 255, 256, 257, 259, 260, 261, 263, 264, 265, 266, 267, 271, 272, 304, 329, 331, 339, 341, 342, 344, 345, 411, 412, 420, 421, 422, 423

Chitosan hydrogel 52, 171, 203, 204, 209, 212, 218, 219, 220, 221, 223, 226, 227, 230, 231, 233, 234, 235, 239, 339

Cupriavidus necator 150, 156, 158, 166, 167, 168, 282, 283

D

Decontamination 26, 27, 28, 193, 369, 370, 371, 373, 376, 377, 378, 380, 383, 384, 390, 392, 393, 400, 438, 443, 474, 479, 510, 511

drug delivery systems 45, 49, 53, 204, 207, 221, 222, 225, 226, 229, 230, 231, 233, 236, 237, 238, 239, 274, 306, 325, 341, 346, 347, 451

E

Eco-Materials 7, 8, 10, 21, 25
emerging pollutant 457
Environmental Materials 1, 10, 20, 21, 152
e-Waste 427, 435, 445, 447, 448, 450

F

fermentation 13, 120, 124, 128, 131, 141, 142, 145, 150, 156, 157, 158, 159, 160, 161, 162, 163, 164, 168, 248, 249, 258, 265, 270, 271, 272, 273, 276, 277, 278, 279, 282, 283, 351, 355, 356, 357, 358, 360, 361, 362, 363, 364, 365, 366, 384
Fibrous Adsorbents 19, 369, 373, 374, 376, 377, 382
Fly ash 11, 18, 19, 22, 23, 24, 26, 27, 28, 29, 30, 31, 372, 387, 389, 393, 394, 426, 487, 488, 489, 490, 491, 492, 496, 497, 498, 499, 500, 501, 502, 503, 504, 505, 506, 507, 508, 509, 510, 511, 512, 513, 514, 515
Food Packaging 15, 32, 43, 56, 137, 143, 146, 149, 154, 170, 176, 177, 178, 186, 187, 188, 190, 191, 192, 195, 198, 201, 302, 304, 311, 315

G

gelation 15, 31, 44, 46, 205, 208, 210, 211, 214, 217, 218, 227, 233, 256, 299, 300, 301, 302, 308, 310, 312, 313, 320, 328, 335, 336, 434, 451, 534
Geopolymer 11, 18, 23, 24, 26, 27, 29, 30, 31, 487, 488, 489, 490, 491, 492, 493, 494, 495, 496, 497, 498, 499, 500, 501, 502, 503, 504, 505, 506, 507, 508, 509, 510, 511, 512, 513, 514, 515

H

Heavy Metal Ions 27, 30, 179, 190, 192, 193, 196, 197, 219, 220, 231, 235, 236, 238, 375, 381, 387, 389, 392, 393, 395, 406, 414, 423, 496, 501, 502, 503, 504, 505, 511, 512, 517, 520, 523, 526

I

Industrial waste 10, 11, 15, 16, 18, 34, 36, 41, 119, 121, 149, 152, 155, 163, 167, 288, 289, 352, 360, 387, 425, 426, 428, 429, 433, 501
Insect Chitin 241, 242, 244, 245, 249, 253, 256, 257, 258, 268
Insect Food Industry 241

L

lactic acid 12, 26, 100, 112, 158, 189, 248, 270, 271, 272, 273, 274, 276, 279, 281, 282, 283, 284, 304, 313, 368

M

Medical Applications 45, 46, 48, 50, 59, 208, 266, 326, 339, 340
Microbe 267, 384, 399
microbeads 61, 62, 63, 64, 65, 66, 67, 68, 69, 70, 71, 72, 73, 74, 91, 92, 94, 171, 192, 233, 316, 338
Mineral Adsorbents 372

N

N-acetylglucosamine 184, 204, 246, 256, 323, 411
nanofiber 28, 29, 30, 32, 61, 74, 79, 109, 110, 112, 138, 139, 141, 144, 145, 146, 151, 184, 247, 266, 339, 347, 373, 394, 397

P

Pectin 12, 15, 24, 25, 26, 28, 29, 31, 34, 57, 120, 162, 176, 183, 193, 194, 210, 285, 286, 288, 289, 290, 291, 292, 293, 294, 295, 296, 297, 298, 299, 300, 301, 302, 303, 304, 305, 306, 307, 308, 309, 310, 311, 312, 313, 314, 315, 316, 317, 318, 319, 320, 321, 329, 348, 408
photocatalysis 26, 388, 457, 460, 461, 462, 463, 464, 465, 466, 468, 471, 472, 474, 475, 477, 478, 481, 482, 484
Phytoremediation 399, 400, 413, 414, 418, 419, 422, 438, 450
Pollutant Remediation 369, 378, 386, 460
Pollutant Soil 425, 444
polyhydroxyalkanoates 100, 109, 155, 156, 167, 168, 281
polylactic acid 7, 153, 269, 270, 281, 282, 283, 284
Polymer Fibers 379, 381, 389
porous geopolymer 499

R

Recycle Materials 8
Recycling-Oriented Society 3, 4, 5, 7, 33, 351, 354, 365, 380, 400, 417, 425, 487

S

Scaffold 31, 47, 48, 49, 56, 59, 174, 179, 187, 219, 230, 266, 307, 347, 348
semiconductor 61, 62, 68, 446, 455, 457, 461, 462, 463, 464, 465, 468, 469, 475, 477, 478, 479, 481
Smart Drug Delivery 221, 325, 336, 338, 340
Sodium hydroxide 40, 64, 249, 411, 432, 492, 501
Soil Contamination 416, 425, 426, 428, 439, 517, 518, 519, 520, 530, 533, 535
Sustainable Natural Resources 533
Sustainable Society 1, 10, 11, 15, 20, 33, 35, 49, 107, 210, 258, 276, 324, 368, 487, 495, 498, 518

U

Ultrasound 16, 19, 25, 26, 31, 49, 54, 55, 57, 105, 204, 221, 225, 229, 231, 232, 233, 235, 239, 250, 262, 290, 292, 308, 311, 312, 313, 314, 315, 319, 321, 325, 326, 327, 332, 333, 334, 340, 341, 342, 343, 344, 345, 346, 347, 348, 425, 428, 429, 430, 431, 432, 433, 434, 435, 436, 437, 438, 439, 440, 441, 442, 443, 445, 446, 447, 448, 449, 450, 451, 452, 453, 454, 460, 480, 495
Ultrasound Washing 26, 425, 428, 439, 442, 448
Upcycle Technology 425

W

Waste products 34, 39, 41
waste utilization 121, 129
water treatment 187, 192, 197, 203, 205, 206, 219, 227, 372, 373, 379, 387, 388, 392, 394, 421, 455, 470, 471, 477, 479, 481